The Evolution Wars

The Evolution Wars:

A Guide to the Debates

Michael Ruse

Grey House
Publishing

PUBLISHER:	Leslie Mackenzie
EDITORIAL DIRECTOR:	Laura Mars-Proietti
MARKETING DIRECTOR:	Jessica Moody

| AUTHOR: | Michael Ruse |

| COMPOSITION & DESIGN: | ATLIS Systems |

Grey House Publishing, Inc.
185 Millerton Road
Millerton, NY 12546
518.789.8700
FAX 518.789.0545
www.greyhouse.com
e-mail: books @greyhouse.com

Publisher's Cataloging-In-Publication Data
(Prepared by The Donohue Group, Inc.)

Ruse, Michael
  The evolution wars: a guide to the debates / Michael Ruse– [2nd ed.]

  p. : ill. ; cm.

  Originally published: Santa Barbara, Calif. :
ABC-CLIO, c2000.
  Includes bibliographical references and index.
  ISBN: 978-1-59237-288-1

1. Evolution (Biology)—History.   I. Title.

QH361 .R874 2008
576.8/09

For Ronald Brooks and Brian Calvert,
and to the memory of Jay Newman

Contents

Acknowledgments

This is the second edition of a book that was written and published about ten years ago. At that time, I expressed the hope that the book could be read with interest and profit by anyone. I added, however, that if I were pressed to name a specific target audience, I would say that the reader I had most in mind is the person who does not know a great deal about evolutionary thinking but who has heard enough to want to find out more. In other words, I have in mind students of all ages. I would like to think, somewhat immodestly, that the call for a second edition means that I have succeeded in my aim.

As I update what I wrote then, I should say that I have been a student and teacher all of my life—for thirty-five years as a professor at the University of Guelph in Canada and now, for almost a decade, at Florida State University in the United States of America. My earlier dedication stands, to three men who taught alongside me, not merely offering me friendship and encouragement but also models of how our job should be done. Now two are retired: Brian Calvert was a fellow member of the Guelph Philosophy Department and Ron Brooks was a fellow member of the Guelph Zoology Department. Sadly, the dedication can now be only to the memory of Jay Newman of the Guelph Philosophy Department. I truly have been their student.

For this edition, I am greatly indebted to the editors at Grey House Publishing, especially for thinking that my ideas still are worth having in the public domain. And finally my family: my wife, Lizzie, and our children, Emily, Oliver, and Edward, have continued to show their usual enthusiastic support for matters Darwinian and for those who write on such subjects.

Prologue

Charles Robert Darwin was laid to rest more than a century and a quarter ago, yet his bones surely do not rest easily, even today. Like none other, he had and has his defenders: passionate defenders. Like none other, he had and has his critics: passionate critics.

Charles Darwin himself is controversial. There are those, scientists particularly, who see in Darwin the ideal researcher—dedicated, persistent, innovative, comprehensive, working patiently and professionally toward his ends, troubled by illness yet not distracted. A man for whom the truth is the only value appropriate for a scientist, himself willing to give and to sacrifice all to this end. At the same time, this Darwin is a man of personal generosity, offering friendship and support to all, close acquaintance and stranger alike. When his lieutenant Thomas Henry Huxley fell sick, it was Darwin who at once passed the hat, making a typically generous personal donation. This was the man he was.

There are others, however, who see a different Darwin. These people, frequently trained professionally in history and related subjects such as cultural studies, see a man who is a classic upper-middle-class Victorian with the prejudices of that class: racist, sexist, chauvinist, capitalist. Their Darwin has a second-rate mind; he was one who stumbled upon ideas that were truly beyond his grasp; and he was a man who quite probably stole most of his discoveries anyway. Rather than a man of genuine warmth and generosity, these critics see a user who concealed a heart of ice behind a facade of congeniality. They see one who used his illness to avoid responsibility, and they find many failures stemming from Darwin's personal inadequacies.

Controversial though Darwin himself may be, this is nothing to the work he produced. At the center is his major book, *On the Origin of Species*. His supporters and enthusiasts regard this work as a paragon of scientific excellence, a model of how to do good science—clear, thorough, balanced, suggestive, innovative. They think Darwin anticipated problems and—the mark of really important science—left work to do for generations to come. Darwin's detractors, however, see a mishmash of ideas and suggestions and hypotheses and half thoughts—half-baked thoughts!—that were strung together without order or reason, not just in the *Origin* but also in a series of secondary writings of genuine Victorian length and tedium. And these were just the first editions. By the time Darwin had written and rewritten his works in the face of criticism, one was left with material that showed as many disparate pieces as a crazy quilt, and with about as much organization. Only those with their own personal agendas to satisfy could find in Darwin that of real worth and value.

Finally, there is Darwin's legacy. His supporters today—neo-, ultra-, or just plain Darwinians—think that he left us one of the most important theories humankind has yet discovered. After the *Origin,* our thinking about the world and about ourselves could never again be the same. Darwin's was a revolution that equaled that of Copernicus. Indeed, one might even say that in the secular realm, Darwin's ideas and influence equal—and perhaps supercede—those of Jesus Christ in the spiritual realm. Never before or again can there be a body of work of this significance. But his detractors think just about the opposite. They appreciate the "dangers" of Darwinism. They argue that Darwin's ideas are overblown, unsubstantiated, and little more than ideology—secular religion—masquerading as disinterested description and explanation. They think that Darwinians are deluded, arrogant, and mischievously influential especially on the young. Destroying the legacy of Charles Darwin must be the aim and obligation of every right-thinking person.

In the last century, several U.S. states banned the teaching of Darwin's ideas, and to this day we find boards of education warning teachers and students against the dangers of accepting his theories. (See Figure 1.) Of course, you might respond, one should always keep an open mind about anything one is told, especially in science. Was it not the great philosopher Karl Popper who warned us that nothing in science is permanent that every idea may and someday probably will fall to the ground? However, which high-school teacher feels the need to caution about the Copernican revolution, telling the class that it may be necessary to revise and revamp, perhaps one future day going back to an earth-centered static universe? None, obviously! But in the case of Darwin, students are told to beware and to take heed. Perhaps one is going to be seduced from the true faith by vile heresies and misrepresentations.

Now, obviously, when people fall out like this, there is something interesting going on. There is no smoke without fire—although what is burning is perhaps another matter. Indeed, trying to find the answer to this puzzle is one of the reasons why I have written this book. I want to introduce you to Charles Darwin and to his ideas, to the people who came before him and the people who came after. I want to see what it is that makes people so passionate, either for or against. And achieving this aim is the reason why this book is structured as it is. I shall go more or less historically, from the past to the present, and each chapter will be introduced by a clash between people or groups: hence my title, *The Evolution Wars.* But although I love a good fight as much as anyone, truly I want to dig out, going behind the arguments and the polemics. I want to see why it is that people disagree and what is at stake and whether there was or is or ever could be a solution to what so divides the antagonists. A word of caution: I am not a social worker or psychiatrist, so frankly I do not care whether a resolution is ever reached or whether anyone feels happier when I have finished. I am a teacher, so I do care very much whether you understand a lot more when I am finished.

Which brings me to my final point, and then we can begin in earnest. You have a right to know where I stand. I think Darwin was a great scientist, and I think his ideas were truly important. Although I think he was often wrong—and I shall be telling you much more about his mistakes—I believe that essentially Dar-

How the States Teach Evolution

The science standards that govern how public schools teach biological evolution — the idea that life in all its forms has evolved over billions of years through mutation and natural selection — have improved in many states since 2000, according to the National Center for Science Education. But most states' standards — New York included — do not explicitly require teachers to explain that humans evolved from earlier life forms.

States where biological evolution is ...

Not mentioned in standards   Mentioned briefly, unclearly   Treated straightforwardly and/or thoroughly

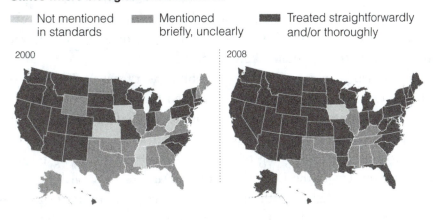

2000   2008

States where human evolution is ...

Not mentioned in standards   Implied, but not mentioned explicitly   Mentioned directly

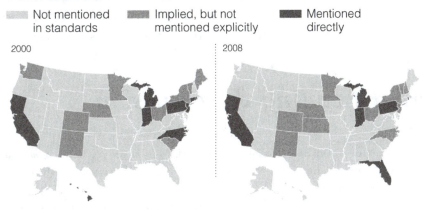

2000   2008

Note: Iowa does not have curriculum standards at the state level.

Sources: National Center for Science Education; Thomas B. Fordham Foundation, "Good Science, Bad Science: Teaching Evolution in the States," by Lawrence S. Lerner

©New York Times Graphics   *Figure 1*

win got it right. This mattered back then, and it matters right now. Darwin told us things of importance about the world and about ourselves. I think Darwin's ideas impinge on other areas of human inquiry and interest. Most importantly they rub up against religion, the Christian religion in particular. And those who say that religion and science can never be in conflict are deluding themselves. Science and religion can be at war, they have been at war, and Darwinism is right in the thick of it. But science and religion can work together; that is the other side to the story. And Darwinism is in the thick of this too. Those who say that religion and science must always be in conflict are likewise deluding themselves.

But I have said that I am a teacher, and I take that responsibility seriously in two ways. First, I am not here to convert you one way or the other. It is my job to give you the information, the tools, and then to let you work on things yourself. I can fault you on your knowledge of the facts, but when it comes to the interpretations, you are on your own. To be honest, I am indifferent as to whether you end up agreeing with me or disagreeing with me. I always tell my students that before I assign their marks, I do not look at their final sentences, in which they give their conclusions. I do care about the arguments they use to get to their conclusions, and I feel the same way about you. Agree or disagree with me as you wish, but show me that I should take you seriously.

And this brings up my second responsibility as a teacher. If, when you have finished, you do not care to argue with me, then I have let you down. Above all, Darwin, Darwinism, the Darwinian legacy, is absolutely fascinating. It is the story of terrific people and terrific ideas. These are important issues, and they matter—to me and to you. I am not going to trivialize, and I am not going to glamorize. You are not about to get the Disney version of Darwin. But I shall be very disappointed if you do not think that this topic is something that makes learning worthwhile. We may be grubby little primates on a grubby little planet, but every now and then we rise above ourselves. We escape the tawdry humdrum of everyday life and make sense of the Christian claim that through our intellect we are made in the image of God. Thinking on the questions raised in this book is one of those times.

Further Reading & Discussion

The standard history of evolutionary thought is Peter Bowler's *Evolution: The History of an Idea,* 3rd ed. (Berkeley: University of California Press, 2003). This is very comprehensive and fair, although to be honest a little bit of a textbook and reads like one. Very different is Robert J. Richards's *The Meaning of Evolution: The Morphological Construction and Ideological Reconstruction of Darwin's Theory* (Chicago: University of Chicago Press, 1992). Short, opinionated, brusque with the views of others, it is fun to read and legitimated by its author's very deep learning and understanding of his subject. My own *Monad to Man: The Concept of Progress in Evolutionary Biology* (Cambridge: Harvard University Press, 1996) is very long and detailed. Only graduate students working on their theses have to read it through from beginning to end; others should read the short introduction and then dip into it as it interests them. It is written in a kind of modular form so you can easily move around from one point to another. You will find that there is a lot of detail about the personalities and ideas of many of the people mentioned in this book.

My *Mystery of Mysteries: Is Evolution a Social Construction?* (Cambridge: Harvard University Press, 1999) covers some of the same ground and is much easier going. It is the best, short, overall introduction to the history of evolutionary thought around at the moment. Also let me recommend the *Dictionary of Scientific Biography* (New York: Scribner, 1970, and supplementary volumes later). There are many excellent articles on the major figures in the history of evolutionary thought and useful guides to further reading.

The Evolution Wars

Part One:
A HISTORY OF EVOLUTION

Friends & Foes: A Scientific Idea is Born and Explored by all Disciplines

Chapter 1
Early Evolutionists in the Debate: The Birth of the Idea

Overview

In this chapter we will explore how an argument of vertebrates vs. invertebrates in 1830s France, 29 years before the publication of *The Origins of Species*, began the Evolution Wars that continue today and remain hotly debated by 21st century academics, religious believers and political leaders. This chapter primarily discusses the life and work of four scientists who not only started the debates, but also set the stage for the wars to continue—Erasmus Darwin, Jean Baptiste de Lamarck, George Cuvier, and Etienne Geoffroy Saint-Hilaire.

Whereas later debates focus heavily on religion vs. science, these early scientists were exploring, debating and disagreeing on evolution with both sides having strong religious beliefs. Early evolutionists like Erasmus Darwin were neither agnostics nor atheists. They tended to be deists, that is, to believe in God as a supreme being who created the physical universe, but who doesn't intervene in its operation. Their God works through unbroken law and there is no need of miracles. Evolution, for them, therefore supports rather than detracts from the belief in God.

Because theorists in the early Evolution Wars did not necessarily see evolution in conflict with their religion, the early debates did not focus on religion vs. science, but on three aspects of evolution itself: There is the very *fact* of evolution: the slow, natural development of all organisms, living and dead, from simple, shared forms, perhaps ultimately from inorganic materials. There is the *path* of evolution: what direction did evolution take; are the birds for instance, descended from the dinosaurs? Then there is the *mechanism* or *cause* of evolution: what drives the process of change?

Charles Darwin's grandfather, late-eighteenth-century English physician Erasmus Darwin was an early evolutionist. What really drove him, rather than any empirical facts, was The Social Doctrine of Progress, the belief that through our unaided (by God) effort we can improve science, technology, and life generally, as evidenced by the Industrial Revolution.

The late-eighteenth-century and early-nineteenth-century botanist and zoologist, French minor aristocrat Jean Baptiste de Lamarck, was the first to write a systematic account of evolution. Like Erasmus Darwin, he too was a deist who believed in progress (hence his success, despite his noble status, during the Revolution). He laid on this the belief that acquired characteristics (like the long neck of the giraffe) can be inherited through parts that respond to use and disuse. This mechanism, known as "Lamarckism," fell out of favor as the modern theory of evolution began to take hold.

The great, early-nineteenth-century French comparative anatomist Georges Cuvier was skeptical about the progress theory. He was a practicing Christian (a Protestant) who disliked the deistic notion of God. Further, he had empirical evidence against evolution, citing the unchanged, mummified animals Napoleon's scientists had brought back from Egypt. But his main objection to evolution was that he could not see that tightly designed, well-functioning organisms that he explored as an anatomist could gradually change from one form to another. To him, this meant that a midpoint organism would be literally neither fish nor fowl and hence could not exist and reproduce.

Cuvier and his one-time friend, another early-nineteenth-century French comparative anatomist, Etienne Geoffroy Saint-Hilaire, clashed over the possibility that there might be connections or significant similarities between vertebrates and invertebrates. To accept these connections also meant accepting the very *fact* of evolution mentioned earlier. This famous clash of two titan personalities in the 19th century European scientific community is the first battle in the Evolution Wars that we will explore.

The Role of the Scientific Community

The work of the following scientists is discussed in this chapter. Short, biographical essays of these individuals appear in **Biographies** on page 607.

Erasmus Darwin (1731–1802)
Jean Baptiste de Lamarck (1744–1829)
Georges Cuvier (1769–1832)
Etienne Geoffroy Saint-Hilaire (1772–1830)

Setting the Stage

It was 1830 and Georges Cuvier was angry. And when Cuvier, the most powerful scientist in France, was angry, he was really livid. Pompous too. And very dangerous. He knew more than anyone else, and he set the standards and judged the results (Coleman 1964). You crossed him at your peril. For thirty years now he had been listening to this stupid, unfounded, dangerous nonsense from his fellow scientists. First there had been Jean Baptiste de Lamarck, and now when finally Lamarck had died and peace was in the offering, Etienne Geoffroy Saint Hillare—an old friend and a man who should have known much better—had taken up the cudgels and was promulgating the same detritus of the intellectual world, pseudoscience if ever there was such a thing. Action had to be taken. No longer could this be a civilized debate between savants of the same stature and learning. Things had to go public (Appel 1987).

No better forum could be found than the chief learned scientific society of France, the Academie des Sciences, of which both Cuvier and Geoffroy were members, and where indeed Cuvier was a Permanent Secretary, one of the chief positions of power and authority. Yet as so often happens when things explode after many years of provocation, the ostensive topic of debate was very minor and arcane. In October 1829, two unknown naturalists, Pierre-Stanislas Meyranx and a Monsieur Laurencet—a man so obscure that no one today knows his first name!—had submitted a memoir to the Academie on the subject of molluscs, a well-known group of marine invertebrates, that is, animals without backbones. They argued that there are significant similarities between the molluscs—they took the cuttlefish as a typical example—and the vertebrates, that is, animals with backbones. At least they argued—for nothing in this world is simple and straightforward—that if you bend a vertebrate backward in a bow, so that its head is virtually sticking up its butt, then you can see similarities. Geoffroy (as a member of the Academie) was asked to make a report on their claim, and his response came in very positively. Rubbing salt into open sores, he quoted (without identifying either source or author) an old paper of Cuvier's that denied forcefully that there could be any similarities between vertebrates and invertebrates. Now, claimed Geoffroy, we see that this kind of zoology is outdated and unneeded.

Incandescent with rage—so much so that the unfortunate authors of the memoir wrote earnestly to Cuvier, denying that their work had any implications whatsoever or that they intended in any way to contradict "the admirable work that you have written and that we regard as the best guide in this matter" (Appel 1987, 147)—Cuvier held forth before the Academie, with charts and tables showing that similarities are absent and that only the truly deluded could think otherwise. At which point, realizing that the best form of defense is attack and that Cuvier had forgotten far more about the invertebrates than he could ever learn, Geoffroy switched topics, arguing now that real similarities across species could best be discerned within the vertebrates (rather than across the vertebrate/invertebrate line). Now his point of argument was focused on the bones in the ears of humans and cats, which although different in size, shape, and number were (according to Geoffroy) essentially similar. Again Cuvier responded, and again his arguments were mixed with scorn and derision. Define your terms, he thundered at Geoffroy. "If our colleague had made a clear and precise response to my requests, that would be a fine point of departure for our discussion." Unfortunately, all he does is introduce one airy-fairy philosophical construction after another. All words and no substance. "It is to say the same thing in other terms, and in much more vague, much more obscure terms" (p. 150).

And so the debate went back and forth, with Geoffroy bobbing and weaving, always changing ground. Chasing him round the ring was Cuvier, flailing away, every now and then landing a good hard punch but never able to strike his opponent on the chin and end the contest. Finally, the fight petered out, with the contestants threatening their opponents with long series of justificatory memoirs. But not before the audience had had a wonderfully good time. Including the aged poet Johann Wolfgang von Goethe, who exclaimed to a friend, "The volcano has come to an eruption, everything is in flames"—an event that he saw as being "of the highest importance for science" (Appel 1987, 1).

But, even accounting for poetic license, could this really be so? An event "of the highest importance for science"? Are we truly talking about the same things: the similarities between a cuttlefish and a vertebrate bent backward until it resembled nothing so much as a participant in a prerevolutionary Cuban sex show? The bones in the ears of humans and of cats? Who cares? Or rather, since some obviously did care, why should we care? To answer these questions, we must go back a hundred years and start our story: then we shall see why it was that two distinguished French scientists did hammer it out in the spring of 1830, to the joy of onlookers then and of historians ever since.

Essay

Defining Evolution

We must not fall into the same trap that Cuvier accused Geoffroy of falling into. We must be careful to define our terms. At least, we must be careful to define one particular term. I realize that at this point you will probably start to groan and fear that I have forgotten already what I said at the end of the Prologue about my duty to be interesting and informative. You will find that I am a professional philosopher, and you will remember that someone once told you that the trouble with philosophers is that they are obsessed with language. They get hold of an important problem, start defining and redefining the pertinent terms, turning them upside down and inside out, and then they end up by announcing triumphantly that there was no genuine problem to begin with!

Georges Cuvier

I cannot deny that there is some truth to this. But terms and language are important, and unless one does take care one can waste an awful lot of time. I expect many of us have gotten into heated arguments about the existence of God, only to find at the end that we are arguing completely at cross purposes. The atheist is denying a God who looks a little bit like Santa Claus in a bed sheet, sitting on a cloud surrounded by angels with wings. The Christian is asserting a God who is the ground of our being or some such thing. The Christian would be appalled to learn that he or she is supposedly defending the odd entity that the atheist is denying. The atheist has never really thought seriously about the being that the Christian is affirming.

So, without further apology, let me turn to the term that is going to be at the heart of this book: *evolution*. And let me tell you that, traditionally, there are three things to which the term *evolution* applies (Ruse 1984). First, there is what we might call the very *fact* of evolution. By this is meant the idea that all organisms—you and I, cats and dogs, cabbages and kings, living and dead—are the end result of a long process of development, from forms vastly different. Usually it is thought that the original forms were very simple and today's forms are rather complex—some of them at least—and that everybody and everything is related in some form through descent. We shall see, however, that there are variations on this. Usually it is also thought that if you go back far enough then you pass from the living to the merely material—chemicals and so forth. In other words, the organic (that is to say the living) came from the inorganic (that is to say the nonliving). We shall have to go into this. And usually evolution is said to be "natural," in the sense that the processes (more on these in a moment) that fuel evolution are simply regular laws of nature—there is no need for divine or any other kinds of interventions. Again this is a matter that will get a lot more attention.

Second there is the *path* or paths of evolution, known technically as *phylogeny* (phylogenies). Here we are dealing with the tracks that evolution takes through time. When did life first occur on earth? When did multicellular organisms evolve from simpler forms? Was the Cambrian explosion one of a kind, or are there many such events? Did the birds come from the dinosaurs or simply from ordinary kinds of reptiles? When did the dinosaurs vanish, and was this associated with any grand terrestrial events? What do we know of human origins? Did humans get up on their legs and then the brains explode in size, or was it the other way around? In many respects, it is this aspect of the idea that most people think of when they think of evolution. "Missing links" is a favorite refrain of the critics of evolutionism, meaning that there are gaps in the fossil record (so the critics claim) where there should be transitions between one major kind (like land mammals) and another major kind (like sea mammals, such as whales). As we shall learn, in various ways the finding and establishing of paths stand somewhat aside from much else in the evolutionary enterprise. How, why, and what this all means will be a matter of some considerable interest.

Third and finally we have the question of the *causes* or *mechanisms* or *theory* of evolution. What makes the whole process go and work? What drives evolution? What is its motive force? In physics, this was Newton's great achievement. He did not discover that the planets go around the sun. This was the job of Copernicus. He did not map the heavens accurately. Tycho Brahe did this. He did not find the planetary motions. Kepler's job. Nor did he work out what happens down here on earth. Galileo. But he did find the law of inverse gravitational attraction and show how everything follows from this—orbiting planets and soaring cannon balls. For this reason alone, we venerate Newton and his genius. Likewise we have such questions in evolutionary biology. Is there a biological equivalent to the force of gravitational attraction and, if so, does it work in the same way? Is there indeed one prime cause, or are there many such forces that collectively make for the overall mechanism? And is the whole thing theoretical, and if so in what sense?

This division of evolution into three is somewhat artificial. Obviously you cannot have a path of evolution or a cause without the fact of evolution. And certainly any thoughts that you have about causes are going to be very much influenced by the paths that you think that evolution took. For instance if you thought—what nobody in fact has ever thought—that trilobites (a form of marine invertebrate that went extinct over three hundred million years ago) gave birth in one step to elephants, you would have a very different theory of evolution from thinking that the trilobite-elephant link (even if it existed) took 500 million years

Etienne Geoffroy Saint-Hilaire

with many, many intermediates. Indeed, if your fact of evolution includes the origin of life itself, then you are probably going to be thinking differently causally than if you think that the question of ultimate origins lies outside your ken. But, for all of the artificialities, it is useful to make a three-part division—fact, path, cause—and it will help us to structure our discussions in this book. Let us use it but not be ruled by it.

Now we are ready to start into our story, so let us go back to the eighteenth century.

Erasmus Darwin

The eighteenth century is called the Age of the Enlightenment, the time when the discoveries in science were consolidated and extended and when in the arts and in literature people started to turn from the past and look to the future. It is the time when we find such great writers and critics as Voltaire; philosophers such as David Hume and (a little later) Immanuel Kant; and the beginnings of social science in the hands of such men as the Scottish political economist, Adam Smith. Physics had had its great revolutions in the two centuries previously. Chemistry was to have its revolution toward the end of the century, thanks particularly to Antoine Lavoisier—whose reward was to be the loss of his head under the guillotine. Biology was still looking forward (Roger 1997). But the way was being prepared, thanks especially to the labors of two men. On the one hand, there was the Frenchman Georges Louis Leclerc, Comte de Buffon, author of the multivolumed *Histoire Naturelle* (from 1749 on), a discursive series of books that covered nature from one end to the next. Then on the other hand, there was the Swedish naturalist Linnaeus (Carl von Linné), whose ever-expanding *Systema Na-*

Erasmus Darwin

turae (first version 1735) introduced the modern system of organic classification, wherein every animal and plant can be fitted into its own unique place in the order of things.

Although neither was entirely successful in holding the dike, essentially both men had static pictures of nature. They had pictures that were, if not directly biblically based, then at least were views of life that might be called "Creationist," in the sense that God had created animals and plants basically in the forms that we see today, subject perhaps to a certain amount of variation, particularly of a degenerative kind. But the Age of the Enlightenment was above all an age of change, both as people saw the world around them and as the leading thinkers of the time saw the course of history. It is true that Christianity is itself a historical religion. One starts back with the Creation in Genesis and works through the Old Testament until the Incarnation, in the form of Jesus Christ. Then one moves forward until some time in the future when God judges us all, for good or for ill. But although our actions are certainly relevant, it is not a history over which we have much control. Indeed, ultimately, our greatest gains "count for naught" and we are dependent on God's grace for our salvation. With the development of science, however, and the advances of literature and philosophy and political economy and more, people began to develop the confidence that not only is there change, but this can be permanent and brought about by us, through our own efforts. Moreover, whatever the naysayers may have claimed to the contrary, this was thought to be change for the good. Progressive change, in short. Such a philosophy, if one

Carl Linnaeus

may so call it so, was bound to have an effect on thinking about the organic world, and now as we shall see it truly did (Ruse 1996).

Erasmus Darwin, the grandfather of Charles, was a physician in the British Midlands in the second half of the eighteenth century (McNeil 1987). Famed for his skill—his diagnostic abilities were formidable—Darwin several times refused the earnest entreaties of poor oft-times mad King George the Third to come south and take on the role of court physician. He was happy in his station in life and particularly in his place in the country, which was just then experiencing the first wave of the Industrial Revolution. Around him enterprising engineers were putting to use the powers of coal and steam in the running of those machines that were to produce finished goods at a rate far more rapid than could ever be achieved by hand. The Midlands and the North of England were the sites of the action, and Darwin was in the thick of it, mixing with industrialists, scientists, engineers, and others, and himself contributing knowledge and advice drawn from his medical studies and experience, not to mention his general grasp of things scientific. A particular interest was the world of agriculture, something that had to experience no less of a revolution than industry, as people moved from the land to the cities, and as population numbers exploded, and hence as there was need to produce far more food with far less remaining available labor.

Erasmus Darwin was a man big in every sense of the word. His appetites were gargantuan. He loved his food so much that it was necessary to cut a semi-circle in his table so that he could get close to the action. Preparing for one of his visits required considerable forethought and expense. Expensive dishes—prefera-bly many of them—were expected and appreciated. But Darwin gave as he received. He was a wonderful conversationalist and a much-loved friend, valued for his sensible advice. Yet, for all that he was fat, was missing his front teeth, and (with or without them) stammered badly, there was a romantic side to Dr. Dar-win. Sexually, he was a man of some considerable action. Three children with a first wife, two with a mistress during a kind of interregnum, and then seven more when, nearing fifty, he married the widow of one of his patients. This last he did in the face of several younger suitors. Intellectually also there was a lighter side to Darwin. In his day, he was one of England's better known and appreciated poets, as well as a writer of prose on many and varied subjects.

One of Darwin's closest friends was the potter Josiah Wedgwood, he who was responsible for the development of the British china trade—cups and saucers, plates and dishes, as well as vases and other objects of great beauty. In this prerail-way age, the chief mode of transportation—especially safe and careful transporta-tion—was by water. Supplementing the sea and the rivers, the eighteenth century was a time of great canal building: something that required an intimate knowledge of geology, especially when there were questions of boring tunnels through mountains. Wedgwood was a major figure in this work, and Darwin was in the midst of this activity, looking and searching and thinking and exclaiming. "I have lately travel'd two days journey into the bowels of the earth, with three most able philosophers, and have seen the Goddess of Minerals naked, as she lay in her in-most bowers" (King-Hele 1981, 43).

Erasmus Darwin was absolutely fascinated by discoveries such as these. Looking back two centuries later, there is no "smoking gun" that proves defini-tively just what it was that tipped him toward evolutionism or (as, in those days, he would have called it) transmutationism. Most probably it was the marine re-mains (shells and fossil fish) found hidden away in that mountain, in the middle of England, where he journeyed with his companions. Certainly, soon thereafter Darwin adopted *E conchis omnia* (Everything from shells) as his personal motto, and to celebrate he had the phrase painted on the door of his own carriage. He did not rush into print, however. Setting a pattern that was to be followed by his grandson Charles, Erasmus Darwin took some 20 years before he felt ready to an-nounce his thinking to the outside world.

His ideas were first written about explicitly in his major medical treatise *Zoonomia*, although one could hardly say that the treatment there was particularly systematic. Darwin made little or no attempt to disentangle the various threads of his thinking. Claims about the *fact* of evolution were mingled with ideas about the *paths* of evolution, and then threaded through the whole discussion were all sorts of hypotheses and speculation about the *causes* of evolution. Quite often he would start a paragraph talking about paths and then end up talking about causes. Or he would start off talking about causes and end up arguing for the general fact. He may have been an innovative thinker; he was no great systematist (Darwin 1794–1796, Vol. 1, 500–505).

Trying our best to disentangle his thinking, we find that probably there were two direct arguments that Erasmus Darwin put forward for the fact of evolution. First of all, he was much impressed by the analogy that he presumed between individual development and group development. If we can transform the individual—"from the feminine boy to the bearded man, and from the infant girl to the lactescent woman"—then why should we not transmute the group? Second, he thought very significant the similarities that he saw holding between the parts of the members of quite different species. These similarities, which today we call "homologies," were taken—as, indeed, they are taken today—to be evidence of common ancestry. Although, as I have just said, it is almost certain that it was fossil discoveries that made Darwin an evolutionist in the first place, he did not really bring in the fossils as a major piece of information in favor of the fact of evolution. They are mentioned but not as an important plank in the evidential foundation.

Today, we would surely want to use the fossils as evidence of pathways. Erasmus Darwin made no move in this direction either, although in fairness he had virtually none of the evidence that today makes the fossil record so important a source of information. As we shall see in a moment, he had an overall vision of the path of evolution, but as far as the specifics are concerned, he said little. In his opinion, the best source of information for actual pathways lay in the natures of living organisms. Take the presumed transition from sea to land. Erasmus Darwin touched on the peculiarities of animals like whales, seals, and frogs. He seemed to think that animals of this kind are somehow representative of those transitional forms that must have existed when life made its move from the sea to the land. Since we have such hybrid types today, it is reasonable to assume that they existed in the past, and these types today give us some clue as to their former nature.

What interested Erasmus Darwin more was the question of causes. He collected and offered all sorts of jumbled anecdotal bits and pieces of information. As you might expect, given that Darwin was living in a particularly important agricultural part of England, many of his suggestions were based on the folklore of animal and plant breeders. Indeed, Darwin spoke explicitly of "the great changes introduced into various animals by artificial and accidental cultivation." He was a strong supporter of the idea that characteristics acquired by an organism in one generation can be passed straight to members of the next generation. He instanced the docking of dogs' tails. Darwin believed that this practice eventually results in the birth of animals without tails at all, and therefore without any need of docking. This inheritance of acquired characteristics is today known as "Lamarckism" after the great French evolutionist of that name, although it should be noted that Lamarck's writings came at least a decade after Darwin put pen to paper. (Actually, the inheritance of acquired characteristics is an idea much older

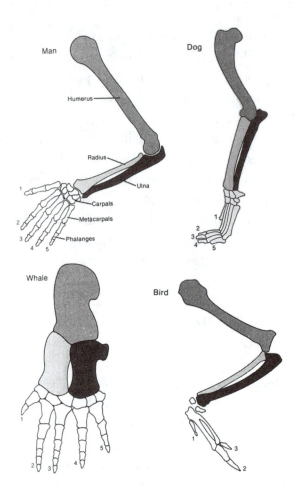

The homology between the forelimbs of vertebrates

than either Erasmus Darwin or Lamarck, although indeed it was Lamarck who made much of the mechanism as a force for evolutionary change.)

Naturally, as a physician, Erasmus Darwin was much interested in the nature of the mind and in the ways in which mental attributes can affect and be affected by physical causes. The popular psychological theory of his day—the brainchild of the eighteenth-century thinker David Hartley—was known as "associationism." In line with the general associationist position, Erasmus Darwin thought that habits and experiences could lead to new beliefs, and that these beliefs could be passed straight on thanks to reproduction. Hence, people's mental attributes could be a result of things having happened in the past to members of earlier generations. From this, there was an easy analogical slide to the physical world: "I would apply this ingenious idea to the generation or the production of the embryo or new animal which partakes so much of the form and propensies of the parent" (p. 480). Also, most interestingly, there was an anticipation of an idea that was promoted by grandson Charles. Erasmus Darwin thought that it was entirely possible that the body throws off small parts; these are carried around, presumably by the blood; and finally they are gathered in and transmitted via the sex organs. This supposedly gave a physiological backing to the already mentioned Lamarckism. The blacksmith's arms get stronger and stronger through use. These newly developed arms cast off modified particles that go down to the sex organs. And so the children of the blacksmith are born with strong arms as part of their biological heritage.

Truly, though, for Erasmus Darwin it was the big picture that counted. He found the nuts and bolts of evolutionism to be rather boring. Later in life, he was much given to poetic expression of his evolutionary vision:

> Organic Life beneath the shoreless waves
> Was born and nurs'd in Ocean's pearly caves;
> First forms minute, unseen by spheric glass,
> Move on the mud, or pierce the watery mass;
> These, as successive generations bloom,
> New powers acquire, and larger limbs assume;
> Whence countless groups of vegetation spring,
> And breathing realms of fin, and feet, and wing.
>   Thus the tall Oak, the giant of the wood,
> Which bears Britannia's thunders on the flood;
> The Whale, unmeasured monster of the main,
> The lordly Lion, monarch of the plain,
> The Eagle soaring in the realms of air,
> Whose eye undazzled drinks the solar glare,
> Imperious man, who rules the bestial crowd,
> Of language, reason, and reflection proud,
> With brow erect who scorns this earthy sod,
> And styles himself the image of his God;
> Arose from rudiments of form and sense,
> An embryon point, or microscopic ens!
> (Darwin 1803, 1, 295–314.)

Understanding the Past

We are going to be looking at a lot of evolutionists before we have finished, so I do not want to linger too long over any one. Fortunately, the overall ideas of Erasmus Darwin are not too hard to follow. We start at the bottom with the most primitive form, what was then often called the "monad," and we work our way up to the most complex and best form, what was then (unself-consciously) known as the "man." From butterfly (monarch) to king (monarch), as he expressed himself on another occasion. From that which is totally without value to that which we value above all else. A progressive rise up the chain of life. Yet, straightforward though this vision may be, I do want to make a couple of points before we move on.

The first is a general point but applied specifically to Erasmus Darwin. It is about the way in which we should treat figures in the past. There is a temptation to go too far in one way or the other, to see too many virtues or too many faults. Either we see the historical figure as a pure genius, with no flaws, and as having anticipated just about everything. Or we see him or her as a real fool, who found his or her way in the history books by chance or default or even fraud. Erasmus

Darwin is a good case in point. On the one hand, he surely did come up with evolution as fact long before a lot of other people. He was right on there. Moreover, he did pick up on some good points. Fossils are important. The similarities between the bone structures of very different organisms are puzzling at the least, and surely suggestive of some hidden links. And embryology? Well, we do develop from primitive beginnings, so why should not the same be true of life itself?

On the other hand, if ever anyone was credulously open to absurd arguments it was Erasmus Darwin. There was no systematic treatment with things properly quantified—the very things that, by the end of the eighteenth century, one took for granted in the physical sciences. Again and again the reader would get something far more suited for Ripley's *Believe It or Not* than for anything with pretensions to being serious science. One prominent anecdote told by Dr. Darwin was of a man who had fathered a dark-eyed daughter in a family of otherwise very fair children. How had this come about? Darwin tells that when the man's wife was pregnant, he (the father) had become totally enamored sexually of the dark-eyed daughter of one of his tenant farmers. Yet, although the man offered the girl money for sex, she would have nothing to do with him. The obsession remained, however, and "the form of this girl dwelt much in his mind for some weeks, and that the next child, which was the dark-eyed young lady above mentioned, was exceedingly like, in both features and colour, to the young woman who refused his addresses" (Darwin 1794, 523–524).

There was much more in this vein. For instance, we are told about the "the phalli, which were hung round the necks of the Roman ladies, or worn in their hair, might have effect in producing a greater proportion of male children" (p. 524). At times, even those who liked Darwin's work showed a tone of regret about the level at which he was writing. "If Dr. Darwin had indulged less in theory and enlarged the number of his facts our satisfaction would have been complete" (McNeill 1987, 174, quoting an anonymous writer in the *Monthly Review* 1800). The simple fact of the matter is that by the end of the eighteenth century, the notion that artificial penises hanging at the ends of chains supposedly affected the sex of future children was just not taken seriously by people who cared about serious science.

What am I trying to tell you? Basically, that at a certain level there was something rather ambiguous or questionable about both the quality and the status of the evolutionary speculations of Dr. Erasmus Darwin. Of course, we today would think this; but the point I am making is that even in the eyes of his contemporaries the ideas of Darwin were somewhat dubious or suspect. Which raises another question. If Darwin was indeed writing at such a loose or unsubstantiated level, why was he driven to do so? He was no fool, nor was he an unsophisticated thinker about technical issues. I told you that he truly had a great and justified reputation as a physician. Why then did he write as he did about evolution, and why was it that others at the time responded favorably to his ideas?

The answer has been given already. Darwin and his followers were absolutely obsessed with the new philosophy of the day: the philosophy or ideology of progress. For Darwin and his supporters, the Industrial Revolution—which was now going ahead at full steam, to use an apt metaphor—was the best thing that had ever happened to rural, sleepy, church-dominated England. What was needed, therefore, was a complete change of worldview. A worldview making central the success of machines and of the men of purpose who devised and drove them. That is to say, a worldview making central the achievements and aims of Darwin himself and of his industrialist friends. Evolution for Darwin, and for his supporters, was very much part and parcel of this philosophy or vision. Darwin (as we saw just above) did not see evolution as a slow, meandering process going nowhere. Rather, he saw it as an upwardly directed, progressive process reflecting the social progress that Darwin thought was now highly desirable. In fact, Darwin himself drew the connection, saying that evolution "is analogous to the improving excellence observable in every part of the creation; such as in the progressive increase of the wisdom and happiness of its inhabitants" (Darwin 1794, 509).

All in all, therefore, the evolutionism of Dr. Darwin was the industrialist's philosophy of action made flesh—or embedded in the rocks! One goes from "an embryon point, or microscopic ens!" to "imperious man, who rules the bestial crowd." At work here is a full-blown circular argument, or perhaps more charitably one might say a feedback argument. You start with the idea of progress, the philosophy of the British industrialist. You read this into nature. And then you read it right back to confirm your philosophy. "All nature exists in a state of perpetual improvement ... the world may still be said to be in its infancy, and continue to improve FOR EVER and EVER" (Darwin 1801, 2, 318).

Is this the philosophy of a man who has turned his back against religion? In a sense, this has to be true. Erasmus Darwin was certainly putting himself in opposition to conventional Christianity. For the Christian, the overall history of the world is one of miraculous creation, of subsequent sin and fall, and of the need for redemption that comes through, and only through, God's grace. Christ's great sacrifice on the cross and his miraculous rising from the dead wash away the sins of us all. For the Christian, therefore, Providence is the key to understanding history and the future. We humans can do nothing, save only with God's help and love. Darwin, as a progressionist, was arguing strongly that we humans are capable of improving our lot ourselves. So, in this sense, quite apart from the fact that as an evolutionist he had no place for the creation story of Genesis, Darwin was putting himself against traditional religion.

However, one should not at once conclude that Darwin was an atheist, or even an agnostic in the sense of having any doubts about God's existence. Darwin was no Christian, but like many intellectuals of his age (including many of the early American presidents), Darwin believed in a God who was an unmoved mover. He believed in a God who has put things in motion and who then stands back and watches how things work out through the agency of unbroken law. To use the technical language of scholars, Darwin was a deist, as opposed to a theist, traditionally a Christian, a Jew, and a Muslim. A deist sees the greatest mark of God's power and forethought in the working out of unbroken law, as opposed to the theist who sees God's power in direct intervention, that is, in miracles.

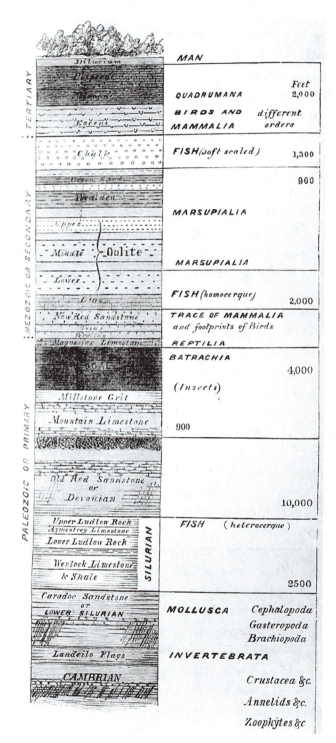

The progressive history of life, as published in 1861

Using a modern metaphor, what one might say is that Darwin's god—the god of the deist—has preprogrammed the world so that he did not have to intervene further. Evolution, therefore, can be seen as the greatest triumph of God. It is the strongest proof of his existence. It is certainly not something that disproves the need for or existence of a Creator or Designer. In Darwin's own words, "What a magnificent idea of the infinite power of *The Great Architect! The Cause of Causes! Parent of Parents! Ens Entium!*" (Darwin 1794, 509)

Jean Baptiste de Lamarck

I am not now sure that you would want to say that Darwin's evolutionism was a religious theory, nor even am I quite sure what that might mean. But this is the first moment at which you should start to realize that the science-religion relationship—the relationship in the context of evolution—is more complex than you might have thought. Those people (and there are many) who seem to think that evolutionists become atheists in the morning and then think up their theories in the afternoon, as a kind of bad joke, could not be more mistaken. Certainly, Erasmus Darwin—the man who can first claim unambiguously the label of "evolutionist"—became an evolutionist as much because of his religious beliefs as despite them. And that is a good point on which to move forward.

Jean Baptiste de Lamarck

As it happens, forward and sideways, for we cross over the Channel to France. Had things been normal, there is no telling what effect Erasmus Darwin might have had. But things were not normal. At the end of the century came the French Revolution, that bloody explosion that destroyed the Old Regime and absolutely terrified the rest of Europe, especially Britain. At once, all radical progressivist ideas came under heavy attack, being seen (with some considerable justification) as one of the major factors that brought on the events in France. Erasmus Darwin, enthusiast for the American and then the French Revolutions (until the latter got out of hand), ardent progressionist, came under particularly bitter attack from the conservatives. Devastating was a brilliant and cruel parody of one of his major poems—where he extolled the love of the plants, his detractors extolled the love of the triangles! The world laughed at him, his reputation sagged, and his evolutionism was crushed beneath the reaction.

But in France, for all of the revolution, evolution proved a more hardy plant. The key figure was Jean Baptiste Pierre Antoine de Monet, chevalier de Lamarck, son of minor nobility, who on being invalided out of the army became a botanist under the patronage of Buffon in the Jardin du Roi (Burkhardt 1995). Scientists did well in the revolution: they represented the kind of forward-looking attitude the leaders cherished (Lavoisier, the obvious exception, lost his head because he was also a tax collector). Although Lamarck found it politic to change his name from the hitherto more aristocratic de la Mark, he found himself in the newly reconstituted Jardin, now called the Museum d'Histoire Naturelle, in charge of (what he himself was to name) the invertebrates.

It is often said that the really revolutionary scientists tend to be young—mathematicians are all washed up by the time they are thirty. You need to have the vitality to move into new fields and not yet to have acquired the vested interests to stay with the old. I am not sure how true this really is—the man who (a few years ago) cracked Fermat's Last Theorem was 40—but Lamarck is certainly an exception to the rule. Although by century end he was 56, it was not until then that he swung from a lifetime's commitment to a static world picture and became an evolutionist. The particular trigger apparently was those invertebrates over which he had just assumed control. There were many fossil specimens for which Lamarck could find no living counterparts, yet since most were marine he could think of no competitors so strong and violent as to make them go extinct without trace. Hence, Lamarck came to the conclusion that they must have changed into other forms, or rather have given birth to other forms, without leaving descendants like themselves. This insight, if we may so call it, was enough to spur Lamarck to further speculation, and before long he was a full-blown evolutionist, a position he articulated fully in his major work, *Philosophie zoologique* (1809).

Lamarck believed unequivocally in the fact of evolution. Complex forms come from older simpler forms. Moreover, for Lamarck there can be no question but that evolution encompasses the production of life from nonlife. He endorsed venerable ideas of "spontaneous generation," believing that heat and lightning (just at that time, electricity was a very trendy phenomenon thanks to the experiments of Franklin and others) and other natural causes would stir up mud and other substances. From this, supposedly, would emerge primitive life-forms—worms and mites and the like. Indeed, not only did Lamarck believe in spontaneous generation, but he thought that it is going on all of the time, in the past and down to and including the present.

It is when we come to the path of evolution that Lamarck starts to get confusing and really quite interesting. Today, thanks to Charles Darwin, we tend to think of life's history in terms of the metaphor of a tree—the tree of life. Primitive forms are down by the roots, and then (going up the trunk with time) we have branching out into the major life-forms, with today's organisms (including us) up at the top, facing up to the sun. In diagrams given in the *Philosophie zoologique*, Lamarck rather gives the idea that this is his position also. But it was not. I have spoken of the main motive force for Erasmus Darwin as being that of progress—the progress, as we have seen, of a British industrialist who thinks that through human effort things will get better and happiness and so forth will be distributed and maximized. Lamarck likewise was a progressionist, but of a distinc-

A medieval rendering of the Chain of Being

tively French variety. For him, from a near feudal country with no major industry, the improvement of progress tended to be intellectual: improvement in the arts and literature and science and philosophy. To this he tied a very old idea—it goes back to the Greek philosopher Aristotle—that all organisms can be put in a line from the simplest to the most complex. This idea, the great Chain of Being or the *scala natura* as it is called in Latin, which was very popular in medieval times, was based on the idea that God would have left no gaps. It would have been incompatible with His Goodness and Greatness that, had it been possible to create intermediate forms, He would have failed to have done so (Lovejoy 1936).

Before Lamarck, the Chain was completely static. It was a way of laying out the living world. It followed, supposedly, from the creative nature of God and as such had no implications about origins. Lamarck fused it with his progressivism—things getting better all of the time—and his evolutionism emerged. One point of immediate interest therefore is that Lamarck had a somewhat ambiguous relationship with the fossil record. It was fossils that made him into an evolutionist. You might therefore think that he was then going to use the fossils to trace out the path of evolution through time, from simpler to more complex. Indeed, he was read that way by later commentators, notably by Charles Lyell the Scottish geologist (who thus gave a distorted picture of Lamarck to a whole generation, including Charles Darwin). But in fact Lamarck (like Erasmus Darwin) was basically uninterested in the record as evidence of the path of evolution. The path—from monad to man—was given to him through the Chain, and no more was needed.

Somewhat connected to all of this is the fact that, appearances to the contrary, Lamarck's evolution was not essentially treelike. Rather it was a series of climbs up the Chain: a staircase no longer but now an escalator. Organisms hopped on at the beginning, thanks to spontaneous generation, and then kept going right up to the top, penultimately as orangutans and then as humans. Thus, rather than a tree, we have parallel upward progressions, as life keeps starting over and over again. For someone who believes in a tree, extinction is forever. The dinosaurs will not reappear. Their branch has come to an end. For Lamarck, however, extinction is always a matter of time. If tigers were wiped out, then it might be a while before they reappeared, but they would—when the next escalator reaches the appropriate point. It is as simple as this.

But then how do you account for the tree-diagrams given in Lamarck's book? Here we need to turn to the third arm of the evolutionary picture: we need to look at causes. Notoriously, Lamarck believed in the inheritance of acquired characteristics—the giraffe's neck is long because ancestral giraffes with short necks stretched and stretched and stretched to reach the leaves high up in tree. Now their descendants are born with necks suited to the job. Although, as we have seen, this mechanism is to be found in Erasmus Darwin—it is indeed part of a cluster of very old ideas, although not previously used in a full-blooded evolutionary context (remember how Jacob tricked Laban by altering his sheep and goats before they were born)—the mechanism is today known as Lamarckism. It is this that Lamarck thought makes for irregularities in the chain of being—some organisms get deflected off the main path—and it was this that Lamarck was trying to show in his pseudotree diagram.

But if Lamarckism is the minor mechanism, what are the main mechanisms? Here things get a little fuzzy, mainly because Lamarck's thinking was itself a little fuzzy! He thought, in a mechanical materialist fashion, that there are bodily fluids (the *sentiment interieur*) that flow through organisms, carving out new paths and constantly complexifying things. Hence, we get a constant movement up the Chain, brought about by purely mechanical causal factors. Yet one might well ask why it is that organisms stay on the path that leads upward to human beings. Here we get no answer from Lamarck, but the impression one has is that in some sense this upward passage is foreordained. In other words, to use the language of the philosophers, Lamarck's is a "teleological" system, meaning that the end point in some sense influences the activities before it is achieved. Whereas normal causation works from back to front, from past to present to future—the banging door (past) made the servant jump (present) and then she dropped the plate (future)—

in teleology (or, as it is sometimes called, "final causes" or "purposeful" or "end-directed" situations), the future somehow reaches back to affect the present.

Now normally, there is nothing terribly mysterious about any of this teleological thinking: what we have are human beings or God thinking about the future, and based on these thoughts (which although referring to the future are in the present) we take action. But some people, Aristotle 2,500 years ago was one and at the beginning of this century the French philosopher Henri Bergson was another, have thought that life itself—even if it is not conscious—has a kind of forward-looking aspect to it. This is said to happen, not through thought but through something analogous to it—a kind of life force or "vital force" as it is often called. (Bergson called it an *élan vital* and his contemporary, the German embryologist Hans Driesch, spoke of an "entelechy." Supporters of such a position are known as "vitalists.") Although Lamarck denied strenuously that he was a vitalist—the opposite position is often known as "materialism," implying that there is nothing but material substance and forces—one has to say that there is a whiff of this about his thinking. Somehow everything fits together just too patly. I am not implying that Lamarck was a hypocrite or deceitful—claiming one thing and doing another—but rather that he was in respects more of a prisoner of his own past than he realized himself.

You might want to say that, given all of this, you do not really want to speak of Lamarck as an "evolutionist" at all. After all, evolution has been defined in terms of unbroken regular laws. But this seems to me to be too strict. What we have to recognize is that his evolutionism was not as unambiguously scientific as one might find in physics and chemistry. Which of course makes us all rather wonder if Lamarck, like Erasmus Darwin, had religious factors at play, driving him in his thinking. And the answer is that he certainly did. In fact, in respects his thinking was very much like that of Erasmus Darwin: Lamarck was no orthodox Christian, but he was a deist, seeing God as working through unbroken laws. It is just that the laws for Lamarck probably included something akin to vital forces. But the important point is that for Lamarck, as for Darwin, together with the vital influence of progress, it is true to say that he was an evolutionist far more because of his religious beliefs than despite them. A god who works through law, rather than through miracle, is a god who creates through evolution rather than in one creative spurt in six days at the beginning of time.

Georges Cuvier

Lamarck had a pretty shaky reputation. People admired and respected him for his taxonomic skills, but he was altogether too given to wild hypotheses. He had some really daft ideas about meteorology, which he suckered the French government into supporting at great expense. Supposedly on one occasion he offered his *Philosophie zoologique* to the Emperor Napoleon, who spurned it with contempt. It turns out that this was less because the Emperor was a creationist than because he thought he was being offered yet more wild and inaccurate weather forecasts! For

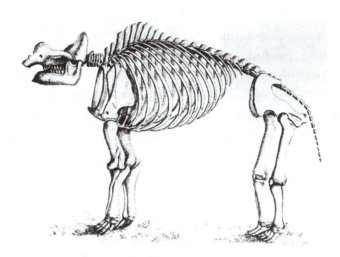

A mastodon as reconstructed by Cuvier

most people, Lamarck's evolutionism had altogether too much of the speculative about it. It was not that they were close minded, but that they had heard much of this kind of guff from Lamarck before.

No one felt more strongly on this subject than Cuvier. So, as I reintroduce him, let me start by stressing that he had every right and authority to feel this way. As a student of the life sciences, Georges Cuvier was head and shoulders above his contemporaries. His anatomical studies were simply outstanding—rightly he is known as the "father of comparative anatomy"—and then he turned to paleontology, taking what was a mess of fragmented ideas and hypotheses and leaving a full-blown scientific discipline. By any standard, this man was a really great scientist. He knew it and his contemporaries knew it. But he did not like evolution: he thought it unnecessary, he deemed it bad science, he found it philosophically offensive, he knew it was socially dangerous, and he found it threatening to him personally. Let us start to unpack these objections.

First, there is the question of the science. Cuvier appreciated that one had to speak to origins. Although he himself was rather inclined to think that the present state of knowledge did not make any real suggestions plausible or convincing, he was not faulting Lamarck for the very attempt to give an explanation. Cuvier's own geological explorations and his work on the fossil record around Paris persuaded him that the earth is subject to violent periodic convulsions—what his English supporters were to deem "catastrophes"—and that life in some sense starts anew after each catastrophic event. He rather inclined to think that new life came in from elsewhere, invading now empty territory, but on this he did not say much. The point is that historical inquiry as such was certainly legitimate. In fact, judged as a historical record, Cuvier was inclined to accept the biblical account of the Flood as the last catastrophe. But, as a sophisticated French scientist, the last thing that Cuvier was going to do was to appeal to the Creation account of Genesis as the beginning and end of inquiry. This was not how one did science. More on this point in a moment.

Why not evolution, especially since it was Cuvier's paleontological inquiries that were first starting to show in a definitive fashion that the fossil record is roughly progressive, leading up from strange and unknown forms to fossil remains not so very different from beings living and breathing today? The record itself, however, was taken as speaking against evolution, especially because, even if progressive, it was not continuously so. One got all sorts of gaps, with abrupt transitions from one distinct form to another. There was no way that this could be the record of continuous change. Better by far to speak of extinctions and then of restockings. In any case, argued Cuvier, drawing on specimens brought back by Napoleon's savants from the ill-fated French incursions into Egypt, the mummified forms of cats and birds and other organisms—beings that lived literally thousands of years ago—are absolutely identical to forms living today. Where then is the evolution, the change, in all of this? If Lamarck be right, we should expect to see some change right before our eyes, and this we do not see.

In a way, though, all of this was surface for Cuvier. He had much deeper reasons for dislike of evolution—reasons that were part scientific, part philosophical. Cuvier, born in a border state between France and Germany, was educated in Germany and clearly felt the influence of the philosophy of the great German philosopher Immanuel Kant. This was reinforced by readings of Aristotle—something Cuvier was able to do when, with enforced leisure during the worst excesses of the revolution, he lived far from Paris in Normandy, tutoring the children of a noble family. Like Aristotle, Kant took a teleological view of living nature—in particular, like Aristotle, Kant thought that one must try to understand organisms in terms of ends or purposes and not just prior causes.

We are not now dealing with the wide sweep of history, so we are not now dealing with vital forces. We are rather dealing with the way in which an individual organism is put together and organized—and for Aristotle and Kant (and Cuvier following them) the secret is that all of the parts are to be understood as seeming as if designed to serve the ends of the organism's well-being. Something like the hand or the eye is not just a piece of an organism, but rather an intricately integrated composition, which serves the end—which has the purpose or function—of the organism's well being. We have teleology, because we are trying to understand the present hand or eye in terms of what we think they will do in the future. Obviously, no one is saying that the hand and the eye are actually caused by the future well-being. What if the organism died young?

Cuvier (1817) spoke of this teleological way of regarding organisms as the "conditions of existence"—these are the kinds of integrative principles that one must have if organisms are to live and work and function. Random collection of bits will not do. He thought (with some justification) that with this approach he had a very powerful way of analyzing organisms, since one knows that many parts must fit together harmoniously with the whole. Apparently proving what he was doing, he was fond of taking some isolated fossil bone and "deducing" the whole of the rest of the organism. A carnivorous tooth, for instance, would imply feet and claws designed for chasing and holding and killing—one could not have the hooves of a horse—as well as a stomach ready to digest huge chunks of raw meat—the digestive system of the cow would not do—and so on and so forth.

One or two cases where he did this inference correctly, working out from fragmentary bone parts the nature of the whole organism (which was later discovered), convinced his fellows that his method was indeed as powerful as he claimed.

The conditions of existence did more than this. Translated into practice, which Cuvier called the "correlation of parts," it gave him a way of classifying organisms in a "natural" manner. If once you have a basic part in place—the backbone for instance—then you cannot have many of the features of an invertebrate—an exoskeleton for instance. Then, if the vertebrate is a meat eater, once you have got the carnivorous teeth in place, you cannot then have the features of a herbivore. With the carnivore, if once you have the features of the cat in place, you cannot then mix them up with the features of the dog. And so forth. Everything has it place—starting (Cuvier thought) with four great divisions, what he called *"embranchements"*: vertebrates, molluscs (like clams), articulates (like insects), and radiates (like starfish).

You can see now why Cuvier had to be, absolutely and completely, against evolution. Moving from one form to another would smash to smithereens his beautiful static picture of the organic world. Nothing would be permanent and every inference would be open to doubt. And you can see now why Geoffroy's attack was so powerful and why it had to be resisted. Geoffroy, by endorsing the analogy between the vertebrate and the mollusc, was suggesting precisely that Cuvier's nice neat system was open to fundamental revision. Ultimately nothing remains the same. Everything is open to change. If you can go from one *embranchement* to another, or if you can find evidence that there are links between one *embranchement* and another, then the game is over. The way is open for evolution to come flooding in and spoil everything.

What about religion? If the evolutionists were all deists, might we infer that Cuvier was not—that in fact he was a theist, a Christian and that this was part of his opposition to evolutionism? As a matter of fact, Cuvier was a Christian, and interestingly a Protestant—a legacy of that border state where he had been born. (It was not in fact incorporated into France until after he was born.) There is no question but that, as their deism influenced the evolutionists, so also Cuvier's theism influenced his antievolutionism. He did believe that God was the Creator and that He had intervened miraculously to place organisms here on earth. They could not have appeared naturally. But, as I have explained, Cuvier was anything but a literalist. It may have been legitimate to use the Bible as a historical record. It could never be a source or substitute for serious scientific research. Genesis should simply not be read that way. It is the story of our moral relationship to God. It is not a scientific text.

Finally, let me raise some social questions (Outram 1984). Cuvier was a powerful scientist, but his was a power circumscribed and defined by his circumstances. He had been trained in Germany, and his training was less as a professional scientist and more as a civil servant—as a bureaucrat. He was big on deference, by him to his superiors, to him by his inferiors. Lamarck and Geoffroy galled precisely because they would not accept his status and thought of him as an equal. Cuvier knew that it was politic for him to serve the ends of his masters, first Napoleon and then the government after the Restoration. And he knew that, given the revolution, his masters were terrified of any social upheaval or of any philosophy that tended that way. Since evolution was so blatantly a tool of change and turmoil, he saw it as his task to oppose it as best he could whenever he could. If he was going to be really useful to the State, this was a place where he could show it. And so, Cuvier did.

But there was more than this, and here the personal factor comes in. As a Protestant, and by no means high-born, in Catholic France—Catholic France, which became increasingly conservative in the early decades of the nineteenth century—Cuvier had to tread carefully. Not only had he to show his personal worth to the state, but he had to be nonthreatening. Here, science was the perfect medium. At least, a science shorn of value and culture and ideology was the perfect medium. Cuvier could, as it were, say to his masters: "Look, give me power and status in science, and feel no threat because science is precisely that area of inquiry where there is no place for culture or value. The fact that I am a Protestant might be worrisome in a sensitive area like education or the like [in fact, Cuvier was put in charge of Protestant education in France], but in science uniquely my religion does not count. Trust me, for my personal ideological and religious commitments are irrelevant."

When people like Lamarck and Geoffroy came along, touting their philosophies and ideologies and religions dressed up as serious science, using their authorities as senior scientists, they threatened to wreck Cuvier's careful social strategy no less than they threatened to wreck Cuvier's careful scientific strategy. No wonder he was drawn into public dispute. And now we can see what hidden depths there were beneath a technical and dry debate about cuttlefish classification, and why Goethe was spot on when he explained to his friend: "I am speaking of the contest, of the highest importance for science, between Cuvier and Geoffroy Saint-Hilaire, which has come to open rupture in the Academy" (Appel 1987, 1).

Further Reading & Discussion

There are several good books on the main characters in this chapter. The aeronautical engineer Desmond King-Hele is somewhat of an Erasmus Darwin buff and has written many books on and around his subject. The latest version is *Erasmus Darwin: A Life of Unequalled Achievement* (London: De La Mer, 1999). Richard Burkhardt has produced the standard biography of Lamarck, *The Spirit of System: Lamarck and Evolutionary Biology (with a New Foreword by the Author)* (Cambridge: Harvard University Press, 1995); and William Coleman wrote a really good scientific biography of Cuvier: *Georges Cuvier Zoologist: A Study in the History of Evolutionary Thought* (Cambridge: Harvard University Press, 1964). Somewhat more technical, but top-quality scholarship, is Toby Appel's account of the dispute between Geoffroy and Cuvier, *The Cuvier-Geoffroy Debate: French Biology in the Decades before Darwin* (New York: Oxford University Press, 1987). She is really sensitive to the science of the day and to the institutional background.

Unfortunately, in a book like the *Evolution Wars*, you do have to be awfully selective, else you just end with a massive encyclopedia that only recommends itself because it leaves no one unmentioned. There was a terrific amount of activity between the disputes at the beginning of the nineteenth century and the controversies that erupted once Charles Darwin had published the *Origin of Species*. A really great book dealing with some of this activity in England in the pre-*Origin* years is Adrian Desmond's *The Politics of Evolution: Morphology, Medicine and Reform in Radical London* (Chicago: University of Chicago Press, 1989). I should tell you that Desmond is an ardent "social constructivist," meaning that he thinks that there is no ultimate truth, that science does not progress in any absolute way, and that evolution is to a great extent less a description of objective reality and more a reflection of the culture of its day. In *Mystery of Mysteries* (Cambridge: Harvard University Press, 1999) I argue strongly against this philosophy of history, but this is not at all to deny Desmond's brilliance as a historian and the deep understanding he brings to the history of evolution. Almost always, I learn more from those with whom I disagree than from those whose thinking parallels my own.

Finally, let me recommend another of my books, *Darwin and Design: Does Evolution have a Purpose?* (Cambridge: Harvard University Press, 2003). It is about the whole question of the design-like nature of the organic world and the consequent problem of explaining in the life sciences. As you might expect, Cuvier has a big role in the book, as I try to explain how in one sense you might think him very wrong to oppose evolution, but in another sense you might think his ideas about purpose were an absolutely fundamental piece of the puzzle as scientists moved toward acceptance of evolution.

Chapter 2

Conflict Before, During and After The Origin of Species: The Legacy of Charles Darwin

Overview

This chapter is a detailed and fascinating look at the step by step development of an idea, the survival of the fittest, as a young, brilliant, well-trained scientist observes the factual world in front of him and integrates these observations with the wide and conflicting theories around him. Unlike some of the prominent scientists of the day, Darwin was not dogmatic, allowing him to meld the factual world he saw with the wide range of theses he delved so deeply into. He readily absorbed the latest treatises and debates on geology, biology, theology, animal husbandry and even sociology and integrated them with his own observations, developing the remarkable theory of natural selection while still a young man. So revolutionary and complete was the theory because of its integration into so many disciplines, that Darwin had concerns regarding its impact on a Victorian scientific community that could be as remarkably progressive as it was rigid. He did not publish *The Origin of Species* for twenty years after its development. Ultimately it was the anticipated publication of a similar theory by Alfred Russel Wallace that forced Charles Darwin's hand and the release of *The Origin of Species*.

There are three main parts to the *Origin*. First, Darwin tries to convince the reader of the reasonableness of natural selection using the analogy of artificial selection, the process by which animal and plant breeders improve their stock. Second, Darwin gives arguments showing first that there is an ongoing struggle for existence, and then, that the struggle for existence leads to natural selection, or survival of the fittest. Third, and for most of the *Origin*, Darwin applies his mechanism to the findings of the biological world—instinct, paleontology, biogeographical distribution, morphology and anatomy, systematics, and embryology. He uses his theory of natural selection to explain these areas and, conversely, the success of the explanations makes natural selection plausible.

Over that span of time he had his own internal religious battles as he moved from a literal interpretation of the bible to a more deist approach that his grandfather had favored. But he found that even this approach, which emphasized the basic beauty of organism design as being God given, gradually gave way as he drifted toward agnosticism.

The thought of a man like his father, Robert Darwin, a man whom Charles loved and venerated above all others, being condemned to eternal damnation because of his lack of religious belief, acted powerfully on Charles Darwin, and moved him toward skepticism about any kind of God.

The Role of the Scientific Community

The work of the following Victorians is discussed in this chapter. Short, biographical essays of these individuals appear in **Biographies** on page 607.

Reverend Archdeacon William Paley (1743–1805)
Reverend Thomas Robert Malthus (1766–1834)
Charles Lyell (1797–1875)
Alfred Russel Wallace (1823–1913)

Setting the Stage

Charles Darwin and Alfred Russel Wallace really liked each other. This was just as well, for their names are forever linked as the two men who discovered the chief cause of evolutionary change. It would have been so easy for them to have quarreled: Darwin resenting Wallace, who came many years later but who yet drove Darwin into action; Wallace resenting Darwin because the older man had beaten him to the punch and then hogged all of the limelight. But although their followers and supporters have tried their best to divide the two, the friendship and respect lasted all of their lives. Wallace admired Darwin for the great scientist that he was; Darwin appreciated Wallace for his genius and his modesty and his firm convictions in the search for the truth.

This said, they rarely agreed about anything. They battled over their jointly parented child in a way that makes today's custody battles look like Quaker meetings. If Darwin had an idea, then Wallace opposed it. If Wallace had a thought, Darwin thought he must be wrong. You thought that cuttlefish classification was a boring topic. Try female bird coat color. Darwin (1859) had an elaborate theory to explain what is known as "sexual dimorphism": the differences between males and females in the same species. Of course, you have got to have some differences. If everyone had a penis you would be as badly off as if no one had a womb. But why do you have the big and visible differences? Why do human males have beards when the women are hairless—at least on their faces and chests and so forth. Why do men go bald, for that matter, and not women? Why are male walruses so much bigger than the females—so much bigger that sometimes the females get crushed to death during copulation? Why do stags have massive heads of antlers and the females go around with little or nothing? And why, why, does the peacock have such a magnificent backside when the female has nothing—magnificent and yet kind of stupid, because who can escape with tail feathers like that when the predator comes calling?

Darwin tended to put the emphasis on the male. Stags have massive heads of antlers because they do combat with each other in the rutting season—winner takes all, and that is why there are the horns. Females hang around passively, not competing, and so they do not need or obtain such appendages. The same sort of thing is true of walruses, who fight like mad for possession of a harem of females. And in the case of the peacocks, it is basically the males' showing off that counts. It is true that the female chooses the male with the most magnificent display, but it is the male who is (literally) the center of attention.

Wallace (1870) felt very uncomfortable about this. He could not deny that the stags fight and so do the walruses and that such an attribute probably is the cause of the differences in those sorts of species. But he disliked intensely the claim that the peacock grew his feathers because the peahen was attracted to beautiful backsides. Rather, he suggested that Darwin had got things bottom backward, as one might say. It is not so much that the males are beautiful and showy

Charles Darwin

as that the females are drab and inconspicuous. Sometimes, Wallace argued, being the center of attention is precisely what one does not want—and this sometimes occurs particularly when one is sitting on eggs, incubating them. Wallace thus claimed that sexual dimorphism is a function of female camouflage, protecting the females from predators, rather than male gaudiness with consequent female preference.

Why the quarrel? Was it just a matter of fact or facts? Well, in a sense it was, as obviously the cuttlefish classification was a matter of fact or facts. Wallace thought he had good evidence of the significance of such things as coloration in mimicry and camouflage, so it was natural to apply his findings to an important question such as dimorphism. But there was a lot more than just that. Today's feminists would at once suspect that prejudice and attitudes were involved—Darwin was excluding the active input of females, whereas Wallace was making this absolutely central. In fact, as we shall see, the feminist would not be so far wrong. Darwin was a bit of a male chauvinist, and Wallace was exceptional in his sensitivity to the significance of the female sex. Yet, there was something even more important, and without now giving away the game completely, let me point out to you that Darwin was claiming that the female peahen's aesthetic sense was very much like a human aesthetic sense. She chooses a feather display for the same reasons as we find it beautiful. And if a peahen's aesthetic sense is like a human's aesthetic sense, then a human aesthetic sense is like a peahen's aesthetic sense. And this was a matter that neither Darwin nor Wallace thought trivial. But let's explore first how they got to their opposing, but not entirely dissimilar, points of view.

Essay

The Making of a Modern Day Scientist in Victorian Times

Charles Robert Darwin was born to a life of upper-middle-class English privilege (Browne 1995, 2002; Desmond and Moore 1992). His father, Robert—oldest son of Erasmus Darwin—was a very successful physician and financier, and his mother was the daughter of Erasmus Darwin's old friend, Josiah Wedgwood. There was simply lots of cash in Darwin's background, and this was augmented when at the age of 30 he married his first cousin, Emma, another grandchild of Josiah Wedgwood. I make this point right at the beginning because it is an absolutely vital key to understanding Darwin's actions and much of his thinking. For instance, it is often said that Darwin never worked for a living, with an implication that he simply was not bright enough to obtain and hold down a proper university professorship. But this is to distort matters entirely. Darwin never worked for money (although he was good with his investments and canny in his dealings with publishers) because he never had to. Not for him were boring department meetings and officious administrators and whining students intent on mark grubbing. He could avoid all of that.

More significantly, because Darwin did so well out of Victorian society, one should not expect to find him a rebel in the sense of repudiating all of his background. Why should he? He was doing very nicely out of it, thank you! This is not in any sense to minimize Darwin's achievements but to point out that Darwin's achievements will most probably involve taking what he has been given and rearranging them into a new pattern. We should not look for Darwin to be the Christian God, making everything out of nothing. Rather, Darwin will be the sculptor or modeler who takes what he has and makes of it something new.

As a boy, Darwin was sent to one of England's famous private schools (misleadingly they are known as "public" schools, but they are anything but). Something of a square peg in a round hole—the main educational diet was Latin and Greek, a terrible bore and burden for the already science-sensitive Darwin—he went next, as had his father and grandfather before him, to the University of Edinburgh to train as a physician. Revolted by the operations and driven to madness by the tedium of the lectures—Darwin hated having to rise on dark Scottish winter mornings to listen to dry old men with incomprehensible accents lecture on dry old topics with incomprehensible significance—by the age of 19 he was back home and at a loose end. Desperate that young Charles not slouch into a life of indolent ease—one son was already going that way—Robert Darwin (himself an

A cartoon of Darwin as a student at Cambridge

atheist) somewhat cynically pushed Charles to the path of an Anglican clergyman, a traditionally safe and respectable position for a young man of wealth and minimal career objectives. This meant getting a degree from an English university, and so, in 1828, Charles Darwin enrolled at Christ's College in the University of Cambridge.

It was a fortuitous move. Although there were then no formal courses in the sciences—Darwin did not get a science degree at Cambridge because there were none—this was just the time when a group of men was starting to take a serious interest in the natural sciences (including geology and biology). Anyone with a like concern, including an untutored undergraduate, was welcome to join in. For three years then, Darwin did formal courses—Latin, Greek, mathematics—and informal courses covering many aspects of the contemporary sciences.

An Invitation to Sail on the Beagle

In 1831, when he graduated, came the big break. The Napoleonic wars now well behind, the Industrial Revolution was starting to get its second breath. Industry demands markets, and some of the biggest were in South America, long settled by Europeans and very wealthy. Ships were going out from the British ports—London, Liverpool, Glasgow—laden down with factory-made goods. There was a need for good naval charts, and so the British Navy was sending a ship down to the southern continent to map the coasts and shoals and waters. The captain of this ship, Robert Fitzroy of H.M.S. *Beagle*, was only 23 and—faced with a long and lonely trip, given that as captain he would be a person apart from the crew—was looking for a gentleman who could be his friend and traveling companion. It had to be someone outside the chain of command, personable and able to pay his own mess bills. Through a friend of a friend, Darwin got the call, he fit the ticket entirely, and so—for all that his father grumbled that he ought to be settling down and starting on the career as a clergyman—he spent the next five years

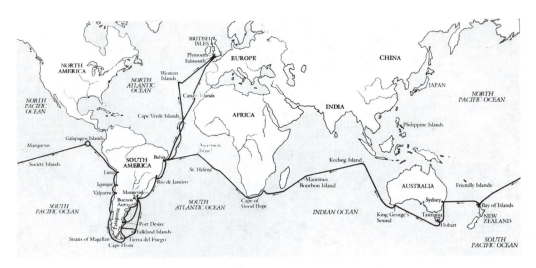

The voyage of the Beagle (1831–1836)

(1831–1836) eventually going around the whole globe as what became, de facto, ship's naturalist, on H.M.S. *Beagle*.

Revealed vs. Natural Religion

Darwin did not become an evolutionist on the voyage, but it was the experiences and discoveries on the voyage that turned him into one shortly after he returned (spring 1837). Since religion is going to play a large role in our account, as a preliminary let me make a distinction that will help our understanding. Students of religion make a division between two kinds of inquiry: revealed religion or theology and natural religion or theology. Revealed religion is the religion of faith—it is what you get when you read your Bible or have direct insights from God or (especially if you are a Catholic) what the Church tells you to believe. So, for the Christian, revealed religion covers such things as Jesus' birth and death, the miracles and the resurrection, and that sort of thing. Natural religion or theology is the religion of reason—it is what you get when you try to get at God through pure thought. If someone says that a good proof for the existence of God is the fact that everything has a cause and so the world must have a cause—call this "God"—they are in the realm of natural religion. (This particular argument is known as the "cosmological" argument.)

There is a lot of debate between theologians as to the significance of the two branches of religion, and their relationships. Here, we need not bother with this. It is enough that they exist and that they will both prove pertinent in the Darwin story.

Darwin's Shift Toward Deism and the Influence of Charles Lyell

Going back now to our hero, it is revealed religion that is first up front and relevant. When he left England, by his own admission Darwin believed in the Bible pretty literally, and this extended to the origins of the earth and of organisms. But his views started to change as the *Beagle* worked its way around South America. It

The frontispiece of Lyell's Principles of Geology. *Lyell is using this picture to show that land sinks (hence the erosion of the pillars) and then rises (hence the pillars out of water), as confirmation of his theory of climate.*

is clear that the major influence—the major influence always on Darwin—was a new book just appearing: *Principles of Geology* (1830–1833), in three volumes by the Scottish-born sometime lawyer Charles Lyell. (Darwin took the first volume with him, and the other volumes were sent out as they appeared.)

The full title to Lyell's work gives the clue to what it was about: *The Principles of Geology, being an Attempt to Explain the Earth's Surface by Reference to Causes now in Operation.* Lyell wanted to counter the catastrophic geology of Cuvier (1813)—a geology that had found much favor in Britain—by arguing that if one has enough time (indefinite time as far as he was concerned) then causes that we see around us today, governed by laws operating today, are quite enough to explain everything: seas, mountains, rivers, canyons, and all else. All one needs is time, and then rain and wind and earthquake and volcano and the rest can do the work. Above all else, one has no need of miracles, in the sense of divine interventions from above mixing things up and creating anew.

Lyell & Deism

This forswearing of miracles and reliance on unbroken law will probably ring a bell and so it should! Could it be that Lyell had inclinations toward deism, away from conventional Christianity (which seemed to fit well with catastrophism)?

The answer is that he did very much—his geological philosophy of "uniformitarianism" was deism in the stones, as it were. And this rang a bell or a chord in Darwin also. For all that he had had a conventional Christian (Anglican) education, intending to be a priest no less, in ways his formal belief sat lightly on him. His mother's family (the Wedgwoods) were practicing Unitarians—people who deny the Trinity, hence the divinity of Christ and the legitimacy of all of his miracles, and thus deists by another name. Before he was long into the voyage, it is clear that Darwin saw himself likewise moving toward deism (a position he was to hold almost the rest of his life), and we know already how that inclines one to views like evolution.

A Grand Theory of Climate

But Lyell had another part to play. Not only was he an enthusiast for unbroken law and causes—causes of a kind and intensity we see around us today—he had a particular theory that was intended to reinforce this uniformitarianism. The catastrophists tended to see the earth as directional, cooling from an original incandescent state down to the temperate state that it has today: this they saw as a background to the progressivism that Cuvier had found in the fossil record. As the world took on the form it has today, so its denizens took on the form they have today. Lyell to the contrary argued that there is no genuine direction to earth history. Yesterday was much like today. Today will be much like tomorrow.

Yet he could not deny some change: there is fluctuation. The fossil plants around Paris are definitely tropical, implying that the climate was warmer. So herein came Lyell's "grand theory of climate": he argued that temporary fluctuations of earth climate are a direct function of the distributions of land and sea around the globe. The Gulf Stream, that body of water that flows up across the Atlantic from the West Indies to Britain, makes for a much more temperate climate in Britain than the latitude would suggest. But, like all else, this will be temporary: the world is in a constant state of rising and falling. As rivers deposit silt at their bottoms, they press down the earth; then, like a gigantic water bed, another part of the earth rises upward. Thus the currents are altered and the local climates are changed. But overall, the general state is one of uniformity. Within limits, nothing changes. There is no direction to earth history.

Darwin bought into this theory all the way down (or up!). Much of the geological work he did on the *Beagle* voyage was devoted to finding evidence that the earth is (and was) in a constant state of rising and falling. But what kind of evidence does count at a time like this? Fossils are helpful, of course, but even more so are the distributions of animals and plants around the globe: what is known as "biogeography" or "biogeographical distribution." Lyell was a bit vague about where he thought organisms come from—for all his deism, he was not keen on evolution because he thought it would downgrade the status of humankind—but he was fairly certain that they come into being on a regular basis and that by and large the new arrivals tend to be fairly similar to those most recently arrived.

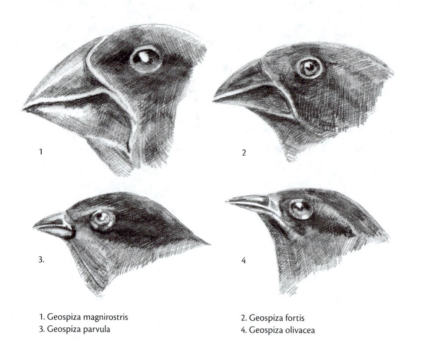

1. Geospiza magnirostris
3. Geospiza parvula

2. Geospiza fortis
4. Geospiza olivacea

The finches of the Galápagos

Thus, if (for instance) we find two groups of animals, very similar, divided by a natural barrier like a river or mountain, we can infer that the barrier is fairly recent. If, on the other hand, the animals are very different, we might infer that the barrier is ancient.

The Creativity that Set Darwin Apart

I explain this all in some detail, because here we are about to see one of the most important aspects of scientific creativity. Finding the answers is easy. It is asking the right questions that is difficult, and important. Once you know where to look, you are on your way. It is finding the right direction that is what counts. The *Beagle* lands in the Galapagos and evolution is on its way to becoming a *fact*. Keyed by Lyell's climate theory, looking intently at biogeographical distributions, Darwin was well primed when the *Beagle* put in (in 1835) at the Galápagos Archipelago, a group of volcanic islands in the mid-Pacific. At first he saw nothing very peculiar, as he collected the birds on the various islands and goggled—as did everyone else—at the giant tortoises that live on the islands. Then, thanks to information furnished by the governor of the archipelago, Darwin realized that from island to island the inhabitants are different. Even on islands within calling distance one has different forms of bird and tortoise. This had to be significant, especially since on the South American mainland (which Darwin had just left), one sometimes found the same animal inhabiting the land from top to bottom, from steamy Brazilian jungle to snowy Patagonian desert.

Back home in England, John Gould, the leading ornithologist of the day, assured the young Darwin that his collections did indeed represent different species

(Darwin was already making enough of a name for himself that the top people were happy to look at his specimens). For someone who was thinking in terms of unbroken law, for someone who was nevertheless trying to fit everything into a scheme that was designed by an understanding and good Creator, to someone who had read his grandfather's works and was well aware of Lamarck's ideas (Lyell conveniently gave a digest in the second volume of the *Principles*, intending to dissuade his readers of the attractions of evolutionism), there was only one answer to the problem. One simply had to argue that the birds and reptiles had come to the Galápagos, and then once there had changed in significant ways as they moved from island to island. Evolution had to be the key! This was evolution as fact.

But straightaway Darwin had his basic picture of the path of evolution. He was thinking of ancestors coming to the islands and then evolving as they moved around. This at once gives a treelike pattern to life's history. Not for Darwin was the upward parallelism of Lamarck—an aspect of the Frenchman's theory that, incidentally, Darwin quite missed. I have mentioned how, in Lyell's discussion of the French naturalist, he had mistakenly presented Lamarck's theory as a response to a progressive fossil record, that is, as a one-off phenomenon no doubt complicated by branching. Yet still there was the question of evolution as cause, and to show how he was far ahead of his grandfather in scientific sophistication—no evolution as pseudoscience for Charles Darwin—we find that the young naturalist now spent some 18 months searching systematically for an answer. His teachers at Cambridge had instilled in him the importance of causal thinking—after all, this was the achievement of the great Newton, and a biologist should aspire to no less (Ruse 1979).

An Evolutionist Because of His Religion

As with earlier evolutionists, it is surely true to say that Darwin became an evolutionist because of his religious beliefs, rather than despite them. The same is true of his path to causal understanding. Darwin was ever a Lamarckian believing in the inheritance of acquired characteristics, but he knew that this alone could not be adequate. One needed some overall *cause*—a kind of force equivalent to a Newtonian power. But it could not be any kind of force. Cuvier may not have been an evolutionist, but his legacy hung over everything anyone thought about the organic world. In particular, one had to pay attention to *function*. This was a given, even for the evolutionist. Not that Darwin wanted to dispute this. By the time of the *Beagle*'s return, he was thinking of himself as a professional scientist, and as such he knew that one might modify and build on Cuvier's legacy, but one ignored it at one's peril. Moreover, his own personal Cambridge theological training had likewise convinced him of the significance of a functional—a teleological—approach to organisms. It is here that natural religion or theology starts to become important.

Archdeacon William Paley

Actually, we have already encountered natural religion or theology at work in Darwin's thinking. When he worried about God's wisdom in creating separate species for each Galápagos island, he was appealing to the kind of Supreme Being that reason would dictate. But now natural religion was to become really important, a direct function of Darwin's having read at Cambridge the classic text on the subject: *Natural Theology* ([1802] 1819) by the Reverend Archdeacon William Paley (an Anglican clergyman). Paley gave the definitive version of the argument from design (for God's existence), also known as the teleological argument. He pointed out that in many respects the mammalian eye is just like a telescope—the lens, the way it focuses images, and so forth. But telescopes, argued Paley, have designers and creators. Hence the eye must have a designer and creator: the Great Optician in the Sky.

Design as Proof of God's Existence

Darwin no longer accepted Paley's belief that this designer had to be a miraculous intervener—all was to happen through unbroken law—but he accepted entirely Paley's premise that the eye seems as if designed. And more generally, Darwin agreed entirely with the theologians that the definitive mark of the living is that organisms seem not have been put together randomly, but that they seem as if they were designed. The features that help organisms to thrive, to survive and re-produce—features that go under the heading of "adaptations"—were for Darwin, as they had been for Cuvier and the natural theologians, things that bore all of the marks of intentionality and forethought.

The point is that any adequate evolutionary mechanism or cause had to be able to speak not just to change, but to change of a particular sort. It had to be able to speak to the evolution of adaptation, meaning it had to be able to show how designlike features come into being, even though—especially though—all was going to be done (by God as Darwin still thought) at remote control through regular laws of nature. The cause of evolution therefore had to produce design.

Darwin soon realized how this could be done in principle. He was perfectly stationed, living with his family in the heart of England, where the rural revolution was still in full swing. It had been necessary to produce such animals as cows and sheep and such plants as vegetables—especially the turnip, crucial for feeding overwintering animals—of far better quality than hitherto. Breeders had come to see that the secret lies in selective breeding: one chooses the animal or plant with the features that one most desires and one breeds from it, discarding all of the others. Fatter cows, shaggier sheep, fleshier turnips appear almost by magic, thanks to the selective skill of the professional breeder.

A Political Economy That Points the Way to Survival of the Fittest

But how is this to occur in nature? Finally, after months of searching, at the end of September 1838, Darwin read a well-known political-economic tract by yet another English Anglican clergyman. This time it was the *Essay on a Principle of Population*, by the Reverend Thomas Robert Malthus, the sixth edition of which (the edition read by Darwin) had appeared a dozen years earlier, in 1826. Here we see in action the precise point made above about Darwin's rearranging parts that he had received from others. Malthus's work was conservative and appealed strongly to the segment of society from which Darwin arose. The *Essay* argued that state welfare schemes are pointless—worse than pointless—because population numbers have always a tendency to outstrip the supplies of food and space. There is bound to be a struggle for existence, which can only get worse if one feeds and coddles the poor and destitute. Better by far to let them suffer at the immediate level: then they will be persuaded to work and support themselves and to practice prudence and temperance and to restrict their family sizes.

This was music to the ears to people like the Wedgwoods, whose manufacturing enterprises depended on low taxes—no large, state welfare bills to pay—and lots of cheap and desperate labor. If it could all be wrapped up in the guise of God's stern unbending laws, so much the better. Charles Darwin, however, took Malthus's ideas, standing them on their head. He generalized from population pressures among humans to population pressures occurring throughout the animal and plant world, arguing that numbers will always have the potential to outstrip food and space. Consequently, there will always be an ongoing struggle for existence (and more importantly, struggle for reproduction).

A cartoonist's vision of what might happen if the struggle for existence is relaxed

But far from this having conservative do-nothing, go-nowhere effects, it is the motive force required to fuel a kind of selection: a lawbound natural kind of selection throughout the living world, which will lead to permanent and significant change. Only a few organisms will be able to get through—to survive and to reproduce—and those that do will tend on average to be different from those that do not. Those that do survive and reproduce (those that later Darwin was to call the "fitter") will do so precisely because they have features that the losers do not have. They will be faster, stronger, sexier, and so forth. In time, this will lead to a full-blown evolution, and moreover, it will be evolution in the direction of adaptive advantage. This new mechanism, that Darwin was to call "natural selection," has the effect precisely of producing the designlike effects of which Cuvier and the natural theologians had made so much.

The Long Delay

Having a bright idea is one thing. Having a full-blown theory that will convince other people, especially doubters and critics, is quite another. In science, no less than in the fast-food business, what counts is the sizzle not just the steak. Darwin realized fully that he was going to have to work to put things together into a fully finished form that would be presentable to others. In the end, it took him 20 years to do this, which is cautious by anybody's standards. If nothing else, it shows just how much science has changed over the past century and a half and how it has become a collaborative big business. No one today could sit on an important idea for 20 years. Jim Watson and Francis Crick discovered the double helical shape of the DNA molecule in 1953. Can you imagine if they had concealed their finding until 1973?! Of course, they could not have done so. Someone would have scooped them. In any case, today most scientists are funded by governments and big business, unlike Darwin, who was supported by the family fortune. If you want to keep the grants coming, you had better come up with a steady stream of results.

In fact, I do not think that Darwin suspected that the delay would be anything like as long as it eventually proved. Within a year or two he had put things together in theory form—he wrote a 35-page outline in 1842, and then a full version of 230 pages in 1844. But a number of factors intervened. One was that Darwin fell very sick from some unknown illness. It slowed him right down. From being a vibrant young man who had braved the elements on the *Beagle* and through South America, he became a near invalid, wracked with headaches and other ailments. He and his increasingly large family spent long periods at spas and other places of treatment as vainly he searched for relief. He became a recluse, totally dependent on his wife for every minor item of everyday life.

This was not a man to take on the scientific community with a daring and dangerous new hypothesis. Especially given that in 1844 there appeared an anonymously authored evolutionary tract: *Vestiges of the Natural History of Creation*. This caused a huge sensation, being wildly popular with the general public, especially women. Almost naturally, all of the Oxbridge science professors who were Darwin's teachers and mentors took a leading role in opposition. Adam Sedgwick (1845, 1850), Cambridge Professor of Geology, evangelical Christian, and ardent catastrophist, led the attack with an 85-page critical review, followed by a 300-page Preface and a 500-page Afterword—all condemnatory of *Vestiges*—added to an inoffensive little 30-page essay on good conduct by undergraduates at university. Having suggested that the anonymous author had such low standards that it had to be a woman, Sedgwick drew back and denied that any member of the fair sex could have penned so vile a work.

In the same vein were the sentiments of David Brewster, Scottish man of science and biographer of Newton: "Prophetic of infidel times, and indicating the unsoundness of our general education, 'The Vestiges ... ' has started into public favour with a fair chance of poisoning the fountains of science, and sapping the foundations of religion" (Brewster 1844, 471). He knew wherein lay the trouble: "The mould in which Providence has cast the female mind, does not present to us those rough phases of masculine strength which can sound depths, and grasp syllogisms, and cross-examine nature" (p. 503).

In the light of all of this, Darwin wisely decided to remain silent. He buried himself in a massive project of barnacle taxonomy, letting a few selected friends in on the great secret. His reputation grew, meanwhile, both as a scientist and as a general man of letters, thanks to a wonderful travel book that he produced from the diary kept on his long journey from England. Darwin was a man known, loved, and respected by the Victorian public, and so it was perhaps no great surprise that in the summer of 1858 a young naturalist and collector, then in the Malay Peninsula, should have sent to Darwin of all people a copy of an essay that he had penned just after recovering from a malarial attack. Shocked beyond belief, Darwin read this piece, by Alfred Russel Wallace (1858), realizing that at last someone had hit on exactly the same ideas as he some 20 years earlier. Material was rushed into print, Darwin wrote frantically, and in the autumn of 1859, *On the Origin of Species by Means of Natural Selection or the Preservation of the Favoured Races in the Struggle for Existence* finally saw the light of day.

Darwin's Origin

Darwin later referred to his work as "one long argument," and this it was. He knew he had a selling job to do. Darwin was never that much interested in the actual path of evolution, although with hindsight we can spot some fascinating speculations in the ostensibly nonevolutionary work on barnacles (published in the early years of the 1850s). But he had to persuade people of the fact of evolution, and he hoped also to convince them of his mechanism or cause for evolution. Running these two tasks together and influenced, I might add, by some of the leading methodologists of his age, Darwin reasoned in two quite distinct ways.

First, he tried to persuade people of evolution through selection by analogy. He thought that if he could introduce people to something they already knew and accepted—in this case the success of breeders in transforming animals and plants through artificial selection—then he might be able to persuade them of something they neither knew nor accepted—full-blown evolution through natural selection. To this end, Darwin trotted out all sorts of examples of the triumphs of animal breeders—with pigeons, with horses, with cows, with sheep, and much more—and then hinted heavily that this is no less than we might expect to find in nature. In a way, therefore, practical agriculture together with the work of those who breed for pleasure (pigeons, fighting cocks, bulldogs) was serving as the experimental evidence for the case that Darwin was building.

When this done, Darwin moved to the second phase of his agreement. First, the struggle leading to selection was introduced and discussed. To this end, Darwin turned to Malthus ([1826] 1914) and argued that as the political economist had claimed there is a struggle for existence in the human world, so likewise there is a struggle for existence in the organic world.

> A struggle for existence inevitably follows from the high rate at which all organic beings tend to increase. Every being, which during its natural lifetime produces several eggs or seeds, must suffer destruction during some period of its life, and

THE ORIGIN OF SPECIES

BY MEANS OF NATURAL SELECTION,

OR THE

PRESERVATION OF FAVOURED RACES IN THE STRUGGLE
FOR LIFE.

By CHARLES DARWIN, M.A.,
FELLOW OF THE ROYAL, GEOLOGICAL, LINNÆAN, ETC., SOCIETIES;
AUTHOR OF 'JOURNAL OF RESEARCHES DURING H. M. S. BEAGLE'S VOYAGE
ROUND THE WORLD.'

LONDON:
JOHN MURRAY, ALBEMARLE STREET.
1859.

The right of Translation is reserved.

The title page of The Origin of
Species

during some season or occasional year, otherwise, on the principle of geometrical increase, its numbers would quickly become so inordinately great that no country could support the product. Hence, as more individuals are produced than can possibly survive, there must in every case be a struggle for existence, either one individual with another of the same species, or with the individuals of distinct species, or with the physical conditions of life. It is the doctrine of Malthus applied with manifold force to the whole animal and vegetable kingdoms; for in this case there can be no artificial increase of food, and no prudential restraint from marriage. (Darwin 1859, 63)

The Transfer of a Social Idea to a Biological Idea

As Darwin himself recognized, strictly speaking what he found in the organic world was not necessarily a struggle for selection or indeed for existence. Rather, there is competition of a kind between organisms for space and food, and this competition centers more directly on reproduction than it does on existence. But either way, what one has is some kind of transference of a social idea from the human realm to the biological realm, where Darwin made of it a biological idea.

Then after the struggle, Darwin moved to say that, given that there is constant new variation in populations, one will get a natural form of the selection practiced by animal and plant breeders. This leads to ongoing change, but change of a particular kind: change in the direction of adaptive advantage.

> Let it be borne in mind in what an endless number of strange peculiarities our domestic productions, and, in a lesser degree, those under nature, vary; and how strong the hereditary tendency is. Under domestication, it may be truly said that the whole organization becomes in some degree plastic. Let it be borne in mind how infinitely complex and close-fitting are the mutual relations of all organic beings to each other and to their physical conditions of life. Can it, then, be thought improbable, seeing that variations useful to man have undoubtedly occurred, that other variations useful in some way to each being in the great and complex battle of life, should sometimes occur in the course of thousands of generations? If such do occur, can we doubt (remembering that many more individuals are born than can possibly survive) that individuals having any advantage, however slight, over others, would have the best chance of surviving and of procreating their kind? On the other hand, we may feel sure that any variation in the least degree injurious would be rigidly destroyed. This preservation of favourable variations and the rejection of injurious variations, I call Natural Selection. (pp. 80–81)

As a substitute for the term *natural selection,* later editions of the *Origin* introduced the term *survival of the fittest.* This was an invention of the English philosopher, social scientist, and biologist Herbert Spencer. The term was urged on Darwin by Wallace as less misleading than *natural selection.* But, whatever name the rose was given, do not think that natural selection was the only causal mechanism endorsed in the *Origin.* Darwin always endorsed secondary mechanisms of evolutionary change. We have seen the acceptance of Lamarckian acquired characteristics. More important—indeed, the most important of all of the alternative mechanisms—was *sexual selection,* a kind of secondary mechanism to natural selection. This corollary, as one might call it, tacked onto Darwin's earliest (private) writings on selection, centers less on the struggle for existence and reproduction and more on the struggle for mates. There is a differential reproduction leading to evolutionary change as features that help in the mating game get selected and refined. Sexual selection clearly came by analogy from the breeders' world, where one selects, on the one hand, for physical characteristics like fleshier meat and shaggier skins (the practical agricultural side, the natural selection equivalent) and, on the other hand, for the kinds of characteristics that organisms have to attract mates and repel rivals (the pleasurable fanciers' side, the sexual selection equivalent). Things brought about by sexual selection include characteristics used for intraspecific fighting, like the antlers of the stag, and characteristics used for sexual attraction, like the peacock's tail. (As you will have realized, it was this sexual selection at the center of the dispute between Darwin and Wallace. Later we will see more on this topic.)

Selection in itself does not explain how a group of organisms might split into two. Most particularly, how one might start with one species and then end up with two species. (This process is known as *speciation*). The model that Darwin had always in mind was the speciation that occurred within the reptiles and the

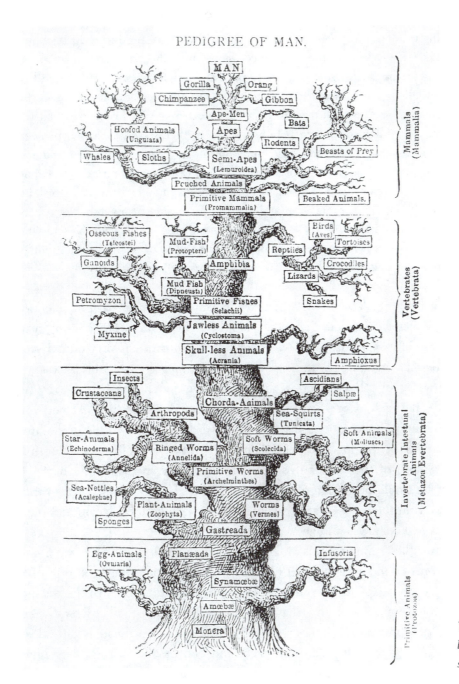

The tree of life (as drawn by Darwin's German supporter Ernst Haeckel)

birds on the Galápagos Archipelago. He needed something that would positively induce selection to tear groups apart, and he thought he had found it in what he called his *principle of divergence*. The essential idea here is that by breaking up into smaller groups, organisms can better exploit their ecological circumstances. Two groups with somewhat different adaptations can do better than one. Big finches can eat big nuts and plants, and small finches can eat seeds or insects.

From this, Darwin was led immediately into his well-known description of life's history, where he drew an analogy with a magnificent tree.

The affinities of all the beings of the same class have sometimes been represented by a great tree. I believe this simile largely speaks the truth. The green and budding twigs may represent existing species; and those produced during each former year may represent the long succession of extinct species. At each period of growth, all the growing twigs have tried to branch out on all sides, and to overtop and kill the surrounding twigs and branches, in the same manner as species and groups of species have tried to overmaster other species in the great battle for life. The limbs divided into great branches, were themselves once, when the tree was small, budding twigs; and this connexion of the former and present buds by rami-fying branches may well represent the classification of all extinct and living species in groups subordinate to groups. … As buds give rise by growth to fresh buds, and these, if vigorous, branch out and overtop on all sides many a feebler branch, so by generation I believe it has been with the great Tree of Life, which fills with its dead and broken branches the crust of the earth, and covers the surface with its ever branching and beautiful ramifications. (Darwin 1859, 129–130)

It hardly needs saying that if Darwin's mechanism of natural selection were to work, then he needed a constant supply of new variations coming into every population of organisms. Otherwise everything runs down very quickly to a ster-ile uniformity. In addition, these new variations must be heritable. If they are not, then however effective selection may be in one generation, it cannot pass on its results to the next. Here Darwin's genius rather deserted him. At best one got a compendium of speculations that would have done credit to his grandfather, and indeed many of the speculations—Lamarckism had a prominent role—were the same as those of the earlier evolutionist. As I have mentioned, Charles Darwin (not in the *Origin* but in later publications) floated ideas about the transmission of particles from the body to future generations, via the bloodstream and the sex or-gans. (This theory was known as *pangenesis*.)

After this somewhat unsatisfactory discussion, Darwin moved to a quick sur-vey of some of the difficulties of his theory, for instance, the evolution of features that are highly adaptive or complex. Then he was able to turn to the second major part of the *Origin*. It was here that Darwin really came into his own as he surveyed the different branches of biology showing how evolution through natural selection throws light on so many different areas and conversely, in turn, is supported by each and every one of these areas. One area of major interest to Darwin was that of instinct and behavior. Like many biologists of his era, he was absolutely fasci-nated by the social insects, particularly the ants and the bees. He was concerned particularly to show how their social characteristics, just as much as anything else, were things that could be explained by natural selection. "No one will dispute that instincts are of the highest importance to each animal. Therefore I can see no diffi-culty, under changing conditions of life, in natural selection accumulating slight modifications of instinct to any extent, in any useful direction" (p.243). As we shall see later, there was a lot more to the story than this, and perhaps Darwin was being overly optimistic in what he wrote. Indeed, he worried a great deal about how organisms can cooperate as tightly as they do in an ant's or bee's nest. However, as we shall see also, it was not until modern theories of heredity had been developed that the full story could be uncovered.

Geology and paleontology naturally got full treatment in The Origin of Species. On the one hand, Darwin was somewhat defensive. Like any evolutionist, he had to face the problem of the incompleteness of the fossil record. He had to show not only why he thought there would be few if any transitional forms but also why the fossil record starts so suddenly. The record does not go very gradually from the most primitive up to the most complex but starts off with a bang with really quite complex and sophisticated forms. (In fact, this is no longer quite true. In a later chapter, we will see that this problem has been remedied somewhat by new discoveries. Darwin, however, was driven to all sorts of speculations about how the early organisms would have lived where there are now seas, and how the weight of the land above them would have squashed their fossils to nothingness, and so forth.)

On the other hand, Darwin happily stressed the positive side to geology and paleontology. For all the problems, the fossil record does have a roughly progressive upward favor, which is what one expects given evolution. "The inhabitants in each successive period in the world's history have beaten their predecessors in the race for life, and are, in so far, higher in the scale of nature; and this may account for that vague yet ill-defined sentiment, felt by many paleontologists, that organisation has on the whole progressed" (p. 267). Moreover, Darwin was able to show that we find the more general and putative linking types of organisms lower down in the fossil record, and hence earlier. Conversely, the more specialized organisms come higher, and therefore later. This is just what one would expect if evolution were true. Darwin stressed also that once an organism has gone extinct it never reappears. Evolution through natural selection would lead one to expect this. On a theory of divine, miraculous, instantaneous creation, it is quite anomalous.

Moving on to geographical distribution, here (as you might expect) Darwin grew positively expansive. This was (and still is) always one of the really strong areas of biological inquiry supporting the evolutionist's case. For Darwin, given his Galápagos experience, it was an area of special importance, and naturally he made much of it. The Galápagos Archipelago itself gets a full treatment, and there is much discussion of oceanic islands in general. Then, following this, Darwin went quickly through a range of topics—classification, morphology, embryology, and rudimentary organs—showing how each and every one of these can be explained by evolution through natural selection, and conversely gives support to the mechanism. Embryology particularly got a detailed and vibrant discussion. Darwin was extremely pleased with his explanation of the fact that often organisms that are very different as adults have embryos that are very, very similar—humans and dogs, for instance. Darwin pointed out, using the analogy of artificial selection, that embryos have much the same selective environment and so are not ripped apart, whereas adults have different environments and so are driven apart by natural selection. In the world of animal breeders, no one cares much how the juveniles look. What really counts are the adults. "Fanciers select their horses, dogs, and pigeons, for breeding, when they are nearly grown up: they are indifferent

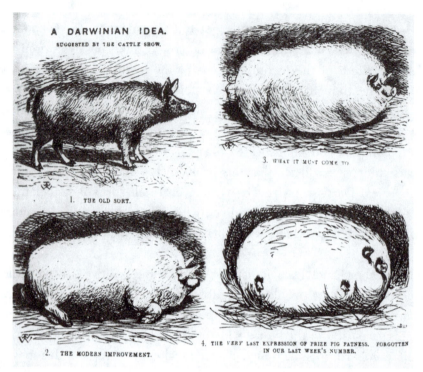

A cartoonist's thoughts on the power of artificial selection

whether the desired qualities and structures have been acquired earlier or later in life, if the full-grown animal possesses them" (p. 446).

And so the case was brought to completion. Truly, Darwin described what he had done as "one long argument" from beginning to end.

> From the war of nature, from famine and death, the most exalted object that we are capable of conceiving, namely, the production of the higher animals, directly follows. There is grandeur in this view of life, with its several powers, having been originally breathed into a few forms or into one; and that, whilst this planet has gone cycling on according to the fixed law of gravity, from so simple a beginning endless forms most beautiful and most wonderful have been, and are being, evolved. (p. 490)

Note incidentally the final word. You will often see it said that Darwin never used the word *evolution* in the *Origin of Species*—as if this tells you something significant, such as that Darwin did not really believe in and argue for evolution. This is misleading nonsense. The word *evolution* only came into the modern use— our use—around the time of Darwin, and he clearly had no strict objection to its use (Richards 1992). The more common language was that of "transmutation" or, Darwin's own preference, "descent with modification."

The argument of this second part of the *Origin* was as deliberate and as structured as was the argument of the first part. A good analogy can be drawn from criminal detection, where we have similar challenges to that faced by Darwin. Suppose you have a crime, let us say a murder. We have a suspect, but there were no eyewitnesses to give testimony. Let us suppose now that we find a similar crime, and there is good evidence that the suspect committed that crime. Now

A cartoon of Darwin showing how the status of humankind was the real problem and fear

the detective would feel much more convinced of the suspect's guilt. This is analogous to Darwin's use of the artificial selection analogy. No one saw evolution occur, but now we have something similar that produces a similar effect. The guilt may not be there, proven absolutely, but the detective/evolutionist feels that we are on the right track.

So what does the detective do next? He or she looks for circumstantial evidence. The search is on for clues. Lord Rake lies dead in the library, a dagger through his heart. The detective pins the guilt on the butler because of the bloodstain (the butler has a rare blood group that is found on the knife), because of the efficient way in which his lordship was killed (the butler was a commando in earlier life), because of the motive (the butler's daughter was seduced by Lord Rake), because of many, many more little bits of information and evidence. The clues point to the guilt and the guilt explains the clues. It is exactly the same for Darwin. The facts of instinct, the paleontological record, the distribution of organisms, morphology, embryology, systematics, all of these point to evolution through natural selection. Conversely, evolution through natural selection explains all of these facts of the biological world. Why the progressive fossil record, why the Galápagos distribution, why the homologies, and so forth. In Darwin's opinion, evolution through selection is proven "beyond reasonable doubt."

After the Origin

The year 1859 really was a watershed in the history of evolutionary thought. Before then, people knew about the idea of evolution (fact, that is) and would speculate about paths, and the Lamarckian inheritance of acquired characteristics was part of general lore (although of course most people did not think that it could cause more than minor effects and certainly did not think it would lead to the change of one species into another). But as a general belief, evolution was looked down upon by serious thinkers, and among professional scientists in particular it was regarded with scorn, and with not a little of the contempt and fear that had marked Cuvier's response and attitude.

If things were to change, something exceptional had to happen. And it did. Darwin was a person with stature: as a scientist and as a general figure in Victorian life. He could and did write extremely well—his travel book showed that and not a few remarked of the *Origin* that it had the same easy and inviting style. The reader is brought into the argument, never condescended to, and seduced by familiar examples and practices. Pigeon breeding—the classic working man's hobby (think of Andy Capp)—who could be scared of that? The main argument was convincing—after all, if not evolution, then how do you explain the Galápagos birds and tortoises, the homologies between the limbs of very different organisms, the facts of embryology? And on top of this, even if the old guard was never going to change, by 1859 Darwin had built up a group of younger scientists and supporters who would see that his ideas got full coverage and fair treatment.

It worked! At least, it worked in part. Virtually overnight people became evolutionists—evolution as fact that is. It was a little like the Hans Christian Anderson story of the Emperor's new clothes. Once the child had spoken—"But Daddy, he doesn't have any clothes on"—everyone said the same. Said they had known it all along! Once Darwin had spoken—"But evolution does occur"—everyone said the same. Said they had known it all along! I do not want to exaggerate. Of course, some of the established scientists and their friends never accepted evolution. Sedgwick went to his grave (in the 1870s as a very old man) denying and denouncing the vile doctrine. Mr. Gladstone, four times Prime Minister, classicist and churchman, never deviated from very old fashioned religious belief. But generally, evolution became the flavor of the decade.

How can one be so certain? The most compelling evidence is from surveys of magazines and newspapers and other such organs—especially religious publications, where one might expect to find opposition (Ellegård 1958). It is quite remarkable how quickly contributors accepted evolution and urged it on their readers. Obviously, liberal writers more quickly than conservative writers, but before long—certainly by 1865, and usually much earlier—evolution was the norm, the orthodoxy. What convinced me personally of the rapidity of the change was looking at the examination papers that students had to attempt at the universities. (In England, examinations are all printed up and copies kept on file.) In 1851, when Sedgwick was an examiner, one question read: "Reviewing the whole fossil evidence, show that it does not lead to a theory of natural development through a natural transmutation of species." But just a few years after the *Origin*, students were being told to assume "the truth of the hypothesis that the existing species of plants and animals have been derived by generation from others widely different" and to get on with discussing causes! When something is part of the standard undergraduate curriculum, you can be fairly sure that it is established truth.

Evolution as fact went over quickly and well. As I have said, Darwin was never really that interested in evolution as path. This was the job for the professional paleontologist, and he never really had aspirations in that direction. But what about evolution as cause—what about natural selection—Darwin's real pride and joy? Here, to be candid, he had a lot less success. No one wanted to deny it outright, but by and large people looked to other mechanisms. There were a number of reasons for this, and in the next chapter we shall be looking at what I think is the most significant of these. But for the moment, let me tell you that it soon became apparent that there were some fairly serious scientific problems with Darwin's theory—Darwinism as it was usually called, and as we shall now call it, meaning not just evolution as fact but the theory that makes natural selection the chief and central mechanism of change. Let me mention two such problems.

First, there were problems with Darwin's thinking about heredity—that is, about the means by which new variations come into populations (the "raw stuff" of evolution, necessary for indefinite change) and even more about the ways in which variations are passed on from generation to generation. It is clear that one must have some such theory of heredity or—as it is known today—of "genetics." Suppose natural selection picks out some feature as especially valuable—say a new predator comes along, and those potential victims that are darker than others are better camouflaged against the background and hence tend to be the "fitter." It does not matter how dark a successful organism may be, if it does not pass this feature on to its offspring. Without some way of preserving and transmitting good characteristics, the clock is put back in each generation and selection goes nowhere.

Unfortunately, at this point, Darwin took a false step. You might think that in each generation characteristics blend in with each other, and certainly this seems to happen sometimes. A black man and a white woman have brown children. Or you might think that in each generation characteristics stay distinct and entire. Eye color, for instance, or sex for that matter. You either have boy features or you have girl features. Depending on the way you go—what you take to be the norm—you explain the other side as anomalous or temporary or some such thing. For instance, if you think that sex is the norm (you take the "particulate" side), then you explain skin color as a temporary manifestation and probably the underlying causes are unaffected. The same obviously if you take the other side ("blending"), thinking skin color the norm and eye color and sex to be explained away.

The point about the two positions is that the particulate side lends itself immediately to a selection position—no matter what happens on the surface, the essential causes remain unchanged and preserved through the generations, always ready to show their effects again. Blending does not so lend itself—however good selection may be, in a generation or two a good new feature gets diluted right down and out to invisibility. One drop of black paint in a gallon of white makes little difference.

Let me be fair to Darwin. None of this is very obvious. There is no clear surface reason why inheritance should be particulate rather than blending, or conversely. It is a judgment call based on the overall background information that you have. But a false judgment is a false judgment, and this is where Darwin faltered: he went the wrong way, thinking that blending is the norm. And the critics pounced, showing how selection simply could not do what was needed. As we shall learn in time, it was in fact not until the twentieth century that the problem was resolved. But that was little consolation to Darwin in the nineteenth century!

The same is true of Darwin's other big scientific problem: although here one has rather more sympathy for Darwin, for he was the victim of the erroneous thinking of others (Burchfield 1975). The physicists, ignorant of the warming effects of radioactive decay, argued strenuously that the earth must be much younger than Darwin needed for the slow processes of natural selection. In the *Origin*, Darwin rather suggested that time was almost infinitely available, and now the physicists cut him down to a hundred million years maximum. This was still huge by what people had believed even a few years earlier. No one in the nineteenth century believed in the 6,000 years since creation that the sixteenth-century Archbishop Ussher had calculated from the genealogies given in the Bible. But one suspects that the catastrophists were thinking in terms of hundreds of thousands of years, a few million at most. The physicists were certainly being generous by earlier standards, but this was not enough for Darwin, who spent years trying to speed things up in the face of criticism. Ultimately, he had to tough it out and hope that something would turn up—which it did, but not until 25 years after he had died. Then and only then was radioactive decay discovered, its warming effects appreciated, the span of earth-life lengthened, and natural selection given full rein. (Today, there is lots of time. The universe is believed to be about 15 to 20 billion years old, the earth is about 4 and a half billion years old, and life started about a billion years later. More on this in due time.)

Genetics and geological history were somewhat technical questions. The big popular question of course was our own species: *Homo sapiens*. It is here that Wallace comes back into our story. In the *Origin*, Darwin made it clear that we are part of the scenario, but he did not want to make too much of this. "Light will be thrown on the origin of man and his history" (Darwin 1859, 488). This silence was deliberate. Darwin wanted to get the main ideas on the table before people got diverted straight into questions of human origins. Darwin knew that, once he published, his theory would be swallowed up by the "monkey question": a prediction that not only proved true but that was reinforced by the almost simultaneous first arrival from Africa, in Victorian England, of the gorilla.

At first, Wallace had been as hard nosed about humans as Darwin—"hard nosed" in a comparative sense, for Darwin certainly thought that God was creating everything including us, if through natural laws—but in the 1860s, the junior evolutionist became enamored with spiritualism. Wallace (1870) started to believe that there are occult forces ruling the world and responsible for our evolution. Selection alone could not do the job, because we humans are fundamentally different from all other organisms. Hence, Wallace's reluctance to allow that the peahen might have the same standards of beauty as humans. There must be something nonmaterial, nonphysical, about human evolution.

Darwin was appalled at Wallace's apostasy. But he realized that there was a challenge here to be met. The younger man had come up with all sorts of human characteristics that he claimed could not have been perfected by natural selection—human hairlessness, big brains, racial differences, and more. Darwin's response, really driving in the wedge between him and Wallace, was to make more and more of sexual selection. In *The Descent of Man*, published in 1871, Darwin argued that it is males competing and women choosing that makes us what we are. The bigger, stronger, brighter men got the pick of the women; the nicer, sexier, more sensitive women got the pick of the men (or picked by the men). My favorite example, if that is the right term, is Darwin's explanation of why Hottentot women have big backsides. Apparently they are lined up, and the warriors crouch down and squint along the line. She who protrudes farthest (*a tergo*) is she who is chosen by the bravest warrior.

Supposedly, all of this tells us not only why there are racial differences but also why there are sexual differences, and even why we humans are different from the brutes. The most beautiful women were desired by the strongest and most intelligent men, and the women in turn were happy to lie down for a good cause. If nothing else, by going with the flow they would then determine that their own sons would have precisely the features that make for male success in the struggle for reproduction. And if this were not enough, for good measure Darwin threw in a defense of capitalism! "In all civilized countries man accumulates property and bequeaths it to his children. So that the children in the same country do not by any means start fair in the race for success. But this is far from an unmixed evil; for without the accumulation of capital the arts could not progress; and it is chiefly thorough their power that the civilised races have extended, and are now everywhere extending, their range, so as to take the place of the lower races" (Darwin 1871, 1, 169).

In the face of this kind of argumentation, Wallace got somewhat short shrift. Apart from the fact that, by the 1860s and 1870s, appeal to spirit forces was simply not acceptable in forward-looking science, his innate assumption that all humans are likewise distinctive in their intelligence and other defining characteristics was much against the temper of the times. The "lower races" were certainly considered human and much above the apes—remember there had just been a bloody civil war in America over this very issue, with forward-looking liberals arguing that slavery is immoral precisely because all humans are in one family. But even the most liberal were generally not about to equate the Negroes and aborigines and Indians and native north Americans along with Europeans, or even southern Europeans and Slavs and Jews with Anglo-Saxons. Darwin's approach, which rested ultimately on competition and differences between peoples and with some coming out ahead of others, fit perfectly with what people already knew (or "knew"). Especially since Darwin made it very clear that the inhabitants of a small island off the coast of Europe are the apotheosis of human development.

It was no wonder the Victorians loved Charles Darwin and ended by burying him in Westminster Abbey. This was a man who spoke a language they could all understand. The coming of evolution was indeed a momentous event, but you should not think that it faced united opposition and hostility. In respects, it lent itself very nicely to the most standard and basic of societal beliefs and prejudices and was welcomed accordingly.

Further Reading & Discussion

There are many books on Charles Darwin, but start with *On the Origin of Species* (London: John Murray, 1859) itself. It is remarkably readable for a "great book." Darwin kept revising and rerevising his work, and by the time he had finished it had rather lost its original, clean, spare form. So try to get hold of the first edition. You can tell if you have found it, because in the fourth chapter where Darwin introduces natural selection, it is only in later editions that he adds Spencer's alternative name of "survival of the fittest." Harvard University Press has produced a facsimile of the first edition, and the Penguin edition is also of the first. This latter has an excellent introduction by the historian John Burrow.

For Darwin himself, the best single volume biography is by Adrian Desmond and James Moore: *Darwin: The Life of a Tormented Evolutionist* (New York: Warner, 1992). This is compulsively readable and simply packed with information about Darwin and his friends and family and the society in which he lived. Be warned, however, that it is written from a Marxist perspective that sees England on the verge of revolution and Darwin as a key figure in precipitating potential trouble. Darwin is portrayed as racked with guilt because, through his promotion of godless evolution, he was thereby betraying his own social class (hence the subtitle of the book). I think this is silly nonsense. I am not convinced that Britain was so very unstable. In any case, Darwin was well liked and secure in his position in society, and although his theory was truly revolutionary it never bothered him that he had it. Moreover, as I shall be telling you in the next chapter, he saw that his basic ideas rapidly became orthodoxy. This was a man who was buried in Westminster Abbey.

The best overall biography of Darwin is Janet Browne's two-volume biography, *Charles Darwin: Voyaging, Volume 1 of a Biography* (New York: Knopf, 1995) and *Charles Darwin: The Power of Place, Volume 2 of a Biography* (New York: Knopf, 2002). The first part particularly is terrific. It is thorough, judicious, well-written with keen insight into Darwin's psyche, and very detailed and knowledgeable about the pertinent science. The second part was bound to be a bit of an anti-climax, if only because any life would be an anti-climax after the *Beagle* voyage, the discovery of natural selection, and the publication of the *Origin*. However, I think that Browne compounds things a bit by simply offering a year-by-year account of Darwin's life. You can only take so much letter writing, spa visiting, and pool playing with the family retainer to while away the winter hours. I would have preferred to have had a more expansive treatment, looking at the reception of the ideas and that sort of thing. Immodestly, therefore, I am going to recommend my own books, which do try to tackle these issues. Start with my own general history of the whole Darwinian revolution: *The Darwinian Revolution: Science Red in Tooth and Claw,* 2nd ed. (Chicago: University of Chicago Press, 1999). It is particularly strong on the religious and philosophical factors in the revolution and has been reissued with a new afterword that discusses findings and interpretations since it first appeared 20 years ago. Then go on to the chapters in *Monad to Man* (Cambridge: Harvard University Press, 1996) that cover things after the *Origin*. Or if you do not have the time for this, read the next chapters of this book!

Chapter 3
Darwinism Explodes onto the Victorian Stage:
Evolution as Religion

Overview

In this chapter we explore how *The Origin*, Darwin's ideas on evolution, influenced not only scientific development but spilled over into the humanities as well. Taken up by historians, social and religious leaders and political theorists not only in Victorian England but throughout the western world, these were ideas seized upon with remarkable fervor by theorists wildly opposed to each other's movements. At times, *The Origin* seemed to be a one-size-fits-all theory used by friend and foe alike in our history of evolution wars.

It is hard to imagine a more perfect place for Darwinism to land in history than in Victorian England. Branches of science were developing at a rapid rate, and explorers and scientists were the era's rock stars. Advances in our understanding of the world, our place in it and our obligations to it, were advancing at a remarkable pace. In this chapter we will touch on the new branches of science (physiology, embryology, morphology to name just a few) that were being developed side by side and often competing with each other for new-found monetary support.

By mid to late 19th century, that evolution had become an accepted fact was yesterday's news in the scientific world. Where evolution continued to wreck havoc was in the secular and religious realms. Darwinism was seen as the keystone of progress, the cause of our ever-increasing upward development as espoused by Herbert Spencer. Evolutionists' belief in this progress and good became, for many, their new religion with worship in museums with the latest displays of fossil progression and skeletal 'missing links,' instead of in churches. It was, indeed, a brave new world.

Some evolutionists felt that this new discovery meant that societies should not help the poor, not establish barriers to trade, and not try to manage economics. This laissez-faire approach meant that the strongest nations with the best goods and strongest people would survive and rule as intended, and societies should not interfere with this progress.

In some cases these theories of an ever-increasing progress led to an interesting marriage of Darwinism and religion. Calvinism (American's Protestantism) belief in predestination meshed with Darwinism and survival of the fittest. It is in this harsh pre-ordained world of Calvinism that the stern laws of nature decided fates for all with God as the redeemer.

A socialist theory also based on Darwinism was on the rise, however, which purported that natural selection could operate for the good of an entire group as well as for individuals. It went on to say that the group could work together to overcome adversity that might befall individuals. This contributed to the development of the socialist movement, which had a profound effect on Russian revolutionary thought. And so, the evolution wars continued with each side taking up the cause for their own purpose.

But more damaging, by far, was the use of Darwinism as a way to justify war. Certain theorists maintained that progress depended on war, proclaiming, "Without war inferior or decaying races would choke growth." Such philosophy and social theories were used to buttress entry into World War I and later, in the mid-20th century, Hitler would twist evolution into a rage against non-Aryans with terrible consequence.

Finally, this chapter explores the age-old question of man and woman on the evolutionary scale. And this war will not be ended any time soon. Many used, and still use, evolution to support the belief that women are inferior to men. On the other hand, there were Victorian-era evolutionists who believed that the only way for the human race to achieve success and salvation was to put our hopes in the hands of the female of the species.

The Role of the Scientific Community

The work of the following theorists is discussed in this chapter. Short, biographical essays of these individuals appear in **Biographies** on page 607.

Samuel Wilberforce (1805–1873)
Herbert Spencer (1820–1903)
Thomas Henry Huxley (1825–1895)
Richard Owen (1804–1892)
Joseph Hooker (1817–1911)
Prince Petr Kropotkin (1842–1921)
Frederick von Bernhardi (1849–1930)

Setting the Stage

I should like to ask Professor Huxley, who is sitting by me, and is about to tear me to pieces when I have sat down, as to his belief in being descended from an ape. Is it on his grandfather's or his grandmother's side that the ape ancestry comes in?" And then taking a graver tone, [Samuel Wilberforce] asserted, in a solemn peroration, that Darwin's views were contrary to the revelation of God in the Scriptures. Professor Huxley was unwilling to respond: but he was called for and spoke with his usual incisiveness and with some scorn: "I am here only in the interests of science," he said, "and I have not heard anything which can prejudice the case of my August client." Then after showing how little competent the Bishop was to enter upon the discussion, he touched on the question of Creation. "You say that development drives out the Creator; but you assert that God made you: and yet you know that you yourself were originally a little piece of matter, no bigger than the end of this gold pencil-case." Lastly as to the descent from a monkey, he said: "I should feel it no shame to have risen from such an origin; but I should feel it a shame to have sprung from one who prostituted the gifts of culture and eloquence to the service of prejudice and of falsehood." (Huxley 1900, 1, 200–201)

A wonderful confrontation. This was the meeting of the British Association for the Advancement of Science, held in Oxford in 1860, the year after the *Origin* was published. Thomas Henry Huxley was clashing with Samuel Wilberforce (son of William Wilberforce, famous in England for leading the fight against slavery), a leader of the "high church" movement in the Anglican Church. This truly was a David and Goliath encounter, for Huxley was young and vigorous, a morphologist and paleontologist and now professor at the London School of Mines—a worthy home but with virtually no status whatsoever—and Wilberforce was old and established and important and occupying one of the most distinguished of bishoprics—at Oxford, no less, the city of the most venerable and powerful university in the realm. The defender of science, the "bulldog" who spoke for the new theory of evolution, battled the champion of the Church of England, speaking for all that was set and important and traditional. And as David slew Goliath, so Huxley's verbal slingshots left the bishop vanquished and speechless.

A wonderful confrontation and a wonderful story, told and retold by generations of evolutionists. I myself first became interested in Darwinism thanks to a graphic reinactment by my history master when I was a schoolboy some forty years ago. I still remember his striding about the room, smashing fist into hand as he made Huxley's rhetorical points. (He was a terrific teacher!) But probably more a myth than true, I am afraid, although (as these things tend to be) a very revealing myth for all that. Let us go back and set the scene, following the story through to the end.

Bishop Samuel Wilberforce of Oxford

Essay

Early 19th Century Britain

Start at the beginning of the century (Ruse 1979). Thanks to the French Revolution, compounded by the rise of Napoleon, Britain was in a conservative phase. But it was a country in tension, with the seeds of change germinated and sprouting. On the one hand, Britain—southern England particularly—was ruled, owned, and controlled by large, generally aristocratic landowners (identified with the Whig party), with the spaces in between belonging to smaller landlords, the squires or gentry (identified with the Tory party). Parliament consisted of an une-

Thomas Henry Huxley

lected house of peers (which included judges and senior bishops of the Anglican church) and an elected house of commons, although this latter was controlled by those who had power over the nomination and election of the memberships of parliament. In an age where the vote was open, tenants knew full well that the wishes of their landlords were paramount. The duke of Norfolk, for instance, was barred from taking his own seat in the House of Lords because he was a Roman Catholic. Nevertheless, through his holdings he controlled the occupancy of several seats in the House of Commons.

Laws tended to be very much in favor of those in power and naturally tended to reflect rural interests. Hunting was given full rein and poaching was heavily prosecuted. Most notorious of all the laws were the so-called Corn Laws, enacted at the end of the Napoleonic Wars (1803–1815). During the wars, thanks to the French navy, imports of corn (the term by which the English referred to wheat, not to the North American maize) had been difficult or impossible; so landlords had done well as their land was used for every last cultivated patch. Now with

*An ironworks, the epitome
of the Industrial Revolution*

supplies coming in and rents dropping, the government enacted laws that specified that imported corn would be subject to restrictive taxes unless locally grown corn reached a certain price. Thus rents were pushed back up again, to the delight of the landowners—which landowners incidentally included Darwin's teachers and mentors at Oxford and Cambridge, for they were all fellows (members) of colleges at those universities, and the colleges got their incomes from rents of very large rural holdings.

Victorian Era in Ushered in

But things were starting to change—things had to change no matter what the authorities in power wanted. I have spoken several times already of the Industrial Revolution. This started in the second half of the eighteenth century, particularly in the North and the Midlands, close to major supplies of coal and water and minerals. This change brought great wealth to many people, including landowners, but the interests of the industrialists and the landowners tended not to be the same. The industrialists, for instance, wanted cheap corn so that they did not have to pay high wages. They did not care if the materials for bread were imported from around the world. And they resented very much that they tended to be outside the corridors of power—seats in the elected house of parliament were not distributed equally. "Rotten boroughs" might have but a handful of voters—all controlled by some powerful interest—whereas a new city might have little or no representation at all.

In any case, not everyone could vote. Not women obviously. And by and large, not workers either. You had to be a property owner. And, not dissenters (Protestant, non-Anglicans) and certainly not Catholics. This latter was a grave injustice, particularly because of the Ireland question, where almost everyone (parts of the north excepted) was a Roman Catholic. At that time there was a united kingdom of Great Britain (England, Wales, and Scotland) and the whole of Ireland. This was a major factor. To give you some idea of how major, look at population numbers. Today, there are 60 million people in Britain, and 5 million in

Ireland. Then, at the beginning of the nineteenth century, there were 10 million people in Britain and already 5 million in Ireland—poor, rural, uneducated, not overly fond of the British, and (in the opinion of those British) appallingly superstitious and priest ridden. Right or wrong, the point is that the United Kingdom had a major fault line running right down it: the Irish Sea.

Most dramatic of all was the population explosion. For reasons that are still not fully understood, numbers started to climb at a high rate, and it was not only an absolute climb but one away from the countryside toward the towns. Just to give you some figures: between 1831 and 1851, London jumped from 1, 900,000 to 2, 600,000 people, Manchester from 182,000 to 303,000, Leeds from 123,000 to 172,000, Birmingham from 144,000 to 233,000, and Glasgow from 202,000 to 345,000. Between 1801 and 1851, Bradford grew from 13,000 to 104,000. With growth like this things have to happen. You have to have food, and law and order, and sewage, and education, and much much more—especially, you have to have the entertainments and occupations of a large, closely packed, urban group rather than the ways and means of traditional small village groups. No longer can you leave charity to the wives of the squire and the vicar. No longer can a couple of old women in the village give out remedies for everything from childbirth to cancer. No longer can literacy be a privilege of the spoiled few. You have to have a more modern society.

Society and its institutions did start to respond—slowly and unwillingly but inexorably. Catholics were emancipated—this did not mean that they could necessarily vote but that their religion did not at once exclude them. Some of the worst rotten boroughs were abolished and parliamentary seats given to major new urban centers. At the same time there were moves to reform education. New universities, starting with London, were being formed—University College started by Radicals and then Kings College started in response by Anglicans. The old universities had reform thrust upon them. They were forced, for instance, to offer science degrees. The Church itself was made to distribute a little of its wealth a little more equitably, with provision for the unchurched, new, urban areas. And as the century went on, reform also came to places like elementary education—something that was always tense given conflicting religious interests. The military and the nursing and hospital and medical professions generally showed that they needed change, especially after the appalling conditions that were revealed during the Crimean war. The civil service also had to start thinking in terms of a meritocracy, as it became clear that connection without talent and industry simply was not enough to run a modern country.

So it went through the nineteenth century, and as the needs arose the men (and sometimes, as in the case of Florence Nightingale, the women) rose up also to tackle and meet the challenges. One such person was Thomas Henry Huxley, born of a mentally distressed schoolteacher, who triumphed over his own personal demons and grew up to be the Cuvier of late Victorian Britain: the most important and influential scientist of his day (Desmond 1994, 1997).

Thomas Henry Huxley, a Professional Scientist

The contrast with Darwin is the most striking. Whereas the author of the *Origin* was born to upper-middle-class security, driven to work only as his ambition dictated, never once in his life ever having to worry about mortgage or school fees or that little extra cash for a house extension, Huxley had to make his own way from the beginning. Apprenticed to medical relatives, he started his rise through brilliant performances at Charing Cross Hospital. He joined H.M.S. *Rattlesnake* as ship's surgeon—significantly, whereas Darwin took his meals with the captain, Huxley ate with the midshipmen—and it was on his journeys through the South Seas that Huxley started to build his scientific reputation and career.

Daily, fishing up delicate marine invertebrates like jellyfish and sponges, Huxley dissected them, showing in wonderful detail their structure and morphology, and most especially the relationships between different forms. No one could be ignorant of or indifferent to teleology, the Cuvierian functional approach; but, from the beginning, Huxley was less interested in ends and workings and more interested in the very ways in which things are put together and how they are transformed from species to species. Hence, from the beginning, he was attracted to a biology that put an emphasis on similarities and isomorphisms: homologies. This of course was the approach of Geoffroy in opposition to Cuvier, but even more (especially by mid-century) it was the approach of a school of German biologists—*Naturphilosophen*—for whom homology was the defining mark of the living, far more than functionality (Gould 1977). Thus we find that, from the beginning, Huxley stood outside the tradition in which Darwin was trained and in which he excelled. It is characteristic of Huxley that, having determined the significance of German thought, he immediately set about teaching himself the language. Darwin was never able to do this.

Returning to England, Huxley rose rapidly through the ranks of science. He became a Fellow of the Royal Society—Britain's premier society for distinguished scientists—and got himself good jobs in London institutions. These did not carry the prestige of an Oxford or Cambridge post, but Huxley saw them as stepping-stones to the control of science as he envisioned it. At the same time he continued to establish himself as a master of the science of living form—morphology—as well as beginning to turn his gaze backward toward paleontology. The most powerful and influential man in these fields in England in the 1850s was Richard Owen, for many years an employee of the Royal College of Surgeons (Rupke 1994). Almost naturally, the touchy older Owen and the pushy younger Huxley fell out, and this set up a lifetime's rivalry. But Huxley had a strong capacity for friendship, and so he forged links with other younger scientists, like him determined to take over British science and convert it into the kind of university-based, professional, government-supported enterprise that they saw as the needed component of a modern forward-looking society. Darwin was linked to these scientists, especially through the botanist Joseph Hooker, and so it was natural that when Darwinism needed a champion at the British Association in 1860, Huxley was there to play the role.

Richard Owen

For a moment longer, however, let us leave evolution on one side. Huxley and his chums did take over British science. By the 1860s and 1870s, they controlled the Royal Society—the presidency, the secretaryship, and the like. They got plum university posts and saw that their students got the same; although Huxley himself never left London and turned down offers from Oxford and Cambridge. They set and marked the examinations. They influenced elementary and secondary education, seeing that science got a firm foothold. They invaded the civil service and insisted that there be a place for science—and for properly trained scientists. They started journals and supported the efforts of others in this direction. Early issues of *Nature* owed much to Huxley. They took over the museums and much, much more. A little dining club started by Huxley and friends—the X-Club—became the very center of the English scientific establishment.

But what sort of science did these men want for their society, or rather what sort of biological science did Huxley want? Here we start to get some very interesting answers, which have a surprising significance for our tale (Ruse 1996).

Joseph Hooker

There were two branches of science particularly that caught Huxley's attention and concern. One was physiology, the study of the workings of organisms: a field in which Huxley himself was not a great practitioner, for it puts a premium on experiential technique and expertise, not something in which he shone. But through his students, especially H. N. Martin in London (and later at Johns Hopkins in Baltimore) and Michael Foster in Cambridge, Huxley supported and encouraged the science. Moreover—and this is absolutely crucial when you are founding and building a professional science—he found cash for its practitioners, and students for its teachers, and jobs for its students. In particular, he persuaded the medical profession—desperate to start curing rather than killing patients, and no less desperate to exclude pretenders—that physiology was just the training required for would-be doctors. And, the message finding very receptive ears, physiology was off and running.

Morphology was the other area of science that found professional favor with Huxley. This was indeed his own field, and here it was the teaching world that was the object of attack and persuasion. Huxley argued that a modern society puts aside such useless subjects as Latin and Greek and takes up science, morphology in particular. He was forever trumpeting the moral virtues of individual empirical experience. There is an intentionally biblical echo to his most famous dictum: "Sit down before fact as a little child, be prepared to give up every preconceived notion, follow humbly wherever and to whatever abysses nature leads, or you shall learn nothing" (Huxley 1900, 1, 219). And, to further his ends, we find the Hux-

H. G. Wells, the novelist who trained under T. H. Huxley to be a schoolteacher

ley and his associates not only taught full courses in morphology but that they started summer schools for teachers, where the message could be passed on, as well as encouraging all that they could to take up cudgels on behalf of the science. H. G. Wells, the novelist, is probably the most famous product of the Huxley system—his fascination with science shows right through his writings—but he was one of many.

Physiology and morphology. Where does evolution fit into all of this? Well, of one thing you can be absolutely certain. Thomas Henry Huxley was a fanatical evolutionist. This was not always the case. Early on he had been against the idea, and when he returned to England one of his first publications was an absolutely savage review of a later edition of the *Vestiges of the Natural History of Creation*, originally published in 1844. (Although he knew that this was not really true, Huxley rather pretended that Owen might have been the author and went at the task with extra zeal!) But then like Saint Paul, also a convert to a belief that hitherto he had rejected, Huxley swung round and became a total fanatic on the subject of evolution. He too had his Romans and his Ephesians and his Corinthians and he preached and wrote accordingly. A brilliant showman and rhetorician, he knew precisely both the weak points of the opposition and all the flashy persuasive ideas that support evolution. Shortly after the *Origin* was published, the first full skeletons of archeopteryx—the reptile-bird—were uncovered in Germany. Given the propaganda value of the find, Huxley brought the discovery to the public podium, complete with diagrams and illustrations (Huxley [1868] 1898). (In those preslide days, it helped mightily that Huxley was a brilliant blackboard artist.)

Fossil finds bridging different kinds of organisms were important. But these bridging fossils—known as *missing links* (until they were no longer missing!)—were not the most important focus of evolutionary studies. This honor was taken by our own species, *Homo sapiens*. Realizing at once that the most pressing question was human evolution, Huxley hammered away incessantly on our likeness to the apes and to our simian ancestry. He wrote a little book on the subject (*Evidence as to Man's Place in Nature*), he lectured on the subject, he discoursed at length on the subject—applying special attention to the recently discovered Neanderthal remains—and all of the time, he constantly inflated the mythic memory of the encounter with the Bishop. Why indeed should not his grandmother and grandfather be lower ape forms? After all, it is as dignified to be modified monkey as modified mud. Huxley's opponents complained, sometimes bitterly, that he misrepresented them and painted them into far more conservative positions than they truly held—in fact, if you look at Bishop Wilberforce's (1860) written review of the *Origin* you will find that although it is critical it is anything but negative. Huxley knew full well, however, that to make an effective positive case you need to portray yourselves as fighting forces of reaction and prejudice and succeeding only against great odds and in spite of gross knavery and trickery.

But what about evolution as a science? Did Huxley think of it as a field like physiology or morphology, that could be developed as a field of professional endeavor? Now it is certainly the case that some people thought this, especially German biologists influenced by Darwin's great supporter and enthusiast in that land, Ernst Haeckel. He promoted evolutionism tirelessly and built around himself at Jena University and elsewhere a group who worked hard to make of the subject a professional discipline. In fact, for all the talk, it was not terribly Darwinian, for no one was much interested in natural selection, and the main emphasis was on that very part of the enterprise that Darwin himself had rather neglected, namely the tracing of paths or phylogenies. Haeckel himself was responsible for the notorious "biogenetic law," which states that ontogeny (the developmental path of the individual) "recapitulates" phylogeny (the developmental path of the group). Using this as a tool, and working with the ever-increasing fossil record, he and his followers strove mightily to map out the details of life's evolutionary history and to start the long and laborious task of filling in the details (Haeckel 1866; Bowler 1996; Richards 2008).

Huxley took some interest in this work, and his younger followers and students—most notably the leading end-of-the-century morphologist E. Ray Lankester—worked hard at this German-inspired activity. But by and large this was not the use or role at all for which Huxley intended evolution. For all that he took proudly the label of "Darwin's bulldog," and for all that I am sure that Darwin himself wanted to see evolution as a thriving professional discipline like morphology and physiology, Huxley was essentially uninterested in promoting evolution in this wise. His lectures to his students, for instance, would be spread over two

Ernst Haeckel

years and would take over a hundred and fifty classes, not to mention practica where one would be dissecting specimens. They were marvels of detail and instruction—Huxley was a brilliant teacher. Evolution would be lucky if it got half a lecture! Natural selection five minutes!

Amazing but absolutely true: "One day when I was talking to him, our conversation turned upon evolution. 'There is one thing about you I cannot understand,' I said, 'and I should like a word in explanation. For several months now I have been attending your course, and I have never heard you mention evolution, while in your public lectures everywhere you openly proclaim yourself an evolutionist'" (Huxley 1900, 2, 428). This was a question by a puzzled student to his great teacher.

Why the silence? There were negative and positive reasons. Negatively, as a scientist himself—as a morphologist, and later as a paleontologist—Huxley really had little need of natural selection or any other cause, and to be quite frank evolution itself (evolution as fact) was not that pressing. His specimens were all dead and on the dissecting table by the time he got to them, so natural selection did not do much for him. Here Huxley was in a very different situation from students working on questions to do with ecology and behavior. You might nevertheless think that evolution as such had to be important, because (in paleontology particularly) one is dealing with change through time. But although developmentalism was certainly crucial, evolution in the sense of a natural (that is, lawbound) connected succession, from one form to another, was not essential. In fact, following Cuvier both in time and commitment, much of the record as known in Huxley's day had been worked out by nonevolutionists (Bowler 1976). They saw change, but they saw change that was God driven, without connected succession. Rather a series of miracles. Now, as we shall see, this was not for Huxley. However, as a scientist, the paleontological succession was all he needed. Even if one invoked embryological analogies thanks to the biogenetic law, well, this approach too had in essence been formulated, pre-Haeckel, by people violently opposed to evolution!

On top of these negative scientific factors, there were the negative social factors. Huxley just could not see how one could find cash support for evolutionary studies as a professional science. Evolution did not help the physician, and in schools it was certainly going to be regarded with suspicion. And it was an absolutely key part of the strategy of Huxley and friends, as they worked to establish power in Victorian society, that they seem even more honest and conventional and moral than anyone else. They were pushing things regarded with doubt and misgiving, so they themselves had to be purer than pure. Huxley knew and liked the novelist George Elliot, but since she lived unmarried with a man, he would not allow her to visit his wife and children at his home.

A place was found for evolution, however, and this starts to push us toward the positive reasons for Huxley's attitude. Museums welcomed evolution into their halls. As we shall see more fully in the next chapter, museums were developing in a major way as places of instruction and entertainment as the century drew to a close—the British Museum (Natural History) and the American Museum of Natural History and others—and evolution had a natural role to play here and could find its support. For what Thomas Henry Huxley wanted positively of evolution was a popular science, a kind of metaphysics, or secular religion if you like—one that could be used to challenge and substitute for the conventional religion of Christianity, which he saw embedded in society and standing in the way of those many reforms he and his fellows were attempting.

Hence, in a fashion, Huxley stood right in the tradition of Erasmus Darwin and Lamarck, except he himself had dropped the deism and evolution was now a respectable doctrine, no longer revolutionary, and being used to reform society rather than break it or overthrow it. Which brings up the question of progress. For the earlier evolutionists, what really counted was that evolution represented progress, in all of its various manifestations: progress against Christian providentialism and progress as a philosophy that represented everything for which the ev-

Herbert Spencer

olutionists stood. Now this was likewise important for Huxley and his friends and associates and students. They too wanted to promote progress, and they too looked to evolution as the ideal vehicle. But here they felt they had to go beyond the work of Charles Darwin. It is true that Charles Darwin himself believed in progress, but it was given only a limited role in the *Origin*. A gap had therefore to be filled, and fortunately there was at hand the man of the hour. For real faith—faith slopping right over into fanaticism—no one could hold the candle to Darwin's and Huxley's fellow Englishman and ardent evolutionist, Herbert Spencer. He lived and breathed and wrote—at very great length in one long volume after another—the subject of upward development, from the simple to the complex, from the blob to the human. Sometimes change takes a break—it achieves a point of "dynamic equilibrium"—but then something disrupts the balance and we are off upward again.

Spencer was truly Mr. Progress and so it was he far more than anyone else who became Mr. Evolution to his countrymen (Richards 1987; Ruse 1996). Even the way in which Spencer wrote of the topic, from the undifferentiated or what he called the "homogeneous" to the thoroughly mixed up or what he called the "heterogeneous" had a very Victorian ring to it. Progress was not just a biological or a social phenomenon: it was an all-encompassing world philosophy.

Now, we propose in the first place to show, that this law of organic progress is the law of all progress. Whether it be in the development of the Earth, in the development of Life upon its surface, in the development of Society, of Government, of Manufactures, of Commerce, of Language, Literature, Science, Art, this same evolution of the simple into the complex, through successive differentiations, hold throughout. From the earliest traceable cosmical changes down to the latest results of civilization, we shall find that the transformation of the homogeneous into the heterogeneous, is that in which Progress essentially consists. (Spencer 1857, 2–3)

Evolution therefore took on the role of a substitute religion for Christianity, and whereas Christians worshipped in churches, evolutionists worshipped in museums, where one found grand displays intending to illustrate and confirm the faith.

Huxley, a close friend of Spencer (they were both members of the X-Club and it was in fact Spencer who in the 1850s first started to persuaded Huxley that evolution as an idea makes good sense), bought entirely into the view of evolution as secular religion. He preached the gospel nonstop, from every public platform he could find—at learned societies, before groups of fellow savants, in workingmen's clubs—traveling far and wide, including a highly successful trip to North America. In every sense of the word, Huxley was the Saint Paul of the movement, although later in his life the press took to calling him "Pope Huxley." There was even a Judas Iscariot of the movement. St. George Mivart (1871), Catholic convert and student of Huxley, was seduced by the Jesuits and became the most bitter critic of the Darwinian establishment: labeled "not quite a gentleman" for an intemperate attack on one of Darwin's sons, he was excluded from positions of power and comfort.

Social Darwinism

A good and full religion has a moral code, directives that it gives to its acolytes. "Love your neighbor as yourself." "Honor thy mother and thy father." "Do not lust after the wives of other men." Evolutionists took very seriously, as part of their system, this need for obligation. This led to the full development of what came to be known as Social Darwinism—a moral code based on evolution—although truly it would be better known as Social Spencerianism. The way in which the directives were obtained were fairly simple and direct. One ferrets out the nature of the evolutionary process—the mechanism or cause of evolution—and then one transfers it to the human realm (if this has not already been done), arguing that that which holds as a matter of fact among organisms holds as a matter of obligation among humans. (There will be much more on this in Chapter 11.)

Take the case of Herbert Spencer. Several years before Darwin published—although some considerable time after Darwin made his own discoveries—Spencer (1852) recognized the significance of the struggle for existence for human population development. He saw clearly that natural urges to reproduce would bring on a differential survival and reproduction of organisms within and between populations, and that this could lead to permanent biological change. Always more interested in humans than in the rest of the organic world, Spencer at once drew the implications for our species. Take, to use his example, the different natures and behaviors of the Irish and the Scots. In true Victorian fashion, Spencer argued that even though the Irish have lots of children, because of their lazy, indolent ways they are going to fail in life's struggles. The far more frugal and hardworking Scots will succeed and thrive, as indeed they do. Change in human nature will ensue.

From this satisfying biological inference, Spencer made an easy transition to economics, arguing that just as biology favors an unrestricted struggle and consequent selective success, so also economically this is the way that one should go for success. In particular, one should promote policies based on extreme laissez-faire socioeconomics. States should stay away from the activities of people following their own self-interest. In no way should politicians try to regulate or otherwise control unrestricted competition. Spencer felt, with some considerable regret, that mid-Victorian Britain was far from the ideal libertarian society, but he thought that if it was to continue and to thrive and to succeed, then it should strive to maximize to the fullest extent its citizens' freedoms to pursue their own interests and ends. The state should be helping people to do what they want to do rather than acting as a deterrent and barrier.

> We must call those spurious philanthropists, who, to prevent present misery, would entail greater misery upon future generations. All defenders of a poor-law must, however, be classed among such. That rigorous necessity that, when allowed to act on them, becomes so sharp a spur to the lazy, and so strong a bridle to the random, these paupers' friends would repeal, because of the wailings it here and there produces. Blind to the fact, that under the natural order of things, society is constantly excreting its unhealthy, imbecile, slow, vacillating, faithless members, these unthinking, though well-meaning, men advocate an interference that not only stops the purifying process, but even increases the vitiation—absolutely encourages the multiplication of the reckless and incompetent by offering them an unfailing provision, and *discourages* the multiplication of the competent and provident by heightening the prospective difficulty of maintaining a family. (Spencer 1851, 323–324)

Spencer could sound positively brutal about those who would help the unfortunate within society: "If the unworthy are helped to increase, by shielding them from that mortality which their unworthiness would naturally entail, the effect is to produce, generation after generation, a greater unworthiness" (Spencer 1873 [1961], 313). And one can find similar sentiments in the writings of Spencer's followers. Listen, for instance, to the turn-of-the-century American sociologist William Graham Sumner, who makes the converse case:

> The facts of human life … are in many respects hard and stern. It is by strenuous exertion only that each one of us can sustain himself against the destructive forces and the ever recurring needs of life; and the higher the degree to which we seek to carry our development the greater is the proportionate cost of every step. For help in the struggle we can only look back to those in the previous generation who are responsible for our existence. In the competition of life the son of wise and pru-

dent ancestors has immense advantages over the son of vicious and imprudent ones. The man who has capital possesses immeasurable advantages for the struggle of life over him who has none. The more we break down privileges of class, or industry, and establish liberty, the greater will be the inequalities and the more exclusively will the vicious bear the penalties. Poverty and misery will exist in society just so long as vice exists in human nature. (Sumner 1914, 30–31)

But there is much more to the story than this. Quite apart from the fact that Spencer had somewhat ambiguous feelings about natural selection—feelings shared by just about everyone else but Darwin—if anything Spencer's ethical theory was due chiefly to his background of Protestant nonconformism, which saw the Poor Laws and the like as keeping much of the population in a state of perpetual poverty and dependency. Spencer (rightly) saw establishment Christianity as serving the ends of the rich and powerful (represented by the Anglican church), who inherit their wealth and status and who have no fear of the threat of competition from the more gifted and industrious (Spencer 1904; Duncan 1908). Spencer's evolutionism certainly moved in to confirm and support his alternative, supposedly secular, Social Darwinian views, but there was no simple deduction of ethics from biology. It was as much a question of one branch of Christianity set against another branch as it was a question of science set against all of Christianity.

Confirming this claim that there were strong Christian elements lurking in even the most ferocious-sounding Social Darwinian systems, it is clear (from statements and from actions) that it was never the intent of Spencer or his followers to deny the importance of individual charity. Take two of Spencer's more notorious disciples. John D. Rockefeller spent the first part of his life building up the vast petroleum company Standard Oil and the second part of his life fighting the federal government as it tried to break up the monopoly he had established over so vital a national resource as fuel oil. From his childhood, Rockefeller had tithed to his church, and he gave seriously and deeply to charity. The University of Chicago would never have become the world institution that it is without Rockefeller munificence.

The same generosity is true of Andrew Carnegie, who came from Scotland and made his fortune by founding and building U.S. Steel. He always claimed that no man should die rich, and he gave huge amounts of money directed toward the founding of public libraries. Carnegie's charity was an immediate function of his reading of Spencer, a reading that stressed the positive rather than the negative side of laissez-faire. Carnegie (like other industrialists) was proud of what he had done, thinking it a credit to his own abilities rather than a black mark against the lesser abilities of others. That poor but gifted children might likewise have the opportunity to develop and use their talents, Carnegie wanted to found public places of instruction and learning where one might go to better oneself. A public library, seeded by Carnegie and then supported by the community, was a perfect outlet for his philanthropic drive (Bannister 1979; Russett 1976).

Alternatives to Laissez-Faire

It is interesting to note just how often the proponents of a Spencerian-inspired Social Darwinism had childhoods that were not only deeply and sincerely Christian but were from that branch of Protestantism indebted to Calvinism, in America particularly the Scottish version known as Presbyterianism. The great sixteenth-century religious reformer John Calvin had given prominent place to the doctrine of predestination: God has ordered things according to His stern, unbreakable laws, and thus the fates of all are decided before we are even born. Some are predestined to be saints, and others sinners. Many Victorians who thought this way embraced evolutionism with enthusiasm—a theory that stressed the force and power of law was music to their ears. Thus Dr. James McCosh, Scottish-born president of Princeton University and one of the most influential churchmen and educators in America in the second half of the nineteenth century, claimed that there are no accidents, all is foreordained:

> It is in the very constitution of things. It is one of the most marked characteristics of the state of the world in which our lot is cast. It is, in fact, the grand means by which the Governor of the world employs for the accomplishment of his specific purposes, and by which his providence is rendered a particular providence, reaching to the most minute incidents and embracing all events and every event. It is the special instrument employed by him to keep man dependent, and make him feel his dependence. (McCosh 1882, 164)

The belief that some are chosen by nature to be successes and some are doomed to failure, that not only are all humans not born equal but that this is a right and proper state of affairs, was to the likes of Rockefeller and Carnegie as much a matter of theology as it was of scientifically based philosophy.

At first this was true also of Thomas Henry Huxley. He spoke of himself as a "scientific Calvinist," meaning that he thought that the stern laws of nature decided the fates of us all, determining some to succeed and others to fail. However, increasingly, as Huxley and his friends succeeded in their aims of changing and reforming Victorian Britain, he was drawn metaphysically toward a position where an individual's own free will and efforts are the true determinants of life (Huxley 1893). Socially, despite his continuing friendship with Herbert Spencer, he pulled away from laissez-faire. For the mature Huxley, ethical success lay not in a conformity with and acquiescence to nature's laws. It lay rather in fighting such laws and the evil consequences to which they lead. At the same time, Huxley saw the virtues of a functioning civil service and of intervention by the state into such things as education and medicine and the military and the like (Huxley 1871 [1893]). (See also Jones 1980.)

One senses that for Huxley there was always a conflict within: his enthusiasm for naked evolutionism, which he always interpreted as based on a brutal struggle, battled with his innate decency and his conviction that it is our ultimate moral obligation to fight those vile personal attributes that come in a package deal as part of our biology. No such worries ever troubled the happy thinking of Alfred

Prince Petr Kropotkin

Russel Wallace. As a boy, he had been taken by one of his older brothers to hear the Scottish mill owner and early socialist Robert Owen (Wallace 1900). He always looked back to this moment as a real turning point and, for the rest of his very long life, Wallace was ever an ardent socialist (Wallace 1905; Marchant 1916). Against Darwin, he believed that selection can operate for the good of the group as well as for the individual, and he thought that evolutionary success would be something that promoted the harmonious whole over the selfish individual.

Similar sorts of views appealed to the exiled Russian Prince Petr Kropotkin. He claimed that there exists between all animals, including humans, a natural sense of sympathy, something that he called *mutual aid*. Kropotkin did differ from Wallace in having little or no time for the state whatsoever. One suspects that his anarchism owed as much to the fact that he hailed from czarist Russia, one of the nineteenth century's most repressive societies, as it did to anything in evolution. But one should not dismiss entirely the influence of the particular spin that Kropotkin put on the evolutionary process.

The terrible snow storms that sweep over the northern portion of Eurasia in the later part of the winter, and the glazed frost that often follows them; the frosts and the snow-storms that return every year in the second half of May, when the trees are already in full blossom and insect life swarms everywhere; the early frosts and, occasionally, the heavy snowfalls in July and August, which suddenly destroy myriads of insects, as well as the second broods of birds in the prairies; the torrential rains, due to the monsoons, which fall in more temperate regions in August and September—resulting in inundations on a scale that is only known in America and in Eastern Asia, and swamping, on the plateaus, areas as wide as European States; and finally, the heavy snowfalls, early in October, which eventually render a territory as large as France and Germany, absolutely impracticable for ruminants, and destroy them by the thousand—these were the conditions under which I saw animal life struggling in Northern Asia. They made me realize at an early date the overwhelming importance in Nature of what Darwin described as "the natural checks to overmultiplication," in comparison to the struggle between individuals of the same species for the means of subsistence. (Kropotkin 1902 [1955], vi–viii)

To survive, it was necessary to work together against the elements.

In the animal world we have seen that the vast majority of species live in societies and that they find in association the best arms for the struggle for life: understood, of course, in its wide Darwinian sense—not as a struggle for the sheer means of existence, but as a struggle against all natural conditions unfavourable to the species. The animal species, in which individual struggle has been reduced to its narrowest limits, and the practice of mutual aid has attained the greatest development, are invariably the most numerous, the most prosperous, and the most open to further progress. ... The unsociable species, on the contrary, are doomed to decay. (Kropotkin 1902 [1955], 293)

War and Peace

By now you may be convinced that I must be exaggerating. In my eagerness to show to you a different, more friendly side to Social Darwinism, in my urge to counter belief in the extreme laissez-faire socioeconomic doctrine that so many people today associate with Herbert Spencer, I have to be ignoring much that is true and pertinent. Surely there was a side to Social Darwinism that stressed conflict and violence. Surely there was a side to Social Darwinism—perhaps more characteristic of continental thought, of German thought in particular—where violence and conflict and, ultimately, all-out warfare were seen to be the right and proper expressions of evolutionary principles. Indeed, can one not say that, in some ways, Social Darwinism was a major motivating force that led to World War I, not to mention the hateful systems that followed in its aftermath? I refer of course to Soviet communism and to German national socialism. This, I might add, is a particularly popular line of thinking among American evangelicals today: Darwin to Hitler in a few easy steps. (See, for example, Weikart 2004.)

In fact, as with social and economic questions, matters rather mixed. Certainly one cannot and should not exonerate Social Darwinism from all responsibility for the monstrous happenings and philosophies of the century that has just passed. One does find people who argued strongly that war and violence are natural states of affairs and who happily expressed their sentiments in evolutionary or pseudoevolutionary language. One enthusiast claimed that war is "a phase in the life effort of the State towards complete self realization, a phase of the eternal nisus, the perpetual omnipresence strife of all beings towards self fulfilment" (Crook 1994, 137). Even though writing like this probably owes as much to the early-nineteenth-century German philosopher Hegel as it does to Charles Darwin, it was not a sentiment expressed by one person alone. Others put matters in similar language: "Man has always been a fighter and his passion to kill animals ... and inferior races ... is the same thing which perhaps in the dark past so effectively destroyed the missing link between the great fossil apes of the tertiary and the

Friedrich von Bernhardi

lowest men of the Neanderthal type. All these illustrate an instinct which we cannot eradicate or suppress, but can best only hope to sublimate" (pp. 143–144).

Perhaps with more direct input, there was General Friedrich von Bernhardi, pushed out of the German army because he was signaling a little too bluntly the General Staff's intentions, and leaving no place for the imagination in his best-selling *Germany and the Next War* (1912). "War is a biological necessity," and hence: "Those forms survive which are able to procure themselves the most favourable conditions of life, and to assert themselves in the universal economy of nature. The weaker succumb." Progress depends on war: "Without war, inferior or decaying races would easily choke the growth of healthy budding elements, and a universal decadence would follow." And, anticipating horrible philosophies of the twentieth century: "Might gives the right to occupy or to conquer. Might is at once the supreme right, and the dispute as to what is right is decided by the arbitrament of war. War gives a biologically just decision, since its decision rests on the very nature of things" (von Bernhardi 1912, 10).

However, countering this kind of writing, one finds that there were many who argued that war and violence are, if anything, the antitheses of evolution, especially inasmuch as one thinks that the course of evolution can and must be progressive. Herbert Spencer himself spoke eloquently to this end. As always with Spencer, the Christian training was never far from the surface—significant here was surely the fact that spicing the nonconformist elements in his intellectual broth was a large pinch of Quakerism—but there were other factors also that led him to deplore militarism. As one who was keen on free trade and open competition, Spencer had little or no tolerance for intersocietal rivalries. Quite properly he saw them as major barriers to such trade. Moreover, he deprecated strongly the arms races that began, at the end of the nineteenth century, to obsess and burden countries like Britain and Germany. He thought that expenditure on such things as ever bigger and more powerful battleships was an appalling waste of money and resources. Far better that these be spent on peaceful things. In these sentiments, Spencer was far from alone. In fact, some of the most important relief work done during World War I came at the hands of evolutionists. They thought that only by trying to ameliorate the appalling consequences of conflict could one have any possible hope of rescuing the desired upward progress of evolution from the degenerate state into which, sadly, it had fallen.

As always, evolutionism's relationship with people's actions and beliefs is ambiguous. This also proves true when we turn to look at the ideologies of the post–World War I period (Mitman 1992). You might think that communism—at least, the nineteenth-century communism of Karl Marx and Friedrich Engels—owed much not just to evolutionism in general but to Charles Darwin's *Origin of Species* in particular. We know that Marx spoke warmly of the *Origin* (Young 1985). At Marx's funeral, Engels went so far as to say that as Darwin had provided great insights about the biological world, so Marx had done likewise for the social world. But although the Soviet system ostensibly responded warmly to these sentiments—after the revolution there were always departments of Darwinism in Russian universities—the English materialistic science of Darwin was not the major influence on Soviet science. This honor was held by the Germanic idealistic philosophy of Hegel. The Soviet bible on scientific methodology was Engels's posthumously published *Dialectics of Nature,* which owes a great debt to *Naturphilosophie* and nothing at all to Darwinism and natural selection. Darwinism, it appears, was more something used to give people's thinking a veneer of intellectual respectability than something that profoundly altered the way that people thought. One amusing side note is that in America the compliment was returned. Early American communism owed more to evolutionism, to Herbert Spencer in particular, than it ever did to Karl Marx (Pittenger 1993). Socialists in the New World found the progressivism of the English evolutionist far more to their taste than the complex dialectic of the German thinkers!

National socialism likewise has a very ambiguous relationship with evolutionism. Darwinism—at least a bastardized form of Darwinism—found its way across the English Channel and ended up in Bismarck's newly unified Germany. Ernst Haeckel, a professor at Jena, preached nonstop "Darwinismus." It is true that, by the time that the German evolutionists had finished converting the Englishman's ideas to their own purposes, the doctrine bore little resemblance to anything to be found either in *The Origin of Species* or *The Descent of Man.* There was for instance a rather heavy bias toward the group as the major unit in evolution—something that Haeckel saw as nicely justifying the strong emphasis on the virtues of the state, a major theme in Bismarck's Germany. But genuinely Darwinian or not, there was much enthusiasm for an evolutionism applied to social and political issues. Moreover, this kind of thinking continued right through the time of the kaiser and resurfaced with the founding of the Third Reich (Gasman 1971). Even in that hotchpotch of half truths and lies that poured into *Mein Kampf,* one can find sentiments that seem on the surface to be strongly influenced by Social Darwinism.

All great cultures of the past perished only because the originally creative race died out from blood poisoning.

Karl Marx

The ultimate cause of such a decline was their forgetting that all culture depends on men and not conversely; hence that to preserve a certain culture, the man who creates it must be preserved. This preservation is bound up with the rigid law of necessity and the right to victory of the best and strongest in this world.

Those who want to live, let them fight, and those who do not want to fight in this world of eternal struggle do not deserve to live.

Even if this were hard–that is how it is! Assuredly, however by far the harder fate is that which strikes the man who thinks he can overcome Nature, but in the last analysis only mocks her. Distress, misfortune, and diseases are her answer.

The man who misjudges and disregards the racial laws actually forfeits the happiness that seems destined to be his. He thwarts the triumphal march of the best race and hence also the precondition for all human progress, and remains, in consequence, burdened with all the sensibility of man, in the animal realm of helpless misery (Hitler 1925, 1, chapter 11).

However, it does not take much to see that there could have been no simple relationship between any philosophy based on evolutionary ideas and the ideology that was so important for the national socialists (Kelly 1981; Richards 2008).

Adolph Hitler

Apart from anything else, evolutionism—Darwinism in particular—stresses the unity of humankind. The Victorians were quite happy to put themselves at the top of the evolutionary tree—others, including Slavs and Jews, came lower down. However, ultimately, we are all part of one family. A consequence like this was anathema to Hitler and his cronies. It is revealing that although Haeckel (like so many of his countrymen at the time) was anti-Semitic, his solution to the Jewish problem was one of assimilation rather than elimination. This was the very opposite of the policy endorsed and enacted by the Nazis. It is no surprise that celebrations in the Third Reich of the anniversary of Haeckel's birth were muted in the extreme. Truly, as scholars have shown, national socialism owed far more to the Volkish movements of the nineteenth century, and particularly to the so-called redemptive anti-Semitism of the group of Wagnerians at Bayreuth, than it did to anything to be found in the writings of the evolutionists (Friedlander 1997). Note that in the passage quoted just above, the real motivation seems to be that of preserving racial purity, rather than the struggle as such. Search as you might, there is nothing in the *Descent of Man* about the dangers of creative races dying out thanks to a kind of pollution—"blood poisoning" thanks to infection by lesser races.

The Nature and Status of Women

Let me conclude this brief survey of the ways in which Social Darwinism was interpreted and molded and used by turning to a discussion of the nature and status of women. This was a matter of as much pressing interest at the end of the nineteenth century as it was at the end of the twentieth century. It is the popular view today that in many respects Social Darwinism was grossly sexist. "Darwin's theories were conditioned by the patriarchal culture in which they were elaborated. … The *Origin* provided a mechanism for converting culturally entrenched ideas of female hierarchy into permanent, biologically determined, sexual hierarchy" (Erskine 1995, 118). It is difficult not to feel sympathy for views like this. If one looks at the writings of Charles Darwin himself, particularly in *The Descent of Man,* there is much to justify the conclusion that Darwinian evolutionary theory was (and perhaps still is) little more than a thinly covered ideology intended expressly for the suppression and demeaning of the female sex. Darwin spoke of "man is more courageous, pugnacious and energetic than woman has more inventive genius" (Darwin 1871, 2, 316). Women in compensation show "greater tenderness or less selfishness" (p. 326). In many respects, one could as easily be reading a novel by Charles Dickens or the most reactionary country vicar as a work of science.

Moreover, one finds that even those evolutionists who claimed to be favorable to the cause of women, notably Huxley, often behaved in ways that rather belied their good intentions. The general sentiment of leading evolutionists was that women simply do not have the intelligence and drive of men and that therefore they ought to be kept out of scientific societies and universities and the like. Lesser-known evolutionists shared these sentiments, albeit the real influence was often German idealism as much as anything written by Darwin or even Spencer.

> In the animal and vegetable kingdoms we find this invariable law—rapidity of growth inversely proportionate to the degree of perfection at maturity. The higher the animal or plant in the scale of being, the more slowly does it reach its utmost capacity of development. Girls are physically and mentally more precocious than boys. The human female arrives sooner than the male at maturity, and furnishes one of the strongest arguments against the alleged equality of the sexes. The quicker appreciation of girls is the instinct, or intuitive faculty in operation; while the slower boy is an example of the latent reasoning power not yet developed. Compare them in after-life, when the boy has become a young man full of intelligence, and the girl has been educated into a young lady reading novels, working crochet, and going into hysterics at the sight of a mouse or a spider. (Allan 1869, cxcvii)

But, going the other way, once again what one finds is that just as Social Darwinians were divided on something like war and peace, so likewise they were divided on the woman question. For every Darwin or Huxley, there was someone on the other side. Take, one more time, Alfred Russel Wallace. So far was he from thinking that women are inferior to men that he came to the opinion that the only way in which the human race will achieve success and salvation is through putting all of our hopes in the hands of the females of the species. What we need

Oliver Twist, the eponymous hero of Charles Dickens's novel, asks for more. The workhouses, within the walls of which Oliver spent his childhood, were made as unpleasant as possible to deter the simply idle from remaining there.

is for young women to come to the fore and to choose only the better-quality young men. Thus, society and civilization will move upward. If we do not do this, then doom and destruction will be our fate.

> In such a reformed society the vicious man, the man of degraded taste or of feeble intellect, will have little chance of finding a wife, and his bad qualities will die out with himself. The most perfect and beautiful in body and mind will, on the other hand, be most sought and therefore be most likely to marry early, the less highly endowed later, and the least gifted in any way the latest of all, and this will be the case with both sexes. From this varying age of marriage, ... there will result a more rapid increase of the former than of the latter, and this cause continuing at work for successive generations will at length bring the average man to be the equal of those who are among the more advanced of the race. (Wallace 1900, 2, 507)

To be honest, you may feel that ideas like this are about as naive as Wallace's already expressed faith in spiritualism. Wallace's daughters and their friends must have been very peculiar and distinctive young women if they were choosing only the better members of the opposite sex. But whether or not Wallace's ideas had any genuine connection with the real world, the simple fact of the matter is that he put forward these ideas in the name of evolution, in the name of Darwinism, even. And surely if anybody in the nineteenth century after Charles Darwin himself has the right to call himself a Darwinian, it was Alfred Russel Wallace. We can properly conclude that, as with other matters, the implications of Social Darwinism for the questions of sex and equality are by no means as straightforward and one-sided as critics today often claim.

Secular Religion

By now you will be starting to get the picture and realizing that secular religions tend to run into the same problems as regular spiritual religions. The problem with Christianity, the love commandment, is that it can mean as many things as there are Christians. Take slavery. Before the Civil War, in the American North, people were adamantly opposed to it on Christian grounds. Quakers and evangelicals led the way. Owning another person can never be harmonized with the teachings of Jesus Christ. In the American South, equally, we find people supportive of slavery on biblical grounds (Noll 2002). When the runaway slave went to Saint Paul, he did not free the slave of his bonds. Paul told him to return to his master! This was no chance decision. Elsewhere Paul had articulated his thinking on the subject. "Servants, be obedient to them that are your masters according to the flesh, with fear and trembling, in singleness of your heart, as unto Christ; Not with eyeservice, as menpleasers; but as the servants of Christ, doing the will of God from the heart; With good will doing service, as to the Lord, and not to men: Knowing that whatsoever good thing any man doeth, the same shall he receive of the Lord, whether he be bond or free." It is true that Paul immediately added: "And, ye masters, do the same things unto them, forbearing threatening: knowing that your Master also is in heaven; neither is there respect of persons with him" (Ephesians 6:5–9). But still! It is little wonder that many in the South thought themselves truer Christians than those in the North. And it is little wonder that those who look at Darwinism and the prescriptions made in its name see significant similarities here with what goes on in Christianity when people start moralizing.

Of course, this is an exaggeration of both Christianity and Darwinism. The two systems do put constraints on behavior, even if they allow much flexibility within these constraints. I cannot imagine either Christians or evolutionists positively welcoming wanton cruelty, at least not if they read their systems correctly. Even though there are Christians who have persecuted Jews quite dreadfully, Adolf Hitler no more truly comes out of Christianity than he does out of Darwinism. The point I would emphasize is the extent to which the two systems, spiritual and secular, have run so parallel in the past. Of course, when we come to reasons or foundations, there are and have been differences. For the Christian, presumably, ultimately all goes back to God and His will. One ought to obey the

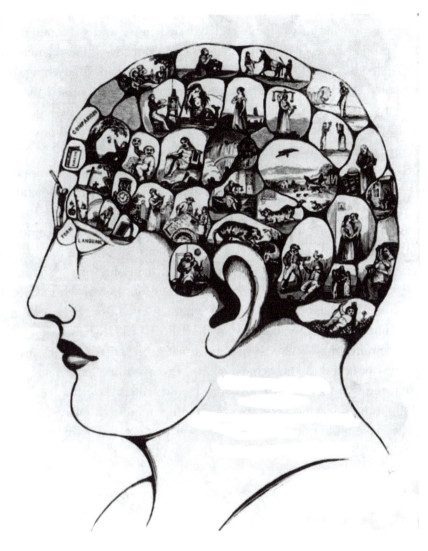

Phrenology explained (phrenology is the belief that you can read mental attributes from the shape of the skull)

love commandment because this is what He wants of us. The evolutionist as evolutionist has no such recourse. Rather we find that, to a person, evolutionists turned to progress to justify the stands that they took. With Herbert Spencer, they would and did argue that evolution makes sense, it has meaning, for it is ever striving and moving upward. The reason why we ought to cherish evolution and its processes is that if we do not, if we let evolution be and perhaps even stop and reverse itself, then progress will end and perhaps decline. And this, by definition, cannot be a good thing. Hence, it is up to us to keep things going. Offering a public library where a poor but bright child can better him or herself is a way of making sure that society keeps up to the mark, perhaps even improving, rather than sinking back to a preenlightened state. The same is true of economic theory and of militarism and of sexual relations. Our various evolutionists prescribed and proscribed as they did because, in this way, they thought that progress could be kept moving right along upward.

I should say that by century's end, not everyone was entirely happy that a full and satisfying world-picture had been sketched out. Huxley himself before his death in 1895 began to have severe doubts about whether evolution is quite a beneficent as the neo-Spencerians preached. He himself was subject to quite wracking depressions and periods of guilt and worried more and more about whether real secular progress is ever possible or desirable. Moreover, the science on which this all depended did not flourish quite as people had hoped. If anything, increased knowledge of the fossil record and of embryology rather increased the problems than reduced their magnitude. The biogenetic law was seen to be a guide—and a rather misleading guide—at best. All too often, ontogeny and phylogeny take very different routes (Bowler 1996).

Increasingly, bright young biologists turned away from evolutionary problems and concentrated on other matters that seemed more tractable and of more immediate value, intellectually and practically. First cytology (the study of the cell) became important, and then heredity (or genetics, as it became known). Evolution, never very Darwinian, never very scientific, took on more and more of the guise of a secular religion, a world picture, than of a forward-reaching professional discipline. *The Origin of Species* lifted evolution up from the pseudoscience status of phrenology and astrology. But whether evolution became quite what Darwin had hoped and expected back in the fall of 1838, when he hit on natural selection, is another matter. My suspicion is that even when he died, in 1882, for all that he was honored for his achievements, Darwin must have been a little disappointed at the way in which things were turning out. But in the realm of ideas, as in real life, our children do not always grow up in quite the way that we had intended. They take on lives of their own. As was certainly the case with evolution.

Further Reading & Discussion

I mentioned one of Adrian Desmond's books as additional reading for Chapter One. A good background source for Chapter Three is Desmond's massive biography of Thomas Henry Huxley: *Huxley: From Devil's Disciple to Evolution's High Priest* (Reading, Mass.: Addison-Wesley, 1997). As always with Desmond's writings, it is very strong on the characterization and the social factors within and without the science. Where it falls down I think is on the whole question of the professionalization of biology in the second half of the nineteenth century. My *Monad to Man* (Cambridge: Harvard University Press, 1996) gives a very different reading of Huxley, for there (and in this book) I see his attitude toward evolution as being very much at odds with the usual picture of an unqualified advocate.

Social Darwinism has a vast literature. Thomas Henry Huxley was one of the most thoughtful writers on the evolution/ethics relationship. His classic essay "Evolution and Ethics" has just been reprinted with a new introduction by me: *Evolution and Ethics* (Princeton, N.J.: Princeton University Press, 2009). This should not be missed. The best discussion of that very peculiar man Herbert Spencer can be found in Robert J. Richards's massive *Darwin and the Emergence of Evolutionary Theories of Mind and Behavior* (Chicago: University of Chicago Press, 1987). This book incidentally, which is even less easy to read at a sitting than *Monad to Man*, contains many fascinating details about evolution's history, backed by formidable scholarship. Follow this up with another great book by Richards, *The Tragic Sense of Life: Ernst Haeckel and the Struggle over Evolutionary Thought* (Chicago: University of Chicago Press, 2008). This gives a wonderful perspective on evolution in Germany in the years after the *Origin* was published.

At the secondary level, two books I rather like are both by the American historian of ideas, Cynthia Eagle Russett. *Darwin in America: The Intellectual Response, 1865–1912* (San Francisco: Freeman, 1976) deals with the social and political issues around evolution in the new world. *Sexual Science: The Victorian Construction of Womanhood* (Cambridge: Harvard University Press, 1989) is a very good account of how biologists and others wrestled with the sex and gender issues in the light of evolution. The whole question of Darwinism and the link with national socialism is very controversial and tense. Two books by Daniel Gasman lay out the case for the prosecution: *The Scientific Origins of National Socialism: Social Darwinism in Ernst Haeckel and the German Monist League* (New York: Elsevier 1971) and *Haeckel's Monism and the Birth of Fascist Ideology* (New York: P. Lang, 1998). More recently, Richard Weikart in *From Darwin to Hitler: Evolutionary Ethics, Eugenics, and Racism in Germany* (New York: Palgrave Macmillan, 2004) argues strongly for the links. Alfred Kelly speaks for the defense: *The Descent of Darwin: The Popularization of Darwinism in Germany, 1860–1914* (Chapel Hill: University of North Carolina Press, 1981). My feeling (as I hint in the chapter) is that simplistic connections are surely wrong but that something had to be responsible for that vile phenomenon and I am not sure that biology is entirely guilt-free. One point I do want to draw your attention to is the importance of not reading selected extracts out of context. If you go back to the Hitler quote and isolate one line—"Those who want to live, let them fight, and those who do not want to fight in this world of eternal struggle do not deserve to live."—it is easy to read him as a straightforward Social Darwinian. But if you look at the sentences I quote around this line, you see at once that really Hitler is talking about the Jews and about not letting them infect pure German blood. This has nothing to do with Darwin, Spencer, the Americans, or almost all others.

Finally, I want to mention another of my books, *The Evolution-Creation Struggle* (Cambridge: Harvard University Press, 2005). There I examine in some detail the claim (to which I subscribe) that in many respects people made of Darwinism a kind of secular religion, one they could put in opposition to and substitution of Christianity. *The Evolution-Creation Struggle* has irritated intensely some of the atheistic Darwinians I shall be discussing later in this book. They don't mind calling others religious. They just hate it when you do the same of them.

Chapter 4
Darwin in America: The New World

Overview

This chapter explores how the evolution debates spilled from the religious, scientific and philosophical worlds it comfortably occupied in Europe into the political and educational worlds in the United States.

We will meet some of the great scientists whose visions in the last half of the 19th century opened the study of science to more students. Thanks to this group of visionary and ambitious men some of our greatest museums were open to the public in a remarkably short time. As a result of museums and well-attended lectures, the idea of evolution easily flowed into the public space without being viewed, initially, as the threat it later became.

One of the first major evolutionary clashes in America remained in the academic realm. Two Harvard professors, the Swiss-born ichthyologist Louis Agassiz, who was never able to accept evolution on scientific grounds, and Asa Gray, the Professor of Botany, who was both a friend of Darwin and a practicing evangelical Christian, battled for funds and recognition. Gray comfortably believed that since God gave man the gift of reason and man developed the evolution theory as a result of his power of reason then evolution was God given. An accommodation idea far removed from the later Evangelical Movement.

Fuel was added to the Evolution Wars with the amazing fossil finds in the American West. The sheer size and numbers of fossil finds was unlike any previous fossil discoveries. Enormous dinosaur bones were shipped back by the crate to eastern museums for further analysis and re-assembly. From these enormous treasure troves the now famous series of horse skeletons was assembled showing the skeletal evolution to the modern horse. But still the question remained among scientists and an interested public as to whether there was simply an upward progression of the many individual species versus the ability of one species to eventually evolve from another.

Because America is more religious than Britain or Europe, these scientific questions became paramount in the religious world as well. Surprisingly, given the later reactions, many faiths openly embraced evolution noting that 'God hath ordained whatsoever comes to pass.' That indeed, evolution was God's plan by which the race should steadily ascend ... that the good in men become mightier than the animal in them.' Others outright rejected evolution as simply atheism. However there was recognition among these theologians that the Bible needed interpretation and could not be used as a scientific text. Their position could not be seen as a blind rejection of science.

As America moved into the twentieth century, however, there began a real campaign for Christians to take a much more literal stand on the readings of the Bible. Highly influential here were a series of pamphlets, *The Fundamentals* (hence the term "Fundamentalism") published early in the new century. This was a paradox because two of the pamphlets contained articles endorsing a form of theistic evolution. But gaps in the fossil record were never the real reason for opposition to evolution. It was always cultural. After World War I Fundamentalism was on the rise. Many Americans associated German militarism with Social Darwinism. In addition, secondary school education was becoming available to more children growing up in the United States. This meant that evolution was no longer contained in museums or the occasional lecture. It was in every child's classrooms and, through textbooks, in every home. The swing to increasing conservatism that led to legislation enacting prohibition, eventually led to laws prohibiting the teaching of evolution. And thus to the famous Scope Trial, an analysis of which ends this chapter.

The Role of the Scientific Community

The work of the following scientists is discussed in this chapter. Short, biographical essays of these individuals appear in **Biographies** on page 607.

Charles Hodge (1787–1878)
John Henry Newman (1801–1890)
Louis Agassiz (1807–1873)
Asa Gray (1810–1888)
Othniel Charles Marsh (1831–1899)
Edward Drinker Cope (1840–1897)
Clarence Darrow (1857–1938)
William Jennings Bryan (1860–1925)

Setting the Stage

Darrow picked up the Bible and began to read: "'And the Lord God said unto the serpent, Because thou hast done this, thou art cursed above all cattle, and above every beast of the field; upon thy belly shalt thou go and dust shalt thou eat all the days of thy life.' Do you think that is why the serpent is compelled to crawl upon its belly?"

"I believe that," William Jennings Bryan responded.

"Have you any idea how the snake went before that time?"

"No sir."

"Do you know whether he walked on his tail or not?"

"No, sir, I have no way to know."

There was a howl of laughter from the crowd.

Suddenly Bryan's voice rose, screaming, hysterical: "The only purpose Mr. Darrow has is to slur at the Bible. ... I want the world to know that this man, who does not believe in a God, is trying to use a court in Tennessee—"

"I object to your statement." Darrow was contemptuous. "I am examining you on your fool ideas that no intelligent Christian on earth believes."

Judge Raulston put an end to the argument by adjourning the court.

That night, at last, it rained. (Settle 1972, 108–109)

"The Scope's monkey trial"! This exchange comes directly from the transcript of a court case in Tennessee in 1925, when a young schoolteacher, John Thomas Scopes, was put on trial for having taught evolution to his class. Prosecuted by three-time presidential candidate William Jennings Bryan—an ardent evangelical Christian—Scopes was defended by well-known lawyer Clarence Darrow—a notorious agnostic and freethinker. In a country that loves a good court case—remember the O.J. Simpson trial—this was entertainment of the highest order. The lawyers put on a wonderful show, dueling openly before the whole American public. And how that public laughed! The best-known and most savagely funny reporter in the country, H. L. Mencken of the *Baltimore Sun,* wrote scathingly of a society that takes seriously "degraded nonsense which country preachers are ramming and hammering into yokel skulls." Which may have been true, but in the end Scopes was found guilty and fined $100.

Evolution in the New World! Let us put back the clock and see how ideas (and people) crossed the Atlantic and what happened when they did.

Essay

The Harvard Clashes

Louis Agassiz was a striking, florid, self-confident man, who could charm dollars out of eager New Englanders like a conjurer with his hat and his rabbit (Lurie

Lousis Agassiz

1960). And a man who could spend those dollars just as rapidly. Born in Switzerland of a father who was a Protestant pastor in a Catholic canton, religion was ever a major factor in Agassiz's world picture. He left his home and country in midlife, crossing the Atlantic in 1846 to a professorship at Harvard. His first wife (from whom he was estranged) conveniently died, he married again, this time into the Boston aristocracy. With the change, he moved also from the piety of his youth to the American Unitarianism of his new spouse. But on one thing Agassiz stood firm all his life: with all his being and with all of his formidable energy, he opposed the vile doctrine of transmutationism. He was against it in his youth and he was against it in his old age, and Agassiz being Agassiz, everybody knew this and the reasons for the opposition.

So let me start by stressing that Agassiz was a great scientist. Toward the end of his life, his energies were given more to institution building, but his achievements were real and important. It was he who established the fact of ice ages in Europe—his coming from Switzerland, where he had first-hand experience of glaciation, played a major role in this achievement, but the triumph was Agassiz's nevertheless. And in the field of biological studies, specifically of ichthyology (fish) both in nature today and as represented in the fossil record, Agassiz was rightly recognized as the world leader. Indeed, his start in this had been fast and precocious, for as a young man he had visited the aged Cuvier in Paris. The French scientist had been so impressed that he had given to Agassiz his

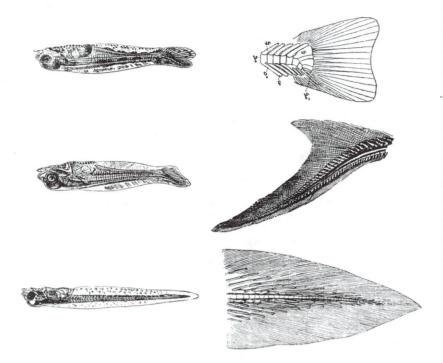

The supposed parallel between the development of the individual (left column) and the chain of living life from the primitive to the complex (right column). Note that the embryo on the left (a flatfish Pleuronectes) *goes from a nonlobed tail (diphycercal) to an asymmetric form of tail (heterocercal) and then to a symmetrical tail (homocercal). On the right we have the same pattern as we go from the primitive* Protopterus *to the middle-range sturgeon to the advanced salmon.*

own notes on fish, that the visiting junior scholar might take them up and use them in his own studies.

This was clearly a defining moment for Agassiz, who felt for the rest of his life (with some justification) that he was carrying the mantle of Cuvier—the torch that had been passed on. He adopted and never relinquished Cuvier's fourfold *embranchement* division, he put his ice age thinking into a context of Cuvierian catastrophism (which was one reason why many people, Lyell and Darwin especially, had trouble with immediate acceptance), and one senses that the lifelong opposition to evolution was in major part very much a paying of debt and homage to Cuvier as Agassiz's mentor. There was an emotional bond struck at once between the two men, something that might have been based in part on their shared religious situations: Protestants in Catholic territory.

But Agassiz was always more than simply a reflection of Cuvier. He was educated in Germany, and at Munich had sat at the feet of two of the greatest of the *Naturphilosophen*: Schelling the philosopher and Oken the biologist (Agassiz 1885). This experience affected him greatly. Schelling was not only a great system builder but a charismatic lecturer, and Oken was little less and (a pattern inherited by Agassiz) was himself a great friend of the students: a real old-fashioned college prof who would drink beer and talk until the small hours of the night. Although the *embranchement* theory meant that Agassiz could never himself be a full-blown *Naturphilosoph*, he adopted many of its ideas, especially the belief that (within *embranchements,* the vertebrate branch particularly) one could see an upward rise from primitive to complex. "One single idea has presided over the development of the whole class, and that all the deviations lead back to a primary plan, so that even if the thread seem broken in the present creation, one can reunite it on reaching the domain of fossil ichthyology" (Agassiz 1885, 1, 241).

Moreover, Agassiz was very much into parallels between life history and embryological development, and indeed he saw a threefold parallelism that included living beings today. "One may consider it as henceforth proved that the embryo of the fish during its development, the class of fishes as it at present exists in its numerous families, and the type of fish in its planetary history, exhibit analogous phases through which one may follow the same creative thought like a guiding thread in the study of the connection between organized beings" (Agassiz 1885, 1, 369–370). With this perspective, Cuvier notwithstanding, one might question why Agassiz would be so strongly against evolution. Especially since he was forced to admit that, just before the appearance of any new form or type in the record, the fossils start to forecast what will come. They are "prophetic types." "It seems to me even that the fishes which preceded the appearance of reptiles in the plan of creation were higher in certain characters than those which succeeded them; and it is a strange fact that these ancient fishes have something analogous with reptiles, which had not then made their appearance" (Agassiz, 1885, 1, 393).

But to ask questions about evolution is to misunderstand the philosophy within which Agassiz was reared. By the time of Haeckel, *Naturphilosophie* had matured into a philosophy that could bear an evolutionary interpretation, but the early beginning-of-the-century version—the version of Schelling and Goethe and Hegel and others—although deeply developmental was no less deeply idealistic. As shown by the passage quoted above from Agassiz, "the same creative thought like a guiding thread in the study of the connection between organized beings," it was the idea that counted, not the reality. No one believed that one could actually move from one species to another—Kant's teleology (which is what Cuvier inherited) precluded the transition between types. But it was not needed anyway. German idealism was just that: idealism. Real evolutionism would have messed things up, and so Agassiz never thought or could think in terms of evolution. And when he went to America, to Boston, the people he mixed with tended less to be scientists of his own stature and more the intellectuals, the philosophers and poets, men like Emerson and Longfellow, who were themselves much taken with German philosophy ("transcendentalism"). So here was reinforcement for the beliefs.

Why did Agassiz want the funds that he sought? Five thousand dollars from this donor, 10,000 dollars from that legislature? His American dream was to build at Harvard a magnificent museum, one that could house collections drawn from the world over and where researchers could study and advance our understanding of the world of animals (Winsor 1991). It would be (in the Cuvierian tradition) a museum of comparative anatomy, where one could draw on many, many specimens and make comparisons and inferences based on the widest possible range of specimens. This project—part of that already mentioned worldwide movement to the building and developing of natural history museums (Richard Owen was just then campaigning to get the British Museum [Natural History] off the ground)—

was finally completed later in the century under Agassiz's son, Alexander: the Museum of Comparative Zoology or Agassiz Museum. But it was Louis Agassiz's dream and that for which he would lecture incessantly in the public forum.

Although he would have denied it vigorously, one senses that Agassiz was not entirely disappointed at the publication of the *Origin*—Darwin sent him a copy with a polite note—for it gave him full opportunity to mount the stage before large audiences and to talk on the topic of the day, making his points of rebuttal and underscoring the need for massive (and expensive) facilities to look at these issues in a full and professional manner. Agassiz's own opposition in itself showed that the matter was not yet resolved once and for all.

All of this was gall and wormwood to another Harvard professor, a native-born scientist, who had risen up from humble roots in upstate New York and through medical training (a route shared by Huxley and other evolutionists) had eased himself into a life of full-time science. By the time the *Origin* was published, he became one of America's leading botanists (Dupree 1959). I refer to Asa Gray, a man so far within the Darwinian circle that he had been let into the great secret some years before the publication of the *Origin*. Gray was by nature your scientist's scientist, a man whose love was the private professional discussion or gathering, where experts in the field could assess data and evaluate hypotheses. He was a man for whom the real respect came from fellow scientists and who thought that ultimately the appeal to the public dimension was a little bit vulgar. The trouble of course with such an attitude is that if you seek out the professional respect and disdain the public forum, professional rather than public respect is precisely what you get. The big funds—monies provided by rich private donors or by enthused state legislatures—just do not come your way. (Agassiz was getting $100,000 from the Massachusetts government alone for his laboratory, while the botany department had to grub around for $10,000 total.) Nor can you attract and support those flocks of students that were flowing toward Agassiz, or any of the other perks of the academic life.

Hence, when the *Origin* was published and Agassiz started declaiming against it, Gray was primed and motivated to start the counterattack, even if it meant going out into the glare of the public arena. And this is precisely what he did, taking on Agassiz and his antievolutionism before audiences at the American Academy of Arts and Sciences. (This was and is a New England–based organization. The National Academy of Sciences had not yet been founded. This was to come later in the decade and owed much to Agassiz's stimulus.) At the same time, Gray took to the pen, reviewing the *Origin* and making sure that it got a fair and full exposition in the American press. His essays were later to be published in a collected volume, *Darwiniana* (1876), which was to prove one of the more appealing and lasting publications from the evolutionary fray.

The Huxley-Wilberforce debate repeated itself across the Atlantic in Boston. But it was not truly the same debate, or at least there were significant differences. For Huxley, whatever he said later about agnosticism, a major motivation was his attack on the Church, not just the dogma but everything it symbolized. You really

Asa Gray

can think of their battle as a clash between science and religion. But in respects Gray was even more devout than Agassiz, who was (it will be remembered) moving from Christianity to Unitarianism. Gray was ever a devoted evangelical Christian—indeed, loyal and attached though he was to Darwin, he never much cared for Huxley whom he thought a rather vulgar man, with little appreciation of or sensitivity toward people's religious beliefs. Gray did believe desperately sincerely in evolution and bound this up with his religious belief: God has given us our powers of sense and reason, to understand His creation, and if He decided in His power and magnificence to create in a developmental evolutionary fashion, then it is for us to accept and glorify. (More on some of these points later in this chapter.)

But precisely because Gray was a Christian, whereas Huxley was indifferent toward natural selection because he was indifferent to design, Gray was so sensitive to and overwhelmed by design that he could not accept that natural selection working on nondirected variation could do the full evolutionary job adequately. To the despair of Darwin, who thought that the move gutted the very principle of which he was so proud, Gray ever supplemented natural selection with divinely guided variations: "We should advise Mr. Darwin to assume, in the philosophy of his hypothesis, that variation has been led along certain beneficial lines" (Gray 1876, 121–122). Interestingly, perhaps connected to the fact that as a Christian he was more committed to Providence than to Progress, Gray was not a great enthusiast for evolutionary progress. But probably the main factor here was that he was a botanist rather than a zoologist. "We have really, that I know of, no philosophical basis for high and low. Moreover, the vegetable kingdom does not culminate, as the animal kingdom does. It is not a kingdom, but a commonwealth; a democracy, and therefore puzzling and unaccountable from the former point of view" (letter to Charles Darwin, 27 January 1863; in Gray 1894, 496).

One hardly need remark that, whatever the motivation, the nonprogressiveness of evolution was a sharp stiletto in the war with Swiss transcendentalism, for whom the ultimate emergence of our species was the very point of God's creative efforts: "The history of the earth proclaims its Creator. It tells us that the object and the term of creation is man. He is announced in nature from the first appearance of organized beings; and each important modification in the whole series of these beings is a step towards the definitive term of the development of organic life" (Agassiz 1859, 103–104).

The Fossil Wars

The general lore is that as Huxley had vanquished Wilberforce, so Gray vanquished Agassiz. I am not sure that this is true of Huxley, and I am certainly not sure that this is true of Gray. All biologists are evolutionists now, so there is a temptation to think that what we believe true today must have been apparent to those working and debating back then. If you combine this with the fact that, for all that he thought humans the culmination of God's creative process, Agassiz held really rather repellent views on the independent creation of human races and hence the independent origin of whites and blacks, whereas as an evangelical Christian Gray was passionately opposed to slavery, the case for Gray over Agassiz seems definitive. But it is not always true that what we find plausible and convincing today was equally plausible and convincing back in the past, and certainly not so here.

For a start, Gray himself held views that would be anathema to today's evolutionists, as indeed they were then to someone back in England. Gray was a friend and he was fighting the good fight, but was the cost too high? "The view that each variation has been providentially arranged seems to me to make Natural Selection entirely superfluous, and indeed takes the whole case of the appearance of new species out of the range of science" (Letter to Charles Lyell, August 1, 1861, Darwin and Seward 1903, 1, 191). But even if you ignore this and allow that Gray won the battle, in the long haul it is more accurate to say that Agassiz won the war. It was he who had and trained the students. Every one of them may have become an evolutionist—they all did, including Agassiz's own son!—but the picture of change traces back to those lectures in Munich rather than to the teaching at Cambridge.

In fact, Agassiz had a somewhat uneasy and difficult relationship with his students. He would welcome them in, make them part of his family, overwhelm them with friendship and advice and instruction. But he could not let them go, nor even could he see that they might be capable of working on their own and thus deserving of public recognition for their efforts. In Hegelian terms, he could not realize that the master-slave relationships he had imbued in Europe do not translate readily into American terms. Eventually, his best and brightest students revolted. Forming the Society for the Protection of American Students from Foreign Professors, they upped and left, with bitter things felt and said on both sides. Yet they took with them Agassiz's teaching: progress, the search for underlying forms or patterns or "archetypes," and an indifference to natural selection.

Paradigmatic of these Agassiz students was Alpheus Hyatt: Maryland-born, military academy–trained, passionate naturalist, and directed by the master to a lifetime's study of marine invertebrate life. Also a man of considerable moral and physical courage, for—breaking with his Confederate-sympathizing family and with Agassiz (who wanted his students to stay out of controversy and turn to science)—he joined the Union army, putting his boyhood training at its service. Yet although he broke with Agassiz personally, not only was his choice of material Agassiz-influenced but so also was his very research program. In particular, Hyatt was fascinated ("morbidly obsessed" might be a better description) with the possibility of degeneration. Could it not be that instead of uniform progress upward, sometimes evolution overtops itself as it were, and starts a slide downward? Perhaps when we reach a certain point indolence and degeneracy set in and instead of going forward, organisms relapse back into a kind of second childhood?

I should say that, although supported by appeal to the fossil record and given a firm Lamarckian (inheritance of acquired characteristics) causal backing—you get to the top and then get slack and so start to slide down, as your organs atrophy through non- or misuse—much of this thinking owed more to Hyatt's reading of social changes than to anything in real biology. As a schoolboy, before he went to Harvard, he was taken by his mother on a trip to Italy. What he saw there impressed and shocked and rather depressed the impressionable lad. All around was the glory that was—the monuments, the buildings, the statues, the pictures—and all around is the filth and decay that is. "The lazzaroni live, beg, starve, make love and shit upon the church steps and along the quai, which last being the most public is the place generally preferred for the last picturesque action" (Hyatt 1857, 6). Was this a universal law of nature? Apparently so. Think of a society that designs and builds huge and beautiful edifices—temples, meeting places, palaces, and more. Eventually, "the nation, having outgrown its strength, would begin to decline. The vast buildings would have to be abandoned, and smaller habitations would arise, in answer to the requirements of a poorer population. The architects, faithful to their inherited canons, but forced into simplicity, would gradually follow the decline, and record it in the structures of the decadence" (Hyatt 1889, 79).

This is an advance on Agassiz, but in basic respects it is right out of Agassiz, for the whole picture is one of internal movement up and then down, something that the Agassiz-like Hyatt found to be mapped in embryological development. Indeed, even the degeneration may have had an Agassiz source, for apparently it was the teacher who first suggested to the student that certain invertebrate groups might be worth of study precisely because they showed evidence of fall after their climb. Of course, Hyatt was an evolutionist, but here also there was something funny. Nobody is quite sure exactly when Hyatt became an evolutionist. You cannot really tell from the crucial papers! You can of course tell once the Lamarckism is added on, but this is old hat and not very Darwinian per se—it is not something that Hyatt was using as a research tool but rather added to make the picture complete. Nor was evolution something that Hyatt was deliberately hiding or anything like that. I have said that Hyatt was a man of courage, and he was not one to conceal his beliefs out of cowardice. The point is that evolution for someone like Hyatt really did not make that much difference to his science at all. It was rather a metaphysical assumption about the working of the world—according to natural law rather than miracle—than a tool of scientific inquiry.

This all ties in precisely with what I have been saying about the English, Huxley-driven situation. Contrast the happenings back in the 1950s after Watson and Crick had discovered the nature of the DNA molecule. Immediately a whole industry sprang up, trying to decipher the genetic code and working out how the information on the DNA molecules gets transferred into the building of the cell. Genetics changed dramatically with the double helix. This was not the way at all for Hyatt and evolution. Evolution was not functioning as a tool of professional scientific research, or at least only in a background sort of way. Evolution was rather a kind of basic way of looking at the world, a sort of secular religion, rather than something to be used as science in itself. And of course all of the stuff about progress and degeneration fit the bill precisely.

Nothing in this chapter so far makes American evolutionism that worthy of note. It is at best all a bit derivative, and even if you avoid simply making judgments based on today's knowledge, American evolutionism all seems a bit uninspiring. Things were not helped by Hyatt's being one of the world's foggiest and most confusing writers. Would that he had learned from Agassiz in this respect. Darwin, for whom Hyatt had terrific admiration and to whom he sent key papers, found him quite incomprehensible. It was not so much a question of not agreeing but simply of not following! But when this is said, let us not forget that no one was doing very much of real note in evolutionary circles in the years after the *Origin*. Other than for one or two students of the insects and similar fast-breeding organisms, natural selection languished unused, and at best people were into the business of spinning phylogenies from embryological analogies. The main function of evolutionary thought was to provide social and moral messages rather than insights about the living world.

Expectedly, the degeneration notion got picked up generally—whether from America or in parallel as it were—and we find that, as the century drew to a close, almost every evolutionist was worrying that social development had peaked and was now in a decline. This was a time when the failures of capitalism—poverty, slums, ill health, and overcrowding—were becoming apparent to all, and when the military arms race now on between countries like England and Germany bode exceedingly bad for the future. Just as religion bends itself to the time—original sin is a Christian notion that has had far higher prominence since such appalling events as the Holocaust—so also evolution proved capable of bending with the time.

But this was not all that there was to the American story, by any means. If you are willing to concentrate on evolution as path, on phylogenies, then fossils become more than just a side interest. They become absolutely central. And it was here, in the second half of the nineteenth century, that America proved its worth in ways that were beyond the wildest dreams of scientists before evolution came to town. It turned out that America, particularly the West (the Canadian West also), was a charnel house for the denizens of the past, and absolutely fabulous fossil finds were there for those who would look. Although it helped also, in those days before government grants, to have a large private fortune to support the large number of assistants and helpers necessary to dig the remains from the soil and rock and to transport the spoils back to the civilized East.

Two men above all others were qualified to go fossil hunting in the West, and they did so with such vigor and enthusiasm and violent personal rivalry and success that their exploits are still today talked of with admiration or censorious disapproval—but always with respect for the abilities and achievements (Shor 1974). The first was Othniel Charles Marsh (1831–1899), who came into a fortune when he came of age, thanks to his maternal uncle George Peabody, a business partner of the great financier Junius Spencer Morgan. Entering college at a much older age than most students, Marsh ended by spending his whole life at Yale, where he was an unpaid professor of paleontology, with the money and leisure to devote all of his time to his science. Not an easy man with whom to work, more difficult and ever more suspicious of the motives of others as he grew older, Marsh was nevertheless a good manager: one who knew how to seize an opportunity when it arose and how to get the most from his employees and underlings. He became president of the National Academy of Sciences for twelve years and wielded very considerable power both within science and on the interface between science and government.

Marsh was not a man to be crossed, a fact that did not at all perturb the second of our fossil hunters, Edward Drinker Cope. Son of a wealthy Quaker, Cope resisted parental entreaties that he become a farmer and opted rather for the life of a vertebrate paleontologist (Osborn 1931). Unlike Marsh, for most of his life Cope had no university affiliation, although toward the end a connection was forged with the University of Pennsylvania. Unlike Marsh also, Cope did not have great managerial skills and he was positively naive when it came to money, eventually losing his great fortune (a quarter of a million dollars) in a mining fraud. He too was a difficult man when dealing with his own generation, and many is the

Othniel Marsh

spiteful or critical tale that one finds in the letters of the day. Part of this no doubt was predicated on Cope's voracious sexual appetite, the quenching of which led to an early death from syphilis. But, for all his faults, there was a manliness about Cope. He was raised a pacifist, but he was no coward, and in the course of his fossil searches would unflinchingly brave Indian territory when more prudent people stayed home. When he lost his fortune, he did not cry or whine, but busied himself with his work, as though nothing had happened. More importantly, he was a brilliant scientist, and although his speculations on evolutionary causes were little different or more imaginative than those of Hyatt—they were known as the American neo-Lamarckian school (Marsh stayed away from all causal speculations)—Cope could reconstruct a long dead animal from the most unpromising of fossil material. Personally, he was warm and charming toward the young, and the

Edward Drinker Cope

next generation of paleontologists felt always that they had learned much from him.

Marsh enters our story first, for it was he who made quite fabulous finds of fossil horses. In 1876, Huxley at long last fulfilled promises to visit the New World and came calling at New Haven. Huxley was committed to giving public lectures on evolution in New York, and in search of dramatic examples had settled on the horse, the evolution of which he and a brilliant Russian student (Vladimir Kovalevsky) had been tracing. Marsh's collections of equine materials quite staggered Huxley, and at once he revised his lecture, fully admitting that (although extinct in North America by the time of human occupancy) the horse had evolved in America and not in Europe.

> At each enquiry, whether he had a specimen to illustrate such and such a point or exemplify a transition from earlier and less specialised forms to later and more specialised ones, Professor Marsh would simply turn to his assistant and bid him fetch box number so and so, until Huxley turned upon him and said, "I believe you are a magician; whatever I want, you just conjure it up." (Huxley 1900, 1, 462)

The result was one of the most famous and most reproduced pictures of the lineage of the horse from the four-toed ancestor to the single-toed living representative. Moreover, it came with a prediction that soon would be unearthed a

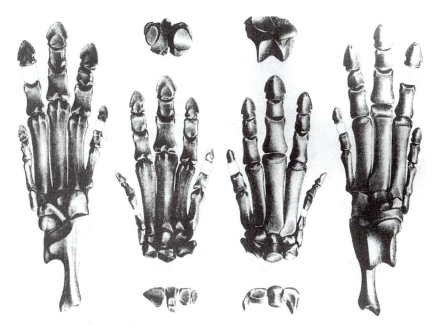

The multitoed feet of the earliest horse, Phenacodus *(ancestral even to* Eohippus*)*

five-toed ancestor, older than all known forms: "In still older forms, the series of the digits will be more and more complete, until we come to the five-toed animals, in which, if the doctrine of evolution is well founded, the whole series must have taken its origin." Within two months, Marsh discovered just such an animal, the famous five-toed Eohippus. Today, we know that the horse record is far, far more complicated than Huxley implied, with masses of branches and extinctions. But it was precisely the forceful simplicity of Huxley's demonstration that impressed. This was evolution that people understood. Evolution that convinced.

Then, when Huxley had returned home, the battle between Marsh and Cope began in earnest. Out from the West, in Colorado and Montana and Wyoming and other states, came reports of fantastical monsters, reptiles of truly gigantic size and shape, buried in the rocks but already poking out into the air. Both Marsh and Cope sent out teams to see and to excavate, and the specimens began arriving back in the East. Digging was frenetic, and on more than one occasion men from the two sides met and clashed and there were reports of fisticuffs. Certainly, neither leader was above subterfuge, concealing results from the other and trying to snatch prize specimens from beneath the nose of the other. Huge sums were spent. Cope put out at least $70,000 of his own riches. Marsh spent $200,000. And this at a time when Asa Gray was thinking himself lucky to get $1,500 a year.

Many specimens were lost through crude and amateurish methods of recovery and transportation. But the overall results were truly magnificent: Allosaurus, Ceratosaurus, Brontosaurus, Camarasaurus, Amphicoelous, Diplodocus, Camptosorus, Stegosaurus, and many many others. And these were just from one period (the Jurassic). Then they went after more recent specimens (from the Cretaceous). Amusing is the discovery of Triceratops, that fabulous monster with (as its name implies) three horns. At first Marsh described them as the horns of an extinct monstrous form of buffalo, *Bison alticornis.* Only later, when more complete specimens were unearthed, was he willing to assign it to the dinosaurs. By the

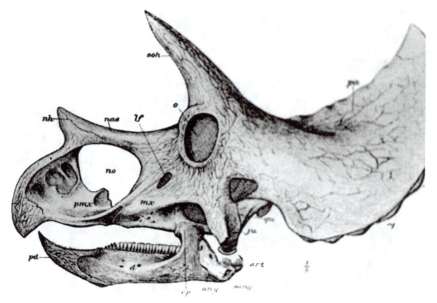

Triceratops

time he had finished—just to give you some idea of the immensity of the labors that were involved—Marsh had fifty specimens of Triceratops. It is true that most of these were represented simply by skulls, but before you start downgrading the achievement, reflect that these on average weighed a ton each, and the biggest was three and a half tons. This was evolution with a vengeance!

It was also evolution—or rather evolutionary evidence—of a kind that people could appreciate. With the move to museums—and by the 1870s and 1880s more are being built, including the already-mentioned American Museum of Natural History in New York (founded by an Agassiz student)—the demand more and more was for visible striking evidence of evolution. People did not want fine-grained experimental fodder. No one was doing experiments anyway! What they wanted were striking, impressive demonstrations of evolution in action. Things to appeal to the emotions as well as—if not more than—the minds. And monstrous reptiles from the past did just that.

All of this was brought together and given its most polished presentation, at the end of the nineteenth century and the first decades of the twentieth, by Henry Fairfield Osborn, for many years director of the American Museum of Natural History (Rainger 1991). Another man of great wealth—his father was one of the great pioneers of the railroad system—Osborn devoted his life to paleontology and to the administration of the halls within which it could occur. Befriending Cope—for whom he had great admiration—Osborn snapped up his collections for the museum when the great fossil hunter fell on hard times. Then he went out and added more of his own, always conscious that there was an end to serve: the dignified amusement and the cultural and social education of the New York public.

I have hinted already at this important point, so let me emphasize it here. The great achievements in museum building at this time—Harvard, London, New York, and elsewhere—did not happen just by chance or simply through civic or

Henry Fairfield Osborn

national pride (although the latter was certainly an important factor). On the one hand, all of those lower- and middle-class urbanites needed distractions and occupations for their free time. They needed distractions and occupations that would be moral and healthy and fulfilling in a spiritual and moral sense. Getting soaked in a gin palace was not the solution. Nor were such things as cock fighting or gambling or other traditional entertainments. Museums were a perfect answer—places of wholesome entertainment, suitable for the whole family, low cost (or free), and easily accessible in the city. On the other hand, all of those people needed—and this was especially true in cities like New York, with a large immigrant population—instruction in moral and cultural norms. At the most basic level they needed instruction in such things as hygiene and nutrition and at the more conceptual level they needed instruction in the proper ordering of the state and of the roles that we all play within it. Museums could offer this teaching—through displays about cleanliness and threats to health such as vermin and microbes. They could also teach about the great heroes of the past and present—the men (and very few women) who had made a contribution to the greatness of the country. They could tell of such things as the wildlife of the country and its other virtues and treasures, thus helping to forge a sense of pride in city, county, and country. And most of all, they could instruct about the nature of society and about the rightful place of those at the top and of those at the bottom. In short, they could tell something of progress and of how some peoples are rightly in control and others are not.

This was the philosophy of Osborn as director. This was the philosophy of his board of trustees. They were all rich men, powerfully established, concerned about immigration and degeneration, supportive of eugenics (the idea that you can improve humankind through selective breeding), wanting to maintain the status quo with the Anglo Saxon elite at the top and the Jews, Irish, Poles, Slavs, Italians, and—above all—the blacks down the ranks and forever staying there. Evolution was the perfect vehicle for their ends. It was interesting, it was fun, it was amazing—all of these things and more. And it preaches a message. Some have succeeded and risen higher than others. That is the way of nature and we must learn to accept it.

A master showman, Osborn put on wonderful displays of horse evolution—he even cadged the skeletons of famous race horses for his ends. And the dinosaurs. Generations of little East Siders were shipped over to Central Park (the American Museum of Natural History is halfway down the West Side), to stand in amazement before these monsters of the past. The dinosaurs blew people's minds away. They still do! A testament to the wonderful ways of nature and to the men who revealed them. Osborn, a student of Huxley in his youth—he had gone to Europe to study and, thrill of thrills, had once been introduced to Darwin—brought evolution as popular science, as secular religion, to its highest point.

Christian Reactions

"As secular religion"? Gray, we know, was an evangelical Christian. Hyatt nearly became a Catholic priest. Although Cope moved from the Quakerism of his youth, he never relinquished a deep conviction that God rules and cares for the world. And Osborn crossed from the Presbyterianism of his childhood to an Episcopalianism of middle age. There were certainly agnostics and atheists in America—agonistics and atheists for whom their evolutionism was an important part of their overall world picture. American Marxists were one such group. And more generally, there were voices who wanted to separate science and religion, arguing that the former looks ahead and the latter looks back. Some of the classic works proclaiming the warfare of science and religion came from American pens in the second half of the nineteenth century. But, America is a religious country—far more so than Europe, including Britain—and at this time we simply do not find the equivalent of Huxley, a major science-based evolutionist for whom evolution is a real religion substitute.

If religion was such a large element in people's lives—especially in American people's lives—what then were the responses by the religious toward evolution in general, and Darwinism in particular? Of course, for any category, there are always exceptions or people who do not fit exactly, but roughly speaking (in Britain and America especially) one sees three basic responses (Moore 1979). The first two came from groups who accepted, even welcomed, evolution in some sense. The first—following custom let us call them the Darwinians—were religious people (Christians) who more or less accepted evolution as is and extended this acceptance to natural selection, that is, to Darwin's own ideas and causal suggestions. Some, like Asa Gray, felt it necessary to modify Darwinism to allow for direction, but others did not even feel this. They happily accepted Darwinism raw, as it were.

And I stress "happily." There were people like John Henry Newman, the great convert to Catholicism (he ended as a cardinal), who were not themselves scientists, but who were appreciative of science and even if they could not go all the way with full-blooded Darwinism, were more than happy to embrace evolution and give Darwin himself great respect. Newman did not accept a literal reading of Genesis. "The Fathers are not unanimous in their interpretation of the 1st chapter of Genesis. A commentator then does not impute untruth or error to Scripture, though he denies the fact of creation or formation of the world in six days, or in six periods. He has the right to say that the chapter is a symbolical representation, for so St Augustine seems to consider" (letter of 1864, in Newman 1971, 266). Then, writing to the conservative Anglican Edward Pusey, in support of Darwin's receiving an honorary degree from Oxford University, Newman mused: "Is this [Darwin's theory] against the distinct teaching of the inspired text? If it is, then he advocates an Antichristian theory. For myself, speaking under correction, I don't see that it does—contradict it" (letter of June 5, 1870, in Newman 1973, 137). As it happens, Newman did not think natural selection could do the evolutionary job, but that was a matter for science and not theology.

Then there were also people who were interested in science and who positively welcomed Darwinism. Interestingly, these were often people of a more conservative or orthodox or high-church bent than otherwise. They were people who were interested in teleology and who saw in natural selection precisely the teleology-producing mechanism that they had been seeking. They were people (who, as I mentioned in the last chapter, were often Calvinists) who took very seriously the facts of cruelty and struggle and pain that are our fate on earth and who saw in natural selection God's way of deciding between sheep and goats. And they were people who saw in the unbroken law of evolution, not deism, but God's constant interest in and sustaining of—His immanence in—the creation.

The late-nineteenth-century Oxford theologian Aubrey Moore—a very high-church Anglican—was one of the more attractive and articulate of this num-

John Henry Newman

ber. He welcomed Darwinism with enthusiasm—with joy, even—seeing in the theory the proof definitive of God's constant care for and action in his creation. The Divine is no remote designer but one always and everywhere present and active.

> Science had pushed … God farther and farther away, and at the moment when it seemed as if He would be thrust out altogether, Darwinism appeared, and, under the guise of a foe, did the work of a friend. It has conferred upon philosophy and religion an inestimable benefit, by showing us that we must choose between two alternatives. Either God is everywhere present in nature, or He is nowhere. He cannot be here, and not there. He cannot delegate His power to demigods called "second causes". In nature everything must be His work or nothing. We must frankly return to the Christian view of direct Divine agency, the immanence of Divine power in nature from end to end, the belief in a God in Whom not only we, but all things have their being, or we must banish Him altogether. (Moore 1890, 73–74)

In America, George Frederick Wright, although later in life to become much more conservative theologically, urged Asa Gray to publication and argued himself that Darwinism threw up no new challenges for the man of god. "The student of natural history who falls into the modern habits of speculation upon his favorite subject may safely leave Calvinistic theologians to defend his religious faith. All the philosophical difficulties which he will ever encounter, and a great many more, have already been bravely met in the region of speculative theology" (Wright 1882, 219). Everyone has had a go at the true faith. Nevertheless, "The Calvinist has stood manfully in the breach, and defended the doctrine that method is an essential attribute of the divine mind, and that whatsoever proceeds from that mind conforms to principles of order; God 'hath foreordained whatsoever comes to pass'. The doctrine of the continuity of nature is not new to the theologian. The modern man of science, in extending his conception of the reign of law, is but illustrating the fundamental principle of Calvinism" (p. 220).

The second group, also in favor of evolution but a lot less directly Darwinian, consists of those generally known as "Darwinistic." Generally, these were people who took a liberal Christian approach, often known as "modernism." They wanted to modify Christianity in directions that they thought more in tune with the modern world and thought. They tended to downplay the stern unforgiving aspects of Calvinism—such things as predestination would be ruled out completely, and original sin would fare little better—and providence tended to get fairly short shrift. Rather they wanted to get on the bandwagon of progress, and their Christianity was going to reflect this. They prided themselves on being more scientific than the scientists, which meant that they simply loved evolution. But of course the evolution they wanted was a user-friendly evolution—one where effort paid off and where the struggle could be played down—and an evolution that was firmly progressive.

In short, what they wanted was a Spencerian type of evolution rather than a Darwinian type of evolution, which was precisely what most Americans of all kinds wanted anyway. Remember how Spencer, in America particularly, was *the* philosopher of evolution. Henry Ward Beecher, brother of the novelist, charismatic preacher, adulterer, liar—religion is not the only great American tradition—put things well:

> If the whole theory of evolution is but a slow decree of God, and if He is behind it and under it, then the solution not only becomes natural and easy, but it becomes sublime, that in that waiting experiment which was to run through the ages of the world, God had a plan by which the race should steadily ascend, and the weakest become the strongest and the invisible become more and more visible, and the finer and nobler at last transcend and absolutely control its controllers, and the good in men become mightier than the animal in them. (Beecher 1885, 429)

It is worth noting that at least one of the reasons why Newman rejected Darwinism is that he saw (correctly or incorrectly) that it was bound up with thoughts of progress, and this was a philosophy that he loathed intensely. He thought that we are tainted by sin and that we need God's help to earn eternal salvation. Alone, we are worthless.

Then, in addition to these two evolution-friendly approaches, there was the third response: that which rejected evolution. Charles Hodge, professor of systematic theology at Princeton Theological Seminary, the leading Calvinist theological school in the United States, had no doubts on the subject. "What is Darwinism?" he asked in one of his books. "It is atheism," came the reply. Hodge exonerated Darwin himself from the charge of deliberate infidelity, but his theory simply could not be held by a believer. "God has revealed his existence and his government of the world so clearly and so authoritatively, that any philosophical or scientific speculations inconsistent with those truths are like cobwebs in the track of a tornado. They offer no sensible resistance" (Hodge 1872, 2, 15). Backed by his reading of Agassiz, "a giant in palaeontology," Hodge had little difficulty in rejecting evolution in any form, especially the Darwinian incarnation. Darwin's "theory is that hundreds or thousands of millions of years ago God called a living germ, or living germs, into existence, and that since that time God has no more to do with the universe than if He did not exist. This is atheism to all intents and purposes, because it leaves the soul as entirely without God, without a Father, Helper, or Ruler, as the doctrine of Epicurus or of Comte" (p. 16).

Hodge, of course, preferred the account of Genesis to the account of the evolutionists. But it is important to note that he did not reject evolution simply because it was science, nor did he accept Genesis simply because it was religion. The point is that (in his opinion) evolution failed as science and hence the way was open for someone to accept the account of Genesis as good history. Here Hodge was very much following the tradition of his church: Calvin, although concerned to stay with a literal reading of the Bible, realized that some interpretative work was needed. To this end, the great reformer had introduced his famous doctrine of "accommodation," one recognizing that the Bible is sometimes written in such a form as to make itself intelligible to scientifically untutored folk who would not have followed sophisticated discourse.

> Moses wrote in a popular style things which, without instruction, all ordinary persons endued with common sense, are able to understand; but astronomers investigate with great labour whatever the sagacity of the human mind can comprehend. Nevertheless, this study is not to be reprobated, nor this science to be condemned, because some frantic persons are wont boldly to reject whatever is unknown to them. For astronomy is not only pleasant, but also very useful to be known: it cannot be denied that this art unfolds the admirable wisdom of God. ... Nor did Moses truly wish to withdraw us from this pursuit in omitting such things as are peculiar to the art; but because he was ordained a teacher as well of the unlearned and rude as of the learned, he could not otherwise fulfil his office than by descending to this grosser method of instruction. ... Moses, therefore, rather adapts his discourse to common usage. (Calvin 1847–1850, 1, 86–87)

Likewise Hodge. He accepted the geologists' claim that the earth must be very old and entertained seriously the rival hypotheses that either there was a very long period of time (unrecorded in the Bible) after the initial creation or that the six days of creation must be understood as six very long periods of time. He himself inclined to the second hypothesis, but the point is that Hodge—and his stand was definitive for many many people—recognized fully with Calvin that the Bible needs interpretation and cannot be used as a scientific text. If Darwinism is to be rejected, it must be on scientific grounds. Hence, even this third position was far from one of blind rejection of science, even though evolution did fail to find favor.

The Scopes Trial

It was not until the end of the nineteenth century and into the twentieth century that things started to get a lot tighter, with full-blown campaigning against evolution (Larson 1997). Now increasingly we find Christians inclined to take a much more literal stand on the meanings of the Bible. This came slowly, not all at once. The definitive conservative Christian position was given in a series of pamphlets published between 1905 and 1915, *The Fundamentals*—hence the term *fundamentalist* for those who take the Bible as the inerrant word of God. But even here opposition to science was not absolute. Some of the pamphlets even endorsed evolution! Not Darwinism, for natural selection alone could not do the job and was clearly atheistic, but some kind of theistic guided evolution. "A new name for 'creation'" (pp. 20–21). Most would not have gone this far, but the interpretation of "days" as long periods or an unmentioned lengthy gap between the creation of heaven and earth and the edenic creation was standard.

However, by now there were hard-line literalists: people who subscribed to a six-day creation some 6,000 years ago. Prominent among these absolutists were the Seventh-day Adventists, a sect starting in the middle of the nineteenth century, strongly committed to the Second Coming and the conflagration that would precede it (Numbers 2006). For them, such a literal reading of the Bible was needed as confirmation of visions of their founder, Ellen G. White, as well as support for their insistence on Sabbath observance (which for them falls on a Saturday). Unless one has a 24-hour day of creation, the biblical support for Sabbath observance becomes less secure. In addition, as people believing in the coming Armageddon, the universal Flood played an important part in their theology as something that, having happened, showed God's ability and willingness to act again in such a way if necessary. A kind of balance to and foretaste of the disruptions to come.

But even if this extremism then attracted no immediate great following, by the end of the 1920s opposition to evolution of all kinds was starting to rise. There were a number of factors here. One, undoubtedly, was World War I. Rightly or wrongly, many associated Germanic militarism (militarism of all kinds, in fact) with Social Darwinism. Hence, evolution was seen as directly implicated in the carnage in Europe—a carnage that many felt was no concern of America's, anyway. A second was the fact that evolution was becoming more of a personal threat to nearly everyone, thanks to an explosion in American secondary education. The numbers of children enrolled in such education shot up from 200,000 in the whole of America in 1890 to 2 million in 1920. Tennessee, of which more in a moment, jumped from 10,000 in 1910 to 50,000 in 1925. Evolution was no

Attornies in the Scopes trial (Clarence Darrow, left; William Jennings Bryan, right)

longer just a faraway phenomenon, concerning only professors. It was now coming into every home, through the textbooks. The evil was right there.

And third, and I suspect most important of all, the evangelicals saw who was the real threat to their way of life and thinking. In America, it was not the agnostics and atheists: these were and are no real threat; they are a minority not to be taken that seriously. The real threat was the liberal Christians. The conservative believers saw this kind of Christianity as representing everything they loathed. It did not help (although it was hardly any surprise) that people like Shailer Mathews, dean of divinity at the University of Chicago and a leader of the liberal Christian wing in America (known as Modernism), had endorsed America's participation in World War I as a Christian duty. And evolution as we know was a centerpiece of this kind of Christianity. The temples of science, places like the American Museum of Natural History, dedicated to evolution, were just the sorts of places that the liberal Christians (Osborn, for instance) were building and endorsing. With reason, evolution was seen as part and parcel of a whole philosophy, a way of life, that was resented and disliked and to be opposed.

With the war over and with the campaign against alcohol brought to a successful conclusion—Prohibition was enacted—attention could be turned to evolution, and so in the 1920s we see attempts to make illegal the teaching of evolution in state-supported schools. Tennessee was one such state, and this led directly to the famous Scopes monkey trial. The interest then was intense—the press coverage was enormous—and it has continued to be so, thanks particularly to an account in a best-selling history—*Only Yesterday: An Informal History of the Nineteen-Twenties*—and then later a wonderful play (1955) and film (1960): *Inherit the Wind*. What gave an edge to the whole affair was that Darrow was denied the opportunity to call his own witnesses in favor of evolution—the case after all was over whether or not Scopes had broken the law, not whether evolution was true. Darrow had therefore put Bryan on the stand as an expert witness on the Bible and had apparently made Bryan into a fool, as the former politician stumbled around trying to keep some semblance of consistency between his religious beliefs and then-standard science. Although evolution may have lost the immediate battle (in fact, the conviction and penalty of $100 were overturned on a technicality on appeal), it won the war. The nation—the world—laughed at Tennessee and at the fundamentalists. Never again was right-wing Christianity to challenge science in such a way.

In fact, as so often happens, real life is not exactly like the myth. The trial took place in Dayton, Tennessee, a town that set up the trial in the first place, thinking that the subsequent publicity would be good for business. Scopes had let himself be prosecuted deliberately, for the American Civil Liberties Union (working hard to define itself as a body needed in America) that organized and bankrolled the defense was looking for a case to test the constitutionality of the law. He was not the regular biology teacher but the physical education teacher who substituted for the biology teacher. There is even some question as to whether he ever did actually teach evolution. Bryan was a big-name figure, but not necessarily everyone's choice for prosecuting attorney. He was not an experienced trial lawyer. Darrow was, but certainly not everyone's choice for defense attorney. His non-Christian views made many on the defense side very uncomfortable, some because they disagreed with him, and others because they thought that his reputation would hurt their cause. And an overturned verdict was precisely not what the defense wanted. They needed a conviction to fight all the way to the Supreme Court and to challenge the law constitutionally. As it happens, in the absence of such a challenge, the Tennessee law remained on the books until the 1960s.

What about the overall effects from the viewpoint of religion and science? Memory, reinforced by book and film, is that fundamentalism went down to defeat, never to rise again. The reporting—especially that of Mencken—was so savage that no one again could take such Christianity seriously. In fact, there is certainly evidence that virulent fundamentalism, linked directly to anti-Darwinism, seems to have peaked with the Scopes trial and did go into decline; although the extent to which this was direct cause and effect is another matter. What is not the case is that Bryan was made quite the fool that he appears in the movie—apart from anything else, the key scene where he is made to look stupid through his subscription to a literal reading of "days" was simply not true. Bryan always believed that the days were periods of time (a fact that upset the Adventists).

> Bryan: I think it would be just as easy for the kind of God we believe in to make the earth in six days as in six years or in 6,000,000 years or in 600,000,000 years. I do not think it important whether we believe one or the other.
> Darrow: Do you think those were literal days?

John Thomas Scopes

Bryan: My impression is that they were periods, but I would not attempt to argue as against anybody who wanted to believe in literal days. (Settle
1972, 80)

General opinion among the fundamentalists after the trial was that Bryan had acquitted himself well and that a good job had been done.

Finally, it should be noted that in some respects the fundamentalists certainly won the war, for the textbooks were immediately gutted of controversial evolutionary material—and then things stayed that way for many years thereafter.
The popular *Civic Biology* by George W. Hunter, used by Scopes, was dropped by
the Tennessee Textbook Commission. More broadly, a six-page section on evolution was dropped from the edition for southern states, and the author set about
revising the text for general use. The explicit discussion of evolution was trimmed
and concealed and material modified, with explicit charts vanishing. The word
evolution itself vanished, and Darwin was no longer described as the "grand old
man of biology." Relatedly, Darwin's "wonderful discovery of the doctrine of

evolution" became "his interpretation of the way in which all life changes" (Larson 1997, 231). This was no unambiguous victory for evolutionism.

Looking back over the years, what should we say? For myself, I cannot say these occurrences—gutting evolution from the textbooks—were right and proper. Indeed, as one who is an ardent evolutionist—a Darwinian even—I deplore them. But equally, I cannot say that I am surprised or that I am entirely unsympathetic to the fundamentalists. Evolution after Darwin had set itself up to be something more than science. It was a popular science, the science of the marketplace and the museum, and it was a religion—whether this be purely secular or blended in with a form of liberal Christianity. I do not think that it had to be, but it was. When believers in other religions turned around and scratched, you may regret the action but you can understand it—and your sympathy for the victim is attenuated.

I do not say that evolution as religion was always a bad thing. Indeed, at the social and moral level, we have seen that it can be entirely admirable. But let us not pretend that it was not what it was and that right and decency was all on one side. As is usually the case in these things, both sides had their saints and their sinners, their people of reason and their people of emotion. Most may have pulled back from the extremes, but the polarization was more than simply one of black and white, chalk and cheese, science and religion. As always when you are dealing with history, when you dig beneath the surface, things become a lot more complex and interesting than appears at first sight.

Further Reading & Discussion

At the conceptual level, James Moore's *The Post-Darwinian Controversies: A Study of the Protestant Struggle to Come to Terms with Darwin in Great Britain and America, 1870–1900* (Cambridge: Cambridge University Press, 1979) is a really detailed account of the science/religion relationship in Britain and (even more) in America in the years after the *Origin*. But the social questions are also important, and especially museums are significant now. Mary P. Winsor's *Reading the Shape of Nature: Comparative Zoology at the Agassiz Museum* (Chicago: University of Chicago Press, 1991) is excellent on the early years of museum building after the *Origin*, and Ronald Rainger's *An Agenda for Antiquity: Henry Fairfield Osborn and Vertebrate Paleontology at the American Museum of Natural History, 1890–1935* (Tuscaloosa: University of Alabama Press, 1991) is not only strong on the museum scene at the end of the century but gives you a real insight into the persona and activities of Henry Fairfield Osborn, the leader in American evolutionism from the late nineteenth century right up to the 1930s. Rainger is particularly good at showing how much evolution back then, palaeontology particularly, was a vehicle for social messages of one sort or another.

Ronald Numbers is the leading authority on science/religion relationships in American history. Having come himself from a fundamentalist background, he was particularly well prepared to write his magisterial work on the history of literalist readings of Genesis. This is now in a revised edition, brought up to date by dealing with modern controversies: *The Creationists: From Scientific Creationism to Intelligent Design* (Cambridge: Harvard University Press, 2006). This work digs into its subject with a vigor and penetrating understanding that one rarely finds in works of scholarship. Highly recommended! At a more specific level, Edward J. Larson (Numbers's student) has given us a detailed and brilliant account of the Scopes monkey trial: *Summer for the Gods: The Scopes Trial and America's Continuing Debate over Science and Religion* (New York: Basic Books, 1997). As I explain in my text, it is quite amazing how much myth has grown up around that event. A major reason for this quasi-fictional status must lie at the feet of *Inherit the Wind*, especially the movie starring Spencer Tracey and Frederick March and with Gene Kelly as a wonderfully cynical newspaper man.

For all that it is fictionalized, the movie—which is easy to find on video—is well worth watching. It does raise most interesting questions about the science/religion relationship. You should be aware that it dates from the height of the Cold War, when Americans were defending the virtues of democracy in the face of the external threat of communism. At the same time—thanks to the witch hunting activities of Senator Joseph McCarthy—they were thinking about themselves, trying to establish to what lengths people should be allowed to dissent in a free society. Why defend democracy if this is equated with total conformity? The writers are using the Scopes monkey trial to explore issues that they found important rather than simply to give a historically accurate account.

Finally let me recommend to you the science fiction novel, *The Time Machine: An Invention*, by the English novelist H. G. Wells. (It was first published in 1895 and is easily available in editions today. There is also a good movie version (1960) with Rod Taylor.) Wells was the student of Huxley and started life as a science teacher. Hyatt was not the only one worried about degeneration. As mentioned in the text, by the end of the nineteenth century, with the rise of militarism and the unsolved problems of industrialism, many were obsessed with the prospect of inevitable degeneration and downward slide. Wells picks up on this worry and explores it in the form of fiction. Thanks to his machine, the time traveler goes forward into the future and finds that humans have sunk into two races or species: the Eloi, warm, friendly, childlike, useless; and the Morlocks, industrious, intelligent, vile, underground-living cannibals. Can this really be the fate of us all? Is there even worse beyond that? These are Wells's themes in a story that is as fresh today as it was then.

Chapter 5
Evolution Denied & Extolled:
The Rise of Creationism, Intelligent Design & Darwinian Religion in America

Overview

This chapter explores how the Fundamentalists in the United States continued to raise alarm at the increasing acceptance of evolution as a *fact* which, combined with long-range ineffectiveness of legislating against the teaching of evolution, opened the doors to a new campaign. This one would mandate the teaching of alternative scientific theories of evolution (or creation) right along side the teaching of evolution: A campaign for inclusion. But what would these alternative theories be called so they could be taught in science classes?

The first, and most successful of these was Creationism. The Creation Science movement was a child of the 1960s. It takes the creation story of Genesis—six days of creation, six thousand years ago, universal flood—absolutely literally. The Institute for Creation Research was founded to develop Bible-based explanations for scientific information resulting from fossil research.

But the educational and political communities continued to resist the integration of, what was to many, religion into sciences. To counter this resistance, Creation Science gradually morphed into a more friendly form, the so-called Intelligent Design Theory.

The key work in Intelligent Design Theory was by Berkeley law professor, Phillip Johnson: *Darwin on Trial*. Effective though the work was, its weakness was that it was mainly a critique of Darwinism with no alternative. In the 1990s, two scholars repaired this deficit: first, biochemist Michael Behe argued that there are examples of organic "irreducible complexity," things that demand intervention by an "intelligent designer"; then, mathematician/philosopher William Dembski argued that statistically one can and should argue for such interventions.

As the Fundamentalists rose against Evolution there were scientists who offered equally strong support for it. In the Naturalism approach it is important to distinguish between methodological naturalism and metaphysical naturalism: the former, an attitude by scientists involving the refusal to use miracles in their scientific explanations; the latter, a philosophy that denies that there is anything beyond the purely material or natural. We shall see that Johnson argued strongly that the former collapses into the latter.

The Intelligent Design Theorists, of course, loathe all kinds of naturalism because they think it associated with the philosophy of modernism: the philosophy that points to a secular world, with a liberal attitude toward society and its denizens. We shall see that there is indeed some truth in this suspicion. Looking at the writings of leading evolutionists today shows that they, like the Intelligent Design Theorists, often have a social and cultural agenda which are more often counter to the values firmly held by the other side.

Many evolutionists today continue their counter attacking, arguing against any kind of religious belief. Critics are particularly scornful of those, like the author, who want to tread a middle line, allowing for the possibility of both religious belief and sincere scientific commitment. We shall see how Richard Dawkins, above all, leads the attack, arguing that evolution shows the lack of necessity for religious belief, and indeed that it destroys such belief by reinvigorating old arguments like the problem of evil.

Finally, let us not deny that some evolutionists, notably Edward O. Wilson, want to go all the way, and create a new religion out of evolution. For them, evolution is a story of origins, a story about the coming and importance of humankind, and a story with a moral message about the need to preserve humans and the world within which they live. In a way, for these people we have come full circle, with the ultimate triumph of Darwinism over Christianity.

The Role of the Scientific Community

The work of the following scientists is discussed in this chapter. Short, biographical essays of these individuals appear in **Biographies** on page 607.

Henry Morris (1918–2006)
Edward O. Wilson (1929–)
Phillip Johnson (1940–)
Richard Dawkins (1941–)
Micheal Behe (1952–)
William Dembski (1960–)

Setting the Stage

Q: **D**r. Ruse, having examined the creationist literature at great length, do you have a professional opinion about whether creation science measures up to the standards and characteristics of science that you have just been describing?

A: Yes, I do. In my opinion, creation science does not have those attributes that distinguish science from other endeavours.

Q: Would you please explain why you think it does not.

A: Most importantly, creation science necessarily looks to the supernatural acts of a Creator. According to creation-science theory, the Creator has intervened in supernatural ways using supernatural forces.

Q: Do you think that creation science is testable?

A: Creation science is neither testable nor tentative. Indeed, an attribute of creation science that distinguishes it quite clearly from science is that it is absolutely certain about all of the answers. And considering the magnitude of the questions it addresses—the origins of man, life, the earth, and the universe—that certainty is all the more revealing. Whatever the contrary evidence, creation science never accepts that its theory is falsified. This is just the opposite of tentativeness and makes a mockery of testing.

Q: Do you find that creation science measures up to the methodological considerations of science?

A: Creation science is woefully lacking in this regard. Most regrettably, I have found innumerable instances of outright dishonesty, deception, and distortion used to advance creation-science arguments.

Q: Dr. Ruse, do you have an opinion to a reasonable degree of professional certainty about whether creation science is science?

A: Yes.

Q: What is your opinion?

A: In my opinion creation science is not science.

Q: What do you think it is?

A: As someone also trained in the philosophy of religion, in my opinion creation science is religion. (Ruse 1988, 304–306)

My moment of glory in Little Rock, Arkansas! It is not often that a philosopher finds himself on national television, and although I no longer dine out on it quite as much as I did, it still brings me pleasure to think of it! It was indeed a moment of glory. In 1981, appearing as an expert witness for the American Civil Liberties Union alongside such evolutionary luminaries as Stephen Jay Gould, I was asked to appear in an attack on the constitutionality of a new law mandating the "balanced treatment" of so-called Creation science with evolution in the publicly financed biology classrooms of the state. And we won! The law was declared unconstitutional, and that was the end of that.

Michael Ruse

But I should have known better. Court cases, particularly in America, are rarely the end of anything. As we have changed from one millennium to another, the science-religion debate, the evolution-creation debate, rages as never before. Let me bring you up to date, show you where we stand now, and offer a few thoughts of my own.

Essay

Creation Science

It really all started with the Russians. In 1957, we were in the depths of the Cold War, and it was then that the Soviet side scored an absolutely massive propaganda victory. Sputnik! They put aloft an unmanned satellite, and then, to rub salt in the wounds, they put up another that (they informed the world) was as big as a Cadillac, the epitome of American opulence and success. In fact, looking back, it was not much more than a propaganda victory. Russia was ahead in rocket technology, partly because at the end of World War II they had grabbed more German rocket engineers than the Americans and partly because they needed long-range missiles. America had its nuclear weapons in Turkey, sitting on the Russian border. The Soviets had needs that the Americans did not, and so they had moved to fill them. It was hardly an unbiased question of the superiority of one world system over another. But America certainly perceived itself as lagging behind, not just in rockets but in science and technology generally.

This meant, among other things, that if parity were to be achieved, then education needed to be upgraded dramatically. One way in which this was done (given that education falls under State jurisdiction) was by the Federal government's sponsoring the writing of good quality, new textbooks, which could then be made available to school boards at attractive prices. Following the chilling effects of the Scopes trial, evolution had become something of a nonsubject in high-school biology texts. Evolution figured in a major way in the new works. As word of this started to filter out, the provocation set the evangelical literalists moving, and so things start to move toward a new confrontation. Aided, I might say, by something that looks suspiciously like Divine Intervention, for just at the moment of crisis, the men of the hour arrived. John C. Whitcomb, a Bible scholar, and Henry M. Morris (who died recently), a hydraulic engineer, jointly authored a book, *The Genesis Flood: The Biblical Record and Its Scientific Implications* (1961), which put once again the whole and full case for a literal Genesis-based account of origins. The case was supposedly supported in its entirety by the best quality modern science. Creation that occurred about six thousand years ago, took just a week, and was miraculous, with humans coming last. At some point after all of this had occurred, there was a massive worldwide flood, which wiped out virtually everything except apparently for a few, carefully chosen survivors.

An alternative to evolution was there for all to see and to adopt. Worried about the fossil record? No need to be. The progressiveness of the record is an artefact of the Flood, with the slowest creatures caught at the bottom and more agile creatures getting up to the tops of mountains before perishing. How else do you explain human footprints found down among the dinosaurs? What did lions eat in Eden? A vegetarian diet obviously, since they could hardly have feasted on other animals. Troubled by the age of the earth question? Be assured that you are less troubled than conventional scientists. "Age measurements by radioactivity are not nearly so precise nor so reliable as most writers imply." Indeed, "the great majority of the measurements have had to be rejected as useless for the desired purpose" (p. 343).

Henry Morris, with a group of like-minded thinkers, founded the Institute for Creation Research. Realizing that the situation had changed from the days of the Scopes trial and that no court was going to stand for the elimination or expulsion of evolution, they campaigned rather for the inclusion of their own beliefs.

Thus, through the 1970s, Morris and the others—notably Duane T. Gish, author of *Evolution. The Fossils Say No!*—wrote and lectured and (very successfully) debated evolutionists on the alternative pictures of origins. At the same time they refined and polished their position—considerable effort had to go into compressing the time-scale down from several billion years to just a few thousand. And also, they took care to see that their position could be presented ostensibly without any reference to biblical matters. In *The Genesis Flood*, for instance, when faced with monstrous-sized human footprints in the fossil record, confident mention is made of the passage in Genesis (6.4) that tells us that there were "giants in the earth in those days" (p. 175). This sort of thing rapidly became unacceptable, at least in "public school editions" of the Creationists' books. The important thing was to offer themselves up as a reasonable, secular alternative to the dominant evolutionarism of the day. Hence, the new name: "Creation science."

One has to say that the Creationists worked hard and succeeded brilliantly in their tactics and aims. They caught evolutionists napping, making them look fools—inarticulate and irrational and prejudiced fools. Working with humor and charm and sincerity, Morris and Gish particularly were masters at the public debate, usually reducing their scientific opponents to choleric rage and intellectual impotence. Moreover, they started to influence state legislatures, and the end result was that early in 1981 Arkansas passed a bill mandating "balanced treatment." I might add that this all happened when Bill Clinton was not in the governor's office, and the bill was signed into law by a man whose unsuitability for the office was equalled only by his surprise at achieving it. And I should say also that the law was a rather unpleasant surprise for many powerful people in the state. The Junior Chamber of Commerce in particular was not happy. It was working flat out to persuade new industry—often high tech, involving electrical engineering or computers—to relocate in the state. The last thing it needed was for a prospective employee, perhaps a newly minted Ph.D. from MIT, to learn that the children would be taught Creationism in the schools. Such a prospective employee would keep on moving until reaching other states—perhaps Arizona, also in the market for the new technology and the people to produce it. Whatever the personal convictions of the leaders of these rival states, they knew enough to maintain a decent hypocrisy of having one set of beliefs for the weekdays and another set for Sundays.

Indeed, the leading Creationists themselves were somewhat torn on the Arkansas law. They knew that once their ideas were made public like this, they would be pilloried in the press and probably defeated in the courts—as indeed they were. Better by far to work at the grassroots level, influencing public opinion, putting pressure on school boards and individual teachers, and like actions. Which is precisely the way the Creationism movement went for 20 years, again with considerable success. A new round of faces was recruited, notable for being much more established academically than the earlier Creation scientists. It is true that Morris and Gish have advanced degrees in science—much is made of this point—but now we find supporters of the movement at leading universities, and not just junior faculty either. Notable are Phillip Johnson, onetime law clerk to former Chief Justice Earl Warren and (retired) professor of law at Berkeley, and Alvin Plantinga, professor at Notre Dame (despite being a Calvinist) and North America's most distinguished philosopher of religion.

With the development of the Creationist side came, perhaps as a kind of counter in reaction, a development of the evolutionist side. There are still many evolutionists, probably the majority, who want nothing to do with the science/religion conflict. They want to get on with their science and leave matters at that. Among the minority who are or were interested in religion, one found Stephen Jay Gould (1999). He had certainly read Genesis: readers of his column knew that he had a biblical knowledge that would challenge any priest or rabbi and knew also that he was prominent among those who think that good fences make good neighbors. Science is science and religion is religion and never the two should meet. He spoke of science and religion as being rival "magisteria"—realms of inquiry and understanding—and advocated what he called the NOMA principle. Science and religion are Non-Overlapping MagisteriA and should stay that way.

But many of those interested in the science/religion interface, ardent in their evolutionism—usually ultra-Darwinism—are among those who have really taken a strong and almost personal dislike to Christianity. Richard Dawkins, author of the 2006 smash best-seller *The God Delusion*, leads the pack, with the philosopher Dan Dennett, *Breaking the Spell*, the graduate student Sam Harris, *The End of Faith: Religion, Terror, and the Future of Reason,* and the journalist Christopher Hitchens, *God is not Great*, close behind. They loathe and detest religion—all religion—and feel very strongly that you cannot serve science and religion at the same time. They argue that Darwinism positively excludes Christianity—not just Creationist Christianity, but any kind. In commenting on a letter favorable to evolution, written by Pope John Paul II (1997), Dawkins (1997a) spoke of a "flabbiness of the intellect" affecting those who turn to religion—and if you are prepared to say that about John Paul II, you are prepared to say it about anybody. In fact, Dawkins is! Because I, someone who has no more religious faith than any of them, am willing to listen seriously to people of religion, I am labeled a craven fool. Dawkins likens me to Neville Chamberlain, the British Prime Minister who tried to appease Adolf Hitler. Dawkins introduces a new norm for journalists, begging them to interview others and get the "real" truth, after they have spoken to me.

What are the pros and cons of the issue? Let us start with the new Creationists.

Intelligent Design Theory

At first, one of the most important things about what I like to call Creationism-lite but what its supporters call Intelligent Design Theory (IDT) is that it is asymmetric. It told you what it did not like but was irritatingly silent on what it did

Archaeopteryx, midway between the reptiles and the birds

like. Phillip Johnson's major book, *Darwin on Trial* (1993), was a paradigm. The new Creationists did not like evolution; especially they did not like Darwinian evolution. So Darwin was put in the dock. But what these Creationists did believe was not specified. Did they believe in a young earth or an old earth? We are not told. Did they believe in a universal Flood or a limited Flood? We were not told. Did they believe that humans necessarily came last (and were Adam and Eve a one-off event or did Eve come later)? We were not told. What we did not know, we cannot criticize—which was a major problem faced by the earlier Creationists. What about Darwinism? Many of the criticisms were familiar—going back to Cuvier, in fact. Natural selection was a favorite target. It cannot do what is required, it is trivial, probably false, and in any case is simply a redescription of what is going on—it is a "tautology," a necessary truth since it tells you that the fittest survive but then the fittest are defined as those that survive! Mutation was also criticized heavily. It is random, and random means random. You cannot get order from randomness. That is the truth. Organisms need something more—they need something in the intelligence line to put them on the road to being. The fossil record speaks eloquently against evolution. Nor do the so-called missing links help. Consider archaeopteryx, the bird-reptile seized on by Thomas Henry Huxley.

> *Archaeopteryx* is on the whole a point for the Darwinists, but how important is it? Persons who come to the fossil evidence as convinced Darwinists will see a stunning confirmation, but skeptics will see only a lonely exception to consistent pattern of fossil disconfirmation. If we are testing Darwinism rather than merely looking for a confirming example or two, then a single good candidate for ancestor status is not enough to save a theory that posits a worldwide history of continual evolutionary transformation. (Johnson 1993, 81)

The molecular evidence for evolution was found no more convincing. In fact, it was all a little bit of a con job. As far as Johnson was concerned, it was a

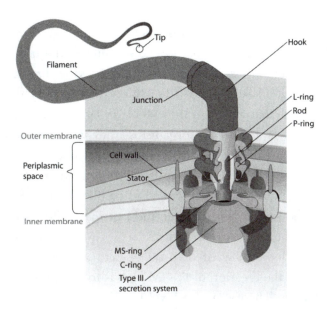

Flagellum

classic case of circular argumentation. We start by assuming that the molecules are important and then, backed by this belief, we set out to prove that they are important! "As in other areas, the objective has been to find confirmation of a theory which was conclusively presumed to be true at the start of the investigation" (p. 101). Obviously, although this is very comforting to the true believer, it is a parody of true scientific methodology and understanding. "The true scientific question—Does the molecular evidence as a whole tend to confirm Darwinism when evaluated without a Darwinist bias?—has never been asked" (p. 101). And the same thing holds again and again elsewhere. Indeed, there is little need to go on, for the main thing that remains to be discussed is the origin of life question, and we can guess on what shaky ground that stands. During the 1990s, Johnson—backed by a conservative "think tank," the Discovery Institute in Seattle Washington—made major efforts to repair the deficiencies in the neo-Creationist position. Two very important figures were recruited to the cause. First there was Michael Behe, a biochemist at Lehigh University in Pennsylvania. He wrote *Darwin's Black Box: The Biochemical Challenge to Evolution* (1996), in which he argued that some phenomena in the living world are "irreducibly complex." By this he means that they cannot function unless they are put together in one fell swoop. Drawing on the analogy of a mousetrap—five parts, all necessary, totally non-functioning unless put together in one creative act—Behe argued that things like the flagellum on bacteria (little whip-like appendages that drive the carrier forward) and the complicated chemical reactions needed for blood to clot (known as a "cascade" because so many sequential processes are needed) simply could not have come about slowly. They could not have come about slowly through a blind process like natural selection. They must therefore have been put together by a thinking being—an "intelligent designer."

Then the mathematician and philosopher of science William Dembski came along to back Behe. In *The Design Inference*, he set himself the task of finding criteria for saying that something is designed and then showing that the sorts of things highlighted by Behe fit these criteria. One thing that marked the work of both Dembski and Behe was that although they were trying to show that intelligence is involved in the origins and nature of organisms, they were not committing themselves at all to the actual nature of this intelligence. It could in theory have been perfectly natural. Hence, they were (supposedly) not moving into the realm of religion. Their position therefore was intended to be like, let us say, a researcher on the origin of life who says: I take as basic the fact that water is made of two hydrogen molecules linked to one oxygen molecule. My job is to go from there. Likewise, Behe and Dembski wanted to say: I take as basic the fact that intelligence was involved in the creation of life. My job is to go from there.

Intelligent Design Theory at one level has been a huge success. It has been adopted by people far and wide. It did receive a nasty jolt a year or two back in the town of Dover, Pennsylvania. The school board decided to insist on some form of IDT being introduced to state-supported biology classes. The end result was similar to that at Arkansas more than twenty years earlier. The judge decisively rejected the ideas (considered as science) and banned them from the classroom. Michael Behe, who was one of the few IDT enthusiasts prepared to stand up for it in court, was made a figure of fun. However, wisely, its supporters took their licks and (like the earlier generation of Creationists before them) vowed to keep up the battle at the less visible level. Recently, several states—including my own state of Florida—have been pressured by IDT supporters to allow some form of anti-evolutionism into biology classes.

This is all at the political level. At the more intellectual level, as you can imagine, the criticisms have rained down on IDT. Behe's analogy of the mousetrap has given evolutionists many happy hours of inventive fiddling, as they make mousetraps with increasingly smaller numbers of parts—from five to four, to three, to two, and even to one. Behe's biological examples have also been laughed to scorn.

Take Behe's claim that the blood-clotting mechanism in vertebrates is too complex to have come through evolution. The world authority on blood clotting (Russell Doolittle of the University of California at San Diego) replies that Behe is just out of date and that the evolution of blood clotting is now well supported. (See Behe 1996 and Miller 1999 for details.) Likewise, Dembski's mathematics has received rough treatment. No one denies that setting out to find criteria of design is a legitimate enterprise; it is just that Dembski's ideas did not work and, even if they did, they do not apply to the biological cases that he highlights.

The ploy of claiming that Intelligent Design Theorists are not talking about religion has been the subject of withering scorn. One must be fair here. Not all of the IDT enthusiasts are Young Earth Creationists like the late Henry Morris or Duane T. Gish. Some are. Philosopher and historian of science Paul Nelson, a very big figure in the IDT movement, believes in a short earth span. Others ac-

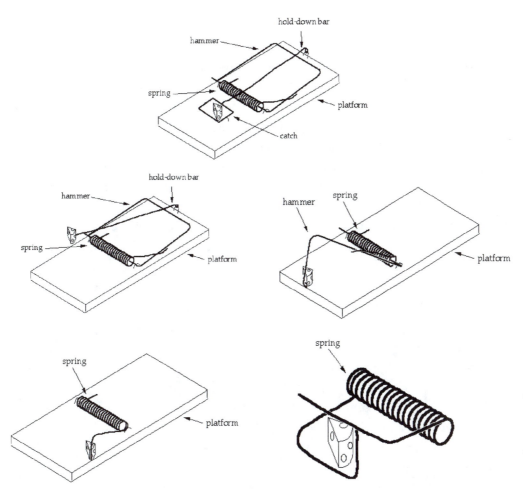

John MacDonald's drawing of a standard five-part mousetrap (top), followed by a series of simpler mousetraps with four, three, two, and finally just one part.

cept conventional dating on the earth, and some—Michael Behe particularly—think that evolution has been important (just not *all* important) in the history of life. (One senses that the Young Earth Creationists are biding their time. In the language of football, they are letting the IDT gang do the blocking for them at this point. If and when they get something into the schools, that will be the time to start divvying up the spoils and to make greater demands. Down the road, I don't think Behe should be looking for much gratitude.)

Fair or not, the simple fact is that the theological push behind IDT is there and thinly concealed—better concealed since they realized that evolutionists were reading their Web pages and circulated e-mails where they were being candid. Time and again we are told that the Intelligent Designer is the Logos of the Gospel of Saint John.

> [1]In the beginning was the Word, and the Word was with God, and the Word was God. [2]The same was in the beginning with God. [3]All things were made by him; and without him was not any thing made that was made. [4]In him was life; and the life was the light of men. [5]And the light shineth in darkness; and the darkness comprehended it not.

But, while important, the science—and even the religion—is only one part of what motivates the new Creationists. (In respects, the old Creationists too, although for them biblical literalism is the key plank.) If one stopped here, one would be missing a very important part of the story. Quite explicitly, there is more to the IDT case than we have thus seen, and here (at first) we start to move to more philosophical questions. Johnson particularly is strong on this matter. In particular, the IDT supporters argue that Darwinism succeeds *faux de mieux*, simply because it is the only game in town. The scene is set so that a position such as IDT is ruled out of court at the beginning, and then Darwinism is declared the winner! The way in which this is done is through an insistence that science—all science—be naturalistic, that is to say something that works according to unbroken law. Then since this is true of Darwinism and is not true of any theistic position which postulates the action of miracle, Darwinism alone qualifies as a proper answer about origins. It wins by default.

In fact, Johnson's position is a little more forceful than this. Not only does he think that Darwinism wins by sleight of hand, but also he thinks that (although some Darwinians may say otherwise) the evolutionary position tips one into atheism. In Johnson's opinion, the classic move made by the Darwinian is to distinguish between so-called methodological naturalism and so-called metaphysical naturalism. A methodological naturalist is one who insists that natural explanations can be given for anything, including organic origins. "Hence all events in evolution (before the evolution of intelligence) are assumed to be attributable to unintelligent causes. The question is not *whether* life (genetic information) arose by some combination of chance and chemical laws, to pick one example, but merely *how* it did so" (Johnson 1995, 208).

Johnson is at pains to allow, indeed to stress, that this does not mean to say that methodological naturalists think that all of the crucial scientific problems have now been solved. Indeed they will agree that this is not the case. But their optimism is that through time and effort the unsolved problems will fall away, dissolved and settled by the scientist—the scientist working purely in a naturalistic mode. "Bringing God or intelligent design into the picture is giving up on science by turning to religion (miracle) and invoking a 'God of the gaps.' The Creator belongs to the realm of religion, not scientific investigation" (p. 208).

Metaphysical naturalism, on the other hand, is a philosophical thesis about the nature of reality. Here the assumption is that what you see is what you get is what there is. There is nothing more to existence than basic particles interacting without end, without purpose. "To put it another way, nature is a permanently closed system of material causes and effects that can never be influenced by anything outside of itself—by God, for example. To speak of something as 'supernatural' is therefore to imply that it is imaginary, and belief in powerful imaginary entities is known as superstition" (pp. 37–38). The position here is "metaphysical" because it is making a claim about ultimate reality, in particular that there is no such reality beyond that within the scope of the scientist. Johnson argues that whatever methodological naturalists may say to the contrary, invariably they find themselves sliding into metaphysical naturalism, and before you know it you have full-blown atheism on your hands.

Now there are two questions that arise here. First, is Johnson's claim well taken? Is methodological naturalism the slippery slope to metaphysical naturalism? Second, are we getting the whole story or is there something else bugging Johnson? Is there an even deeper level of concern? Let us start with the first question. Is it truly the case that, if once you have accepted methodological naturalism, you are on the slippery slope to atheism? I am not at all convinced. I will agree that if you accept methodological naturalism (and I would think of myself as being one who does, incidentally, so you know where I stand), then you are almost certainly going to be an evolutionist. I suppose logically you could think that all of the world's organisms are as old as the universe and that therefore there was no evolution, but we know that this is empirically false. The evidence points to evolution—I myself would say that the evidence points to Darwinism—so it is certainly true that, as things are, descent with modification is a consequence of methodological naturalism.

But does this now mean that the whole god question is ruled out? It would surprise the Pope and it rather surprises me. Let us suppose, for I do not want to get an easy victory by unfair definition, that you are a Christian and as such you think that the Bible must be true. Obviously if you insist on a literal reading, that is an end to matters. Evolution is out, and you might as well agree at once to a denial of methodological as well as metaphysical naturalism. But—and here I am not making things up but simply reporting fact—it has never been part of orthodox Christianity, Catholic or Protestant, that the Bible must be taken literally, word for word (McMullin 1985). We have seen already the most sincere of Christians, people like Cuvier, knew that this is not the way to go. Literalism is a nineteenth-century American invention. In fact, as I told you in an earlier chapter, the insistence that the days of creation are of twenty-four-hour duration comes out of that sect known as Seventh-day Adventism who, keen as they were to insist on the Sabbath (Saturday) as the day of rest, wanted the other days to be of the same length so as to reinforce their special beliefs about the seventh day. One could hardly insist on people taking long periods of time off to rest, which would seem the consequence if the six days are understood metaphorically. (George McCready Price, who inspired the authors of *Genesis Flood*, was a Seventh-day Adventist.)

But if Genesis is not literally true, but only metaphorically true, what price God then? Can you be an evolutionist—a genuine one, not the Asa Gray variety who goes in for guided mutations—and yet take in the essential heart of the Bible? The answer of course depends on what you take to be the "essential heart" of the Bible. At a minimum we can say that, to the Christian, this heart speaks of our sinful nature, of God's sacrifice, and of the prospect of ultimate salvation. It speaks of the world as a meaningful creation of God (however caused) and of a foreground drama that takes place within this world. I refer particularly to the original sin, Jesus' life and death, and His resurrection and anything that comes after it. And clearly at once we are plunged into the first of the big problems, namely that of miracles—those of Jesus himself (the turning of water into wine at the marriage in Cana), his return to life on the third day, and (especially if you are a Catholic) such ongoing miracles as transubstantiation and those associated, in response to prayer, with the intervention of saints (Ruse 2001).

The metaphysical naturalist would reject all of these. But what about the methodological naturalist? There are a number of options. You might simply say that such miracles occurred, that they did involve violations of law, but that they are outside your science. People do not usually rise from the dead three days after being crucified, but on one occasion someone did. You cannot explain the event scientifically, but this does not mean that it did not happen. And the same is true of other miracles. They are simply exceptions to the rule. End of argument. A little abrupt, but I am not sure that this is an impossible option. You simply say that God laid the salvation history on top of the normal course of events. The world goes by law, and then Jesus and the saints worked their ways on top of this. In fact, turning an apparent weakness into a strength, you say that what makes the biblical miracles particularly miraculous and wonderful is the fact that they are so uncommon. If miracles happened on a daily basis, the resurrection would be disvalued. Precisely because people do not rise from the dead three days after being crucified, the fact that Jesus did makes it truly significant.

Or you might say that miracles occur but that they are compatible with science, or at least not incompatible. Jesus was in a trance and his rising on the Third Day involved no breaking or lifting of law. Likewise, the cure for cancer after the prayers to Saint Bernadette is according to rare, unknown, but genuine laws. This position is less abrupt, although I will admit that I worry whether it is truly Christian, in letter or in spirit. It seems to me a little bit of a cheat to say that the Jesus taken down from the cross was truly not dead, and the marriage in Cana (when Jesus turned water into wine) starts to sound like outright fraud. Did he bring a barrel of Chardonnay and not tell anybody, or were the guests so drunk that they could not tell what they were drinking? You start stripping away at more and more miracles, downgrading them to regular occurrences blown up and magnified by the Apostles, but in the end this rather defeats the whole purpose.

The third option is simply to refuse to get into the battle at all. You argue that the law/miracle dichotomy is a false one. Miracles are just not the sorts of things that conflict with or confirm natural laws. This is not such a strange or ad hoc suggestion. Christians already accept that some miracles fall into this category. Take for instance transubstantiation—the miracle accepted by Catholics that in the Mass there is a turning of the bread and the wine into the literal body and blood of Christ. This miracle (or if you prefer, this purported miracle) is simply not something open to empirical check. You cannot disconfirm religion or prove science by doing an analysis of the host. Likewise one might say that the same is true even of the resurrection of Jesus. After the Crucifixion, his mortal body was irrelevant. The point was that the disciples, downcast and dispirited, suddenly felt

Jesus in their hearts and were thus emboldened to go forth and preach the gospel. Something real happened to them, but it was not a physical reality—nor, for instance, was Paul's conversion a physical event, even though it changed his life and those of countless after him. Today's miracles also are really more a matter of the spirit than the flesh. Does one simply go to Lourdes in hope of a lucky lottery ticket to health or for the comfort that one knows one will get, even if there is no physical cure? Surely the latter at least as much as the former. Miracles are matters of feeling and meaning, not of transgressions of nature. In the words of the philosophers, it is a category mistake to put miracles and laws in the same set.

It seems to me that there are at least these options for the would-be Christian who wants also to be an evolutionist. I myself am not equally keen on each and every one, but there is here surely enough to satisfy the would-be believer. I recognize that not every one would be acceptable to every Christian. Protestants, for instance, do not accept transubstantiation, and although they do not have a shared alternative, many (probably most) think that the Eucharist (the ceremony involving the bread and the wine) simply is symbolic of Jesus' last supper with his disciples. The same is true of the other miracles and their possible explanations. Taking the resurrection metaphorically or in spirit only is certainly not accepted by all or even most Christians. But the point is that these options are all accepted by some Christians, and by no means indifferent or careless believers. Indeed, some of the most passionate and devout go for these alternatives.

Johnson (1995), however, sneers that such options are not "intellectually impressive" (p. 211). He adds: "Makeshift compromises between supernaturalism in religion and naturalism in science may satisfy individuals, but they have little standing in the intellectual world because they are recognized as a forced accommodation of conflicting lines of thought" (p. 212). Which of course is absolutely true. Johnson is right. Makeshift compromises rarely do having much standing in the intellectual world. But are the sorts of options I have listed of this nature? To the contrary, the very difficulties I have been discussing—having to take miracles on faith despite the evidence against them or having to admit that there are no physical miracles at all—are taken by some very significant theologians of our age to be the very crux of what it is to be a Christian (Barth [1949] 1959; Bultmann 1958; Gilkey 1985). They believe that, if we can get a guarantee on all of the answers, then commitment is devalued. Faith without difficulty and opposition is not true faith. "As the Danish philosopher Soren Kierkegaard … taught us, too much objective certainty deadens the very soul of faith. Genuine piety is possible only in the face of radical uncertainty" (Haught 1995, 59).

Such thinkers, often conservative theologically—revealingly they are known as the "neo-orthodox"—are inspired by the Jewish philosopher Martin Buber (1937) to find God in the center of "I-Thou" personal relationships. For them there is something degrading in the thought of Jesus as a miracle man, a sort of fugitive from the Ed Sullivan show. What happened with the 5,000? Some hocus-pocus over a few loaves and fishes? Was the Redeemer no more than a high-class caterer? Or did Jesus fill the multitude's heart with love, so there was a spontane-

Phillip Johnson

ous outpouring of generosity and sharing, as everyone in the crowd was fed by the food brought by a few? Surely this is what truly happened. This is what Christianity is really about.

Part of the problem when dealing with matters to do with Christianity and science, evolution in particular, is that so many people believe in so many things. For instance, the position I have just been sketching—that faith is only genuine faith in the face of uncertainty—would be denied by Catholics. They believe that one can in fact prove the existence and nature of God through reason. Although, especially given the Pope's position on evolution, this certainly does not mean that they would now swing round and think that Johnson is right. If anything, Catholics tend to be more opposed to biblical literalism than Protestants. But at this point we can honorably pull back from the details. It is enough to show, and this surely has been shown, that the whole science/religion relationship is more complex than allowed by people like Phillip Johnson. More complex, and I would say more interesting and more fruitful.

Modernism

In one sense, I do want to agree with Johnson somewhat. It is true that evolution, Darwinism in particular, is identified with what is usually known as "modernism." This term has various uses, including reference to a liberal kind of theology. My sense, although obviously all of the senses are linked, is cultural and social, meaning a kind of liberal attitude to society and its denizens. That was the case in the nineteenth century, it was the case in the twentieth century, and it is still true today.

The fact is that when you read Johnson, and when you read the other ID theorists as well, very quickly we start to leave the realm of science and religion and get into the realm of morals, of social behavior. People like Johnson are absolutely appalled at what they think is the dreadful turn that has been taken by American society. (The same is true of critics of evolution elsewhere.) They take Darwinism to be emblematic of everything that is wrong. It is not so much Darwinism in itself but Darwinism as a symbol, as a flag, as the kind of ideology that has been pushed by people from Thomas Henry Huxley on. If you doubt me, then ask yourself what this passage from Johnson has to do with gaps in the fossil record.

> "A responsible society is based first and foremost on responsible parents who fulfill their obligations to each other and to their children. Probably the most important thing that most adults do is to prepare the next generation for the joys and responsibilities of life. To do this they must ensure to the best of their ability that their children are born healthy. Following birth, children must be nurtured and educated in moral behavior by loving parents, preferably *two* parents. That is one reason it is important for lovers to regard marriage as a sacred bond, rather than as a contractual arrangement to be terminated at the convenience of either party. That is also why mothers in a rational society regard their children, born and unborn, as a sacred trust rather than primarily as an encumbrance that men impose on women in order to make them unhappy and impede their pursuit of wealth, power and pleasure. Similarly, fathers in a rational society regard their offspring from the beginning of pregnancy as their own flesh, so that they become enthusiastic providers and conurturers rather than the unwilling objects of child-support orders." (Johnson 1995, 150-1)

Paradoxically, I think it has everything to do with gaps in the fossil record. Johnson and fellows see science generally, evolution specifically, as being bound up with a philosophy of life that promotes abortion on demand, homosexual marriage, teen out-of-wedlock pregnancy, and more. Johnson also obsesses about cross-dressing, a somewhat strange fixation until you realize that it is connected to his anti-feminism — women wearing pants and that sort of thing.

William Hamilton is generally considered the evolutionary genius of the second half of the twentieth century. He was responsible for major innovations that we shall encounter in *Chapter 9: Human Sociobiology*, innovations explaining intricacies of animal social behavior. He made breakthroughs on problems that had puzzled evolutionists since the days of the *Origin of Species*. In the words of Richard Dawkins: "Those of us who wish we had met Charles Darwin can console ourselves: we may have met the nearest equivalent that the late twentieth century had to offer" (Hamilton 2001, xi). Listen to Hamilton on the family.

> "One of the ways in which I think backing plus curbing of the hypocrisies of individualism will come about will be through a greater measure of *family* responsibility that political parties will see it as a necessary measure to impose."

Hamilton believes that individuals rather than society should be those facing consequences of decisions made (say) about handicapped children, and if groups in society (for instance, church organizations) want to get involved in advocating various practices — by example, insisting on a total prohibition of abortion even though it is known that the fetus is dreadfully damaged — then they too should be prepared to offer support. In fact, society should be relieved of any obligations.

> "In general along such lines, it will be a great step in the equitable running of modern society if a sincerity tax comes to be imposed on all propaganda — what you say you believe in you must show you believe in through hard cash and sacrifice; as an example again, there should be no option but that your child attends the idealistic comprehensive school you say you believe in." (Hamilton 2001, xlviii).

I suspect that most of us would not want to go this far. I certainly don't. My personal feeling is that if we are not prepared to force behaviors on people — say compulsory abortions — and I am not, then we as a group have a responsibility to any and all children. I would go so far as to say that we have a responsibility to the parents with whom we disagree. My guide here is *Meditation XVII* of the great English poet of the seventeenth century, John Donne:

> "All mankind is of one author, and is one volume; when one man dies, one chapter is not torn out of the book, but translated into a better language; and every chapter must be so translated...As therefore the bell that rings to a sermon, calls not upon the preacher only, but upon the congregation to come: so this bell calls us all: but how much more me, who am brought so near the door by this sickness....No man is an island, entire of itself...any man's death diminishes me, because I am involved in mankind; and therefore never send to know for whom the bell tolls; it tolls for thee."

I don't think you have to be religious to see the force of what is being said here. We are all part of one family and that brings responsibilities for all. But this is not really my point here. My point — actually my two points — is that first Hamilton is trying to make his case on biological grounds and as such thinks that this means we must approach matters from a perspective that reflects the workings of biology, and second — and this is what is really pertinent here — what I find really interesting is the fact that Hamilton, like Johnson, wanted to talk about the family. There is a real clash here. People are not talking past each other. They are talking at each other. And this I suspect is much of what is at stake when the Creationists — full strength or light — start bashing Darwinism. It is about the way to live. We shall see confirmation of this point shortly.

Darwinian Atheism

Swing around now and look at the other side. Let us focus in on Richard Dawkins. In *The God Delusion*, he does not mince words: "The God of the Old Testament is arguably the most unpleasant character in all fiction: jealous and proud of it; a petty, unjust, unforgiving control-freak; a vindictive, bloodthirsty ethnic cleanser; a misogynistic, homophobic, racist, infanticidal, genocidal, filicidal, pestilential, megalomaniacal, sadomasochistic, capriciously malevolent bully." (31) Dawkins is not much more friendly to the God of the New Testament either, writing of "his insipidly opposite Christian face, 'Gentle Jesus meek and mild'." Dawkins makes no bones about the immorality of giving a child a Christian education. "Once, in the question time after a lecture in Dublin, I was asked what I thought about the widely publicized cases of sexual abuse by Catholic priests in Ireland. I replied that, horrible as sexual abuse no doubt was, the damage was arguably less than the long-term psychological damage inflicted by bringing the child up Catholic in the first place." (317)

This is nothing new. Let me quote a couple of paragraphs from an interview that Dawkins gave a few years ago.

> I am considered by some to be a zealot. This comes partly from a passionate revulsion against fatuous religious prejudices, which I think lead to evil. As far as being a scientist is concerned, my zealotry comes from a deep concern for the truth. I'm extremely hostile towards any sort of obscurantism, pretension. If I think somebody's a fake, if somebody isn't genuinely concerned about what actually is true but is instead doing something for some other motive, if somebody is trying to appear like an intellectual, or trying to appear more profound than he is, or more mysterious than he is, I'm very hostile to that. There's a certain amount of that in religion. The universe is a difficult enough place to understand already without introducing additional mystical mysteriousness that's not actually there. Another point is esthetic: the universe is genuinely mysterious, grand, beautiful, awe inspiring. The kinds of views of the universe which religious people have traditionally embraced have been puny, pathetic, and measly in comparison to the way the universe actually is. The universe presented by organized religions is a poky little medieval universe, and extremely limited.
>
> I'm a Darwinist because I believe the only alternatives are Lamarckism or God, neither of which does the job as an explanatory principle. Life in the universe is either Darwinian or something else not yet thought of. (Dawkins 1995a, 85–86)

These paragraphs are very revealing, showing the emotional hostility that Dawkins feels toward religion, including (obviously) Christianity. I am sure the reader will not be surprised to learn that Dawkins has characterized his move to atheism from religious belief as a "road to Damascus" experience (Dawkins 1997c). Saint Paul would have recognized a kindred spirit. But my purpose in quoting Dawkins's words here—and I could equally quote Dennett or Harris or Hitchins—is not so much to pick out the emotion, as to point to the logic of Dawkins's thinking. This comes through particularly in one of the passages just quoted. It is clear that for Dawkins we have here an exclusive alternation. Either you believe in Darwinism or you believe in God, but *not both*. For Dawkins—as for Phillip Johnson on the other side—there is no place for what philosophers call an inclusive alternation, that is to say either a or b or possibly both. (The third way mentioned is Lamarckism, the inheritance of acquired characteristics. But neither Dawkins nor anybody else today thinks that this is a viable evolutionary mechanism.)

Why not simply slough off Christianity and ignore it? At the purely intellectual level (if, after the passages just quoted, we are ever capable of finding this level again), things are not this simple: as we saw in earlier chapters, Dawkins—like any good Darwinian including Charles Darwin himself—recognizes that the Christian religion poses the important question, namely that of the designlike nature of the world (Dawkins 1986). Moreover, Dawkins believes that until Charles Darwin no one had shown that the God hypothesis, that is to say the God-as-designer hypothesis, is untenable: more particularly, Dawkins argues that until Darwin no one could avoid using the God hypotheses.

In this context, Dawkins is fond of telling a story about a conversation he once had with a well-known philosopher. (Although Dawkins never tells us in print who it is, he himself has told me that it was the late Sir Freddy Ayer, a fellow Oxford professor and a notorious atheist.) Apparently the conversation took place at one of those famed Oxford college feasts, where the food is abundant (in my experience, usually pretty dreadful) and the wine even more abundant (in my experience, always very good). Probably everyone was indulging well if not wisely, and finally the philosopher—in a rather sneering and condescending way—challenged the biologist. Surely, he asked, there is nothing in the living world that demands special explanation. All of this nonsense by Christians and biologists alike about the special nature of animals and plants is silly make-believe, pretending that things are more significant and interesting than they really are.

In reply, the aroused biologist demanded an explanation of the complexity that we see around us. Asked Dawkins, does this not require some special understanding? Not at all, replied the philosopher. The living world is as it is and simply exists. That is all and that is enough. But it is not enough, replied Dawkins then, and to this question that continues to haunt him, he still replies sternly. The living world is special. "Paley knew that it needed a special explanation; Darwin knew it, and I suspect that in his heart of hearts my philosopher companion knew it too" (Dawkins 1986, 6).

It is true—as Dawkins concedes—that David Hume made devastating criticisms of the argument from design. In his *Dialogues Concerning Natural Religion* (first published in 1777), Hume showed, for instance, that the living world might as reasonably have had a team of gods making it as having one unaided creator. He showed that if the world is designed by God or by gods, then it is reasonable to think that there were many previous attempts and trials. There must exist somewhere a whole series of cruder or botched earths, which were the forerunners of our earth—or we must have been formed out of them. Indeed, it may be the case that we ourselves are not living in the final and perfected world. Hume showed in fact that we might as well think that the world is as much like a giant vegetable as like an object of design!

But in Dawkins's opinion there is still a gap requiring a filling. "What Hume did was criticize the logic of using apparent design in nature as *positive* evidence for the existence of a God. He did not offer any *alternative* explanation for apparent design, but left the question open" (p. 6). Dawkins continues: "An atheist before Darwin could have said, following Hume: 'I have no explanation for complex biological design. All I know is that God isn't a good explanation, so we must wait and hope that somebody comes up with a better one'" (p. 6). But, in Dawkins's opinion, this is not enough. "I can't help feeling that such a position, though logically sound, would have left one feeling pretty unsatisfied, and that although atheism might have been *logically* tenable before Darwin, Darwin made it possible to be an intellectually fulfilled atheist" (Dawkins 1986, 6).

At this point, some of the sorts of questions asked of Johnson start to seem pertinent. (Stay for the moment at the intellectual level.) Why should we not say that Dawkins is certainly right in stressing the designlike nature of the organic world, but he is wrong in thinking that it is either Darwinism or God, but not both? At least, even if he is not wrong, he has failed to offer an argument for this. Perhaps the designlike nature of the world testifies to God's existence. It is simply that God created through unbroken law. Indeed, as we have seen, people in the past would argue that the very fact that God creates through unbroken law attests to his magnificence. Such a God is much superior to a God who had to act as Paley's watchmaker would have acted, that is, through miracle.

In fairness, I think that at this point Dawkins does have a second argument up his sleeve. It is the venerable argument based on the problem of evil. But for Dawkins it is more than just the traditional argument (which is in itself not particularly evolutionary). What Dawkins would argue is that not only does evolution intensify the problem of evil, but Darwinism in particular makes it an overwhelming barrier to Christian belief. This argument is expressed most clearly in one of Dawkins's books published a few years ago: *River Out of Eden: A Darwinian View of Life* (1995b). In a chapter entitled "God's Utility Function," he starts by pointing to the fact that many adaptations require that other organisms suffer, sometimes greatly. "A female digger wasp not only lays her egg in a caterpillar (or grasshopper or bee) so that her larva can feed on it but ... she carefully guides her sting into each ganglion of the prey's central nervous system, so as to paralyze it *but not kill it*. This way, the meat keeps fresh. It is not known whether the paralysis acts as a general anesthetic, or if it is like curare in just freezing the victim's ability to move. If the latter, the prey might be aware of being eaten alive from inside but unable to move a muscle to do anything about it" (p. 95). All of this sounds pretty dreadful and cruel, but Dawkins's conclusion is that speaking of cruelty in such a situation is no better than speaking of beneficence and kindness. "Nature is not cruel, only pitilessly indifferent. This is one of the hardest lessons for humans to learn. We cannot admit that things might be neither good nor evil, neither cruel nor kind but simply callous—indifferent to all suffering, lacking all purpose" (Dawkins 1995b, 95–96).

Then, Dawkins goes on to reinforce this point. He talks about organisms being excellent examples of designlike engineering. If we tried to unpack the engineering principles involved in organisms, the problems of pain and evil would

Cheetah

come to the fore. Meaning by the notion "utility function" the purpose for which an entity is apparently designed, Dawkins asks about God's Utility Function when it comes to carnivores and their prey. Consider the cheetah, a beautiful piece of design if anything is. We can work backward, "reverse-engineering," trying to ferret out the way it which it was put together and the purposes of its various adaptations. We can probably be fairly successful in our labors, for the problem posed by cheetahs is relatively easy. "They appear to be well designed to kill antelopes. The teeth, claws, eyes, nose, leg muscles, backbone and brain of a cheetah are all precisely what we should expect if God's purpose in designing cheetahs was to maximize deaths among antelopes" (p. 105). The same is true of the cheetah's prey. "If we reverse-engineer an antelope we find equally impressive evidence of design for precisely the opposite end; the survival of antelopes and starvation among cheetahs. It is as though cheetahs had been designed by one deity and antelopes by a rival deity" (p. 105). Or if we want to suppose that there was one designer responsible for both cheetahs and for gazelles, then legitimately we might ask about His intentions. "Is He a sadist who enjoys spectator blood sports? Is He trying to avoid overpopulation in the mammals of Africa? Is He maneuvering to maximize David Attenborough's television ratings?" (p. 105).

This is silly of course. No one would draw such a conclusion as this. The point at best seems to be that if there be a God, then He is one who certainly is nothing like the Christian God. He is unkind and unfair or, more likely, totally indifferent. And indeed, this is the point at which Dawkins ends the discussion of this chapter. We simply have to accept that natural selection works by and through pain, pain, and more pain. All of the time, animals are dying: from starvation, from disease, from being eaten by prey, and from many other horrible causes. Things may ease up for a minute or two, but then trouble and pain reappear, even brought on by the pauses. If the predator number is reduced, the prey increase, and then there are more predators in turn and yet more killing than average.

In Dawkins's opinion, this is pointing to an appalling theology, unless we simply stop and realize that our argument is entirely on the wrong track. There is neither a good god nor a bad god. There is simply no god. You ask about human tragedy. Why expect an answer for there is no answer. There is simply nothing.

> In a universe of blind physical forces and genetic replication, some people are going to get hurt, other people are going to get lucky, and you won't find any rhyme or reason in it, nor any justice. The universe we observe has precisely the properties we should expect if there is, at bottom, no design, no purpose, no evil and no good, nothing but blind, pitiless indifference. As that unhappy poet A.E. Houseman put it:
>   For Nature, heartless, witless Nature
> Will neither know nor care.
>   DNA neither knows nor cares. DNA just is. And we dance to its music. (1995b, p. 133)

This is powerful stuff, and whether you agree with Dawkins or disagree, I have no time for anyone who trivializes it. The problem of evil is the biggest of all the obstacles to Christian belief, and Dawkins is absolutely right to point out—to stress—that Darwinism brings it right to the fore. The way in which Darwinian evolution works is through pain and suffering and cruelty and hardship and deprivation and much, much more. You cannot get away from this fact, nor should you pretend to do so. But my suspicion is that Dawkins himself provides the answer! This is a paradox, but true nevertheless. The ardent Darwinian, the Richard Dawkins (or Michael Ruse for that matter), believes above all that the mark of the organic world is its designlike nature. Animals and plants are adapted. We are with Archdeacon Paley and Georges Cuvier and Charles Darwin on this. But how can one produce this design? If the only way that is possible is through natural selection, then one can argue that God did what He did because He had to. There was no choice. And so the pain follows naturally. It is not God's fault for not preventing it. He could not prevent it. It has always been stressed in Christian theology that God's power—His omnipotence—never meant doing the impossible. God cannot make $2 + 2 = 5$.

Might it not be that, God having decided to create, did then create—perhaps His choice, perhaps not—in an evolutionary fashion? And this being so, might it not be that He was now locked into a path that would necessarily lead to physical evil? It comes with the method employed. The theologian Bruce Reichenbach (1976) makes this objection against the suggestion that God might have used better laws of nature, that is, laws that do not lead to physical evil. At first sight it seems easy for God to have done a better job, making a universe without all of the pain and suffering that we find throughout. But would it really have been all that easy? "For example, what would it entail to alter the natural laws regarding digestion, so that arsenic or other poisons would not negatively affect my constitution? Would not either arsenic or my own physiological composition or both have to be altered such that they would, in effect, be different from the present objects which we now call arsenic or human digestive organs?" (Reichenbach 1976, 185) And this is just the beginning. Think of such everyday things as fire and electricity and the solidity of wood—things that we would be most loathe to relinquish. But in a nonpainful world all sorts of changes would be needed. "Fire would no longer burn or else many things would have to be by nature non-combustible; lightning would have to have a lower voltage or else a consistent repulsion from objects; wood would have to be penetrable so that clubs would not injure" (p. 185).

More than this. Suppose we accept that the world evolved. How could we prevent pain and suffering from occurring? Either we are going to have to change the laws of nature in significant ways, or we are going to have to alter the initial conditions of the universe so that different results come about. Both alternatives raise major problems. If we alter the laws themselves, then at a minimum we will have to alter humans, and this might entail unpleasant or unacceptable theological conclusions. We human are sentient beings, part of nature. That is to say we have a natural physiology, we work according to fixed laws of nature, we see and sense generally because we function like the rest of the world. It comes with the territory that we will encounter unpleasant phenomena—pain and the like—and that we will be conscious of it and not like it. From a biological point of view, if we did not have pain and did not dislike it, we would not function properly. But to alter all of this, we would have to be removed from nature in the sense that we now know it. We would have to be immune from the ways of the world. But if this comes about to be the case, do we now have a being that would be loving and giving (or hating and hurtful)—in other words, would we still have a being of a kind that is supposed to be at the center of God's creation and on which a religion such as Christianity claims that He lavishes so much care and love?

The other alternative suggests that the initial conditions might be altered, thus avoiding unwanted painful conclusions. But what would this mean and entail in fact? If the Big Bang story is right, way back at the near beginning everything was hydrogen. Altering the initial conditions would presumably therefore mean altering the nature and functioning of hydrogen, and probably consequently all of the other elements. But where do you stop and what guarantee do you have that things will now turn out better? In particular, we do not know if humans would have evolved and if they did evolve whether the things that make for pain and so forth would have failed to have evolved alongside. "Whether humans would have evolved but no infectious virus or bacilli, or whether there would have resulted humans with worse and more painful diseases, or whether there would have been no conscious, moral beings at all, cannot be discerned. Given a change in initial conditions, it is possible that this world would not have had any less natural evil while not preserving free moral activity" (Reichenbach 1976, 192–193). All in all therefore, we seem to be in as much trouble after we have made these moves as before. Clearly there is here no devastating argument against the person who believes in a caring and loving God.

I have stressed that the key aspect of organic form is (as we have seen) its adaptedness, and it is this that (as we have also seen) is addressed by natural selection. Physical or natural evil is a result of the causes or a consequence of this selective process. But could one not have got adaptedness by a physical process much nicer than selection? Here we return to the paradox mentioned just above. It is Dawkins (1983) himself who comes to the aid of the theist, for he more than anyone argues strenuously that selection and only selection can do the job. Most putative processes simply do not lead to adaptation: saltationism, evolution by jumps, for instance. Indeed, in Dawkins's opinion, there is a general principle in biology that adaptive complexity always comes through small, gradual processes rather than from big, sudden, incremental changes. And those rivals to selection that address adaptation and that might do things gradually—notably Lamarckism—are known to be false. So it is selection or nothing.

> My general point is that there is one limiting constraint upon all speculations about life in the universe. If a life-form displays adaptive complexity, it must possess an evolutionary mechanism capable of generating adaptive complexity. However diverse evolutionary mechanisms may be, if there is no other generalization that can be made about life all around the Universe, I am betting that it will always be recognizable as Darwinian life. The Darwinian Law ... may be as universal as the great laws of physics. (p. 423)

God had no choice but to take the option that He chose.

In the end therefore, for all that so many people think that a true Christian could never be an evolutionist—certainly could never be a Darwinian—it turns out that, for the Christian, Darwinism is to be welcomed positively at this point. Physical evil exists, and Darwinism explains why God had no choice but to allow it to occur. He wanted to produce designlike effects—without producing these He would not have organisms, including humankind—and natural selection is the only option open. Natural selection has costs—physical pain—but these are costs that must be paid. What more need be said?

Darwinian Religion

I am arguing what history has shown: there is really no reason why a Christian should not be a Darwinian, and there is really no reason why a Darwinian should not be a Christian. I am not saying that you should be a Christian, and I am not really saying that you should be a Darwinian, but I am saying that the one does not preclude the other. But is this not all a bit redundant? We have seen that Darwinism has been used as a kind of secular religion—Religion without Revelation. Should we not all be going that way now? One more time, let us take up the thinking of Richard Dawkins. I argued at the end of my treatment of the Creationists that more was at stake than simple science and religion (assuming that these things are ever simple) and that there was a battle over the very way you run your life. For the anti-evolutionists, it is modernism in some sense that is the real anti-Christ. Quoting Hamilton, I suggested that this was not a purely one-way phenomenon. Many evolutionists feel the same way as does Johnson—except in reverse! They want to embrace modernism. And this of course is what is motivating someone like Dawkins through and through. For him, religion is not just wrong. It is abusive. It leads to great ills. It does not take a genius to see that the spate of atheistic books like his have arrived as soon as they could have been written after the terrible events in New York City on September 11, 2001. For Dawkins, Harris, Hitchins, and all of the rest, religion is not an option. That is why someone like me is labeled an "appeaser." It is like tangling with Hitler. Now is not the time to be open and fair. Now is the time to fight with all of your might. We are engaged in a great moral crusade.

At times like this, I always think of the story of Noah and the Flood. I don't think it is about boat-building at all, or about the geological effects of too much water. I think it is a caution against simplistic solutions to complex problems. God is faced with a world gone wrong. So at one fell swoop he washes it all away except for Noah and his family. But what happens next—the part that you are usually not encouraged to read when you are a child? Noah gets blind drunk and one of his sons laughs at the old man in his nakedness. Evil is still there. All of God's efforts were for nothing. The same is true of saying you are going to sweep away Christianity and all of the other religions. First, you are never going to do it. If the Enlightenment meant anything it meant the end of Christianity. Yet look at America today! Second, even if you do, still nastier things lie waiting to take its place. Dawkins goes to great efforts to suggest that truly Hitler and Stalin and company were Christians and so Christianity is responsible for the Holocaust and gulags and so forth. Which is about as plausible as the claim that the earth is six thousand years old. Only someone truly on a mission could think that the evils of Mao Zedong stemmed exclusively from a reading of the Sermon on the Mount. Religion has many faults, but to think that you are going to solve the problems of the world by getting rid of religion is about on a par with planning another big flood. So for or against modernism, I am not about to take Richard Dawkins or his friends as my guide on what to think and do.

There is a difference between Dawkins and Edward O. Wilson. The former says it is morally wrong to be a religious believer. The latter certainly does not say that. Nevertheless, he does himself reject Christianity and wants to embrace a kind of secular religion based on evolution. As we know, he believes we have an upward rise to humankind, yielding moral prescriptions—telling us what we should do— and for him this is where it begins and ends. He (1978) writes:

> But make no mistake about the power of scientific materialism. It presents the human mind with an alternative mythology that until now has always, point for point in zones of conflict, defeated traditional religion. Its narrative form is the epic: the evolution of the universe from the big bang of fifteen billion years ago through the origin of the elements and celestial bodies to the beginnings of life on earth. The evolutionary epic is mythology in the sense that the laws it adduces here and now are believed but can never be definitively proved to form a cause-and-effect continuum from physics to the social sciences, from this world to all other worlds in the

visible universe, and backward through time to the beginning of the universe. (p. 192)

And, in fact, Wilson goes even further than this. He thinks that biology now can explain religion, as something that is needed for group cohesion or some such thing. This means that religion is on the way out. In the future, it will at best be seen as a consequence of a more powerful, more adequate, world picture. Theology will no longer survive as an autonomous subject.

Well, perhaps! But what if you do not share Wilson's vision of progress up to humankind? What if you think that any progress you see in evolution is something that you have read into the process rather than found and read out? What if you do not share Wilson's materialism? What if, like me, you think that (in this quantum age) materialism is slightly silly and that even if you extend your understanding of the term, it still does not follow that evolutionism equals materialism? What if you think you can be an evolutionist and a nonmaterialist? And most particularly, what if you think that whether or not evolution can explain religion, nothing is said about what if anything is more important or basic? After all, if evolution be true, at some level everything we know or understand has to come from evolution. But does this tell us about ontological status or importance? I feel hunger and there are good evolutionary reasons. Does this make my hunger any less real? I feel sexual pangs and there are good evolutionary reasons. Does this make my love any less real or genuine or worthwhile? Does this reduce to nothing all of the poetry that has been written? If evolution be true, I fully expect there to be good evolutionary reasons for religion. But does this mean that God does not exist?

I am sure you know by now how I am going to answer these questions. Frankly if God is going to create in an evolutionary fashion, He would be tempting fate if He then made all belief in Him and all religious practice into things that went against our evolved nature. The fact that religion may have an evolutionary base—a selective base even—tells us nothing about the nonreality of religion. For that, we need an additional argument that there are reasons, perhaps evolutionary reasons, to think that our biology is deceiving us over the religion matter. These may exist, but they are not forthcoming. As it is, one has a feeling here that the Creationists may have a good point. The philosophy is being fed in at the beginning of the paragraph, and then triumphantly at the end of the paragraph it is being produced as proven. There is no secret to success in hide and seek that beats first hiding the prizes yourself.

None of what I have just said is to stop someone making a religion of evolution if they so wish. Edward O. Wilson is my friend, and I am proud to acknowledge our relationship. He is a good and gentle man, generous to a fault, with a real moral concern for the world's ills, for problems of biodiversity and ecological preservation in particular. If he wants to take evolution as the new myth, something replacing Christianity, I am happy for him to do so. I see only good coming from this move. Those who condemn the man because they do not share his beliefs are bigots and worse. But I do not see that his fellow evolutionists have to follow him into making a religion of our shared science. This has nothing to do with whether or not we want to opt for some other religion, Christianity for instance, or if we have no religion at all—if perhaps we find no ultimate meaning to life, other than that of everyday living and the joys and troubles with that. The point is that just as being an evolutionist neither compels nor denies Christian belief, so also being an evolutionist neither forces one into nor, for that matter, prevents one from being a member of the Church of Darwin. And that is my final (well, almost final) word on the subject.

Further Reading & Discussion

Spearheading the New Creationist attack on evolution are Phillip Johnson's *Darwin on Trial,* 2d ed. (Downers Grove, Ill.: InterVarsity Press, 1993) and his follow-up work *Reason in the Balance: The Case against Naturalism in Science, Law and Education* (Downers Grove, Ill.: InterVarsity Press, 1995). Although his mentor, Supreme Court Chief Justice Earl Warren, must now be revolving in his grave at his protégé's behavior, Johnson is a brilliant man and these are clever and skilfully written books. I hope you are not convinced by them but do not underestimate them.

Whatever else you might want to say about Michael Behe, I am sure he is a terrific teacher. *Darwin's Black Box: The Biochemical Challenge to Evolution* (New York: Free Press, 1996) is indeed a great read and very persuasive. Behe has a great ability to make a difficult point clear through a simple but appropriate example. I think he is wrong, wrong, wrong, but do not take my word for it. Rather turn to *Finding Darwin's God: A Scientist's Search for Common Ground between God and Evolution* (New York: Cliff Street Books, 1999) by biologist (and practicing Christian) Kenneth Miller. He knocks down both Johnson and Behe with great skill, drawing on a deep and profound understanding of modern biology, both evolutionary and those parts more directed toward physiology and the molecular realm. Miller like Behe has the ability to pick on the right and illuminating example, and I am sure he is also a great teacher. Some people just have a gift for communication. William Dembski and I co-edited a volume: *Debating Design: From Darwin to DNA* (Cambridge: Cambridge University Press, 2004) in which we brought together Darwinians and Intelligent Design Theorists to present our different world pictures.

Balance your reading of the Creationists by reading works by the "new atheists": Richard Dawkins, *The God Delusion* (New York: Houghton Mifflin, 2006); Daniel Dennett, *Breaking the Spell: Religion as a Natural Phenomenon* (New York: Viking, 2006); Sam Harris, *The End of Faith: Religion, Terror, and the Future of Reason* (New York: Free Press, 2004); and Christopher Hitchens, *God is not Great: How Religion Poisons Everything* (New York: Hachette, 2007). I am on record as saying that they make me ashamed to be an atheist (because I think the arguments are so bad), but please don't let my opinion prejudice you.

Judged as a scientist, Edward O. Wilson is today's leading evolutionist. He is also the leading spokesman for a religious-type reading of evolutionary thought. This comes through strongly in his Pulitzer Prize-winning work, *On Human Nature* (Cambridge: Harvard University Press, 1978). Turn also to a recent book, *The Creation: A Meeting of Science and Religion* (New York: Norton, 2006), penned as a letter to a hypothetical Southern Baptist minister. (Wilson is from Alabama and he was raised as a Baptist.) In the book, Wilson lays out his humanist philosophy, but in a way as different from Dawkins as it is possible to do. He wants to build bridges not to burn them. I would also recommend Wilson's autobiography, *Nat-*

uralist (Washington, D.C.: Island Press/Shearwater Books, 1994). It too is inspirational, but the bit I like best is about how miffed was Jim Watson of double helix fame, when he was beaten by Wilson in the race to get tenure at Harvard. More seriously, Wilson's book gives great insight into the life and mind of a scientist—the dedication necessary for real success is rather frightening.

But Is It Science? The Philosophical Question in the Creation/Evolution Controversy (Buffalo, N.Y.: Prometheus), is an edited volume that brings together many different readings on and about the debate over evolution and Creationism. There is some historical material as well as a good selection dealing with the 1981 Arkansas creation trial, not to mention criticisms from my fellow philosophers over the kind of performance I gave in the witness box. It originally appeared in 1988 but a new edition (2008) has just appeared, co-edited with Robert Pennock, who was one of the evolution witnesses at Dover. It is now up to date on Intelligent Design Theory. Finally, let me recommend a trilogy of my books dealing with the science and religion relationship. *Can a Darwinian Be a Christian? The Relationship between Science and Religion* (Cambridge: Cambridge University Press, 2000) tries to look seriously not only at the pertinent science in the evolution/creation debate but also at the relevant theology. *The Evolution-Creation Debate* (Cambridge: Harvard University Press, 2006) tries to show, through history, how many things the two sides share. *Making Room for Faith: Christianity in an Age of Science* (Cambridge: Cambridge University Press, 2009) tries to show how there may still be a place for religion, no matter what the successes of science. I don't have a religion myself, but I don't see why other people should not have one.

Part Two:
EVOLUTION MATURES

New Sciences Emerge:
Disciplines Develop, Independent of Religion

Chapter 6
Darwinism and Genetics:
A New Frontier Opens

Overview

As we shall see in this chapter, and those that follow, Darwin's theory of adaptation through natural selection became an important lynch pin in many of the new 20th century sciences. If Darwin's theory had not been developed a century earlier, it would have had been now, as more and more hard science backed-up his observations.

But Darwin left a big gap in evolutionary thought. What is the nature of the mechanism responsible for passing information on from one generation to the next and why do new variations keep appearing in each generation? The clue came from the thinking of Darwin's virtually unknown contemporary, the Moravian monk Gregor Mendel. When Mendel's experiments in heredity were rediscovered at the beginning of the twentieth century, they were quickly up-dated with even newer discoveries regarding cells and combined to create the 'classical theory of the gene.' But this group of scientists downplayed the influence of natural selection influencing what they saw as the slow but sure progression and change that occurred as a result of occasional mutation. These changes or mutations were seen as ongoing in jumps or steps and not as a continual process. On the other front, we have the biometricians who statistically felt that natural selection was the operating mechanism for change. Genetic changes might occur, but natural selection is what helps make it a permanent part of the population.

It was not until around 1930 that theoreticians melded Darwinian selection and Mendelian genetics to make one unified theory. This "population genetics"—devised by Ronald A. Fisher and J.B.S. Haldane in Britain and Sewall Wright in America—could now serve as the foundation of an invigorated evolutionary theory, "neo-Darwinism" (as it was called in Britain) or the "synthetic theory" (as it was called in America).

A number of thinkers in Britain and America then began experimenting and enhancing the genetic theoretical skeleton. In America, the key figure was Theodosius Dobzhansky, who had left his homeland of Russia in the late 1920s. He was backed by a number of other thinkers, most crucially ornithologist Ernst Mayr, paleontologist George Gaylord Simpson, and botanist G. Ledyard Stebbins.

But work still had to be done to create a fully fledged scientific discipline, with students, organizations, jobs, a journal, grant money and the like. This was the task of people like Mayr, who proved to be a brilliant organizer. As the topic of evolutionary studies was professionalized, effort had to be made to break with the past, when evolution was primarily a vehicle for promoting people's social views. At least, if these views were still to be promoted, that activity had to be in strictly popular venues.

The big row between Dobzhansky and Nobel Prize winner H.J. Muller was over variation in populations. Dobzhansky wanted to argue that selection could act to preserve variation and that hence, when organisms need it due to changed circumstances, it is always waiting there for exploitation. Muller, who was an ardent eugenicist, thought that there are species ideals, and that normally selection wipes out all variation.

It was impossible to solve this problem with conventional techniques, but then came molecular biology—something initially regarded as a threat by conventional biologists. In 1953, the most important single event in the history of twentieth-century biology occurred with the discovery of the double helix structure of the DNA molecule. This too, set off Evolution Wars as the discovery at first was seen to be too narrow to effect the larger issues at hand. What did occur, as a result of these advances, however, was that evolution at last could no longer be seen as a secular religion. It was a true science and could now be separated from the religious world.

The Role of the Scientific Community

The work of the following scientists is discussed in this chapter. Short, biographical essays of these individuals appear in **Biographies** on page 607.

Gregor Mendel (1822–1884)
Julian Huxley (1887–1975)
Sewall Wright (1889–1988)
Hermann J. Muller (1890–1967)
Ronald A. Fisher (1890–1962)
J.B.S. Haldane (1892–1964)
Trofim Lysenko (1898–1976)
Theodosius Dobzhansky (1900–1975)
George Gaylord Simpson (1902–1984)
Ernst Mayr (1904–2005)
G. Ledyard Stebbins (1906–2000)
Richard Lewontin (1929–)

Setting the Stage

I t was too bad you couldn't be at Princeton, where we had a kind of gladiatorial combat from which both sides finally emerged apparently uninjured, so far as each side thought of itself, but demolished, so far as each side thought of the other. At the end, Dobzhansky held out his hand for me to shake and I grasped it firmly, saying "I think you may in time come around after all," at which everybody laughed and the meeting broke up. (Beatty 1987, 289)

Thus a letter from the Nobel Prize–winning geneticist Hermann J. Muller to a student, about an encounter with the leading American evolutionist, the Russian-born Theodosius Dobzhansky. Through the 1950s, they had battled in an increasingly bitter fashion, over the nature of evolution and particularly over the amount of heritable variation one might expect to find in any wild population of organisms. Dobzhansky had rallied his forces, his own students mainly, and had run experiment after experiment on populations of fruit flies (Drosophila), subjecting them to radiation and trying to assess the effects. Muller, who had won his prize precisely for his work on the effects of radiation, had responded through his students, critiquing the work of the Dobzhansky group and devising experiments of his own that proved precisely the opposite of what his opponents claimed!

There was a fair amount of name calling here, for while Dobzhansky supported (what we shall see was) the fairly straightforwardly described "balance" hypothesis, he succeeded in getting Muller's option labeled the "classical" hypothesis, with the connotations that it was something old-fashioned and outdated. But underneath were some deep convictions, far more than mere science. We were now in the frozen depths of the winter of the Cold War, and this affected the real positions, as did absolutely fundamental convictions about the nature of humankind and its future. But to unpack all of this, we must go back to the beginning of the century and to the birth of genetics.

Essay

Population Genetics

We know that a major scientific problem with the theory of the *Origin* was the lack of an adequate theory of heredity. Darwin opted for a blending view of characteristics and their causes. What we now see was needed was a particulate theory, where the units causing organic features, passed on from generation to generation, remain unchanged, no matter in what individual combinations they may appear, in any particular organism. Unknown to all, the right approach was even at that time being formulated by a Moravian monk, Gregor Mendel, but the work was not noticed and he died obscure. Then, at the beginning of the twentieth cen-

Gregor Mendel

tury, with renewed interest in heredity, his work was brought to light and developed. Together with discoveries in the nature of the cell (cytology), the basic ideas of heredity (now called "genetics") were fused together in a satisfying overall picture, thanks particularly to the work of the American biologist Thomas Hunt Morgan and his students (one of whom was Muller) at Columbia University in the second decade of this century (Allen 1978).

The "classical theory of the gene" located the units of heredity (the genes) on threadlike entities (the chromosomes) in the center (the nuclei) of the basic building blocks of organisms (the cells). The genes are the units of heredity, that is to say, they are the units that are passed on in each generation, in sexual organisms via the sperm and the ovum, carrying the blueprint as it were for the new organism. It is believed that the chromosomes come in pairs and that the genes are matched across chromosomes—the particular place on the chromosome (common to all members of the species) being known as the "locus," and a gene form that can occupy a particular locus being known as an "allele." The genetics is Mendelian in the sense that each parent contributes equally to the new offspring, one and only one allele from each locus being transmitted. Which particular allele is transmitted is random, not in the sense of being uncaused but in the sense of being equiprobable and the choice not being a function of the efforts of the organism or the nature of the mate or the needs of the possessors or whatever.

How do you account for blending—skin color? How do you account for nonblending—sexuality? Genes are the units of function as well as of heredity—it is the genes, in combination with the environment, that cause the grown individual. The genes of the individual are known as the "genotype," the genes of the species are known as the "gene pool," and corresponding to the genotype we have the physical organism, the "phenotype." (There is no such corresponding term for the

Thomas Hunt Morgan

gene pool.) Paired alleles can be identical (this is called a "homozygote" with respect to that locus, or "homozygous") or different ("heterozygote," "heterozygous"). Sometimes the effect of one allele swamps the other allele, so that the heterozygote looks like homozygote of the swamping allele. In this case, the swamping allele is said to be "dominant" over the swamped allele, which is in turn "recessive." What this all means is that a characteristic might be hidden for generation, only reappearing when identical recessive alleles get mated up again. What it also means is that a characteristic can be very rare indeed but will persist in the population so long as selection does not eliminate it. It will not get swamped out. Early geneticists did not realize this, but two mathematicians showed that original ratios will always stay the same as will the proportion of genotypes (the two homozygotes and the heterozygote), so long as no other factors are disrupting things. This simple ratio is known as the Hardy-Weinberg law, after those who found it.

The genes are very stable. From generation to generation, they change only rarely. Such changes as do occur are not uncaused. Muller won his Nobel Prize for showing how radiation can bring on changes. But they too are random in the sense that you can only say statistically how many changes there will be, and they do not occur according to the needs of the possessor. Most gene changes affecting the phenotype are harmful or deleterious. The gene is therefore the unit of change or "mutation." The early Mendelians somewhat naturally concentrated on large differences, so the idea grew up that significant changes are always largish, which led to the Mendelians seeing evolution as going in jumps or steps, from one varia-

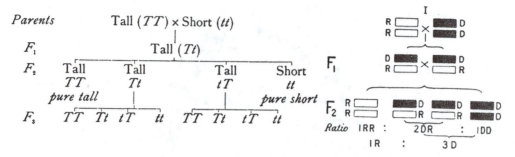

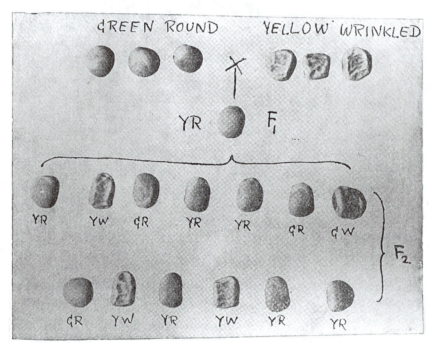

Mendelism as illustrated at the beginning of the twentieth century by William Bateson, the first British champion of the new science of heredity (the main point is that breeding may mask characters for a generation or two but does not destroy or blend them away)

tion to the next. They were therefore "saltationists" (as people like Huxley had been before them) and much inclined to play down the significance of natural selection. In this they were opposed by another turn-of-the-century group, the biometricians, who were working with statistical techniques trying to calculate the variations one finds in natural populations and who were the first group after Darwin actually to start taking natural selection seriously as a mechanism of change (Provine 1971).

In fact, although the debate between the Mendelians and biometricians was fierce and deadly—the leading biometrician dropped dead at 45 from stress—it did not last long. By about 1910, people were starting to realize that mutations can have very small effects as well as large ones, and it was understood that Mendelian genetics and Darwinian selection can be complements making the whole picture rather than rivals or contradictories. But as I have explained to you, these were not good days for evolutionary studies, at least not as a practicing professional science. People had ideas and intuitions but generally did not follow them up. It was to be another 20 years, around 1930, before mathematically inclined evolutionists put together genetics and Darwinism in one integrated theory: Mendelian genetics generalized to populations with the effects of such causes as muta-

Trofim Lysenko

tion and selection factored in. With reason, this subject is usually known as "population genetics."

Three names are usually associated with this major advance in evolution's history. In England there were Ronald Fisher, one of the greatest statisticians of all time, and J. B. S. Haldane, a biochemist and mathematician. In America, one had Sewall Wright, an agricultural geneticist who had worked extensively on the blood lines of cattle, but who by 1930 was on the faculty at the University of Chicago. Today, we know that there was related work going on elsewhere, in Russia particularly. After the Soviet Revolution, people were looking for practical, low-cost science, and genetics fit the bill entirely. Unfortunately at the time this work was little known elsewhere, although it had an effect on Dobzhansky, who left his homeland in the 1920s, never to return. Even more unfortunately, by the 1930s, Stalin had fallen under the spell of the charlatan Trofim Lysenko, who promised quick and easy—and totally fallacious—ways of obtaining favorable genetic results, and that was the end of that. Russian Mendelian approaches crashed never to rise again (Joravsky 1970).

Formally, the English and the American population geneticists produced identical work. They used different techniques—Fisher, for instance, used powerful classic mathematics, whereas Wright invented his own pragmatic techniques for problem solving—but they got the same answers from the same premises. However, as historians now realize, in intent there were major differences. For the moment, I will concentrate on the American picture, although in later chapters I will swing back to some of the English ideas. Fortunately, not only for my exposition but also for the men who followed him and who did not possess his mathematical skills, although Wright first presented his theory in formal style, at its heart was a pictorial metaphor. This was the famous "adaptive landscape."

Wright (1932) invited us to think of an area of land, with hills and valleys and plains. It is three dimensional: left and right, forward and back, up and down. Think of the surface as made up of points that could be occupied by different genotypes. Two organisms, very similar and just different in one or two alleles, would be next to each other, whereas two organisms with many differences would be far apart. Now the third dimension, up and down, is the kind of Darwinian dimension of fitness—if an organism was very much better at surviving and reproducing than another it would be higher, and if not, then lower. One would expect fairly smooth curves, because one or two allele changes would make only small differences to survival and reproduction ability.

This metaphor of the landscape—with organisms occupying spots on the surface—was the heart of Wright's theory. Initially, one would expect to find organisms clustered around the peaks of the landscape—not all of the peaks, but some of them. How then does evolution take place? Here, Wright's background in animal breeding became very important. He knew that the way that breeders have optimal success is not by trying to change the whole group at one time. Rather, you look out for features that you think particularly desirable, and you isolate them, trying to get them confined and spread through a small subgroup. When once you have done this, then you start to try to spread it through the whole group, by selective mating and choosing. But fragmentation and isolation are the initial key.

In real life, Wright thought that species tend to be divided into small subgroups. Obviously, however, such fragmentation and isolation on its own cannot do everything. Somehow one has got to get change taking the subgroups away from their shared, uniform past. Here Wright introduced what is now known as "genetic drift" or the "Sewall Wright" effect. If populations are fragmented into small subgroups, then one can show that within these subgroups selection might not be effective (even though it is at work), because the random factors of breeding might overwhelm it. In other words, change might come about through chance, and new features that have a higher fitness could appear. In terms of the metaphor, groups might wander down the sides of hills under the influence of drift, and then shoot up other hills thanks to selection—to peaks that were higher than the ones they left. Then these could thrive and perhaps swamp out everyone else.

Wright called his theory the "shifting balance theory" of evolutionary change, and this name often puzzles people because it is not obvious to what the term *balance* applies. In Wright's opinion, what you have always is a balance between forces that are leading to fragmentation and differentiation (drift and the like) and forces leading to recombination and uniformity (selection and so forth cleaning up afterward). In other words, everything is in a state of fluid balance or,

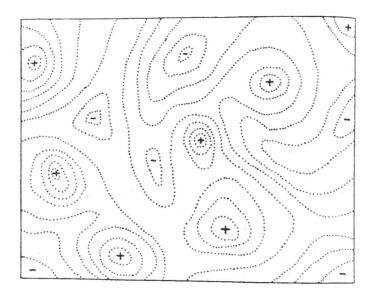

Sewall Wright's adaptive landscape

to coin a phrase, "dynamic equilibrium," with the forces toward heterogeneity squaring off against the forces toward homogeneity.

Now of course we have heard this kind of talk before—the terms are those of Herbert Spencer. But why should one be surprised? We know that Herbert Spencer was by far the most important evolutionist in North America, and in fact Wright was a student of a man who was an ardent Spencerian. (I refer to L. J. Henderson, one of Wright's professors at Harvard, where Wright was a graduate student around 1915.) And frankly, for all of the talk about selection and fitness, there is really nothing very Darwinian about the adaptive landscape metaphor. The chief force of change, certainly the most creative force of change, is genetic drift, which is about as non-Darwinian as you can get: something that Fisher, who was an ardent Darwinian, kept pointing out nonstop. Moreover, for Wright as for Spencer, what really counted was progress. You might think that the landscape is a bit like a Lyellian water bed—as one peak rises, so another falls. But this was not really how Wright saw things. He thought that the landscape was pretty rock-like and that over time real progress will occur. Certainly he did not think that the arrival of humans was pure chance, and in fact he had some pretty funny personal ideas about how everything is evolving upward so eventually we will all be part of one eternal mind.

But there was a big difference between Spencer and Wright. The Victorian had worn his values and his culture and his ideology in a very public fashion. He was preaching a doctrine, a secular religion, and it was there in full view for all who would read and listen. Wright was trained as a careful, professional scientist—one did not study science at Harvard to become a theologian, secular or otherwise. Furthermore, significantly, Wright was in a rather insecure branch of science. Today, genetics is a pretty top-dog sort of science—millions were spent on the Human Genome Project and some of the brightest minds go into the molecular biological business. Almost every year it gets a Nobel Prize or two. But back in the early years of the last century, genetics was new with promises but no triumphs. The people who did it and supported it were agriculturalists—Wright spent the first 10 years of his career at the U.S. Department of Agriculture—and we all know where they tend to stand in the pecking order of academia. Above education and sociology, but not by much.

Hence, for all that he had a whole parade of private, Herbert Spencer–type values, Wright was absolutely not going to let these come to the surface of his professional work. In the tradition of Cuvier (of whom I suspect that Wright had never heard), for his own private subjective reasons, he was intent on pushing forward his science as objective. So whatever depths there may have been beneath the adaptive landscape, and I suspect that there were many and that they went down a long way, the population genetics of Sewall Wright represented a way of doing evolutionary biology that had not been seen hitherto. It was science of a much more professional standard.

The Synthetic Theory of Evolution

In the 1930s and 1940s, things now really started to move forward (Cain 1993). The key figure was Theodosius Gregorievitch Dobzhansky, to give him his full name. As a youth he read the *Origin of Species* (in Russian translation) and was at once converted to evolutionism, with a strong sympathy for Darwin's ideas. He trained as a biologist, making great trips across (prerevolutionary) Russia, specializing in that common little insect, the ladybug. But in the 1920s, Dobzhansky moved on a scholarship to America (never to return to his homeland) and, being located in the laboratory of Thomas Hunt Morgan, switched to the study of chromosomal variations in fruit flies. Clearly destined for big things, in 1936 Dobzhansky was invited to give a prestigious series of lectures in New York City, and the following year these were written up as *Genetics and the Origin of Species*.

Unlike his American colleagues, who tended to be city types, Dobzhansky knew from his early training that there are simply masses of variation in wild populations—one finds differences between individuals and differences between groups. The standard uniform type is a fiction of the laboratory geneticist's imagination. Dobzhansky knew moreover that you get gradations from group to group: rarely if ever do you get abrupt changes. And he was keenly aware of adaptation, realizing that since Lamarckism is false, natural selection is really the only game in town. "A biologist has no right to close his eyes to the fact that the precarious balance between a living being and its environment must be preserved by some mechanism or mechanisms if life is to endure. No coherent attempts to account for the origin of adaptations other than the theory of natural selection and the theory of the inheritance of acquired characteristics have ever been proposed" (Dobzhansky 1937, 150).

But when the time came, it was not Darwin who really provided the inspiration and foundation for Dobzhansky. In 1932, at an international genetics congress, Dobzhansky saw a poster display of Wright's shifting balance theory. Although he was himself completely devoid of any mathematical ability whatso-

Theodosius Dobzhansky

ever, Dobzhansky knew a good idea—a good picture—when he saw one. Wright's adaptative landscape, with its peaks and valleys, with groups of organisms either sitting on the tops of the peaks or subject to factors that were moving them from one peak to another, was the causal theory that Dobzhansky needed and within which he could place his knowledge of organisms in the wild as well as of experimental subjects in the laboratory.

> Each living species or race may be thought of as occupying one of the available peaks in the field of gene combinations. The evolutionary possibilities are twofold. First, a change in the environment may make the old genotypes less fit than they were before. Symbolically we may say that the field has changed, some of the old peaks have been levelled off, and some of the old valleys or pits have risen to become peaks. The species may either become extinct, or it may reconstruct its genotype to arrive at the gene combinations that represent the new "peaks." The second type of evolution is for a species to find its way from one of the adaptive peaks to the others in the available field, which may be conceived as remaining relatively constant in its general relief. (p.187)

We have, thought Dobzhansky, a group (like a species) "exploring" the slopes of a mountain peak, working in some way through "trial and error" until at last it escapes from its home base and moves across a valley and shoots up the side of a neighboring mountain.

The theory in Dobzhansky's book is therefore a funny synthesis. It is in essence Wright's Spencerian shifting balance theory, but to this is added both a deep knowledge of the real world of organisms and at the same time a keen appreciation of key Darwinian ideas, especially those of adaptation. What Dobzhansky does not offer is a synthetic unifying vision, as we find in the *Origin of Species*—he makes no effort to cover the wide range of topics that Darwin thought essential to his case. There is no mention of paleontology in *Genetics and the Origin of Species*, nor embryology, nor many of the other subjects that interested Darwin and that he thought so important. The real emphasis is on speciation, not a subject on which Darwin dwelt at length. Expectedly, given Dobzhansky's time in Morgan's lab, there was much discussion of such factors as chromosomal variation and of how it can and cannot become important when groups split and new reproductively isolated groups (species) are formed. And this led straight to a crucially vital underlying assumption of Dobzhansky's whole case, namely, that the way to understand major evolutionary changes is through the study of minor changes— changes so minor that you might not normally notice them or think them significant. "Experience seems to show … that there is no way toward an understanding of the mechanisms of macro-evolutionary changes, which require time on a geological scale, other than through a full comprehension of the micro-evolutionary processes observable within the span of a human lifetime and often controlled by man's will" (p. 12).

Strange hybrid though it may have been, *Genetics and the Origin of Species* proved to be an absolutely seminal publication. Here was an attractively written and reasoned work on evolution that all could understand—the mathematics was kept to a minimum!—and that could inspire a young researcher and offer a program leading to a career as a professional evolutionist. But there was more than just this, for at this point Dobzhansky showed himself to be a master at organization: if he did not himself want to cover the spectrum of evolutionary topics, then he was ready and very willing to bring others into the arena, urging them to work alongside him, filling out the picture of evolutionary change—a picture that went back ultimately to Wright's adaptive landscape metaphor. First there was Ernst Mayr. An immigrant like Dobzhansky, trained as an ornithologist and systematist (classifier), Mayr left his native Germany, traveling west to the American Museum of Natural History (still in the early 1930s under the directorship of Osborn), where he became curator of birds. Drawing on a vast knowledge of nature's denizens and combining this with a sensitivity to geographical conditions and variation, in 1942 Mayr produced his masterwork, *Systematics and the Origin of Species*. Most dramatic of all of the instances on which Mayr drew were the so-called rings of races" where interbreeding subpopulations of organisms circle the globe, finally touching but unable to interbreed at such meeting points. Here was natural, gradual variation before one's very eyes—not just evolution (evolution as fact, that is) in the making, but the process of evolution (evolution as cause) showing itself. It simply was not possible to think that evolution proceeds by jumps, saltations: the touching subpopulations blended one into another without a break or a step, even though the end populations were genetically isolated from each other. More causal speculation than this was beyond Mayr's scope of inquiry, but it was the landscape metaphor that was assumed and that was in turn confirmed.

More theoretical was the paleontologist of the group, George Gaylord Simpson. He was unique among the "synthetic theorists" (as their theory came to be

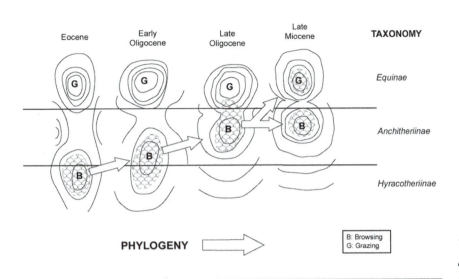

Eocene | Early Oligocene | Late Oligocene | Late Miocene | **TAXONOMY**

Equinae

Anchitheriinae

Hyracotheriinae

PHYLOGENY

B: Browsing
G: Grazing

The evolution of the horse as envisioned by George Gaylord Simpson

known) in having a facility with figures. He could go behind Wright's pictures, understanding the mathematics, and thus modifying what was a theory for populations in action over a few short generations into a theory that dealt' with populations (at the physical or phenotypic level, for nothing was known of genes) over long periods of time—millions of years in fact. Particularly noteworthy in Simpson's *Tempo and Mode in Evolution* (published in 1944, having been delayed somewhat because of the war) was his treatment of horse evolution. Simpson showed how the mammals had been bush and tree browsers, then they started to move toward a grazing lifestyle (somewhat incidentally because of other changes, particularly toward a larger overall body size), then at some crucial point the horses had split, with some going right back to browsing and others moving across the valley and right up the path of Mount Grazing. As it happens, at some point after this the browsers went extinct, but this was an event only after all the exciting action had occurred.

Finally there was botany. Dobzhansky had deliberately canvassed the field, looking for someone to write on plants from the perspective that he and his friends were exploiting. His first choice let him down, and when he found a substitute, G. Ledyard Stebbins, Dobzhansky had Stebbins stay in his own home. The geneticist fed the botanist pertinent information until Dobzhansky was convinced that Stebbins would complete the task, and complete it properly. And so, in 1950, *Variation and Evolution in Plants* made its appearance. There were of course major differences in Stebbins's work from that of the animal evolutionists. For a start, in botany you do get jumps, as new species are formed by chromosomal events virtually unknown in the animal world. For a second, hybridization (where members of different species breed and produce fertile offspring) is a well-known and common and important method of change. But, for all of these differences and more, the underlying story is the same. Peaks and valleys, and organisms

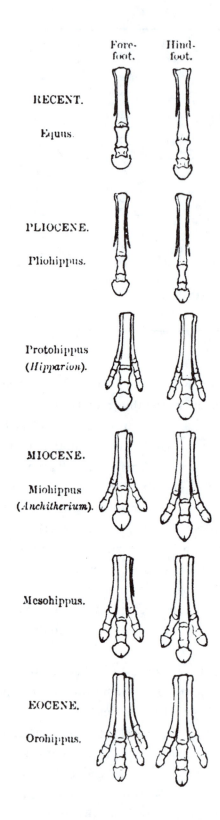

Fore-foot. Hind-foot.

RECENT.

Equus.

PLIOCENE.

Pliohippus.

Protohippus
(*Hipparion*).

MIOCENE.

Miohippus
(*Anchitherium*).

Mesohippus.

EOCENE.

Orohippus.

T. H. Huxley's reconstruction of the evolution of the horse foot (based on specimens of O. Marsh)

struggling up the sides and sitting triumphantly on the top, or being displaced and having to start the evolutionary process all over again. The landscape model triumphed.

Forging a Discipline

These were some of the major intellectual moves that were made by evolutionists after Darwinian selection (together with other bits and pieces such as Spencerian progress) had been fused with Mendelian genetics. But there was more to be done than this. Thomas Henry Huxley had done it for physiology and morphology. He had failed to do it for evolution. Make a professional discipline of it, that is. Now finally, there was a group who felt (with some good reason) that they had an adequate theory—one that was ready for development through experiment, natural observation, theoretical amplification, and more—and who wanted to spend their time as professionals working on it. They did not want to spend their time as museum-based priests of a secular religion. So Dobzhansky and his coworkers deliberately set about making a professional science of evolutionary biology—not just a second-rate enterprise tracing hypothetical phylogenies with too little information and too much imagination, but a real discipline that was causally based and that took seriously experiment and theory.

The theologian and mathematician Blaise Pascal once asked about belief in the existence of God. He concluded that one has an asymmetrical situation: if God exists then you had better believe in Him, and if God does not exist then not believing in Him does not really matter. Hence, the sensible thing is to believe in Him. And if you complain that you cannot believe in Him, then go through the motions and you will be surprised how faith will come. This is known as "Pascal's Wager"—Pascal's branch of expertise was probability theory and he is offering you a bet or a wager you really ought not refuse. Founding a scientific discipline is a bit like Pascal's Wager. If you go through the motions, then you and others will start believing in it. And to this end, you need good university jobs, you need students, you need journals (preferably with lots of esoteric language), you need associations (that you and your pals are in and others are not), you need grants and other monies, you need supporters, and you need to shove it to your enemies and detractors.

The synthetic theorists achieved all of these ends and more. For a start, they moved into plum university posts and once there brought their friends in too. Dobzhansky got a job at Columbia. Stebbins got a job at Berkeley. Mayr got a job at Harvard, and before long he was campaigning (successfully) for Simpson. Dobzhansky had masses and masses of students and postdoctoral fellows that he treated like his children, supporting, guiding, encouraging, scolding. They worshipped him—there was no talk here of societies for the preservation of native Americans from foreign professors—and fanned out to carry the word. A journal, *Evolution*, was started, with Mayr as the first editor. Firm guidelines were put in place. The obvious esoteric language was mathematics, and even though Dobzhansky and Mayr would not have known a symbol if their sisters had married one, care was taken to see that their students were properly trained, and associates with mathematical skills were dragooned into coauthoring papers. Dobzhansky wrote a whole series of Drosophila articles with Sewall Wright: articles of which he understood the first lines and the last lines and absolutely nothing in between.

As importantly, causal, experimental work was encouraged. Path tracing was shown the door, or rather the pink rejection slip: "Your manuscripts have been scrutinized by two readers and both of them report that they consider them unsuitable for publication in *Evolution*. I have tried to get some detailed criticism for you (as you asked) but there seems to be nobody in this country now who is interested in phylogenetic speculations" (letter from Mayr to F. Raw, 2 August 1949; *Evolution* Papers, American Philosophical Society). I should add on a personal note that I dug this letter out of the meticulous files that Mayr, as editor, kept for those early years of the journal. Very atypically, I could find no record of the negative referees' reports. I challenged him on this, suggesting that the two referees might have been Ernst Mayr in the morning and Ernst Mayr in the evening, at which he just smiled.

An association was started: The Society for the Study of Evolution. President: G. G. Simpson; Secretary: E. Mayr; Council members: S. Wright and Th. Dobzhansky. And then there was the question of grants. Fortunately by the time things really got going, World War II was over and the U.S. government was realizing that it needed to subsidize basic science. The National Science Foundation was begun, and the evolutionists were right up at the head of the line with their begging bowls outstretched. They were not always successful, but they got some nice juicy grants, and at the same time had the satisfaction of seeing that others were denied. Simpson (like Mayr) squirreled away every piece of paper, including all of the referee's reports he wrote on grant applications. Of one man, at that time the editor of *Evolution*, for all that he and Simpson were at personal loggerheads (Simpson had troubles with personal relationships, to put matters euphemistically), the report reads: "This application is first-class in every respect." Of another, a brilliant systematist who had been rather rude about the synthetic theorists: "His approach is narrow-minded and shows consistent lack of thought into biological, as distinct from strictly mathematical, aspects of the problems considered." Merit rating: "Questionable"!

What about the Cuvier tactic, to which we have seen Sewall Wright sensitive? What about the need to come across as serious, objective scientists? The Scopes trial was not that long before and even if the evangelicals were now taking a low profile, there were still many happy to make of evolution a lot more than mere science—the spirit of Thomas Henry Huxley still roamed the land. Indeed, rather more than this—the genes of Thomas Henry Huxley were still active. His grandson Julian Huxley (brother of the author Aldous Huxley) was a prominent British evolutionist, and although he wrote more sober works—his *Evolution: The Modern Synthesis* appeared in 1942—his forte was inspirational material, blending science and value in an unabashed, neo-Spencerian fashion. He was an ardent humanist; you do not need to turn many pages of his 1927 *Religion without Revelation* to guess his intent and lifelong convictions: evolution as progress, evolution as popular science, evolution as religion. Although an ardent atheist, in the 1950s Julian Huxley became enthused with the speculations of the Jesuit priest-cum-paleontologist Pierre Teilhard de Chardin and, sensing a kindred spirit, wrote the foreword to the English translation of Teilhard's major opus, *The Phenomenon of Man* (1959).

As you can imagine this kind of activity did not sit well with sober scientists—the English Nobel Laureate Peter Medawar wrote a scathing review of Teilhard and of Huxley's involvement (Medawar 1961, 71–81). As you can imagine also, the synthetic theorists were tense about all of this and determined not to be tarred by the same brush. This does not mean that they did not have values—even that they did not have yearnings to treat evolution as a secular religion. There was not one of them who had not turned to evolution in the first place to find the meaning of life, hoping especially to discover the implications of an evolutionary approach for our own species. There was not one of them who was not a gung-ho progressionist, and Dobzhansky—a deeply committed Christian—was no less supportive of Teilhard than was Huxley. (Simpson, the paleontologist, knew and liked Teilhard, although as a sometime Presbyterian disapproved of Teilhard's womanizing—or rather of the hypocrisy of someone who took a vow of chastity and then broke it, flagrantly.)

But the synthetic theorists knew that, if they were to upgrade their science, they had to keep their evolution-as-professional-science separate from their evolution-as-secular-religion (not so very secular in Dobzhansky's case). So what they did was to write two series of books! The first series was the professional series: lots of talk about models and causes and quantification and so forth. Not a whiff of culture or social values. Then there was the second series: openly written for the general reader (if it did not say this on the title page, then it said it in the preface), with the same science as before supported with lots of nice illustrations and the mathematics removed (not much work here!), and with a couple of final chapters on life and its total meaning.

Simpson was the paradigm. His *Tempo and Mode in Evolution* (1944) is sufficiently pious and straight-faced in its serious intent that it could pass for a church sermon. Then in 1949 comes *The Meaning of Evolution: A Study of the History of Life and of Its Significance for Man*, a book that Simpson admitted meant much to him, with a subtitle that tells all. Four years later (1953), we are back with a revision of *Tempo*, retitled *The Major Features of Evolution*, as strict and straight as you could ever wish. But, oh, how those evolutionists did let it all hang out when they were allowed to! For the paleontologist Simpson when writing in popular mode there were two major social directives. First, there was the need to improve and promote knowledge—knowledge in itself, as a good.

> The most essential material factor in the new evolution seems to be just this: knowledge, together, necessarily, with its spread and inheritance. As a first proposition of evolutionary ethics derived from specifically human evolution, it is submitted that promotion of knowledge is essentially ... both the acquisition of new truths or of closer approximations to truth (metaphorically the mutations of the new evolution) and also its spread by communication to others and by their acceptance and learning of it (metaphorically its heredity). (Simpson 1949, 311)

Next we have personal responsibility, leading to individualization and thus to integrity and dignity.

> The responsibility is basically personal and becomes social only as it is extended in society among the individuals composing the social unit. It is correlated with another human evolutionary characteristic, that of high individualization. From this relationship arises the ethical judgment that it is good, right, and moral to recognize the integrity and dignity of the individual and to promote the realization or fulfillment of individual capacities. It is bad, wrong, and immoral to fail in such recognition or to impede such fulfillment. This ethic applies first of all to the individual himself and to the integration and development of his own personality. It extends farther to his social group and to all mankind. (p. 315)

Fully to understand Simpson's thinking, especially his high valuing of responsibility and dignity, we must take note of the context and time within which he was writing. The Cold War was frozen in a seemingly endless winter. Worse, science in the Soviet Union was subject to dreadful pressures. I have mentioned that, even in the 1930s, biology was firmly under the thumb of charlatans led by the agriculturalist Trofim Lysenko, promoting neo-Lamarckian pseudotechniques intended to raise wheat harvest yields to dramatic new levels. Using his friendship and connections with Stalin, Lysenko continued right through the 1950s to harass and persecute those who dared raise a voice against him. For all of his difficulties with personal relationships, Simpson was a man of the highest moral ideals: he had volunteered for dangerous action in World War II, despite the fact that he was over the enlistment age. It was natural, therefore, for him to take the crusade for democracy and freedom as a personal issue, and you will not be surprised to learn that, having established (as he thought) a biological basis for dignity and responsibility, he moved straight to a vehement condemnation of the totalitarian systems that spread from Eastern Europe right across Asia. At the same time, he made it clear that there was an alternative: the society within which he and his fellows were able to work so freely. "Democracy is wrong in many of its current aspects and under some current definitions, but democracy is the only political ideology which can be made to embrace an ethically good society by the standards of ethics here maintained" (1949, 321). And then added to this was a repeat affirmation of the significance of evolutionary biology for providing foundations and standards: "It bears repeating that the evolutionary functioning of ethics depends on man's capacity, unique at least in degree, of predicting the results of his actions. A system of naturalistic ethics then demands acceptance of individual responsibility for those results, and this in fact is the basis for the origin and function of the moral sense" (1949, 145–146).

Material similar to that of Simpson can be found in the popular writings of the other synthetic theorists. Stebbins (1969), during the late 1960s, while a faculty member at Berkeley, even went so far as to argue that it is good biologically to have a small group of radicals upsetting an otherwise complacent population!

The Classical-Balance Dispute

The mark of really good science, productive fertile science, is that it gives you lots to do. With every problem you solve you get a couple more questions thrown up. The last thing you need is just to sit polishing a theory like a prized antique. Fortunately, the synthetic theory showed its worth and more. All sorts of exciting problems and questions and anomalies kept coming to the surface. This really was dynamic stuff. For Dobzhansky (and hence for the others) the main move was, in the 1940s, to a much more selective stance than hitherto. This change came about because, for all that it might have pretended otherwise, Wright's shifting balancing theory was never very Darwinian. The key causal mechanism is genetic drift, supposedly occurring when a small population is fragmented from its fellows and subject to the vagaries of mutation and breeding. It is claimed that this is enough to fuel a slide down the side of an adaptive peak until, reaching the bottom, the group can then move up the side of a neighboring mountain. Although he was always much keener on selection than was Wright, in the first edition of *Genetics and the Origin of Species* (1937) Dobzhansky had accepted (without critical comment or question) much that Wright argued. But then, just a few years after the first edition was published, Dobzhansky became increasingly convinced that drift is far less significant than he had thought previously. Correspondingly, selection must be yet more important (Lewontin et al. 1981). Drift leads one to expect that there will be differences between populations, both at the genetic level and at the level of the chromosomes. In these respects, isolated populations will move all over the place. What drift does not lead one to expect, and in fact what drift leads one positively not to expect, is any kind of systematic variation in genes or chromosomes. One does not, for instance, expect to find that there are cyclical differences in chromosomal variation, from one season of the year to another and then repeating itself in subsequent years. Systematic change of this kind is a sign that natural selection is at work rather than purely random factors.

Studies of natural populations of fruit flies in the American West showed Dobzhansky that there are indeed such cyclical variations tied to the seasons, and he concluded that natural selection must be at work. Then his initial observations were backed by experimental studies that showed that variation in temperature and humidity and the like are just what is needed to bring about systematic variation in populations. Some chromosomal variations are clearly better adapted to some conditions than are others: there is no absolute perfect form, suitable for all climates. It is all a question of the relative conditions. The question for Dobzhansky now was how one is to explain this variation in populations, and how in particular such variation is held from one generation to another. Why does selection not simply wipe out all variations so a population is relatively uniform throughout? But then, how could selection operate to raise the levels of first one form and then the levels of another form? Mutations provide the raw material necessary for variation, but simply waiting on required variations when selective needs arise is hardly adequate. One needs some theory or mechanism showing how variation can be held within populations, so that there is always material to work on when new selective needs arise.

After considerable thought, Dobzhansky finally decided that selection itself keeps up the required variation. The key mediating mechanism is "balanced superior heterozygote fitness" (Dobzhansky and Wallace 1953; Dobzhansky and Levene 1955). This supposes that the heterozygote is different from both homozygotes and that, for various reasons, the heterozygote is favored by selection over the homozygotes. (More technically: that the heterozygote is fitter than either homozygote.) When one has a situation like this, then it can be shown easily that the heterozygote keeps reappearing in populations. Indeed, under favorable conditions, there will be a balance between the ratios of heterozygotes and of homozygotes. And what this means is that the different genes or alleles (or chromosome variations) persist in the populations and are therefore always ready to be used when selective needs arise. But it is selection itself that is keeping the variation present in the first place.

To use a metaphor, the conventional view would have one waiting on the occasional mutation, which will probably be no use in any case. This is akin to having to write an essay for an instructor where the only source of material you have is monthly offerings of the Book of the Month Club. You can be fairly certain that whatever comes your way through the mail will rarely, if ever, be the material that you need for your essay. Suppose, however, that you have the balanced heterozygote fitness situation. This is akin to having a whole library at your disposal. Pretend that your instructor demands that you write an essay on dictators. If, when you go to the library, you find there is nothing on Hitler, there may well be something on Napoleon. And if there is nothing on Napoleon, then one can look further down the alphabet for something on Stalin. You can be certain that there will be something on some dictator that you can use for your essay. Reverting back to real life, suppose a new predator appears, threatening a population of organisms. There is a whole new set of selective pressures. The balanced heterozygote fitness position suggests that if there is no variation within the population allowing some members to escape through camouflage, then perhaps there will be something allowing some to escape through a more effective defense like a thick shell or skin. Or, if neither of these defenses exists, then perhaps there will be the capacity for a different kind of behavior, perhaps behavior inclining one to get out of the area and to move somewhere else that is predator free (Ruse 1982).

Balanced superior heterozygote fitness is a neat way of solving the variation problem, but is it true? Well, there are certainly some cases where it seems to be in action. The most famous case occurs in humans, particularly among inhabitants of certain parts of West Africa where malaria is a major health threat. There is a certain allele, the sickle-cell allele, which confers a natural immunity on its possessors when it is in the heterozygote state (in other words, the bearer is fitter than the homozygote without the gene) but is lethal when it is in the homozygote state. There is a balance between the heterozygote's good qualities and the inferior qualities of the homozygotes. But as Aristotle was wont to say, one swallow does not make a summer. Because balanced heterozygote happens occasionally, it does not follow that it is a near universal phenomenon as Dobzhansky and fellow supporters of the balance hypothesis supposed. It was at this point that controversy started to grow. To prove his case, Dobzhansky and his students (particularly Bruce Wallace) ran experiments subjecting fruit flies to radiation. This would cause mutation, and because the mutations would generally occur only in a few alleles, the overall effect would be to increase the heterozygosity in populations—by and large, one would not expect members of new generations to be at once homozygotes for the newly mutated genes. And Dobzhansky and associates claimed that, overall, the radiated populations were indeed fitter than the nonradiated populations, supposedly showing that the increased number of heterozygotes pushed the populations farther up their adaptive peaks. In other words, the balance hypothesis was vindicated.

I hardly need say that all of this is a bit inferential, to put it mildly. One is not exactly measuring heterozygosity directly, nor is one doing anything very much about individual organisms and their fitnesses. Muller disliked the balance hypothesis intensely—more on the reasons in a moment—and found little reason to accept Dobzhansky's findings. Muller (1949) wanted to argue (the classical hypothesis) that organisms in a population are more or less uniform and at their adaptive peak. Mutations generally are deleterious, heterozygosity is therefore discouraged (certainly heterozygosity that has any phenotypic effect), and the only variation one finds generally are those very few new mutant genes that prove fitter than the old types and that are therefore moving through the population to become the norm. Muller therefore argued that the Dobzhansky experiments were flawed, and when he and his students (particularly the Israeli Raphael Falk) ran their experiments, they failed to replicate the Dobzhansky results. As seems to be almost the norm in these cases, the experiments that they ran were slightly different—they used stronger radiation rates, for instance—and as is certainly the norm in these cases, Dobzhansky and company complained that the Muller results were irrelevant and meant nothing!

Now, why was everybody so tense about all of this? Why not just sit back and wait for some good results to come in? Well, one reason is that good results do not come from sitting back and waiting. The way to get things done is by pushing, and this is what everyone was doing. More seriously, there were serious scientific issues—intuitions if you will—at stake here. As we have seen, Dobzhansky desperately needed a supply of variation above single mutations, or else selection as he saw it could not function properly. Unlike Muller moreover, he was a field naturalist, and so he really knew that, whatever the theory might say, variation (however caused) does truly exist in natural populations. And obviously, he was properly sensitive to the workings of external causes, including selection. Muller, on the other hand, from a lifetime's experience of experimentally inducing mutations through radiation, knew how horrendous were the effects of most mutations. He just could not see how mutations could remain positively in populations.

I look upon mutation as a random process, so far as the nature of its effects are concerned, much as Brownian movement is random, and I therefore find it a necessity to conclude that the effects are on the whole disintegrating rather than integrating, just as the effects of Brownian movement are to decrease rather than increase the free energy of a system. In other words, I regard the principle as being merely a logical extension of the second law of thermodynamics and I would as soon expect to see the average or usual effect of irradiation on later generations to be an improvement as to believe a man who said that he had perfected a perpetual motion machine. (Beatty 1987, 299)

These factors were important in driving Muller and Dobzhansky apart, but one suspects that there was more than this, and there was. We have seen already how the evolutionists were affected by the Cold War, and here was another point of impact—a point entwined with another factor already encountered—the need to find funds to support research. Both Dobzhansky and Muller had reason to feel tense about the Cold War, and they were indeed. The reason for Dobzhansky's state of mind is obvious: as a native-born Russian with a deep attachment to his homeland and an emotional identification with the United States and a hatred of the Soviet system, he was very torn about what was then happening. The reason for Muller's tenseness is hardly less obvious: as a former communist who in the 1930s had gone to live in Russia and barely escaped with his life, he felt very ambivalent about a military arms race that seemed bent on destruction of the whole world.

One of the main features of the Cold War was the buildup and stockpiling of nuclear weapons: weapons that required testing, in those days massive amounts of atmospheric testing with the consequent radiation fallout. People were very worried about the effects, short term and long term, on the human population, and here Dobzhansky and his school stepped in, prepared to use fruit flies as models for humans. The Atomic Energy Commission accepted the offer, and so through the 1950s we find that Dobzhansky and his students were getting monies for their work from this organization. Although, in principle, Dobzhansky declared himself against testing—or at least worried by its effects—one need hardly say that his results, showing that radiation far from being deleterious can be beneficial, went well with his paymasters. They were happy to support research like that! Conversely, Muller, who was regarded with suspicion by military and civil authorities, precisely because of his communist background, tended in turn to be wary of anything that spoke well for these powers and was certainly very uncomfortable with results that supposedly negated the bad effects of a military arms race. Interestingly, he was not against nuclear testing as such: he had no love of the Soviet system, nor did he have any illusions about it. But he did worry about bad science's glossing over the ill effects of such testing.

There was yet another factor dividing Dobzhansky and Wallace, perhaps the most important of all. *Eugenics* refers to the belief and practice of altering the genetic composition of the human species for supposedly good ends. It is a topic fraught with tension, if only because of the Nazis' efforts in that direction: efforts based on their pseudoscientific understanding of human biology and motivated by their vile racist ends. For good and ill, eugenics is a topic that has fascinated geneticists since the birth of their subject—indeed, motivation and finance for early genetics often owed almost as much to eugenics as it did to agriculture.

Both Dobzhansky and Muller had strong opinions on the subject, and it is clear that these opinions influenced thinking on the balance/classical dispute. For Muller, there was an ideal, and it is our obligation to see that we do not fall beneath this. Bad mutations are bad mutations and we should get rid of them. Most mutations, especially most new mutations, are bad mutations, and so eugenical efforts should be directed toward their elimination. For Dobzhansky, there is no ideal. Populations are varied, and this is the way nature intended them. Just as we have fruitfly geneticists, so also we have hewers of wood and drawers of water—we are better off this way and should not tamper with things. There is no such thing as a bad mutation in itself. It is all a question of context

Parenthetically, I need hardly say that Dobzhansky's Christian commitments were working flat out here—we may well be genetically nonidentical, but in the eyes of God and of Theodosius Dobzhansky we are of equal worth. No wonder he felt so strongly about the balance hypothesis. Even though it was deep in the heart of his professional science, for Dobzhansky it was an absolutely key element in his religiously infused vision of humankind.

Molecular Biology

The most important single event in the history of twentieth-century biology was the discovery in 1953 by James Watson and Francis Crick of the double helical structure of the deoxyribonucleic acid (DNA) molecule. At once it was seen that this molecule was the underlying foundation of the Mendelian gene and that the information of heredity was carried along its back, coded by the order of its many sequential parts. At first, the synthetic theorists tended to be suspicious of, if not outrightly hostile to, the work and results of the molecular biologists. As individuals, these scientists were not much loved either. The new science and its practitioners were bright, pushy, and contemptuous of old-fashioned biology, and willing and able to say just this on every available occasion. Striking back, the synthetic theorists—Ernst Mayr particularly—claimed that molecular biology is narrow and limited, that its technical achievements conceal its spiritual aridity. They claimed that whole organism biology—evolutionary biology centrally—necessarily looks at questions and issues that a molecular approach misses. Extreme "reductionism," trying to explain everything in terms of the very small, is a waste of time. Distracting from the real problems, in fact.

In a way, looking back over a half century, this dispute seems about on a par with the dispute half a century earlier, between the biometricians and the Mendelians. And about as worthless. We see now that molecular biology was no enemy of evolutionary studies but a really good friend. Whether it has made any difference to the actual theory of evolutionism is a matter we shall discuss in Chapter

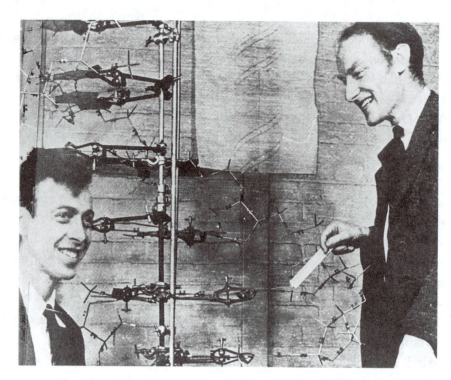

James Watson (left) and Francis Crick (right) demonstrate their model of the DNA molecule

11. We shall see that it has given systematists powerful new tools in their attempts to discern phylogenies. And, what cannot be denied and is pertinent here is that it has certainly given conventional evolutionists wonderful new techniques to cut through problems that were hitherto intractable. Nowhere more so than over the balance/classical dispute. In the 1960s, Dobzhansky's prize student, Richard C. Lewontin, was one of a number who developed the method of "gel electrophoresis." Essentially this is a technique that uses the different electrostatic charges on molecules to discern differences in the molecules themselves, and through this it was possible at last to measure directly the genetic differences between organisms. And through this one can shortcut all of the laborious methods used by Dobzhansky and Muller in their efforts to find genetic differences and similarities, going straight to the genes themselves for information about their nature.

The results were really quite incredible. The balance hypothesis was vindicated to a degree that not even Dobzhansky could have envisioned. Absolutely massive amounts of variation were being held in populations. Take the organism so favored by population geneticists: the fruit fly *Drosophila pseudoobscura*. Lewontin discovered that, from one organism to the next, there were simply huge genetic differences. As many as one-third of the genes that were investigated using gel electrophoretic techniques proved to have different forms. (They were "polymorphic" for different "alleles.") Taking any locus on a chromosome, on average 12 percent of the allele pairs were heterozygotes (Lewontin and Hubby 1966, 608). All in all, therefore, Dobzhansky apparently was as right as one could possibly be and Muller apparently was as mistaken as one could possibly be.

But, as I am sure you are now starting to realize, as in love nothing in science is quite that simple. The variation could not be denied. But the balance hy-

Richard Lewontin

pothesis also gives us a cause of this variation, namely that it is held in populations because of heterozygote fitness. This was still denied—no longer by Muller, for he died in 1967, but by his students and associates. Perhaps, they argued, there are other causes, or perhaps no cause at all! Perhaps, because it has no effect on the phenotype, much of the variation at the genetic level quite escapes the effects of selection. Gel electrophoretic techniques pick out molecular differences in genes, or more precisely they pick out differences in the substances the genes code for immediately. But perhaps these differences make no difference to the finished organism, in which case they would be quite neutral with respect to natural selection. Molecular variation might therefore just "drift" up and down in populations, giving the results obtained through gel electrophoresis but without the immediate supportive consequences for the balance hypothesis. The classical hypothesis rides again!

This issue is still not resolved completely, and I shall return to it. For now we can pull away. We have seen how Darwinian selection theory was integrated with genetics to make a full and satisfying evolutionary theory and how in addition this was then used as a base to build a professional scientific discipline. The cost was that all of those secular religion features that so many found so attractive had to be dropped—or at least, corralled off to one side. But note how this was done, not so much from conviction that evolution as secular religion was a bad thing, but because unless it was done the desired status of professional science would not be achieved.

Further Reading & Discussion

William Provine is *the* historian of population genetics. He is not always easy to read but well repays the effort. His first book, *The Origins of Theoretical Population Genetics* (Chicago: University of Chicago Press, 1971), is an excellent account of the coming of Mendelian genetics and its integration into evolutionary studies. Then his second book, *Sewall Wright and Evolutionary Biology* (Chicago: University of Chicago Press, 1986), is an absolutely massive intellectual biography of one of the last century's giants in evolutionary studies. What is really valuable about this work is that it treats not only of its central subject but also of others who interacted with Wright, most especially Theodosius Dobzhansky. The story of the always-fruitful, sometimes-uncomfortable relationship between Wright and Dobzhansky is itself worth the price of the volume. Then third there is a book Provine coedited with the senior evolutionist Ernst Mayr: *The Evolutionary Synthesis: Perspectives on the Unification of Biology* (Cambridge: Harvard University Press, 1980). The book is based on two conferences nearly forty years ago, when the leading figures in the making of neo-Darwinism were still alive and willing to give their recollections of the main events and moves. (It was reissued in paperback in 1998, with a very short new preface.) Much more work is still needed on the development of twentieth-century evolutionism—some can be found in the pertinent chapters of *Monad to Man* (Cambridge: Harvard University Press, 1996)—but thanks to Provine we now have a good grasp of the really important ideas and their fates.

The Lysenko affair will no doubt be an ongoing focus of scholarly interest. With the opening of archives in Russia, we will surely be learning much more about what happened and why. But already we have several accounts about what went on in Russian genetics and biology generally under the Stalin and Khrushchev dictatorships. David Joravsky's *The Lysenko Affair* (Cambridge: Harvard University Press, 1970) is a very solid discussion based on a great deal of evidence. You should supplement it with the work of Loren Graham, who is the authority on science in Russia under the Soviets. His *Science, Philosophy, and Human Behavior in the Soviet Union,* 2d ed.(New York: Columbia University Press, 1987) is excellent for its insights into the trials and tribulations of Russian biology as well as its triumphs (such as Oparin on the history of life, a topic to be discussed in chapter seven).

Finally, let me draw your attention to the autobiography of Edward O. Wilson, the great ant specialist (of whom more appears in later chapters). *Naturalist* (Washington, D.C.: Island Press, 1994) gives a fascinating account of what it was like to be a whole-organism biologist at Harvard in the 1950s when James Watson was also in the same department and intent on recreating the life sciences in a molecular image. It was not very pleasant!

Chapter 7
Life: The Early Years
In the Beginning

Overview

This chapter explores the idea of spontaneous generation, how it became scientifically unacceptable, and how life took root on this planet of ours.

In France, around the time of the publication of *The Origin of Species*, there was a major row about the origin of life, with the skeptical Louis Pasteur firmly countering Felix Pouchet who was a proponent of "spontaneous generation." This had been an idea under attack at least since the seventeenth century, but until Pasteur it was by no means universally rejected. Many people, particularly those with inclinations towards evolutionary ideas, thought that there might be something to the idea. The "nature philosophers" or *Naturphilosophen,* at the beginning of the nineteenth century, saw patterns linking the inorganic to the organic and so for them such generation was by no means a silly idea.

As it happens, even after Pasteur there were many who still thought that life could arise spontaneously from non-life. Particularly the evolutionists, notably Thomas Henry Huxley in Britain and Ernst Haeckel in Germany, looked for such jumps. All the possible candidates fell flat, however, and after a couple of decades everyone realized that there were no simple solutions to the origin of life.

In the 1920s, English biologist and chemist J. B. S. Haldane and Russian biologist Aleksandr Oparin independently proposed that life came naturally through a series of small events, one after another. There was to be no instantaneous appearance of life.

There was great excitement in the 1950s when it seemed that the first stages of this life process had been replicated in the laboratory, with some of the basic building blocks of life synthesized under conditions believed to duplicate life's beginning. Thus a new science, microbiology, was introduced to the world stage.

Since then, however, the task of making life artificially has proven much more difficult than expected.

The more we learn about the early stages of life, from its supposed beginning 3.7 billion years ago, the more it seems that it was a natural process fueled by natural selection. We see evidence of what we would expect—the gradual growth of more and more complex forms of life. The fossil record (and related geological record) tells a very consistent story and modern-day microbiologists are able to look well beyond the fossils, into the cellular make-up and beyond. How did they make life? What role did oxygen play in respiration or fermentation and what came first? Increasingly the new sciences go deeper into the origins of life than we ever imagined even a few decades ago, and these origins are no longer quite the mystery they once were.

The Role of the Scientific Community

The work of the following scientists is discussed in this chapter. Short, biographical essays of these individuals appear in **Biographies** on page 607.

Felix Archimede Pouchet (1800–1872)
Louis Pasteur (1822–1895)
Ernst Haeckel (1834–1919)
Aleksandr Oparin (1894–1980)
Lynn Margulis (1938–)

Setting the Stage

Back across the English Channel and back in time to 1859. This was the year of the *Origin*, but in France it made no waves. At least, this was the case for a few years until a translation was published, a translation with an inflammatory preface by radical thinker Clémence Roger—positivist, materialist, atheist, and female to boot—that may or may not have stirred her countrymen but as sure as anything sent shudders of discomfort and disapproval through a reclusive naturalist across the Channel in the little Kent village of Downe.

But that year did see a major controversy in France. This was between Félix Pouchet, director of the natural history museum at Rouen, and the great Louis Pasteur, discoverer of the means of the prevention of rabies and many other diseases as well as seminal investigator into the preservation of foodstuffs (pasteurization) and into much, much else (including important studies on viticulture, especially on the properties of yeast).

The controversy was over "spontaneous generation," the one-step appearance of living organisms from nonliving (inorganic or organic) material: flies, worms, grubs, or smaller beings coming (by natural causes) from mud or slime or whatever (so long as it was itself inert). Pouchet affirmed his belief in the venerable doctrine of spontaneous generation. In a major work, *Heterogenie ou traite de la generation spontanée*, he argued that the eggs of microscopic-sized organisms are produced in single steps from other organic materials. And so effective was he in his case that the French Academie des Sciences put up prize money for further studies on the subject. Which brought in Pasteur, who was determined to refute what he saw as a thoroughly false doctrine. To this end, Pasteur was led to perform a dazzling series of experiments—experiments of such a caliber that today they are often highlighted as the paradigm of scientific excellence—supposedly showing that spontaneous generation of a kind promoted by Pouchet is simply impossible. There is just no way in which life can be derived manually from nonlife.

And this would seem to be the end of the matter. Although then you might well ask how this can possibly be so. If indeed evolution be true, that is, if the fact of evolution—the rise of today's organisms from primitive forms—be true, then surely one is forced back to ask about the origin of those first primitive forms. From where did they come? Was the earth seeded from outer space, or is the living part of the basic fabric of the universe and as old as time itself, or (most obviously) did the living come naturally from the nonliving? It is to these and other questions—about the beginnings of life and about the history of its early forms—that we turn in this chapter. As always, we shall find that the story is more complex and interesting than it appears on the surface. For a start, you might think that Pouchet was in favor of spontaneous generation precisely because he was an evolutionist. In fact, he was nothing of the sort. And, until the end of his life, Pasteur—he who goes into every elementary textbook as the final assassin of the

Louis Pasteur

spontaneous generation position—thought it was an open question! But we get ahead of ourselves, so let us start back at the beginning.

Essay

Spontaneous Generation

The belief that life can come naturally and spontaneously from nonlife is indeed a venerable doctrine. It goes back at least to the ancient Greeks and to Aristotle. Down to the time of the Scientific Revolution (the sixteenth and seventeenth centuries), people were torn on the subject. On the one hand, it seemed so clearly true and backed by the evidence of the senses. Meat left unattended rots and in a very short time is swarming with maggots—remember, these were the days before refrigeration! Where could these living beings have come from, except from the nonliving? On the other hand, what was the biblical evidence for such ongoing creation of life? Surely, God had finished His work on the sixth day, and that was that? Was not man himself the culmination of His creative outpouring? Yet, if man was the final act of creation, did this mean that there were maggots in Eden? And parasites and other ugly and unwanted organisms? Surely not!

It was all a puzzle, but with the coming of modern science—particularly with the coming of instruments of magnification—at least some of the issues were resolved. Or rather some of the boundaries were drawn in and made clearer. The Jesuit-trained physician Francesco Redi, for instance, was able to show how many putative instances of spontaneous generation were the results of insects laying eggs, which later hatch. If meat is covered with muslin or some such cloth, so that flies cannot get at it, then the maggots simply do not appear. "Although it be a matter of daily observation that infinite numbers of worms are produced in dead bodies and decayed plants, I feel, I say, inclined to believe that these worms are all generated by insemination" (Redi 1688 [1909], 27).

However, showing that for every step forward there seem to have been two steps back, Redi went on to say that he did not thereby intend to deny the possibility of spontaneous generation. Indeed, he was quite prepared to accept that as a caterpillar can be transformed into a butterfly, so also a piece of the flesh interior to an organism can be transformed into a parasite. He denied that the inorganic can be turned into the living (this is known as "abiogenesis"), but he asserted that the organic can be turned into the living (this is known as "heterogenesis"). More particularly, he asserted that something that is living can be turned into something new and quite different. (This was not evolution, in the sense of transformation from one form to another, but something stronger, where a piece of a living organism is turned into a new animal.) "If the thing is alive, it may produce a worm or so, as in the case of cherries, pears, and plums; in oak glands, in galls and welts of osiers and ilexes worms arise, which are transformed into butterflies, flies, and similar winged animals. In this manner, I am inclined to believe, tapeworms and other worms arise, which are found in the intestines and other parts of the human body" (p. 116).

Theological arguments, however, continued strong against spontaneous generation. To revealed theological objections (that is, to objections based on the Bible) were added natural theological objections (that is, objections based on reason). John Ray (1691), the English clergyman-naturalist, was adamant that the designlike nature of the living world altogether precludes the appearance in one step by natural causes of living organisms. Laws lead to randomness and to things not functioning. (Murphy's Law, that if something can go wrong it will, is a metaphysical fact of nature and not just an engineer's joke.) God designed the world to work, and there is simply no way that things could spring into immediate being and work as well as anything else.

But now came a powerful counter, giving spontaneous generation a whole new lease on life as the eighteenth century moved toward its climactical events, first in North America and then in the final years in France. This was the Age of the Enlightenment, when philosophers and others challenged the old ways of thinking, going counter to the repressive religion that had dictated so much to so many for so long. New confidence was found in the power of unaided reason—philosophical, political, social, literary, and scientific—the latter most particularly Newtonian scientific, as the great English thinker's ideas spread and were developed and applied. With the new strains of thought, theological worries (particularly revealed theological worries) about spontaneous generation started to abate, and at the same time Newtonian undercurrents gave the doctrine a whole new lease on life.

Newton's theory of gravitation (which, incidentally, scholars now trace to some very strange "alchemic" views about subtle powers pervading all of creation and being the key to the transformation of base metals into gold) suggests that the

Georges-Louis Leclerc,
le Compte de Buffon

universe is not what it seems on the surface (Westfall 1980). Bodies can affect each other at a distance, even though nothing is touching, and other odd effects likewise are to be considered part of everyday science. Could it be therefore that there are special powers akin to gravity, affecting and animating living creatures? Even though it may be that nothing living can come from the inorganic (abiogenesis), it may well be that the living can come from the organic (heterogenesis). Such at least was the opinion of the most powerful and influential naturalist of the age, Georges Leclerc, le Comte de Buffon, in France. He believed in a kind of internal living force, the *moule interieur*, which animates and forms the shape and function of organisms, and for him it was quite plausible to think that a primitive life-form could thus be formed from the matter of other such forms. Nor were he and his followers convinced otherwise by experimentation, for instance, by the celebrated work of the Italian Lazzaro Spallanzani, who boiled various broths and the like in sealed flasks, showing that once life had been destroyed it never reappears. The Buffonians' counterargument was that these experiments proved precisely their case, because Spallanzani's boiling had destroyed not just life-forms but the very potential for life that existed hitherto! The *moule interieur* is itself something that must be protected, or else it cannot perform its intended functions.

This all leads us into the nineteenth century, where we start to see a polarizing of positions that lasted to the Pouchet-Pasteur debate, and indeed beyond right down to the present day in respects. On the one hand, we have the successors of Buffon, who thought that spontaneous generation is a constant and common fact of life. In fact, there were two groups of importance. First, there were the evolutionists, notably Erasmus Darwin and Lamarck (who had, incidentally, been a protégé of Buffon). They saw the natural origins of life as part and parcel of their overall developmental position. And for them, natural origins meant spontaneous generation. Thus Darwin, in verse:

> Then, whilst the sea at their coeval birth,
> Surge over surge, involv'd the shoreless earth;
> Nurs'd by warm sun-beams in primeval caves
> Organic life began beneath the waves ... .
> Hence without parent by spontaneous birth
> Rise the first specks of animated earth
> (Darwin 1803, 1, 231–248)

Lamarck had similar views. He thought that worms and the like are produced from warm mud by the action of electricity and heat and so forth. Remember, this was just after the time when Franklin had performed his spectacular experiments with electricity, so it was in itself a rather fashionable possible force of natural change.

Second there were the *Naturphilosophen*, those German thinkers who saw and stressed patterns or isomorphisms throughout the living world—things that we today interpret in an evolutionary fashion and call "homologies." These people were (like the French and English evolutionists) also developmentalists, seeing life going from blobs to complex forms, although in the early part of the century they tended to understand things in an idealistic fashion rather than in terms of actual organic change. They were primed to see patterns not only in the organic world but also in the inorganic world—crystals were a particular focus of attention—and it was but a simple move to link patterns in the inorganic with the organic. Combined with the otherwise apparently inexplicable appearance of living forms where none had existed before—parasites were the big favorite in this respect—it was an easy and ready move to argue that here the links truly did exist and that life sprang spontaneously from nonlife. "Those who defend the doctrine of spontaneous generation do so through experience; when one sees an organized being born without being able to discover either a germ, or any way by which the body had been able to reach the place of formation, one admits that nature has the power to create an organized being with heterogeneous elements" (Burdach 1832, 1, 8).

Empirical experience was clearly an important factor in the support for spontaneous generation at the beginning of the nineteenth century. But we know that there was much more than experience driving people—whether full-blooded evolutionists or not—in their enthusiasm for developmentalism. It would not be fair to say that the main factor was materialism in any simple sense—Erasmus Darwin and Lamarck were both deists, and the *Naturphilosophen* saw life force or spirit in some sense pervading the whole of nature. What was important was the upward-driving force that they saw throughout nature—the force of progress, moving teleologically (purposefully) up from the primitive blob, the monad, to the most complex, the man. Spontaneous generation was part and parcel of this, whether one believed (as did Darwin) that life was created at the beginning and then not again, or whether one believed (as did Lamarck) that life was being created all of the time, in a continuous fashion. And this all went with a progressivist radical view on life—one identified with the forces and philosophies of change and of reform. Progessionists saw themselves as set against the conservative or reactionary elements in society—aristocracy, inherited wealth, the church, and much much more. Spontaneous generation therefore becomes a symbol of a radical progressive view of life, as well as a mark of an empirically justified scientific stance.

All of this on the one hand. On the other hand, we have those who opted for no less of an amalgam of scientific claim and justification together with religious and philosophical urges that supported a conservative, religion-supportive view of life. Most obviously and definitively one finds this in the life and thought of our old friend, Georges Cuvier. He loathed spontaneous generation and thought it quite unsupported by the empirical facts. "Life has always arisen from life. We see it being transmitted and never being produced" (Cuvier 1810, 193). If evolution was judged impossible, you can imagine how much less likely would be spontaneous generation. But clearly, Cuvier's opposition was more than just this. At one level, he hated both Lamarck's speculations and *Naturphilosophie* because they represented the kind of speculative and sloppy science that he thought so threatening to the neat, tight, objective work he was promoting, both for its own sake and for his own sake as an important (although minority-religion) scientist in a conservative society.

At another level he hated evolution and *Naturphilosophie* because they violated his teleologically inspired view of the living world—one that, thanks to the conditions of existence (and its corollary the correlation of parts), gave Cuvier (as he thought) a predictive science, as one tries to fit parts of organisms into an end-driven purposeful functioning whole. And at a third level, he hated evolution and *Naturphilosophie* precisely because they represented radical and revolutionary elements in society, and everything for which he himself stood was on the side of stability, and the status quo, and the establishment. All of these levels and more came into play against developmentalism and hence were focused even more on spontaneous generation. To Cuvier, it was a false doctrine in every possible way.

The Opposition Continues

These two positions—part empirical, part metaphysical, part political—dominate thinking from here on through the century. In the 1840s, the supporters of spontaneous generation were dealt a heavy blow when finally people worked out the basic facts of parasitism. It was shown how organisms such as tapeworms take on several different forms, according to the hosts in which they are embedded: far from being generated at one fell swoop in one set of hosts, they come from other hosts where they pass unrecognized because they do not yet have their final forms. Although a person may never eat food containing a tapeworm, as is found in humans, these foodstuffs do actually have parasites that are transmitted to humans: parasites that are indeed tapeworms in potentiality if not yet in actuality. This discovery was obviously a severe thrust against the doctrine of spontaneous generation, although apparently it was not the death knell. Supporters still continued to have faith in its existence.

> I am at the very first struck by the great a priori unlikelihood that there can have been two modes of Divine working in the history of Nature—namely, a system of fixed order or law in the formation of globes, and a system in any degree different in the peopling of these globes with plants and animals. Laws govern both: we are left no room to doubt that laws were the immediate means of making the first; is it to be readily admitted that laws did not preside at the creation of the second also? (Chambers 1846, 19–20)

Enter Pasteur. Pouchet was desperate not to be labeled a radical because of his belief in spontaneous generation. He thought it was proven by the facts. Hence, his adamant denial of evolutionism. He had no desire to be tarred by that particular brush. But to no avail. Pasteur was determined to roll right over such ongoing claims about the empirical plausibility of spontaneous generation. In a series of celebrated experiments, he boiled sugared yeast to kill off the live contents, and then showed that the treated material remained sterile unless and until it was recontaminated. He showed that these results hold in different conditions. Even when the material is open to the outside air, so long as the openings are such that contaminants cannot enter (through being long and thin and curved), there is no appearance of life. Potential breeding grounds remain sterile. But as soon as air is allowed to enter freely, or other nonsterile substances are permitted to infect the inner material, fermentation begins almost at once. Only when life is introduced does life multiply. If life is barred, it seems to remain forever absent.

Celebrated then and celebrated now. Pasteur was a brilliant experientialist, and he and others recognized this fact. He did truly strike a heavy empirical blow against spontaneous generation. But the joy of his countrymen at his successes far exceeded the mere empirical. France in the 1860s was a deeply conservative society, with the monarchy, the aristocracy, the church in full force. The last particularly was setting its face resolutely against change or modernity. In a Papal Encyclical of 1864, Pope Pius IX denied explicitly that "the Roman Pontiff can and must make his peace with progress, liberalism, and modern civilization and come to terms with them" (Error 80, quoted by Farley 1977, 95). All of these sorts of factors played a major role in the reception of Pasteur's work and the canonical status it achieved. The committee set up by the French Academie des Sciences was deeply conservative, quite determined to find in Pasteur's favor, and no less willing to preach to one and all that the fatal blow had been struck against the radical doctrine of spontaneous generation. And when Pouchet and friends complained against the bias of the judges, the Academie set up a committee even more conservative and predetermined to find in Pasteur's favor! By this time, France had received the *Origin* with its radical new introduction, and the forces of the French scientific establishment felt (with reason) that the counterattack must be mounted and supported with every possible weapon. Pasteur's work was just what was needed, and so his results were trumpeted far and wide.

Brilliant though Pasteur's work truly was, the existence of extraneous non-scientific factors in its reception—a reception the success of which echoes down to the present—is amply attested by the fact that (as I mentioned at the beginning of this chapter) Pasteur himself did not truly believe that he had forever disproved spontaneous generation. He did not believe that one could get life from living material or even from just plain organic matter (heterogenesis)—his experiments showed this and his conviction was part of his overall thinking on fermentation. But work he had done early in his career on crystallization rather disposed him toward the possibility of life from the inorganic (abiogenesis). "Life is the germ and the germ is life. Now who may say what might be the destiny of germs if one could replace the immediate principles of these germs (albumin, cellulose, etc.) by their inverse asymmetric principles. The solution would constitute in part the discovery of spontaneous generation, if such be in our power" (Pasteur [1883] 1922, 1, 375).

Pasteur, however, kept quiet about these speculations until late in his career, when—a new, more liberal government being in power after the disastrous Franco-Prussian war—he felt free to make them public. Although, paradoxically, by this time the rest of the world was moving on beyond spontaneous generation. Such had not been the case when, at the beginning of the 1860s, Pasteur began his assault on Pouchet. In Britain and Germany, the rapid rise in the respectability of evolutionism, thanks to Darwin's *Origin*, led to an immediate enthusiasm for spontaneous generation that had not engulfed those countries to such a degree ever before. At once, people saw that evolutionism demanded answers about the ultimate origins of life, and the old ideas were brought out and polished to shine more brightly than they had ever done in earlier times. Pasteur was ignored, if indeed he was noted at all.

Interestingly and paradoxically, Darwin himself made no contribution to this enthusiasm for spontaneous generation. In the *Origin*, he said virtually nothing about origins, merely talking of the rise of life from "one or a few forms." I am not quite sure about the reason for this silence. Part, I strongly suspect, is that Darwin realized that it was a topic surrounded by controversy, associated with radical thinking. Naturally cautious by nature and determined to push his ideas in a nonthreatening manner (because then they would be more likely to be accepted), Darwin stayed away from ultimate origins because speculation would and could only harm his case. If there is a nasty gap in your knowledge, then your best policy is to say nothing and to say it firmly! Also, I suspect his silence was in part because—a problem that plagues evolutionists to this day—although evolution seems to demand answers about ultimate origins, there is little reason to think that the evolutionist is in any way capable of answering them. I do not mean that no answers can be given, or that the evolutionist as such is a bad or inadequate scientist, but rather that the problems and answers lie outside of his or her professional domain. The evolutionist is a biologist. Origins require chemistry, biochemistry in particular, and lots of it. There is really no reason why the evolutionist as evolutionist should be able to answer these questions, even though the evolutionist's work points to these questions as demanding answers. So here was another reason for Darwin's silence.

Also at this point remember that Darwin was trying to promote evolution as a potential professional science, and so here again was reason to stay away from speculation. And remember that his supporters—Huxley in England particularly, and Ernst Haeckel in Germany also—had other ends in view. They wanted broad metaphysical speculations, grounds for the secular world philosophies or religions that they were spinning and endorsing and promoting. For them, evolution was the popular science par excellence, and they had no hesitation in pushing its limits to the ultimate and beyond. They had no need of caution on origins or on hypotheses about spontaneous generation. Indeed, in the 1860s Huxley and friends even thought that they had new empirical evidence of its truth. Mud dredged from the sea was taken to be full of life or life-potential forms or particles.

> I conceive that the granule-heaps and the transparent gelatinous matter in which they are imbedded represent masses of protoplasm. Take away the cysts … [it would] very nearly resemble one of the masses of this deep sea "Urschleim," which must, I think, be regarded as a new form of these simple animated beings … described by Haeckel. … I propose to confer upon this new "Moner" the generic name *Bathybius*, and to call it after the eminent Professor of Zoology in the University of Jena, *B. Haeckelii*." (Huxley 1868, 212)

Unfortunately the euphoria did not last and neither did *Bathybius haeckeli*. In 1876, it was discovered to be inorganic—a precipitate of sulfate of lime—and that was the end of that. Huxley withdrew his claim. And despite the fact that there were those who still wanted to defend spontaneous generation, although by now evolution had conquered almost all that lay before it, one senses that the days of spontaneous generation were coming to an end. On the one hand, just too many things that had been hailed as evidence for the belief had by now been shown explicable by other means or simply not supportive of the doctrine. On the other hand, and probably more important overall, people were now starting to dig further and further into the elements of the organisms—first cells, and then cell parts, and then parts of these parts. With each move to a yet-smaller level, the intricate complexity of the stuff of life was reinforced even more strongly than before.

And with this, whether or not one put all or any of this down to God's design and activity, the improbability of things having come together spontaneously became less and less. It was not so much that people like Pasteur had given definitive proof that spontaneous generation was impossible—we have seen that Pasteur himself had hardly done that because he thought that it was always an open question—but rather that the weight of evidence about the nature of the living world made spontaneous generation less and less plausible as an explanation. In an era of cell biology, it simply did not make sense.

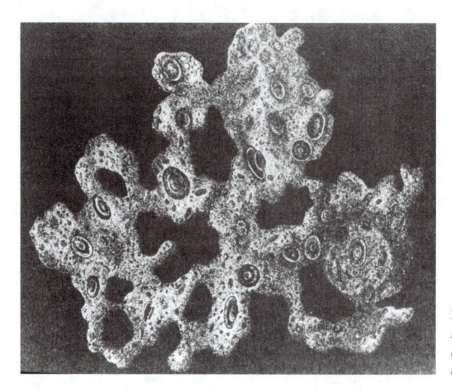

Bathybius haeckelii,
*supposedly an example of
early life but in fact a
chemical precipitate*

The Oparin-Haldane Hypothesis

But still an answer must be sought. If not spontaneous generation, then what? If
evolution be true, then the Darwinian strategy of silence can only last so far. One
must return at some point to the issue of origins. As indeed Darwin himself did in
a private letter.

> It has often been said that all the conditions for the first production of a living or-
> ganism are now present which could ever have been present. But if (and oh! what a
> big if!) we could conceive in some warm little pond, with all sorts of ammonia and
> phosphoric salts, light, heat, electricity, etc., that a protein compound was chemi-
> cally formed ready to undergo still more complex changes, at the present day such
> matter would be instantly devoured or absorbed, which would not have been the
> case before living organisms were formed. (Darwin 1887, 3, 18; letter written in
> 1871)

This gradual (but natural) appearance of life was also a position endorsed by
Herbert Spencer. He set himself entirely against spontaneous generation. "That
creatures having quite specific structures are evolved in the course of a few hours,
without antecedents calculated to determine their specific forms, is to me incredible" (Spencer 1864, 1, 480). As Spencer pointed out, reasonably, in a way spontaneous generation threatens to undercut the whole evolutionary enterprise. If
primitive-yet-complex organisms can appear in one fell swoop, what is to stop
more sophisticated organisms appearing in like fashion? And if these, then where
does such generation end? Why bother with evolution at all? Yet, Spencer did not
want to deny the natural appearance of life. It had to appear gradually, that is all.
"The evolution of specific shapes must, like all other organic evolution, have re-

J. B. S. Haldane

sulted from the actions and reactions between … incipient types and their environments, and the continued survival of these which happened to have specialities best fitted to the specialities of their environments" (p. 481).

But it was to be another fifty years before people started to make a research program out of such speculations. Two people particularly, in the 1920s, are credited with the ideas that started things moving in the direction of gradual development of life from nonlife: first the Russian biochemist Aleksandr Ivanovich Oparin ([1924] 1967), and then the English biochemist and theoretical population geneticist J. B. S. Haldane (1929). Like Darwin and Spencer before them, although in somewhat more concrete terms given the advances in chemical understanding, they postulated the emergence of the living from more simple inert substances, through natural evolutionary-type laws. Not that the program that they started has yet been brought to full fruition. There are still major questions about how life might have come naturally, even if (especially if) it takes a very long time. But for all that, work in the twentieth century on the origin of life became almost a full-time industry, and the researchers themselves certainly think that they have made significant advances, if not in the traditional direction of spontaneous generation.

I should say that whatever the nature of the advances—the merits of which we shall consider in a moment—some of the controversy and questioning that still swirls around the question is to a certain extent self-imposed. Famously, or notoriously, both Oparin and Haldane were Marxists—Oparin especially so, for he was a key figure in Soviet science and a major backer of the agricultural genetics charlatan Trofim Lysenko. You might think therefore that even though you no longer have to subscribe to an outmoded philosophy like German *Naturphilosophie*, if you are today going to take a naturalistic stand on origins, you must endorse a philosophy that many (perhaps including yourself) find thoroughly objectionable. Fortunately, although Oparin particularly was given to tying in his theorizing with dialectical materialism, the links are at best tenuous (Graham 1987). In fact, in the 1920s, neither Oparin nor Haldane was yet a Marxist, so the strongest possible connections simply are not there. And when in the 1930s, Oparin did start to put things in Marxist terms (Haldane never did so, for his contribution to the question was confined to a suggestive essay), much that he said could be translated at once into nontheoretical language.

Marx—or rather his coworker Engels, who was more interested in natural science than Marx himself—postulated a number of laws (or "laws") that supposedly govern the workings of nature. One is the "law of quantity into quality," as when cooled water does not simply get colder but turns into something new, namely ice. Oparin took this law to incorporate the fundamental truth of his whole approach to the origin of life question, inasmuch as he was suggesting that nonlife turns into life. But this is hardly Marxist, as such, for it was also Darwin's view—and famously, although the Englishman received from the author a copy of *Das Kapital*, he never cut the pages and read it! The same holds true of the supposed connection of the origin of life question to Engels's "law of the negation of the negation" (as with Newton's law, that to every action there is an equal and opposite reaction). Oparin took this as proof that life once started makes impossible the further creation of new life. But again this was Darwin's view, and so not Marxist per se. (As I have said earlier, in a way, I see Engels's work going back to that very German idealism the Marxists thought they were refuting!)

Having disposed of this ideological red herring—"red" in more senses of the word than one, and a great favorite of the evangelical Christian opponents to any scientific approach to the question of origins—we can turn now to contemporary thought on the beginnings of life. It is customary and useful to break down the Oparin-Haldane hypothesis (as it is generally known) into a number of steps. First, there is need of the right conditions for the (natural) creation of organic molecules—those that make the ultimate building blocks of life—from nonorganic molecules (from a warm prebiotic soup, as is sometimes said). If conditions were like they are today, with a 20 percent oxygen atmosphere, then (as Haldane pointed out) the molecules simply could not have formed and persisted. One needed a very different sort of atmosphere. But as it happens, this fits precisely with what students of the subject think in fact might have been the case. The earth is believed to be about four and a half billion years old and initially in a molten

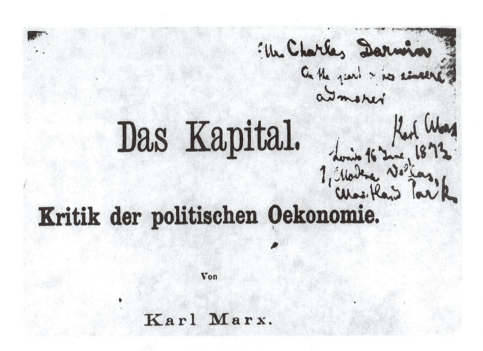

The title page of the presentation copy of Das Kapital *from Marx to Darwin*

state. As it cooled over the next billion years, the oceans formed. It was one rather heavy in such gases as methane (CH_4), ammonia (NH_3), carbon dioxide (CO_2), and hydrogen sulphide (H_2S). Moreover, it would have been an atmosphere permitting the passage of much ultraviolet radiation, for there would have been no ozone (O_3) layer to block it.

Next, there is the making of these elementary organic molecules. Even today, nearly a half century after it was first achieved, this is the most celebrated part of the chain. Back in 1953 a number of chemists—notably Stanley Miller, then just a graduate student—set up a relatively simple apparatus showing how, under what were presumed original conditions, complex molecules ("amino acids," the components of proteins, chainlike molecules that make the structure of the cell) would form quite rapidly. A little electricity (simulating lightning) or radiation would turn inorganic substances (such as methane and ammonia) into organic molecules of the required sort. Since those first experiments, even more successes (including the creation of the composites of the nucleic acids, the templates of life) have been formed. It should be added that there are now doubts about how oxygen-free the early atmosphere truly was. It might have been impossible for the required reactions to have occurred above ground. However, there is now reason to believe that the reactions could have taken place deep in the sea, close to vents where magma (molten rock) bubbles to the earth's surface.

Third (in the Oparin-Haldane sequence), one has to get the individual organic molecules to link together into the chains that are needed for the maintenance of life—proteins and nucleic acids. In fact, the joining is no great problem. It is just that such chains tend to break apart rapidly before the job is finished. Here it is thought that naturally occurring clays may be significant causal factors. Organic molecules adhere to clays and can build up chains while at the same time resisting the urge to break apart. Already, experimenters have shown how quite long chains can be formed in such (presumably analogous to natural) conditions.

The experimental apparatus of Stanley Miller, through which he made amino acids naturally

The fourth step is a lot more tricky. Now you have to get the long chain molecules to replicate themselves. In cells today, deoxyribonucleic acid (DNA, the modern molecular equivalent of the gene) reads off itself to make copies for new cells. Also, ribonucleic acid (RNA) reads off the DNA, and then this RNA acts as an information template to make chains of amino acids, proteins. But at the beginning you have a bit of a chicken and egg situation. Without the superstructure of the cell, made of proteins, it is hard to see how the nucleic acids could function. But without the functioning nucleic acids, you get no proteins! Perhaps, suggest some workers, the mineral clays continue to play a significant role. Crystals repeat themselves, building copies on templates. And sometimes errors get incorporated into the crystal patterns and get repeated. Could it be that originally it was crystals that were reproducing (no one says that they were alive), with organic molecules as it were piggybacking on them? Then the organic molecules themselves started to take over reproduction, and eventually they dropped their mineral supports (Cairns-Smith 1982, 1986).

Other workers are suspicious of this hypothesis—their trouble is that it contains the claim that today there is no confirmatory evidence of what happened originally! They rather prefer to think that the organic chains themselves may have gone directly to reproduction—most likely through the medium of RNA, which is needed for proteins and which is in some organisms the only nucleic acid and thus capable of acting as a template for itself without need of DNA. Of course, there are questions about how this it to be done. Suffice it to say here that biochemists are trying with some success to get RNA molecules on their own to replicate themselves. No one has succeeded in getting this to work properly yet, but already one can get an RNA molecule to add bits of chains like itself to other such molecules. Of such tiny steps are great edifices built!

Even if this all works out nicely, more steps remain. One has to tuck everything away in a nice globular cell, for instance. It was here that Oparin put much of his energies, as also did a number of American workers, notably Sidney Fox (1988) and his colleagues. They strove with some success to show how some organic molecules can be made to form self-contained spheres (like the outer shells of cells); how they can maintain themselves, even budding off to form other spheres; how such shells can keep and even promote differences between the inside and the outside, even selecting as it were certain compounds to cross over from the outside to the inside (while barring others); and how some of these compounds can be precisely the kinds of molecules (like ribonucleic acid molecules) that one would expect to have been preserved and cherished in new or protocells.

Enough! There is more, much more. Cells are very complex entities, certainly the cells of higher organisms. (A distinction is drawn between "prokaryotes" that roughly speaking are the cells of simple organisms and "eukaryotes" that roughly speaking are the cells of more evolved organisms.) These latter, eukaryotes, contain not just the centers (the nuclei) where you find the DNA, but also other bodies (like mitochondria and ribosomes) that have various functions. It has been suggested by Lynn Margulis (1970) and others, with some considerable plausibility, that perhaps the eukaryotic cells were formed by incorporating prokaryotic cells: not so much cannibalism but in a form of symbiotic relationship. And then after that, you have the development of sexuality, something that is thought to be closely linked to the appearance of eukaryotic cells: virtually all major groups of organisms with eukaryotic cells have sexuality—and those organisms without sexuality in such groups are thought to have been sexual and then (for various selective reasons) to have lost it.

There are a lot of steps here and almost all of them are tentative—they require a measure of faith, not necessarily "faith" in the religious sense but in the sense that one might say one is making a gamble or prediction on what one thinks are reasonable grounds. But are they "reasonable grounds," or rather if you add everything up together are they reasonable grounds? One or two steps you might swallow. But six or seven or eight steps? Surely this is all a bit like gambling on all of the winners on a day's racing card. You might pick two or three winners, but would anyone want to put their money on picking all of the winners for that day—however good the odds? One prominent critic of all things evolutionary— Alvin Plantinga, noted already as North America's most distinguished philosopher of religion—is so contemptuous of the work thus far performed on the origin of life question that he cannot even bring himself to write on it. He speaks of hypotheses about the origin of life as "the most part mere arrogant bluster," adding that "given our present state of knowledge, I believe it is vastly less probable, on our present evidence, than is its denial." Indeed, so contemptuous is he of such claims that he finds that he cannot bring himself to "summarize the evidence and the difficulties here" (Plantinga 1991).

Can things really be this bad? Obviously the stand you take here is going to depend on a number of things, and most of them are not going to be purely scientific. If (with most scientists) you are firmly committed to the belief that natural explanations can be found for all physical (including organic) phenomena, then you are going to think that natural origins of life are reasonable, no matter what the gaps. If you are firmly committed to the significance of supernatural forces—divine interventions or miracles—then I suspect that you are probably going to think that God had a role here. Perhaps you will think this even if you are an evolutionist. If you are somewhere in the middle, then presumably you are going to end up somewhere in the middle!

But is it reasonable to be a naturalist? Are not scientists turning their heads away from the truth—out of ignorance or prejudice or whatever? (Do not discount the powers of indoctrination and prejudice. It would be a brave scientist indeed today to admit that he or she was going to invoke miracles. I can just imagine the comments on the next grant application.) Let me make two remarks. First, even though it is surely true that the scientist's assumption that everything can be given a natural explanation is an assumption that goes beyond the evidence—how could it be otherwise?—it does not follow that it is an unwise or irrational or even a risky move to make. The fact of the matter is that time and again things that have seemed incredibly puzzling, surely defying scientific explanation, have succumbed to constant pressure and investigation. Think of the wonders of physics—the planets, for instance, and how they shine bright when they loop the loop in the heavens ("retrogress")—for many years a monstrous puzzle, but then explicable in terms of Copernican theory. Think of the strange distributions of animals and plants around the globe. For years people wondered about their causes, and now we know that it is because the continents move around the globe on massive plates. Think of diseases that seemed beyond doubt to be acts of God, but that now are almost commonplace phenomena, explicable through microbes or viruses or whatever. When acquired immunodeficiency syndrome (commonly known as AIDS) was first reported, no one had any idea of its cause, but investigation soon brought the answer in the form of the human immunodeficiency virus (HIV).

My point is that being a scientific naturalist is a good strategy because again and again it brings results, even in the most unpromising situations. It may not be a logically sure bet—there could always be a miracle around the next corner—but pragmatically it is a very sensible way to go. It is not just a "leap of faith" in the sense of going against the evidence. It is the opposite. It is going with the evidence and precisely what we mean by being "reasonable." But what about the particular case of the origin of life? Is this not a special case that makes the naturalistic approach highly unpromising? I do not see that this is the case at all. Remember, until about a hundred years ago, people were trying to shortcut the whole process with spontaneous generation. Eventually, that fell to the ground, but it has only been in the past 60 or 70 years that people have really tried to crack the problem in a way that (even in principle) stands any chance of succeeding.

And it is even more recently that researchers of the subject have had at hand some of the really relevant tools of the trade—detailed knowledge about DNA and RNA and proteins and such things, for instance. It is a massive problem that faces the origin of life researchers—indeed, part of their advance is to realize precisely how massive a problem it is. I for one would be suspicious (especially given the history of the all-too-slick spontaneous generation idea) if the claim were that the problem was now licked or close to being so. Big problems require lots of effort—time too—to get at big solutions. I would not expect more than a progress report from the battlefield, which is precisely what we get.

But there are reports and then there are reports. Not all progress reports are progressive in the sense of reporting on genuine advance. I would say that here, however, we are given just such a report. The researchers are making advances—on the self-synthesizing of nucleic acids, for instance—and feel confident that more such advances lie ahead. One is not just given a whole heap of questions, with no one having the slightest idea about how to crack any one of them. One does not simply have awe and mystification and nothing else. One has real work and real results. I simply do not see that one could ask for or expect any more at this point. It is a fallacy to think that, because there are many links to be filled and most or all are thus far not connected, this means that collectively the case is hopeless. The point is that the links are open to study and investigation and that they are yielding to pressure.

For this reason, while not wanting to pretend that more has been done than actually has been done, I would suggest that the researchers' faith that answers will come—naturalistic answers will come—is not misplaced. It is simply silly (and a sign of almost wanton ignorance) to say that the work thus done is for "the most part mere arrogant bluster."

Early Life

About 600 million years ago, there was a huge increase in the number of life-forms on this planet. After the "Cambrian explosion," the earth teemed with life and nothing was ever again quite the same. But what about the time before the Cambrian? If the first life came over three and a half billion years ago, then we have a vast period for which to account—a period five or six times as long as after the start of the Cambrian and the much more familiar modern era. Is there any evidence of past life? What about the paths that it took? Can we say something about causes? Even if you agree that natural selection was the chief motivating factor, is there anything to show why certain things happened at certain times?

I have mentioned in an earlier chapter how the pre-Cambrian was something that worried Darwin a lot, for at the time of writing the *Origin*, there was simply nothing at all in the record to show that there had been life.

> If my theory be true, it is indisputable that before the lowest Silurian [today, called the Cambrian] stratum was deposited, long periods elapsed, as long as, or probably far longer than, the whole interval from the Silurian age to the present

day; and that during these vast, yet quite unknown periods of time, the world swarmed with living creatures.

To the question why we do not find records of these vast primordial periods, I can give no satisfactory answer. ... The case at present must remain inexplicable; and may be truly urged as a valid argument against the views here entertained. (Darwin 1859, 307–308)

As it happens, not long after the *Origin* was published, a number of strange objects were unearthed in the pre-Cambrian rocks of Canada. These were identified by Sir William Dawson, sometime principal of McGill University and doyen of Canadian geologists, as primitive life-forms, and they were given the name *Eozoon canadense*, the "dawn animal of Canada." (O'Brien 1970). Darwin picked up on this at once, and the discovery duly made its way into later editions of the *Origin*, supposedly filling the acknowledged major gap in life's history. Paradoxically, although perhaps by now the kind of move you are coming to expect, Dawson—a lifelong opponent of evolution—also focused on *E. canadense*, making it the linchpin of his case against evolution! He argued that since it occurred isolated from all of its fellows, it must have been created miraculously and placed in position by Divine intervention.

As it happens, everybody's house was built on sand—metamorphic sand. It was soon discovered that *E. canadense* is no genuine organism but an artifact of great heat and pressure on limestone. The pre-Cambrian therefore seemed as empty as ever before, and this was the way that things lasted right down and well into the twentieth century. But then the record started to open up in a major way, taking us back virtually to the (presumed) beginning of life, over three and a half billion years ago. Moreover, what is really exciting is that the most primitive organisms seem to be the oldest and that the most sophisticated, those edging close to Cambrian forms, come in the last of the pre-Cambrian deposits. Nothing is out of order. What happened where, and who evolved into whom—working out the path of pre-Cambrian evolution, that is—is of course another matter, and the fossil record is hardly good enough for that. Here one needs to turn to cellular and molecular traces, trying to infer past connections and phylogenies. And at least some of this seems to be possible. One quest has been toward finding the latest common ancestor of all living organisms (the "cenancestor"). It obviously came fairly early on in the story, and today it is believed that it might have been as long ago as three and a half billion years, although (showing how crude things still are, as yet) it might be as recent as two billion years.

One major question about early life history—perhaps the major question—is about the move from life that is exclusively prokaryotic to life that is also eukaryotic. We have seen already that part of what was going on here was probably the symbiotic coupling of prokaryotes to make eukaryotes. Is there more evidence of the evolutionary emergence of eukaryotes from prokaryotes, and is there evidence of when and why it happened? There are several suggestive lines of evidence bearing on these questions, with the key factor being oxygen. First of all, there are the ways in which the two kinds of cells obtain energy: their metabo-

Relative Durations of Major Geologic Intervals	Era	Period	Epoch	Duration in Millions of Years (approx.)	Millions of Years Ago (approx.)
Cenozoic	Cenozoic	Quaternary	Recent	Approx. last 5,000 years	
			Pleistocene	2.5	2.5
Mesozoic		Tertiary	Pliocene	4.5	7
			Miocene	19	26
			Oligocene	12	38
Paleozoic			Eocene	16	54
			Paleocene	11	65
	Mesozoic	Cretaceous	**Events** Last Dinosaurs First Primates First Flowering Plants	71	100 / 136
		Jurassic	Dinosaurs First Birds	54	150 / 190
		Triassic	First Mammals Therapside Dominant	35	200 / 225
	Paleozoic	Permian	Major Marine Extinction Pelycosaurs Dominant	55	250 / 280
		Pennsylvanian (Carboniferous)	First Reptiles	45	300 / 325
		Mississippian (Carboniferous)	Scale Trees, Seed Ferns	20	345 / 350
		Devonian	First Amphibians Jawed Fishes Diversity	50	395 / 400
		Silurian	First Vascular Land Plants	35	430 / 450
		Ordovician	Burst of Diversification in Metazoan Families	70	500
		Cambrian	First Fish First Chordates	70	550 / 570
		Ediacarian	First Skeletal Elements First Soft-Bodied Metazoans First Animal Traces (Coelomates)	130	600 / 650 / 700

Life's history

lisms. Both kinds of cells get energy from glucose, but whereas prokaryotes work by fermenting their foodstuffs, simply breaking down the glucose, and getting energy that way, eukaryotes work by respiration, burning the glucose in oxygen, and thus releasing energy. The second mechanism, respiration, is far more effective than fermentation, which in itself is suggestive—one has a presumed move from the less to the more efficient. But more significantly, fermentation and respiration are not two completely different mechanisms. The one metabolism, respiration, follows on the other, fermentation, by adding on more steps: an oxygen-using phase. This is just what one would expect were evolution at work: building on what you have rather than starting anew.

Second, what about the coming of oxygen? We have seen that free oxygen in the atmosphere cannot have been present when life first formed, for it would have had a devastating effect on the beginning organic molecules. One would expect to find therefore, that although eukaryotes need oxygen, the prokaryote story would be different. As indeed it is. Some prokaryotes can tolerate and even need oxygen (which is what one might expect if the oxygen-needing eukaryotes are to evolve

Pre-Cambrian life

from them), whereas for other prokaryotes oxygen is a poison (which is what one might also expect). But where would the oxygen have come from, if there was none free at first? Presumably, as today, it would have come through photosynthesis, where organisms free up oxygen from carbon dioxide, thanks to the energy provided by the sun. And expectedly, we find that some prokaryotes can perform photosynthesis. This is within the power of the blue-green algae (known as "cyanobacteria"), which interestingly and significantly seem to have a metabolism halfway between that of fermentation (like regular prokaryotes) and respiration (like regular eukaryotes). Surely pertinently, the cyanobacteria function most efficiently when oxygen levels are around 10 percent, that is, about half of today's levels. One presumes that they evolved at a time when the oxygen level was not as high as it is today, but that they paved the way for the evolution of higher oxygen level–using organisms.

Finally, what about evidence of the time of the arrival of the eukaryotes and the rise in the level of oxygen? There are factors relevant to both of these questions, and they come together with coinciding answers, suggesting that we can claim to know the whole picture. Larger fossils of a kind one would associate with eukaryotes are to be found from about one and a half billion years ago, and this is a point somewhat after the evidence points to the rise in oxygen levels. For instance, uraninite (UO_2) oxidizes (to U_3O_8) in the presence of more than a 1 percent oxygen atmosphere. Predictably, in rocks older than two billion years, uraninite is to be found, but it is absent from younger deposits. Conversely, iron rusts in oxygen. In deposits less than two billion years old, we find iron oxides. These are missing from earlier deposits.

Some of the most compelling evidence for oxygen scarcity on the early Earth comes from gravel and sand deposited by ancient rivers as they meandered across Archean and earliest Proterozoic coastal plains. Pyrite [FeS_2—fool's gold] is com-

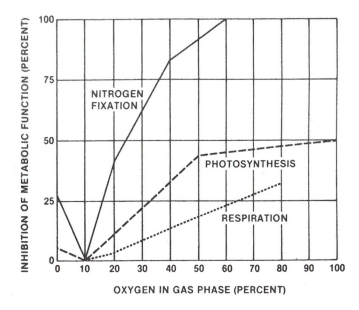

Dramatic evidence that cyanobacteria function best when the atmospheric oxygen levels are half what they are now, suggesting that cyanobacteria evolved when oxygen was at half its present level

mon in organic-rich sediments, forming below the surface where H_2S produced by sulfate-reducing bacteria reacts with iron dissolved in oxygen-depleted groundwaters....

The same is true of two other oxygen-sensitive minerals: siderite (iron carbonate, or $FeCO_3$ and uraninite (uranium dioxide, or UO_2). Neither of these minerals is found today among the eroded grains that make up sediments on coastal floodplains, but both occur with pyrite grains in river deposits older than about 2.2 billion years.... (Knoll 2003, 97)

Conversely, after 2.2 billion years, we get the deposition of minerals that can form only in the presence of iron. The bright red sandstones of the Grand Canyon are a stunning example. "These rocks—called red beds, in the button-down parlance of geologists—derive their color from tiny flecks of iron oxide that coat sand grains. The iron oxides form within surface sands, but only when the groundwaters that wash them contain oxygen. Red beds are common only in sedimentary successions deposited after about 2.2 billion years ago." Before this date, there cannot have been more oxygen than about 1% of today's levels; after this date, there was at least 15% of today's levels. The way was being prepared for the rise of the eukaryotes.

Lots of questions still remain, but the answers are starting to come in. And importantly, the answers fit together. They are consistent and coherent. A unified picture of life's early history is starting to shine through. (See the accompanying pictures and diagrams.) Charles Darwin, who admitted to so much ignorance, would have been pleased. We should feel the same way also. Origins are no longer quite the mystery that they were once.

Further Reading & Discussion

I cannot speak sufficiently highly of John Farley's *The Spontaneous Generation Controversy from Descartes to Oparin* (Baltimore: Johns Hopkins University Press, 1977). His knowledge of the science is deep and profound, and his ability to move from work in one language to the next is simply staggering. He is truly superb on the developments in our thinking about life's origins from the seventeenth century right down to the near present. I have used his translations in my discussion. A more recent book is a great complement to this older treatment. Iris Fry's *The Emergence of Life on Earth: A Historical and Scientific Overview.* (New Brunswick, N.J.: Rutgers University Press, 2000) is strong not only on the science but also on the underlying philosophical elements that enter into people's thinking about such a difficult and challenging subject as life's origins. She is balanced and fair without being in the least boring.

As I have intimated in the text, the problem with origin of life studies is that they carry you into areas of science, areas of chemistry particularly, where a knowledge of regular evolution is not much help. These areas tend to be fairly complex and thus discussions are difficult to follow unless one has had some training. An excellent up-to-date overview on the origin of life question can be found in an encyclopedia to evolution, edited by me and Joseph Travis, *Evolution: The First Four Billion Years* (Cambridge: Harvard University Press, 2009). You will be amazed at the amount of work that is being devoted to the topic. A superb account of our understanding of life before the Cambrian comes from the pen of Harvard professor Andrew Knoll, *Life on a Young Planet: The First Three Billion Years of Evolution on Earth* (Princeton, N.J.: Princeton University Press, 2003).

A topic for which I have no room in my main text but which certainly impinges on the subject of the origin of life is that of life elsewhere in the universe. Did life start here uniquely on earth, or has it occurred again and again throughout the depths of space? Did life perhaps start somewhere else, and was our planet seeded from outside? These have been topics of fascination to scientists, philosophers, and theologians ever since the Greeks. An excellent trilogy of works covers the field. First there is Steven Dick's *Plurality of Worlds: The Origins of the Extraterrestrial Life Debate from Democritus to Kant* (Cambridge: Cambridge University Press, 1982), taking us up to the end of the eighteenth century. Then Michael J. Crowe carries the story through to the beginning of the twentieth century: *The Extraterrestrial Life Debate, 1750–1900: The Idea of Plurality of Worlds from Kant to Lowell* (Cambridge: Cambridge University Press, 1986). Finally Dick again, in a truly magnificent work, deals with the whole extraterrestrial issue in the twentieth century just gone: *The Biological Universe: The Twentieth Century Extraterrestrial Life Debate and the Limits of Science* (Cambridge: Cambridge University Press, 1996). A popular version is *Life on Other Worlds: The Twentieth Century Extraterrestrial Life Debate* (Cambridge: Cambridge University Press, 1998). This is a really wonderful coverage of science and fiction and speculation and much, much more. It is interesting how the astronomers are so eager to argue for extraterrestrials, including humanlike forms, whereas the evolutionists are much more skeptical. Of course, the astronomers have an interest in seeing all of those rockets shot off into space. This is not an interest much shared by evolutionists.

Chapter 8

Two New Sciences at War: Placing Ancestors in Time

Overview

This chapter discusses how the emergence of two new 20th century sciences—paleoanthropology and molecular biology—helped to answer the question of when humans first appeared.

Even before the *Origin* was published, people were getting excited about the natural origins of humans. There were all sorts of arguments about the place of the newly discovered Neanderthals in human development. In the first half of the 19th century, before new sciences emerged, 'placement' arguments and discussions could take decades to resolve, if resolution was even possible. Early in the 20th century, however, a new branch of anthropology—paleoanthropology—looked at skeletal remains and comparative development in order to catalogue "missing links". Later in the 20th century, another new science—molecular biology—developed techniques for assessing absolute dates with just a few samples in a laboratory.

Remains of a clear "missing link" was unturned in Java in the late 19th century, and the first specimen of Australopithecines was discovered in the early 20th century. However, for many years things were thrown off course by the greatest hoax in the history of science, the Piltdown man, supposedly a human-ape type being, but truly bits of different species put together. Today, back on track, we have a lot of evidence of human evolution, backed by molecular findings.

What were the causes of human evolution? In the course of answering this question, beware of thinking that there would be an inevitable progression up to beings with big brains. Apart from anything else, big brains demand lots of protein which in the past meant meat, and the need to hunt. No doubt the move to sociality was a major factor in our past, with the ability to work together fueling success in food searching and feeding and, in turn, promoting more social behavior.

There are still major disputes about human prehistory. One centers on the Neanderthals. Did modern humans come out of Africa more than a hundred thousand years ago and wipe out the Neanderthals, as is claimed by British paleoanthropologist Chris Stringer? Or was there a kind of parallel evolution all over the world relating humans to the Neanderthals, who were part of this picture, as is claimed by Michigan researcher Milford Wolpoff? This chapter discusses both possibilities and, as we shall see, the jury is still out.

But in the end man is more than a collection of bones for the paleoanthropologist and chemicals for the microbiologist. What about the evolution of language? And consciousness? We can trace a bit of language development using hard science to watch the necessary skull formation, but language and communication quickly fall out of the abilities of these sciences. As we progress through this chapter, controversies are uncovered, although no one denies the adaptive significance of language. Consciousness is still very controversial, due in major part to the fact that there has not yet been a good philosophical theory about the nature of consciousness. It is hard to explain its evolutionary significance with such a gap in our understanding. That consciousness has some major value is the assumption of almost all evolutionists, but how exactly it occurs and functions is still shrouded in mystery. So even with the development of new 20th century sciences, Darwin's theory of evolution can still lead us to unanswered questions as much today as it did well over a hundred years ago.

The Role of the Scientific Community

The work of the following scientists is discussed in this chapter. Short, biographical essays of these individuals appear in **Biographies** on page 607.

Milford Wolpoff (1942–)

Chris Stringer (1947–)

Setting the Stage

In 2003, a group of Australian and Indonesian researchers—their field is known as "paleoanthropology" because they are students of human evolutionary history—were searching on the island of Flores, part of the Indonesia archipelago. They were looking for evidence of human (*Homo sapiens*) migration from Asia on to Australia. To their incredible surprise, they found specimens in a cave of little creatures (about half as tall as us humans) with small brains (about 400 cm², around ape size, compared to human brains of about 1,400 cm² for men and rather less for women), but with evidence of sophisticated tool use and hunting ability. (There were remains of dwarf elephants that had been killed and eaten.) And if this were not enough, radioactive dating put these creatures at about 18,000 years old. In other words, it seems that at one point, not that very long ago by evolutionary standards, we modern humans had little cousins who survived and flourished. It was natural, given that this was just the point when the movies based on the *Lord of the Rings* trilogy were appearing to great acclaim, that the little creatures, whose official name became *Homo floresiensis*, were quickly nicknamed the "hobbit" (Brown et al. 2004; Morwood et al 2004; Morwood et al 2005).

Is this the most exciting discovery in evolutionary history since the digging up of the huge dinosaurs in the American West during the second half of the nineteenth century, or is this the biggest mistake since Georges Cuvier declared that evolution is impossible? There has been no shortage of advocates for both positions. In favor of the hobbit's special status are the original discoverers, backed by a number of leading specialists, most notably Dean Falk (a woman) who is the world's foremost expert on fossil brains (she is known as a "paleoneurologist"). Of course, brains by and large do not stay around to get fossilized. It is the skulls that remain. But Falk is an expert at filling up the skulls with rubbery material, which when hardened can be extracted. Then you can start to look at the various

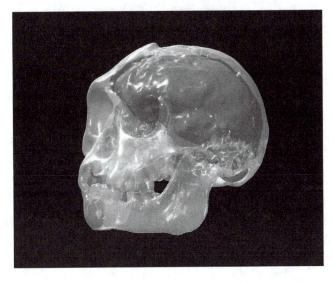

Hobbit

Map of Indonesia

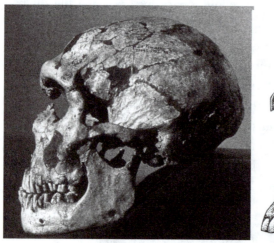
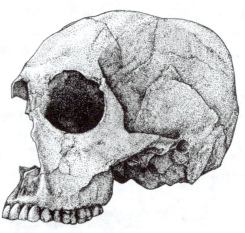

Modern Homo sapiens *(right) and Neanderthal man (left)*

contours and make inferences about the parts and their functions. More recently, the paleoneurologists have availed themselves of advances in medical science, most notably Magnetic Resonance Imaging (MRI) and computerized axial tomography (CAT) scanning, which enables them to photograph the insides of skulls and to make their inferences.

This was all just as well in the case of the hobbit. There are the remains of nine individuals thus far, but only one has a really good skull ready for study. Luckily Falk got her hands on a CAT scan of this skull and on the basis of this decided unambiguously that *Homo floresiensis* is a species different from *Homo sapiens* (Falk et al. 2005). Indeed, although she was very cautious on this, she suspected (as she still suspects) that it might be very different indeed and that our joint ancestors might range quite far back. I speak of "luck" because unfortunately extra-scientific factors started quickly to intrude into the hobbit story, perhaps no great surprise given its potential significance and the fact that it was discovered in a Third World country, naturally jealous of its status and possibly a tad resentful that First World outsiders (Australians and then Americans) were hogging the limelight. A leading Indonesian palaeontologist, Teuku Jacob, essentially kidnapped the specimens, appropriating them for his own use and study. They were returned, but in appalling condition, with cut marks (where rubber casts had been removed) and, even worse, key bones snapped and then glued together in altogether misleading fashion. Therefore, the anguished cry, quoted at the beginning of this chapter, by Michael Morwood, one of the key figures in the discovery of the hobbit. (This comment was reported in the *Sydney Morning Herald* on March 5, 2005.) Hence, Falk's luck at getting information on the specimen before it was damaged.

Where do things stand now, five or so years after the discovery? There is some very solid evidence that the hobbit is what its enthusiasts claim of it. Falk thinks that the brain is very distinctive. It is very small, admittedly, but it has features that identify it with significant cognitive skills. It may be a brain the size of a chimpanzee's, but Falk stresses that it is *not* the brain of a chimpanzee. Areas and parts are developed that one associates with powerful mental abilities. For instance, a part of the brain known as the dorsomedial prefrontal cortex, closely associated with self-awareness, is as developed in the hobbit (despite its generally much smaller brain) as in humans. In addition to the brain, other parts of the body also suggest that the creature was significantly different from humans. The arms and legs are very non-human. The wrist bones, for instance, are much more ape-like than human-like. In the words of one expert, Matthew W. Tocheri of the Smithsonian's National Museum of Natural History, the bones are "basically indistinguishable from an African ape or early hominin-like wrist." (A "hominin" is a member of the group Hominini, which contains only humans and chimpanzees and their ancestors back to the point at which they broke from other lines. There will be more on this, later in the chapter.)

However, there are many critics who argue that the hobbit is basically little more than a crippled human. It has been suggested, for instance, by people at the Field Museum in Chicago, that the hobbits were microcephalic, that is to say people with (generally genetically caused) very small brains, usually associated with various degrees of mental retardation. Falk has challenged this strongly, doing a comparative study of known microcephalic skulls, arguing that the hobbit brain is very different. This has not stopped the critics. More recently the claim has been made that the creature suffered from something known as Laron syndrome. This is a genetic disorder caused by insensitivity to growth hormone (GH). Needless to say, this suggestion has not gone down well with supporters of the special and significant nature of *Homo floresiensis*!

There is more to the story, including all sorts of tantalizing tales by the indigenous people of the islands about tiny people flourishing in the past, possibly surviving to the modern period, perhaps but a century or two ago. Could they even be hiding out in unknown caves today?! However, we must pull back here for the point that we need to make has now been made. *Homo sapiens* is not just another species. It is *our* species. Hence any discussion about us and our evolution is going to be exciting and interesting. It is also going to be fraught with tension. If the hobbit really is something new, think of the theological implications for a start! Who was really made in God's image: us or them? I am not going to answer this. It is left an exercise for the reader! I want to turn now to what evolutionists know about our prehistory. I also want to caution that this is not an area where it is easy for cool logic and hard evidence to prevail. Let us start our story by going back to the beginning of the nineteenth century.

Essay

The Antiquity of Man

Cuvier set the background position. Although in the eighteenth century there had been talk of "pongos" and "jockos" and other fabulous creatures, he found no fossil evidence of humans or humanlike creatures (Greene 1959). We are modern and appear in Europe after the last catastrophe, which you may remember that Cuvier identified with Noah's Flood. This was a conclusion that was welcomed by all, especially by those who had no intention of admitting anything to the vile evolutionary doctrines. People like Adam Sedgwick, Darwin's old teacher and friend, could allow that the earth is old with previous now-extinct inhabitants. Humankind came after all of this, and it is them that the Bible describes and discusses and explains. Christianity is a story about our relationship with God, and what happened before is irrelevant with respect to faith and those things that really matter. Pre-Adamite men are no more supported by science than they are welcomed by religion.

The first break in this picture came in 1847, when the French customs officer Jacques Boucher de Perthes described stone tools (axes) found in northern France in deposits also containing the remains of now-extinct animals (Oakley 1964). This all rather implied that humans go back some considerable time and that (as he saw it) we may not indeed be the first humanlike species. Boucher de Perthes's work, *Antiquites celtiques et antediluviennes*, attracted little attention for over a decade. Then, in 1858, the trained English geologists William Pengelly and Hugh Falconer explored Brixham Cave near Torquay in Devon, also finding tools and the bones of extinct animals, and rapidly opinion swung toward recognition of the Frenchman's achievements. Popular books, including Charles Lyell's *Antiquity of Man* (1863), together obviously with the acceptance of evolutionism that was just then occurring at a rapid pace, completed the demolition of Cuvier's conservative rejection. Mention has already been made of the fact that Thomas

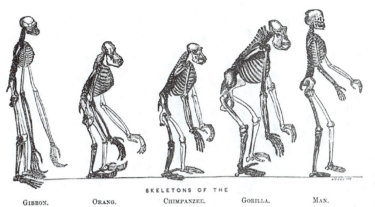

SKELETONS OF THE

GIBBON. ORANG. CHIMPANZEE. GORILLA. MAN.

Photographically reduced from Diagrams of the natural size (except that of the Gibbon, which was twice as large as nature), drawn by Mr. Waterhouse Hawkins from specimens in the Museum of the Royal College of Surgeons.

The frontispiece of Huxley's Evidence as to Man's Place in Nature

Henry Huxley at once went to the heart of the evolutionary issue, and, in his *Evidence as to Man's Place in Nature* (1863), on comparative grounds he argued strongly for our simian ancestry. It is true that there is more gap between humans and the nearest ape than between successive apes themselves, but there is more gap between the highest and lowest apes than between humans and apes. There is no question but that, as Lyell wrote worriedly in a private notebook, picking up on the penultimate organism in Lamarck's evolutionary scheme, we simply must "go the whole orang." We humans are part of the primate evolutionary picture.

But what about the "missing link"? Everybody knew what link this referred to, and everybody knew how important its discovery was going to be. Humans had evolved, there was no question about that: how and where and when were the key questions. Darwin and Huxley rather favored an African origin for humankind. The great apes live now in Africa, and the homologies between them and us were precisely what these two men were stressing in their efforts to convince people of the facts of human evolution. The more apelike we could be made or the more humanlike they could be made, the tighter the conceptual links and the greater the case for evolution as fact. But when it came to paths, most other people had different ideas. The racist progressionism of the late nineteenth century saw white European humans as clearly superior to other races, especially to blacks. It was argued—by Spencer and his followers particularly—that the colder climates required more effort to survive than did the warmer climates, and hence protohumans advanced more rapidly (through Lamarckian inheritance following effort) in the colder climates than in the warmer climates. So, it was thought by many that Africa could not have been the home of human evolution—certainly it could not have led the way.

Asia became the favored origin of humankind—all those grassy steppes seemed tailor-made for the evolution of humans out of the trees and up onto two legs. Most influential were Ernst Haeckel's writings and, inspired by them, toward the end of the century the Dutch doctor Eugene Dubois found "Java man," pieces of a skull with a smaller cranial capacity than today's humans and yet appar-

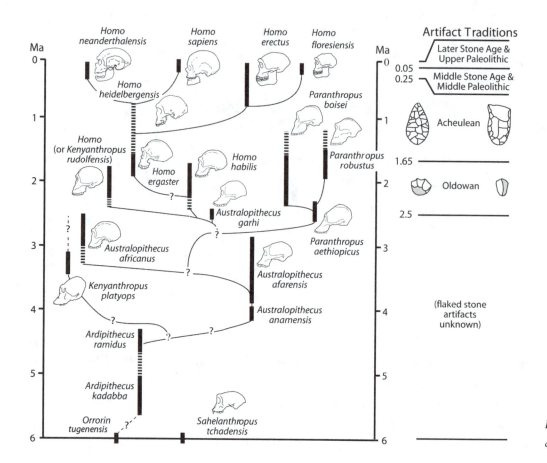

Human evolution
diagram

ently, on the basis of a thighbone also found in the deposit, an upright walker. He named this being *Pithecanthropus erectus*, although today we put it in the same genus as ourselves, *Homo erectus* as opposed to us, *Homo sapiens*. Haeckel seized at once on the significance of Dubois's discovery, and as can be seen from the diagram given in a little book he penned (revealingly entitled *The Last Link: Our Present Knowledge of the Descent of Man*), he had no doubt but that it represented the very piece of evidence long awaited (Haeckel 1898).

If one were just giving a rational reconstruction of the history of the discovery of human ancestral fossil remains—that is, if with hindsight one were just looking back at what happened or what one thinks ought to have happened if everyone were rational (that is, as rational as oneself!)—then one might expect that the next moves would have been directed to the finding of humanlike fossils even older than Java man. After all, the chimpanzee in Haeckel's picture is one of today's organisms, and no one claims that it was also our ancestor. Rather, the picture is intended to hint that our ancestor was chimpanzee-like in significant respects. And indeed, history seems to fit this reconstruction rather exactly, for the next major discovery in the 1920s in South Africa was of precisely an organism with upright stance but a far smaller brain. Raymond Dart, newly established professor of anatomy at Witwatersrand University (in Johannesburg), discovered and described this animal, informally known as Taung baby (because it was a juvenile) and officially classified in a different genus from humans, *Australopithecus africanus*.

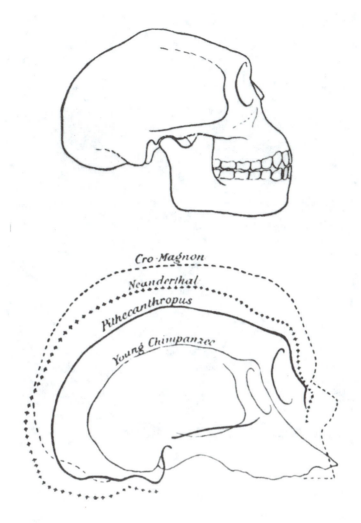

Haeckel's illustration comparing the brain of Cro-Magnon man (an early specimen of modern humans) with other brains

But, primed by the story of the hobbit, we are now ready to realize that real history has a nasty way of taking on a life of its own and of not following the path that rational people think that it should. Dart's discovery was opposed right from the beginning, as not at all significant for the story of human evolution. It was not until the 1940s, when the climate started to change (and there was the discovery of more fossils, including parts pointing unambiguously to upright walking), that Dart's *Australopithecus* was recognized for the significant finding that it really was. Why the delay? One major factor clearly was that Dart's animal came from Africa, and most informed people were looking to Asia. The thought that we humans might have evolved in Africa was altogether too horrendous to contemplate. Taung baby just did not fit in, and anyone who knows anything at all about science will realize that prior convictions and expectations are a far more significant factor in observation than any thing out there in the real world. (That is an exaggeration, but not too much of one. No one denied that Taung baby existed or that it had the features that it had. The question was rather about what these features represented.)

Also there is no doubt but that a lot of people simply did not like the smooth upward rise that Haeckel showed and that Taung baby would seem to confirm. There was agreement, of course, that we had evolved and that ultimately

Raymond Dart with Taung Baby

we had evolved from beings with small brains. But there was a desire to push this back as far as possible. People did not want to be too closely associated with the apes—even the Neanderthals were now out of favor and portrayed as highly brutelike and not all respectably human. Indeed, white people (who were after all making the running in paleoanthropology) did not want to be too closely associated with their fellow darker humans and wanted many years of evolution independent from other groups. Fortunately, there was what was thought to be good evidence for this position of long-time separation—evidence that we know now to be one of the most notorious scientific frauds of all time.

I refer of course to the Piltdown man, or Piltdown Hoax, as it has been known since it was uncovered in the early 1950s. In southern England, around 1912 (the exact date of first discovery is clouded in mist), an amateur archaeologist, Charles Dawson, unearthed pieces of skull and jaw that seemed to confirm that precisely the required sorts of humans had lived and thrived, long before the present. These were humans with massive brains—virtually as big as ours in fact—and yet clearly primitive in other respects, particularly in the lower face and jaw. Conferring authenticity, Arthur Smith Woodward, a curator at the British Museum (Natural History), became involved in the discoveries, as well as the then-young French priest/paleontologist Pierre Teilhard de Chardin. Quelling doubters, a year or two later some really major pieces of evidence came to light.

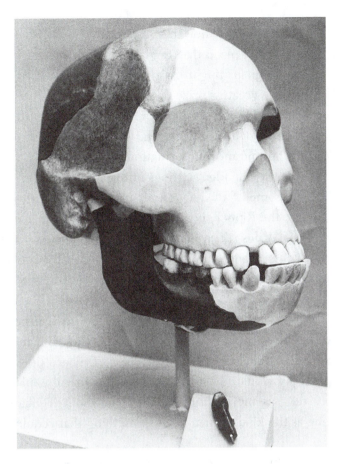

Piltdown Man

(Supposedly Dawson found these new fossils in 1915, although they were not announced by Woodward until after Dawson's death in 1916.)

We now know that it is hardly surprising that Piltdown man had a brain as big as ours, since the key skull was in fact a human skull! Nor was it surprising that the lower face was primitive and apelike—the jaw and teeth that were recovered came from an orangutan. The pieces were suitably shaped and stained, and then the awkward bits (precisely those bits that would cast doubt on the brain and jaw being from the same animal) were broken off and thrown away. As I said earlier, "anyone who knows anything at all about science will realize that prior convictions and expectations are a far more significant factor in observation than anything out there in the real world. (That is an exaggeration, but not too much of one.)" Exactly.

The remarkable thing about Piltdown man was not that the fraud was eventually uncovered but rather that it lasted as long as it did. It really was quite a crude job. As soon as anyone looked, you could see all sorts of file marks and such things, including evidence of staining rather than weathering through time. And this was apart from physicochemical methods of dating materials. It ought to have been spotted early on, and indeed to their credit some people did always feel that it was highly and uncomfortably anomalous. But it fit precisely what most people were after—almost too patly one might say (especially when more relevant bits appeared almost to order)—and there are none so blind as those determined to see. And people were nothing if not this, especially English people, who were highly sensitive to the proud place that England now possessed in the search for human ancestors. The Germans might have those nasty Neanderthals, but fair Albion has been home to the greatest prize of all.

The possible identity of the perpetrator of Piltdown has filled more books than has the quest for the identity of Jack the Ripper—with about as much success. On the Internet, I found more information on the topic than I truly need for one lifetime. To be honest, the identity does not really matter, which I suppose is part of the attraction. Some of the suggested suspects rather boggle the imagination—although, unlike the Ripper, no one yet has suggested that the Piltdown hoaxer was the Prince of Wales. (The hoax may not have been a great work of art, but it required more energy and gumption than one generally associates with British royalty in the twentieth century.) One far-out suggestion is Sir Arthur Conan Doyle, the author of the Sherlock Holmes stories. He was a keen spiritualist and had a keen dislike of scientists who regarded his enthusiasms with contempt. Hoaxing them all like this would have been very satisfying. But motive alone does not make for criminal action, nor does opportunity. Teilhard de Chardin has been fingered by Stephen Jay Gould (1980b). However, as I shall explain later, the accusation probably tells us more about Gould than about Teilhard, who simply does not strike one as the kind of man to do something that required such systematic deception.

The most recent purported culprit is one Martin Hinton, a curator at the British Museum (Gee 1996). He has been indicted on grounds of bits and pieces of supposedly incriminating evidence discovered in his effects after his death. But, it appears that he cannot have been the sole perpetrator—he was simply not around at some of the required times—and the evidence may not be what it seemed. (Particularly suggestive was a discovery in Hinton's effects of various chemicals that were needed for "aging" the orangutan jaw, but Hinton's chemicals do not match exactly the chemicals used on Piltdown.) General suspicion has always centred on Dawson, who had a bit of a reputation for being shifty, and probably this is not far off the mark. Woodward may well have been a dupe—it is interesting to note that his speciality was fish rather than humans.

The story continues and no doubt will continue to continue, so let us return to the main thread of our tale. Since the acceptance of *Australopithecus*, the last half century has seen massive efforts, richly rewarded, in tracing human origins—centered now almost exclusively in Africa. Thanks to the labors of fossil hunters at least the equal of the dinosaur hunters of the last century, we now have a reasonably good pattern of human evolution back for the last five million years. Our earliest-known, direct ancestor seems to have been *Australopithecus afarensis*, represented dramatically both by more complete skeletons than we normally expect—notably "Lucy," the woman from Ethiopia—and by footsteps in drying volcanic ash in Tanzania (Johanson and Edey 1981). The animal was about half our height or a bit more, with a small ape-size brain of less than 500 cubic centimeters as compared to a human brain of around 1,400 cubic centimeters for a male and a

Donald Johanson and friend

bit less for a female. (The brain size of *Australopithecus afarensis* was ape size, but internal casts suggest that it was already not an ape brain. Nor was the hobbit brain the same as Lucy's. The hobbit in many respects was far more advanced.)

Most exciting of all, Lucy was undoubtedly and unambiguously bipedal. She walked up on her own two feet—she did not run around on all fours nor was she a knuckle walker like the great apes (who can run around very quickly, using their knuckles for support). Yet at the same time—terrific music in the ears of the evolutionist—it seems clear that *Australopithecus afarensis* was not as efficient a walker as are we humans. This does not mean that Lucy was an unstable hybrid, neither fish nor fowl. To assume so is to fall into the same kind of progressionist thinking as held sway at the beginning of the century. She was not an item on a directed line to humans. Had another meteor wiped out mammals two million years ago, she would still have been just fine. It was just that she was not fully human. And in fact, slight curvature of the bones of hands and feet suggest that she would have been much better at tree climbing than we tend to be.

After *Australopithecus afarensis*, the line split—some australopithecines went one way, evolving into more robust forms and eventually to extinction. Others, more graceful or delicate, went on to turn eventually into the human line, and down through several species of *Homo* to our own *Homo sapiens*. More on us in a moment, but first what should we believe about life before the Australopithecines? Here, as is well known, have come some of the most dramatic discoveries and changes of perspective. Until about 20 years ago, the firm conviction of paleoanthropologists was that we humans are a long way from the apes, comparatively speaking. It was thought that, probably, one needed to go back about 15 million years or so before one would find a common ancestor with the chimpanzees and gorillas and orangutans. Humans may or may not have evolved together with other groups, but we surely have evolved apart from the rest of creation.

The molecular biologists would have none of this (Pilbeam 1984). They had developed new techniques for assessing absolute dates, and by comparing the macromolecules of apes and men, they came to the conclusion that the ape-human break had to be much more recent—as recent, indeed, as five million years ago, which is really quite astounding when you think that Lucy is nearly four million years old. Expectedly these results—offered less as tentative suggestions and more as firm corrections—did not sit well with people who had spent their lives finding and interpreting fossils. How dare rank outsiders presume to tell them their business?! Listen to an eminent physical anthropologist, writing just a decade ago—nearly fifty years after Watson and Crick discovered the double helix. "Unfortunately there is a growing tendency, which I would like to suppress if possible, to view the molecular approach to primate evolutionary studies as a kind of instant phylogeny. No hard work, no tough intellectual arguments. No fuss, no muss, no dishpan hands. Just throw some proteins into the laboratory apparatus, shake them up, and bingo!—we have the answers to questions that have puzzled us for at least three generation" (Wolpoff and Caspari 1997, 112). It just isn't fair! One can hear the plaintive cries of rejection and dismissal.

Fair or not, the molecular biologists won. Now it is accepted that although the break may be a little older than five million years, it is that order of magnitude. Moreover, although the evidence is still ambiguous, it could easily be that we humans are more closely related to the chimpanzees than we are to the other apes, the gorillas in particular. (Hence the already-mentioned grouping of Hominini.) Although to our eyes chimps and gorillas may look more alike than chimps and humans, it could be that we have gone off on our own and the apes (in those similar-looking respects) have stood comparatively still. Recent fossil findings certainly suggest that around five or six million years ago, the human-ape line was probably one. *Sahelanthropus tchadensis*, which gets its name because it was found in Chad (Central Africa), is almost seven million years old, and (thus far, we have only a skull) combines ape-like features (brain case) with some more human-like features (specifically, teeth and the shape of the lower face) (Brunet et al. 2002). From Kenya, at about six million years old, we have *Orrorin tugenensis*. Its bones suggest strongly that it was bipedal and walked, yet it has ape-like upper features

Milford Wolpoff

(the slant of the neck, particularly) (Pickford et al. 2002). *Ardipithecus ramidus kadabba* (the third term of this trinomial designates the subspecific classification) was found in Ethiopia (close by the home of Lucy). It has toe bones that are intermediate between those of humans (upward tilt to joint surface) and apes (long and downward curving). It flourished over five million years, although other specimens have been found (and put in a different subspecies, *A. r. ramidus*) and they are nearly a million years younger.

Not everything has been put together in a way that satisfies everyone. Nevertheless, no matter what the details, we are a lot closer to the rest of the animal world than anyone dared think just a few years ago.

Causes

So far I have been talking more about the path of evolution, about phylogenies, than about causes—something one does rather dread broaching, for the discussion goes right off the subjectivity-emotion index, time and again. Indeed, one enterprising scholar has likened the causal tales told by students of human evolution to fairy stories, in a rather literal sense. Misia Landau (1991) draws on analyses of folk tales to show that common patterns keep reappearing. The hero starts in a happy initial situation that is disrupted by external forces—death or famine or the like. The hero then sets out on a journey to find salvation or the golden fleece or something similar and along the way has to struggle with forces and the elements, sometimes falling but eventually triumphing. So with the story of human evolution. We were happy apes up the trees in darkest Africa, minding our own business and happily surviving and reproducing. Then something happened. A drought is a favourite causal factor, and the home we loved was no more. We had to leave the trees and come down on the plains or savannah. But we were hardly suited to this, so we had to start evolving in a big way. We needed to be able to run around on the plain, so we developed bipedalism, jettisoning the now no-longer-needed adaptations for tree life. At the same time, things were tough out there on the plain—far more dangerous than up trees. We had to learn to cooperate, to get along with each other. What better way than through intelligence? So we humans (or protohumans) started the path up to full-time thinking ability. And now, finally, we have won. We have conquered the tasks set before us and achieved the goal, full humanhood. Our journey is ended.

Of course, you can run variations on all of this, depending on various factors. If, for instance, you incline to the view that encephalization (large brainedness) preceded bipedalism (two leggedness), then you might well look for external factors other than drought as the stimulae for the initial evolution. Perhaps, for instance, it was all a question of new or different predators. And some writers are going to be more daring in their hypotheses than are others. They are going to be more inventive about the challenges supposedly faced and the solutions supposedly found. But the fairy tale—hero makes epic journey, conquering through trial, and arriving eventually at the promised land—persists. And a moral tale, too, especially as one can tie in some strands about the white race having had to travel farther and struggle more decisively than the members of other races—with the expected results.

> But if we know nothing of the wonderful story of Man's journeying toward his ultimate goal, beyond what we can infer from the flotsam and jetsam thrown upon the periphery of his ancient domain, it is essential, in attempting to interpret the meaning of these fragments, not to forget the great events that were happening in the more vitally important central area—say from India to Africa—and whenever a new specimen is thrown up, to appraise its significance from what we imagine to have been happening elsewhere, and from the evidence it affords of the wider history of Man's ceaseless struggle to achieve his destiny. (Elliot Smith 1924, 79)

Of course, today's paleoanthropologists deny vigorously that such approaches to causal factors are faults of which they are guilty. Although all of this may have happened in the past—undoubtedly did happen in the past—it is no longer true of today's work. It is far more objective and value free and so forth. After all, we are all Darwinian evolutionists now, so talk about "achieving destinies" is simply ruled right out of court. Darwinian organisms do not achieve destinies. If they are lucky, they survive and reproduce—for a time.

To which response—that paleoanthropology has changed and that with the coming of the synthetic theory it has become more scientific, and less simply a vehicle for telling one's favorite story—one can say that there have certainly been changes but that whether they are as absolute as some seem to think might be doubted. There is no question but that more attention is paid to fundamental biological principles and that new techniques have thrown up all sorts of new ways of finding pertinent information. But at the same time, values and culture still play a major role in the pictures painted and stories told by students of the human fossil past. Let me not exaggerate. We do know some things now that were not known before. Thanks to the fossil evidence, we know now that humans came down out of the trees and that only then did the brain start to explode up to three times its original size. And we know that there had to be some large selective pressures at work here, if only because brains take a huge amount of energy to run. Selective advantages that cows and horses, or chimpanzees and gorillas for that matter, have not found in their interests (or within their abilities) to follow or satisfy. So any pictures of human evolution that do not fit in with these constraints have to be false. Of this we can be certain.

But after this, there is huge scope for variation and inventiveness. Probably climate did have something to do with our leaving the trees and becoming denizens of the plains, but there are major questions as to why it all happened. Why, for instance, did the other apes not come down to the ground like us? And what was it on the plains that made it so attractive to be bipedal? Was it foodstuffs, and if so what kind? Was it seeds, as has been suggested, and did our hands evolve to pick and eat these seeds? Or was it the need to move around the plains to find food that was less evenly distributed than it was in forests? Walking is an efficient way of traveling—certainly, if the option is going on your knuckles all day, walking has its virtues (Lewin 1989, 68). One attractive hypothesis (due to Dean Falk 2004) is that being upright protects us from the sun. There are fewer rays that hit the body if we are vertical than if we crouch or otherwise stand with major parts of the body exposed to the rays from above.

Move on to about two and a half million years ago. *Homo* was making its appearance now, and here we get the first human-made tools as well as the beginning of the really massive expansion of the brain. What is the cause? In the 1960s, the popular hypothesis was that of "man the hunter." Little groups of early members of our genus would set out hunting with their tools; catch, kill, and cut up their prey; and then eat it. Brains were needed for this exercise, for obviously we had to depend on skill for the hunt, not being fast and furious like other mammals, and with the coming of a meat or partially meat diet, brains could grow that much bigger because meat is a very rich food and can support organs that are high energy cost. By the end of the 1970s, this hypothesis, at least in its crude form, was starting to fall right out of favor. In its place was coming the hypothesis of man as scavenger—early humans just followed around behind big animals, and when they got into trouble, or when something else killed them, we would move in to grab our share and more. We were a kind of primate jackal.

In addition, there was now a lot more emphasis on food sharing, and females started to take a more prominent active role. The hunters (as in modern societies) were taken almost universally to be male—females therefore had a passive or noneffective role in early human life. Shades of Charles Darwin! With scavenging, and with associated food gathering, it was a whole family activity. Why the new perspective? At least some of this change of viewpoint was fact driven. Increasingly sophisticated studies of teeth and of bones, for instance, could tell that there had to be much more to diet than meat—vegetable matter was very significant. Then there was work being done on the possible modes of travel and life of the early hominids—archaeological studies of where they cut up their meat, for instance, trying to work out lifestyles. Did one kill and eat? Or did one kill and transport and eat? And if the latter, did one kill and cut and transport and eat, or did one kill and transport and cut and eat? Lots of questions like these, which were being tackled using molecular and microscopical and comparative and other studies.

At least some of the change was derived by changes elsewhere in evolutionary biology. The whole question of cooperation was becoming a big thing in Darwinian studies, and these undoubtedly slopped over into paleoanthropology. I shall be looking at cooperation in the next chapter so need say nothing here, except to remark that (as one might expect) if things get hot in one area of evolutionary biology one expects fully that workers in other areas will take note and see if there is anything in it for them. And some of the change was simply driven by ideology. The 1970s was the time when the feminist movement really got up a head of steam, and in an area like paleoanthropology—which has its full share of women workers—one could have predicted that "man the hunter" would get little sympathy. Which it did not! "Woman the gatherer" was an almost perfect counter—here, if anything, females were doing all the real work of collecting seeds and other small food stuffs, and men basically parasites, as always (Zihlman 1981).

Scavenging and gathering, a gender-reciprocal, food-sharing hypothesis, was a natural outcome from this polarization, appealing to those who wanted to acknowledge the significance of the female role in human life and evolution but yet did not want to relinquish entirely the important role of males in this picture. Here now we had a happy balance, with both men and women providing foodstuffs and sharing. What could be nicer? On the one hand, all of that aggressive stuff about hunting now takes a back seat—at best, we men have a rather low role in the meat-gathering business. Although one that requires intelligence. A perfect job for professors, as one might say, rather than for he-men in plaid shirts. On the other hand, the new male now takes his place along with his mate (there was also some stuff brought in about the virtues of sexual fidelity), sensitively sharing his bounty with hers. Those who do not think that such an approach is drenched with social values are as naive as the people writing it. I am not saying that it is bad. I am not saying that I could or would want to do better. I am saying that this approach was the way that it was and looks fair to being for the future.

The Neanderthal Controversy

Let us pick up now on recent human history and on the Neanderthal question. *Homo habilis* goes back about two and a half million years; *Homo erectus* appears about one and a half million years ago and lasts until about 500,000 years ago, at which time we start to get the appearance of *Homo sapiens*, or rather a group of *H. sapiens*–like organisms often known informally as "archaic sapiens." So far, so good. Now we put the Neanderthals into the mix. In August 1856, in the once-peaceful Neander valley in Germany, people unearthed the first identified specimen of what came to be known as Neanderthal man. (Thal is the old German term for valley.) At once there was controversy about the meaning of the find. These beings slept for a long time but they have not slept since. Are they human? Some have portrayed them as respectable citizens, hardly distinguishable from the chap next to you on the bus or subway. Others—including most gloriously the cartoonist Gary Larson—paint them as hairy hunched monsters, stupid and criminal, like something from a Boris Karloff movie in the 1930s. Some have seen them as obvious ancestors, because they are so similar. Others have seen them as too different and stupid to be other than extinct. And yet others have seen them as ancestors (of others!) precisely because they are degenerate: "Ferocious gorilla like living specimens of Neanderthal man are found not infrequently on the west coast of Ireland, and are easily recognized by the great upper lip, bridgeless nose, beetling brow with low growing hair, and wild and savage aspect. The proportions of the skull which give rise to this large upper lip, the low forehead, and the superorbital ridges are certainly Neanderthal characters" (Grant 1916, 95–96).

What we do know is that Neanderthal man appeared about 150,000 years ago and that he lasted until around 35,000 years ago—found mainly in Europe but with some in the Middle East. All told we have about 200 specimens, beginning with that first identified discovery a year or two before the *Origin*. (I say "identified" because we now know that there were unappreciated specimens found in Belgium in 1829 and on Gibraltar in 1848.) Modern humans, that is, *Homo sapiens* like us, were at one point thought all to come after Neanderthals, but now the thinking is that our remains date back almost as far, and there is evidence in some places that modern humans lived together with Neanderthals without interbreeding—or at least without interbreeding enough to wipe out differences. (About a decade ago there was the discovery of a new skeleton, apparently a modern human/Neanderthal hybrid [Duarte et al. 1999]. This does not prove that hybridization was common or that the offspring were fertile. And indeed the meaning of the discovery itself has been hotly disputed. Other factors pertinent to interbreeding will be discussed in a moment.)

How and in what respects were Neanderthals different from us? This question reveals much of the difficulty of the whole Neanderthal problem: those who want to argue that we are descended from Neanderthals tend to minimize differences, whereas those who argue that we are not descended from Neanderthals tend to emphasize differences.

His thick neck sloped forward from the broad shoulders to support the massive flattened head, which protruded forward, so as to form an unbroken curve of neck and back, in place of the alternation of curves which is one of the graces of the truly erect Homo sapiens. The heavy overhanging eyebrow-ridges and retreating forehead, the great coarse face with its large eye-sockets, broad nose and retreating chin, combined to complete the picture of unattractiveness, which it is more probable than not was still further emphasized by a shaggy covering of hair over most of the body. (Elliot Smith 1924)

At least some of this is pure fancy. Why on earth should Neanderthal man be covered by shaggy hair like a gorilla? Only in the author's imagination does this occur, but once done the Neanderthal comes out that much more apelike and different from us.

There are differences, and to be candid if anything these differences are such as to give rise to the ape-connection perspective. The Neanderthal is more robust and stronger than we and more significantly does have a face—the lower face particularly—which sticks out more. However, before you pack up and go home, thinking that the Neanderthals are definitely more apelike and could not possibly be our ancestors, I should also mention that if anything their brains tended to be larger than ours. Hence if brain size is a mark of progress, if anything we represent a step backward—although, as you can imagine, a good number of people have jumped in to warn against easy and facial identifications of brain size with intelligence. Often these have been precisely the same people who have been happy to accept and stress the significance of the difference in size between the human male and human female!

With all of these various issues and prejudices floating around, it is no great surprise that students of the subject have divided into two major camps. On the one side, championed particularly by the University of Michigan's Milford Wolpoff, we have the "multiregional evolution" model or hypothesis. This sees *Homo erectus* as having evolved in Africa—the fossil findings on this seem to be definite—and then it was this species that traveled far from home, spreading at least through the old world, into Europe the one way and then toward Asia and up into China the other way. Once *Homo erectus* was in place, *Homo sapiens* emerged about 500,000 years ago or a little earlier—a significant point is that there is going to be no sharp dividing line and *Homo erectus* blends gradually into *Homo sapiens*. Then *Homo sapiens* kept on evolving, up through time to the present. By and large the separate populations kept separate, but there was a certain amount of gene flow—interbreeding between populations—thus ensuring that the populations did not go off and evolve into separate species and that there would be a substantial degree of continuity and uniformity in the form that this evolution took. You fit into this picture all of the fossil discoveries that have been made, and of course part of the picture is the Neanderthals being shown as the immediate ancestors of Europeans. Most Neanderthals are found in Europe, so most Neanderthals are now represented by modern-day Europeans. There may well be—there surely will be—some Neanderthal genes in today's Australian aborigines, but most Neanderthal contributions end up right in the places where we find their remains.

On the other hand, championed particularly by the British Museum's Chris Stringer (2002, 2003), we have the "out of Africa" hypothesis. The beginning part is the same as the multiregional hypothesis. We start with the origins of humans in Africa—this was the home of *Homo erectus*. Moreover, it is agreed that *Homo erectus* went traveling around the Old World—Java man shows that that was the case.

*Christopher
Stringer*

And these populations did go on surviving and evolving, but gene flow was insignificant or nonexistent and so there were different populations, perhaps well on the way to speciation. Meanwhile, back in Africa about the 500,000 or a bit more years ago mark, *Homo erectus* was evolving into *Homo sapiens*. Then at some point, around the 100,000 year mark, this population (or perhaps species now) starting moving out—at least some did, although others stayed at home. This group, *Homo sapiens*, spread around the world, and as it did it wiped out the populations of hominids already living there. How it did so is not in itself a matter of great moment—it was not necessarily through violence but could have been through disease or some such thing. Superior technology may have been involved. The point is that *Homo sapiens* did take over, and specifically in Europe this meant the end of the Neanderthals. They did not evolve into us, they are at closest related to us through *Homo erectus*, and they are now extinct.

These are very different hypotheses, starkly so, and would seem to lend themselves readily to test and comparison. One might think one is going to have a textbook case of science in action here; but, although one does in fact have a textbook case, it is rather one that shows just how difficult it can be to test and compare rival models, even when they seem unambiguously clear and different. Most obviously, one has the physical facts, that is, the remains of Neanderthals and the remains of modern *Homo sapiens*, and their relationships or nonrelationships. But as I have pointed out already several times in this chapter, people tend to interpret things in the ways that accord with their own hypotheses. Stringer has started with a number of modern techniques for classification—initially a statistical process known as "multivariate analysis" and more recently a newly refined form of systematics known as "cladistics"—and he finds clear differences between us and Neanderthals. He argues that we are not the same, that transitions are rare or nonexistent (although note the recent hybrid discovery mentioned above), and moreover (and expectedly) the real differences come between European Neanderthals and those Neanderthals found in the near-East and (very rarely) farther afield. Moreover, using increasingly sophisticated methods of dating, he argues that we do not find the Neanderthals giving way gracefully as it were to us, but rather that there is overlap and if anything the two groups evolve in different ways. Instead of converging as one might expect, the two groups stay apart or even move farther away from each other.

Wolpoff will have nothing of this. His philosophical remarks quoted earlier are a warm-up to a knife through the heart of multivariate analysis, something that we learn he himself had used and discarded (or learnt to regard with suspicion) long ago. We are told: "Multivariate techniques are attractive because they seem to give the data an opportunity to speak for themselves. However, there are many problems with the incautious use of these techniques that stem from a variety of sources" (Wolpoff and Caspari 1997, 353). And then: "The danger of using multivariate analyses to address the human origins issue is that the analysis presupposes the solution. When you plug your data into a statistical program, you will get an answer, whether you are using the appropriate statistics or not. It's like adding up the diameters of apples and oranges and taking the average. There *is* an average, but what is it an average of?" (p. 354). So much for that!

In this molecular age, can one use something from that kind of biology to throw light on the two hypotheses? Stringer thinks one can and in fact turns to one of the flashier (that does not seem an inappropriate term) scientific hypotheses of recent years. I refer to the so-called mitochondrial Eve hypothesis formulated by Allan Wilson and others at Berkeley (Cann, Stoneking, and Wilson 1987). Mitochondria are parts of the cell, outside the nucleus. They contain genes (DNA) and are passed on in reproduction. However, the peculiarity is that one gets all of one's mitochondria from one's mother and none from father. By comparing mitochondria in different people and by working out the rate of mutation (mitochondrial DNA mutates up to 10 times faster than nuclear DNA), one can work back to how long it has been since people shared the same great- great- and so on grandmother (the source of the original mitochondria). The amazing finding was that this female—immediately christened "Eve"—the uniting link for all humans on earth, seems to have lived less than 200,000 years ago. Now note what this hypothesis does not say and what it does say. It does not say that at one point there was just one human or hominid female on earth. It does not even say that the human species went through a major bottleneck with just a few members. It does say that, although we are all no doubt descended from many people, we are all descended from this female. (A good analogy is to think of surnames, in a case in which women took their husbands' names and so did their children. Think of four people: Jim White, Mary Brown, Fred Green, and Ann Black. Jim marries Mary and they have two sons. Fred marries Ann and they have two daughters. Sons marry daughters and there are four grandchildren. No bottleneck, all four original people equally related to the grandchildren, but all of the grandchildren with the name White. Hence, Jim is the Eve equivalent.)

Stringer seizes on this hypothesis and argues that it proves his point. Around 200,000 years ago we all had a shared ancestor, which means that we all come from one shared population—just what his hypothesis demands. It was after this that the migrations around the world began. Wolpoff is not convinced, contending that the quality of the work is a bit like the engineer's classical way of finding a solution: "Think of a number and double it. The answer you want is half the total." In any case, in his opinion, the Eve hypothesis is irrelevant to the debate. The multiregional hypothesis admits—insists on—gene exchange between populations. Eve could come at any time. "*Only if human groups were isolated after Eve's time would her age be of importance.* The finding that human populations were connected by low levels of genic exchanges means any age for Eve could be compatible with Multi regional evolution because her DNA type could potentially spread throughout the world at any time" (Wolpoff and Caspari 1997, 309).

What about the archaeological evidence? This is the really dramatic stuff, although by its very nature it is the most tantalizing—how much is lost, how do we interpret it, and so forth. The more you get away from human beings themselves, the more subjective things all become. But, the fact is that the evidence from archaeology—artifacts and so forth—really is very striking and does prima facie tell strongly for the out of Africa hypothesis. Tools start coming in with the arrival of *Homo.* For a long time, these are all pretty crude stone hand tools. What is remarkable is how little change there is for so long. Then with "archaic sapiens," we start to get a significant move in the direction of sophistication. But this is nothing to what we get 100 thousand years or so ago, and increasing as time goes on, intensifying 50 or so thousand years ago. Tools, materials, decorations, and so forth are just levels of magnitude above what they were before. It is very tempting to link this to the arrival of modern humans and to argue that even if there is a little bit of this among the Neanderthals it is because they copied us. It is even more tempting, if we can locate the earliest modern complex tools in Africa, because then there would seem to be some sort of casual connection between tool use and the subsequent migrations and successes of *Homo sapiens* (us).

Not that Wolpoff will accept any of this: "Africa may differ from other areas, but if it does so it is in the extent of its marked regionalization" (Wolpoff and Caspari 1997, 327).This means that you cannot expect to find, and do not in fact find, one culture swamping the human population and taking off from there. The most sophisticated "technologies are local" and moreover "on the whole they do not seem to reflect particularly more progressive behaviours. These and other similarities to much later industries and technologies are short-lived and disappear, hardly the pattern we would expect if they were heralding a new superior, pattern of behavior" (p. 327).

It is starting to be clear that nothing, simply nothing, is going to shift the protagonists at this point. One move has been to try to extract the DNA of Neanderthals and, after sequencing it, to compare it to our DNA (Krings et al. 1997). This would seem surely to give definitive answers. Unfortunately, not quite so. The Neanderthal Genome Project has been wonderfully successful—to a quite ambiguous end! We humans are at least 99.5% genetically similar to the Neanderthals. Edward Rubin, director of both the Joint Genome Institute and the U.S. Department of Energy's Lawrence Berkeley National Laboratory's Genomics Division nevertheless concluded: "While unable to definitively conclude that interbreeding between the two species of humans did not occur, analysis of the nuclear DNA from the Neanderthal suggests the low likelihood of it having occurred at any appreciable level." Erik Trinkaus of Washington University to the contrary argued that Neanderthals are extinct because they have bred themselves out of existence with modern humans: "Extinction through absorption is a common phenomenon." In other words, the debate about whether we humans drove the Neanderthals to non-being by killing them off or by loving them out of existence is otiose. "From my perspective, the replacement vs. continuity debate that raged through the 1990s is now dead." (For details see Green et al. 2006; Noonan et al. 2006; Trinkaus 2006.)

In short, the impasse over the Neanderthals continues. But probably this is nothing very exceptional in science. The number of times that one side simply collapses and admits that it is wrong is rare indeed in science, as it is rare in real life. Perhaps the revolution in geology in the early 1960s, when people swung from thinking that the earth is stable and the continents unmoving over to thinking that the continents slide around the globe on big plates ("continental drift") is one such case—although there was really no question of two sides persisting. Rather, almost everyone switched over. In conflicts with two sides such as we have over the Neanderthals, we get more the persistence of the debate until people get tired or one side drops out (through retirement and death) or points on both sides are brought into an amalgamation in the middle.

Language and Consciousness

We have gone this far in the chapter without yet mentioning what many people— every philosopher!—would think are the most distinctive and important aspects of our species: the facts that we can talk and that we are conscious. To a certain extent this is cowardice, or perhaps prudence, on my part. Language and thought tend not to get caught in the fossil record, so one had best be silent. But we cannot be completely silent, nor need we be. As you can imagine, language particularly has got caught up in the Neanderthal debate, with the out of Africa proponents arguing that the key difference between us and the Neanderthals is language—we have it and they did not, or at least not to the same extent—and the

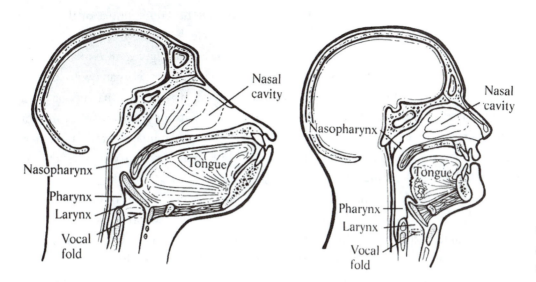

Comparison of ape and human vocal tracts

multiregional proponents arguing that this is unproven, untrue, and not needed anyway!

At least, let me modify things somewhat. No one today who takes evolution seriously wants to deny that human language is a deeply biological phenomenon, and no one who is not in some sense a Darwinian wants to deny that language has adaptive value in communication and so forth and that is why it evolved. Since the work of Noam Chomsky in the 1950s, it has been realized that languages are related with a shared "deep structure" and that they are not rational phenomena, but rather jerry-built, reflecting the constraints of biology and the vagaries of history. It is true that Chomsky himself opposes Darwinism for language, but his students and followers have shown precisely how language is the sort of thing put together by selection (Pinker 1994). But from here on we have difference and debate.

There are at least two ways in which you can approach the question of language. First there is the brain itself. This seems to imply that the growth of language has been a fairly gradual process, at least it does if you equate brain size with language ability. However, if you take organization into account, that is, the parts of the brain actually used in language, the traces left on the insides of skulls suggest that language may have come in a bound or leap, early on—certainly with the arrival of *Homo erectus*, and perhaps even with *Homo habilis* over two million years ago. Whatever else seems clear, by the time you get to *Homo sapiens*, and this includes Neanderthals, language was in play. It had evolved.

If, second, you go with archaeology, then the implications seem to be that language came in leaps and bounds. As we have seen, you get the development of some tools with the first hominids, *Homo habilis*, a jump with *Homo erectus*, another bigger jump with *Homo sapiens*, and then things go wild with the arrival of modern humans. The implication that has been drawn, especially since this does not reflect brute brain size, is that it reflects developments in language ability. What really marks us off from others, including the Neanderthals, is the fact that we have full and complex language abilities, which we use. Of course, you cannot use this claim as a piece of evidence independent from others in support of the out of Africa hypothesis. Already above we have seen appeal to the archaeological evidence in support of the hypothesis. But one can say that it gives an explanation of what was happening—why it was, in particular, that modern humans were able to succeed so well culturally and the Neanderthals were not. It was that we had language, or rather sophisticated language in a way that they did not. (Note this point. The claim is not that Neanderthals did not have language—although there have been those who have claimed just this—but that they did not have the sophisticated language ability that we have.)

However you interpret the role of language, every evolutionist agrees that the explosion of the brain in size had to be essentially adaptive. Whatever the cost, hominids with bigger brains are better adapted than hominids with smaller brains. But why? Exactly how the brain works and functions has always been a matter of significant debate and dispute. Many people today think that computers are a good analog for brains, and without necessarily making a simple identification—the brain is a computer made of meat, as one joker has said—these people feel that functioning of the brain is much like the functioning of a computer, as the brain operates somewhat akin to a calculator in processing and using information. Extremely popular is the hypothesis that, as with computers, the brain is built on a somewhat modular pattern. This means that there is no one central mechanism doing everything all at once, in a generic sort of way, but rather there are different parts or units that are put together to perform different tasks. Rather like the components of a Swiss Army knife, they are connected together to make the whole.

English archeologist Steven Mithen (1996) in a clever synthesizing hypothesis, ties in the modular theory and the growth of the brain with tool use. He builds on the fact that the growth of brain size was not smooth but jerky, with significant spurts about two million years ago and then about half a million years ago. Mithen suggests that before these events we had a general intelligence (possessed also by the higher apes), and then came those modules (also possessed by the apes) for special skills. These focused on social abilities and navigating and understanding the environment. Nothing at this point was very well integrated. With the first spurt (to *H. habilis*), came a new module for technical ability. We can infer this from the existence of the first primitive tools (chipped stones forming hand axes and the like). What is nevertheless striking is that even though the tools arrived, there was (as noted above) basically no subsequent innovation despite their being really very limited in scope—no one used bone and antler, for instance, even though these substances have virtues that stone does not have. The second spurt (taking us to *H. sapiens*) brought far more integration of the various modules. Now and only now was it possible for sophisticated language and tool use and culture generally to take off, although even this did not really happen until (as we saw above) within the last fifty thousand years, or even later.

Of course, none of this addresses the ultimate question, namely, that of consciousness. As you might expect, there are divided opinions on this matter. There

The mind as a cathedral

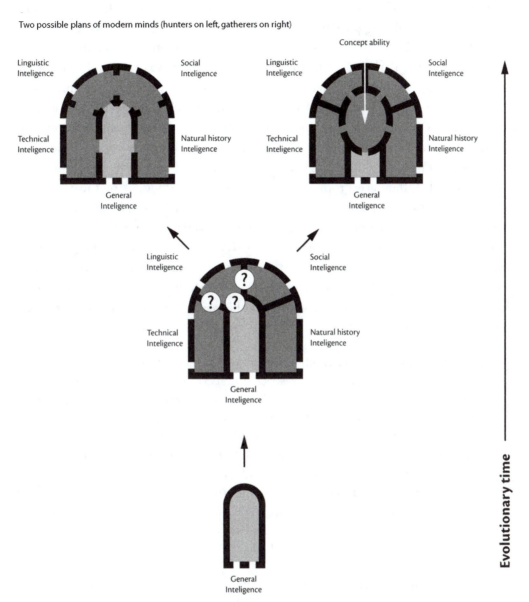

Two possible plans of modern minds (hunters on left, gatherers on right)

Concept ability

Linguistic Inteligence — Social Inteligence

Technical Inteligence — Natural history Inteligence

General Inteligence

Linguistic Inteligence — Social Inteligence

Technical Inteligence — Natural history Inteligence

General Inteligence

Linguistic Inteligence — Social Inteligence

Technical Inteligence — Natural history Inteligence

General Inteligence

General Inteligence

Evolutionary time

Steven Mithen's hypothesis

are those who, even today, want to deny that consciousness has any great biological significance. Others, relatedly, feel that consciousness is something very recently acquired, and so it cannot have been a major factor in human evolution. The average evolutionist, however, particularly the average Darwinian, feels extremely uncomfortable with such a dismissive attitude. Consciousness seems a very important aspect of human nature. Whatever it may be, consciousness is so much a part of what it is to be human that Darwinians are loath to say that natural selection had no or little role in its production and maintenance.

Whatever position is taken on evolution, no one is denying that consciousness is in some sense connected to or emergent from the brain. The question—at least the question that concerns Darwinians—is whether, over and above the brain, consciousness has some biological standing in its own right. General opinion (my opinion!) is that somehow, as brains got bigger and better during animal evolution, consciousness started to emerge in a primitive sort of way. Brains developed for calculating purposes and consciousness emerged and, as it were, got dragged along. Most Darwinians think that at some point, consciousness came into its own right. Perhaps, then, the causal connection was reversed, and brains were now dragged along, in order to make bigger and better conscious animals.

This raises the question of what consciousness actually does. Why should we not just have a nonthinking machine, which does everything? Is consciousness little more than froth on the top of the electronics of the brain? Is consciousness just an epiphenomenon, as philosophers would say? Slowly but positively, brain scientists do feel that they are groping toward some understanding of the virtues of consciousness, over and above the operation of blind automata. It is felt that consciousness may act as a kind of filter and a guide—coordinating all the information thrown up by the brain. Consciousness helps to prevent the brain from getting overloaded, as happens all too often with computers. Consciousness regulates experience, sifting through the input, using some and rejecting some and storing some. One important brain scientist, referring to this aspect of consciousness as access consciousness, writes as follows:

> Any intelligent agent incarnated in matter, working in real time, and subject to the laws of thermodynamics must be restricted in its access to information. Only information *relevant* to the problem at hand should be allowed in. That does not mean that the agent should wear blinkers or become an amnesiac. Information that is irrelevant at one time for one purpose might be relevant at another time for another purpose. So information must be *routed*. Information that is always irrelevant to a kind of computation should be permanently sealed off from it. Information that is sometimes relevant and sometimes irrelevant should be accessible to a computation when it is relevant, insofar as that can be predicted in advance. This design specification explains why access-consciousness exists in the human mind and also allows us to understand some of its details. (Pinker 1997, 138)

Still, you might complain that this does not explain consciousness in itself. Why do we have "sentience," as we might call it? Why do we have the capacity of self-awareness? To what was the seventeenth-century French philosopher René Descartes referring when he spoke of the *cogito,* as when he said, "I think, therefore I am"? Why is it that what is essentially no more than a bunch of atoms should have thinking ability? Why is it that I am able to write now and to think about what I am doing, and you are able to read what I have written: perhaps agreeing, perhaps disagreeing, perhaps liking what I say, perhaps disliking what I say, but certainly reacting in some fashion or another? I am afraid that at this point, we start to run out of answers. The Darwinian qua Darwinian is reduced to silence. This is not to deny the existence of consciousness. Anything but! "Saying that we have no scientific explanation of sentience is not the same as saying that sentience does not exist at all. I am as certain that I am sentient as I am certain of *anything*, and I bet you feel the same. Though I concede that my curiosity about sentience may never be satisfied, I refuse to believe that I am just confused when I

René Descartes

think I am sentient at all!" (Pinker 1997, 148). The point is that as a Darwinian, that is to say as a scientist and an evolutionist, there seems to be no answer. At least, no answer at the moment.

The psychologist David Chalmers refers to this as the "hard question."

> What makes the hard problem hard and almost unique is that it goes *beyond* problems about the performance of functions. To see this, note that even when we have explained the performance of all the cognitive and behavioral functions in the vicinity of experience—perceptual discrimination, categorization, internal access, verbal report—there may still remain a further unanswered question: *Why is the performance of these functions accompanied by experience?* A simple explanation of the functions leaves this question open. (Chalmers 1997, 12)

At this time, perhaps it is best to turn to philosophy. Certainly, philosophers have thought much about the problem. Simplifying somewhat, we find two main approaches. On the one hand, there are the dualists. This group includes the great Greek philosopher Plato as well as Descartes, mentioned just above. They argue that consciousness is something altogether different from physical matter. They speak of it as being a substance in its own right: in Descartes's language it was *res cogitans* (thinking substances) as opposed to *res extensa* (material or physical substances). As the language implies, these people take thought or thinking as the mark of the substance of consciousness, as opposed to extension, which is the mark of the material or physical world. On the other hand, there are the monists. Most famously, there was the seventeenth-century Dutch philosopher Benedict Spinoza. He argued that when thinking of consciousness, there is no reason to

Benedict Spinoza

think that one is considering a separate substance. Consciousness, in some way, is simply a manifestation of the physical world. Spinoza and his modern-day followers do not want to say that consciousness does not exist, or that it is simply material substance in a traditional way. Consciousness is obviously not round, or red, or hard, or anything like that. Rather, consciousness in some sense is emergent from or an aspect of material substance. In other words, the notion of material substance has to be extended, from red and round and hard, to include consciousness.

Most philosophers and scientists today are inclined to monism rather than to dualism. There have been relatively recent defenses of dualism by philosophers and scientists, notably by the philosopher Karl Popper and his friend the brain scientist John Eccles (1997). More recently Chalmers (1996) has endorsed a version of dualism. Since both of these people would have thought of themselves not only as evolutionists but also as Darwinians, clearly one can hold both positions (dualism and Darwinism) at the same time. But there are serious problems with dualism, particularly about how one gets connections between material and thinking substance. Having distinguished them so firmly, it is hard to reconnect the two. For this reason, most Darwinians who think about these sorts of things are inclined to some kind of monism, or (as it is often known today) to some kind of identity theory. They think that body and mind are manifestations of the same thing, and that as selection works on one it affects the other, and as it works on the other it affects the former.

I hardly need say that all of these suggestions raise as many questions and problems as they solve. Philosophers and scientists are working hard toward answers and resolutions. But perhaps this is a point at which we might pull back from the discussion. The important thing from our perspective is that consciousness is a real thing. We are sentient beings. Moreover, consciousness is surely something subject to the forces of evolution, to natural selection in particular. More than this perhaps we need not say, or argue. As with the physical world, take it as a given. It is something wonderful, but commonplace, mysterious, yet familiar. All of these things and a great deal more. We must recognize that all inquiry must start at some point, and perhaps here is one such point. No one ever said that a scientific theory has to explain everything. Although some of my readers will now themselves be inspired to take up the quest. It will be an honorable task to set oneself.

Further Reading & Discussion

A popular account of the hobbit by one of the discoverers is *A New Human: The Startling Discovery and Strange Story of the "Hobbits" of Flores*, Indonesia by Mike Morwood and Penny van Oosterzee [New York: HarperCollins Publishers, 2007].

There are lots of good books on human evolution. *The Smithsonian Intimate Guide to Human Origins* (New York: HarperCollins, 2007) by science writer Carl Zimmer is excellent. You can also rely on anything written by Roger Lewin, including *Human Evolution: An Illustrated Introduction* (New York: Wiley-Blackwell, 2004). A wonderfully opinionated account of the discovery of Australopithecus afarensis is *Lucy: The Beginnings of Humankind* by Donald Johanson and Martin Edey (New York: Simon & Schuster, 1981). It would seem that you need a massive ego to be a successful paleoanthropologist (student of human origins). Don Johanson, the man who discovered Lucy, has that and more. The same is also true of Chris Stringer, who with Peter Andrews is the author of *The Complete World of Human Evolution* (London: Thames and Hudson, 2005). Steven Pinker is not only a good psychologist but also a great writer. His *How the Mind Works* (New York: W. W. Norton, 1997) is detailed, informative, and at times very funny. Earlier he had taken on the question of human language in *The Language Instinct: How the Mind Creates Language* (New York: William Morrow, 1994). Later he takes on everyone in *Blank Slate: The Modern Denial of Human Nature* (New York: Viking, 2002).

In the text I make a somewhat exasperated comment about the World Wide Web, but truly for human evolution it really is invaluable. It is great on such topics as Piltdown man, Neanderthals, as well as detailed claims like the "Out of Africa" and "Multiregional" hypotheses about human evolution. And finally, if you get tired of bones and egos and disagreements, let me recommend something very different. The English novelist Angus Wilson wrote a terrific story inspired by Piltdown: *Anglo-Saxon Attitudes* (London: Secker & Warburg, 1956). He transforms the fraud into one about archaeology, but his novel is not only a great read but a penetrating insight into how a fraud might have started as a joke and then taken on a life of its own. I very much suspect that that is what must have happened back there at Piltdown. If I had done it, my first emotion would have been joy at having pulled it off; then horror at the damage I was doing to the subject I loved so much; and finally rank fear that someone might finger me.

Chapter 9
Human Sociobiology: Genetic Determinism

Overview

This chapter explores social behavior. Darwin himself realized that adaptation could not just be of physical characteristics, but also had to take behavior into account. Most interesting and challenging was social behavior, where organisms seem to help others at the expense of themselves. To Darwin, this was obviously something that applied particularly to the human species. But how it worked with natural selection was not fully explored, if at all, until 100 years after the *Origin* was published. The exploration of social behavior then became one of the most exciting areas of evolutionary inquiry.

After Darwin, thanks particularly to the rise of the social sciences, there was a long period when social behaviour, especially as applied to human beings, was ignored and downplayed. The rise of the Nazis, with their vile doctrines of genetic behavior, made people very unwilling to discuss such matters. But the work of the English graduate student William D. Hamilton transformed things. He devised a number of sophisticated models that helped to explain the whole basis of the evolution of social behavior, a field now known as "sociobiology." As soon as people started working in the field, a new set of methods were devised. Particularly important were the ideas of game theory, developed by the English evolutionist, John Maynard Smith.

A wonderful empirical example of sociobiology in action was furnished by the English evolutionist Geoffrey Parker who worked on dung flies. Aristotle once told us that we should never look down on the most humble of organisms, but see interest and beauty in them all. He was right!

All of these ideas were put together in the most important book on evolution in the second half of the twentieth century—Edward O. Wilson's *Sociobiology: The New Synthesis*. This surveyed the field; but, at the same time, tried to carry ideas further and made provocative suggestions about the importance of looking at humans from a Darwinian evolutionary perspective.

Social scientists felt deeply threatened by Wilson's work. Marxists were incandescent at the thought that biology might matter, rather than simply social conditions, in the development of social behavior. So the battle commenced, probably generating more heat than light, a fact that the human sociobiologists would have expected.

Critics or not, sociobiology moves forward rapidly today. In the animal realm, people like Nicholas Davies work on the mating relationships between birds; in the human realm Canadian researchers Martin Daly and Margo Wilson show interesting and important implications by using human sociobiology for understanding human homicide rates and practices.

The Role of the Scientific Community

The work of the following scientists is discussed in this chapter. Short biographical essays of these individuals appear in **Biographies** on page 607.

Konrad Lorenz (1903–1989)
John Maynard Smith (1920–2004)
William D. Hamilton (1936–2000)
Geoffrey Parker (1944–)
Sarah Hrdy (1946–)
Nicholas Davies
Martin Daly
Margo Wilson

Setting the Stage

In 1978, the eminent Harvard biologist Edward O. Wilson—the world's leading authority on the ants—was giving a talk at the annual meeting of the American Association for the Advancement of Science. Suddenly from the audience, a man carrying a glass of water dashed up to the podium, emptying it over Wilson's head. "There, Professor Wilson," he screeched to the noisy approval of a bunch of supporters, "now everyone can see that you really are all wet!" Even Cuvier, at his most combative, never thought of doing anything like that.

This was but an episode in a war that had now been going on for three years, pitting Wilson and his team against the opponents, several of whom were eminent evolutionists in Wilson's own department of organismic biology at Harvard. They were fighting over something that Wilson had labeled "sociobiology": more particularly, they were fighting over the implications of this sociobiology for our own human species. Wilson thought it was the most important move in evolutionary biology since the *Origin*. His critics, many of whom were Jewish and who loathed and feared any attempt to seek biological factors in human behavior and understanding, thought it bad science, morally reprehensible, and politically dangerous. If a little cold water could show the world the evil of Wilson's ways, then so be it.

Let us go back to Darwin and pick up the story there, bringing it down to the present and to the implications of sociobiology for understanding ourselves.

Essay

Social Behavior

Charles Darwin always recognized that behavior is as important a part of an animal's being as is its physical form. Biologically speaking, there is little point in having the physique of Tarzan if the only thing you are interested in is philosophy! Right from the beginning, in the *Origin*, Darwin acknowledged the significance of behavior and thought it as much an adaptation formed by natural selection as is any physical feature such as the eye or the hand. Indeed, the very first example that Darwin gives of selection at work in the *Origin* is of wolves hunting deer, and how the different strategies and behaviors might well lead to different physical features. Moreover, Darwin recognized that some of the most interesting and intriguing examples of behavior involve what one might call social behavior, where instead of working flat out to deprive or otherwise harm a competitor or fellow struggler for existence, one works to aid or help one's fellow, especially one's fellow species member. He was particularly interested in the hymenoptera (the ants, the bees, and the wasps), the paradigm of social animals, and in fact devoted a whole chapter to their study.

Edward O. Wilson

Now why should social behavior, adaptive social behavior, that is, be particularly interesting and challenging? A mother feeds her offspring. Surely there is no real problem here. If a mother does not feed her offspring, they will die. Although the mother may survive, her reproduction is as truncated as if she were sterile in the first place. Nor is there any real problem when you start to extend the range of social behavior. In a nest of ants, you find the workers (always female) helping the group by feeding the young, or going foraging for food, or acting as soldiers by defending the nest, or a number of other activities. The workers are after all helping their siblings by raising them: also aiding their mother, who is the queen of the nest. Why should there be any worry here? Or indeed, why should there be any worry when an organism helps any fellow species member? After all, surely selection has the good of the group at heart?

But this is precisely the problem. As we saw in an earlier chapter, in the eyes of Wallace, the codiscoverer of natural selection, the mechanism did work for the group. Characteristics, physical and behavioral, work for the group (meaning the species) as much as they work for the individual. In the eyes of Darwin, however, characteristics are adaptively directed toward the individual only, and the group not at all (Ruse 1980). The struggle for existence pits lion with lion and human with human. Group benefits can never come at the expense of individual benefits. One can circumvent this only if, in some sense, the social behavior—behavior, that is, that requires cooperation and working with others and perhaps even giving to others—benefits the individual. There is little point, for instance, in a mother harming her daughter—taking all of the food for herself—because then the mother harms herself. Her own reproduction is blocked. What then of the social insects, where one finds that cooperation has been driven to such a degree that the workers are sterile, giving their whole lives to the nest? How can they benefit, who have no offspring of their own?

Darwin was little worried about the sterility per se, for his knowledge of the agricultural world had shown him how selection can (as it were) work sideways, promoting desirable features for nonreproductive animals (geldings and oxen and porkers) through their fertile relatives. Artificial selection is done for our ends. Who benefits when there is no conscious intention involved? Eventually, Darwin decided that one could treat the whole hymenopteran nest as a kind of supraorganism, with the sterile members as parts of the whole: they exist rather as hands and eyes exist, not for their own sakes but for the sakes of the whole. Darwin was never really comfortable with this, however. But nothing more could be done on the problem, especially in ignorance of the proper principles of genetics (Richards 1987).

One thing that Darwin did always realize is that a significant—and to us humans by far the most interesting—social animal is *Homo sapiens*. We humans have made sociality our speciality. And Darwin was never loath to get right in there and speculate. It is true that there is little on this topic in the *Origin*, but the very first records that we have of Darwin's discussing selection (in a private notebook in the late fall of 1838) has him thinking about human evolution and about how some people are brighter than others thanks to natural selection! I do not think that Darwin became an evolutionist because he was obsessed with human beings—unlike quite a few other prominent evolutionists—but there is no doubt but that he thought that human evolution is an important part of the overall story. The *Descent of Man*, published in 1871, was written to deal with human evolution—with, it will be remembered, a particularly significant causal role being given to sexual selection.

Darwin made it very clear that our social nature just as much as our physical nature (the two of course are very much combined) is the result of a selection-

THE

DESCENT OF MAN,

AND

SELECTION IN RELATION TO SEX.

By CHARLES DARWIN, M.A., F.R.S., &c.

IN TWO VOLUMES.—Vol. I.

WITH ILLUSTRATIONS.

LONDON:

JOHN MURRAY, ALBEMARLE STREET.

The title page of The Descent of Man

driven evolution. Some races (Europeans particularly) come out over others because they did better in the struggle: generally because the winners had a harder time in the struggle, thanks to the more difficult conditions in Europe than elsewhere, as in Africa. Males differ from females because of the different selective forces: not only do males have different physical characteristics but that they have different emotional and behavioral characteristics. The classes are stratified because of selective pressures. Remember how Darwin gives a long discussion of the virtues of capitalism—just what you would expect from the grandson of Josiah Wedgwood! And there is much more along the same lines. The Darwinian man is a social man is a biological man, and that means evolution through natural and sexual selection. We may have come out on top—Darwin thought that we did—but we are still part of the whole. In fact, for Darwin, coming out on top is precisely a matter of being, like everything else, part of the organic world: there was a race and we won. In this sense, the Darwinian picture is very much part of that progressivist world vision, set off against the Christian providentialist world vision, which latter judges us to have won because we were never part of the race in the first place. For the believer, we humans are the top because God made us that way, in His image.

The Long Hiatus

Move the clock forward rapidly, through a hundred years. By the time of the *Origin*'s centenary in 1959, evolutionary theory in general had made major strides forward. Except in the area of behavior, social behavior in particular. It is true that a number of European workers, the "ethologists," were working on such issues as mate recognition and honey bee activity, but compared (say) to the activity in population genetics or systematics, the area was one of neglect. There were a number of reasons for this. Most obviously, behavior is much more difficult to study than something like morphology. If you are interested in anatomy, you can kill your subject, pop it into formaldehyde, and then pull it out and chop it up when you are ready. Behavior has to be studied on the job, as it were. You can try experimenting, but it is well known that experimental conditions can affect even the most basic of activities—consider how difficult it can be to get animals to breed in zoos (or to stop the breeding of other animals, quite reproductively isolated in the wild). And if you try to study in the wild, then costs and difficulties arise. It is one thing to study the eye color of Drosophila in the lab and quite another to measure breeding activity in a jungle or a desert.

Then again, going against the study of social behavior, there was the rise of the social scientists. They were young, insecure, and jealous of their territory. They were terrified that evolutionary biologists might come down, take over, and hang out a new shingle: "Evolutionary biology (sociology division)." So they resisted any attempt at a takeover or even collaboration. Studies were done on white mice or rats, generalized to other animals, and then it was declared that the uniformity showed that there was no need for a comparative approach! Learning behavior, for instance, was considered quite outside the evolutionary context. An animal could learn to avoid or welcome anything, in any way, at any time. The thought that perhaps one might be more receptive to learning in certain periods and not others was considered slightly silly. (I write now with some bitterness as one who was first introduced to foreign languages at the age of eleven, just the point at which we are now assured the biological door closes firmly shut.)

Then finally there was the human question. Here all sorts of factors worked against an evolutionary approach. Freud, for instance, was himself quite receptive to evolutionary ideas (Sulloway 1979). In his seminal works on human sexuality, he started by stating simply that some people are as they are because of their biology—no need of protective mothers and hostile fathers to do the work. Biology is self-sufficient. But his followers, from personal ignorance (not trained as was he in biology) or from arrogance (who needs biologists?) or from avarice (how can one justify high fees listening to moaning about mother when the genes did it in the first place?), cut out the biological component almost completely. Then the social scientists were full of all sorts of progressivist ideas about changing society, so long as we do the right things. The peak of self-deception was achieved by Margaret Mead, who, so eager was she to show that our Western sexuality has no reflection in innate human nature, allowed herself to be the butt of schoolgirl jokes about Samoan sexuality. Thanks to the influence of such studies (if one might so dignify them) as these, it became accepted wisdom that human beings are infinitely plastic—it is all a matter of the environment.

And as the century went on, hanging over everything was the terrible example coming out of Nazi Germany. In that land, there was the claim that humans are different because of their biology, and from this belief stemmed the most terrible actions and injustices. Jews, gypsies, homosexuals, Slavs, the insane, and more and more groups were judged biologically inferior and subjected to oppression and the lack of liberty and ultimately the final punishment, death. Who could think that a biological approach toward humankind could have any merit whatsoever? Even if it be true that biology might play some role, it has to be minor, and the risks raised by studying it far outweigh any potential benefits. There are some things that are simply best left alone.

Sociobiology

But things did start to change, and what began as a trickle soon swelled right out into a torrent. First perhaps came the theory, and this had both a critical side and a positive side. On the critical side, the 1960s started to see a significant shift toward a Darwinian approach to (what became known as) the level of selection, as older assumptions were subjected to withering analysis. At the beginning of that decade, with very few exceptions, the automatic assumption of evolutionists was that natural selection could and did work at all levels—for the benefit of the individual, the group, or the species. An adaptation therefore might help you personally, or it might be of no value whatsoever to you as possessor but of great worth to other members of your species. The ethologists never doubted that this might be the case. Konrad Lorenz (1966) wrote a whole book on aggression, arguing that in fights between species members, constraint is always shown because otherwise the species would suffer. A dog will never knowingly kill another dog, because this would be bad for doggyhood in general. And others thought the same. A major work on animal population numbers argued that they are regulated by individuals because otherwise one might have overpopulation—bad for the group (Wynne-Edwards 1962). You might benefit from one or two more children, but what if everyone did the same? (There was often an interesting subcurrent, to the effect that humans uniquely seem not to obey group rules, to the detriment of all. We have no means of restraining aggression toward fellow humans, and clearly we cannot contain our sexual passions. We are the naked ape with blood-stained jaws.)

This was now seen as totally fallacious reasoning (Williams 1966). Natural selection has no forethought. It acts only in the present. If an organism benefits in this generation, then so be it, however disastrous the long-term consequences. Consider two species members, the one of which acts purely selfishly by having lots of offspring and the other of which acts purely altruistically by having but few offspring. In the next generation, there will be far more of the selfish member's offspring than the altruistic member's offspring, and so on down the line. Even

Konrad Lorenz

though some 10 or more generations hence it might be better for all were the altruist to prevail, by then it would be too late. The selfish member's offspring would be the populational norm. The point is that, as Darwin realized, group selection simply cannot work. (In fact, one can show that under certain special circumstances, a group effect can overwhelm an individual effect, but such cases are few and far between.)

On the positive side, the early 1960s was just the time when theoreticians were starting to produce models, showing how individual selection can work and how in fact one can throw light on interesting problems, hitherto insoluble. For the point is that social behavior does occur, and animals do show altruistic inclinations and actions toward one another—usually (although not necessarily always) toward fellow species mates. If one cannot explain this directly through group selection and must therefore rely on individual selection, the question arises as to how this is to be done. And the answer, obviously, is that one must just show that in helping others one is helping oneself. Indeed, one must show that one helps oneself more by helping others than if one did nothing.

Now, in a way, one can follow through fairly directly on this insight. Mothers care for their offspring. Why? Obviously, because the offspring carry on the mother's line. (None of this necessarily happens at the conscious level. Rather, our genes make us do it.) Or let us put matters another way. Natural selection is a matter of making sure that one's units of heredity, one's genes, are represented in future generations. I am fitter than you if a higher percentage of my genes get through rather than yours. But it is hardly a question of my genes as such. Rather, it is a question of copies of my genes. And thus understood, we can say that a mother cares for her offspring because, by so doing, she is ensuring that copies of her genes are transmitted. If the offspring all die without issue, then the genes are stopped dead.

You can generalize this idea, which is precisely what was done by the English, then-graduate student William Hamilton (1964a, b). He reasoned that altruism—helping other organisms—will always pay if those organisms are bearers of the same genes as oneself. One is helping one's own genes in the struggle for existence, vicariously as it were. Or rather, he reasoned that altruism would pay if one could do more for one's genes through such altruism rather than otherwise. A distant cousin will have only a very small proportion of genes in common with you. Hence, there is little point in forgoing one's own reproduction for that cousin, unless you can have very few offspring yourself and that cousin can have many more offspring than otherwise. And this indeed suggests a simple little formula that governs the altruism relationship: essentially, altruism kicks in only when the benefit through help, or altruism, exceeds the reciprocal of one's blood connection to the beneficiary. As the blood relationship falls away, so it is necessary that the benefits rise accordingly.

Genius is not always recognized at once. Hamilton's thesis supervisor thought so little of his student's insight that he urged Hamilton not to use it in his thesis, for fear of failing! But slowly it was seen for the brilliant move that it is. And what did start the realization of its importance was that Hamilton applied his idea (known now as "kin selection") to that very problem that had stymied Darwin. How is that hymenopteran workers devote their whole lives to the good of others, without breeding themselves? Hamilton pointed out that (as was well known) the hymenoptera have a funny mating system. Whereas females have both mothers and fathers (they are diploid, meaning that they have the usual paired set of chromosomes), males have only mothers (they are haploid, having only one set of unpaired chromosomes). A queen is inseminated but keeps the sperm, sometimes for many years. If an egg is fertilized then a daughter is born, but if an egg is not fertilized a son is born.

What this means (as you can see from the diagram) is that although mothers and daughters have the usual genetic relationship of 50 percent (just like humans), sisters are more closely related than normal (75 percent as opposed to the usual 50 percent). This implies, from a selective viewpoint, that female hymenoptera are better employed raising fertile sisters than fertile daughters. The altruism that workers show in the nest is preserved and cherished by natural selection, even though the workers are sterile! In the case of males, they are 50 percent related to mothers and to daughters (they have no sons), and so there is not the same urge to help. Notoriously, male hymenoptera are "drones," good only for breeding

W. D. Hamilton

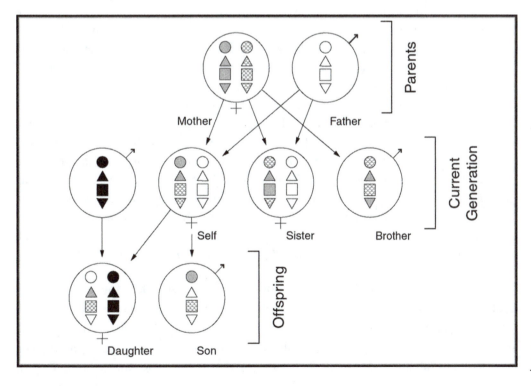

The relationships within a hymenopteran family

purposes. Interestingly (and expectedly), you do sometimes find that "sterile" workers will lay unfertilized eggs that hatch into drones—this is a move that one would expect given natural selection.

To give you some idea of how this thinking first met opposition and then conquered all before it, let me quote to you the full and generous account that Wilson gives of his first encounter with Hamilton's ideas. I do not know of quite anything that gives such a sense of the excitement of scientific ideas or of the way in which science is no respecter of status, only brilliance. Wilson explains that the year was back in 1965, and he was on a train carrying him south from his home in Boston to his field station work in Florida. Keep in mind that Wilson was a Harvard professor, in the same department as Jim Watson of double helix fame (Wilson got tenure before Watson!), and with good reason thinking of himself as the great man in insect biology, before whom all must defer.

> I picked Hamilton's paper out of my briefcase somewhere north of New Haven and riffled through it impatiently. I was anxious to get the gist of the argument and move on to something else, something more familiar and congenial. The prose was convoluted and the full-dress mathematical treatment difficult, but I understood his main point about haplodiploidy and colonial life quickly enough. My first response was negative. Impossible, I thought; this can't be right. Too simple. He must not know much about social insects. But the idea kept gnawing away at me early that afternoon, as I changed over to the Silver Meteor in New York's Pennsylvania Station. As we departed southward across the New Jersey marshes, I went through the article again, more carefully this time, looking for the fatal flaw I believed must be there. At intervals I closed my eyes and tried to conceive of alternative, more convincing explanations of the prevalence of hymenopteran social life and the all-female worker force. Surely I knew enough to come up with something. I had done this kind of critique before and succeeded. But nothing presented itself now. By dinnertime, as the train rumbled on into Virginia, I was growing frustrated and angry. Hamilton, whoever he was, could not have cut the Gordian knot. Anyway, there was no Gordian knot in the first place, was there? I had thought there was probably just a lot of accidental evolution and wonderful natural history. And because I modestly thought of myself as the world authority on social insects, I also thought it unlikely that anyone else could explain their origin, certainly not in one clean stroke. The next morning, as we rolled on past Waycross and Jacksonville, I thrashed about some more. By the time we reached Miami in the early afternoon, I gave up. I was a convert, and put myself in Hamilton's hands. I had undergone what historians of science call a paradigm shift. (Wilson 1994, 319–320)

In the spirit of Hamilton, other models were devised showing how sociality could be preserved given individual selection. The American Robert Trivers (1971) came up with "reciprocal altruism": this is a case of "you scratch my back and I will scratch yours." Here animals cooperate because they both benefit. The interesting thing about this kind of situation is that it can cross species boundaries, and Trivers gave interesting examples drawn from fish, where predatory species will nevertheless refuse to attack other fish that specialize in cleaning them of parasites. The predators get cleaned and the cleaners get a good meal. Both sides

John Maynard Smith

benefit, which would not be the case if the predators immediately ate the cleaners (or ate the cleaners after a cleaning).

And the English evolutionist John Maynard Smith (1982) systematized much of our thinking about social situations by making heavy use of game theory. He showed how selection can promote certain equilibrium situations, where everyone gets the most that is possible, given that everyone else is trying to do the same. These Evolutionarily Stable Strategies are what one finds when one has mixed populations, with different members trying to achieve their ends by different means (or where every member has alternate means to achieve the same end). Most famously, we have a species consisting of hawks and doves (that is to say, some members show hawklike behavior and other members show dovelike behavior, where these translate as fighting as opposed to fleeing). A population of hawks would just tear each other apart, and so a dove would be selectively favored. A population of doves would never threaten, so a hawk would be selectively favored. But given costs and gains (if the cost of fighting is slight, then being a hawk is better than if the cost of fighting is heavy), one can show that the population will achieve a stable equilibrium at certain ratios—different behavior will be held in the population by (individual) selection.

Dung Flies

This kind of theoretical thinking stimulated the empiricists: experimentalists and naturalists. Realizing that one would have to spend much more time in the wild or more care over experimentation than previously, the new models nevertheless inspired people to try to see if one could measure behavior in action and draw solid conclusions. One of the most successful workers was the English evolutionist Geoffrey Parker (1978), who made his mark through a series of papers stemming from his thesis project: the behavior of one of nature's less prepossess-

Geoffrey Parker

ing members, the dung fly, *Scatophaga stercoraria* (Ruse 1996, 1999). Parker spent many long hours in fields, surrounded by herds of cattle, following the brutes around and waiting for them to defecate. He knew that the flies' reproductive behavior is focused on the waste that the mammals expel and leave behind, and he soon found that there are standard behavioral patterns followed by male and female flies. First, it is the males who fly in, looking about for fresh cow pats. Then the females arrive and are seized by the males, who mate with them vigorously. After this, the now fertilized females fly onto the pats and lay their eggs. Sometime later, the larvae hatch and bury down into the cow feces, thus able to feed abundantly from the rich nutrients within which they find themselves embedded. It was in the variations and elaborations on these standard patterns that Parker found much scope for scientific investigation: an opportunity that he exploited to the full, with diligence and intelligence.

To do science successfully, you need hypotheses to build models. With the interest in reproductive behavior, Darwin's mechanism of sexual selection seems the obvious tool of inquiry. Today, were one to suggest this, there would be no great surprise. But even forty years ago, this was not so. For the century after the *Origin* and the *Descent,* for all that Darwin himself had championed sexual selection, it had never been a great success. We saw Wallace's unsympathetic reaction, and while few biologists shared Wallace's enthusiasm for spiritualism (which lay ultimately behind his rejection of aspects of the mechanism), ever fewer wanted to credit Darwin with having found in sexual selection a significant factor in evolutionary change. Indeed, it seems fair to say that for the first two-thirds of the twentieth century, sexual selection (if considered at all) was thought but a minor and not significant form of the general mechanism of natural selection. But with the move to a more individual-based perspective on the working of selection, sexual selection—which is an individual-versus-individual form of selection par excellence—started finally to come into its own. So perhaps after all it was no great surprise that, for all that he was working in the late 1960s, Parker's focus was very much on sexual selection, particularly on the competition between males, who in the dung flies outnumber the females by four or five to one.

Particularly interesting and significant was the distribution of the males, who had to choose a site carefully—fresh pats of dung tend to be far too liquid for safety—where they could be reasonably sure of finding a female and yet able to defend themselves against the needs and desires of other males. "Males should be distributed between zones in such a way that all individuals experience equal expectations of gain. Hence the proportion of females captured in a given zone should equal the proportion of males searching there, assuming that all females arriving are equally valuable irrespective of where they are caught" (Parker 1978, 219–220). What made Parker's work so exciting was the fact that his predictions about spacing held so exactly: observation and theory differed not at all in any significant way. Although the work could not be ended with just one set of findings, for Parker soon discovered that he was dealing with a fluid situation. As the first round of mating comes to an end, successful males must now balance their labors between guarding their females from other males and going off in search of new females. Hitherto unsuccessful males, meanwhile, must move from trying to find mates in their own right to trying to pry females away from successful (copulating or postcopulating) males.

As time goes by, the cow pats form a skin and thus are less hazardous for the flies—in particular, females can start moving toward the pats in order to lay their eggs. One expects therefore that the males will move from a general wide distribution around a field toward the cow pats. Parker found here that theory and findings were close but not quite as close as before. Perhaps the smells of new droppings crowd out the smells of older droppings and the males have to adopt strategies to allow for this: "However, this information about new droppings may be obtained by spending time in the grass upwind" (p. 225). One important assumption in all of this is that the females are able to sustain and use multiple matings—a one-time mating with a male does not exhaust a female's supply of unfertilized eggs. In fact, turning now to experiment in the laboratory, Parker found that the last male in any mating succession was by far the most successful from an evolutionary perspective. By sterilizing selected males, by encouraging multiple matings, and by counting the fertile eggs that females laid, Parker (1970) discovered that an amazing 80 percent of the eggs laid by any particular female were fertilized by the sperm of the last male to mate with her.

It is indeed truly the case that it pays a male to take over a female or—if he already has a female—to protect her from intruders and competitors. Apparently, there is a balance between protecting the female one has already and finding another female where one will be the final male: "In conditions of high male density during reproduction and with mating followed immediately by oviposition, in *S.*

stercoraria evolution seems to have favored the optimum active *copula* duration with inhibition of separation so that pairing is extended for guarding the female during oviposition" (Parker 1970, 785).

The Call to Arms

Work like this—theoretically ambitious and predictively fertile and successful—convinced evolutionists that their theory was moving forward rapidly. It is not surprising that people began to think in terms of synthesis, and in 1975 Edward O. Wilson attempted just this. But Wilson's book, *Sociobiology: The New Synthesis*, was more than just a compilation. It was a manifesto. A call to arms. Speaking of Hamilton's work as revolutionary, *Sociobiology* is a flamboyant, oversized tome with lots of pictures. The title of the first chapter, "The Morality of the Gene," sets the tone, and the opening words continue in the same vein:

> Camus said that the only serious philosophical question is suicide. That is wrong even in the strict sense intended. The biologist, who is concerned with questions of physiology and evolutionary history, realizes that self-knowledge is constrained and shaped by the emotional control centers in the hypothalamus and limbic systems of the brain. These centers flood our consciousness with all the emotions—hate, love, guilt, fear, and others—that are consulted by ethical philosophers who wish to intuit the standards of good and evil. What, we are then compelled to ask, made the hypothalamus and limbic system? They evolved by natural selection. That simple biological statement must be pursued to explain ethics and ethical philosophers, if not epistemology and epistemologists, at all depths. (p. 3)

Although the pace of the book never slackens, as Wilson warms to his task the melodramatic language and imagery do recede somewhat. Having first shown how he sees sociobiology as a natural outgrowth of evolutionary ecology, Wilson turns to a detailed and comprehensive discussion of the causal factors behind animal sociality. We get a basic discussion of the principles of evolution and of genetics, coverage of the sorts of models introduced earlier in this chapter (kin selection, reciprocal altruism, and so forth), and—an area where Wilson himself is a world expert—much attention paid to methods of animal communication, especially chemical communication between insects using so-called pheromones (p. 231).

Then, after brief overviews of such topics as aggression, dominance, caste systems, sexuality, parental care, and the like, Wilson turns to what he obviously considers the real meat of the book: a survey moving upward through the animal social world from colonial microorganisms through insects and lower mammals right up to our own species: "Man: From Sociobiology to Sociology." And here we find (as we have surely been led to suspect all along) that the inclusion of *Homo sapiens* is no last-minute decision, something done for completeness, as it were. We humans in a way are the raison d'être of the whole book.

To visualize the main features of social behavior in all organisms at once, from colonial jellyfish to man, is to encounter a paradox. We should first note that social systems have originated repeatedly in one major group of organisms after another, achieving widely different degrees of specialization and complexity. Four groups occupy pinnacles high above the others: the colonial invertebrates, the social insects, the nonhuman mammals, and man. Each has basic qualities of social life unique to itself. Here, then, is the paradox. Although the sequence just given proceeds from unquestionably more primitive and older forms of life to more advanced and recent ones, the key properties of social existence, including cohesiveness, altruism, and cooperativeness, decline. It seems as though social evolution has slowed as the body plan of the individual organism became more elaborate. (p. 379)

A paradox, but one that is a challenge rather than a barrier (p. 382). Tearing into the "culminating mystery of all biology," namely just how it is that humans have been able to stem the flow away from social integration, we learn that as humans evolved away from the apes, they reached a threshold. Arguing consciously with metaphors drawn from cybernetic thinking, Wilson reasons that at such a point a kind of feedback situation kicks in. There is suddenly an incredibly rapid and significant form of evolution, where it is appropriate to apply a kind of autocatalytic (self-driving) model of change. In a two-stage process, first humans got up on their hind legs and walked, thus freeing hands for tool use, and then sequentially there was an explosion of brain size with corresponding increase in mental power. This opened the way to a kind of cultural evolution, which in some sense takes us humans up and beyond our biology—although only in a sense, for Wilson makes it very clear that in other senses our biology remains (and always will remain) very important. If biology does not control the course of culture directly, then culture feeds back into the biology so that the genes in some fashion track the social. Either way, today and forever, much that we think and do is under genetic control—training and the environment are important but never all-important.

Had sociobiology—as, from now on, we can call the study of the evolution of social behavior—simply confined itself to the nonhuman part of the animal world, then although it would have been celebrated in biological circles, one doubts that it would have been heard of elsewhere. After all, dung flies do not have the sex appeal of dinosaurs. But with the move to humans, even though (or perhaps especially though) this was following in the grand tradition of Charles Darwin himself, it was bound to be controversial. And matters were not helped by works that followed up on Wilson's *Sociobiology*. First, there was a popular account of the whole new rising discipline, an account coming from the pen of a young English student of the evolution of social behavior. *The Selfish Gene* by Richard Dawkins (1976) was as provocative as it was flamboyant as it was compulsively readable. Through a brilliant use of metaphor—who can take group selection seriously after genes have been thus labeled "selfish"?—Dawkins brought home the moves and developments of this new branch of science in ways more vivid and

Richard Dawkins

compelling than would have been achieved by thick volumes of normal academic prose. In fact, Dawkins himself said little about the application of sociobiology to the human realm. Introducing the idea of a "meme"—a kind of unit of culture akin to a gene, a unit of heredity—Dawkins's discussion of the subject rather suggested that cultural evolution is something apart from biological evolution. We shall be talking more later about "memetics" (the science of memes). Here it is enough to note that Dawkins's examples in the animal world spoke about things in the same tenor as did Darwin and (as we shall see in a moment) Edward O. Wilson. You know perfectly well what he thinks of male/female differences after you learn that females have two choices in the battle of the sexes: either they can take the "he-man" strategy, trying to get themselves the strongest and sexiest male, or they can take the "domestic bliss" strategy, trying to get themselves a mate by providing the best home life.

If all of this was not enough, Wilson himself then reentered the scene with a more popular book of his own. *On Human Nature* (1978), a work for which Wilson won the Pulitzer Prize, is an extension of the discussion of the last chapter of *Sociobiology*, given to exploring precisely how it is that biology yet impinges on human consciousness and action. In the case of sexuality, for instance, we learn that male animals tend toward aggression whereas females toward being "coy" and to looking for males who will remain and help with child-rearing. "Human beings obey this biological principle faithfully" (p. 125). Nor is alternative sexuality overlooked. Perhaps, for instance, homosexuals are like worker ants: they themselves might not do so very well in the reproductive stakes, so their efforts are diverted into helping close relatives raise more offspring. Although, of course, all humans are into some forms of help or altruism: "Individual behavior, including seemingly altruistic acts bestowed on tribe and nation, are directed, sometimes very circuitously, toward the Darwinian advantage of the solitary human being and his closest relatives" (pp. 158–159). And so we come to religion. This is no afterthought but is central to Wilson's conception of the functioning human: "The highest forms of religious practice, when examined more closely, can be seen to confer biological advantage. Above all they congeal identity" (p. 188). In belonging to a group, we find meaning in our lives. At the same time, we further individual self-interest.

Critical Reaction

Enough! Although Wilson was genuinely surprised at the reactions his work invoked, one might say that whatever his other faults (real and imaginary), he was being dreadfully naive if he thought there would be no response at all. Social scientists surely were going to be made tense, and those for whom any kind of biological approach to humankind was highly suspect (especially Jews) were going to react negatively. And this is precisely what did happen, especially in America where these things were felt somewhat more deeply. Sociobiology, especially the human variety, was accused of just about every sin under the sun. What gave the debate—if one can thus dignify an all-out war of words and personalities—a particularly keen edge is the fact that among the most prominent critics of Wilson's vision of sociobiology were several of his colleagues at Harvard, including at least two in his own department: the molecular geneticist Richard Lewontin and the paleontologist and soon-to-be-famous popularizer of things evolutionary, Stephen Jay Gould. They were candid about what drove them. If Wilson's program works, then we are right back in the 1930s or earlier.

> Just as theories of innate differences arise from political issues, so my own interest in those theories arises not merely from their biological content but from political considerations as well. As I was growing up, Fascism was spreading in Europe, and with it theories of racial superiority. The impact of the Nazi use of biological arguments to justify mass murders and sterilization was enormous on my generation of high school students. The political misuses of science, and particularly of biology, were uppermost in our consciousness as we studied genetics, evolution, and race. That consciousness has never left me, and it has daily sources of refreshment as I see, over and over again, claims of the biological superiority of one race, one sex, one class, one nation. I have a strong sense of the historical continuity of biological deterministic arguments at the same time that my professional mature research experience has shown me how poorly they are grounded in the nature of the physical world. I have had no choice, then, but to examine with the greatest possible care questions of what role, if any, biology plays in the structure of social inequality. (Lewontin writing in Schiff and Lewontin 1986, xiii)

Human sociobiology was accused of being false. How can one argue for the significance of the genes when culture clearly changes at rates that far exceed the speed at which genes can take effect? For instance, the rise, triumph, and fall of

Islam took less than a thousand years, a mere blink in the evolutionary life of the genes. There is simply no way in which biology can have been significant in this event. In any case, there is nothing but ignorance on the part of the human sociobiologists in their speculations. Who is to say that there are "gay genes," making people into homosexuals? And is there any evidence that homosexuals do in fact help their relatives to have and raise more offspring?

Human sociobiology was accused of being unfalsifiable—a charge not entirely consistent with the one that it is false, but no matter. There is simply no way in which its flabby claims can be put to check. All exits are covered:

> When we examine carefully the manner in which sociobiology pretends to explain all behaviors as adaptive, it becomes obvious that the theory is so constructed that *no tests are possible*. There exists no imaginable situation which cannot be explained; it is *necessarily confirmed by every observation*. The mode of explanation involves three possible levels of the operation of natural selection: 1. classical individual selection to account for obviously self-serving behaviors; 2. kin selection to account for altruistic or submissive acts toward relatives; 3. reciprocal altruism to account for altruistic behaviors directed toward unrelated persons. All that remains is to make up a "just-so" story of adaptation with the appropriate form of selection acting. (Allen et al. 1977, 24)

The *Just So Stories* were the fantastical stories made up by the English author Rudyard Kipling to account for the elephant's nose and other strange features of the living world. The critics claim that, just as it is silly to take seriously Kipling's claim that the nose resulted from a crocodile's pulling on a normal nose, so it is equally silly to take seriously the sociobiologist's claim about such things as sexuality and religion. No matter what counterevidence you produce, the sociobiologist will have an answer.

Sociobiology was (and is) accused of being sexist, racist, classist. It is argued that it is just not true that men are naturally aggressive and women naturally coy and retiring. This is all in the imagination of the evolutionists, and then read into nature—at which point it is read right back out and triumphantly held up as objectively validated! Even if it be true of the nonhuman animal world, the point about humans is that we are flexible and can escape our biology. No one can deny that many societies, including our own, treat women as inferior and that even women internalize this treatment and behave as if they are second in major, desirable characteristics to men. But this is culture and biology has no part in it. From the viewpoint of our genes, it is a level playing field.

Racism is another point of contention. Here, Lewontin—drawing on his expertise as a population geneticist—has had much to say. And bluntly, the conclusion must be that there is simply no evidence for the broadscale differences supposed by the sociobiologists. "Of all human genetic variation, 85% is between individual people within a nation or tribe" (Lewontin 1982, 123). Indeed we can put matters more strongly than this. Suppose there were a world holocaust and only Africans survived. We would have lost only 7 percent of human variation. In fact, "if the cataclysm were even more extreme and only the Xhosa people of the

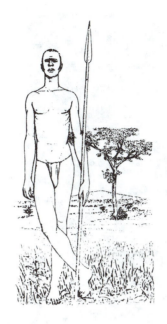

An African (left) and an Eskimo compared

southern tip of Africa survived, the human species would still retain 80% of its genetic variation" (p. 123). Lewontin can hardly deny that some differences between peoples may have an adaptive basis. Take body shape, a plausible candidate if anything is. There are (as we know from other animals, including birds) good adaptive reasons for minimizing surface area in cold regions. "Typically, the Eskimo has a large, chunky torso and short limbs, whereas the Dinka of Africa is tall and thin with very long arms and legs" (p. 128). Yet even this gets guarded treatment by Lewontin: "Although these trends seem to make good sense, there is no actual demonstration that they subserve greater survival and reproduction" (p. 128). And generally, Lewontin has nothing but contempt for those who would tie a strong link between human traits and personalities and abilities and our biology, our Darwinian adaptively shaped biology in particular. Not only can you not separate out genetic factors and environmental pressures but the very attempt is founded on a mistaken view of the way biology and nature interact. Genetic causes and environmental causes are truly "inseparable" (p. 68).

Classism (thinking social classes are found, not made) also is a major problem with human sociobiology. "Since the seventeenth century we seem to have been caught up in this vicious cycle, alternately applying the model of capitalist society to the animal kingdom, then reapplying this bourgeoisified animal kingdom to the interpretation of human society" (Sahlins 1976, 101). Darwin himself led the way, arguing that capitalism is a good thing, because then there will be people freed from toil and strife and able to devote their time to other things. Of course, this was grossly self-serving, and for every Darwin who did work there are a hundred parasites who do nothing for their livings. All that sociobiology does is give false justification to evil social and civil iniquities. "What is inscribed in the theory of sociobiology is the entrenched ideology of Western society: the assurance of its naturalness, and the claim of its inevitability" (p. 101).

Sarah Hrdy

There was more, but this will give you a good flavor of what things were about. Looking back, some thirty-plus years later, it all has a bit of a quaint look about it. There is nothing academics like more than a good fight. After all, this is what we are paid to do! And I cannot say—I ought not say, since I myself was right in the thick of it—that there were no good points made or matters of real issue. There is no doubt about it that some of the work produced by the sociobiologists was sexist, at least as judged by the exacting conditions of political correctness that prevail in universities today. Wilson and his friends did rather assume that males are naturally superior and that females like it that way, and leave matters at that. They did jump way ahead of their evidence, and then congratulate themselves on a hard empirical slog well done. And they were determined not to let a little counterevidence stand in their way. To be candid, they were determined not to let a massive amount of counterevidence stand in their way.

On the other hand, they were by no means as guilty as the critics would claim. Negatively, no one then was always that sensitive about male/female differences. Positively, there has long been a tradition in evolutionary studies of taking a feminist stand, from Alfred Russel Wallace on. Sure enough, sociobiology produced its own feminist counter. Sarah Blaffer Hrdy argued (in *The Woman That Never Evolved* [Hrdy 1981]) that females conceal ovulation, so males do not know exactly when the females are fertile. Hence, males cannot be assured of the paternity of their social offspring unless they stay around and help. In other words, it is females who make the running in the battle of the sexes and the males who are led along on a string. Far from men being on top, it was the women.

In other respects also sociobiologists could clear themselves. It is true that a lot of their speculations about homosexuals were based on very little evidence, but how else do hypotheses start? As it happens, a number of people went out and worked hard on this very issue, and now there is some considerable evidence pointing to the fact that there may indeed be genes coding for homosexual orientation (LeVay 1996). And as far as racism is concerned, it is simply not true that sociobiologists went in for the "blacks got rhythm" sort of thinking. In any case, as they pointed out, in the twentieth century more harm has been done by those who think that you can change human nature through social engineering than through any beliefs in genetic engineering. Certainly, no sociobiologist thought we were unvariably "genetically determined" to do what we do, as the critics often claimed.

As it happens, Lewontin has been accused (by A. W. F. Edwards [2003], R. A. Fisher's last Cambridge student) of making a gross mistake about statistics—so much so that it is now referred to as "Lewontin's fallacy." It is true that there is not that much variation between groups, but the variation within groups is clustered, some genes usually go with other genes, and so one can in fact make fairly sophisticated and reliable judgments about people's groupings given knowledge of their genetic makeups. For instance, a recent major study of human variation, looking at 1,056 individuals from 52 populations, reported: "Of 4,199 alleles present more than once in the sample, 46.7% appeared in all major regions represented: Africa, Europe, the Middle East, Central/South Asia, East Asia, Oceania, and America. Only 7.4% of these 4,199 alleles were exclusive to one region; region-specific alleles were usually rare, with a median relative frequency of 1.0% in their region of occurrence." Putting things another way and seemingly confirming Lewontin: "Within-population differences among individuals account for 93 to 95% of genetic variation; differences among major groups constitute on 3 to 5%" (Rosenberg et al. 2002, 2381).

Nevertheless, correlation does play a major factor. Group differences emerge if you run a cluster analysis across the large sample. People sort into groups that correspond to ethnic sortings. Specifically, geographic Europeans come out as one genetic cluster and Africans come out as another genetic cluster. And as you start to factor in more and more genetic information, the clusters hold up and divisions get ever finer, continuing to map ever finer ethnic and geographical groups. By example, the analysis picks out as anomalous a group in Northern Pakistan. These are the somewhat isolated Kalash, who are believed (by oral tradition) not to be of the same ethnic background as the rest of their countrymen, but to have a European or Middle-Eastern origin. This is confirmed by the study. In short: "Genetic clusters often corresponded closely to predefined regional or population groups or to collections of geographically and linguistically similar populations" (p. 2384). All of this rather makes a mockery of Richard Lewontin's (1972) claim: "Human racial classification is of no social value and is positively destructive of social and human relations. Since such racial classification is now seen to be of virtually no genetic or taxonomic significance either, no justification can be offered for its continuance." At least, even if there are no social reasons for making such classifications, it does not follow that it cannot be done on a sound scientific basis. And even the social reasons have come under fire in recent years. With the completion of the Human Genome Project, we are learning more and more about the genetic basis of many diseases and how some of these diseases are far more common in some groups than in others.

Explicit recognition of the groupings can therefore have immediate and significant results for detection and prevention. Extra special efforts are made to detect prostate cancer in African-American men for example. (That some groups do have atypical potentials for specific genetic diseases is a fact long known. Tay-Sachs disease, an appalling neurological disorder that leads to very early death, is far more common in Ashkenazi Jews than in Sephardic Jews or the general population. Detection procedures have long been in place. What was not known was just how common were group specific diseases.)

There are still those who want to keep fanning the flames of controversy. Philosophers have never been very good at accepting the fact that humans are animals rather than the special creation of a Good God on the Sixth Day. As a hundred years ago there was a steady stream of works showing that life is a mysterious force that can never be the subject of physico-chemical inquiry, so today there is a steady stream of works showing that human sociobiology is impossible. But for the rest, things have rather subsided now. Indeed, those who keep arguing are looked upon more with embarrassment than with respect. Quarrels grow old— they may not be solved, but they get boring. It is true, perhaps trying to distance themselves from past controversies and reflecting discipline affiliations, human sociobiology tends these days to be called "biological anthropology" or "evolutionary psychology" or some such thing. So let us now, in concluding this chapter, turn to the most important question of them all. Criticism and countercriticism, where stands sociobiology today? Where stands, whatever it may now be called, human sociobiology today?

The Contemporary Scene

Animal sociobiology has never really been in question, except it has been attacked as a support for human sociobiology. Let me simply make reference to one celebrated piece of research, a study of the dunnocks or hedge sparrows, small birds that live in the hedgerows and bushes of the English towns and countryside. The British ornithologist Nicholas Davies (1992) has discovered that they have the most remarkable set of sexual customs, something that would not be out of place within the covers of *Playboy*. They have breeding arrangements that go all the way from polygyny (where one male will have two or three mates) through monogamy to polyandry (where one female will have two or three mates) and even to a form of polygynandrous relationship (the polite name for group sex, where several males mate up with several females).

Why? Because then there are selective advantages, given the particular circumstances. If the situation is such that a male can service two or three mates, and the food stuffs are there such that the females can benefit from having an alpha male (or the females and other males cannot prevent the male from acting as he

does), then we get polygyny. And corresponding reasons for the other sexual arrangements. How can Davies be so sure that he is right? Because he has used the most modern of molecular techniques, so-called DNA fingerprinting (the very same that is now used in murder trials), to trace genetic relationships. He finds that he can track, just about exactly, the time that individual birds spend helping with offspring with the genetic relatedness of the males to these offspring. A male who has fathered the whole brood puts in the time to help at the nest—all of his time if it is his only brood, and proportionately if there are others. Conversely, other males give no help—except if there was a chance that they contributed to the brood. Just what one expects given individual selection.

Davies also goes on to discuss the question of parasitism.

The dunnock is a favourite host of the cuckoo in Britain, with about 2% of nests being parasites. Individual female cuckoos specialize on one host species. Experiments with variously coloured model cuckoo eggs show that the degree of host-egg mimicry exhibited by the different cuckoo gentes [*Gens,* plural *gentes,* means a particular group or race related by descent.] reflects the degree of egg discrimination shown by their respective hosts. Unlike other gentes, dunnock-cuckoos do not lay a mimetic egg, as expected from the fact that, in contrast to other hosts, dunnocks show no egg discrimination.

Nevertheless, dunnock-cuckoos still lay a distinctive egg, different in shade from the other cuckoo gentes. Experiments provide no support for predation as an important selective pressure. Either selection by secondary hosts, or by cuckoos themselves (for an egg which is cryptic in the nest) may be involved.

It is unlikely that dunnocks accept nonmimetic eggs because rejection is peculiarly costly for them or of less benefit than for other hosts. Experimental parasitism of species which have no history of interaction with cuckoos shows that before parasitism occurs hosts exhibit no rejection of eggs unlike their own. Dunnocks may, therefore, be recent victims of the cuckoo, lagging behind in their counteradaptations to a new selective pressure. (Davies 1992, 234)

You can see how questions are asked and solved, using natural selection, in a way that would have altogether delighted Darwin. Here is an extension of evolutionary thought—selection-based evolutionary thought—of the most exciting and fertile kind.

But what about humans? Do we really have any significant scientific advances, or is it all a question of hypothesis and supposition and wishing? Do we get anything more than "just so" stories? Let me tell you about one case where the sociobiological approach really does seem to have paid major dividends. It concerns murder or, as the authors call it, "homicide." Two Canadian psychologists, Martin Daly and Margo Wilson (1988), have made an extensive study of homicide: because they are Canadian, they are particularly interested in the differences between homicide in Canada and homicide in the United States. What fascinates them—what fascinates Canadians particularly—is that here we have two countries, with very similar lifestyles, running right next to each other, and yet they have dramatically different homicide rates. The American rates are four to five times higher, or even more.

There are some fairly obvious reasons for this, the most prominent being the availability of guns. By and large, Canadians do not have access to guns, certainly not to handguns, the means by which so many Americans kill each other off (and, to be fair, themselves also). But when it comes to certain kinds of killing, even if the proportions are different, the patterns between Canada and the United States are similar, chillingly similar. In particular, Daly and Wilson concerned themselves with cases of parents killing children. This should not happen in the best-ordered Darwinian worlds—you are stopping your genes in their path. The psychologists hypothesized that perhaps what was happening was that stepparents were doing the killing—especially stepfathers (who are the ones more likely to be living with someone else's children). And the data proved their hypothesis in an incredibly strong fashion. "Daly and Wilson found that step parenthood is the strongest risk factor for child abuse ever identified. In the case of the worst abuse, homicide, a stepparent is forty to a hundred times more likely than a biological parent to kill a young child, even when confounding factors—poverty, the mother's age, the traits of people who tend to remarry—are taken into account" (Pinker 1997, 434). Why is this? "Stepparents are surely no more cruel than anyone else. Parenthood is unique among human relationships in its one-sidedness. Parents give, children take. For obvious evolutionary reasons, people are wired to want to make these sacrifices for their own children, but not for anyone else." The answer is obvious. "The indifference, even antagonism, of stepparents to stepchildren is simply the standard reaction of a human to another human. It is the endless patience and generosity of a biological parent that is special" (p. 434).

These are incredible findings. Moreover no one can accuse Daly and Wilson of twisting the facts to their own end, and even less can you claim that the findings are "obvious" and that you hardly needed a sociobiological perspective to find what they found. So remarkably strong were the biases of social science—supporting the belief that biology has nothing to do with family relationships—that neither in the United States nor in Canada did the authorities keep track of biological versus social parental connections. They simply did not know whether stepparents were more likely to commit violence. Hence, Daly and Wilson had to go out and gather their own data: data that did indeed prove precisely what they predicted. Moreover, the findings about stepparental abuse are backed by other findings about the nature of homicide and the people who do commit it far more often than others. For instance, it turns out that the real killers are young males, who have little to lose and much to gain by violence—precisely what sociobiology predicts. The new enthusiasm for locking people up for long periods of time may indeed have a significant effect on violent crime statistics. It is not that the perpetrators are cured by imprisonment or deterred by the threat of punishment. It is rather that when they get out, they are no longer all that young, with all of that testosterone pumping through their systems.

One example cannot be definitive. There are other equally stunning pieces of research and interpretation. For instance, continuing with the topic of the killing of children, Sarah Hrdy (1999) has looked hard at systematic infanticide, a practice common in the animal world and also it appears more practiced than many suppose (or want to acknowledge) in the human world. She points out that the sex of a child can be very significant when it comes to these things. In India, for instance, among some castes it was very rare indeed for a baby girl to survive. Again one asks: Why would this be so? Isn't there something very non-Darwinian about killing off your own children? Hrdy points out that the answer to that question very much depends on several factors. In particular, there is a well-known (and solidly supported theorem) about the animal world, the Trivers-Willard hypothesis (1973), that states that high-status females tend to have male offspring and low-status females have female offspring. The reason is simple. Females almost always reproduce, and there tends not to be a huge variation in numbers that any one female has (there is some variation, but not by orders of magnitude), but males compete and often just a few have offspring, but these successful males often have very many offspring (by orders of magnitude). Reproductively, a good strategy is to have offspring who are going to succeed. If you are low status your sons will probably not be great successes, so the better strategy is to go female. If you are high status, then your sons have better chances of succeeding so the better strategy is to go male. The coypu, a South American guinea pig-like creature now overrunning parts of Britain, confirms this theorem precisely. High-status females abort female fetuses. Low-status females abort male fetuses. This is all done chemically. In the case of humans, Hrdy argues that we do the same sort of thing through conscious choice. It is always high-status Indians who practice female infanticide. Low-status females look after their daughters and let the sons fend for themselves. The same is true in other parts of the world, notoriously China but also even in Europe among more rural and less sophisticated peoples.

Of course, one or two swallows do not make a summer, but studies like these are the tip of an iceberg. Human sociobiology, evolutionary psychology, can and does work and can throw incredible light on human nature and behavior. Nor is it easy to see that it is infected with all of the faults that critics found endemic of early exercises in human sociobiology. It is certainly not sexist, for instance, and neither is it racist—the figures seem to hold whatever the ethnic group—or classist—the figures are not affected by poverty, for instance. The work is falsifiable and as far as one can see, true not false. In short, a paradigm of good work on problems of social science. Only time can tell whether it will prove to be one of a very few such studies that really work or whether it will prove to be the norm. But for the time being, the future looking promising—one might even say "bright"—for human sociobiology.

Further Reading & Discussion

A good place to start on the general theory of sociobiology is Richard Dawkins's sparkling book, *The Selfish Gene* (Oxford: Oxford University Press, 1976). I would call it a popularization but really it is more than that. The metaphors he uses, especially that of the title, have entered into the scientific discourse and stimulate researchers into looking at problems in altogether new ways. The human side of things is the subject of a provocative essay by Edward O. Wilson: *On Human Nature* (Cambridge: Harvard University Press, 1978), a work for which Wilson deservedly won a Pulitzer Prize. I myself wrote a quick survey of the field dealing not only with the science but with many of the philosophical undercurrents: *Sociobiology: Sense or Nonsense?* 2d ed. (Dordrecht, Holland: Reidel, 1986).

I believe that some of the most interesting and significant implications of sociobiology will be for my own discipline of philosophy, especially trying to answer questions in what is known as epistemology ("What can I know?") and ethics ("What should I do?"). In my *Taking Darwin Seriously: A Naturalistic Approach to Philosophy,* 2nd ed. (Buffalo: Prometheus, 1998), I explore some of these avenues in a preliminary sort of way. This has been a somewhat controversial book—as I have already noted in this chapter, most philosophers are not keen on the idea that evolutionary biology might be the key to unlocking the secrets of their inquiry. So for somewhat different perspectives, turn first to Daniel Dennett's racy (albeit overly long) *Darwin's Dangerous Idea: Evolution and the Meanings of Life* (New York: Simon & Schuster, 1995), a book that managed to offend just about everyone (except me and Richard Dawkins), so it must be saying something right. Then look at *Unto Others: The Evolution and Psychology of Unselfish Behavior* (Cambridge: Harvard University Press, 1998), a work co-authored by philosopher Elliott Sober and biologist David S. Wilson. This is a book that tries to resuscitate the notion of group selection over individual selection, a project in my opinion on a par with King Canute's trying to stop the tide from entering. (Unlike Sober and Wilson, Canute knew that what he was doing was futile and was simply trying to show his sycophantic courtiers that he was not capable of miracles.) You may end by thinking that Sober and Wilson are right and Ruse and Dennett are wrong, but what I want you to see is how modern philosophers of very different convictions are nevertheless turning to evolutionary biology for insight into their philosophical problems. I have gathered together many pertinent discussions in a collection, *Philosophy After Darwin* (Princeton, N.J.: Princeton University Press, 2009).

And in a related way, novelists are also looking at evolutionary biology for insights. *Enduring Love* (Toronto: A. A. Knopf Canada, 1997), by the English Booker Prize–winning novelist Ian McEwan, is a fascinating exploration of sociobiological ideas. The hero is a science writer who is obsessively tracked by a young man who suffers from a form of homoerotic obsession known as de Clérambault's syndrome. The story is the account of how the hero reacts to this pressure: not very well in fact, for he ends up losing his girlfriend and shooting (not fatally) his stalker. But in the course of the account, McEwan explores the ways in which we are all in a sense prisoners of our biology, leading half lives midway between reality and illusion, and how escaping from this state can be dangerous for ourselves and destructive on our relationships. At the same time, however, McEwan shows how this escape can move us to acts of true nobility, beyond our animal natures, and how real love can be achieved. (The title comes from Saint Paul's First Epistle to the Corinthians. "Love bears all things, believes all things, hopes all things, endures all things.")

Finally, I should mention that I discuss McEwan and other creative writers who have turned in some way to evolutionary thinking for insight in my *Darwinism and its Discontents* (Cambridge: Cambridge University Press, 2006).

Chapter 10

Philosophy: Evolution & Thinking About Knowledge & Morality

Overview

In addition to stimulating and interacting with new scientific disciplines—genetics, molecular biology, paleoanthropology, and other areas discussed in earlier chapters—evolutionary thinking impacts on philosophy. Charles Darwin himself realized this, seeing that the mechanism of natural selection has deep and lasting implications both for what is known as "epistemology"—What can I know?—and for what is known as "ethics"—What should I do? This chapter explores how evolution through natural selection leads to new understanding about human nature and about how a process that seems to be focused just on survival and reproduction nevertheless tells us much about precisely those things that make us uniquely human.

There are traditional ways of trying to link evolutionary biology both to our theory of knowledge (epistemology) and to our theory of morality (ethics). In the realm of knowledge, the most obvious way simply argues that the units of knowledge—the ideas or concepts that make up our thinking—are in some sense akin to or analogous to the genes or to individual organisms. Hence just as the latter struggle with some proving fitter than others, so ideas struggle and some prove fitter than others. This ultimately is what truth is all about. In the nineteenth century, the most prominent supporters of this kind of thinking were the American Pragmatists. As the twentieth century drew to its close, the best known thinker of this ilk has been Richard Dawkins, biologist, popular science writer, and new atheist, who argues that culture divides into "memes," units akin to the biologist's "genes."

In ethics, there was likewise an analogical transference of ideas from the biological to the cultural. We have seen that so-called Social Darwinians argued that just as there is a struggle in the world of organisms, so there is a struggle in the social world. We saw also that this did not necessarily translate out into all-out combat (although it could), because people had different ideas about how humans can struggle successfully. Again in the twentieth century we had people who wanted to argue from the biological analogically to the social, finding ethical norms in the process. One who thought this way was Julian Huxley, the grandson of Thomas Henry Huxley. He wanted to promote large technological enterprises in the name of evolution. Another today is Edward O. Wilson. On the basis of evolution, he argues for what we now call an ecological ethic.

Beneath both traditional evolutionary epistemology and evolutionary ethics lie major assumptions about the progressive nature of evolutionary change. It is not something meandering meaninglessly, but directed, going from the simple to the complex, from the valueless to the value-full. Stephen Jay Gould was one of many who find this assumption about the evolutionary process to be very dubious and ill-supported.

Progress supporters fight back vigorously. Some, like Julian Huxley and Richard Dawkins, invoke the idea of an arms race. Perhaps organisms are caught in ongoing battles, with ever-more sophisticated adaptations arising in response to the attacks of opponents. Thus progress occurs as a result. Others, like the Cambridge paleontologist, Simon Conway-Morris, argue that, through Darwinian-fueled evolution, organisms climb up into ever-higher niches and thus progress occurs. As we shall see both of these approaches have obvious weaknesses.

Perhaps a more profitable approach to a Darwinian-based philosophy starts with considering the human brain as a product of natural selection. Could it be that our very ways of thinking are themselves adaptive and thus promoted by natural selection? A number of evolutionary psychologists think just this, although the whole approach has been severely criticized by the Christian philosopher Alvin Plantinga. He thinks the entire program is flawed and he would have us go back to God for justification.

A number of philosophers, including the author, think that Darwin's theory can likewise be applied to moral thinking. We are moral simply because this is adaptive. Humans are social animals and need such an adaptation to get on with each other. This possibly means that ultimately there is no justification for morality, but it does not mean that we can immediately go out and do bad things. Psychologically this is not possible. Our nature, as shaped by natural selection, saves us from the implications of our skeptical philosophy.

The Role of the Scientific Community

The work of the following scientists is discussed in this chapter. Short, biographical essays of these individuals appear in **Biographies** on page 607.

Alvin Plantinga (1932–)
Daniel Dennett (1942–)
Simon Conway-Morris (1951–)

Setting the Stage

"**N**onsense."

Well, that tells you in no uncertain terms what the well-known philosopher Daniel Dennett thinks of an attempt by Edward O. Wilson and me at explaining human moral thinking and behavior in terms of evolutionary principles! Fortunately, we have the courage of our convictions or, if you prefer another interpretation, we are totally insensitive to well-founded criticism. Either way, we remain convinced that evolutionary thinking has great implications for our understanding of humankind—specifically those aspects of human nature that traditionally have attracted the attentions of philosophers. I refer to the theory of knowledge, known technically as "epistemology," and the theory of morality, otherwise known as "ethics." In this chapter I want to look at some of the thinking on these topics, as is my custom, using history to bring us to the present. Now we have reached the stage of the discussion where I am a fairly active participant. I am not going to conceal my views, but the chief aim is to introduce you to the field.

Essay

Traditional Perspectives

Charles Darwin was no philosopher, but thanks to the upper-middle-class education that he received, he was well versed in philosophical issues. He had read Plato as an undergraduate, and by the time he was working on his theory of evolution, he knew the works of the British empiricists like Locke and Hume, not to mention some of the continental thinkers. His older brother, Erasmus, was a man-about-town in London when Charles returned from the *Beagle* voyage. Through his brother, Charles met literary and philosophical figures of the day, encouraging him to dig more deeply into the great issues. It is hard to say how much he kept up with these sorts of things through the years. More and more, Darwin became science-obsessed, admitting that he was leaving literature and culture generally behind. But when things were important, he was prepared to swing out and read more widely. Certainly by the time he came to write *The Descent of Man* in 1871, Charles Darwin's reading was broad enough to include Immanuel Kant's *Metaphysics of Morals.*

Philosopher or not, Darwin never deviated from the rock-solid conviction that his theory of evolution was important for an understanding of both epistemology and of ethics. "He who understands baboon would do more toward metaphysics than Locke." (Notebook *M* 84, Barrett et al, 1987) As it happens, he left little more than a few suggestions about the relevance of evolution for thinking about epistemology. In the *Descent,* he did treat ethics at some length, although

Daniel Dennett

(perhaps expectedly) he was more interested in the evolution of the ethical sense itself rather than in the foundational questions that more philosophical thinkers tend to focus on. For this reason, rather than staying with Darwin for his own sake, it will make more sense to offer a more general discussion of the themes, referring back to Darwin as and when pertinent.

This approach being adopted, what can be said is that there are (probably expectedly) parallels between evolutionary approaches to epistemology and evolutionary approaches to ethics. In particular, both in epistemology (theory of knowledge) and in ethics (theory of morality) we see two different ways of tackling the issues, and in both cases one way is more metaphorical and one way is more literal. As it happens, I am inclined to think that in both cases the metaphorical is rather less satisfactory than the literal, although I rush to qualify by saying that (unlike some, for instance philosopher Jerry Fodor) my objections are not to using metaphor as such but because of other issues that I shall detail and examine. Since the metaphorical way is the more traditional, I shall start there. I will follow the usual pattern of starting with epistemology; although, as it happens (the great Scottish philosopher of the eighteenth century, David Hume, being the notable example), however philosophers may present their results, often their interests are sparked by moral issues and it is only later that they work backwards to problems of knowledge.

Science as a Struggle

The traditional or metaphorical way of applying evolutionary theory to an understanding of knowledge is to regard the ideas of the subject as if they were organisms, and then to bring Darwinian selection to bear on the topic. Basically, one sees a struggle for existence between ideas and the winner emerges. That is what truth is all about. Darwin did not develop this at all, but he certainly adopted a variant of it in the *Descent*:

> The formation of different languages and of distinct species, and the proofs that both have been developed through a gradual process, are curiously the same. But we can trace the origin of many words further back than in the case of species, for we can perceive that they have arisen from the imitation of various sounds, as in alliterative poetry. We find in distinct languages striking homologies due to community of descent, and analogies due to a similar process of formation. The manner in which certain letters or sounds change when others change is very like correlated growth. We have in both cases the reduplication of parts, the effects of long-continued use, and so forth. The frequent presence of rudiments, both in languages and in species, is still more remarkable. The letter *m* in the word *am*, means *I*; so that in the expression *I am*, a superfluous and useless rudiment has been retained. In the spelling also of words, letters often remain as the rudiments of ancient forms of pronunciation. Languages, like organic beings, can be classed in groups under groups; and they can be classed either naturally according to descent, or artificially by other characters. Dominant languages and dialects spread widely and lead to the gradual extinction of other tongues. A language, like a species, when once extinct, never, as Sir C. Lyell remarks, reappears. The same language never has two birthplaces. Distinct languages may be crossed or blended together. We see variability in every tongue, and new words are continually cropping up; but as there is a limit to the powers of the memory, single words, like whole languages, gradually become extinct. As Max Müller has well remarked: "A struggle for life is constantly going on amongst the words and grammatical forms in each language. The better, the shorter, the easier forms are constantly gaining the upper hand, and they owe their success to their own inherent virtue."
>
> To these more important causes of the survival of certain words, mere novelty may, I think, be added; for there is in the mind of man a strong love for slight changes in all things. The survival or preservation of certain favoured words in the struggle for existence is natural selection. (Darwin 1871, 1, 60)

This is less to do with knowledge as such and more with language. Others in the nineteenth century who made more of the analogy, explicitly extending it to knowledge, included the American pragmatist Chauncey Wright not to mention Herbert Spencer. (Spencer's writings are so voluminous and varied that it would be odd if one did not find, somewhere within them, some ideas on anything and everything.) It has been in the twentieth century however that people have made much more of this kind of thinking. I should say that the general tendency has been to focus almost exclusively on scientific knowledge. This is probably a legitimate thing to do. Religious questions aside—these are the topic of the next and

Max Müller

final chapter—scientific knowledge with good reason is usually held to be the firmest and most reliable knowledge that we have. If evolutionary ideas do not apply here, then they probably do not apply anywhere. The English-born philosopher Stephen Toulmin has been at the forefront of this kind of thinking:

> Science develops... as the outcome of a double process: at each stage a pool of competing intellectual variants is in circulation, and in each generation a selection process is going on, by which certain of these variants are accepted and incorporated into the science concerned, to be passed on to the next generation of workers as integral elements of the tradition.
>
> Looked at in these terms, a particular scientific discipline—say, atomic physics—needs to be thought of, not as the contents of a textbook bearing any specific date, but rather as a developing subject having a continuing identity through time, and characterized as much by its process of growth as by the content of any one historical cross-section... Moving from one historical cross-section to the next, the actual ideas transmitted display neither a complete breach at any point—the idea of absolute 'scientific revolutions' involves an over-simplification—nor perfect replication, either. The change from one cross-section to the next is an *evolutionary* one in this sense too: that later intellectual cross-sections of a tradition reproduce the content of their immediate predecessors, as modified by those particular intellectual novelties which were selected out in the meanwhile—in the light of the professional standards of the science of the time. (1967, 465–6)

Another who subscribed to the position was the Austrian-born philosopher, long a professor at the London School of Economics, the late Karl Popper. He is well known for his claim that the mark of genuine science is that it be falsifiable,

Karl Popper

that is to say that it leave itself open to check and to possible refutation. Areas like Freudian psychoanalytic theory, argued Popper, will never let the evidence show them wrong. Hence they are not genuine scientific theories. Theories of physics and chemistry will let themselves be shown wrong—look at what happened to Newtonian mechanics—hence they are genuine science. Popper argued that this theory of falsifiability is essentially Darwinian. You start with a problem, you offer a tentative solution to this problem, a bold conjecture, you open it up to check and if need be rigorous refutation, and then you find yourself with this solution or more likely a modified problem on your hands.

$$P_1 \rightarrow TS \rightarrow RR \rightarrow P_2$$

It is easy to see how things could get even more Darwinian if you offer two tentative solutions TS_1 and TS_2 to the same problem and then let them fight it out—let us say Darwin's theory of pangenesis and Mendel's theory to explain heredity.

A number of historians of science have seized on this evolutionary philosophy with some enthusiasm. David Hull, for instance, has written a deeply insightful book, *Science as a Process,* about a major clash that occurred over biological classification in the 1970s. On the one side, were ranged the traditionalists like Ernst Mayr, the so-called "evolutionary taxonomists," who argued that classification should recognize evolution in all of its facets. Although it may be historically that birds and crocodiles are close, it would be silly to put them together. Birds and crocs have gone off in such different ways. On the other side were the phylogenetic taxonomists, better known as "cladists." (Brief mention was made of them in the discussion of human evolution.) Followers of the German taxonomist Willi Hennig, author of *Phylogenetic Systematics,* they argued that history is all. They devised certain quasi-empirical techniques for making divisions and groupings and went from there. (I say quasi-empirical because, although the techniques were

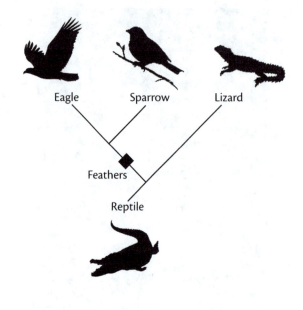

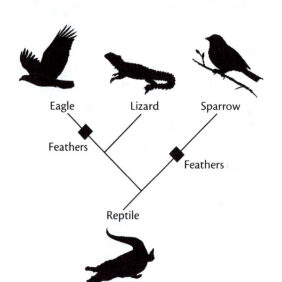

Cladism in action. The upper classification, putting Eagle and Sparrow together, is preferred because it hypothesizes only one move to feathers.

based on reality, they were then formalized into rules allowing no exceptions. For instance, the only information allowed about different species was when one group split into two. Change within a line was not acknowledged nor was it allowed that a split might be three ways or more. Generally, change did come with splitting and in most cases splitting was into two, but this was legislated rather than confirmed empirically.)

The cladists won, decisively. Like an Egyptian plague, they left nothing in their path. Today, if you want to do taxonomy, you do phylogenetic taxonomy. This is the way that things are. Mayr is dead, metaphorically as well as literally. Hull shows in great detail how this happened. There were some things that helped. The coming of computers in an easy-to-use fashion was very important. Cladism lends itself to numerical techniques, counting characteristics and so forth, and computers let you grind up the information in large quantities, do the boring calculations, and spew forth the results quickly. An intuitive assessment, based on years of experience, is helpless before a graduate student with a print out. The cladists also helped themselves. They got into positions of power—in a way taking a leaf from the book of people like Mayr, who in the 1940s and 1950s had got themselves plum university posts as part of the campaign to upgrade evolutionary studies—and used that power ruthlessly. They took over the journals like *Systematic Zoology* and made sure that theirs were the voices heard. And they did so much more. If you want to see nature red in tooth and claw, read Hull's book—a work I might add that was published at considerable personal sacrifice because after it appeared several of the main figures in the story, formerly friends of the author, appalled that their behavior was now public, immediately cut off all social and professional interaction.

There are variants on this approach to epistemology. One recent form that has garnered much attention is that of "memetics," based on an earlier-mentioned suggestion by Richard Dawkins at the end of *The Selfish Gene*. Could it not be, wondered Dawkins, that culture has something equivalent to the units of heredity in biology? Is it possible that, corresponding to the genes, culture might contain its units of heredity, things that Dawkins called "memes?"

> The gene, the DNA molecule, happens to be the replicating entity that prevails on our planet. There may be others. If there are, provided certain other conditions are met, they will almost inevitable tend to become the basis for an evolutionary process.

> But do we have to go to distant worlds to find other kinds of replicator and other, consequent, kinds of evolution? I think that a new kind of replicator has recently emerged on this very planet. It is staring us in the face. It is still in its infancy, still drifting clumsily about in its primeval soup, but already it is achieving evolutionary change at a rate that leaves the old gene panting far behind.

> The new soup is the soup of human culture. We need a name for the new replicator, a noun that conveys the idea of a unit of cultural transmission, or a unit of *imitation*. 'Mimeme' comes from a suitable Greek root, but I want a monosyllable that sounds a bit like 'gene'. I hope my classicist friends will forgive me if I abbreviate mimeme to *meme*. If it is any consolation, it could alternatively be thought of as being related to 'memory', or to the French word *même*. It should be pronounced to rhyme with 'cream'.

> Examples of memes are tunes, ideas, catch phrases, clothes fashions, ways of making pots or of building arches. Just as genes propagate themselves in the gene pool by leaping from body to body via sperms or eggs, so memes propagate themselves in the meme pool by leaping from brain to brain via a process that, in the broad sense, can be called imitation. If a scientist hears, or reads about, a good idea, he passed it on to his colleagues and students. He mentions it in his articles and his lectures. If the idea catches on, it can be said to propagate itself, spreading from brain to brain... When you plant a fertile meme in my mind you literally parasitize my brain, turning it into a vehicle for the meme's propagation in just the way that a virus may parasitize the genetic mechanism of a host cell. And this isn't just a way of talking—the meme for, say, "belief in life after death" is actually real-

ized physically, millions of times over, as a structure in the nervous systems of individual men the world over. (Dawkins 1976, 206–207)

One who has taken up memetics with enthusiasm is the scourge of Wilson and Ruse, Daniel Dennett. Like Dawkins, Dennett dislikes religion intensely, and argues that it is a kind of parasite that infects the brain. Memes have existences of their own and move from mind to mind, like a virus that invades the physical body. In his most recent book, *Breaking the Spell: Religion as a Natural Phenomenon*, Dennett introduces the reader to the lancet fluke (*Dicrocelium dendriticum*). This is an evolutionary marvel (I say this without sarcasm), a parasite that corrupts the brain of an ant, causing it to strive to climb blades of grass, at which point this host gets eaten by a sheep or cow. The fluke is thus able to complete its life cycle before its offspring are excreted and take up again with ants.

> Does anything like this ever happen with human beings? Yes indeed. We often find human beings setting aside their personal interests, their health, their chances to have children, and devoting their entire lives to furthering the interests of an *idea* that has lodged in their brains. The Arabic word *islam* means "submission," and every good Muslim bears witness, prays five times a day, gives alms, fasts during Ramadan, and tries to make the pilgrimage, or *hajj*, to Mecca, all on behalf of the idea of Allah, and Muhammad, the messenger of Allah. Christians and Jews do likewise, of course, devoting their lives to spreading the Word, making huge sacrifices, suffering bravely, risking their lives for an idea. So do Hindus and Buddhists. (p. 4)

Is there a kernel of good sense, a glimpse of the truth, in all of this, or—as is surely the case with someone like Dennett—are we merely turning to evolutionary theory as a fancy cover for expressing our already formed and accepted prejudices? I have nothing against the fact that this is all rather metaphorical. And I certainly want to say that some good things have come from this approach. Right or wrong in his philosophy, Hull was led to write a deeply insightful account of the taxonomic wars. Those named in the dispute found the account upsetting, although, one might grumble that subjectivity and fighting is all one would expect in a field like taxonomy. Even if there are some objective standards and facts—whales really are mammals and not fish—no one could ever think that any major classification was truly objective. Hence, there is going to be a place for pure power politics. If Hull could do for physics what he did for classification, then he might be on firmer ground.

Of course the basic worry one has about this philosophy—the very point its supporters would say we must recognize and accept—is that it reduces all knowledge ultimately to some kind of power politics. If you say that some solutions work better than others, on their own merits, then already you are appealing to something else. As it happened, Popper did precisely this. He was well known for being a "realist," thinking that there is a real world that exists when no one is around, and thus for him solutions work because they correspond in some way to this world (Popper 1972). But whether or not this is true, it is surely to miss the whole point of the Darwinian analogy. Darwinism doesn't care about right or wrong. It doesn't care about getting in touch with the real world for the sake of getting in touch with the real world. It cares about *winning*. It cares about winning to the total exclusion of anything else. For this reason, my suspicion is that if you are going to attack (or defend) this particular kind of epistemology, you are going to have to do more than simply express your prior convictions or prejudices. If you are going to try to show that knowledge is not just a matter of winning, then you are going to have to show that adopting this philosophy challenges other deeply held views that we have. You are going to have to suggest that the costs of the philosophy might be more than people want to bear. I will try to do this in a moment, but first let us turn to traditional evolutionary ethics.

Social Darwinism (redux)

Much earlier in this book, we had one encounter with traditional evolutionary ethics, or as it is usually called "Social Darwinism." Remember that the key move is to suggest that moral worth emerges from a struggle for existence. Humans struggle in society and this is a good thing because better things emerge at the end than if there was no struggle. The metaphor is that what is biologically good is to be taken as what is morally good—at least I regard it as a metaphor, although most of its enthusiasts seem to think it literally true. (Of course, there comes a time when any much-used metaphor is literally true in a sense. Magnets literally attract iron filings, although the use of the word *attraction* obviously comes from the human world of love and friendship. If enough people were to adopt traditional evolutionary ethics then as a matter of language what we mean by "good" would in one sense mean biologically good.)

There has been lots of debate about whether Darwin himself was ever a Social Darwinian. The answer is that sometimes he was a bit and sometimes he wasn't. Others however have gone right down the path. Spencer, Sumner, and company to name a few, although, as we also saw, all sorts of different things were claimed under the same banner—for and against capitalism, for and against militarism, for and against feminism. This tradition continued right into and through the twentieth century. People pushed moral prescriptions in the name of evolution, and these moral prescriptions had a funny way of tallying with their own particular moral beliefs and with the themes and needs and proposed solutions of the age. Take, for instance, Julian Huxley, already introduced as the evolutionary humanist grandson of Thomas Henry Huxley. Although born in the nineteenth century, he was in his prime in the 1930s and 1940s, writing, publishing, broadcasting, and lecturing. Although he was never very much of a scientist himself—one senses he soon got bored with the details and the need for daily slog—he became the public spokesperson for science. And naturally he was led to relate this interest, this passion, to the big problems of the day—first the Great Depression, then World War II, and after that the needs of the Third World. As a matter of fact, one senses a certain insensitivity at the personal level—as a young man Huxley had spent two years on the faculty at Rice University in Houston, and he always had a somewhat condescending attitude towards African Americans and their abilities, and if his behavior towards his wife was any measure, he fell far short of the ideal family man—yet he was a man driven by moral passions, and a consistent pattern or theme emerges. It was the general domain that really excited him.

> All claims that the State has an intrinsically higher value than the individual are false. They turn out, on closer scrutiny, to be rationalizations or myths aimed at securing greater power or privilege for a limited group which controls the machinery of the State.
>
> On the other hand the individual is meaningless in isolation, and the possibilities of development and self-realization open to him are conditioned and limited by the nature of the social organization. The individual thus has duties and responsibilities as well as rights and privileges, or if you prefer it, finds certain outlets and satisfactions (such as devotion to a cause, or participation in a joint enterprise) only in relation to the type of society in which he lives. (Huxley 1931, 138–9)

The key moral principle seems to have been the necessity of planning in running the state and, above all, the application of *scientific* principles and the results of such planning and its implementation. You simply cannot (or should not) leave things to chance or intuition—the implication being that this is precisely where your average politician does leave things—but rather you should bring the trained scientific mind to bear on life's problems. Again and again Huxley returned to this theme. For instance, in a book that he wrote in the inter-war years, *If I Were Dictator*, he stressed the need for science in the running of an efficient state and that such science would need to be of the social variety as well as physico-chemical and biological. During World War II, he wrote a highly laudatory essay on the Tennessee Valley Authority, that marvel of the Rooseveltian New Deal, whereby the federal government built and ran a massive system of river damming and irrigation in what had hitherto been one of the more desolate parts of the United States. Then, after World War II, it was Huxley who insisted on "Scientific" being added to UNESCO (United Nations Educational, Scientific, and Cultural Organization), and he wrote a vigorous polemic arguing that the organization had to be run on evolutionary lines—lines demanding lots of science. So vigorous was his polemic indeed, that he upset his masters and he was refused a full four-year term as director general.

And this was all done in the name of evolution! Thomas Henry Huxley may have had doubts about evolutionary ethics, but not his grandson. Invited to speak in the same lecture series at Oxford in which his grandfather had expressed his doubts about linking evolution and ethics, Julian held forth at length about how all of our moral directives stem from the process of development and change. Our task simply is to continue the journey, leaving things better than when we found them.

Julian Huxley

In the broadest possible terms, evolutionary ethics must be based on a combination of a few main principles: that it is right to realize ever new possibilities in evolution, notably those which are valued for their own sake; that it is right both to respect human individuality and to encourage its fullest development; that it is right to construct a mechanism for further social evolution which shall satisfy these prior conditions as fully, efficiently, and rapidly as possible. (Huxley and Huxley 1947, 136)

Julian Huxley was not the last in his line. Edward O. Wilson espouses exactly the same philosophy. He too thinks that we should act morally and what is moral is what is dictated by evolution. I am not now about to denigrate someone with whom I have written a much-anthologized essay (Ruse and Wilson 1985) on evolution and ethics—always reproduced as a dreadful example to students of how **not** to do philosophy. I will note with some slight amusement that today one of the biggest moral dilemmas that we have is how to stop the destruction of the Brazilian rainforests, how to preserve biological diversity generally, and it just so happens that one of the world's greatest naturalists believes that action in this direction is mandated by the ways of evolution! Wilson argues that we humans have evolved in symbiotic relationship with the rest of nature and that in a world of plastic, quite literally, we would die. This is more than just a practical matter. Wilson argues that we need the biodiversity supplied by the rain forests for practical reasons. Who knows what medicines and other needed products these forests might yield in the future? But he sees our need for biodiversity as an aesthetic, almost a spiritual, thing. In a book published a few years back, he writes: "a sense of genetic unity, kinship, and deep history are among the values that bond us to the living environment. They are survival mechanisms for us and our species. To conserve biological diversity is an investment in immortality" (Wilson 2002, 133).

There is much to praise in the moral prescriptions of both Julian Huxley and Edward O. Wilson. I am not sure that I have quite the enthusiasm for scientific solutions that Huxley had and I am not sure that I have quite the delight in nature that Wilson has. I joke with him, as he sets off in the early morning for a day in the steamy swamps looking for wildlife, that one turtle a summer is quite enough for me. Then I head back to camp for a beer and a good read of a detective story. However, overall their intentions are admirable. I confess nevertheless that I am still with grandfather Huxley, and I am not yet convinced that what is biologically good is necessarily morally good. To go back to Dennett's good friend, the lancet fluke, biologically it is a marvel. Socially, however, it leaves much to be desired. I agree with Huxley and Wilson that there is much to be said for keeping the human race going, but even that is surrounded with qualifications. Suppose the Nazis had won and systematically they set about killing everyone they disliked: not just Jews and Gypsies and gays and the mentally handicapped, but stroppy philosophy professors and promiscuous teenagers and all of the other misfits as judged by National Socialism. Is this a group that should be preserved and cherished no matter what? I hardly think so.

Of course here Huxley and Wilson are going to have exactly the same response as the evolutionary epistemologists. They will tell us that the whole point about the evolutionary approach is that we have got to throw out our prejudices and convictions and think things anew. To which the response, as before, must be: Perhaps so, but before we have a wholesale cleansing, let us see if there are some assumptions being made by the evolutionary ethicists that are not necessarily those to which we truly want to make a commitment. Are there depths thus far unexplored, and when we start looking are things not quite as clear cut and obvious?

Progress

I do not mean to be coy. There are depths and they are obvious. Traditional evolutionary epistemology and traditional evolutionary ethics are united by one underlying assumption. We have been hinting (or more) about it many times in this book, so now let us bring it right out into the open. It is our old friend progress. The belief is that evolution is progressive—it is not a slow meandering process going nowhere but is directed, from the simple to the complex, from the blob to the fully functioning, from the monad to the man. We saw how deeply this idea was embedded in the philosophies of people like Herbert Spencer. It is still with us today. The assumption certainly lies behind the metaphorical approach to evolutionary epistemology. By any measure, science is progressive. Mendel knew more than Darwin; Morgan knew more than Mendel; Watson and Crick knew more than Morgan. If this is not progress, then what is? These epistemologists feel confident in their philosophy because although it is winning that counts, ultimately they believe that winning adds up to something more. And that something more is better—is truer—than what went before. The same is true in the realm of ethics. The reason why people believe that their position does really give the answers is because they believe that evolution yields value. Things are better at the end than they were at the beginning. Huxley was unambiguous.

> When we look at evolution as a whole, we find, among the many directions which it has taken, one which is characterized by introducing the evolving world-stuff to progressively higher levels of organization and so to new possibilities of being, action, and experience. This direction has culminated in the attainment of a state where the world-stuff (now moulded into human shape) finds that it experiences some of the new possibilities as having value in or for themselves; and further that among these it assigns higher and lower degrees of value, the higher values being those which are more intrinsically or more permanently satisfying, or involve a greater degree of perfection.
>
> The teleologically-minded would say that this trend embodies evolution's purpose. I do not feel that we should use the word purpose save where we know that a conscious aim is involved; but we can say that this is the *most desirable* direction of evolution, and accordingly that our ethical standards must fit into its dynamic framework. In other words, it is ethically right to aim at whatever will promote the increasingly full realization of increasingly higher values. (Huxley 1942, 137)

Wilson believes exactly the same thing: "the overall average across the history of life has moved from the simple and few to the more complex and numerous. During the past billion years, animals as a whole evolved upward in body size, feeding and defensive techniques, brain and behavioral complexity, social organization, and precision of environmental control—in each case farther from the nonliving state than their simpler antecedents did" (Wilson 1992, 187). Adding: "Progress, then, is a property of the evolution of life as a whole by almost any conceivable intuitive standard, including the acquisition of goals and intentions in the behavior of animals." From here it is one easy step to arguing that we humans ought to cherish humans and to do this we must preserve the environment.

The question therefore must be: What about progress? What about evolutionary progress, that is? Stephen Jay Gould (1989) was not very keen on it. In a way, it was somewhat paradoxical that he should have been quite as negative as he was. As we learn later in Chapter 12, he toyed with Marxism, had a fondness for *Naturphilosophie*, and (perhaps indeed) there were Spencerian elements in his thinking. Hence, one might have thought him quite favorably disposed to the idea. All three of these philosophies are deeply progressive, with humankind as the culmination at the top. Yet, Gould spent twenty years arguing against precisely this. How could this be? Two points are relevant. First, Gould was not always an opponent of progress. In fact, up to and including the writing of his *Ontogeny and*

Phylogeny in 1977, he was in favor of progress, a process apparently triumphing with our own species. Then Gould swung round against the idea. Which brings in the second point, namely that this was just at the time of the heated human sociobiology debate, a matter on which Gould was as committed negatively as was his colleague Richard Lewontin. Gould saw Wilson's science as being a terrible travesty of the way in which real science should be performed. He saw it as a threat to all that he held sacred and something to be opposed with all his might. Most particularly, Gould saw sociobiology as mixed up with notions of progress. He thought that the idea of and hope for biological progress had been behind many moves to oppress African Americans and Jews and women and others. If there is progress, then some have to be higher than others, and this would be a natural and proper state of affairs. But obviously such a conclusion is unacceptable. Hence, biological progress must be a false hope.

There were various ways that Gould set about opposing biological progress. Some were less than subtle. I have told how Gould fingered Teilhard de Chardin for the Piltdown hoax (Gould 1980b). It is hardly contingent that Teilhard has been one of the last century's greatest boosters of biological progress. If Teilhard could be removed from the scene as a hoaxer, then there surely would be a trickle-down effect against progress. More openly in his campaign against biological progress, Gould was led to write a book, *The Mismeasure of Man* (1981), detailing the ways in which biological approaches to humankind have had a long and ugly history of prejudice and bias. One should expect no more from human sociobiology. (It is worth remembering at this point that Gould himself was Jewish. Much of the book focuses on the ways that, in the early part of the twentieth century in America, IQ tests were used against Jewish immigration.)

Another move by Gould, taken with Lewontin, was to attack the foundations of Darwinism. Since sociobiology is so deeply Darwinian, so deeply adaptationist, then a general attack on this is a particular attack on human sociobiology. Looking ahead to Chapter 12, the papers he wrote were attempts to show that there is more to evolution than adaptation, and hence the very project of human sociobiology—so thoroughly adaptationist—is misconceived. And the point is that inasmuch as one discredits pure Darwinism, one discredits sociobiology, and inasmuch as one discredits sociobiology one discredits biological progress. It is as simple as that. Gould was quite open that he believed in the possibility of social progress and that he saw one of its greatest barriers to be thoughts of biological progress, which latter he took to be deeply Darwinian and very much part of human sociobiology. Hence the real reason for a notorious paper (see Chapter 12) likening many organic features to the functionally useless triangles (spandrels) atop columns of medieval churches.

Arms races and niches

Surely Gould was mistaken in some respects here? If anything, it would be hard to imagine a more non-progressive theory than Darwin's—meaning a theory less likely to support thoughts of biological progress. Darwinism does say that what wins is what wins, and all else is decoration. The lancet fluke is a very unpleasant animal, however is a huge biological success. The great apes are marvels of construction and behavior and much else. They teeter on the edge of extinction. To which, of course, Gould could and would have replied that if this is so, why then have so many Darwinians been so keen on progress? Why do people like Julian Huxley and Edward O. Wilson persist so long in their beliefs? Are they simply schizophrenic, refusing to see the implications of the science with which they work? Such a supposition is not impossible but surely not very likely.

As it happens, some really do not care that much and just plow ahead. Wilson is one. He thinks the progressive nature of the evolutionary process is so obvious that it is not really something to be defended. One suspects that mixed in with his Darwinism are elements of Spencerianism, which certainly took deeper roots in North America than they did in Britain. Others however, hard-line Darwinians, are prepared to offer some arguments for their commitment to progress. One such is Richard Dawkins. He picked up on an idea of Julian Huxley about the prevalence of what today we call "arms races." (Huxley was writing before this term was introduced.) In his first book, *The Individual in the Animal Kingdom*, written before World War I, Julian Huxley drew an analogy between the course of biological evolution and the results of the competition (that Spencer so deplored) between nations in preparation for war. Germany and Britain were competing on the sea, leading Huxley to write: "The leaden plum-puddings were not unfairly matched against the wooden walls of Nelson's day." He then added that today "though our guns can hurl a third of a ton of sharp-nosed steel with dynamite entrails for a dozen miles, yet they are confronted with twelve-inch armor of backed and hardened steel, water-tight compartments, and targets moving thirty miles an hour. Each advance in attack has brought forth, as if by magic, a corresponding advance in defence" (Huxley 1912, 115–116).

Curiously Huxley himself, for all of his enthusiasm for progress, never picked up on this idea as something that would be of use in the search for causes—he opted for a near vitalistic force pushing things upwards—but Richard Dawkins has been unsparing in his enthusiasm for and use of the idea. Obviously arms races lead to a kind of comparative progress, and in Dawkins's view, overall this leads to a kind of absolute progress. He pushes the analogy hard. According to Dawkins, the history of arms races in the last century is highly instructive in this regard. Military strategy depended less on sheer brute force and more on sophisticated weaponry using high-tech electronic equipment. Such artifacts are analogous to the development of organisms' on-board computers, better known as brains. To make his case, Dawkins turns to Harry Jerison's (1973) notion of an Encephalization Quotient (EQ). This is a sort of universal animal IQ, that works from brain size and subtracts the gray matter simply needed to get the body functioning—whales require bigger brains than shrews because they have bigger bodies. The important measure is what is left when you take off the body-functioning portion. Through the lens of this kind of thinking, humans win hands down, leading Dawkins (1986, 189) to reflect: "The fact that humans have an EQ of 7 and

British Battleship

hippos an EQ of 0.3 may not literally mean that humans are 23 times as clever as hippos!" But, he concludes, it does tell us "something."

Dawkins has had more to say about progress elsewhere, invoking the notion of the "evolution of evolvability." Sometimes, you just get evolutionary breakthroughs—like the eukaryotic cell—that have more potential, and hence evolution has made a jump to a new dimension.

> There really is a good possibility that major innovations in embryological technique open up new vistas of evolutionary possibility and that these constitute genuinely progressive improvements (Dawkins 1989; Maynard Smith and Szathmáry 1995). The origin of the chromosome, of the bounded cell, of organized meiosis, diploidy and sex, of the eucaryotic cell, of multicellularity, of gastrulation, of molluscan torsion, of segmentation—each of these may have constituted a watershed event in the history of life. This is not just in the normal Darwinian sense of assisting individuals to survive and reproduce, but watershed in the sense of boosting evolution itself in ways that seem entitled to the label "progressive." It may well be that after, say, the invention of multicellularity, or the invention of metamerism, evolution was never the same again. In this sense, there may be a one-way ratchet of progressive innovation in evolution. (Dawkins 1997, 1019–1020)

As always, computer technology provides the analogy.

Computer evolution in human technology is enormously rapid and unmistakably progressive. It comes about through at least partly a kind of hardware/software coevolution. Advances in hardware are in step with advances in software. There is also software/software coevolution. Advances in software make possible not only improvements in short-term computational efficiency—although they certainly do that—they also make possible further advances in the evolution of the software. So the first point is just the sheer adaptedness of the advances of software make for efficient computing. The second point is the progressive thing. The advances of software open the door—again I wouldn't mind using the word "floodgates" in some instances—open the floodgates to further advances in software. (Ruse 1996, 469, from a presentation given in Melbu, Norway, in 1989)

Evolution is cumulative, for it has "the power to build new progress on the shoulders of earlier generations of progress." And brains, especially the biggest and best brains, are right there at the heart, or (perhaps we should say) end: "I was trying to suggest by my analogy with software/software coevolution, in brain evolution that these may have been advances that will come under the heading of the evolution of evolvability in [the] evolution of intelligence."

Coming from a very different perspective—a right-wing Christian trying to make the case for the inevitability of humans given the evolutionary process—Cambridge paleontologist (famous for his work on Cambrian fossils found in the Canadian Burgess Shale) Simon Conway Morris, in his *Life's Solution: Inevitable Humans in a Lonely Universe* (2003), has tried another tack, although he too wants to stay strictly within the Darwinian compound. His basic starting position is that only certain areas of potential morphological space are going to be capable of supporting functional life. To this, he adds the assumption that selection is always pressing organisms to look for such spaces. Hence, sooner or later they will be occupied—probably sooner rather than later, and probably many times. Conway Morris argues that this is not just wishful thinking because life's history shows an incredible number of instances of convergence—instances where the same adaptive, morphological space has been occupied again and again. The most dramatic perhaps is that of saber-toothed, tiger-like organisms, where the North American placental mammals (real cats) were matched item for item by South American marsupials (thylacosmilids). It is beyond doubt that there existed a niche for organisms that were predators—predators of a particular kind, with cat-like abilities and shearing/stabbing-like weapons—and natural selection found more than one way to enter it. Indeed, it has been suggested, long before the mammals, the dinosaurs might also have found this niche.

Conway Morris claims that this sort of thing happens repeatedly. Hence, one must conclude that the historical course of nature is not random but strongly selection-constrained along certain pathways and to certain destinations. From this, Conway Morris infers that movement up the order of nature, the chain of being, is bound to happen. The appearance of some kind of intelligent being (what has been termed a "humanoid") is no chance. It had to emerge. Our own existence is the best possible proof that a kind of cultural adaptive niche exists—a niche where intelligence and social abilities are the defining features. And we know full well that this niche is no freak, in the sense that we arrived at it by chance and were then trapped. Many other organisms have (with greater or lesser success) aspired to occupy this niche. We know of the kinds of features (like eyes and ears and other sensory mechanisms) that have been used by organisms to enter new niches; we know that brains have increased as selection presses organisms to ever new

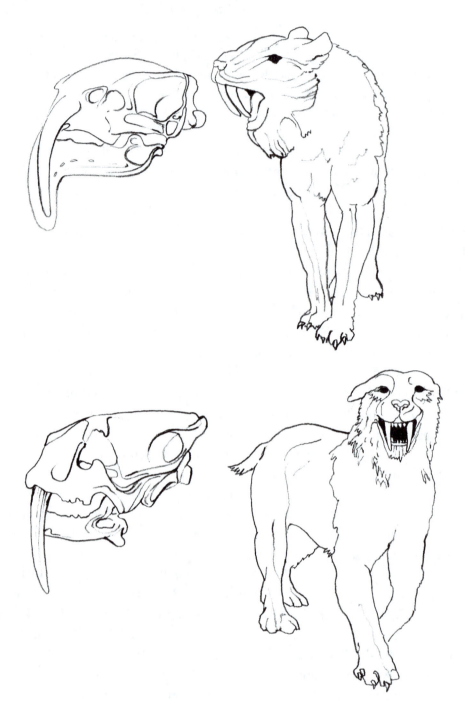

Saber-toothed Tigers—marsupial, top; placental, bottom

and empty niches; and we know that, with this improved hardware, have come better patterns of behavior and so forth (more sophisticated software). Is not the conclusion staring us in the face?

If brains can get big independently and provide a neural machine capable of handling a highly complex environment, then perhaps there are other parallels, other convergences that drive some groups towards complexity. Could the story of sensory perception be one clue that, given time, evolution will inevitably lead not only to the emergence of such properties as intelligence, but also to other complexities, such as, say, agriculture and culture, that we tend to regard as the prerogative of the human? We may be unique, but paradoxically those properties that define our uniqueness can still be inherent in the evolutionary process. In other words, if we humans had not evolved then something more-or-less identical would have emerged sooner or later. (p. 196)

What does one say about these arguments? Obviously they are not stupid, but there are gaps in them through which one could drive a Sherman tank. Some of the worries are empirical. Although some Darwinians think that arms races are ubiquitous and important and effective, others are not quite so sure. For instance, at the empirical level, there are serious questions based on the fossil evidence about the classic, supposed arms race—predators and prey engaged in arms races for greater speed. Is it really the case that lions get faster and so also do antelopes? The evidence is not definitive. At the more conceptual level, some have questioned whether niches just sit there waiting to be occupied as Conway Morris supposes. To a certain extent—some would say to a great extent—organisms create the niches they occupy. Beavers, for instance, build dams so they can occupy lakes—but before the beavers there are often no lakes. Most important, obviously, one runs into the questions about what constitutes "better" in any absolute way. We have been here before, so I will not linger. The simple fact of the matter is that big brains are very good in many circumstances, but they are expensive—you need lots of protein to keep them functioning—and often there are other adaptations that would be much more sensible. Adaptations are always relative to needs, and needs are never absolute. It is as simple as that. Ask the lancet fluke. Ask the chimpanzee.

In a way, the interesting question is why biological progress has such a hold on people's imaginations. I doubt Gould was right putting it all down to the desire to oppress peoples of those groups to which one does not belong. More likely I suspect is the fact that we are humans and we want to find ourselves winners. In the case of scientists, the simple fact that if as a scientist you do not believe in progress of one sort or another you will never get anywhere. Unless you think there are answers and you can find them and do better than others, you might as well take up poetry or philosophy. I am sure that thoughts of progress in one realm slop over to thoughts of progress in others. And, of course, there is at work something like Descartes asking whether he exists or not. As soon as he asks it, he realizes that he must exist. *Cogito, ergo sum.* If we ask if there is progress, at once we know we are at the end of the line, the (or a) final product of evolution, and that we have the ability to ask, is there progress? What more does one need to tip one into thinking that one has won?

Taking Darwin Literally

Is this the end of the story? Is there no more to be said about evolution and the theory of knowledge and the theory of morality? Well, obviously not in one sense. At the very least, one can mount an empirical inquiry, looking into the ways in which we come to gain knowledge and why, looking also into the ways in which we interact socially and how morality fits into all of this. In other words, one can do the sort of thing that Darwin did for ethics in *The Descent of Man*.

In the past twenty or thirty years, a huge amount of effort has been directed to these ends. The evolutionary psychologists particularly have been interested in knowledge claims. Darwin had the insight, but he never followed it up: "Plato... says in Phaedo that our '*imaginary ideas*' arise from preexistence of the soul, are not derivable from experience. ...read monkeys for preexistence." (Notebook M. 121, Barrett et al 1987, 558) More recently, people have been looking at knowledge claims and how we come to make them. John Tooby and Leda Cosmides have been pioneers here (Cosmides 1989; Tooby et al. 2005). The basic idea is really quite simple. Assuming that there is a real world out there, generally speaking one is better off if one knows about it rather than otherwise. If it is hot, it is silly to think that it is cold and conversely. If a tiger is bearing down on you, it is silly to think it is a panda bear, and conversely. If meat stinks to high heaven, it is silly to think it smells fresh, and conversely. Of course, much knowledge is more than a simple matter of observation. It demands reasoning. Two tigers went into the cave. One came out. Should you go on in? You go down to the water hole at the end of the day, and see brushes crumpled, footsteps in the sand, growling in the undergrowth. Do you say, "Tigers, let's leave." Or do you say, "Tigers, just a theory not a fact."

Tooby and Cosmides follow these sorts of things right through, arguing that things get interesting when life is not straightforward, and when for various reasons your thinking plays tricks on you. Cosmides especially has been interested in those puzzles that so delight psychologists. Particularly pertinent are those paradoxes where humans perform well on one task and badly on another task, even though formally they are identical. Take for example the Wason test: Given four cards, with a number on one side and a letter on the other, and the distribution D, F, 3, 7, which cards must you turn over to see if the following rule holds true: "If a card has a D on the one side, it must have a 3 on the other"? Now try this one: "Given cards corresponding to four drinkers in a bar—beer, lemonade, 25 years old, 16 years old—and if the bar bans under 18 year old drinking, which equivalent cards must you turn over to see that no one is breaking the law?" Everyone gets the second problem right but most people flunk the first. Why? Simply because, in everyday life, we much more commonly encounter the boozing-type situation than the abstract number-letter situation, and so are better at solving it. In other words, Cosmides argues (what we might expect) the brain is not a simple all-purpose computer but one that reflects the needs of its possessors as they strive for success in life's struggles.

This is work at the empirical level. But we did set out to ask philosophical questions. Suppose that the science of Tooby and Cosmides and fellow workers all succeeds and gives us genuine information about human nature. Suppose, for instance, that those pro-humans who said $2-1=1$ rather than $2-1=0$ survived and reproduced, and those that did not did not. Suppose that those who took circumstantial evidence seriously survived and reproduced, and those that did not did not. Suppose that those humans who could work out who should be drinking in a bar survived and reproduced, and those who could not did not. Suppose that it really did not matter at all whether we could solve the abstract game with letters and numbers, and so it made no difference to survival and reproduction. What does this tell us about knowledge? Does it tell us that most of the time we really are in touch with a real world, and those times that we are not there are good reasons why. Probably that is true. But what if someone says, "Yes, but how do you know you are not being systematically deceived? How do you know that the world is really that way?" If selection does not care about you being deceived once, perhaps it does not care about you being deceived a great many times. So long as it all works pretty well, that is enough for selection.

The philosopher Alvin Plantinga (1991) expresses a worry like this. Imagine we are in a factory making those mythical objects known as widgets, and suppose that these widgets are all red. If a supervisor were to tell us that the widgets seem red because, to find cracks and other defects, the factory is bathed in red light, then obviously you will think yourself deceived. The widgets are not really red. They are cream-colored, say. This is hardly troublesome in the sense of upsetting to someone trying to use knowledge of our evolution to arrive at estimates of the reliability of our knowledge claims. But suppose now that supervisor's boss tells you that the supervisor is a liar or hallucinating. Then you really do get stuck about the redness of the widgets. An observer "doesn't know what to believe about those alleged red lights." Ultimately "she will presumably be agnostic about the probability of a widget being red, given that it looks red; she won't know what the probability might be; for all she knows it could be very low, but also, for all she knows, it could be very high." We are in the same position with respect to evolutionary theory. Perhaps the whole thing is deceiving us, and even the basic beliefs against which we judge false or misleading beliefs are themselves unreliable. Then we really are in a skeptical mess. Plantinga makes reference to David Hume, arguing that: "What we really have [is] one of those nasty little dialectical loops to which Hume draws our attention." And he quotes Hume: "'Tis happy therefore, that nature breaks the force of all skeptical arguments in time, and keeps them from having any considerable influence on understanding."

For Plantinga, a deeply committed Christian who loathes and detests Darwinian evolutionary biology, that is an end to matters. Bring on God to guarantee our beliefs about the real world. But what if (even though you might be a Christian) you would prefer not to bring in God too quickly? Then it seems that you are probably going to have to agree to some kind of pragmatic solution. Ultimately we can say why we believe in certain things, but ultimately equally we cannot give any absolute guarantees. The great American philosopher of the twentieth century, Willard Van Orman Quine, wrote about the problem of induction, the problem of why it is reasonable to think that the future will be like the past:

> One part of the problem of induction, that part that asks why there should be regularities in nature at all, can, I think, be dismissed. *That* there are or have been regularities, for whatever reason, is an established fact of science; and we cannot

Alvin Plantinga

ask better than that. *Why* there have been regularities is an obscure question, for it is hard to see what would count as an answer. What does make clear sense is this other part of the problem of induction: why does our innate subjective spacing of qualities accord so well with the functionally relevant groupings in nature as to make our inductions come out right? Why should our subjective spacing of qualities have a special purchase on nature and a lien on the future?

There is some encouragement in Darwin. If people's innate spacing of qualities is a gene-linked trait, then the spacing that has made for the most successful inductions will have tended to predominate through natural selection. Creatures inveterately wrong in their inductions have a pathetic but praise-worthy tendency to die before reproducing their kind. (Quine 1969, 162)

That is all you can say, and incidentally that is about all Hume would have had us say. Ultimately, philosophical inquiry leads to skepticism, from which fortunately our psychology rescues us.

The intense view of these manifold contradictions and imperfections in human reason has so wrought upon me, and heated my brain, that I am ready to reject all belief and reasoning, and can look upon no opinion even as more probable or likely than another. Where am I, or what? From what causes do I derive my existence, and to what condition shall I return? Whose favour shall I court, and whose anger must I dread? What beings surround me? And on whom have, I any influence, or who [has] any influence on me? I am confounded with all these questions, and begin to fancy myself in the most deplorable condition imaginable, environed with the deepest darkness, and utterly deprived of the use of every member and faculty.

Most fortunately it happens, that since reason is incapable of dispelling these clouds, nature herself suffices to that purpose, and cures me of this philosophical melancholy and delirium, either by relaxing this bent of mind, or by some avocation, and lively impression of my senses, which obliterate all these chimeras. I dine, I play a game of backgammon, I converse, and am merry with my friends; and when after three or four hours' amusement, I would return to these speculations, they appear so cold, and strained, and ridiculous, that I cannot find in my heart to enter into them any farther. (Hume 1978, 1, 7)

Darwin incidentally used to play endless games of backgammon with his wife, Emma, keeping score over many years.

Justice as Fairness

What about ethics? Really, given evolution through natural selection and the intention to apply this theory literally, the basic approach is much as with epistemology. One is going to look at moral behavior and thinking from an adaptive perspective. Getting on with your fellow humans is a good thing rather than otherwise. This was very much Darwin's approach in the *Descent*.

> It has often been assumed that animals were in the first place rendered social, and that they feel as a consequence uncomfortable when separated from each other, and comfortable whilst together; but it is a more probable view that these sensations were first developed, in order that those animals which would profit by living in society, should be induced to live together, in the same manner as the sense of hunger and the pleasure of eating were, no doubt, first acquired in order to induce animals to eat. The feeling of pleasure from society is probably an extension of the parental or filial affections, since the social instinct seems to be developed by the young remaining for a long time with their parents; and this extension may be attributed in part to habit, but chiefly to natural selection. With those animals which were benefited by living in close association, the individuals which took the greatest pleasure in society would best escape various dangers, whilst those that cared least for their comrades, and lived solitary, would perish in greater numbers. With respect to the origin of the parental and filial affections, which apparently lie at the base of the social instincts, we know not the steps by which they have been gained; but we may infer that it has been to a large extent through natural selection. (1, 80)

One thing that did worry Darwin was (as we know already) answering questions about the level at which selection operates. Morality seems to be such a group sort of thing. If I sacrifice my life for yours, I am really not doing much to further my own self-interests, my genes as we would say. In part in the end he wondered if here group selection might be at work. But immediately he suggested that we might have a case of what we have seen called "reciprocal altruism." You scratch my back and I will scratch yours.

Today, biological anthropologists and evolutionary psychologists show much interest in the evolution of morality. They argue that essentially morality is much as you would expect if it is an adaptation. It pays to get on with others and so we do, but not as suckers or infinite givers. We expect something in return, not necessarily consciously but instinctively. Quine's colleague at Harvard, ethical philosopher the late John Rawls (1971), used to argue that the supreme principle of

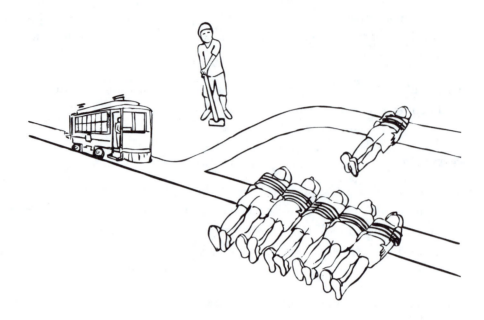

Trolley Problem

morality was "justice as fairness." We ought to be just and to be just is to be fair. This is not to say that we treat everyone identically, but that we treat others as we might want to be treated ourselves. If I am handicapped then I need and expect more help than otherwise, but also I feel the obligation to do or offer the same to others. Again, as with the Wason test in epistemology, we get those nagging paradoxes in ethical behavior that the empirical researchers think cast important light on the nature and origin of moral thinking. Consider a conundrum much discussed these days by moral philosophers—the trolley problem. Suppose you are down a mine by a rail track and you see a laden, unmanned truck coming your way. Down the track beyond you stand five people who will die if you do not do something. Fortunately you are standing by some points and can switch the rails so the trolley is diverted to a sideline, where it will kill only one person. Most people unhesitatingly say you should switch. Suppose now, however, the situation is the same except there is no sideline and you are standing next to a large overweight person. You are so small that if you threw yourself on the tracks it would make no difference, but your neighbor is big enough so that if you pushed him, he would stop the trolley, albeit at the cost of his life. Would you do this? Most people say no, even though the situations are formally the same. One life sacrificed would save five. Why are we willing to pull the switch but not to push the neighbor? Ethicists now suggest that in our evolution we have been primed to care about neighbors because then they will care about us, but abstract reasoning is another matter. In other words, there is no logical or rational reason for our behavior. It is all emotion, as designed by natural selection. Interestingly it turns out that we use different parts of the brain to make these conflicting decisions, grist for the Darwinian mill. (See Singer 2005 for a fascinating discussion of these points.)

Finally, let us ask the same question we asked about epistemology. What does any of this have to do with truth and falsity? Suppose it is the case that thanks to my evolution through natural selection I evolved the sentiment that killing humans for fun is wrong. Does that mean that it really is wrong to kill humans for fun? Or is this just some sort of convention put in place by our genes? Although by now you will realize that this is not really the position of Edward O. Wilson, this was the position I had rather dragooned him into saying in one of our jointly authored papers and it was to this that Daniel Dennett crisply responded: Nonsense! So perhaps I had better bring this chapter to an end by leaving the answering of the question as an exercise for the reader. My personal feeling is that once you have the Darwinian explanation of morality out on the table, there is not much more to be said. Ultimately I don't think there is any more to prohibitions against rape and pillage. This does not mean that I am urging you to go out at once and start raping and pillaging, or that I think you can. Apart from anything else, others will have something to say on the matter. More importantly, evolution has made us so that we want to be moral—most of the time anyway (and when we are eyeing someone else's manservant or maidservant, there are good reasons for that too)—and breaking with morality makes us feel very uncomfortable. It is known as conscience.

The great Russian novelist Fyodor Dostoevsky, although hardly an evolutionary ethicist, knew all about these things and portrayed them dramatically in *Crime and Punishment*. The student Raskolnikov wants to be a Napoleon of crime, beyond morality, and kills an old woman and her sister. The police detective knows full well that he is the culprit but waits until Raskolnikov can stand it no longer and confesses. David Hume also knew all about these things. There is no reason to ultimate moral commitments. "It is not contrary to reason to prefer the destruction of the whole world to the scratching of my finger" (Hume 1978, 2, 3, 3). It is all a matter of feelings or emotions. But, as the Darwinian will tell you, that's what being human is all about.

Further Reading & Discussion

In my *Taking Darwin Seriously: A Naturalistic Approach to Philosophy,* 2nd ed. (Buffalo: Prometheus, 1998) I explore some of these philosophical avenues in a preliminary sort of way. This has been a somewhat controversial book—most philosophers are not keen on the idea that evolutionary biology might be the key to unlocking the secrets of their inquiry. I have gathered together many pertinent discussions in a collection: *Philosophy After Darwin: Classic and Contemporary Readings* (Princeton: Princeton University Press, 2009). This really covers the spectrum, from Darwin to the very latest findings in evolutionary psychology and their applications to the perennial problems of philosophy. On the side of epistemology, I recommend Karl Popper's autobiography (worth reading in its own right), *Unended Quest: An Intellectual Autobiography* (La Salle, Ill.: Open Court, 1976). I also strongly recommend a book I talked about in this chapter, David Hull's account of the taxonomic wars, *Science as a Process* (Chicago: University of Chicago Press, 1988). For a somewhat different perspective turn to Daniel Dennett's racy (albeit overly long) *Darwin's Dangerous Idea: Evolution and the Meanings of Life* (New York: Simon & Schuster, 1995), a book that managed to offend just about everyone (except me and Richard Dawkins), so it must be saying something right.

On the ethics side, Peter Singer has written a couple of really good, clear books: *The Expanding Circle: Ethics and Sociobiology* (New York: Farrar, Straus, and Giroux, 1981) and then a short work trying to show that the critics of sociobiology who claim that it is bound to be right-wing are just plain wrong: *The Darwinian Left: Politics, Evolution, and Cooperation* (New Haven: Yale University Press, 2000). A corresponding book arguing that Darwinism justifies right-wing thinking (and a good thing too!) is Larry Arnhart's *Darwinian Conservativism* (Exeter, U.K.: Imprint Academic, 2005). Also look at *Unto Others: The Evolution and Psychology of Unselfish Behavior* (Cambridge: Harvard University Press, 1998), a work co-authored by philosopher Elliott Sober and biologist David S. Wilson. This is a book that tries to resuscitate the notion of group selection over individual selection, a project in my opinion on a par with King Canute's trying to stop the tide from entering. (Unlike Sober and Wilson, Canute knew that what he was doing was futile and was simply trying to show his sycophantic courtiers that he was not capable of miracles.) You may end by thinking that Sober and Wilson are right and Ruse and Dennett are wrong, but what I want you to see is how modern philosophers of very different convictions are nevertheless turning to evolutionary biology for insight into their philosophical problems.

Chapter 11
Evolutionary Development: Minimizing Natural Selection

Overview

This chapter explores how, as scientists were increasingly able to refine their genetic research, new discoveries resulted in the emergence of a new field of biology called Evolutionary Development. These biologists argue that evolution, or organism change, happens at the embryonic level or deeper and can happen without natural selection. Once again Darwinism finds itself at war with a new science.

Well before Darwin, embryology was important and controversial in evolutionary thinking. It continued so after *The Origin of Species* was published in 1859. For Darwin himself, embryology was important because he could show the importance of natural selection in the evolutionary process. Embryonic similarities of organisms that then grow into very different adults are due to the fact that the selective pressures on the young are similar, while these pressures on adults are different. Germanic-type thinkers, however, used embryology to discern pathways, particularly through Ernst Haeckel's so-called biogenetic law, "ontogeny recapitulates phylogeny."

Although we will see, in the following pages, that the "original" biogenetic law had too many exceptions to be really successful , embryology has roared back into evolutionary studies in recent years, thanks to work at the molecular level. Most stunning are discoveries that there are molecular homologies between genes controlling development in both humans and fruitflies. It is clear that organisms are built on the Lego principle—the same parts and processes are put together in different ways to make different organisms.

Is this the clue to something deeper and more profound? Could it be that organisms are built by purely physico-chemical processes like the evolutionary development field claims, without the help or need of natural selection? In the tradition of the Scottish morphologist of the early twentieth-century, D'Arcy Wentworth

Thompson, there are those who argue just this. Nature yields "order for free." Darwinism is unneeded.

Expectedly, Darwinians are unimpressed. They agree that we are going to find constraints on development, but they disagree that this means the end of natural selection. They agree that a phenomenon like phyllotaxis, the patterns revealed by many plants as they grow, is indeed something that is ruled by fairly complex mathematical formulae, but they disagree that this means the end of selection, claiming that, as they strive for adaptive excellence, organisms must still succeed in the struggle for existence—selection at work.

In fact, Darwinians argue that today we have more and more evidence of natural selection at work. The deservedly celebrated, long-term study of Peter and Rosemary Grant on the beak-size of finches on the Galapagos Archipelago makes it clear that adaptation is the key to understanding.

Today's Darwinians go beyond direct studies and look at nature in more subtle ways. Very important are "optimality models," where one assumes adaptive excellence and then studies nature in this light. Although there are critics of this practice, the work of people like Edward O. Wilson on the caste distributions among the leaf-cutter ants is highly instructive and shows the great success of the models. Optimal models are part of the overall reason why today's Darwinians think the evidence for natural selection and its importance has never been stronger.

The Role of the Scientific Community

The work of the following scientists is discussed in this chapter. Short, biographical essays of these individuals appear in **Biographies** on page 607.

D'Arcy Wentworth Thompson (1860–1948)
Brian Goodwin (1931–)
Jerry Fodor (1935–)
Stuart Kauffman (1939–)
Peter and Rosemary Grant

Setting the Stage

The homologies of process within morphogenetic fields provide some of the best evidence for evolution—just as skeletal and organ homologies did earlier. Thus, the evidence for evolution is better than ever. The role of natural selection in evolution, however, is seen to play less an important role. It is merely a filter for unsuccessful morphologies generated by development. Population genetics is destined to change if it is not to become as irrelevant to evolution as Newtonian mechanics is to contemporary physics. (Gilbert, Opitz, and Raff 1996, 368)

Oh my goodness! Here we go again! These three writers are at the forefront of a new subdiscipline in the evolutionary spectrum known as "evolutionary development" or "evo-devo" for short. No sooner does a new biological idea or discovery arrive on the scene—Mendelian genetics, newly refurbished paleontology, you name it—and the Darwin-bashing index is pushed up one more notch. Philosopher Jerry Fodor joins in the fun. Talking about what structures the physical form of organisms (their "phenotypes"), Fodor suggests that natural selection producing adaptively fine-tuned beings may be the wrong answer. It could all be a matter of development as brought on by the genes.

> External environments are structured in all sorts of ways, but so, too, are the insides of the creatures that inhabit them. So, in principle at least, there's an alternative to Darwin's idea that phenotypes 'carry implicit information about' the environments in which they evolve: namely, that they carry implicit information about the endogenous structure of the creatures whose phenotypes they are. This idea currently goes by the unfortunate soubriquet 'Evo-Devo' (short for 'evolutionary-developmental theory'). Everybody thinks evo-devo must be at least part of the truth, since nobody thinks that phenotypes are shaped directly by environmental variables. Even the hardest core Darwinists agree that environmental effects on a creature's phenotype are mediated by their effects on the creature's genes: its 'genome'. (Jerry Fodor, "Why Pigs Don't Have Wings", London Review of Books, 18 October, 2007, 29(20), 19–22)

Fodor continues, suggesting that it is really the genes that do the donkey work, not something from outside:

> Indeed, in the typical case, the environment selects a phenotype by selecting a genome that the phenotype expresses. Once in place, this sort of reasoning spreads to other endogenous factors [that is factors that come from *inside* the organism]. Phenotypic structure carries information about genetic structure. And genotypic structure carries information about the biochemistry of genes. And the biochemical structure of genes carries information about their physical structure. And so on down to quantum mechanics for all I know. It is, in short, an entirely empirical question to what extent exogenous variables [variables that come from *outside* the organism] are what shape phenotypes; and it's entirely possible that adaptationism is the wrong answer. (p. 21)

As is our wont, let us start with some history and move toward the present.

Essay

Embryology

Embryology and history were an item long before Darwin wrote on the topic in *The Origin of Species*. Remember Louis Agassiz with his ideas about the three-fold parallel: between the history of life, the history of the individual, and the spectrum of organisms as it exists today? He was not the first to think in these terms. Even in the eighteenth century we start to find people who argue that life's history has the same kind of developmental momentum as we find in the individual. The philosopher Hegel spoke for many:

> Nature is to be regarded as a *system of stages*, one arising necessarily from the other and being the proximate truth of the stage from which it results: but it is not generated *naturally* out of the other but only in the inner Idea which constitutes the ground of Nature. *Metamorphosis* pertains only to the Notion as such, since only its alteration is development. But in Nature, the Notion is partly only something inward, partly existent only as a living individual: *existent* metamorphosis, therefore, is limited to this individual alone. (Hegel 1970, 21)

Others, perhaps including the aged Goethe, probably crossed the evolutionary (in our sense of the term) divide. But the real point for everyone was that there is a parallel between the individual and the group, and both are, as it were, caught in an inevitable thrust upwards, to the completed individual or to the completed (meaning modern, complex) group. There is more than a hint of teleology about all of this, as nature unfolds and takes its course to a much better and developed end than from which it set out.

Although it is true that Darwin's writings sometimes reflect this kind of thinking—perhaps *acknowledge* would be a better word—essentially he broke entirely from this way of thinking about the past. Darwin wanted nothing to do with any sort of world spirit bringing all to fruition, of any kind of inner momentum, so beloved of Goethe and other *Naturphilosophen*. For Darwin, embryology was (as we have seen) very important, but it was to play a key role in *his* theory, not in the theories of others. Embryology was to be an essential lynchpin in the case for natural selection. And so it was in the *Origin*. First there is the most striking fact that embryos of organisms widely different as adults are frequently very similar if not virtually identical:

> It has already been casually remarked that certain organs in the individual, which when mature become widely different and serve for different purposes, are in the embryo exactly alike. The embryos, also, of distinct animals within the same class are often strikingly similar: a better proof of this cannot be given, than a circumstance mentioned by Agassiz, namely, that having forgotten to ticket the embryo of some vertebrate animal, he cannot now tell whether it be that of a mammal, bird, or reptile. The vermiform larvae of moths, flies, beetles, & c., resemble each other

Goethe

much more closely than do the mature insects; but in the case of larvae, the embryos are active, and have been adapted for special lines of life. (p.439)

Then comes the explanation. The adults are ripped apart by natural selection. The young feel no such pressures, protected as they so often are, and hence they stay similar. In an adaptationist passage that would no doubt send shudders down Jerry Fodor's spine, were he ever to read this far into the *Origin*, Darwin pointed out that animal breeders only care about the adult forms, and hence expectedly the juveniles are often very similar even though the adults are very different.

> As the evidence appears to me conclusive, that the several domestic breeds of Pigeon have descended from one wild species, I compared young pigeons of various breeds, within twelve hours after being hatched; I carefully measured the proportions (but will not here give details) of the beak, width of mouth, length of nostril and of eyelid, size of feet and length of leg, in the wild stock, in pouters, fantails, runts, barbs, dragons, carriers, and tumblers. Now some of these birds, when mature, differ so extraordinarily in length and form of beak, that they would, I cannot doubt, be ranked in distinct genera, had they been natural productions. But when the nestling birds of these several breeds were placed in a row, though most of them could be distinguished from each other, yet their proportional differences in the above specified several points were incomparably less than in the full-grown birds. (p.445)

Jerry Fodor

One is hardly surprised that Darwin was particularly pleased with this explanation. It was an absolutely beautiful application–not just of the idea of evolution but of the causal concept of natural selection.

Unfortunately, as we learnt earlier in this book, Darwin's thinking at this point fell on deaf ears. His great supporters, Thomas Henry Huxley in particular, were far more interested in pushing Darwinism as a world picture than in using natural selection as a precise tool of causal inquiry. Big hypothetical pictures of life's past were the order of the day, not precise little experiments showing selection in action. It is no wonder therefore that, for all people may have thought and said otherwise, truly it was the *Naturphilosoph* vision that triumphed. This was particularly thanks to Ernst Haeckel's appropriation of the thinking and his incorporation of it into his "biogenetic law": "ontogeny recapitulates phylogeny." For the rest of the nineteenth century, spurred on particularly by all of those fabulous fossil finds, people happily used embryology to speculate about the paths of the past. Selection was irrelevant. Parallels were everything.

By century's end, the anomalies and exceptions and consequent contradictions were becoming too great to ignore. Increasingly, bright young people turned from evolutionary speculations and into areas of inquiry where one could do good experimental science with prospects of firm and important results. As an old man remembering the days when he started out, William Bateson, an English biologist and early enthusiast for the genetics of Mendel, wrote that he and his fellows all became morphologists because they thought that here lay the way to unlock the secrets of evolution—and for morphology, embryology was at the cutting edge. "Therefore every aspiring zoologist was an embryologist, and the one topic of professional conversation was evolution...." But it did not work. It went nowhere. "Discussion of evolution came to an end because it was obvious that no progress was being made. Morphology having been explored in its minutest corners, we turned elsewhere" (Bateson 1928, 390–1). People like Bateson wanted to do productive fruitful work in the life sciences, as trained professionals–experimenting, observing, predicting, explaining, and all else that goes with such activity (Allen 1978). German-inspired tracing of histories, phylogenies, was not enough. Thus, as these young men turned to other fields, up rose cytology (the study of the cell), genetics (the study of heredity), and experimental embryology, where one looks at development in its own right and forgets about the evolutionary implications, real or apparent.

These scientists made good career decisions. They put biology on its modern foundations. At the same time, however, when the evolutionists did finally start to get their act together—first the theoreticians like Fisher and Haldane and Wright, and then the experimentalists and naturalists like Dobzhansky and Mayr—the young turks had become old codgers. Often they were the very ones who blocked at every turn the new evolutionary speculations and the move of evolution to professional status. Ernst Mayr (with some delight) used to tell the story of the founding of the journal *Evolution*, in the late 1940s. He and his fellows went cap in hand to the American Philosophical Society in Philadelphia. (Founded by Benjamin Franklin, it uses the word "philosophy" in the old sense of natural scientist.) The committee, with one dissenter, gave them $500. The opposition came from Edwin Grant Conklin, a very eminent Princeton embryologist who wrote and lectured extensively on evolution, but who was convinced that it could never be a real science and could only be a metaphysical background to real empirical inquiry!

There were some among the new breed of evolutionists of the 1930s who were interested in development and embryology—C. D. Darlington (1932) for one did important work on the origins of the chromosomes. But generally embryology was not a subject included in the professional affiliations of the synthesizers, and one gets the strong sense that this suited them fine and dandy. The embryologists did not want any part of them. They did not want any part of embryology. In fact, this was easy enough to do. The theoretical models of the population geneticists all worked simply with genes. One started with something like the Hardy-Weinberg law, a formula dealing with the distributions of genes in large populations, and then one introduced different causal factors like natural selection and one watched what the gene ratios would do. If one turned to fruitflies or finches or dinosaurs or plants, one rather treated the organisms as black boxes. There was the level of the genes–the genotype, and the level of the physical characteristics–the phenotype, and don't ask too many questions about what goes on in between. Rather like making sausages. The pigs go in the one door. The links come out another. And don't be too nosy about what goes on inside!

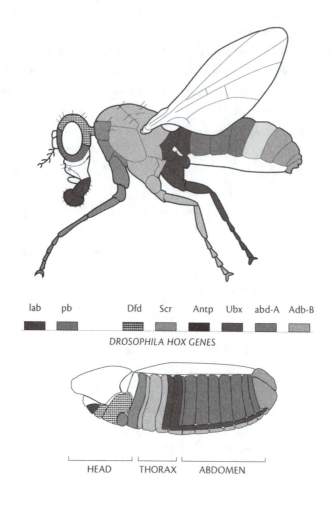

Fly Dfd	P KRQRTAYTRHQI LELEKEFHYNRYLTRRRR I EI AHTLVLSERQUKIWFQNRRMKWKKDN	KLPNTKNVR
AmphiHox4	TKRSRTAYTRQQVLELEKEFHNRYLTRRRR I EI AHSLGLTERQI KIWFQNRRMKWKKDN	RLPNTKTRS
Mouse HoxB4	P KRSRTAYTRQQVLELEKEFHYNRYLTRRRRVEI AHALCLSERQI KIWFQNRRMKWKKDH	KLPNTKIRS
Human HoxB4	P KRSRTAYTRQQVLELEKEFHYNRYLTRRRRVEI AHALCLSERQI KIWFQNRRMKWKKDH	KLPNTKIRS
Chick HoxB4	P KRSRTAYTRQQVLELEKEFHYNRYLTRRRRVEI AHSLCLSERQI KIWFQNRRMKWKKDH	KLPNTKIRS
Frog HoxB4	AKRSRTAYTRQQVLELEKEFHYNRYLTRRRRVEI AHTLRLSERQI KIWFQNRRMKWKKDH	KLPNTKIKS
Fugu HoxB4	P KRSRTAYTRQQVLELEKEFHYNRYLTRRRRVEI AHTLCLSERQI KIWFQNRRMKWKKDH	KLPNTKVRS
Zebrafish HoxB4	AKRSRTAYTRQQVLELEKEFHYNRYLTRRRRVEI AHTLRLSERQI KIWFQNRRMKWKKDH	KLPNTKIKS

The fruit-fly—human gene homology

Evolutionary Development

Let me say straight off that today there is no area of evolutionary inquiry, experimental and theoretical, which is more exciting. Forty years ago, if you were an aggressive young evolutionist looking for an area to conquer, it would have been sociobiology. Today it is evo-devo. And let me add that there is good reason for this. Some of the results have been absolutely stunning. Among the most incredible are those of the kind to which my three critics at the head of this chapter are referring. Homologies, the structural similarities between organisms, have been with us since Aristotle and (as we know) they have been the stock in trade of evolutionists since Erasmus Darwin. But there are homologies and homologies. It is one thing to look for similarities between horses and humans. It is quite another to look for similarities between humans and fruitflies. Indeed, in one of his books about fifty years ago, Ernst Mayr (1963) raised that possibility only to laugh it to scorn. That is the kind of quest that makes a mockery of evolutionary studies. Any similarities between humans and fruitflies are "purely accidental," as they say at the beginning of detective stories.

Well, guess what! There are human and fruitfly genes that are virtually identical, in the same order, doing exactly the same things. In particular, the genes that control development in animals are carbon copies of each other, and they lay down the developing organism in exactly the same ways. (See figures.) There is no great mystery about any of this, incredible though the findings surely are. Nature is economical. Having found a good solution, she uses it over and over again. Why keep on reinventing the wheel when the ones you have work just fine? In particular, it turns out that organisms are built on the Lego principle. Those same Legos can be put together to make the White House or the Creature from the Black Lagoon. Similarly the same little bits and pieces of DNA can be used to build a fruitfly or a human. It is as simple as that! It's as astounding as discovering that Elvis is alive and well and living in a retirement home in Florida. But it has the all-important difference of being true!

How would Ernst Mayr feel about this discovery? How would Charles Darwin feel about this discovery? I know the answer. They would have been overjoyed. Mayr would have told one of his incredibly convoluted, not-very-funny German jokes. Darwin would have taken another large pinch of snuff and challenged his butler to a game of billiards. But should they have been so pleased? The discoverer of natural selection in the nineteenth century? The cherisher of natural selection in the twentieth century? Should they have welcomed this and other findings of evo-devo? Scientists do not take personally being proven wrong on particulars. They do not necessarily get upset at having bigger pictures overthrown. They often get very excited. Biblical Creationists frequently make the mistake of thinking that the worst thing that can happen to a scientist is finding that he or she has committed to a false idea or theory. That's just not true. It's being committed to a boring idea or theory that is the kiss of death. Finding a better idea or theory is very much akin to the soldier and the tinderbox going into the next room, his pockets full of copper, and discovering that this room is filled with silver (and then gold in the third room). Don't regret the past. Celebrate the present. Anticipate the future. When plate tectonics arrived around 1960, a whole generation of geologists who had spent their professional lives opposing moving continents swung around and happily spent their final days speculating about the ways in which the earth's surface slips around the globe. Likewise, I am sure that Darwin is hugging himself in his grave in Westminster Abbey. Likewise, I am sure that Mayr would be happy to take time off from arguing with Gould about punctuated equilibrium to celebrate the homologies between humans and fruitflies.

The question I raise now is the deeper one about natural selection itself. Before plate tectonics no one really had much idea about how to make sense of the globe. It was a bit of hypothetical land-bridging here and a bit of ignoring there. The new theorem filled a vacuum. It is true that Einstein's theory showed that

Newton's theory—the most successful in the history of science—is false, but only in a sense. Most of Newton's ideas could be incorporated readily into the new theory. But what the critics like Gilbert and company, together with their supporters like Fodor, are arguing is that the Darwinians just got completely wrong. They were simply off base from day one. In reply to respondents who complained that he did not know enough science, Fodor (2008) made reference to Thomas Kuhn's celebrated *The Structure of Scientific Revolutions*: "I am, to be sure, in danger of having insufficient 'acquaintance with the biological theory that [I aspire] to replace'; but I'm prepared to risk it. A blunder is a blunder for all that, and it doesn't take an ornithologist to tell a hawk from a handsaw. Tom Kuhn remarks that you can often guess when a scientific paradigm is ripe for a revolution: it's when people from outside start to stick their noses in."

So what is going on here? Why is there the feeling that evo-devo shoves Darwinism out in the cold? Two things I think. First, there is the assumption by evo-devo enthusiasts that what really counts in evolutionary change is the nature of the raw building blocks—the variations or (in the language of genetics) the mutations. Everything else is periphery. This of course is a feeling sparked and reinforced by such findings as the homologies between humans and fruitflies. Now we can see how organisms are put together and how they develop from the egg to the adult. We see how the parts are made and work together. Hence the assumption is that change from one organism to another is no more than a matter of new parts and (most particularly) of rearranging old parts. It is all a question of structure, or (in the old language) of form. This assumption has been around a long time, well before Darwin, and persisted after Darwin—it was clearly the position of Thomas Henry Huxley—and it has thriven right up to (as we shall see) Stephen Jay Gould. The feeling is that if you are, say, going to turn a reptile into a mammal, then what really counts are those variations that make for hotbloodedness and hairiness and all of the other features associated with mammals and not with reptiles. Let us say variation (or mutation) A, variation B, and variation C. Reptiles don't have them. Mammals do. End of story.

The second thing driving the evo-devo scientists is the belief that Darwinism is trying to solve a pseudo-question, namely that of function or adaptive excellence, and once that is seen the need for Darwinism's (meaning natural selection's) existence or invocation is much reduced, to virtual non-being. Darwin's challenge, Darwin's revolution, was to say: Stop for a moment. Structure alone is not the whole story. Change is not the whole story. Organisms have to be functional. They have to be adaptive. If they aren't, they will die without reproducing. (True, this was not Darwin's insight. The natural theologians believed this. Cuvier believed this. But Darwin's genius was to put it into an evolutionary context and to come up with natural selection.) So it can't be just variation A and variation B and variation C. It has to be these variations as possessed by organisms that are working, that are living and surviving and reproducing. The Darwinian says that you cannot just consider the variations in isolation. The evo-devo people want to challenge this and to say: Oh yes, you can consider the variation in isolation. We do so and do so successfully. That is why traditional population genetics, something that has selection as a major causal force, must be changed drastically or even rejected.

Self-organization

Why do the evo-devo people feel this way? In part, I guess, because they are human. They are doing terrific science and they want to show that their work has big implications. What bigger implication than that they are no longer just part of the evolutionary synthesis, but that they *are* the evolutionary synthesis?! We could call this the Stephen Jay Gould syndrome. As we shall see, much of his life was devoted to promoting the importance of paleontology. Make your field really important. Given the biogenetic law and its non-evolutionary predecessors, there has long been a connection between embryology and paleontology—that is what Gould's *Ontogeny and Phylogeny* was all about—so it is really no surprise that in his last book, *The Structure of Evolutionary Theory*, Gould waxed enthusiastic about evo-devo.

In part, the attitudes of evo-devo people come from the fact that they tend to be bench scientists and not naturalists. Like Thomas Henry Huxley, they work inside the laboratory. What excites them is structure. Function enters into their thinking only secondarily. In particular, these people do not encounter living organisms having to fight and survive and reproduce. They are not in a position of someone like Edward O. Wilson who once said to me, "Mike, without adaptation, my work grinds to a halt." And as with Huxley, so with the evo-devo workers. Without adaptation, natural selection becomes otiose.

In part, the evo-devo people are responding to other, non-Darwinian currents. There has long been a tradition, linked to the formalists, of seeing the laws of physics and chemistry as determining structure and hence of being the all-important factors in evolutionary change. The evo-devo people, who obviously insist that the Lego pieces must be joined together properly—less metaphorically, who insist that bodily components must fit together properly and who, accordingly, have to know a lot more physics and chemistry than they have to know biology—are part of this tradition. In the twentieth century, the most important figure in this movement was the Scottish morphologist D'Arcy Wentworth Thompson, author of *On Growth and Form* (1917). In that book, Thompson—interestingly and significantly a great hero of Gould (1971)—argued that most organic form has little to do with selection and much to do with physics. A favorite example is of a jellyfish that is shaped exactly like an ink drop falling in water. More generally, Thompson delighted in showing how different organisms (fish particularly) could be generated by simple mathematical functions, suggesting that shapes are all a matter of accidental changes in the mechanisms that control form and have little to do with natural selection. "To seek not for ends but for antecedents is the way of the physicist, who finds 'causes' in what he has learned to recognise as fundamental properties, or inseparable concomitants, or unchanging laws, of matter and of energy. In Aristotle's parable, the house is there that men may live in it;

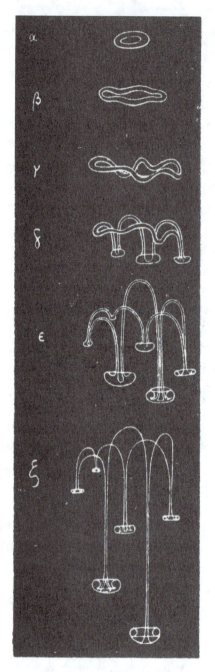

A: Ink drops falling in water

B: Jelly Fish

Similarities between patterns show that jelly fish's shape is due solely to physical laws and has nothing to do with adaptation and natural selection (W. D'A. Thompson, On Growth and Form*)*

but it is also there because the builders have laid one stone upon another" (Thompson 1948, 6). Continuing: "Cell and tissue, shell and bone, leaf and flower, are so many portions of matter, and it is in obedience to the laws of physics that their particles have been moved, moulded and conformed.... Their problems of form are in the first instance mathematical problems, their problems of growth are essentially physical problems, and the morphologist is, *ipso facto*, a student of physical science" (p.10). Thus: "We want to see how, in some cases at least, the forms of living things, and of the parts of living things, can be explained by physi-

Stuart Kauffman

cal considerations, and to realise that in general no organic forms exist save such as are in conformity with physical and mathematical laws" (p.15).

In recent years, Thompson-type thinking has been picked up by a number of people, often theoretical biologists whose main connection with the real world is mediated through the glowing monitors before which they spend their working days devising ever more subtle algorithms to try out their ideas. They argue that form carries everything forward. Nature obeys its laws inexorably and from this emerges structure. "Order for free" is the catchy slogan. Canadian-residing, American theoretician Stuart Kauffman writes: "The tapestry of life is richer than we have imagined. It is a tapestry with threads of accidental gold, mined quixotically by the random whimsy of quantum events acting on bits of nucleotides and crafted by selection sifting. But the tapestry has an overall design, an architecture, a woven cadence and rhythm that reflect underlying law—of self-organization" (1995, 185).

Self-organization! The so-called Beloussov-Zhabotinsky reaction yields an example. This phenomenon, discovered by a Russian team in Moscow in the 1950s, has a mixture of organic and inorganic substances on a flat plane (as in a Petri dish) making concentric rings, moving out from the center and vanishing as they encounter other such rings. (See figure on next page.) Their significance is that these kinds of rings are seen also in nature. In particular, the cellular slime mold goes through a phase in which it simulates the Beloussov-Zhabotinsky reac-

Wave patterns in aggregating slime mold amoebas showing expanding concentric circles associated with the (purely chemical) Beloussov-Zhabotinsky reaction.

tion very precisely. Usually, such slime molds are from colonies of free-living amoebas, eating bacteria. However, if and when food supplies become scarce, they begin to aggregate. "Cells start to signal to one another by means of a chemical that they release. This initiates a process of aggregation: the amoebas begin to move toward a center, defined by a cell that periodically gives off a burst of the chemical that diffuses away from the source and stimulates neighboring cells in two ways: (1) cells receiving the signal themselves release a burst of the same chemical; and (2) they move toward the origin of the signal" (Goodwin 2001, 46). What is truly significant is that, as these amoebas begin to move together—at which point, combining into a mullticellular organism that can fruit and reproduce, making another crop of independent amoebas—the patterns exhibited are identical to those of the Beloussov-Zhabotinsky reaction. It is important to stress that, in fact, the molecules in the chemical state and the living state are quite different. Nevertheless, the underlying process is similar. Linking the two cases, we find that substances are produced in increasing amounts until other processes take over to inhibit the production of these substances. Then, the whole system exists in an unstable condition of oscillation, as the various processes switch on and off.

The new formalists seize on the similarities. Here we have a case where an organism (or group of organisms, depending on how you count the slime molds) uses a self-generating, chemical process for its own biological ends. The overt pattern was not shaped by selection, but emerged spontaneously from the way that the non-living world works. Defining a field as "the behaviour of a dynamic system that is extended in space," Canadian-born, British biologist Brian Goodwin

Brian Goodwin

(2001) writes: "a new dimension to fields is emerging from the study of chemical systems such as the Beloussov-Zhabotinsky reaction and the similarity of its spatial patterns to those of living systems. This is the emphasis on self-organization, the capacity of these fields to generate patterns spontaneously without any specific instructions telling them what to do, as in a genetic program. These systems produce something out of nothing." Continuing: "There is no plan, no blueprint, no instructions about the pattern that emerges. What exists in the field is a set of relationships among the components of the system such that the dynamically stable state into which it goes naturally—what mathematicians call the generic (typical) state of the field—has spatial and temporal pattern" (pp. 51–2). To be honest, in some of the earlier writings, particularly that of Thompson, it is not always clear whether the claim is that physics and chemistry make organisms that are adaptive, or whether the claim is that physics and chemistry make organisms and the matter of adaptation is simply irrelevant. With people like Kauffman and Goodwin one senses the latter is the case. Adaptation is not an issue in the biological world.

Is Natural Selection Irrelevant?

What can the Darwinian say in response? In part, the evo-devo people do have some good points. It may be that we all need to look more at the nature of variation and of how this plays out in development. If changes, say, are usually less a matter of coming up with something completely fresh and more a matter of taking what you have and reorganizing it, then we should all be aware of this and think about its implications. Richard Dawkins (2007) uses the analogy of a Jumbo Jet, a Boeing 747. You are not going to build a functioning Jumbo Jet out of parts found in a scrap yard. That is just not possible. But you can take a Jumbo Jet and stretch it, in one step making a longer, bigger plane, and still have a functioning method of transporting people through the sky. If, in nature, these kinds of changes not only occur but are common, or are involved in key evolutionary events, then Darwinians should know about them and take note of them. There is no question about that.

Relatedly, whatever may be the particular scientific or philosophical ends of the order-for-free brigade, they too make points worth considering by the Darwinian. In particular, physics and chemistry do matter when it comes to building organisms. They set constraints on what you can and what you cannot do. Why do you never have a cat as big as an elephant? Simply because, as the length increases, the volume goes up by the cube. To bear their weight, elephants have massive, tree-trunk-type legs. Cats are agile and if they were the size of elephants, their legs would break immediately. On the other hand, one needs to be careful about using this argument. Critics of Darwinism often think that an appeal to constraints surprises Darwinians—which is simply not true because they have known about them all along—or forces Darwinians to abandon their theory—which again is simply not true. The developmental morphologist Rudolf Raff, one of Darwinism's critics quoted at the head of this chapter, raises the issue of genome size. "Having a large genome has consequences outside of the properties of the genome per se. Larger genomes result in larger cells. Because cells containing large genomes replicate their DNA more slowly that cells with a lower DNA content, large genomes might constrain organismal growth rates. Cell size will also determine the cell surface-to-volume ratio, which can affect metabolic rates" (1996, 304). Raff notes that salamanders often have large genome sizes. Hence, if we do find constraints in action, we might expect to find them here. And there does seem to be some evidence of their operation. "Roth and co-workers have observed that in both frogs and salamanders, larger genome size results in larger cells. In turn, larger cells result in a simplification of brain morphology. Thus, quite independently of the demands of function, internal features such as genome size can affect the morphology and organization of complex animals. Plethodontid salamanders share the basis vertebrate nervous system and brain, but they have very little space in their small skulls and spinal cords" (p.305, referring to Roth et al [1994]).

Having said this, however, Raff is too good a biologist not to admit that if there are constraints at work, they apparently do not make much difference. The salamanders can do some pretty remarkable things—remarkable salamander things, that is—seeming not at all to be functionally constrained. "These salamanders occupy a variety of caverniculous, aquatic, terrestrial, and arboreal habitats. They possess a full range of sense organs, and most remarkably, a spectacular insect-catching mechanism consisting of a projectile tongue that can reach out in ten milliseconds to half the animal's trunk length (snout to vent is the way herpetologists express it)." They have pretty good depth perception too. And indeed, their slow metabolic rate brought on by large genome size may even be of adaptive advantage. "Plethodontids are sluggish, and the low metabolic rates introduced by

Salamander

large cell volume may be advantageous to sit-patiently-and-wait hunters that can afford long fasts. Vision at a distance is reduced to two handbreadths, but since these animals are ambush hunters that strike at short range, that probably doesn't affect their efficiency much" (p.306). In other words, far from refuting Darwinism, it seems if anything to help the cause. Moreover, if nature demands, apparently the salamanders can start to bring down their genome sizes. The constraints are just not that strong.

But what about the stronger claim? After all, constraint talk does still imply that selection can do its work, if only within limits. What about the kind of claim of someone like Brian Goodwin, who maintains that really physics and chemistry do everything? Let us look at another example that is a favorite of people like him. Phyllotaxis refers to a very common phenomenon in the plant world, where many identical elements are packed together. A sunflower shows this very dramatically, for the seeds on the head form a highly characteristic pattern of clockwise and counterclockwise spirals. One sees this also in pinecones and even in cauliflowers as you tear them apart. All told, more than 80% of the quarter-million higher plants show it in one form or another. This pattern, phyllotaxis, is produced by the leaves appearing at the center (the "growing apex") and then, as it were, being pushed outwards (Mitchison 1977). The appearing leaves follow a spiral (known as the "genetic spiral") and, given constant growth, the angle between successive leaves is also constant. The spirals that catch one's eye are known technically as "parastichies." Botanists long ago discovered that one could express phyllotaxis in mathematical form by means of a formula discovered by thirteenth-century Italian mathematician Leonardo Fibonacci, made popular by the bestseller *The Da Vinci Code*. Rather boringly, unlike the hero of the novel, he was not fighting some secret Catholic society, but looking for a way to calculate

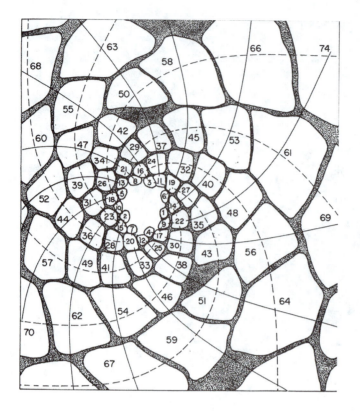

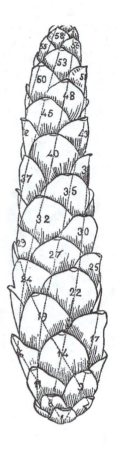

Phyllotaxis: left, Monkey Puzzle tree Aravearia excelsa *(8,13); right,* Pinus strobus *(5,8).*

the growth of the offspring of a pair of rabbits. He thus discovered the series formed by adding together the previous two members of the series, starting with zero and one. The series thus being 0, 1, 2, 3, 5, 8, 13, and so on, or more generally, $n_j = n_{j-1} + n_{j-2}$. Botanists have found that the numbers of parastichies, one set clockwise and one set counterclockwise, on a particular species of plant are always related by being consecutive numbers of the Fibonacci series. In the stylized picture given in the diagram, the example is of an 8, 13 phyllotaxis. As can also be seen from the diagram, another way of calculating the measure is by using the order of production of the "contact" leaves on the same spiral. This is not a measure based on the order of production of the leaves but on the pattern itself, and refers to those leaves, along shared paratischies, that will be touching. Using examples furnished by Asa Gray in the sixth edition of his textbook, *Structural Botany* (1881), the American larch produces a cone that is 2,3; holly is 3,5; and the cone of *Pinus strobus* is 5,8.

Why do plants show this pattern? Darwinians are obviously going to think in terms of selection and adaptation. Indeed, shortly after the *Origin* was published, the American pragmatist Chauncey Wright argued that the arrangement gives the best way of exposing each leaf to the light, without undue overlap from its fellows. With this end in view, the differences between the various phyllotactic arrangements are so minute as to not really matter that much. "To realize simply and purely the property of the most thorough distribution, the most complete exposure to flight and air around the stem, and the most ample elbow-room, or space for expansion in the bud, is to realize a property that exists separately only in abstraction, like a line without breadth" (Wright quoted in Gray 1881, 125).

The critics of Darwinism will have none of this. D'Arcy Thompson listed one objection after another. The differences between the arrangements are indeed significant, and the Darwinian teleological intent is something which "cannot commend itself to a plain student of physical science." Thompson argued that there are all sorts of other ratios that would do the job as well, and that the plant could have taken other and better paths to exposing the leaves to sunlight, and much more. "We come then without more ado to the conclusion that while the Fibonacci series stares us in the face in the fir-cone, it does so for mathematical reasons; and its supposed usefulness, and the hypothesis of its introduction into plant-structure through natural selection, are matters which deserve no place in the plain study of botanical phenomena" (Thompson 1948, 953).

Calling back over his shoulder as he walked away, Thompson accused the Darwinian of "harking back to a school of mystical idealism." There is a certain irony to this given the fascination that mathematics exerts over the order for free supporters. When faced with phyllotaxy, Brian Goodwin (2001) sounds like a follower of Pythagoras, such is his fascination with the underlying numerology. He begins with the happy observation that the vulgar-fraction series formed from dividing consecutive members of the Fibonacci series homes in on the irrational number 0.618. This is no casual finding for it is what the ancient Greeks called the Golden Mean or Section—the ratio of the sides of a rectangle, where the rectangle left after removing the biggest possible square is of the same proportions as the original rectangle. This is but a beginning. You can get the Golden Mean out of circles also, if you divide up the perimeter in an appropriate way. This yields the major angle of 137.5 degrees, and—a finding so wonderful you begin to think you are seeing the hand of God—this is just about the angle you tend to get with successive leaves on the genetic spiral. "So plants with spiral phyllotaxis tend to locate successive leaves at an angle that divides the circle of the meristem in the proportions of the Golden Section. Plants seem to know a lot about harmonious properties and architectural principles" (p. 127). (The meristem is the growing tip of a plant, such as stems and roots. The connections, of course, are not arbitrary, but follow mathematically from the properties of lattices, which is what we have here.)

We are not yet finished. Now is the time for a little experiment. Drop a ferrofluid (a fluid with magnetic properties) slowly into the center of a polarized film of oil (Douady and Couder 1992). The drops repel each other and move away from the center. If this is done sufficiently slowly, each drop is affected only by the previous drop. If done more rapidly, wonderful things start to happen. "As the rate of adding drops (equivalent to the rate of initiation of leaves in a meristem) is increased, a new drop experiences repulsive forces from more than one previous drop, and the pattern changes: the initial simple symmetry of the alternate mode gets broken, and a spiral pattern begins to appear. It takes a while for the system to settle on a steady pattern, the duration of this transient depending on the rate of adding drops. If this is rapid, so that there is strong interaction between drops, then a stable pattern emerges rapidly and successive drops quickly settle into a divergence angle of 137.5° , the spirals obeying the normal Fibonacci series" (Goodwin 2001, 127–8). Once again we get self-organization, and just as one can get different patterns by altering the rates at which the oil drops, so also in plants the different patterns reflect simply the rates at which the plants grow and generate leaves. In short, "the frequency of the different phyllotactic patterns in nature may simply reflect the relative probabilities of the morphogenetic trajectories of the various forms and have little to do with natural selection" (p. 132). Or as Kauffman (1995) puts it: "Like the snowflake and its sixfold symmetry, the pinecone and its phyllotaxis may be part of order for free" (p. 151).

Let's go back to the Darwinians. By now, you will not expect them to be convinced. The formalists overlook the "obvious possibility" that "natural selection may universally favor close packing by phyllotaxis over alternative arrangements" (Reeve and Sherman 1993, 21). Obviously, however, this can be only a partial response. After all, the flowers must pack in their parts in some way, and if not phyllotaxis then what? And if phyllotaxis, then it is impossible to get away from the mathematics underlying the phenomenon. A measured response has been to recognize the nature of phyllotaxis and how and why it comes about—denying it would be a bit on a par with denying Pythagoras's theorem itself—but to point out that there still remains lots of scope for selection. Plant scientist Karl Niklas (1988) writes: "Computer simulations indicate that phyllotaxy can influence the quantity of light intercepted by leaf surfaces. Model plants constructed with equal total leaf area and number differ significantly in flux, even when leaf-divergence angles are very similar.... Nonetheless, computer simulations indicate that a variety of morphological features can be varied, either individually or in concert, to compensate for the negative aspects of leaf crowding resulting from "inefficient" phyllotactic patterns. Internodal distance and the deflection ("tilt") angle of leaves can be adjusted in simulations with different phyllotactic patterns to achieve equivalent light-interception capacities" (p. 566). Niklas would rather speak of phyllotaxy as a limiting factor rather than as something that involves constraint. It is not anti-adaptation. It is rather to be thought of as background, something in which adaptation is embedded. It is indeed that which makes adaptation possible. "The distinction between a 'constraint' and a limiting factor is important, because it reflects a measure of plasticity within the developmental repertoire."

Making the Positive Case

It's *déjà vu* all over again. We are caught again in one of those indeterminable disputes, like that over the Neanderthals, that have no end or resolution. To every suggestion or criticism, there is a counter suggestion or criticism. However, through and through Darwinians make the same point and ultimately there is no gainsaying it. If you don't function, you don't live. If you don't live, you don't reproduce. If you don't reproduce, then you are out of the evolutionary game. It is as simple as that. Darwinians make this point in terms of the analogies of others. Go back to the Lego example. It is a terrific metaphor or analogy for thinking about the way in which organisms are put together. But like all metaphors or analogies it has the capacity to mislead. Suppose you start with a simple form back in the Cambrian, let us say a Lego outhouse. You are going to work up to lions or apes today, Lego White Houses or Pentagons. If you are a Lego enthusiast, you look at the outhouse, you spot some good principles of design, and then you keep them in mind while you are building your big modern building. But between the outhouse and the White House you have pieces of Lego all over the carpet, lying doing nothing, waiting to be picked up, or put together in smaller units waiting to be picked up and used in the whole. Nature is not like that. Between the Cambrian pre-vertebrate and the lion or ape, you have to have a continuous series of functioning organisms. There can be no gaps—ever. It takes only one little instance of non-functionality to spoil the whole sequence. Darwinians point out that you must ignore this to make the critic's point, and once you stop ignoring it, then suddenly you have a different perspective on variation and mutation. However good it may be potentially, some new change can never get away from selection. You cannot think of variation except in this context. And we all know that the bigger the change, the less likely is it going to be fully functional right from the start. Without the hand of God to direct things, Murphy's Law steps in. If it can go wrong, it will. Moreover, talk about order for free is not very helpful, unless you can show that this order for free—this self-organization—kicks in right at the moment when new variations are introduced into the mix. Even if it happens sometimes, there is nothing in evo-devo to say that it happens when organisms shuffle their Lego pieces. New variations are not snowflakes. That is why calling for a new population genetics is just plain silly unless you are taking into account the things that today's population genetics takes very seriously.

Darwinians also want to make the case in their own terms. They want to say that they have lots of positive evidence for the workings and significance of natural selection. They do not simply spend their time answering the objections of others. Let me give just one example: namely the already-classic work done by Peter and Rosemary Grant on Darwin's finches, those little birds of the Galapagos that so excited Darwin himself when he visited the archipelago (Grant 1986, 1991; Grant and Grant 1989, 1995). Looking at the finches was nothing new. The English ornithologist David Lack led the way for those who did extensive studies of these organisms in the 1930s and 1940s. Famously, having first endorsed a non-selection-based position, Lack then swung around and wrote a highly adaptationist account. But it is the Grants and their associates, starting in the 1970s, who have most carefully studied the birds and shown the action of selection in the wild. They worked on an islet (Daphne Major) only a few hundred meters each way, focusing on a population of medium ground finches, *Geospiza fortis*. On average there are about 1,200 specimens, and the Grants have caught and ringed them all. Given that the birds can live for up to sixteen years, and given also that they have a generation time of about four and one-half years, there is plenty of death and

Darwin's finches

destruction going on. The Grants asked whether this death and destruction is systematic and, if it is, whether it has selective effects.

Their answers were unambiguously positive. Confining attention particularly to the finches' most distinctive feature, the Grants asked about beak size and shape. First, are these features heritable? If there is no genetic causal connection, then selection could work away indefinitely without effect. In fact, this was a fairly easy question to answer. By measuring parents and offspring, it was seen that beaks in shape and size are strongly under the control of the genes. Big-beaked parents have big-beaked offspring and so forth. Second, what is the significance of beak size and shape? Another question readily answered. The birds eat nuts and fruits and the like. Big-beaked birds are going to be able to crack bigger and harder fruits and nuts than are small-beaked birds. Smaller-beaked birds, however, are going to be able to eat smaller seeds and the like. The implications are obvious. If there are mainly big and hard nuts and fruits, then the bigger-beaked birds are going to be at a selective advantage. If there is lots of everything, then probably the smaller-beaked birds are going to be at an advantage.

Fortuitously for the researchers, if not for their subjects, there was a horrendous drought in 1977. There was no reproduction in that year. Hanging on was the aim of the game. Food supplies dried up, and the advantage shifted to those

Daphne Major

individuals who could exploit rarer or harder-to-access resources—namely big and hard nuts and fruits. The Grants found not only that the dead and emaciated birds tended significantly to be those with smaller and more refined beaks, but that the average beak size shifted strongly over the next year to bigger and coarser. There really was a gene-based shift, and it was in an adaptive direction that favored those birds able to access the more scarce resources. However, things even out in the long run. God tempers the climate to the starving finch, and, in the years after the 1970s, there were many times of plenty. These times favored smaller-beaked birds, able now to take full advantage of the abundant seeds and small-sized fruits. Not that one should assume that fluctuations of this kind imply that nothing of significance happens over the long run. The birds rarely, if ever, return to exactly their original starting points. There are always subtle modifications. On average, over the past thirty years, the birds tend to be smaller and with sharper beaks. Natural selection really does leave its mark.

Optimality Models

Straightforward methods of testing like these are not the only resources open to Darwinians. Another method much favored today is the building of "optimality models" showing just how and where selection might have worked. Here, instead of going out to find selection, we work backwards rather as one does in Darwin's theory of the *Origin*. He assumed natural selection brought on evolution, and then he went out to test this assumption in areas like palaeontology and biogeographical distribution. The success of these explanations is taken as truth of the hypothesis. In the language of philosophers, Darwin had made "an inference to the best explanation." Working at a more limited level, our hypothesis—our best explanation—is that natural selection has been at work. What would this mean? In the best of all possible worlds, this would mean that selection has brought about perfect adaptation—it has "optimized" the situation—and that from here we can work out what is going on and why. Let us therefore build "optimality models" to explore cases of putative adaptation (Orzack and Sober 1994, 2001). The entomologists George F. Oster and (our old friend) Edward O. Wilson (1978) explicitly think of themselves as construction workers, as people making things that work. "In order to employ engineering optimization models the biologist tries to interpret living forms as in some sense the 'best.'" Of course, the trouble is with precisely what one means by "best" in a situation like this. "In effect the biologist 'plays God': he redesigns the biological system, including as many of the relevant quantities as possible and then checks to see if his own optimal design is close to that observed in nature." From then on, it is all rather a matter of trial and error–putting the theoretical design model against the empirical findings. "If the two correspond, then nature can be regarded as reasonably well understood. If they fail to correspond to any degree (a frequent result), the biologist revises the model and tries again. Thus, optimization models are a method for organizing empirical evidence, making educated guesses as to how evolution might have proceeded, and suggesting avenues for further empirical research" (Oster and Wilson 1978, 294–5).

Now I should say that some people sneer at this way of doing things. For Gould, these models fall into his category of "Just So Stories." For his sometime co-author Richard Lewontin, by "allowing the theorist to postulate various combinations of 'problems' to which manifest traits are optimal 'solutions', the adaptationist programme makes of adaptation a metaphysical postulate, not only incapable of refutation, but necessarily confirmed by every observation. This is the caricature that was imminent in Darwin's insight that evolution is the product of natural selection" (quoted in Maynard Smith 1978; reprinted in Sober 1994, 99). Philosopher Robert Brandon and biologist Mark Rausher speak even more strongly: "The attraction of optimality models is clear—they allow one to avoid history and genetics. Years ago in a discussion about number theory, Bertrand Russell said, 'The method of "postulating" what we want has many advantages; they are the same as the advantages of theft over honest toil. Let us leave them to others and proceed with our honest toil'... These are exactly our thoughts with respect to optimality models and the rigorous test of adaptationism" (Brandon and Rausher 1996, 200).

Well, having poured water on the altar, let us look at someone using optimality models, and who better than Edward O. Wilson himself? As you know, he is an expert on the social insects. In a series of papers, written at the beginning of the 1980s, he focused specifically on the caste system in certain groups of the ants. Using the metaphor of a division of labor, Wilson was concerned to find why and how it is that the ants have so many different forms: ranging from tiny workers within the nest to large soldier ants outside the nest, protecting their siblings from attackers of all kinds. Wilson worked exclusively on the so-called leaf-

Leaf-cutter ants

cutter ants, a genus known as *Atta*. They have a very complex social system. First, they send out forgers from the nest to search for vegetation, leaves, and the like. Once they have spotted something, these foragers proceed to cut their bounty into small pieces, which they can then carry back into the nest. At this point, another caste takes over. Its members cut up the leaves into even smaller pieces and treat them with enzymes on which they grow a kind of fungus. Finally, yet another caste takes the fungus and feeds it to the young. "The fungus-growing ants of the tribe Attini are of exceptional interest because, to cite the familiar metaphor, they alone among the ants have achieved the transition from a hunter-gatherer to an agricultural existence" (Wilson 1980a, 153).

Wilson is an ardent Darwinian, so his working assumption was that, from the viewpoint of morphology as well as from behavior, we should find that the ants have been shaped by natural selection. We should find that their body shapes and behavior are about as good (optimized) as it is possible to be. Taking this assumption as a tool of research, as much as an established empirical hypothesis, Wilson turned first to the question of the whole overall caste pattern and distribution to be found in *Atta*. Striving to show that there is indeed a division of labor, Wilson's work here was as much descriptive as experimental. First and most obviously, one finds that the soldiers (who take on the roughest work) are bigger and stronger than any of the others—a hundred times bigger than some of their nest mates. Then one finds that those out foraging are in the middle range. Finally, back home in the nest, one finds that here is the place of the most minute and delicate ants.

Why does one have this division? "The elaborate caste system and division of labor that are the hallmark of the genus *Atta* are an essential part of the specialization on fresh vegetation. And, conversely, the utilization of fresh vegetation is the raison d'être of the caste system and division of labor" (Wilson 1980a, 150). And how does this all come about? Wilson was able to show that from a biological point of view, it is done fairly easily. It is a question of relative growth or allometry, combined with a degree of behavioral flexibility. In primitive species, nest members are not differentiated and anyone can and does do any task. "Most of the monomorphic attines utilize decaying vegetation, insect remains, or insect excrement as substrates, in other words, materials ready made for fungal growth" (Wilson 1980a, 153). In the *Atta,* with specialization, some members of the nest do some tasks and other members do other tasks. But body forms are not radically different; rather they are developed proportionately to their ends.

The point is that if one is going to have a kind of specialization that the *Atta* have developed, namely, the ability to feed on fresh leaves and to grow fungus on them, one needs much more specialization than one finds with primitive monomorphic forms. But can one then show experimentally that there are adaptive reasons behind this? "Is the colony as efficient in its basic operations as natural selection can make it, without some basic change in the ground plan of anatomy and behavior?" (Wilson 1980b, 157). In what way is one to answer this question?

> The ideal way in which to test the natural selection hypothesis and to estimate the degree of optimization is to first write a list of all conceivable optimization criteria, deduced a priori from a knowledge of the natural history of the species. The next step is to conduct experiments to determine which of the criteria has been most closely approached, and to what degree. Finally, with the results in hand, the theoretician can alter behavioral and anatomical parameters in simulations in order to judge whether the species is capable of still further optimization by genetic evolution. If the approach actually taken by the species cannot be significantly improved by the simulations, we are justified in concluding that the species has not only been shaped in this particular part of its repertory by natural selection, but that it is actually on top of an adaptive peak. (Wilson 1980b, 158)

One question that interested Wilson centered on the nature of the ants that would be most efficient for going out foraging, cutting up leaves, and bringing them back. Why should one find that the middle-range ants do this? Why not bigger ants, who could also act as soldiers, or smaller ants, who could also act as nest tenders? Wilson's hypothesis was that the middle-range ants are the best adapted to their allotted task–it is they who make optimal use of the energy resources of the nest. To test this hypothesis, Wilson ran a number of experiments using the so-called pseudomutant strategy. Wilson removed foragers under certain circumstances and saw whether the other castes, who were left in the nest, were more efficient at foraging or whether the foraging dropped off. For instance, was the nest better off with smaller foragers or larger foragers? Or was it truly the case that something in between, as one has at the moment, offers the best solution? (Wilson took note of the fact that in natural conditions, the vegetation available to the *Atta* is of a particular kind. In the rain forests, the vegetation is tough. One must therefore recognize that an ant that is good at cutting up rose petals might not function at all well in nature. One needs an ant at least capable of cutting up rhododendron leaves.)

Wilson showed that his hypothesis and research strategy pay off. "What *A. sexdens* has done is to commit the size classes that are energetically the most efficient, by both the criterion of the cost of construction of new workers ... and the criterion of the cost of maintenance of workers" (Wilson 1980b, 164). More than this, Wilson found that the nests are adapted more to the kind of vegetation that they would experience in the wild than to any general range of vegetation. One has natural selection working flat out, most efficiently. The ants are adapted in such a way as to optimize the overall behavior of the nest. In other words, the colony "sits atop an adaptive peak."

In my opinion, good science like this answers all of the critics. It is ludicrous to speak of the work as a "Just So Story." The ideas are checked against the evidence in the most minute detail. There is far more than a simply metaphysical fabric spun from ideas and fantasies. And as for the snarky comments about avoiding honest toil, one recoils at the chutzpa of a philosopher of all people speaking in such terms. The fact is that Darwinism—meaning evolution through natural selection, explaining the adaptive nature of the living world—is a successfully functioning theory and its critics can only be measured against it.

Further Reading & Discussion

Stephen Jay Gould has a chapter on evo-devo in his *The Structure of Evolutionary Biology* (Cambridge: Harvard University Press, 2002). It is long enough to be almost a book in itself. I disagree with just about every idea that Brian Goodwin has ever had, but I really enjoyed *How the Leopard Changed its Spots* (Princeton, N.J.: Princeton University Press, 2001). It is clearly written and makes its case strongly and forcibly. From the Darwinian side, you really should read something by or about the Grants. Peter Grant's *Ecology and Evolution of Darwin's Finches* (Princeton, N.J.: Princeton University Press, 1986) is a good place to start, and then follow with Peter and Rosemary on *How and Why Species Multiply: The Radiation of Darwin's Finches* (Princeton, N.J.: Princeton University Press, 2007). A work on them, their work, and their lives that deservedly won the 1995 Pulitzer Prize in general nonfiction is Jonathan Weiner's *The Beak of the Finch: A Story of Evolution in Our Time* (New York: Knopf, 1994).

Another Pulitzer Prize-winning work (his second, in 1991 for general nonfiction) is Wilson's big book on ants, co-authored with his colleague Bert Hölldobler, *The Ants* (Cambridge: Harvard University Press, 2000). With the same publisher in 1994, this team wrote a more popular book: *Journey to the Ants: A Story of Scientific Exploration*. Finally a couple of my books are certainly relevant to these issues: *Darwin and Design: Does Evolution have a Purpose?* (Cambridge: Harvard University Press, 2003) looks at the whole question of adaptation, and includes a discussion of the relevance of evo-devo to the debate as well as a discussion of optimality models. *Darwinism and its Discontents* (Cambridge: Cambridge University Press, 2006) looks at the positive evidence for evolution through natural selection.

Chapter 12
New Evolutionary Theories:
Thickening the Plot of Natural Selection

Overview

This chapter explores the thinking of a number of 20th century evolutionists, including Stephen Jay Gould. Before he died early in the new millennium, Gould, a Harvard paleontologist, was the most famous evolutionist of his age. Yet his thinking was disliked intensely by many leading professional evolutionists, including the late John Maynard Smith, the doyen of English biologists and an ardent Darwinian. Why?

In the early 1970s, Gould together with fellow paleontologist, Niles Eldredge, proposed a new theory of evolution: "punctuated equilibrium." This claimed that evolution goes in jumps, moves from one form to another, with periods between of non-change, "stasis." The theory was opposed to traditional Darwinism, which supposedly stresses gradual evolution, "phyletic gradualism." Over the years, Gould broadened his critique of Darwinism, arguing that it over-stresses the extent to which organisms are adapted and hence over-relies on natural selection. Gould argued that many organic features are like the spandrels at the tops of columns in medieval churches, apparently needed but truly without essential function.

The important thing to keep in mind when dealing with Gould's thinking is that there are so many layers at work here. Most obviously, there are scientific questions, specifically about whether the fossil record really does show a jerky history for life or if this is truly an artifact of incomplete fossilization. But more subtly, there are issues about the very status of paleontology as a science, with Gould wanting to argue for a heightened status for his discipline. Then lurking are philosophical questions, about whether function is really the key to understanding organisms, or if other viewpoints are equal or even better.

Emerging from the particular controversies started by Gould and his sympathizers are some really important questions about the nature of the evolutionary process.

In particular, do we need a kind of layered series of theories, dealing with events at different levels of magnitude? It is clear that in some sense we do. Most evolutionists now agree that change at the molecular level may well be non-Darwinian in important respects. But whether we need different theories at higher levels is still highly controversial.

Most importantly, do we need new theories when discussing changes over very long periods of time? There is really no definitive answer to this question yet, because only now are we starting to get really good and reliable surveys of what actually happened. The work of the late John J. Sepkoski, applying ideas of ecology formulated by Robert MacArthur and Edward O. Wilson, is highly informative on and suggestive about these issues.

The Role of the Scientific Community

The work of the following scientists is discussed in this chapter. Short biographical essays of these individuals appear in **Biographies** on page 607.

Stephen J. Gould (1941–2002)
John J. Sepkoski (1948–1999)

Setting the Stage

In the past half-century there have been few science writers more read and honored—and loved—than the late Stephen Jay Gould, of Harvard University (1941–2002). His books were devoured and discussed by millions. By profession, Gould was a paleontologist, and he could write about this in a fascinating way. But his range was far wider–across biology, across science, and across culture, of today and of the past. For some thirty years he wrote a monthly column, "This View of Life," in the American Museum of Natural History's journal, *Natural History*, and his pieces were collected in one sparkling collection after another: *Ever Since Darwin*, *Bully for Brontosaurus*, *The Flamingo's Smile*, and many more. He wrote also full-length works, on the perils of IQ testing—*The Mismeasure of Man*, on obscure organisms from the past found in a deposit in Western Canada—*Wonderful Life*, on science and religion—*Rocks of Ages*, and again many more. And if you thought that a Harvard professor simply has to be an egghead, concerned only with the higher verities, not only did Gould have a deep love of choral singing, he was also a great fan of American baseball. Like a true aficionado, it was the history, the statistics, that really excited him. There was even an entire book, *Full House: The Spread of Excellence from Plato to Darwin*, spurred by Joe DiMaggio's record of 56 continuous games with a hit.

But if Gould had been looking for glory and praise from his fellow evolutionists, he would have been out of luck. You expect that philosophers will sometimes turn a little nasty. That comes with the job. The less we connect with the real world, the more choleric we become. But you do not expect such bile of the leading evolutionary game theorist, John Maynard Smith, a man whose boyhood years at England's leading private school (Eton College) reflect in the courtesy and charm he showed in conversation and in writing. Yet, writing in the *New York Review of Books*—a place, admittedly, where unbalanced emotion is the norm rather than the exception—he suddenly swung from his allotted task (a mild review of something on another topic) and started declaiming against Gould and his false and sloppy thinking.

Stephen Jay Gould

Gould occupies a rather curious position, particularly on his side of the Atlantic. Because of the excellence of his essays, he has come to be seen by non-biologists as the preeminent evolutionary theorist. In contrast, the evolutionary biologists with whom I have discussed his work tend to see him as a man whose ideas are so confused as to be hardly worth bothering with, but as one who should not be publically criticised because he is at least on our side against the creationists. All this would not matter, were it not that he is giving non-biologists a largely false picture of the state of evolutionary biology. (Maynard Smith 1995, 46)

Sometimes a dignified silence, however difficult, is a strategy preferable to all-out counterattack. Such was not Gould's way. Labeling people like Maynard Smith and Richard Dawkins as "Darwinian fundamentalists," Gould lamented that although Maynard Smith has "written numerous articles, amounting to tens of thousands of words" about Gould's work, whereas those were "always richly informed, now alas he has been seduced into adaptationist fanaticism."

He really ought to be asking himself why he has been bothering about my work so intensely, and for so many years. Why this dramatic change? Has he been caught up in apocalyptic ultra-Darwinian fervor? I am, in any case, saddened that his once genuinely impressive critical abilities seem to have become submerged within the simplistic dogmatism epitomized by Darwin's Dangerous Idea [i.e., all-powerful natural selection], a dogmatism that threatens to compromise the true complexity, subtlety (and beauty) of evolutionary theory and the explanation of life's history. (Gould 1997, 37)

As we shall learn, there are different levels to this quarrel, but let us start with the most obvious level—that of the science.

Essay

Punctuated Equilibrium

The year of the centenary of the *Origin*, 1959, was the heyday of Darwinian natural selection. After years of neglect and denial, finally the significance of selection as a mechanism was being recognized, in America as well as Britain. Great and long were the celebrations, with honorary degrees being handed out like candy to all of the major figures in the field. It is therefore no surprise that, when Stephen Jay Gould began his career in the mid-1960s as a paleontologist, specializing in the evolution of snails (Gould 1969), he was an orthodox Darwinian. Confirming this, an earlier review paper on problems of relative growth showed how things considered nonadaptive can be fitted readily into a selectionist framework. (Gould 1966). But in a sense, American Darwinism was always skin deep—remember how Spencer had been a far greater influence—and, for all that George Gaylord Simpson labored in Darwinian fields, paleontology was always on the edge of the pasture. The fact of the matter is that paleontology cannot use selection directly,

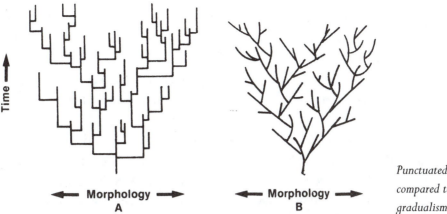

Punctuated equilibria (left) compared to phyletic gradualism (right)

as can the student of today's organisms, such as the sociobiologist. Selection is not a tool of research where you can go out and discover and test and come up with results. You are working at a distance—a very long distance—with evidence (fossils) that is spotty and incomplete and very, very dead. You are always having to take somebody else's exciting ideas and see if they do anything for you.

Those who knew of the self-confident personality of Stephen Jay Gould—he was not about to take a back seat to anyone—could have predicted that he would not tolerate this. Before long, Gould would be moving forward to make his own mark on evolutionary studies. This mark would make paleontology a central focus of attention, arguing that the evolutionist needs paleontology not just for establishing the fact of evolution and for ferreting out the path of evolution but also for discovering the true nature and full extent of the causes of evolution. Expectedly, in the early 1970s, this prediction came true. Together with a former fellow graduate student, Niles Eldredge, Gould began pushing forward a supposedly all-new perspective on the paleontological record—a perspective that Gould and Eldredge somewhat inelegantly labeled "punctuated equilibrium"

The two young palaeontologists started with the fact that traditionally, the course of evolution is seen to be one of smooth, gradual change. This is something that comes about simply because natural selection makes sudden change highly improbable. The only way in which organisms can stay in adaptive harmony with their surroundings is by changing only minutely in each generation. Therefore, any apparently sharp breaks in the fossil record should not be explained in terms of major jumps from one form to another but should be put down to the incompleteness of the record and so forth. What Eldredge and Gould argued, to the contrary, was that the paleontological record is in fact much better and stronger than most people allow, and that hence a causal explanation must be found to explain this. One must accept that there are long periods of relatively little evolutionary change—periods of equilibrium, or stasis—broken, or punctuated, by rapid moves from one form to another. "The history of life is more adequately represented by a picture of 'punctuated equilibria' than by the notion of phyletic gradualism. The history of evolution is not one of stately unfolding, but a story of equilibria, disturbed only 'rarely' (i.e., rather often in the fullness of time) by rapid and episodic events of speciation" (Eldredge and Gould 1972, 84). Although there is no official position on this, the position today is generally known as the "theory of punctuated equilibrium." Informally, friends and critics often call it punk eck! The theory of punctuated equilibrium supposedly explains the phenomena of punctuated equilibria, that is periods where there are no significant changes.

The controversial and exciting part of the Gould-Eldredge thesis was that an explanation can indeed be found. And interestingly, at this point, far from wanting to break from conventional (American) neo-Darwinism or the synthetic theory, Gould and Eldredge argued that it is precisely this theory itself that has the resources to explain the paradox! To make their case, the paleontologists turned to the ideas of Dobzhansky's associate, the major ornithologist and systematist Ernst Mayr. Some years previously, in order to explain speciation (the fact and causes behind new species), Mayr (1954) had proposed what he termed the "founder principle." According to Mayr, speciation results from a small group of organisms getting broken off or isolated from the main species population. Simply because of the new circumstances in which they find themselves, the members of this subpopulation start to evolve rapidly away from the parental form. In addition, argued Mayr, given the masses of genetic variation that occur naturally in any population, any small subpopulation will necessarily be atypical with respect to the whole group. There will therefore be a kind of shaking down as the members get used to each other and learn to do with much reduced genetic resources. Within the "founder population," there will be what one might call a "genetic revolution."

Mayr certainly thought of himself as being fairly orthodoxly Darwinian in his claims about speciation, although with hindsight one can see that what he was proposing was something much more in the spirit of Sewall Wright's shifting balance theory than Darwin's theory of the *Origin*. (Sewall Wright thought it was the shifting balance theory!) Mayr was arguing that a certain randomness, which occurs because of the breaking off of the subpopulation, is the crucial factor in the forming of new species. One has, as it were, a kind of genetic drift writ large. But whatever the true lineage of Mayr's ideas, this hypothesis was highly congenial to Eldredge and Gould. It suggested that new species will form very rapidly, not in the neighborhood of their immediate ancestors, but in new areas. You have species A in one place and then, almost overnight as it were, you have species B somewhere else. This could just be the kind of jerky fossil record that Gould and Eldredge thought was the true story to be read from the rocks. "If new species arise very rapidly in small, peripherally isolated local populations, then the great expectation of insensibly graded fossil sequences is a chimera. A new species does not evolve in the area of its ancestors; it does not arise from the slow transformation of all its forebears" (Eldredge and Gould 1972). In addition, the two paleontologists liked the way that Mayr was making the dynamics of populations (rather than the dynamics of isolated individuals) absolutely central to the evolutionary process. In the eyes of these paleontologists, factors operating over large periods of time, involving groups of organisms, yield the crucial causal keys needed for a full understanding of evolutionary processes. Here, for all that they drew on Mayr, Gould and Eldredge were starting to stand against population geneticists in the Dobzhansky tradition: scientists who looked at microevents often involving just a few individuals.

Yet at this point, although Gould was starting to embrace some ideas with but a loose connection to real Darwinism, he was not presenting himself as a dramatic revolutionary. This was to change in the next decade as Gould began to take a stronger and stronger position, setting himself more and more in opposition to prevailing orthodoxy. Why did he do this? There were a number of reasons. Undoubtedly, one was the fact that in the 1970s Gould immersed himself in a huge reading program in the history of biology. This was in preparation for *Ontogeny and Phylogeny*, his major work that appeared in 1977. Part history and part science, *Ontogeny and Phylogeny* argued that traditional links between embryology and phylogeny are better taken than people in the twentieth century had been prepared to recognize. At the same time—and perhaps in major part because of his reading program—Gould was growing increasingly sympathetic to elements of German evolutionism. He responded particularly warmly to that tradition going back, through Haeckel, to the morphology of the early nineteenth century that had so upset Cuvier: *Naturphilosophie*. Gould embraced with enthusiasm the *Naturphilosophen*'s emphasis on form rather than function, their insistence that what really counts when studying organisms is the architectural nature of the underlying ground plan, or *Bauplan* (Russell 1916). He liked the turn to homology and the retreat from what the German thinkers regarded as a rather superficial cherishing of selection-caused functionality.

From this, it was but an easy step for Gould to move right into an attack on all-embracing adaptationism. Notoriously, in 1979, writing with a colleague in the department of organismic biology at Harvard, the population geneticist Richard C. Lewontin, Gould produced an article arguing that much to be found in the organic world bears little or no direct connection to adaptive advantage (Gould and Lewontin 1979). Gould, with Lewontin, argued that there are significant constraints on development: these constraints forming and molding organisms in nonadaptive ways. And, simply as part of developmental processes, even when selection is at work there are bound to be a great many nonadaptive by-products. Much that seems to have purpose probably exists for no end-related reason whatsoever. With Lewontin, Gould drew attention to the triangular areas at the tops of pillars in medieval churches, things that they labeled *spandrels* (although it turns out that the true technical name is *pendentive*). These triangles—one finds them in St. Marks Church in Venice, as well as on the roof of King's College, Cambridge—are often used as vehicles for wonderful mosaics or carvings. They seem therefore to have a direct adaptive function. But, indeed, they really are simply part and parcel of the architectural constraints that were involved in medieval church building.

Gould and Lewontin argued that, analogously, many organic characteristics have a no true adaptive significance. The human chin, for instance, seems to be

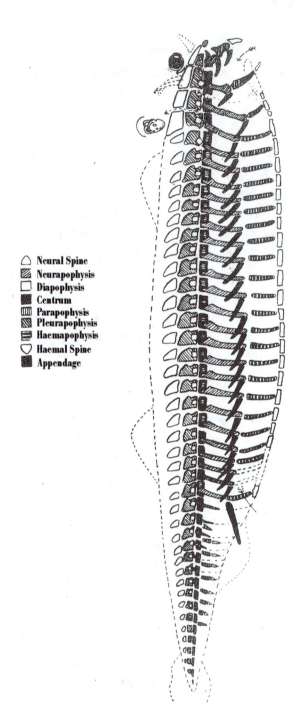

Legend:
- △ **Neural Spine**
- ▨ **Neurapophysis**
- ☐ **Diapophysis**
- ▓ **Centrum**
- ▥ **Parapophysis**
- ▧ **Pleurapophysis**
- ⊟ **Haemapophysis**
- ▽ **Haemal Spine**
- ▨ **Appendage**

Richard Owen's picture of the vertebrate
Bauplan *(which he called an "archetype")*

something with a purpose. Surely, if naught else, it is part of the design of the face for sexual attractiveness. But, in fact, detailed study shows that the chin is really something that comes about simply as a result of trying to put together other adaptive facial features: the jaw and the teeth and so forth. Seeming purpose should never be equated simplistically with genuine purpose.

In King's College Chapel in Cambridge, for example, the spaces contain bosses alternately embellished with the Tudor rose and portcullis. In a sense, this design represents an "adaptation," but the architectural constraint is clearly primary. The spaces arise as a necessary by-product of fan vaulting; their appropriate use is a sec-

The spandrels of San Marco

ondary effect. Anyone who tried to argue that the structure exists because the alternation of rose and portcullis makes so much sense in a Tudor chapel would be inviting the same ridicule that Voltaire heaped on Dr. Pangloss: "things cannot be other than they are ... Everything is made for the best purpose. Our noses were made to carry spectacles, so we have spectacles. Legs were clearly intended for breeches, and we wear them." Yet evolutionary biologists, in their tendency to focus exclusively on immediate adaptation to local conditions, do tend to ignore architectural constraints and perform just such an inversion of explanation. (Gould and Lewontin 1979, 583)

We are now at the end of the decade (1980). Gould was on a roll. He was mounting an all-out assault on the synthetic theory of Theodosius Dobzhansky and his colleagues. Gould (1980a) went so far as to argue that the synthetic theory is "effectively dead." At the same time, punctuated equilibrium theory—which was now becoming more and more identified with Gould alone—was breaking entirely from any connections with conventional evolutionary thought. In particular, it was being presented now as an outright saltationary theory, that is to say as a theory where large jumps (presumably brought about by macro mutations) are the key factors in evolutionary change. There was an expressed likeness for "hopeful monsters": organisms that take phylogenies directly from one form to another form. Drawing on his deep knowledge of evolution's history, Gould was bringing forward evolutionists from the past who were supportive of saltationism: evolutionists who, so Gould maintained, had been unfairly belittled or denied credit simply because they were out of tune with the ideology of the then prevalent Darwinism. The synthetic theory, so he claimed, was little more than an extension of nineteenth-century liberalism, with its fondness for gradual change rather than revolution.

As you might have expected, conventional evolutionists—those working on fast-breeding organisms and concerned more with microevolution than with macro changes—started to get very tense. Here was a very public evolutionist—Gould's *Ever since Darwin*, published the same year (1977) as *Ontogeny and Phylogeny*, was a runaway best-seller—telling the world that their theory was not true science but merely washed-up Victorian ideology. G. L. Stebbins, the botanist member of the cohort who put together the synthetic theory, together with Dobzhansky's student Francisco Ayala, wrote an influential paper pointing out that natural selection is sufficiently powerful to bring about all of the so-called saltationary changes that Gould was demanding (Stebbins and Ayala 1981). In addition, these critics argued that although selection may seem fairly leisurely in the eyes of an individual human, from the perspective of geological time it is more than sufficiently rapid to bring about any conceivable macro changes: both those recorded and those not recorded directly in the fossil record. In other words, as Darwin and his followers had always argued, the gaps in the record are as much artifactual as genuinely representative of things that truly happened.

Continuing their counterresponse, these doughty defenders of tradition pointed out that no Darwinian has ever claimed that the course of evolution is always as smooth and gradual as is implied by Gould's caricature of their theory. It has always been recognized that the pace of evolution is something that speeds up and slows down, according to many different factors. There are impinging conditions imposed both from without the organic world, geological factors, for instance, and impinging conditions imposed from within the organic world, competitors and the availability of desirable ecological niches, for instance. It is true that Darwinism demands that immediate change be gradual—there is indeed no place for hopeful monsters—but over the time scales recorded in the fossil record, there is no reason at all to expect uniformity. "Living fossils" such as horseshoe crabs have persisted over hundreds of millions of years. Other organisms have evolved very rapidly. And in any case, the saltationists of the past, worthy scientists though they may have been in their time, are now simply outdated and wrong.

Gould was never one to acknowledge directly that he was mistaken or even that he was walking on dangerous ground. There was certainly to be no dramatic retraction of any of the claims that he had made when he was writing at his most vehement level. However, over the next decade—that is to say, through the 1980s—in many respects, Gould did start to pull back from the more extreme positions that he had taken or floated. Not entirely accurately, he now denied that he had ever made the extreme claims ascribed to him. In particular, he denied strongly that he had ever been an outright saltationist. Gould (1982) now started to argue that he was not so much against Darwinism as such, but that what he had been advocating and would continue to push for was a kind of expanded Darwinism. This would be a vision where natural selection and adaptation are indeed very important aspects of organic life and of the evolutionary process. A vision, however, where there is a perceived need for the supplementation, sometimes dramatically, of selection by other processes.

More specifically, in Gould's opinion what one has now (at least, what one needs now) is less a single-level theory—as apparently was true of the synthetic theory—and more something that is hierarchical. The image here is of the Catholic church, with its different levels from the parish priest right up to the pope. Likewise in evolutionary theory, argued Gould, we need a layered perspective, going from bottom to top. Neo-Darwinism is good and right, as far as it goes, but it speaks only to a kind of midlevel to the hierarchy. Beneath natural selection working on individual organisms, one has a microlevel that involves molecular biology. Here, at this molecular level, it is pertinent to note that a number of theoretical biologists, particularly Japanese population biologists, have argued that there is ubiquitous randomness: what came to be known, naturally, as molecular drift. It is a well-known fact that at this molecular level, there is a great deal of redundancy. Different molecules encoding the DNA produce the same cellular products. Hence, there is every reason to think that these differences lie below the forces of natural selection and simply drift from one form or ratio to another. (The classic statement of this thesis can be found in Kimura 1983.)

Then, argued Gould, above the microlevels of individual selection, one has macrolevels involving vast periods of time. Here, other new forces come into play. And here, at this macrolevel, the expertise of the paleontologist comes into its own. One sees that individual selection makes no major difference and that such things as constraints on development start to be the major determining factors. Perhaps some of the ideas raised in the spandrels paper are important here. Initially, a certain *Bauplan* is the all-important constraint on what an organism (or a group of organisms) is and must be. A threshold is reached, and there is a rapid change from one *Bauplan* to another—a change that has nothing to do with natural selection, being rather a shuffling of the internal structure (morphological, biochemical, whatever) of the organism. Then selection comes back into play, refining and elaborating on the new form that has been produced. It is all rather as if a kaleidoscope had been shaken, and a new picture emerges from parts that had been fragmented and reassembled.

Crucial to this whole way of looking at things was the belief that what is going at this upper level simply cannot be explained in terms of the lower levels. Gould (like Lewontin) was long an ardent critic of what he labeled "reductionism": the assumption that the key to understanding the upper levels of reality lies in delving ever more deeply into the lower levels of reality. Gould did not deny

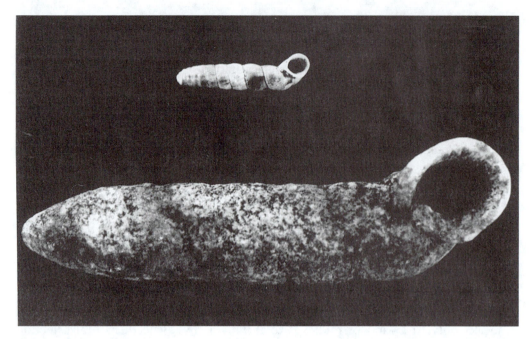

The usual Cerion *height to width ratio is less than 3. At extreme sizes (dwarf and giant)* Cerion *occur with higher height:width ratios. Gould calls these "smokestacks."*

that this assumption can be the basis of very fruitful inquiry—in ecology, it may well be the vital method of investigation—but he was adamant that it is very dangerous if taken as an all-determining metaphysical principle. Sometimes one can and should try for an understanding at an emergent level—at a higher hierarchical level. And here the higher simply cannot be reduced to or explained away at the lower level. Specifically with respect to evolution at the macrolevel, one has things happening that cannot be explained at microlevels. Dobzhansky and his fellows were just plain wrong. Genetics, the science of the micro, must be supplemented by paleontology, the science of the macro. To argue otherwise is to slip into the dreadful sins of Panglossianism or the building of "Just So" stories (things encountered in this and the last chapter).

For the last two decades of his life—he died of cancer in 2002—Gould refined his position, trying to build on and develop his own ideas, while at the same time wearing down the opposition: wearing down the Darwinian opposition, that is. One paper dealt with the shapes of snail shells, showing that certain atypical forms of the shells—so-called smokestack shells—are a function of constraints on growth, rather than Darwinian selection as the synthetic theory would argue.

> Evolution is a balance between internal constraint and external pushing to determine whether or not, and how and when, any particular channel of development will be entered. Natural selection is one prominent mode of pushing, but most engendered consequences of any impulse may be complex, nonadaptive sequelae of rules in growth that define a channel. Most changes must then be prescribed by these channels, not by any particular effect of selection. Natural selection does not always determine the evolution of morphology; often it only pushes organisms down a preset, permitted path. (Gould 1984, 191–192)

Another paper, coauthored by Gould, dealt with the replacement in the same ecological niche of one organic form by another (Gould and Calloway 1980). Gould's claim was that such a replacement might as well be nonadaptive as anything fueled by selection. We may have "ships that pass in the night." To assume otherwise is simply to make a dogma of Darwinism. And yet a third paper dealt with specific forms of nonadaptive characteristics, things that Gould has labeled "exaptations" (Gould and Vrba 1982).

A major contribution to the cause was Gould's (already-mentioned) *Wonderful Life: The Burgess Shale and the Nature of History*, a book published in 1989. On the surface, this is a work about soft-bodied organisms (dating back to the Cambrian) found fossilized in the in the Rockies of Western Canada. There are all sorts of strange forms, truly sparking one's imagination and seeming to defy orthodox classification. But the telling of the tale is only one part of what Gould was about. Truly, indeed, this was a work with a mission. Gould used the Burgess Shale to launch an attack on what he saw as an incorrect picture of the history of life, an incorrect picture that had been brought illicitly into evolutionary studies by enthusiastic Darwinians. A particular bugbear of Gould was the idea of evolutionary progress—our old friend of upward change, from monad to man. He thought this is a truly false picture of history, which is rather one of randomness and chance and lack of any significant direction. Certainly, humans came last. If they did not, we would not be around now to tell the tale. But we are not the finest culmination of a directed process. Like everything else, we just happened. And the fossils of the Burgess Shale show that this is so. There are all sorts of weird and wonderful forms, all now extinct with very few exceptions (one of which may be a vertebrate predecessor), and any one of these might have been the progenitor of today's organisms. It was just chance that it all went one way rather than any other. Life has no ultimate meaning and history shows this. Those who think otherwise, Darwinians particularly, are just plain wrong.

This continued as Gould's theme. Another book, already mentioned, was *Full House: The Spread of Excellence from Plato to Darwin* (1996). This may ostensibly have been about baseball statistics, and the probabilities of anyone ever repeating DiMaggio's feat, but truly it was about the nature of history and how seeming direction can be simply a function of chance:

> If one small and odd lineage of fishes had not evolved fins capable of bearing weight on land (though evolved for different reasons in lakes and seas), terrestrial vertebrates would never have arisen. If a large extraterrestrial object—the ultimate random bolt from the blue—had not triggered the extinction of dinosaurs 65 million years ago, mammals would still be small creatures, confined to the nooks and crannies of a dinosaur's world, and incapable of evolving the larger size that brains big enough for self-consciousness require. If a small and tenuous population of protohumans had not survived a hundred slings and arrows of outrageous fortune (and potential extinction) on the savannas of Africa, then *Homo sapiens* would never have emerged to spread throughout the globe. We are glorious accidents of an unpredictable process with no drive to complexity, not the expected results of evolu-

*The discoverer of Burgess Shale,
Charles Doolittle Walcott*

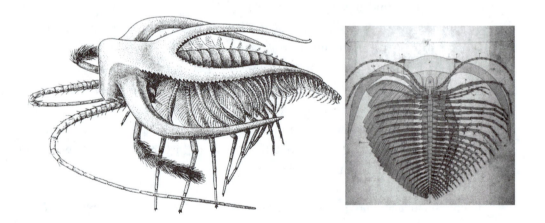

*A denizen of the Burgess Shale (*Marrella*)*

tionary principles that yearn to produce a creature capable of understanding the mode of its own necessary construction. (Gould 1996, 216)

Just as he lay dying, Gould published yet another book, *The Structure of Evolutionary Theory* (2002). This was a truly gargantuan compendium of all of Gould's thinking on just about everything. Certainly, if you could take only one Gould book with you to a desert island, this would be it. But really Gould's genius lay in the short, pithy piece. Love him or loathe him—and at (different) times I felt both emotions in the course of our twenty-five year relationship—the guy really could write a terrific opinion piece.

Taking Things Apart

Whether or not evolution has many different levels or layers, Gould's arguments most certainly do. So let us take them apart and see what we get. At one level, the most basic level, you may say that we have a scientific argument. Was he right that the fossil record is as jerky as he claimed and that this is something that a Darwinian cannot handle or explain? Whatever else, one can certainly say that Gould drew attention to the question of the rates of evolution. Intense effort has been expended on the path or course of evolution as revealed through the fossil record and on its putative support for the theory of punctuated equilibria. And the answer, I am afraid, is one of extreme ambiguity! Indeed, perhaps by now you might have been expecting that I would say this, because several times before in this book when we have come up to crucial points of decision, I back away and say that the facts cannot decide! Although an exaggeration, there is some truth in this. But I think it probably tells you more about science than it does about me. (Although my father used to complain that I could not open my mouth without telling you something about me.)

The truth is that when scientists hold different positions, it rarely is simply one of the physical facts. Both sides can summon up facts to suit their respective causes: Cuvier points to function, Geoffroy to form; Dobzhansky to heterozygosity, Muller to homozygosity; Stringer to Neanderthal differences, Wolpoff to Neanderthal similarities. The facts are not irrelevant, anything but, yet they are not decisive. And certainly this is the case here. There are cases where evolutionary change seems to have been very rapid indeed—so rapid, that it surely qualifies as sudden or jerky in the terms demanded by punctuated equilibrium theory. It seems likely that the evolution of fish (cichlids) in East African lakes qualifies here—one can show that speciation has been so rapid an event that even if there were fossilization, it would be invisible in the record. (Williamson 1985). There are cases where evolutionary change seems not to have been so very rapid—slow enough, in fact, that the changes do come through in the fossil record. This seems true of the evolution of certain mammals, for instance. And there are cases where, depending on your inclination, you can interpret the record one way or the other. The human fossil trail seems to fall into this camp. It is not that punctuated equilibrium theory is wrong and that the Darwinian alternative (what Gould calls "phyletic gradualism") is right, or conversely. Rather it is that the fossil record simply is not decisive.

But this is not the end of the argument, for there are other levels of debate. Just as I am always arguing that the facts are not decisive, so I am also always arguing that philosophical differences really count. I will not disappoint your expectations, for I do think that they are very important here. One thing that may seem important is Marxism. Notoriously, Gould boasted of his connections to this philosophy—we are told that he learnt it "at his daddy's knee"—and he certainly drew attention to the way in which the Marxist view of world history is one of rapid revolutionary change, rather than gradualism (which Gould linked with the liberal philosophy that was Charles Darwin's). Also, the antireductionism—seeing different processes at work at different levels is Marxist—is a translation of Engels's law of quantity to quality. (Lewontin, who coauthored the spandrels paper, is an ardent Marxist. See also the comments in Gould and Eldredge 1977.)

But although I am sure that this is important, I doubt that it ever was all-important, even if we discount the fact that after Stephen Jay Gould became *the* Stephen Jay Gould, he rather backtracked on his earlier influences and enthusiasms. In line with what we have seen, more pertinent to Gould's thinking, I suspect, was that whole Germanic approach to biology (which was, naturally, shared by Marx and Engels). It is the approach of the *Naturphilosoph*, who thinks that form takes precedence over function, who thinks in terms of hierarchy, whose philosophy of history is one of dialectic, swinging from one pole to another. Add to this a good swig of Herbert Spencer—the very name, "punctuated equilibrium," reeks of the old man. More seriously, the obsession with equilibrium is very much a Spencerian concern for evolution (as opposed to Darwinian, where it plays no essential role whatsoever). And certainly in some of his writings Gould showed a liking for the notion of "homeostasis," an idea developed on Spencerian lines in the 1930s by the physiologist Walter B. Cannon (1931), supposing that organisms get themselves into a kind of balance and have a natural tendency to stay or return to the beginning point.

There is one final item that should be added and then the case will be complete. Gould was a paleontologist. In the eyes of the general public, this is what evolution is all about: fossils, dinosaurs, Lucy, and all of that. But as you must now realize, this is not at all the way that professional evolutionists see things. To them, paleontology is just the thing that they have had to escape in order to raise the status of their science. To get out of the museums and away from a quasi-religious system of hypothetical phylogeny building, they have had to turn to tight, mathematical, experimental, causal studies of fast-breeding organisms like fruitflies. I would hardly want to say that dinosaurs are an embarrassment, but even now there are echoes of the past. The past decade or so, for example, has seen a very public and indecisive debate about the origins of the birds (Feduccia 1996)—are they descended from the dinosaurs or from other nondinosaur reptiles? If you cannot answer something as basic as this, what hope of a real quality science?

It is symptomatic of the state of affairs that when, in the early 1980s, Gould began suggesting that one must invoke one-step changes in organisms to account for the fossil record, he was slapped down and into place by the geneticists. Paleontology must do what it is told by the geneticists, rather than conversely. But now, with punctuated equilibrium theory, the case is changed—at least, such was the hope of Gould and others in his field. Geneticists must sit up and take notice.

Not only does paleontology have its own level or levels of understanding—levels that cannot be eliminated (reduced away) by slick appeals to genetics—but there are dimensions where paleontology can actually tell genetics what it can and cannot do. Equality is now in sight. And anyone who thinks that something like this was not of extreme interest to a person with the ego of Stephen Jay Gould, simply did not know the man. Upgrading his subject was always high on Gould's list of things to do. No one wants to spend their professional lives in a subject that is regarded with disdain, if not contempt—the sociology of the life sciences. If Gould's program succeeds, if people do accept the need for an expanded Darwinism, then at long last paleontology will come into its own. It can stand shoulder to shoulder with genetics rather than lurk unobtrusively in the background, coming forward only when called. The title of a talk Gould gave back in 1983 tells all: "Irrelevance, Submission and Partnership: The Changing Role of Paleontology in Darwin's Three Centennials, and a Modest Proposal for Macroevolution." (The three centennials were for the birth of Darwin in 1908, the publication of the *Origin* in 1959, and the death of Darwin in 1982. We Darwinians like centennials.)

It is this, as much as anything, that accounts for the bitter note in John Maynard Smith's criticism quoted at the beginning of this chapter. The trouble is that by the time that Maynard Smith wrote, people were starting to take Gould seriously, and that rankled. It rankled also that Gould did not fight his battles just in the professional journals, where only professional scientists would take notice. He got into the public arena, with his monthly column in *Natural History*, and then in collections and monographs, as well as many other places, notably the influential *New York Review of Books*. For Maynard Smith, geneticist and sociobiologist, this was all the wrong way around. Gould should be judged against the standards set by Maynard Smith and his fellows and should not try to get around difficult points with philosophy and rhetoric. He should have been more respectful of and appreciative toward the ideas that have been developed and inherited. And he should not have reminded the world of the shaky status of so much evolutionary theorizing for so long. It was not just that Gould's ideas are wrong. It was that they are presented as position of reason and tolerance and common sense, and the outside world believed him. That really irritated.

It still irritates even though Gould is now gone. The already-mentioned well-known American philosopher Jerry Fodor (2007) has embraced the spandrels argument with gusto. Terrified that we humans might be part of the animal world, he attacks adaptationism with the frenzied enthusiasm of the true believer. "History might reasonably credit Stephen J. Gould and Richard Lewontin as the first to notice that something may be seriously wrong in this part of the wood. Their 1979 paper, 'The Spandrels of S. Marco and The Panglossian Paradigm: A Critique of the Adaptationist Programme', ignited an argument about the foundations of selection theory that still shows no signs of quieting." Apparently it was all part of a mistaken analogy on Darwin's part, moving from the conscious design of the animal and plant breeders to the supposed mindless design of natural selection.

The present worry is that the explication of natural selection by appeal to selective breeding is seriously misleading, and that it thoroughly misled Darwin. Because breeders have minds, there's a fact of the matter about what traits they breed for; if you want to know, just ask them. Natural selection, by contrast, is mindless; it acts without malice aforethought. That strains the analogy between natural selection and breeding, perhaps to the breaking point. What, then, is the intended interpretation when one speaks of natural selection? The question is wide open as of this writing. (Fodor 2007, 20)

Adding a twist to the Gouldian argument—a twist that might or might not have been appreciated by one of the twentieth century's masters of metaphor—Fodor finds much of the trouble in the metaphor of design as we find it in biology. The function of the eye, the purpose of the heart, and that sort of thing, brought about by selfish genes and the like. "Metaphors are fine things; science probably couldn't be done without them. But they are supposed to be the sort of things that can, in a pinch, be cashed. Lacking a serious and literal construal of 'selection for', adaptationism founders on this methodological truism" (p. 20).

Needless to say, none of this has gone unchallenged. Philosophers and biologists have responded with vigor spliced with a certain amount of irritation. Daniel Dennett (2007) for one points out that the whole point of the Gould-Lewontin argument about spandrels is that some things are adaptive—the pillars in churches, the roofs they support—and that, in the wake of this adaptation, non-adaptive things like spandrels are likely to emerge. "I won't bother correcting, one more time, Fodor's breezy misrepresentation of Gould and Lewontin's argument about 'spandrels', except to say that far from suggesting an alternative to adaptationism, the very concept of a spandrel depends on there being adaptations: the arches and domes are indeed selected for, and they bring spandrels along in their wake. No 'perfectly reasonable biologist' has claimed that the hugely various and exquisitely tuned sense organs of animals, or the superbly efficient water-conserving methods of desert plants, are spandrels, even if they spawn spandrels galore." I am sure that somewhere, looking down or looking up, Stephen Jay Gould is enjoying every moment of this!

Hierarchy Theory

Let us pull away from the motives and countermotives, charges and countercharges. The really important question is whether Darwinism—an ultra Darwinism, which pushes selection without hesitation or apology—is enough, or whether one really wants and needs more to get a full understanding of the evolutionary process. Start with the level below the physical characteristics (the phenotype), the molecular level. At this level (as Gould noted), it has been hypothesized that selection can have only a minimal effect. Even if natural selection produces the hand and the eye, the molecules making everything up are another matter entirely. Selection may (for instance) decide between a blue eye and a brown eye, but suppose there are two ways (with different molecular patterns) of making a blue eye. Selection could not decide between them. Some biologists, extending this possibility, think that in real life there is a huge amount of molecular redundancy, and it was suggested by a leading Japanese population geneticist, Motoo Kimura (1983), that at this level the molecules just drift up or down to total fixation or to elimination. In populations where the effects of selection can make themselves known, drift can operate only on small populations. But where selection is absent or minimalist, drift can (in theory) have major effects on large populations.

But is this true? Well, possibly in some cases. This is the basis for the very successful notion of a "molecular clock," where one judges the time since different organisms had a common ancestor by the amount of genetic molecular difference there is between them. Drift, unlike selection, is presumed to be something that produces change at a fairly standard or regular rate. But in other cases, molecular drift certainly does not hold. Where one is dealing with nonfunctional chunks of DNA (pseudogenes), no doubt drift is the player that counts. But overall the amount of drift at the molecular level has been subject to various experiments, some of which suggest strongly that selection is sifting through the molecules, choosing some and rejecting others. For instance, there has been detailed study of the molecular gene replacements in closely related species of fruitfly (Drosophila). If the genes are drifting up or down, irrespective of selection, then one ought to find the same orders of magnitude of differences between species as one finds within species. Everything is going according to random patterns, so interbreeding and like phenomena should make no difference. In fact, they did make major differences. Between species, one finds significant differences in the molecular genes, but within species although there is some variation, there is far less. This all rather suggests that within the species selection is acting in a positive way to cherish some genes and to eliminate others. A counter to the neutral theory (McDonald and Kreitman 1991).

Move next to the physical level: the phenotype. It is here that selection is supposed to reign supreme. But does it? The Darwinian—the ultra-Darwinian like Richard Dawkins—thinks selection is very, very important. But all important? In fact, no one has ever wanted to claim that selection works in a perfect fashion, forever producing adaptations at their "optimized" peak. One might for instance be dealing with something that had an adaptive function but that now no longer serves such an end. It could be that circumstances have changed, and selection simply has left the feature in place—perhaps selection is unable to reduce the feature. Paradoxically, one ultra-Darwinian has suggested that human sexuality might fall into this category (Williams 1975). Although this is a controversial issue and not all would agree, there are reasons to think that sexuality is really only of adaptive advantage to fast-breeding organisms in unstable environments. For humans, who breed slowly and who stabilize their environments, sexuality may be positively disadvantageous—a single female could do the work herself (as is the case in many mammals and to be candid many human families). But our anatomy and physiology have now become so specialized that we cannot relinquish sexuality. We are stuck with it, for all its problems.

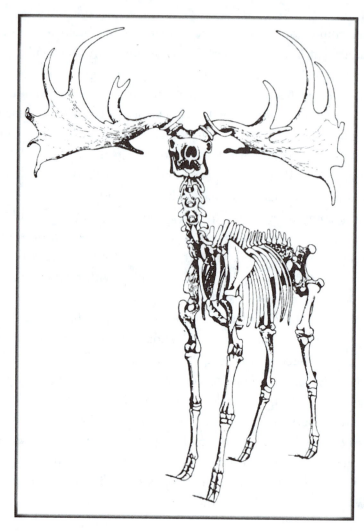

The Irish elk

Another reason for the nonoptimality of adaptation is relative growth (allometry). Sometimes features are linked together, with one part growing faster than other parts. In fact, this is a well-known and studied phenomenon, and it turns out that the usual relationship is logarithmic—the fast-growing part grows at a very much faster rate than the other parts. It could be that such a fast-growing part is of crucial importance in breeding, but unfortunately it then peaks and goes over into nonadaptive status as the rest of the organism matures and reaches full size. It is thought that possibly the massive horn-growth of the extinct "Irish elk" (actually a deer) could have come through such a process. For early breeding purposes, big horns are a decided advantage. But then the horns just keep growing even though they are maladaptive. Unfortunately by this stage the damage is done—the next generation have the potential for big horns—and so the adult maladaptation is perpetuated.

Something similar occurs when one has sexual selection working against natural selection. Big tails are sexually desirable in the peacock, but from a natural selection viewpoint—escaping from predators—they are no good at all. Such characteristics are adaptive in one sense and maladaptive in others. And then finally let me mention pleiotropy. Sometimes more that one characteristic is produced by one gene. If the one characteristic is very valuable in the struggle for life, then it can balance other characteristics that are less valuable or even harmful. In a way, this is an individual phenomenon somewhat akin to the group effect that you get with balanced superior heterozygote fitness, where the homozygotes are less fit than the heterozygote—where, indeed, a homozygotes may be so unfit as to be absolutely lethal.

So you can see that, although they are all selection connected, the Darwinian certainly sees a place for nonadaptive features. Moreover, turning now to Gould's counterarguments, the Darwinian would challenge many of Gould's supposed examples of nonadaptive characteristics. It would be argued that these features are indeed rooted in adaptive advantage, as brought about by natural selection. Take the key example of vertebrate limb number. Gould suggests that the fact that vertebrates have four limbs rather than six (as insects have) is purely a matter of contingency or constraints on building vertebrates or some such thing. Having four rather than three or five can be explained through adaptive advantage—five legs would be lousy for running, although I suppose the kangaroo, with two legs and a tail, might make one pause about three—but why four rather than six or even eight (like arachnids)? However, Maynard Smith (1981) has seized on this example as precisely one where selection does count! He points out that the early vertebrates were sea creatures, with the need to go up and down rapidly in the water. This, as with airplanes in the air, is best effected by two wings or limbs fore and two wings or limbs aft. In fact, there were vertebrates with other numbers of limbs, but selection favored the four-limbed variety. Today, we live with the relict of this need. It may be that we could get by with a different number—snakes, whales, and chickens obviously do—but that is not to deny the fact that four is rooted in selection, contra Gould's claim. And as we have seen, it is certainly not part of the Darwinian case that all features must have maximum adaptive value right now, and always. The point is that such features are connected to selection in some way, at some point in time.

Macroquestions

Finally, what about the upper level of the hierarchy? No one is going to deny that you are going to get effects at the macrolevel—that is, over long periods of time—that are more than just microeffects stitched together. I am not sure that there is anything mysterious or "holistic" about this, but the fact is that the course of history over millions of years simply does not follow from the changes in a fruitfly cage. If this is what antireductionism means, then we are all antireductionists. One does not need an Engels to tell us as much. Take, for instance, the whole question of extinction. Not only do individual species go extinct, but sometimes you get a whole range of species going extinct at the same time: "mass extinction." Such events saw out the Devonian, the Permian, and most famously (when the dinosaurs went) the Cretaceous. No one could have inferred these extinctions from microevents, but then no one would ever have pretended to.

Clearly, some other factors—possibly random and possible not, possibly extraterrestrial and possibly not—were involved. Everybody knows that the popular hypothesis for the end of the Cretaceous is that an asteroid or some such thing hit the earth, causing a great dust cloud and blocking out of the sun, and that as a consequence there was cooling and paucity of food and that this put paid to the dinosaurs (as opposed to the mammals who were just weedy little runt-sized nocturnal animals). This is not something that could have been predicted by population genetics, but it is something that is part of the causal story of life's history (Alvarez et al. 1980).

It has to be granted then that the macroevolutionist—the paleontologist—will tell us something about evolution as path that we cannot get from elsewhere. And this will surely lead into discussion of evolution as cause, as one tries to understand and explain the path. The causes might not be directly biological, but they are part of the picture. I am not sure that there is any question of downward causation—of the paleontologist teaching and instructing the geneticist—but there is certainly some measure of autonomy to the macrolevel. But can one go on from here? Are there biological patterns at the macrolevel that would not be expected from the microlevel? Can the macroevolutionist show and explain biologically fueled events that do not appear at smaller levels with shorter times?

In principle there seems no reason why not, and in fact we do find that some workers have tried to provide explanatory models of this nature. By example, let me take a problem that has long puzzled students of life's history, namely the so-called Cambrian explosion. Nearly 600 million years ago, life suddenly started to explode in diversity and number. From fairly sparse numbers and types, at least as revealed in the fossil record, huge numbers and varieties made their appearance, almost overnight as it were. Now there are a number of questions that you can ask—for instance, about why the explosion happened at all. And some of the answers will surely be framed in terms of adaptive advantage. For instance, it may be that the seawater was carrying much more oxygen, thanks to photosynthesis caused by algae, and this then made possible the sustenance of many more and more complex life-forms than previously.

But what about the actual pattern of the explosion? John J. Sepkoski Jr. (a student of both Edward O. Wilson and Stephen Jay Gould) collected huge amounts of data about the numbers of different kinds of organism that have been recorded as living back then at the time of the explosion. He found, plotting numbers on a graph, that the picture is roughly s-shaped (sigmoidal)—a rapid rise up, and then a flattening out. To explain this, Sepkoski turned to a well-known ecological hypothesis about the colonization of islands by organisms, formulated in the 1960s by Princeton biologist Robert MacArthur and Harvard entomologist Edward O. Wilson. The island biogeography hypothesis specifies that organism species numbers will reach equilibrium (a function of distance from the mainland and island size) after a period of (exponential) growth—new species arriving on an island will equal the old species leaving or going extinct. Reasoning that colonizing in time is much like colonizing in space, Sepkoski (1976) was able readily to

John J. Sepkoski Jr.

show that one can model the sigmoidal rise of organisms in the Cambrian using the MacArthur-Wilson hypothesis.

The first models produced by Sepkoski were understandably crude, but they were sufficiently promising to stimulate him to further effort. He worked diligently to expand his database—for technical reasons he focused on marine animals—and as the material piled up, he found that he needed to refine his theory. Instead of a nice smooth upward rise, a sigmoidal curve carrying one through the Cambrian and beyond, there is a midlevel break as the growth pauses before picking up again to continue the movement upward (Sepkoski 1979, 235). Tantalizingly, those organisms that seem most successful during the Cambrian reach their peak at the time of this pause, before they start into a long, slow decline.

Tantalizing but suggestive. Surely what is needed is a second set of equations, superimposed on the first, with a second curve therefore taking off on the back of the first. The Cambrian organisms (marine fauna) reach their peak halfway up and then start to decline. But in the meantime, rather like a second-stage rocket that takes over when the first stage is exhausted and is now falling down to the sea, the next batch of organisms has taken over and is rising up through the Paleozoic. "The two-phase kinetic model … seems to provide an adequate description of the fundamental patterns observed in the early Phanerozoic diversification of marine metazoan families" (p. 242). This is just a description of what is happening, but the temptation is strong to speculate on causes, and some hypotheses come at once to mind. Could the earlier organisms be rather "generalized" in some sense, good for flourishing and increasing when there is lots of empty ecological space, and could the later organisms be rather "specialized" in some sense, good for flourishing and increasing when the ecological space is much more crowded? Are we looking at the replacement of organisms that have "relatively

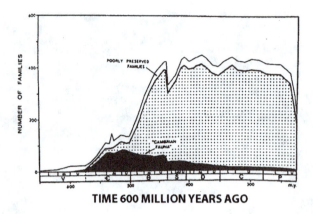

The history of life showing the different phases of the growth of living beings

TIME 600 MILLION YEARS AGO

broad feeding and habitat adaptations" by organisms that "might be expected to exhibit lower rates of speciation and extinction and, as a result, lower rates of diversification but higher equilibria" (p. 243)? Are these replacing organisms better at utilizing crowded or restricted environments, so that we end with "more finely divided and stable ecosystems which can be described as having high equilibrial diversities" (p. 243)?

We are not done yet. As more data flowed in, Sepkoski discovered that the new, replacing organisms ran out of steam at some later point, peaking and then going into a slow decline. But now he knew just what to do! A third set of equations yielded a third curve, with a new set of organisms taking off on the back of the second set. After the great extinction at the end of the Permian, life picked up again, increased in diversity , and grew with some force and speed right up to the present (Sepkoski 1984). Humans, of course, are messing things up at the end. The ways in which we are destroying habitats and the denizens thereof has a major impact. But the overall picture of life's history makes good sense.

Moreover, perhaps we can even try our hand at predictions. Humans aside, we seem to be in a bit of a lull right now. Could it be that there is a fourth group of organisms waiting in the wings, ready to take off on the backs of today's animals, ready to scale yet higher peaks? It seems improbable but cannot be discounted entirely. In the plant world, with the arrival of the Cretaceous, we got a new fourth kind of flora, the angiosperms (the flowering plants). Could not the same be true of the animal world? "By analogy to the plant record, we can speculate that one or more unpredictable innovations of importance comparable to angiosperms might appear among future marine animals, leading to major changes in faunal composition and driving diversity to yet higher levels" (p. 264).

The point is made! Sepkoski was certainly not against Darwinism, meaning explaining evolution through selection. Rather, he was interested in somewhat different questions. To be honest, if I were looking for a predecessor, someone in whose shoes he stands, I would opt for Herbert Spencer rather than Darwin. All of the talk about moving up to a plateau and then a period of stability or equilibrium sounds very much like the synthetic philosophy updated. Which would fit in with the influences under which Sepkoski fell. Remember that he was a student of Gould as well as Wilson, and both of these men are Spencer-influenced: the fact that they fell out bitterly is almost what you expect from family members. Moreover, since in the case of Sepkoski we have a paleontologist who came into evolution via an intense interest in computers—he never took biology courses as an undergraduate—there is really no reason to seek for strong naturalist influences. I mean that we should not expect to find, nor do we indeed find, influences leading to a fondness for selection.

But however you analyze Sepkoski—on content or on influences—the fact is that he worked at a level that is above and beyond that of the Darwinian working on selection-related problems, trying to understand features of and changes in today's organisms. In this sense, Gould was truly right to think of evolutionary theorizing as hierarchical. Darwinism is not the washed-out, inadequate theory he pretends it to be, but there may well be more to the history of life and to our understanding than ultra-Darwinians sometimes claim. In the end, Steve Gould was much like the rest of us. Sometimes he was wrong. And sometimes he was right!

Further Reading & Discussion

As I noted in the text, for many years, in the magazine *Natural History*, Gould wrote a monthly column: "This View of Life." He ranged over many topics, on and around the life sciences. One month you got an account of an old volume on natural history Gould had discovered tucked away in a secondhand book store. The next month you learnt of the mating practices of some tropical bird. But through the diversity of topics and friendly, almost folksy, style, the reader senses that there is not just a keen intelligence but a burning moral passion. Gould's view of life was fun. Gould's view of life was serious. As also noted, the essays were collected in published volumes. The first and still the best is *Ever Since Darwin* (New York: Norton, 1977).

Although I am more passionate about the shorter pieces, the best of Gould's full-length books has also already been mentioned. Ostensibly, *Wonderful Life: The Burgess Shale and the Nature of History* (New York: Norton, 1989) is a discussion of the marvelous finds of soft-bodied, fossilized invertebrates, in a place up in the Rockies between the Canadian provinces of Alberta and British Columbia. These fossils give us an insight into the nature of life just before it exploded up into the rich diversity that marks living beings as we know them, an event that happened during the Cambrian period some half a billion or more years ago. Gould used the fossils and their interpretation as a vehicle to discourse on the process of scientific discovery and theorizing, as well as on nature and form of life itself, drawing conclusions about the paths and causes of evolution, the status of humankind, and the ways of scientific reasoning. I myself disagree with just about every one of his conclusions, but I have rarely enjoyed a book so much. It is simply science writing at its best—clear, informative, provocative.

Wonderful Life was also, I believe, a rather clever pastiche on books about what was, along with the fossils, Gould's other great passion: baseball. See how Gould wrote of his characters as if they were managers and players in America's National Pastime. Judge, for instance, how he treated the British paleontologist Simon Conway Morris, brought up from the minors by a manager who saw real talent in his unpolished character and who took him on to win a Cy Young award of science, fellowship in Britain's Royal Society. Conway Morris, I should say, did not entirely appreciate the honor of Gould's analysis and responded with a somewhat waspish book of his own: *The Crucible of Creation: The Burgess Shale and the Rise of Animals* (Oxford: Oxford University Press, 1998). Unfortunately, although he may be a better scientist, Conway Morris is nowhere like as good a writer.

Given Gould's engaging prose, what more could you ask of anyone or anything? Well, how about a different perspective, from someone who writes as well yet who is as committed to Darwinism as Gould is questioning. I refer to Richard Dawkins, as English as Gould is American. Dawkins's exposition of the ideas and achievements of Darwinism, *The Blind Watchmaker* (New York: Norton, 1986), is just superb. Dawkins has long been something of a computer buff, and he uses his knowledge and skill to great effect, especially when he is dealing with all of those worries so often expressed about whether so simple a mechanism as natural selection can truly generate the complexity we find distinctive of the living world. How can you generate a line of a Shakespeare play in just a few moves, if all you have is the random processes of nature, akin to a monkey striking randomly on the keys of a typewriter? Read Dawkins and find out. Find out this and much more as you are taken from one dazzling chapter to the next, each one proving that those who think that the scientific study of nature in some way impoverishes our sensibility are themselves the ones truly lacking and blind in spirit.

Nearly a century ago, the embryologist E. S. Russell wrote a book trying to trace what he saw as the two conflicting tendencies in biological understanding, between those (like the *Naturphilosophen*) who emphasize the form of organisms and those (like Cuvier) who emphasize the functional nature of organisms. You know that I think this difference is reflected in the opposed thinking of Gould and Dawkins. Read *Form and Function, a Contribution to the History of Animal Morphology* (London: John Murray, 1916) to see that they are the end points of a long tradition of difference. In my *Darwin and Design: Does Nature have a Purpose?* (Cambridge: Harvard University Press, 2003). I try to bring these issues up to date in the light of a century of work on evolutionary questions since Russell wrote his great book. In my *Mystery of Mysteries: Is Evolution a Social Construction?* (Cambridge: Harvard University Press, 1999) I discuss Gould as well as Geoffrey Parker and Jack Sepkoski (as he was always known). I look also at Edward O. Wilson, Richard Lewontin, and Richard Dawkins.

Finally, one question I will ask you is whether, given that the form and function divide predates Darwin and continues still today, this means that in some sense the Darwinian revolution is less revolutionary (for all that Darwin established evolution through selection) than many have assumed. Is evolution just a surface dance on philosophical issues of much greater depth? In my *The Darwinian Revolution: Science Red in Tooth and Claw,* 2d ed. (Chicago: University of Chicago Press, 1999) I try to answer this question myself, and I return to it in two more recent books: *Darwinism and its Discontents* (Cambridge: Cambridge University Press, 2006) and *Charles Darwin* (Oxford: Blackwell, 2008). One suspects that I am not quite sure of the answer.

Epilogue

Charles Darwin's body lies a mould'ring in his grave, but his soul goes marching on. And it marches in the form of the wonderful theory that he bequeathed to us. I am not now particularly interested in whether you are a Darwinian or even an evolutionist—although I do hope that you are certainly the latter if not the former. I am certainly not interested in whether or not you are a Christian or a subscriber to some other faith. I do not mean to be rude: I expect and hope that your religious beliefs are important to you as mine are to me. But I am not about to convert you to or from Christianity or any other religion. That is for you to decide.

What I do care is that, at the very least, you are now able to stand back and appreciate what Darwin and his fellow evolutionists did and still do. I want you to recognize that these were magnificent achievements, even if in the end you decide you cannot accept them. I want you to see that the work of these scientists is the real miracle of life. That grubby little primates should be able to work out all of these things is something one should respect and admire. If God exists—certainly if the Christian God exists—then we have in this life an intellectual challenge as much as a moral challenge. It is our job to discern and understand this wonderful creation, and to give thanks and praise. That is what the evolutionists have been doing. And whether you are a Christian or not, history and logic dictate that you can accept evolution—Darwinism even—for what it is. A wonderful scientific theory—no more but certainly no less.

And with this I come to an end. I have spent a lot of my life working on and around Darwin and his achievements. I have had a lot of fun doing so. If I have passed on to you some of my enthusiasm, then this is a good reason for my having written this book and for your having read it. What more can either of us say?

Part Three: DOCUMENTS

Chapter 11

Chapter 12

Chapter 1

Introduction

My first document is an extract from Jean Baptiste de Lamarck's *Philosophie Zoologique* (1809). Lamarck's theory of evolution was of an upwardly moving escalator, with organisms somehow being carried along rising ever higher until they evolved into humans. On this, as a kind of secondary mechanism, he laid the force for which he is best known, the inheritance of acquired characteristics: through use and disuse features are developed or lost, and then they are passed on to future generations. This is the idea presented in the first document. Note how Lamarck presents his thinking in the form of laws of nature. He wants to come across as a genuine scientist like those in the physical sciences. Georges Cuvier of course would have none of this, and in the second document he takes on Lamarck's ideas, arguing that they are simply not borne out in nature: breeders can never cross the species barrier and the forms of ancient animals are those that live today. Note that whatever his ideological objections to evolutionism, Cuvier the scientist knew that he had to oppose it on scientific grounds.

The third document of this chapter contains extracts from *In Memoriam* by the English poet Alfred Tennyson. Published in 1850, the poem is dedicated to the memory of a friend (Arthur Hallam) who had died some twenty years before. It was long in conception and writing as Tennyson wrestled with his troublesome and worrying thoughts, especially those sparked by his reading of the science of his day. As a young man Tennyson had read Charles Lyell's *Principles of Geology,* and in the first of the two extracts we see the poet recoiling from the endless, mindless picture of nature that the geologist sketched. Nothing seems to make sense. Later Tennyson read a detailed review of Robert Chambers's *Vestiges of the Natural History of Creation,* that evolutionary work published in 1844. It claimed that all of nature is in an evolutionary progression up to humankind and perhaps beyond to something superior. In the second of the extracts, Tennyson recovers hope by supposing that Hallam was a precursor of this future "crowning race," doomed because he was born before his time. I offer these passages here to show that even before the *Origin* people were getting ready for evolutionary ideas. *In Memoriam,* with its evolutionary underpinnings offering hope for the future, was the Victorians' favorite poem. It was a source of great comfort to many, not the least the queen after her husband, Prince Albert, died suddenly in 1860. You should not think that evolutionary ideas faced simple, complete, absolute opposition. The story is more complex than that, with many having more reasons to accept some form of evolution than to reject it outright.

Zoological Philosophy

J. B. Lamarck

In every locality where animals can live, the conditions constituting any one order of things remain the same for long periods: indeed they alter so slowly that man cannot directly observe it. It is only by an inspection of ancient monuments that he becomes convinced that in each of these localities the order of things which he now finds has not always been existent; he may thence infer that it will go on changing.

Races of animals living in any of these localities must then retain their habits equally long: hence the apparent constancy of the races that we call species,—a constancy which has raised in us the belief that these races are as old as nature.

But in the various habitable parts of the earth's surface, the character and situation of places and climates constitute both for animals and plants environmental influences of extreme variability. The animals living in these various localities must therefore differ among themselves, not only by reason of the state of complexity of organisation attained in each race, but also by reason of the habits which each race is forced to acquire; thus when the observing naturalist travels over large portions of the earth's surface and sees conspicuous changes occurring in the environment, he invariably finds that the characters of species undergo a corresponding change.

Now the true principle to be noted in all this is as follows:

Every fairly considerable and permanent alteration in the environment of any race of animals works a real alteration in the needs of that race.

Every change in the needs of animals necessitates new activities on their part for the satisfaction of those needs, and hence new habits

Every new need, necessitating new activities for its satisfaction requires the animal, either to make more frequent use of some of its parts which it previously used less, and thus greatly to develop and enlarge them; or else to make use of entirely new parts, to which the needs have imperceptibly given birth by efforts of its inner feeling; this I shall shortly prove by means of known facts.

Thus to obtain a knowledge of the true causes of that great diversity of shapes and habits found in the various known animals, we must reflect that the infinitely diversified but slowly changing environment in which the animals of each race have successively been placed, has involved each of them in new needs and corresponding alterations in their habits. This is a truth which, once recognised, cannot be disputed. Now we shall easily discern how the new needs may have been satisfied, and the new habits acquired, if we pay attention to the two following laws of nature, which are always verified by observation.

First Law.

In every animal which has not passed the limit of its development, a more frequent and continuous use of any organ gradually strengthens, develops and enlarges that organ, and gives it a power proportional to the length of time it has been so used: while the permanent disuse of any organ imperceptibly weakens and deteriorates it, and progressively diminishes its functional capacity, until it finally disappears.

Second Law.

All the acquisitions or losses wrought by nature on individuals, through the influence of the environment in which their race has long been placed, and hence through the influence of the predominant use or permanent disuse of any organ; all these are preserved by reproduction to the new individuals which arise, provided that the acquired modifications are common to both sexes, or at least to the individuals which produce the young.

Here we have two permanent truths, which can only be doubted by those who haw never observed or followed the operations of nature, by those who have allowed themselves to be drawn into the error which I shall now proceed to combat.

Naturalists have remarked that the structure of animals is always in perfect adaptation to their functions, and have inferred that the shape and condition of their parts have determined the use of them. Now this is a mistake: for it may be easily proved by observation that it is on the contrary the needs and uses of the parts which have caused the development of these same parts, which have even given birth to them when they did not exist, and which consequently have given rise to the condition that we find in each animal.

If this were not so, nature would have had to create as many different kinds of structure in animals, as there are different kinds of environment in which they have to live; and neither structure nor environment would ever have varied.

This is indeed far from the true order of things. If things were really so, we should not have race-horses shaped like those in England; we. should not have big draught-horses so heavy and so different from the former, for none such are produced in nature; in the same way so fleet of foot, nor water-spaniels, etc.; we should not have fowls without tails, fantail pigeons, etc.; finally, we should be able to cultivate wild plants as long as we liked in the rich and fertile soil of our gardens, without the fear of seeing them change under long cultivation.

A feeling of the truth in this respect has long existed; since the following maxim has passed into a proverb and is known by all, *Habits form a second nature.*

Assuredly if the habits and nature of each animal could never vary, the proverb would have been false and would not have come into existence, nor been preserved in the event of any one suggesting it.

If we seriously reflect upon all that I have just set forth, it will be seen that I was entirely justified when in my work entitled *Recherches sur les corps vivants* (p. 50), I established the following proposition:

"It is not the organs, that is to say, the nature and shape of the parts of an animal's body, that have given rise to its special habits and faculties; but it is, on the contrary, its habits, mode of life and environment that have in course of time controlled the shape of its body, the number and state of its organs and, lastly, the faculties which it possesses."

If this proposition is carefully weighed arid compared with all the observations that nature and circumstances are incessantly throwing in our way, we shall see that its importance and accuracy are substantiated in the highest degree.

Time and a favourable environment are as I have already said nature's two chief methods of bringing all her productions into existence: for her, time has no limits and can be drawn upon to any extent.

As to the various factors which she has required and still constantly uses for introducing variations in everything that she produces, they may be described as practically inexhaustible.

The principal factors consist in the influence of climate, of the varying temperatures of the atmosphere and the whole environment of the variety of localities and their situation, of habits, the commonest movements, the most frequent activities, and, lastly, of the means of self-preservation, the mode of life and the methods of defence and multiplication.

Now as a result of these various influences, the faculties become extended and strengthened by use, and diversified by new habits that are long kept up. The conformation, consistency and, in short, the character and state of the parts, as well as of the organs, are imperceptibly affected by these influences and are preserved and propagated by reproduction.

These truths, which are merely effects of the two natural laws stated above, receive in every instance striking confirmation from facts; for the facts afford a clear indication of nature's procedure in the diversity of her productions.

Theory of the Earth

G. Cuvier

Nature appears also to have guarded against the alterations of species which might proceed from mixture of breeds, by influencing the various species of animals with mutual aversion from each other. Hence all the cunning and all the force that man is able to exert is necessary to accomplish such unions, even between species that have the nearest resemblances. And when the mule-breeds that are thus produced by these forced conjunctions happen to be fruitful, which is seldom the case, this fecundity never continues beyond a few generations, and would not probably proceed so far, without a continuance of the same cares which excited it at first. Thus we never see in a wild state intermediate productions between the hare and the rabbit, between the stag and the doe, or between the martin and the weasel. But the power of man changes this established order, and contrives to produce all these intermixtures of which the various species are susceptible, but which they would never produce if left to themselves.

The degrees of these variations are proportional to the intensity of the causes that produce them, namely, the slavery or subjection under which those animals are to man. They do not proceed far in half-domesticated species. In the cat, for example, a softer or harsher fur, more brilliant or more varied colours, greater or less size—these form the whole extent of the varieties in the species; the skeleton of the cat of Angora differs in no regular and constant circumstances from the wild cat of Europe.

In, the domesticated herbivorous quadrupeds, which man transports into all kinds of' climates, and subjects to various kinds of management, both in regard to labour and nourishment, he procures certainly more considerable variations, but still they are all merely superficial. Greater or less size; longer or shorter horns, or even the want of these entirely; a hump of fat, larger or smaller, on the shoulder; these form the chief *differences* among particular races of the *bos taurus,* or domestic black cattle; and these differences continue long in such breeds as have been transported to great distances from the countries in which they were originally produced, when proper care is taken to prevent crossing.

The innumerable varieties in the breeds of the *ovis aries,* or common sheep, are of a similar nature, and chiefly consist in differences of their fleeces, as the wool which they produce is a very important object of attention. These varieties, though not quite so perceptible, are yet sufficiently marked among horses. In general the *forms of* the bones are very little changed; their connections and articulations, and the form and structure of the large grinding teeth, are invariably the same. The small size of the tusks in the domesticated hog, compared with the wild boar, of which it is only a cultivated variety, and the junction of its cloven hoofs into one solid hoof, observable in some races, form the extreme point of the differences which man has been able to produce among herbivorous domesticated quadrupeds.

The most remarkable effects of the influence of man are produced upon that animal which he has reduced most completely under subjection. Dogs have been transported by mankind into every part of the world, and have submitted their actions to his entire direction. Regulated in their sexual unions by the pleasure or caprice of their masters, the almost endless variety of dogs differ from each other in colour; in length and abundance of hair, which is sometimes entirely wanting; in their natural instincts; in size, which varies in measure as one to five, amounting, in some instances, to more than an hundred fold in bulk; in the forms of their ears, noses, and tails; in the relative length of their legs; in the progressive development of the brain in several of the domesticated varieties, occasioning alterations, even in the form of the head; some of them having long slender muzzles with a flat forehead; others having short muzzles, with the forehead convex, &c. insomuch that the apparent differences between a mastiff and a water spaniel, and between a greyhound and a pug-dog, are even more striking than between almost any of the wild species of' a genus. Finally, and this may be considered as the maximum of known variation in the animal kingdom, some races of dogs have an additional claw on each hind foot, *with* corresponding bones of the tarsus; as there sometimes occur in the human species some families that have six fingers on each hand. Yet, in all these varieties, the relations of the bones with each other remain essentially the same, and the form of the teeth never changes in any perceptible degree, except that in some individuals one additional false grinder occasionally appears, sometimes on the one side, and sometimes on the other.*

It follows from these observations, that animals have certain fixed and natural characters, which resist the effect of every kind of influence, whether proceeding from natural causes or human interference; and we have not the smallest reason to suspect that time has any more effect upon them than climate.

I am well aware that some naturalists lay prodigious stress on the thousands of years which they can call into action by a dash of their pens. In such matters, however, our only way of judging as to the effects which may be produced by a long period of time, is by multiplying, as it were, such as are produced by a shorter known time. With this view I have endeavoured to collect all the ancient documents respecting the forms of animals; and there are none equal to those furnished by the Egyptians, both in regard to their antiquity and abundance. They have not only left us representations of animals, but even their identical bodies embalmed and pre served in the catacombs.

I have examined with the greatest attention the engraved figures of quadrupeds and birds upon the numerous obelisks brought from Egypt to ancient Rome; and all these figures, one with another, have a perfect resemblance to their intended objects, such as they still are in our days. On examining the copies made by Kirker and Zoega, we find that, without preserving every trait of the original in its utmost purity, they have yet given us figures which are easily recognised. We readily distinguish the ibis, the vulture, the owl, the falcon, the Egyptian goose, the lapwing, the land rail, the asp, the cerastes, the Egyptian hare with its long ears, even the hippopotamus; and among the numerous remains engraved in the great work on Egypt, we sometimes observe the rarest animals, the algazel, for example, which was known in Europe only a few years ago.

My learned colleague, M. Geoffroy Saint Hilaire, convinced of the importance of this research, carefully collected in the tombs and temples of Upper and Lower Egypt as many mummies of animals as he could procure. He has brought home the mummies of cats, ibises, birds of prey, dogs, monkeys, crocodiles, and the head of a bull; and after the most attentive and detailed examination, not the smallest difference is to be perceived between these animals and those of the same species which we now see, any more than between the human mummies and the skeletons of men of the present day. Some slight differences are discover able between ibis and ibis, for example, just as we now find differences in the descrip-

---

* See, in the Annals of the Museum, XVIII. 338., a memoir by my brother on the varieties of dogs, which he draw up at my request, from a series of skeletons of all the varieties of dogs, prepared by me expressly on purpose.

tions of' naturalists; but I have removed all doubts on that subject, in a Memoir on the Ibis of the ancient Egyptians, in which I have clearly shewn that this bird is precisely the same in all respects at present that it was in the days of the Pharaohs.** I am aware that in these I only cite the monuments of two or three thousand years back: but this is the most remote antiquity to which we can resort in such a case.

From all these well-established facts, there does not seem to be the smallest foundation supposing, that the new genera which I have discovered or established among extraneous fossils, such as the *paloetherium, anoplotherium, megalonyx, mastodon, pterodactylis,* &c. have ever been the sources of any of our present animals, which only differ so far as they are influenced by time or climate. Even if it should prove true, which I am far from believing to be the case, that the fossil elephants, rhinoceroses, elks, and bears; do not differ farther from the presently existing species of the same genera, than the present races of dogs differ among themselves, this would by no means be a sufficient reason to conclude that they were of the same species; since the races or varieties of dogs have been influenced by the trammels of domesticity, which these other animals never did, and indeed never could experience.

In Memoriam

A. Tennyson

> 54.
> Oh yet we trust that somehow good
> Will be the final goal of ill,
> To pangs of nature, sins of will,
> Defects of doubt, and taints of blood;
> That nothing walks with aimless feet;
> That not one life shall be destroy'd,
> Or cast as rubbish to the void,
> When God hath made the pile complete,
> That not a worm is cloven in vain;
> That not a moth with vain desire
> Is shrivell'd in a fruitless fire,
> Or but subserves another's gain.
> Behold, we know not anything;
> I can but trust that good shall fall
> At last—far off—at last, to all,
> And every winter change to spring.
> So runs my dream: but what am I?

** In that dissertation, the ibis of the ancient Egyptians is shewn to be a species of *numenius,* or curlew, denominated by Cuvier *numenius ibis:* the same bird described in Bruce's Travels under the name of *abu-hannes*—Transl.

An infant crying in the night:
An infant crying for the light:
And with no language but a cry.

55.
The wish, that of the living whole
No life may fail beyond the grave,
Derives it not from what we have
The likest God within the soul?

Are God and Nature then at strife,
That Nature lends such evil dreams?
So careful of the type she seems,
So careless of the single life;

That I, considering everywhere
Her secret meaning in her deeds,
And finding that of fifty seeds
She often brings but one to bear,

I falter where I firmly trod,
And falling with my weight of cares
Upon the great world's altar-stairs
That slope thro' darkness up to God,

I stretch lame hands of faith, and grope,
And gather dust and chaff, and call
To what I feel is Lord of all,
And faintly trust the larger hope.

56.
'So careful of the type?' but no.
From scarped cliff and quarried stone
She cries, 'A thousand types are gone:
I care for nothing, all shall go.

'Thou makest thine appeal to me:
I bring to life, I bring to death:
The spirit does but mean the breath:
I know no more.' And he, shall he,

Man, her last work, who seem'd so fair,
Such splendid purpose in his eyes,
Who roll'd the psalm to wintry skies,
Who built him fanes of fruitless prayer,

Who trusted God was love indeed
And love Creation's final law—
Tho' Nature, red in tooth and claw
With ravine, shriek'd against his creed—

Who loved, who suffer'd countless ills,
Who battled for the True, the Just.
Be blown about the desert dust,
Or seal'd within the iron hills?

No more? A monster then, a dream,

A discord. Dragons of the prime,
That tare each other in their slime,
Were mellow music match'd with him.
O life as futile, then, as frail!
O for thy voice to soothe and bless!
What hope of answer, or redress?
Behind the veil, behind the veil.

. . .

A soul shall draw from out the vast
And strike his being into bounds,
And, moved thro' life of lower phase,
Result in man, be born and think,
And act and love, a closer link
Betwixt us and the crowning race
Of those that, eye to eye, shall look
On knowledge; under whose command
Is Earth and Earth's, and in their hand
Is Nature like an open book;
No longer half-akin to brute,
For all we thought and loved and did,
And hoped, and suffer'd, is but seed
Of what in them is flower and fruit;
Whereof the man, that with me trod
This planet, was a noble type
Appearing ere the times were ripe,
That friend of mine who lives in God,
That God, which ever lives and loves,
One God, one law, one element,
And one far-off divine event,
To which the whole creation moves.

Introduction

Chapter 2 is on Darwin. I could have given you something from the *Origin* itself but have decided, for the sake of completeness and to suggest a sense of the excitement of the event, to present as the sole document for this chapter the extracts given by Darwin, read for him by his friends Lyell and the botanist Joseph Hooker, at the Linnean Society meeting in London in early July 1858. The first extract is from a book Darwin was writing (and never published) on natural selection when the arrival of Wallace's essay interrupted his plans. It deals with the struggle for existence leading to natural selection, which gives rise to adaptation. The second extract, talking of something that Darwin called his "principle of divergence," is from a letter that Darwin sent to his American friend the botanist Asa Gray. Here Darwin is trying to show how his evolutionary ideas lead to branching, to new species, and ultimately to a tree of life.

On the Variation of Organic Beings in a State of Nature: On the Natural Means of Selection; on the Comparison of Domestic Races and True Species

Charles Darwin

De Candolle, in an eloquent passage, has declared that all nature is at war, one organism with another, or with external nature. Seeing the contented face of nature, this may at first well be doubted: but reflection will inevitably prove it to be true. The war, however, is not constant, but recurrent in a slight degree at short periods, and more severely at occasional more distant periods; and hence its effects are easily overlooked. It is the doctrine of Malthus applied in most cases with tenfold force. As in every climate there are seasons, for each of its inhabitants, of greater and less abundance, so all annually breed; and the moral restraint which in some small degree checks the increase of mankind is entirely lost. Even slow-breeding mankind has doubled in twenty-five years; and if he could increase his food with greater ease, he would double in less time. But for animals without artificial means, the amount of food for each species must, *on an average,* be constant, whereas the increase of all organisms tends to be geometrical, and in a vast majority of eases at an enormous ratio. Suppose in a certain spot there are eight pairs of birds, and that *only* four pairs of them annually (including double hatches) rear only four young, and that these go on rearing their young at the same rate, then at the end of seven years (a short life, excluding violent deaths, for any bird) there will be 2048 birds, instead of the original sixteen. As this increase is quite impossible, we must conclude either that birds do not rear nearly half their young, or that the average life of a bird is, from accident, not nearly seven years. Both checks probably concur. The same kind of calculation applied to all plants and animals affords results more or less striking, but in very few instances more striking than in man.

Many practical illustrations of this rapid tendency to in crease are on record, among which, during peculiar seasons, are the extraordinary numbers of certain animals; for instance, during the years 1826 to 1828, in La Plata, when from drought some millions of cattle perished, the whole country actually *swarmed* with mice. Now I think it cannot be doubted that during the breeding season all the mice (with the exception of a few males or females in excess) ordinarily pair, and therefore that this astounding increase during three years must be attributed to a greater number than usual surviving the first year, and then breeding, and so on till the third year, when their numbers were brought down to their usual limits on the return of wet weather. Where man has introduced plants and animals into a new and favourable country, there are many accounts in how surprisingly few years the whole country has become stocked with them. This increase would necessarily stop as soon as the country was fully stocked; and yet we have every reason to believe, from what is known of wild animals, that *all* would pair in the spring. In the majority of cases it is most difficult to imagine where the check falls—though generally, no doubt, on the seeds, eggs, and young; but when we remember how impossible, even in mankind (so much better known than any other animal), it is to infer from repeated casual observations what the average duration of life is, or to discover the different percentage of deaths to births in different countries, we ought to feel no surprise at our being unable to discover where the check falls in any animal or plant. It should always be remembered, that in most cases the checks are recurrent yearly in a small, regular degree, and in an extreme degree during unusually cold, hot, dry, or wet years, according to the constitution of the being in question. Lighten any check in the least degree, and the geometrical powers of increase in every organism will almost instantly increase the average number of the favoured species. Nature may be compared to a surface on which rest ten thousand sharp wedges touching each other and driven inwards by incessant blows. Fully to realize these views much reflection is requisite. Malthus on man should be studied; and all such cases as those of the mice in La Plata, of the cattle and horses when first turned out in South America, of the birds by our calculation, etc., should be well considered. Reflect on the enormous multiplying power *inherent and annually in action* in all animals; reflect on the countless seeds scattered by a hundred ingenious contrivances, year after year, over the whole face of the land; and yet we have every reason to suppose that the average percentage of each of the inhabitants of a country usually remains constant. Finally, let it be borne in mind that this average number of individuals (the external conditions remaining the same) in each country is kept up by recurrent struggles against other species or against external nature (as on the borders of the arctic regions, where the cold checks life), and that ordinarily each individual of every species holds its place, either by its own struggle and capacity of acquiring nourishment in some period of its life, from the egg upwards; or by the struggle of its parents (in short-lived organisms, when the main check occurs at longer intervals) with other individuals of the *same* or *different* species.

But let the external conditions of a country alter. If in a small degree, the relative proportions of the inhabitants will in most cases simply be slightly changed; but let the number of inhabitants be small, as on an island, and free access to it from other countries be circumscribed, and let the change of conditions continue progressing (forming new stations), in such a ease the original inhabitants must cease to be as perfectly adapted to the changed conditions as they were originally. It has been shown in a former part of this work, that such changes of external conditions would, from their acting on the reproductive system, probably cause the organization of those beings which were most affected to become, as under domestication, plastic. Now, can it be doubted, from the struggle each individual has to obtain subsistence, that any minute variation in structure, habits, or instincts, adapting that individual better to the new conditions, would tell upon its vigour and health? In the struggle it would bare a better *chance* of surviving; and those of its offspring which inherited the variation, be it ever so slight, would also have a better *chance*. Yearly more are bred than can survive; the smallest grain in the balance, in the long run, must tell on which death shall fall, and which shall survive. Let this work of selection on the one hand, and death on the other, go on for a thousand generations, who will pretend to affirm that it would produce no effect, when we remember what, in a few years, Bakewell effected in cattle, and Western in sheep, by this identical principle of selection?

To give an imaginary example from changes in progress on an island: let the organization of a canine animal which preyed chiefly on rabbits, but sometimes on hares, become slightly plastic; let these same changes cause the number of rabbits very slowly to decrease, and the number of hares to increase; the effect of this would be that the fox or dog would be driven to try to catch more hares: his organization, however, being slightly plastic, those individuals with the lightest forms, longest limbs, and best eyesight, let the difference be ever so small, would be slightly favoured, and would tend to live longer, and to survive during that time of the year when food was scarcest; they would also rear more young, which would tend to inherit these slight peculiarities. The less fleet ones would be rigidly destroyed. I can see no more reason to doubt that these causes in a thousand generations would produce a marked effect, and adapt the form of the fox or dog to the catching of hares instead of rabbits, than that greyhounds can be improved by selection and careful breeding. So would it be with plants under similar circumstances. If the number of individuals of a species with plumed seeds could be increased by greater powers of dissemination within its own area (that is, if the check to increase fell chiefly on the seeds), those seeds which were provided with ever so little more down, would in the long run be most disseminated; hence a greater number of seeds thus formed would germinate, and would tend to produce plants inheriting the slightly better-adapted down.[1]

Besides this natural means of selection, by which those individuals are preserved, whether in their egg, or larval, or mature state, which are best adapted to the place they fill in nature, there is a second agency at work in most unisexual animals, tending to produce the same effect, namely the struggle of the males for the females. These struggles are generally decided by the law of battle, but in the ease of birds, apparently, by the charms of their song, by their beauty or their power of courtship, as in the dancing rock-thrush of Guiana. The most vigorous and healthy males, implying perfect adaptation, must generally gain the victory in their contests. This kind of selection, however, is less rigorous than the other; it does not require the death of the less successful, but gives to them fewer descendants. The struggle falls, moreover, at a time of year when food is generally abundant, and perhaps the effect chiefly produced would be the modification of the secondary sexual characters, which are not related to the power of obtaining food, or to defence from enemies, but to fighting with or rivalling other males. The result of this struggle amongst the males may be compared in some respects to that produced by those agriculturists, who pay *less* attention to the careful selection of all their young animals, and more to the occasional use of a choice male.

Abstract of a Letter from Charles Darwin to Professor Asa Gray, 5 September 1857

It is wonderful what the principle of selection by man, that is the picking out of individuals with any desired quality, and breeding from them, and again picking out, can do. Even breeders have been astounded at their own results. They can act on differences inappreciable to an uneducated eye. Selection has been *methodically* followed in *Europe* for only the last half century; but it was occasionally, and even in some degree methodically, followed in the most ancient times. There must have been also a kind of unconscious selection from a remote period, namely in the preservation of the individual animals (without any thought of their offspring) most useful to each race of man in his particular circumstances. The 'roguing', as nurserymen call the destroying of varieties which depart from their type,

---

[1] I can see no more difficulty in this, than in the planter improving his varieties of the cotton plant. C.D. 1858.

is a kind of selection. I am convinced that intentional and occasional selection has been the main agent in the production of our domestic races; but however this may be, its great power of modification has been indisputably shown in later times. Selection acts only by the accumulation of slight or greater variations, caused by external conditions, or by the mere fact that in generation the child is not absolutely similar to its parent. Man, by this power of accumulating variations, adapts living beings to his wants—may be said to make the wool of one sheep good for carpets, of another for cloth, etc.

Now suppose there were a being who did not judge by mere external appearances, but who could study the whole internal organization, who was never capricious, and should go on selecting for one object during millions of generations; who will say what he might not effect? In nature we have some *slight* variation occasionally in all parts; and I think it can be shown that changed conditions of existence is the main cause of the child not exactly resembling its parents; and in nature geology shows us what changes have taken place, and are taking place. We have almost unlimited time; no one but a practical geologist can fully appreciate this. Think of the Glacial period, during the whole of which the same species at least of shells have existed; there must have been during this period millions on millions of generations.

I think it can be shown that there is such an unerring power at work in *Natural Selection* (the title of my book), which selects exclusively for the good of each organic being. The elder de Candolle, W. Herbert, and Lyell have written excellently on the struggle for life; but even they have not written strongly enough. Reflect that every being (even the elephant) breeds at such a rate, that in a few years, or at most a few centuries, the surface of the earth would not hold the progeny of one pair. I have found it hard constantly to bear in mind that the increase of every single species is checked during some part of its life, or during some shortly recurrent generation. Only a few of those annually born can live to propagate their kind. What a trifling difference must often determine which shall survive, and which perish!

Now take the ease of a country undergoing some change. This will tend to cause some of its inhabitants to vary slightly—not but that I believe most beings vary at all times enough for selection to act on them. Some of its inhabitants will be exterminated; and the remainder will be exposed to the mutual action of a different set of inhabitants, which I believe to be far more important to the life of each being than mere climate. Considering the infinitely various methods which living beings follow to obtain food by struggling with other organisms, to escape danger at various times of life, to have their eggs or seeds disseminated, etc., I cannot doubt that during millions of generations individuals of a species will be occasionally born with some slight variation, profitable to some part of their economy. Such individuals will have a better chance of surviving, and of propagating their new and slightly different structure; and the modification may be slowly increased by the accumulative action of natural selection to any profitable extent. The variety thus formed will either coexist with, or, more commonly, will exterminate its parent form. An organic being like the woodpecker or mistletoe, may thus come to be adapted to a score of contingencies—natural selection accumulating those slight variations in all parts of its structure, which are in any way useful to it during any part of its life.

Multiform difficulties will occur to every one, with respect to this theory. Many can, I think, be satisfactorily answered. *Natura non facit saltum* answers some of the most obvious. The slowness of the change, and only a very few individuals undergoing change at any one time, answers others. The extreme imperfection of our geological records answers others.

Another principle, which may be called the principle of divergence, plays, I believe, an important part in the origin of species. The same spot will support more life if occupied by very diverse forms. We see this in the many generic forms in a square yard of turf, and in the plants or insects on any little uniform islet, belonging almost invariably to as many genera and families as species. We can understand the meaning of this fact amongst the higher animals, whose habits we understand. We know that it has been experimentally shown that a plot of land will yield a greater weight if sown with several species and genera of grasses, than if sown with only two or three species. Now, every organic being, by propagating so rapidly, may be said to be striving its utmost to increase in numbers. So it will be with the offspring of any species after it has become diversified into varieties, or subspecies, or true species. And it follows, I think, from the foregoing facts, that the varying offspring of each species will try (only few will succeed) to seize on as many and as diverse places in the economy of nature as possible. Each new variety or species, when formed, will generally take the place of, and thus exterminate its less well-fitted parent. This I believe to be the origin of the classification and affinities of organic beings at all times; for organic beings always *seem* to branch and sub-branch like the limbs of a tree from a common trunk, the flourishing and diverging twigs destroying the less vigorous—the dead and lost branches rudely representing extinct genera and families.

This sketch is *most* imperfect; but in so short a space I cannot make it better. Your imagination must fill up very wide blanks.

Chapter 3

Introduction

Chapter 3 has seven documents. The first two are by Thomas Henry Huxley, beginning with an autobiographical account of Huxley's becoming an evolutionist. See how he too had read *Vestiges* (not just a review like Tennyson!). But far from being convinced of the truth of evolution, as a professional biologist Huxley was simply irritated. Nor did the pleading of Huxley's friend Herbert Spencer change his mind. Yet the idea would not go away, and under the influence of Charles Darwin not only was Huxley introduced to a more scientifically plausible evolutionism but eventually he was convinced of its truth. Note however that Huxley was less enthusiastic about natural selection, thinking that it was not yet proven by experimental study, and this was a skepticism that he carried to his grave.

The second document is from a letter Huxley wrote to the Reverend Charles Kingsley, known to us today as the author of the popular children's book *The Water Babies*. Huxley's son had just died of scarlet fever at the age of 4, and Kingsley wrote to Huxley offering Christian consolation. This spurred Huxley to respond, stating his own unbelief—he was to coin the term *agnostic*—but also his belief in the power and worth of science. Note how explicitly Huxley is regarding this as a religious experience: to "sit down before the fact as a little child" is akin to surrendering to the will of God. It is small wonder that a man like this should have seized on evolutionism and regarded and treated it as a form of secular religion: something to give meaning to life.

The third document is from an essay by Alfred Russel Wallace, and it expands on the theme mentioned in my main text, the argument that future human progress depends on the power and influence of women. Supposedly they will select the best men and thus there will be a general increase in human well-being and happiness. Note that Wallace's is a biological position—explicitly he downplays the lasting potential and influence of education. Note also how Wallace's position belies the traditional view of Social Darwinism—that is, a view of a philosophy that endorses a stem procapitalism economics, with success to the strong and failure for the weak. Wallace's socialism comes through strongly in our extracts. A major influence in the later part of Wallace's life, from which this document is taken, was the novel *Looking Backward* by the American author Edward Bellamy. It is a tale of a man who sleeps until he awakens in the future, about now, in fact. All of the stuff about female power is lifted directly from the novel.

The paradox is that generally Wallace was (as we saw) uncomfortable with the notion that selection involves female choice: he thought the peahen is relatively dowdy not because she has chosen bright males but because she needs camouflage. There really is no paradox, for, as I pointed out also, Wallace's real worry was that Darwin was making the peahen with her choice seem more humanlike. Humans are already humanlike, so they can do all the choosing that they will!

The fourth and fifth documents continue the exploration of using Darwinian ideas to promote social theories. The fourth is an extract from a very popular philosophy of science text (*The Grammar of Science*) by the statistician and social theorist Karl Pearson (he was English, for all that he was so enthusiastic about things German that he changed his name from Carl to Karl). Pearson was an ardent socialist, as the extract shows. He was also a keen eugenicist and as the extract also (rather unfortunately shows) an appalling racist, who clearly thought (as did many of his fellow countrymen at the time of writing this work, 1900) that the British Empire, with a small island of white people dominating a third of the world (many of the denizens being non-white), was a very good thing. Next (the fifth piece) is by Prince Petr Kropotkin. He was an anarchist, and anarchists have a deservedly bad reputation for their willingness to attack institutions and people in the cause of their philosophy. In theory, however, Kropotkin was far more pacifistic, believing that humans are bound by a sense of mutualism. As the main text argues, this was in major part based on the Russian take on evolution, which downplayed considerably the struggle for existence between humans. The extract shows vividly why one might think of the causes of evolution in this way.

The sixth document is a passage from Friedrich von Bernhardi's *Germany and the Next War*, showing (by contrast with Wallace) how very different things could be claimed in the name of Darwinism and also how there really was reason for people to see Darwinism as having had a causal influence on the militarism of the early twentieth century, especially in Germany. Of course, how much of this really came from Darwin is a matter of speculation. One could as easily say that the character of Thrasymachus in Plato's *Republic*, a man who preached that "might is right," was just as great an influence. As I say in the text, one should be very wary of drawing a simple, straight line from Darwin to Hitler.

There were many more writers on the supposed links between Darwinian evolutionary biology and social issues. I am leaving American writers on the topic until the next section of the Documents. Here, summing up, the seventh and final document in this section is a short essay by me, on the extent to which one can truly say that Darwin's thinking has functioned as a secular religion. I should say that the Creationists have had a field day with this piece, claiming triumphantly that an ardent Darwinian (me!) now agrees that Darwinism is nothing but a religion, just like their own position. Apart from the fact that this now puts the Creationists on very shaky ground—a major plank in their argument is that Creationism is not a religion, and hence is in the USA is legitimately taught in state-supported school biology classes—note that at the end of the essay I carefully stress that my claim is NOT that all evolutionary thinking is religious. It is rather that it can be and sometimes it has been. There is and long has been an entirely secular science of evolutionary studies, based on the thinking of Charles Darwin.

From *The Reception of the "Origin of Species"*

T. H. Huxley

I think I must have read the V*estiges* before I left England in 1846; but, if I did, the book made very little impression upon me, and I was not brought into serious contact with the "Species" question until after 1850. At that time, I had long done with the Pentateuchal cosmogony, which had been impressed upon my childish understanding as Divine truth, with all the authority of parents and instructors, and from which it had cost me many a struggle to get free. But my mind was unbiassed in respect of any doctrine which presented itself, if it professed to be based on purely philosophical and scientific reasoning. It seemed to me then (as it does now) that "creation," in the ordinary sense of the word, is perfectly conceivable. I find no difficulty in conceiving that, at some former period, this universe was not in existence; and that it made its appearance in six days (or instantaneously, if that is preferred), in consequence of the volition of some preexisting Being. Then, as now, the so-called *a priori* arguments against Theism, and, given a Deity, against the possibility of creative acts, appeared to me to be devoid of reasonable foundation. I had not then, and I have not now, the smallest *a priori* objection to raise to the account of the creation of animals and plants given in *Paradise Lost,* in which Milton so vividly embodies the natural sense of Genesis. Far be it from me to say that it is untrue because it is impossible I confine myself to what must be regarded as a modest and reasonable request for some particle of evidence that the existing species of animals and plants did originate in that way, as a condition of my belief in a statement which appears to me to be highly improbable.

And, by way of being perfectly fair, I had exactly the same answer to give to the evolutionists of 1851–58. Within the ranks of the biologists, at that time, I met with nobody, except Dr. Grant of University College, who had a word to say for Evolution—and his advocacy was not calculated to advance the cause. Outside these ranks, the only person known to me whose knowledge and capacity compelled respect, and who was, at the same time, a thorough-going evolutionist, was Mr. Herbert Spencer, whose acquaintance I made, I think, in 1852, and then entered into the bonds of a friendship which, I am happy to think, has known no interruption. Many and prolonged were the battles we fought on this topic. But even my friend's rare dialectic skill and copiousness of apt illustration could not drive me from my agnostic position. I took my stand upon two grounds:—Firstly, that up to that time, the evidence in favour of transmutation was wholly insufficient; and secondly, that no suggestion respecting the causes of transmutation assumed, which had been made, was in any way adequate to explain the phenomena. Looking back at the state of knowledge at that time, I really do not see that any other conclusion was justifiable.

In those days I had never even heard of Treviranus' *Biologie*. However, I had studied Lamarck attentively, and I had read the *Vestiges* with due care; but neither of them afforded me any good ground for changing my negative and critical attitude. As for the *Vestiges,* I confess that the book simply irritated me by the prodigious ignorance and thoroughly unscientific habit of mind manifested by the writer. If it bad any influence on me at all, it set me against Evolution; and the only review I ever have qualms of conscience about, on the ground of needless savagery, is one I wrote on the V*estiges* while under that influence.. . .

But, by a curious irony of fate, the same influence which led me to put as little faith in modern speculations on this subject as in the venerable traditions recorded in the first two chapters of Genesis, was perhaps more potent than any other in keeping alive a sort of pious conviction that Evolution, after all, would turn out true. I have recently read afresh the first edition of the *Principles of Geology;* and when I consider that this remarkable book had been nearly thirty years in everybody's hands, and that it brings home to any reader of ordinary intelligence a great principle and a great fact,—the principle that the past must be explained by the present, unless good cause be shown to the contrary; and the fact that so far as our knowledge of the past history of life on our globe goes, no such cause can be shown,—I cannot but believe that Lyell, for others, as for myself, was the chief agent in smoothing the road for Darwin. For consistent uniformitarianism postulates Evolution as much in the organic as in the inorganic world. The origin of a new species by other than ordinary agencies would be a vastly greater "catastrophe" than any of those which Lyell succesfully eliminated from sober geological speculation.

Thus, looking back into the past, it seems to me that my own position of critical expectancy was just and reasonable, and must have been taken up, on the same grounds, by many other persons. If Agassiz told me that the forms of life which have successively tenanted the globe were the incarnations of successive thoughts of the Deity, and that He had wiped out one set of these embodiments by an appalling geological catastrophe as soon as His ideas took a more advanced shape, I found myself not only unable to admit the accuracy of the deductions from the facts of paleontology, upon which this astounding hypothesis was founded, but I had to confess my want of any means of testing the correctness of his explanation of them. And besides that, I could by no means see what the explanation explained. Neither did it help me to be told by an eminent anatomist that species had succeeded one another in time, in virtue of "a continuously operative creational law." That seemed to me to be no more than saying that species had succeeded one another in the form of a vote-catching resolution, with "law" to catch the man of science, and "creational" to draw the orthodox. So I took refuge in that "thätige Skepsis" which Goethe has so well defined; and, reversing the apostolic precept to be all things to all men, I usually defended the tenability of the received doctrines when I had to do with the transmutationists; and stood up for the possibility of transmutation among the orthodox—thereby, no doubt, increasing an already current, but quite undeserved, reputation for needless combativeness.

I remember, in the course of my first interview with Mr. Darwin, expressing my belief in the sharpness of the lines of demarcation between natural groups and in the absence of transitional forms, with all the confidence of youth and imperfect knowledge. I was not aware, at that time, that he had then been many years brooding over the species-question; and the humorous smile which accompanied his gentle answer, that such was not altogether his view, long haunted and puzzled me. But it would seem that four or five years' hard work had enabled me to understand what it meant; for Lyell, writing to Sir Charles Bunbury (under date of April 30, 1856), says :—

> "When Huxley, Hooker and Wollaston were at Darwin's last week, they (all four of them) ran atilt against species—further, I believe, than they are prepared to go."

I recollect nothing of this beyond the fact of meeting Mr. Wotlaston; and except for Sir Charles's distinct assurance as to "all four," I should have thought my *outrecuidance* was probably a counterblast to Wollaston's conservatism. With regard to Hooker, he was already, like Voltaire's Habbakuk, *capable de lout* in the way of advocating Evolution.

As I have already said, I imagine that most of those of my contemporaries who thought seriously about the matter were very much in my own state of mind—inclined to say to both Mosaists and Evolutionists, "a plague on both your houses !" and disposed to turn aside from an interminable and apparently fruitless discussion, to labour in the fertile fields of ascertainable fact. And I may therefore suppose that the publication of the Darwin and Wallace paper in 1858, and still more that of the "Origin" in 1859, had the effect upon them of the flash of light which, to a man who has lost himself on a dark night, suddenly reveals a road which, whether it takes him straight home or not, certainly goes his way. That which we were looking for, and could not find, was a hypothesis respecting the origin of known organic forms which assumed the operation of no causes but such as could be proved to be actually at work. We wanted, not to pin our faith to that or any other speculation. but to get hold of clear and definite conceptions which could be brought face to face with facts and have their validity tested. The "Origin" provided us with the working hypothesis we sought. Moreover, it did the immense service of freeing us for ever from the dilemma—Refuse to accept the creation hypothesis, and what have you to propose that can be accepted by any cautious reasoner? In 1857 I had no answer read), and 1 do not think that anyone else had. A year later we reproached ourselves with dulness for being perplexed with such an inquiry. My reflection, when I first made myself master of the central idea of the "Origin" was, "How extremely stupid not to have thought of that!" I suppose that Columbus' companions said much the same when he made the egg stand on end. The facts of variability, of the struggle for existence, of adaptation to conditions, were notorious enough; but none of us had suspected that the road to the heart of the species problem lay through them, until Darwin and Wallace dispelled the darkness, and the beacon-fire of the "Origin" guided the benighted.

Whether the particular shape which the doctrine of Evolution, as applied to the organic world, took in Darwin's hands, would prove to be final or not, was to me a matter of indifference. In my earliest criticisms of the "Origin" I ventured to point out that its logical foundation was insecure so long as experiments in selective breeding had not produced varieties which were more or less infertile; and that insecurity remains up to the present time. But, with any and every critical doubt which my sceptical ingenuity could suggest, the Darwinian hypothesis remained incomparably more probable than the creation hypothesis. And if we had none of us been able to discern the paramount significance of some of the most patent and notorious of natural facts, until they were, so to speak, thrust under our noses, what force remained in the dilemma—creation or nothing? It was obvious that hereafter the probability would be immensely greater, that the links of natural causation were hidden from our purblind eyes, than that natural causation should be incompetent to produce all the phenomena of nature. The only rational course for those who had no other object than the attainment of truth was to accept "Darwinism" as a working hypothesis and see what could be made of it. Either it would prove its capacity to elucidate the facts of organic life, or it would break down under the strain. This was surely the dictate of common sense, and, for once, common sense carried the day.

From The Life and Letters of Thomas Huxley

T. H. Huxley

14 Waverley Place, Sept. 23, 1860.

My Dear Kingsley—I cannot sufficiently thank you, both on my wife's account and my own, for your long and frank letter, and for all the hearty sympathy which it exhibits—and Mrs. Kingsley will, I hope, believe that we are no less sensible of her kind thought of us. To myself your letter was especially valuable, as it touched upon what I thought even more than upon what I said in my letter to you. My convictions, positive and negative, on all the matters of which you speak, are of long and slow growth and are firmly rooted. But the great blow which fell upon me seemed to stir them to their foundation, and had I lived a couple of centuries earlier I could have fancied a devil scoffing at me and them—and asking me what profit it was to have stripped myself of the hopes and consolations of the mass of mankind ? To which my only reply was and is—Oh devil! truth is better than much profit. I have searched over the grounds of my belief, and if wife and child and name and fame were all to be lost to me one after the other as the penalty, still I will not lie.

And now I feel that it is due to you to speak as frankly as you have done to me. An old and worthy friend of mine tried some three or four years ago to bring us together—because, as he said, you were the only man who would do me any good. Your letter leads me to think he was right, though not perhaps in the sense he attached to his own words.

To begin with the great doctrine you discuss. I neither deny nor affirm the immortality of man. I see no reason for believing in it, but, on the other hand, I have no means of disproving it.

Pray understand that I have no *a priori* objections to the doctrine. No man who has to deal daily and hourly with nature can trouble himself about *a priori* difficulties. Give me such evidence as would justify me in believing anything else, and I will believe that. Why should I not ? It is not half so wonderful as the conservation of force, or the indestructibility of matter. Whoso clearly appreciates all that is implied in the failing of a stone can have no difficulty about any doctrine simply on account of its marvellousness. But the longer I live, the more obvious it is to me that the most sacred act of a man's life is to say and to feel, "I believe such and such to be true." All the greatest rewards and all the heaviest penalties of existence cling about that act. The universe is one and the same throughout; and if the condition of my success in unravelling some little difficulty of anatomy or physiology is that I shall rigorously refuse to put faith in that which does not rest on sufficient evidence, I cannot believe that the great mysteries of existence will be laid open to me on other terms. It is no use to talk to me of analogies and probabilities. I know what I mean when I say I believe in the law of the inverse squares, and I will not rest my life and my hopes upon weaker convictions. I dare not if I would.

Measured by this standard, what becomes of the doctrine of immortality ?

You rest in your strong conviction of your personal existence, and in the instinct of the persistence of that existence which is so strong in you as in most men.

To me this is as nothing. That my personality is the surest thing I know— may be true. But the attempt to conceive what it is leads me into mere verbal subtleties. I have champed up all that chaff about the ego and the non-ego, about noumena and phenomena, and all the rest of it, too often not to know that in attempting even to think of these questions, the human intellect flounders at once out of its depth.

It must be twenty years since, a boy, I read Hamilton's essay on the unconditioned, and from that time to this, ontological speculation has been a folly to me. When Mansel took up Hamilton's argument on the side of orthodoxy(!) I said he reminded me of nothing so much as the man who is sawing off the sign on which he is sitting, in Hogarth's picture. But this by the way.

I cannot conceive of my personality as a thing apart from the phenomena of my life. When I try to form such a conception I discover that, as Coleridge would have said, I only hypostatise a word, and it alters nothing if, with Fiehte, I suppose the universe to be nothing but a manifestation of my personality. I am neither more nor less eternal than I was before.

Nor does the infinite difference between myself and the animals alter the case. I do not know whether the animals persist after they disappear or not. I do not even know whether the infinite difference between us and them may not be compensated by *their* persistence and *my* cessation after apparent death, just as the humble bulb of an annual lives, while the glorious flowers it has put forth die away.

Surely it must be plain that an ingenious man could speculate without end on both sides, and find analogies for all his dreams. Nor does it help me to tell me that the aspirations of mankind—that my own highest aspirations even—lead me towards the doctrine of immortality. I doubt the fact, to begin with, but if it be so even, what is this but in grand words asking me to believe a thing because I like it.

Science has taught to me the opposite lesson. She warns me to be careful how I adopt a view which jumps with my preconceptions, and to require stronger evidence for such belief than for one to which I was previously hostile.

My business is to teach my aspirations to conform themselves to fact, not to try and make facts harmonise with my aspirations.

Science seems to me to teach in the highest and strongest manner the great truth which is embodied in the Christian conception of entire surrender to the will of God. Sit down before fact as a little child, be prepared to give up every preconceived notion, follow humbly wherever and to whatever abysses nature leads, or you shall learn nothing. I have only begun to learn content and peace of mind since I have resolved at all risks to do this.

There are, however, other arguments commonly brought forward in favour of the immortality of man, which are to my mind not only delusive but mischievous. The one is the notion that the moral government of the world is imperfect without a system of future rewards and punishments. The other is: that such a system is indispensable to practical morality. I believe that both these dogmas are very mischievous lies.

With respect to the first, I am no optimist, but I have the firmest belief that the Divine Government (if we may use such a phrase to express the sum of the "customs of matter") is wholly just. The more I know intimately of the lives of other men (to say nothing of my own), the more obvious it is to me that the wicked does *not* flourish nor is the righteous punished. But for this to be clear we must bear in mind what almost all forget, that the rewards of life are contingent upon obedience to the *whale* law—physical as well as moral—and that moral obedience will not atone for physical sin, or *vice versa*.

The ledger of the Almighty is strictly kept, and every one of us has the balance of his operations paid over to him at the end of every minute of his existence.

Life cannot exist without a certain conformity to the surrounding universe—that conformity involves a certain amount of happiness in excess of pain. In short, as we live we are paid for living.

And it is to be recollected in view of the apparent discrepancy between men's acts and their rewards that Nature is juster than we. She takes into account what a man brings with him into the world, which human justice cannot do. If I, born a bloodthirsty and savage brute, inheriting these qualities from others, kill you, my fellow-men will very justly hang me, but I shall not be visited with the horrible remorse which would be my real punishment if, my nature being higher, I had done the same thing.

The absolute justice of the system of things is as clear to me as any scientific fact. The gravitation of sin to sorrow is as certain as that of the earth to the sun, and more so—for experimental proof of the fact is within reach of us all—nay, is before us all in our own lives, if we had but the eyes to see it.

Not only, then, do I disbelieve in the need for compensation, hut I believe that the seeking for rewards and punishments out of this life leads men to a ruinous ignorance of the fact that their inevitable rewards and punishments are here.

If the expectation of hell hereafter can keep me from evil-doing, surely a *a fortiori* the certainty of hell now will do so? If a man could be firmly impressed with the belief that stealing damaged him as much as swallowing arsenic would do (and it does), would not the dissuasive force of that belief be greater than that of any based on mere future expectations.

And this leads me to my other point.

As I stood behind the coffin of my little son the other day, with my mind bent on anything but disputation, the officiating minister read, as a part of his duty, the words, "If the dead rise not again, let us eat and drink, for tomorrow we die." I cannot tell you how inexpressibly they shocked me. Paul had neither wife nor child, or he must have known that his alternative involved a blasphemy against all that was best and noblest in human nature. I could have laughed with scorn. What! because I am face to face with irreparable loss, because I have given back to the source from whence it came, the cause of a great happiness, still retaining through all my life the blessings which have sprung and will spring from that cause, I am to renounce my manhood, and, howling, grovel in bestiality? Why, the very apes know better, and if you shoot their young, the poor brutes grieve their grief out and do not immediately seek distraction in a gorge.

Kicked into the world a boy without guide or training, or with worse than none, I confess to my shame that few men have drunk deeper of all kinds of sin than I. Happily, my course was arrested in time—before I had earned absolute destruction—and for long years I have been slowly and painfully climbing, with many a fall, towards better things. And when I look back, what do I find to have been the agents of my redemption? The hope of in, mortality or of future reward? I can honestly say that for these fourteen years such a consideration has not entered my head. No, I can tell you exactly what has been at work. *Sartor Resartus* led me to know that a deep sense of religion was compatible with the entire absence of theology. Secondly, science and her methods gave me a resting-place independent of authority and tradition. Thirdly, love opened up to me a view of the sanctity of human nature, and impressed me with a deep sense of responsibility.

If at this moment I am not a worn-out, debauched, useless carcass of a man, if it has been or will be my fate to advance the cause of science, if I feel that I have a shadow of a claim on the love of those about me, if in the supreme moment when I looked down into my boy's grave my sorrow was full of submission and without bitterness, it is because these agencies have worked upon me, and not because I have ever cared whether my poor personality shall remain distinct for ever from the All from whence it came and whither it goes.

And thus, my dear Kingsley, you will understand what my position is. I may be quite wrong, and in that case I know I shall have to pay the penalty for being wrong. But I can only say with Luther, "Gott helle mir, Ich kann nichts anders."

I know right well that 99 out of 100 of my fellows would call me atheist, infidel, and all the other usual hard names. As our laws stand, if the lowest thief steals my coat, my evidence (my opinions being known) would not be received against him.[1]

But I cannot help it. One thing people shall not call me with justice and that is—a liar. As you say of yourself, I too feel that I lack courage; but if ever the occasion arises when I am bound to speak, I will not shame my boy.

I have spoken more openly and distinctly to you than I ever have to any human being except my wife.

If you can show me that I err in premises or conclusion, I am ready to give up these as I would any other theories. But at any rate you will do me the justice to believe that I have not reached my conclusions without the care befitting the momentous nature of the problems involved.

And I write this the more readily to you, because it is clear to me that if that great and powerful instrument for good or evil, the Church of England, is to be saved from being shivered into fragments by the advancing tide of science—an event I should be very sorry to witness, but which will infallibly occur if men like Samuel of Oxford are to have the guidance of her destinies—it must be by the efforts of men who, like yourself, see your way to the combination of the practice of the Church with the spirit of science. Understand that all the younger men of science whom know intimately are *essentially* of my way of thinking. (I know not a scoffer or an irreligious or an immoral man among them, but they all regard orthodoxy as you do Brahmanism.) Understand that this new school of the prophets is the only one that can work miracles, the only one that can constantly appeal to nature for evidence that it is right, and you will comprehend that it is of no use to try to barricade us with shovel hats and aprons, or to talk about our doctrines being "shocking."

I don't profess to understand the logic of yourself, Maurice, and the rest of your school, but I have always said I would swear by your truthfulness and sincerity, and that good must come of your efforts. The more plain this was to me, however, the more obvious the necessity to let you see where the men of science are driving, and it has often been in my mind to write to you before.

---

[1] The law with respect to oaths was reformed in 1869.

If I have spoken too plainly anywhere, or too abruptly, pardon me, and do the like to me.

My wife thanks you very much for your volume of sermons.—Ever yours very faithfully,

T. H. Huxley.

Studies, Scientific and Social

A. R. Wallace

Social Advance Will Result in Improvement of Character.

It is my firm conviction, for reasons which I shall state presently, that, when we have cleansed the Augean stable of our existing social organization, and have made such arrangements that *all* shall contribute their share of either physical or mental labour, and that all workers shall reap the *full* and equal reward of their work, the future of the race will be ensured by those laws of human development that have led to the slow but continuous advance in the higher qualities of human nature. When men and women are alike free to follow their best impulses; when idleness and vicious or useless luxury on the one hand, oppressive labour and starvation on the other, are alike unknown; when all receive the best and most thorough education that the state of civilization and public opinion is set by the wisest and the best, and that standard is systematically inculcated on the young; then we shall find that a system of selection will come spontaneously into action which will steadily tend to eliminate the lower and more degraded types of man, and thus continuously raise the average standard of the race. I therefore strongly protest against any attempt to deal with this great question by legal enactments in our present state of unfitness and ignorance, or by endeavouring to modify public opinion as to the beneficial character of monogamy and permanence in marriage. . . .

We have now to consider what would be the probably effect of a condition of social advancement, the essential characteristics of which have been already hinted at, on the two great problems—the increase of population, and the continuous improvement of the race by some form of selection which we have reason to believe is the only method available. In order to make this clear, however, and in order that we may fully realize the forces that would come into play in a just and rational state of society, such as may certainly be realized in the not distant future, it will be necessary to have a clear conception of its main characteristics. For this purpose, and without committing myself in any way to an approval of tall the details of his scheme, I shall make use of Mr. Bellamy's clear and forcible picture of the society of the future, as he supposes it may exist in America in little more than a century hence.[2]

The essential principle on which society is supposed to be—is that of a great family. As in a well-regulated—family the elders, those who have experience of the—the duties, and the responsibilities of life, determine—general mode of living and working, with the fullest—consideration for the convenience and real well-being of—younger members, and with a recognition of their—independence. As in a family, the same comforts—joyments are secured to all, and the very idea of—any difference in this respect to those who from mental or physical disability are unable to do so much as other, never occurs to any one, since it is opposed to the essential principles on which a true society of human brotherhood is held to rest. As regards education all have the same advantages, and all receive the fullest and best training, both intellectual and physical; every one is encouraged to follow out those studies or pursuits for which they are best fitted, or for which they exhibit the strongest inclination. This education, the complete and thorough training for a life of usefulness and enjoyment, continues in both sexes till the age of twenty-one (or thereabouts), when all alike, men and women, take their place in the lower ranks of the industrial army in which they serve for three years. During the latter years of their education, and during the succeeding three years of industrial service, every opportunity is given them to see and understand every kind of work that is carried on by the community, so that at the end of the term of probation they can choose what department of the public service they prefer to enter. As every one—men, women, and children alike—receive the same amount of public credit—their equal share of the products of the labour of the community, the attractiveness of various pursuits is equalized by differences in the hours of labour, in holidays, or in special privileges attached to the more disagreeable kinds of necessary work, and these are so modified from time to time that the volunteers for every occupation are always about equal to its requirements. The only other essential feature that it is necessary to notice for our present purpose is the system of grades, by which good conduct, perseverance, and intelligence in every department of industry and occupation are fully recognized, and lead to appointments as foremen, superintendents, or general managers, and ultimately to the highest offices of the state. Every one of these grades and appointments is made public; and as they constitute the only honours and the only differences of rank, with corresponding insignia and privileges, in an otherwise equal body of citizens, they are highly esteemed, and serve as ample inducements to industry and zeal in the public service.

At first sight it may appear that in any state of society whose essential features were at all like those here briefly outlined, all the usual restraints to early marriage as they now exist would be removed, and that a rate of increase of the population unexampled in any previous era would be the result, leading in a few generations to a difficulty in obtaining subsistence, which Malthus has shown to be

---

[2]*Looking Backward*. See specially chapters vii., ix., xii., and xxv.

the inevitable result of the normal rate of increase of mankind when all the positive as well as the preventive checks are removed. As the positive checks—which may be briefly summarized as war, pestilence, and famine—are supposed to be non-existent, what, it may be asked, are the preventive checks which are suggested as being capable of reducing the rate of increase within manageable limits? This very reasonable question I will now endeavour to answer.

Natural Checks to Rapid Increase.

The first and most important of the checks upon a too rapid increase of population will be the comparatively late average period of marriage, which will be the natural result of the very conditions of society, and will besides be inculcated during the period of education, and still further enforced by public opinion. As the period of systematic education is supposed to extend to the age of twenty-one, up to which time both the mental and physical powers will be trained and exercised to their fullest capacity, the idea of marriage during this period will rarely be entertained. During the last year of education, however, the subject of marriage will be dwelt upon, in its bearing on individual happiness and on social well-being, in relation to the welfare of the next generation and to the continuous development of the race. The most careful and deliberate choice of partners for life will be inculcated as the highest social duty; while the young women will be so trained as to look with scorn and loathing on all men who in any way wilfully fail in their duty to society—on idlers and malingerers, on drunkards and liars, on the selfish, the cruel, or the vicious. They will be taught that the happiness of their whole lives will depend on the care and deliberation with which they choose their husbands, and they will be urged to accept no suitor till he has proved himself to be worthy of respect by the place he holds and the character he bears among his fellow-labourers in the public service.

Under social conditions which render every woman absolutely independent, so far as the necessaries and comforts of existence are concerned, surrounded by the charms of family life and the pleasures of society, which will be far greater than anything we now realize when *all* will possess the refinements derived from the best possible education, and all will be relieved from sordid cares and the struggle for mere existence, is it not in the highest degree probable that marriage will rarely take place till the woman has had three or four years' experience of the world after leaving college—that is, till the age of 25, while it will very frequently be delayed till 30 or upwards? Now Mr. Galton has shown, from the best statistics available, that if we compare women married at 20 with those married at 29, the proportionate fertility is about as 8 to 5. But this difference, large as it is, only represents a portion of the effect on the rate of increase of population caused by a delay in the average period of marriage. For when the age of marriage is delayed the time between successive generations is correspondingly lengthened; while a still further effect is produced by the fact that the greater the average age of marriage the fewer generations are alive at the same time, and it is the combined effect of these three factors that determines the actual rate of increase of the population.[3]

But there is yet another factor tending to check the increase of population that would come into play in a society such as we have been considering. In a remarkable essay on the *Theory of Population* Herbert Spencer has shown, by an elaborate discussion of the phenomena presented by the whole animal kingdom, that the maintenance of the individual and the propagation of the race vary inversely, those species and groups which have the shortest and most uncertain lives producing the greatest number of offspring; in other words, individuation and reproduction are antagonistic. But indification depends almost entirely on the development and specialization of the nervous system, through which, not only are the several activities and coordinations of the various organs carried on, but all advance in instinct, emotion, and intellect is rendered possible. The actual rate of increase in man has been determined by the necessities of the savage state, in which, as in most animal species, it has usually been only just sufficient to maintain a limited average population. But with civilization the average duration of life increases, and the possible increase of population under favourable conditions becomes very great, because fertility is greater than is needed under the new conditions. The advance in civilization as regards the preservation of life has in recent times become so rapid, and the increased development of the nervous system has been limited to so small a portion of the whole population, that no general diminution in fertility has yet occurred. That the facts do, however, accord with the theory is indicated by the common observation that highly intellectual parents do not as a rule have large families, while the most rapid increase occurs in those classes which are engaged in the simpler kinds of manual labour. But in a state of society in which all will have their higher faculties fully cultivated and fully exercised throughout life, a slight general diminution added to that caused by the later average period of marriage would at once bring the rate of increase of population within manageable limits. The same general principle enables us to look forward to that distant future when the world will be fully peopled, in perfect confidence that an equilibrium between the birth and death rates will then be brought about by a combination of physical and social agencies, and the bugbear of over-population become finally extinct.[4]

How Natural Selection Will Improve the Race.

There now only remains for consideration the means by which, in such a society, a continuous improvement of the race could be brought about, on the assumption that for this purpose education is powerless as a direct agency, since its effects are not hereditary, and that some form of selection is an absolute necessity.

---

[3]See Inquiries into Human Faculty and Its Development, p. 321; and Hereditary Genius, p. 353.
[4]A theory of Population deduced from the General Law of Animal Fertility. Republished from the Westminster Review for April, 1852.

This improvement I believe will certainly be effected through the agency of female choice in marriage. Let us, therefore, consider how this would probably act.

It will be generally admitted that, although many women now remain unmarried from necessity rather than from choice, there are always a considerable number who feel no strong inclination to marriage, and who accept husbands to secure a subsistence or a home of their own rather than from personal affection or sexual emotion. In a society in which women were all pecuniarily independent, were all fully occupied with public duties and intellectual or social enjoyments, and had nothing to gain by marriage as regards material well-being, we may be sure that the number of the unmarried from choice would largely increase. It would probably come to be considered a degradation for any woman to marry a man she could not both love and esteem, and this feeling would supply ample reasons for either abstaining from marriage altogether or delaying it till a worthy and sympathetic husband was encountered. In man, on the other hand, the passion of love is more general, and usually stronger; and as in such a society as is here postulated there would be no way of gratifying this passion but by marriage, almost every woman would receive offers, and thus a powerful selective agency would rest with the female sex. Under the system of education and of public opinion here suggested there can be no doubt how this selection would be exercised. The idle and the selfish would be almost universally rejected. The diseased or the weak in intellect would also usually remain unmarried; while those who exhibited any tendency to insanity or to hereditary disease, or who possessed any congenital deformity would in hardly any case find partners, because it would be considered and offence against society to be the means of perpetuating such diseases or imperfections. . . .

When we allow ourselves to be guided by reason, justice, and public spirit in our dealings with our fellow-men, and determine to abolish poverty by recognizing the equal rights of all the citizens of our common land to an equal share of the wealth which all combine to produce—when we have thus solved the lesser problem of a rational social organization adapted to secure the equal well-being of all, then we may safely leave the far greater and deeper problem of the improvement of the race to the cultivated minds and pure instincts of the men, and especially of the Women of the Future.

Individualism, Socialism, and Humanism From *The Grammar of Science*

Karl Pearson, M.A, F.R.S.

We may fitly conclude this chapter on *Life* by a few remarks on the extent to which Individualism, Socialism, and Humanism respectively describe the features of human development. The great part played in life by the self-asserting instinct of the individual does not need much emphasising at the present time. It has been for long the over-shrill keynote of much of English thought. All forms of progress, some of our writers have asserted, could be expressed in terms of the individualistic tendency. The one-sided emphasis which our moralists and publicists placed upon individualism at a time when the revolution of industry relieved us from the stress of foreign competition, may indeed have gone some way towards relaxing that strict training by which a hard-pressed society supplements the inherited social instinct. This emphasis of individualism has undoubtedly led to great advances in knowledge and even in the standards of comfort. Self-help, thrift, personal physique, ingenuity, intellect, and even cunning have been first extolled and then endowed with the most splendid rewards of wealth, influence, and popular admiration. The chief motor of modern life with all its really great achievements has been sought—and perhaps not unreasonably sought—in the individualistic instinct. The success of individual effort in the fields of knowledge and invention has led some to of our foremost biologists to see in individualism as the sole factor of evolution, and they have accordingly propounded a social policy which would place us in the position of the farmer who spends all his energies in producing prize specimens of fat cattle, forgetting that his object should be to improve his stock all round.

I fancy science will ultimately balance the individualistic and socialistic tendencies in evolution better than Haeckel and Spencer seem to have done. The power of the individualistic formula to describe human growth has been overrated, and the evolutionary origin of the socialistic instinct has been too frequently overlooked. In the face of the severe struggle, physical and commercial, the fight for land, for food, and for mineral wealth between existing nations, we have every need to strengthen by training the partially dormant socialistic spirit, if we as a nation are to be among the surviving fit. The importance of organising society, of making the individual subservient to the whole, grows with the intensity of the struggle. We shall need all our clearness of vision, all our reasoned insight into human growth and social efficiency in order to discipline the powers of labour, to train and educate the powers of mind. This organisation and this education must largely proceed from the state, for it is in the battle of society with society, rather than of individual with individual, that these weapons are of service. Here it is that science relentlessly proclaims: A nation needs not only a few prize individuals; it needs a finely regulated social system—of which the members as a whole respond to each external stress by organised reaction—if it is to survive in the struggle for existence.

If the individual asks: Why should I act socially? There is indeed, no argument by which it can be shown that it is always to his own profit or pleasure to do so. Whether an individual takes pleasure in social action or not will depend upon his character (pp. 47, 125)—that product of inherited instincts and past experience—and the extent to which the "tribal conscience" has been developed by early training. If the struggle for existence has not led to the dominant portion of a given community having strong social instincts, then that community, if not already in a decadent condition, is wanting in the chief element of permanent stability. Where this element exists, there society will itself repress those whose conduct is anti-social and develop by training the social instincts of its younger members. Herein lies the only method in which a strong and efficient society, capable of holding its own in the struggle for life, can be built up. It is the prevalence of social instinct in the dominant portion of a given community which is the sole and yet perfectly efficient sanction to the observance of social, that is moral, lines of conduct.

Besides the individualistic and socialistic factors of evolution there remains what we have termed the humanistic factor. Like the socialistic it has been occasionally overlooked, but at the same time occasionally overrated, as, for example, in the formal statements of Positivism. We have always to remember that, hidden beneath diplomacy, trade, adventure, there is a struggle raging between modern nations, which is none the less real if it does not take the form of open warfare. The individualistic instinct may be as strong or stronger than the socialistic, but the latter is always far stronger than any feeling towards humanity as a whole. Indeed the "solidarity of humanity," so far as it is real, is felt to exist rather between civilised men of European race in the presence of nature and of human barbarism, than between all men on all occasions.

"The whole earth is mine, and no one shall rob me of any corner of it," is the cry of the civilised man. No nation can go its own way and deprive the rest of mankind of its soil and its mineral wealth, its labour-power and its culture—no nation can refuse to develop its mental or physical resources—without detriment to civilisation at large in its struggle with organic and inorganic nature. It is not a matter of indifference to other nations that the intellect of any people should lie fallow, or that any folk should not take to its part in the labour of research. It cannot be indifferent to mankind as a whole whether the occupants of a country leave its fields untilled and its natural resources undeveloped. It is a false view of human solidarity, a weak humanitarianism, not a true humanism, which regrets that a capable and stalwart race of white men should replace a dark-skinned tribe which can neither utilize its land for the full benefit of mankind, nor contribute its quota to the common stock of human knowledge.[1] The struggle of civilized man against uncivilised man and against nature produces a certain partial "solidarity of humanity" which involves a prohibition against any individual community wasting the resources of mankind.

The development of the individual, a product of the struggle of man against man, is seen to be controlled by the organization of the social unit, a product of the struggle of society against society. The development of the individual society is again influenced, if to a less extent, by the instinct of a human solidarity in civilised mankind, a product of the struggle of civilisation against barbarism and

---

[1] This sentence must not be taken to justify a brutalising destruction of human life. The anti-social effects of such a mode of accelerating the survival of the fittest may go far to destroy the preponderating fitness of the survivor. At the same time, there is cause for human satisfaction in the replacement of the aborigines throughout America and Australia by white races of far higher civilisation.

against inorganic and organic nature. The principle of the survival of the fittest, describing by aid of the three factors of individualism, socialism, and humanism the continual struggle of individuals, of societies, of civilisation and barbarism, is from the standpoint of science and the sole account we can give of the origin of those purely human faculties of healthy activity, of sympathy, of love, and of social action which men value as their chief heritage.

From *Preface to* Mutual Aid

Prince Petr Kropotkin

Two aspects of animal life impressed me most during the journeys which I made in my youth in Eastern Siberia and Northern Manchuria. One of them was the extreme severity of the struggle for existence which most species of animals have to carryon against an inclement Nature; the enormous destruction of life which periodically results from natural agencies; and the consequent paucity of life over the vast territory which fell under my observation. And the other was, that even in those few spots where animal life teemed in abundance, I failed to find—although I was eagerly looking for it—that bitter struggle for the means of existence, among animals belonging to the same species, which was considered by most Darwinists (though not always by Darwin himself) as the dominant characteristic of struggle for life, and the main factor of evolution.

The terrible snow-storms which sweep over the northern portion of Eurasia in the later part of the winter, and the glazed frost that often follows them; the frosts and the snow-storms which return every year in the second half of May, when the trees are already in full blossom and insect life swarms everywhere; the early frosts and, occasionally, the heavy snowfalls in July and August, which suddenly destroy myriads of insects, as well as the second broods of the birds in the prairies; the torrential rains, due to the monsoons, which fall in more temperate regions in August and September—resulting in inundations on a scale which is only known in America and in Eastern Asia, and swamping, on the plateaus, areas as wide as European States; and finally, the heavy snowfalls, early in October, which eventually render a territory as large as France and Germany, absolutely impracticable for ruminants, and destroy them by the thousand—these were the conditions under which I saw animal life struggling in Northern Asia. They made me realize at an early date the overwhelming importance in Nature of what Darwin described as "the natural checks to over-multiplication," in comparison to the struggle between individuals of the same species for the means of subsistence, which may go on here and there, to some limited extent, but never attains the importance of the former. Paucity of life, under-population—not over-population—being the distinctive feature of that immense part of the globe which we name Northern Asia, I conceived since then serious doubts—which subsequent study has only confirmed—as to the reality of that fearful competition for food and life within each species, which was an article of faith with most Darwinists, and, consequently, as to the dominant part which this sort of competition was supposed to play in the evolution of new species.

On the other hand, wherever I saw animal life in abundance, as, for instance, on the lakes where scores of species and millions of individuals came together to rear their progeny; in the colonies of rodents; in the migrations of birds which took place at that time on a truly American scale along the Usuri; and especially in a migration of fallow-deer which I witnessed on the Amur, and during which scores of thousands of these intelligent animals came together from an immense territory, flying before the coming deep snow, in order to cross the Amur where it is narrowest—in all these scenes of animal life which passed before my eyes, I saw Mutual Aid and Mutual Support carried on to an extent which made me suspect in it a feature of the greatest importance for the maintenance of life, the preservation of each species, and its further evolution.

And finally, I saw among the semi-wild cattle and horses in Transbaikalia, among the wild ruminants everywhere, the squirrels, and so on, that when animals have to struggle against scarcity of food, in consequence of one of the above-mentioned causes, the whole of that portion of the species which is affected by the calamity, comes out of the ordeal so much impoverished in vigour and health, that no progressive evolution of the species can be based upon such periods of keen competition.

Consequently, when my attention was drawn, later on, to the relations between Darwinism and Sociology, I could agree with none of the works and pamphlets that had been written upon this important subject. They all endeavoured to prove that Man, owing to his higher intelligence and knowledge, may mitigate the harshness of the struggle for life between men; but they all recognized at the same time that the struggle for the means of existence, of every animal against all its congeners, and of every man against all other men, was "a law of Nature. " This view, however, I could not accept, because I was persuaded that to admit a pitiless inner war for life within each species, and to see in that war a condition of progress, was to admit something which not only had not yet been proved, but also lacked confirmation from direct observation.

On the contrary, a lecture "On the Law of Mutual Aid," which was delivered at a Russian Congress of Naturalists, in January 1880, by the well-known zoologist, Professor Kessler, the then Dean of the St. Petersburg University, struck me as throwing a new light on the whole subject. Kessler's idea was, that besides the law of Mutual Struggle there is in Nature the law of Mutual Aid, which, for the success of the struggle for life, and especially for the progressive evolution of the species, is far more important than the law of mutual contest. This suggestion—which was, in reality, nothing but a further development of the ideas expressed by Darwin himself in The Descent of Man—seemed to me so correct and of so great an importance, that since I became acquainted with it (in 1883) I began to collect materials for further developing the idea, which Kessler had only cursorily sketched in his lecture, but had not lived to develop. He died in 1881 ....

Consequently I thought that a book, written on Mutual Aid as a Law of Nature and a factor of evolution, might fill an important gap. When Huxley issued, in 1888, his "Struggle-for-life" manifesto (Struggle for Existence and its Bearing upon Man), which to my appreciation was a very incorrect representation of the facts of Nature, as one sees them in the bush and in the forest, I communicated with the editor of the Nineteenth Century, asking him whether he would give the hospitality of his review to an elaborate reply to the views of one of the most prominent Darwinists; and Mr. James Knowles received the proposal with fullest sympathy. I also spoke of it to W. Bates. "Yes, certainly; that is true Darwinism," was his reply. "It is horrible what 'they' have made of Darwin. Write these articles, and when they are printed, I will write to you a letter which you may publish. "Unfortunately, it took me nearly seven years to write these articles, and when the last was published, Bates was no longer living.

After having discussed the importance of mutual aid in various classes of animals, I was evidently bound to discuss the importance of the same factor in the evolution of Man. This was the more necessary as there are a number of evolutionists who may not refuse to admit the importance of mutual aid among animals, but who, like Herbert Spencer, will refuse to admit it for Man. For primitive Man—they maintain—war of each against all was the law of life. In how far this assertion, which has been too willingly repeated, without sufficient criticism, since the times of Hobbes, is supported by what we know about the early phases of human development, is discussed in the chapters given to the Savages and the Barbarians.

The number and importance of mutual-aid institutions which were developed by the creative genius of the savage and half-savage masses, during the earliest clan-period of mankind and still more during the next village-community period, and the immense influence which these early institutions have exercised upon the subsequent development of mankind, down to the present times, induced me to extend my researches to the later, historical periods as well; especially, to study that most interesting period—the free medieval city republics, of which the universality and influence upon our modem civilization have not yet been duly appreciated. And finally, I have tried to indicate in brief the immense importance which the mutual-support instincts, inherited by mankind from its extremely long evolution, play even now in our modem society, which is supposed to rest upon the principle: "every one for himself, and the State for all," but which it never has succeeded, nor will succeed in realizing.

It may be objected to this book that both animals and men are represented in it under too favourable an aspect; that their sociable qualities are insisted upon, while their anti-social and self-asserting instincts are hardly touched upon. This was, however, unavoidable. We have heard so much lately of the "harsh, pitiless struggle for life," which was said to be carried on by every animal against all other animals, every "savage" against all other "savages," and every civilized man against all his co-citizens—and these assertions have so much become an article of faith—that it was necessary, first of all, to oppose to them a wide series of facts showing animal and human life under a quite different aspect. It was necessary to indicate the overwhelming importance which sociable habits play in Nature and in the progressive evolution of both the animal species and human beings: to prove that they secure to animals a better protection from their enemies, very often facilities for getting food and (winter provisions, migrations, etc.), longevity, therefore a greater facility for the development of intellectual faculties; and that they have given to men, in addition to the same advantages, the possibility of working out those institutions which have enabled mankind to survive in its hard struggle against Nature, and to progress, notwithstanding all the vicissitudes of its history. It is a book on the law of Mutual Aid, viewed at as one of the chief factors of evolution—not on all factors of evolution and their respective values; and this first book had to be written, before the latter could become possible.

I should certainly be the last to underrate the part which the self-assertion of the individual has played in the evolution of mankind. However, this subject requires, I believe, a much deeper treatment than the one it has hitherto received. In the history of mankind, individual selfassertion has often been, and continually is, something quite different from, and far larger and deeper than, the petty, unintelligent narrow-mindedness, which, with a large class of writers, goes for "individualism" and "self-assertion." Nor have history-making individuals been limited to those whom historians have represented as heroes. My intention, consequently, is, if circumstances permit it, to discuss separately the part taken by the self-assertion of the individual in the progressive evolution of mankind. I can only make in this place the following general remark:—When the Mutual Aid institutions—the tribe, the village community, the guilds, the medieval city—began, in the course of history, to lose their primitive character, to be invaded by parasitic growths, and thus to become hindrances to progress, the revolt of individuals against these institutions took always two different aspects. Part of those who rose up strove to purify the old institutions, or to work out a higher form of commonwealth, based upon the same Mutual Aid principles; they tried, for instance, to introduce the principle of "compensation," instead of the lex talionis, and later on, the pardon of offences, or a still higher ideal of equality before the human conscience, in lieu of "compensation," according to class-value. But at the very same time, another portion of the same individual rebels endeavoured to break down the protective institutions of mutual support, with no other intention but to increase their own wealth and their own powers. In this three-cornered contest, between the two classes of revolted individuals and the supporters of what existed, lies the real tragedy of history. But to delineate that contest, and honestly to study the part played in the evolution of mankind by each one of these three forces, would require at least as many years as it took me to write this book.

The Right to Make War:
From Germany and the Next War

Friedrich von Bernhardi

The struggle for existence is, in the life of Nature, the basis of all healthy development. All existing things show themselves to be the result of contesting forces. So in the life of man the struggle is not merely the destructive, but the life-giving principle. "To supplant or to be supplanted is the essence of life," says Goethe, and the strong life gains the upper hand. The law of the stronger holds good everywhere. Those forms survive which are able to procure themselves the most favourable conditions of life, and to assert themselves in the universal economy of Nature. The weaker succumb. This struggle is regulated and restrained by the unconscious sway of biological laws and by the interplay of opposite forces. In the plant world and the animal world this process is worked out in unconscious tragedy. In the human race it is consciously carried out, and regulated by social ordinances. The man of strong will and strong intellect tries by every means to assert himself, the ambitious strive to rise, and in this effort the individual is far from being guided merely by the consciousness of right. The life-work and the life-struggle of many men are determined, doubtless, by unselfish and ideal motives, but to a far greater extent the less noble passions—craving for possessions, enjoyment and honour, envy and the thirst for revenge—determine men's actions. Still more often, perhaps, it is the need to live which brings down even natures of a higher mould into the universal struggle for existence and enjoyment.

There can be no doubt on this point. The nation is made up of individuals, the State of communities. The motive which influences each member is prominent in the whole body. It is a persistent struggle for possessions, power, and sovereignty, which primarily governs the relations of one nation to another, and right is respected so far only as it is compatible with advantage. So long as there are men who have human feelings and aspirations, so long as there are nations who strive for an enlarged sphere of activity, so long will conflicting interests come into being and occasions for making war arise.

"The natural law, to which all laws of Nature can be reduced, is the law of struggle. All intrasocial property, all thoughts, inventions, and institutions, as, indeed, the social system itself, are a result of the intrasocial struggle, in which one survives and another is cast out. The extrasocial, the supersocial, struggle which guides the external development of societies, nations, and races, is war. The internal development, the intrasocial struggle, is man's daily work—the struggle of thoughts, feelings, wishes, sciences, activities. The outward development, the supersocial struggle, is the sanguinary struggle of nations—war. In what does the creative power of this struggle consist? In growth and decay, in the victory of the one factor and in the defeat of the other! This struggle is a creator, since it eliminates."[B]

That social system in which the most efficient personalities possess the greatest influence will show the greatest vitality in the intrasocial struggle. In the extrasocial struggle, in war, that nation will conquer which can throw into the scale the greatest physical, mental, moral, material, and political power, and is therefore the best able to defend itself. War will furnish such a nation with favourable vital conditions, enlarged possibilities of expansion and widened influence, and thus promote the progress of mankind; for it is clear that those intellectual and moral factors which insure superiority in war are also those which render possible a general progressive development. They confer victory because the elements of progress are latent in them. Without war, inferior or decaying races would easily choke the growth of healthy budding elements, and a universal decadence would follow. "War," says A. W. von Schlegel, "is as necessary as the struggle of the elements in Nature."

Now, it is, of course, an obvious fact that a peaceful rivalry may exist between peoples and States, like that between the fellow-members of a society, in all departments of civilized life—a struggle which need not always degenerate Into war. Struggle and war are not identical. This rivalry, however, does not take place under the same conditions as the intrasocial struggle, and therefore cannot lead to the same results. Above the rivalry of individuals and groups within the State stands the law, which takes care that injustice is kept within bounds, and that the right shall prevail. Behind the law stands the State, armed with power, which it employs, and rightly so, not merely to protect, but actively to promote, the moral and spiritual interests of society. But there is no impartial power that stands above the rivalry of States to restrain injustice, and to use that rivalry with conscious purpose to promote the highest ends of mankind. Between States the only check on injustice is force, and in morality and civilization each people must play its own part and promote its own ends and ideals. If in doing so it comes into conflict with the ideals and views of other States, it must either submit and concede the precedence to the rival people or State, or appeal to force, and face the risk of the real struggle—i.e., of war—in order to make its own views prevail. No power exists which can judge between States, and makes its judgments prevail. Nothing, in fact, is left but war to secure to the true elements of progress the ascendancy over the spirits of corruption and decay.

It will, of course, happen that several weak nations unite and form a superior combination in order to defeat a nation which in itself is stronger. This attempt will succeed for a time, but in the end the more intensive vitality will prevail. The allied opponents have the seeds of corruption in them, while the powerful nation gains from a temporary reverse a new strength which procures for it an ultimate victory over numerical superiority. The history of Germany is an eloquent example of this truth.

[B] Clauss Wagner, "Der Krieg als schaffendes Weltprinzip."

Struggle is, therefore, a universal law of Nature, and the instinct of self-preservation which leads to struggle is acknowledged to be a natural condition of existence. "Man is a fighter." Self-sacrifice is a renunciation of life, whether in the existence of the individual or in the life of States, which are agglomerations of individuals. The first and paramount law is the assertion of one's own independent existence. By self-assertion alone can the State maintain the conditions of life for its citizens, and insure them the legal protection which each man is entitled to claim from it. This duty of self-assertion is by no means satisfied by the mere repulse of hostile attacks; it includes the obligation to assure the possibility of life and development to the whole body of the nation embraced by the State.

Strong, healthy, and flourishing nations increase in numbers. From a given moment they require a continual expansion of their frontiers, they require new territory for the accommodation of their surplus population. Since almost every part of the globe is inhabited, new territory must, as a rule, be obtained at the cost of its possessors—that is to say, by conquest, which thus becomes a law of necessity.

The right of conquest is universally acknowledged. At first the procedure is pacific. Over-populated countries pour a stream of emigrants into other States and territories. These submit to the legislature of the new country, but try to obtain favourable conditions of existence for themselves at the cost of the original inhabitants, with whom they compete. This amounts to conquest.

The right of colonization is also recognized. Vast territories inhabited by uncivilized masses are occupied by more highly civilized States, and made subject to their rule. Higher civilization and the correspondingly greater power are the foundations of the right to annexation. This right is, it is true, a very indefinite one, and it is impossible to determine what degree of civilization justifies annexation and subjugation. The impossibility of finding a legitimate limit to these international relations has been the cause of many wars. The subjugated nation does not recognize this right of subjugation, and the more powerful civilized nation refuses to admit the claim of the subjugated to independence. This situation becomes peculiarly critical when the conditions of civilization have changed in the course of time. The subject nation has, perhaps, adopted higher methods and conceptions of life, and the difference in civilization has consequently lessened. Such a state of things is growing ripe in British India.

Lastly, in all times the right of conquest by war has been admitted. It may be that a growing people cannot win colonies from uncivilized races, and yet the State wishes to retain the surplus population which the mother-country can no longer feed. Then the only course left is to acquire the necessary territory by war. Thus the instinct of self-preservation leads inevitably to war, and the conquest of foreign soil. It is not the possessor, but the victor, who then has the right. The threatened people will see the point of Goethe's lines:

"That which them didst inherit from thy sires, In order to possess it, must be won."

Perceptions in Science:
Is Evolution a Secular Religion?

Michael Ruse

A major complaint of the Creationists, those who are committed to a Genesis-based story of origins, is that evolution—and Darwinism in particular—is more than just a scientific theory. They object that too often evolution operates as a kind of secular religion, pushing norms and proposals for proper (or, in their opinion, improper) action. Evolutionists dismiss this argument as merely another rhetorical debating trick, and in major respects, this is precisely what it is. It is silly to claim that a naturalistic story of origins leads straight to sexual freedom and other supposed ills of modern society. But, if we wish to deny that evolution is more than just a scientific theory, the Creationists do have a point.

The history of the theory of evolution falls naturally into three parts (1). The first part took place from the mid-18th century up to the publication of Charles Darwin's theory of natural selection as expounded in his *Origin of Species* published in 1859. Up until then, evolution was little more than a pseudo-science on a par with mesmerism (animal magnetism) or phrenology (brain bumps), used as much by its practitioners to convey moral and social messages as to describe the physical world. At the end of the 18th century, Charles Darwin's grandfather, Erasmus, wrote evolutionary poetry, hymning the progress of life from the monad to man—or, as he put it, from the monarch (the butterfly) to the monarch (the king). He derived this notion of biological progress from the successes of the Industrial Revolution and then used it in a circular fashion to justify the cultural progress of the Britain of his day. For example, in his *Temple of Nature* (2), Erasmus Darwin wrote:

> *Imperious man, who rules the bestial crowd,*
> *Of language, reason, and reflection proud,*
> *With brow erect who scorns this earthy sod,*
> *And styles himself the image of his God;*
> *Arose from rudiments of form and sense,*
> *An embryon point, or microscopic ens!*

The same sort of stuff can be found in the writings of other early evolutionists, notably in the *Philosophie Zoologique*, published in 1809 by the Frenchman Jean-Baptiste Lamarck.

Charles Darwin, a serious full-time scientist, set out to change all of this. First, he wanted to give an empirically grounded basis for belief in the fact of evolution. Second, he wanted to persuade his readers of a particular mechanism of

The author is in the Department of Philosophy, Florida State University, Tallahassee, FL 32306-1500, USA. E-mail: mruse@fsu.edu evolution, the natural selection of the successful brought on by the struggle for existence. In his first aim, Darwin was spectacularly successful. Within a decade of the publication of his *Origin of Species*, thinking people were convinced of the fact of evolution. However, regarding his second aim to convince folk about natural selection, Darwin had less success. Most people went for some form of evolution by jumps (saltationism), inheritance of acquired characteristics (Lamarckism), or some other mode of change. Darwin failed in another respect, too. He hoped to upgrade the study of evolution to a respectable, professional science—the kind offered in lectures at universities, with dedicated students and well-funded research. It was not to be. A kind of bastardized Germanic evolution did make it into academia—but it was concerned less with mechanisms and more with hypothesizing about histories, being more connected to Ernst Haeckel's biogenetic law ("ontogeny recapitulates phylogeny") than anything to be found in the *Origin of Species*. As a mature professional research area, evolution was a flop. It simply did not materialize.

Why was this? Darwin himself was an invalid from the age of 30, and any profession building had to be done by his supporters, in particular by his "bulldog," Thomas Henry Huxley. In many respects, Huxley played to Darwin the role that Saint Paul played to Jesus, promoting the master's ideas. But just as Saint Paul rather molded Jesus' legacy to his own ends, so also Huxley molded Darwin's legacy. At the time that the *Origin of Species* was published, Britain was a country desperately in need of reform, as revealed by the horrors of the Crimean War and the Indian Mutiny. Huxley and others worked hard to bring about change, trying to move public perceptions into the 20th century. They reformed education, the civil service, the military, and much else. Huxley's own work was in higher education, and he succeeded best in the areas of physiology and morphology. He realized that to improve and professionalize these fields as areas of teaching and research, he needed clients (a must in all system building). Huxley sold physiology to the medical profession, just then desperate to change from killing to curing. Huxley's offer of a supply of students, ready for specialized medical training, with a solid background in modern physiology was gratefully received. Morphology, Huxley sold to the teaching profession, on the grounds that hands-on empirical study was much better training for modern life than the outmoded classics. Huxley himself sat on the new London School Board and started teacher training courses. His most famous student was the novelist H. G. Wells.

Evolution had no immediate payoff. Learning phylogenies did not cure belly ache, and it was still all a bit too daring for regular schoolroom instruction. But Huxley could see a place for evolution. The chief ideological support of those who opposed the reformers—the landowners, the squires, the generals, and the others—came from the Anglican Church. Hence, Huxley saw the need to found his own church, and evolution was the ideal cornerstone. It offered a story of origins, one that (thanks to progress) puts humans at the center and top and that could even provide moral messages. The philosopher Herbert Spencer was a great help here. He was ever ready to urge his fellow Victorians that the way to true virtue lies through progress, which comes from promoting a struggle in society as well as in biology—a laissez-faire socioeconomic philosophy. Thus, evolution had its commandments no less than did Christianity. And so Huxley preached evolution-as-world-view at working men's clubs, from the podia during presidential addresses, and in debates with clerics—notably Samuel Wilberforce, Bishop of Oxford. He even aided the founding of new cathedrals of evolution, stuffed with displays of dinosaurs newly discovered in the American West. Except, of course, these halls of worship were better known as natural history museums.

As with Christianity, not everyone claimed exactly the same thing in the name of their Lord. Yet, moral norms were the game in town, and things continued this way until the third phase, which began around 1930. This was the era during which a number of mathematically trained thinkers—notably Ronald Fisher and J. B. S. Haldane in England, and Sewall Wright in America—fused Darwinian selection with Mendelian genetics, and thus provided the conceptual foundations of what became known as the synthetic theory of evolution or neo-Darwinism. Rapidly, the experimentalists and naturalists—notably Theodosius Dobzhansky in America and E. B. Ford in England—started to put empirical flesh on the mathematical skeleton, and finally Darwin's dream of a professional evolution with selection at its heart was realized. But there is more to the story than this. These new-style evolutionists—the mathematicians and empiricists—wanted to professionalize evolution because they wanted to study it full time in universities, with students and research grants, and so forth. However, like everyone else, they had been initially attracted to evolution precisely because of its quasi-religious aspects, regardless of whether these formed the basis of an agnostic/atheistic humanism or something to revitalize an old religion that had lost its spirit and vigor. Hence, they wanted to keep a value-impregnated evolutionism that delivered moral messages even as it strived for greater progressive triumphs.

This all meant that by the 1940s and 1950s the study of evolution was of two sorts. There was serious empirical work, very professional, containing few or no direct exhortations to moral or social action. Along with this, almost all of the leading evolutionists were turning out works of a more popular nature, about progress and the ways to achieve it. By the 1950s, evolutionary works, such as those by the Darwinian paleontologist G. G. Simpson, discussed democracy and education and (increasingly) conservation. In 1944, Simpson published *Tempo and Mode in Evolution*: straight science about natural selection and the fossil record. Then, in 1949, he published *The Meaning of Evolution*: science for the general reader, packed with all sorts of stuff about the virtues of the American way over communism. (Remember, the Cold War was then settling into its long winter, and Trofim Lysenko was destroying Russian biology.) Finally, in 1953, came Simpson's *The Major Features of Evolution*, and we were back to straight science.

Things have continued in much the same way to the present. There is professional evolutionary biology: mathematical, experimental, not laden with value statements. But, you are not going to find the answer to the world's mysteries or to societal problems if you open the pages of *Evolution* or *Animal Behaviour*. Then, sometimes from the same person, you have evolution as secular religion, generally working from an explicitly materialist background and solving all of the world's major problems, from racism to education to conservation. Consider Edward O. Wilson, rightfully regarded as one of the most outstanding professional evolutionary biologists of our time, and the author of major works of straight science. In his *On Human Nature*, he calmly assures us that evolution is a myth that is now ready to take over Christianity. And, if this is so, "the final decisive edge enjoyed by scientific naturalism will come from its capacity to explain traditional religion, its chief competition, as a wholly material phenomenon. Theology is not likely to survive as an independent intellectual discipline" (3). An ardent progressionist, Wilson sees moral norms emerging from our need to keep the evolutionary process moving forward. In his view, this translates as a need to promote biodiversity, for Wilson believes that humans have evolved in a symbiotic relationship with nature. A world of plastic would kill us humans, literally as well as metaphorically. For progress to continue, we must preserve the Brazilian rainforests and other areas of high organic density and diversity (4).

So, what does our history tell us? Three things. First, if the claim is that all contemporary evolutionism is merely an excuse to promote moral and societal norms, this is simply false. Today's professional evolutionism is no more a secular religion than is industrial chemistry. Second, there is indeed a thriving area of more popular evolutionism, where evolution is used to underpin claims about the nature of the universe, the meaning of it all for us humans, and the way we should behave. I am not saying that this area is all bad or that it should be stamped out. I am all in favor of saving the rainforests. I am saying that this popular evolutionism—often an alternative to religion—exists. Third, we who cherish science should be careful to distinguish when we are doing science and when we are extrapolating from it, particularly when we are teaching our students. If it is science that is to be taught, then teach science and nothing more. Leave the other discussions for a more appropriate time.

References and Notes

1. M. Ruse, *Monad to Man: The Concept of Progress in Evolutionary Biology* (Harvard Univ. Press, Cambridge, MA, 1996).
2. E. Darwin, *The Temple of Nature* (J. Johnson, London, 1803), vol. 1, canto 1, lines 309-314.
3. E. O. Wilson, *On Human Nature* (Harvard Univ. Press, Cambridge, MA, 1978), p. 192.
4. See, for example, E. O. Wilson, *Biophilia* (Harvard Univ. Press, Cambridge, MA, 1984) and E. O. Wilson, *The Diversity of Life* (Harvard Univ. Press, Cambridge, MA, 1992).

Chapter 4

Introduction

Chapter 4 has six documents. The first two are exchanges of ideas between Asa Gray and Charles Darwin. Gray, an evangelical Christian, thought that selection could not bring on all of the adaptation—which he took as evidence of God's design—that he saw about him in nature. Hence, in the essay from which the extract printed here is taken he argued that God must have been standing directly behind evolutionary change. For Darwin this would be to make selection redundant, and he expressed himself forcefully in the extracts given here, taken from his letters to Gray. Darwin thought science should keep direction and purpose and unseen forces out of its explanations. He wanted no truck with guided variations bring on adaptation. For him it was natural selection or nothing. Note however that Darwin himself seems to allow that there is (or may be) something or Someone behind everything. It is just that he does not want this guiding force to be active on the job. Rather, leave it all to normal unguided laws of nature.

The next three documents, picking up again on themes from the last chapter, give a range of the kinds of views that were being expressed in America on Social Darwinian issues and a sense of how widely spread were these views among the general public of the day. First (third in this section) is an extract from an essay ("The Challenge of Facts") by the sociologist William Graham Sumner, perhaps Spencer's greatest supporter, where he argues that Darwinism supports laissez faire economics and gives the lie to socialism. Next (fourth in the section) comes part of an essay ("The Gospel of Wealth") by the great industrialist (and another Spencer enthusiast) the Scottish-born Andrew Carnegie. In his life, his actions, his thinking, Carnegie shows how paradoxical can be the whole Social Darwinian thrust and how careful one must be in making judgements. Carnegie was a brutal businessman—he had to be, to build US Steel, that back then completely dominated the landscape of Pittsburgh and had great influence through the land. He put down strikes with cruel ruthlessness. Yet, he never valued wealth for its own sake and put into practice his famous dictum that, if one refuses to give away one's wealth, "The man who dies thus rich dies disgraced." I myself owe much to this philosophy for as a child in industrial England I went regularly to our local Carnegie library. no one should die rich. Finally (fifth in the section), I give you a passage from Jack London's novel, *The Call of the Wild*. The title of the chapter, "The law of club and fang"—a riff on Tennyson's "nature red in tooth and claw" (itself pre-Darwinian and reflecting Charles Lyell's discussion, in his *Principles of Geology*, of the struggle for existence)—tells you everything you need to know. I note, without comment, that checking on amazon.com, I found thirty six editions, most still in print, not to mention teachers' guides and much more.

The final and sixth document comes from the end of our period. Henry Fairfield Osborn, whom we met in the text as director of the American Museum of Natural History in New York, had been chosen as one of the expert witnesses to appear on behalf of evolution in the Scopes trial. As it happened, the judge would not admit any of this evidence, but this did not stop Osborn—a sincere Christian—from expressing his views in print. In the passage given, from a collection revealingly entitled *The Earth Speaks to Bryan,* Osborn is affirming his faith in evolution. He is also giving his decided opinion on the primitive nature of Neanderthals, whom he contrasts unfavorably with a later race, Cro-Magnon man. Osborn was treading a careful line here, intending his history to convey a powerful message for the present. As a modern, enlightened American from the North, he thought that all humans—including Negroes—were the same species. His mother had read *Uncle Tom's Cabin* to him as a child. But as an upper-class, white, Protestant American he had no doubt that not all humans were equal and that blacks, Jews, Catholics, Irish, Italians—in short, anyone not of his own narrow group—were inferior to good old Anglo-Saxon stock. Osborn was keen on eugenics and supported moves to restrict immigration to the United States. The Neanderthal/Cro-Magnon divide was intended to show that these differences were a natural state of affairs among humans and that his own position was merely a reflection and appreciation of the history of humankind. Neanderthals were human but, like recent immigrants, not the best kinds of humans. As I have argued in the text, and as I have tried to show through the readings in this section (as well as the last section) of the documents, evolution was proving a perfect vehicle for social messages of a kind that one associates with an ideology or religion.

Darwiniana

A. Gray

However that may be, it is undeniable that Mr. Darwin has purposely been silent upon the philosophical and theological applications of his theory. This reticence, under the circumstances, argues design, and raises inquiry as to the final cause or reason why. Here, as in higher instances, confident as we are that there is a final cause, we must not be overconfident that we can infer the particular or true one. Perhaps the author is more familiar with natural-historical than philosophical inquiries, and, not having decided with which particular theory about efficient cause is best founded, he meanwhile argues the scientific questions concerned—all that relates to secondary causes—upon purely scientific grounds, as he must do in any case. Perhaps, confident, as he evidently is, that his view will finally be adopted, he may enjoy a sort of satisfaction in hearing it denounced as sheer atheism by the inconsiderate, and afterward, when it takes its place with the nebular hypothesis and the like, see this judgment reversed, as we suppose it would be in such event.

Whatever Mr. Darwin's philosophy may be, or whether he has any, is a matter of no consequence at all, compared with the important questions, whether a theory to account for the origination and diversification of animal and vegetable forms through the operation of secondary causes does or does not exclude design; and whether the establishment by adequate evidence of Darwin's particular theory of diversification through variation and natural selection would essentially alter the present scientific and philosophical grounds for theistic views of Nature. The unqualified affirmative judgment rendered by the two Boston reviewers, evidently able and practised reasoners, "must give us pause." We hesitate to advance our conclusions in opposition to theirs. But, after full and serious consideration we are constrained to say that, in our opinion, the adoption of a derivative hypothesis, and of Darwin's particular hypothesis, if we understand it, would leave the doctrines of final causes, utility, and special design, just where they were before. We do not pretend that the subject is not environed with difficulties. Every view is so environed; and every shifting of the view is likely, if it removes some,. difficulties, to bring others into prominence. But we cannot perceive that Darwin's theory brings in any new kind of scientific difficulty, that is, any with which philosophical naturalists were not already familiar.

Since natural science deals only with secondary or natural causes, the scientific terms of a theory of derivation of species—no less than of a theory of dynamics—must needs be the same to the theist as to the atheist. The difference appears only when the inquiry is earned up to the question of primary cause—a question which belongs to philosophy. Wherefore, Darwin's reticence about efficient cause does not disturb us. He considers only the scientific questions. As already stated, we think that a theistic view of Nature is implied in his book, and we must charitably refrain from suggesting the contrary until the contrary is logically deduced from his premises. If, however, he anywhere maintains that the natural causes through which species are diversified operate without an ordaining and directing intelligence, and that the orderly arrangements and admirable adaptations we see all around us are fortuitous or blind, undesigned results—that the eye, though it came to see, was not designed for seeing, nor the hand for handling—then, we suppose, he is justly chargeable with denying, and very needlessly denying, all design in organic Nature; otherwise, we suppose not. Why, if Darwin's well-known passage about the eye[1]—equivocal though some of the language be—does not imply ordaining and directing intelligence, then he refutes his own theory as effectually as any of his opponents are likely to do. He asks:

> "May we not believe that under variation proceeding long enough, generation multiplying the better variations times enough, and natural selection securing the im-

---

[1]Page 188, English edition.

provements a living optical instrument might be thus formed as superior to one of glass as the works of the Creator are to those of man ?"

This must mean one of two things: either that the living instrument was made and perfected under (which is the same thing as by) an intelligent First Cause, or that it was not. If it was, then theism is asserted; and as to the mode of operation, how do we know, and why must we believe, that, fitting precedent forms being in existence, a living instrument (so different from a lifeless manufacture) would be originated and perfected in any other way, or that this is no the fitting way? If it means that it was not, if he so misuses words that by the Creator he intends an unintelligent power, undirected force, or necessity, then he has put his ease so as to invite disbelief in it. For then blind forces have produced not only manifest adaptations of means to specific ends—which is absurd enough—but better adjusted and more perfect instruments or machines than intellect (that is, human intellect) can contrive and human skill execute—which no sane person will believe.

On the other hand, if Darwin even admits—we will not say adopts—the theistic view, he may save himself much needless trouble in the endeavor to account for the absence of every sort of intermediate form. Those in the line between one species and other supposed to be derived from it he may be bound to provide; but as to an infinite number of other varieties not intermediate, gross, rude, and purposeless, the unmeaning creations of an unconscious cause," born only to perish, which a relentless reviewer has imposed upon his theory—rightly enough upon the atheistic alternative—the theistic view rids him at once of this scum of creation." For, as species do not now vary at all times and places and in all directions, nor produce crude, vague, imperfect, and useless forms, there is no reason for supposing that they ever did. Good-for-nothing monstrosities, failures of purpose rather than purposeless, indeed, sometimes occur; but these are just as anomalous and unlikely upon Darwin's theory as upon any other. For his particular theory is based, and even over-strictly insists, upon the most universal of physiological laws, namely, that successive generations shall differ only slightly, if at all, from their parents; and this effectively excludes crude and impotent forms. Wherefore, if we believe that the species were designed, and that natural propagation was designed, how can we say that the actual varieties of the species were not equally designed? Have we not similar grounds for inferring design in the supposed varieties of species, that we have in the case of the supposed species of a genus? When a naturalist comes to regard as three closely-related species what he before took to be so many varieties of one species, how has he thereby strengthened our conviction that the three forms are designed to have the differences which they actually exhibit? Wherefore, so long as gradatory, orderly, and adapted forms in Nature argue design, and at least while the physical cause of variation is utterly unknown and mysterious, we should advise Mr. Darwin to assume, in the philosophy of his hypothesis, that variation has been led along certain beneficial lines. Streams flowing over a sloping plain by gravitation (here the counterpart of natural selection) may have worn their actual channels as they flowed; yet their particular courses may have been assigned; and where we see them forming definite and useful lines of irrigation, after a manner unaccountable on the laws of gravitation and dynamics, we should believe that the distribution was designed.

To insist, therefore, that the new hypothesis of the derivative origin of the actual species is incompatible with final causes and design, is to take a position which we must consider philosophically untenable. We must also regard it as highly unwise and dangerous, in the present state and present prospects of physical and physiological science. We should expect the philosophical atheist or skeptic to take this ground; also, until better informed, the unlearned and unphilosophical believer; but we should think that the thoughtful theistic philosopher would take the other side. Not to do so seems to concede that only supernatural events can be shown to be designed, which no theist can admit—seems also to misconceive the scope and meaning of all ordinary arguments for design in Nature. This misconception is shared both by the reviewers and the reviewed. At least, Mr. Darwin uses expressions which imply that the natural forms which surround us, because they have a history or natural sequence, could have been only generally, but not particularly designed—a view at once superficial and contradictory; whereas his true line should be, that, his hypothesis concerns the *order* and not the *cause*, the *how* and not the *why* of the phenomena, and so leaves the question of design just where it was before.

Letter to Asa Gray

Charles Darwin

One word more on "designed laws" & "undesigned results." I see a bird which I want for food, take my gun & kill it, I do this *designedly*.—An innocent & good man stands under tree & is killed by flash of lightning. Do you believe (& I really sh^d like to hear) that God *designedly* killed this man? Many or most persons do believe this; I can't & don't.—If you believe so, do you believe that when a swallow snaps up a gnat that God designed that that particular swallow sh^d snap up that particular gnat at that particular instant? I believe that the man & the gnat are in same predicament.—If the death of neither man or gnat are designed, I see no good reason to believe that their *first* birth or production sh^d be necessarily designed. Yet, as I said before, I cannot persuade myself that electricity acts, that the tree grows, that man aspires to loftiest conceptions all from blind, brute force.

. . .

Yesterday I read over with care the third Article; & it seems to me, as before, *admirable*. But I grieve to say that I cannot honestly go as far as you do about Design. I am conscious that I am in an utterly hopeless muddle. I cannot think that the world, as we see it, is the result of chance; & yet I cannot look at each separate thing as the result of Design.—To take a crucial example, you lead me to infer (p. 414) that you believe "that variation has been led along certain beneficial lines."—I cannot believe this; & I think you would have to believe, that the tail of the Fan-tail was led to vary in the number & direction of its feathers in order to gratify the caprice of a few men. Yet if the fan-tail had been a wild bird & had used its abnormal tail for some special end, as to sail before the wind, unlike other birds, everyone would have said what beautiful & designed adaptation. Again I say I am, & shall every remain, in a hopeless muddle.—

. . .

Your question what would convince me of Design is a poser. If I saw an angel come down to teach us good, & I was convinced, from others seeing him, that I was not mad, I sh^d believe in design.—If I could be convinced thoroughily that life & mind was in an unknown way a function of other imponderable forces, I sh^d be convinced.—If man was made of brass or iron & in no way connected with any other organism which had ever lived, I sh^d perhaps be convinced. But this is childish writing.—

I have lately been corresponding with Lyell, who, I think, adopts your idea of the stream of variation having been led or designed. I have asked him (& he say he will herafter reflect & answer me) whether he believes that the shape of my nose was designed. If he does, I have nothing more to say. If not, seeing what Fanciers have done by selecting individual differences in the nasal bones of Pigeons, I must think that it is illogical to suppose that the variations, which Nat. Selection, preserves for the good of any being, have been designed.

From *The Challenge of Facts and Other Essays*

William Graham Sumner

Socialism is no new thing. In one form or another it is to be found throughout all history. It arises from an observation of certain harsh facts in the lot of man on earth, the concrete expression of which is poverty and misery. These facts challenge us. It is folly to try to shut our eyes to them. We have first to notice what they are, and then to face them squarely.

Man is born under the necessity of sustaining the existence he has received by an onerous struggle against nature, both to win what is essential to his life and to ward off what is prejudicial to it. He is born under a burden and a necessity. Nature holds what is essential to him, but she offers nothing gratuitously. He may win for his use what she holds, if he can. Only the most meager and inadequate supply for human needs can be obtained directly from nature. There are trees which may be used for fuel and for dwellings, but labor is required to fit them for this use. There are ores in the ground, but labor is necessary to get out the metals and make tools or weapons. For any real satisfaction, labor is necessary to fit the products of nature for human use. In this struggle every individual is under the pressure of the necessities for food, clothing, shelter, fuel, and every individual brings with him more or less energy for the conflict necessary to supply his needs. The relation, therefore, between each man's needs and each man's energy, or "individualism," is the first fact of human life.

It is not without reason, however, that we speak of a "man" as the individual in question, for women (mothers) and children have special disabilities for the struggle with nature, and these disabilities grow greater and last longer as civilization advances. The perpetuation of the race in health and vigor, and its success as a whole in its struggle to expand and develop human life on earth, therefore, require that the head of the family shall, by his energy, be able to supply not only his own needs, but those of the organisms which are dependent upon him. The history of the human race shows a great variety of experiments in the relation of the sexes and in the organization of the family. These experiments have been controlled by economic circumstances, but, as man has gained more and more control over economic circumstances, monogamy and the family education of children have been more and more sharply developed. If there is one thing in regard to which the student of history and sociology can affirm with confidence that social institutions have made "progress" or grown "better," it is in this arrangement of marriage and the family. All experience proves that monogamy, pure and strict, is the sex relation which conduces most to the vigor and intelligence of the race, and that the family education of children is the institution by which the race as a whole advances most rapidly, from generation to generation, in the struggle with nature. Love of man and wife, as we understand it, is a modern sentiment. The devotion and sacrifice of parents for children is a sentiment which has been developed steadily and is now more intense and far more widely practiced throughout society than in earlier times. The relation is also coming to be regarded in a light quite different from that in which it was formerly viewed. It used to be believed that the parent had unlimited claims on the child and rights over him. In a truer view of the matter, we are coming to see that the rights are on the side of the child and the duties on the side of the parent. Existence is not a boon for which the child owes all subjection to the parent. It is a responsibility assumed by the parent towards the child without the child's consent, and the consequence of it is that the parent owes all possible devotion to the child to enable him to make his existence happy and successful.

The value and importance of the family sentiments, from a social point of view, cannot be exaggerated. They impose self-control and prudence in their most important social bearings, and tend more than any other forces to hold the individual up to the virtues which make the sound man and the valuable member of society. The race is bound from generation to generation, in an unbroken chain of vice and penalty, virtue and reward. The sins of the fathers are visited upon the children, while on the other hand, health, vigor, talent, genius, and skill are, so far as we can discover, the results of high physical vigor and wise early training. The popular language bears witness to the universal observation of these facts, although general social and political dogmas have come into fashion which contradict or ignore them. There is no other such punishment for a life of vice and self-indulgence as to see children grow up cursed with the penalties of it, and no such reward for self-denial and virtue as to see children born and grow up vigorous in mind and body. It is time that the true import of these observations for moral and educational purposes was developed, and it may well be questioned whether we do not go too far in our reticence in regard to all these matters when we leave it to romances and poems to do almost all the educational work that is done in the way of spreading ideas about them. The defense of marriage and the family, if their sociological value were better understood, would be not only instinctive but rational. The struggle for existence with which we have to deal must be understood, then, to be that of a man for himself, his wife, and his children.

The next great fact we have to notice in regard to the struggle of human life is that labor which is spent in a direct struggle with nature is severe in the extreme and is but slightly productive. To subjugate nature, man needs weapons and tools. These, however, cannot be won unless the food and clothing and other prime and direct necessities are supplied in such amount that they can be consumed while tools and weapons are being made, for the tools and weapons themselves satisfy no needs directly. A man who tills the ground with his fingers or with a pointed stick picked up without labor will get a small crop. To fashion even the rudest spade or hoe will cost time, during which the laborer must still eat and drink and wear, but the tool, when obtained, will multiply immensely the power to produce. Such products of labor, used to assist production, have a function so peculiar in the nature of things that we need to distinguish them. We call them capital. A lever is capital, and the advantage of lifting a weight with a lever over lifting it by direct exertion is only a feeble illustration of the power of capital in production. The origin of capital lies in the darkness before history, and it is probably impossible for us to imagine the slow and painful steps by which the race began the formation of it. Since then it has gone on rising to higher and higher powers by a ceaseless involution, if I may use a mathematical expression. Capital is labor raised to a higher power by being constantly multiplied into itself. Nature has been more and more subjugated by the human race through the power of capital, and every human being now living shares the improved status of the race to a degree which neither he nor anyone else can measure, and for which he pays nothing.

Let us understand this point, because our subject will require future reference to it. It is the most shortsighted ignorance not to see that, in a civilized community, all the advantage of capital except a small fraction is gratuitously enjoyed by the community. For instance, suppose the case of a man utterly destitute of tools, who is trying to till the ground with a pointed stick. He could get something out of it. If now he should obtain a spade with which to till the ground, let us suppose, for illustration, that he could get twenty times as great a product. Could, then, the owner of a spade in a civilized state demand, as its price, from the man who had no spade, nineteen-twentieths of the product which could be produced by the use of it? Certainly not. The price of a spade is fixed by the supply and demand of products in the community. A spade is bought for a dollar and the gain from the use of it is an inheritance of knowledge, experience, and skill which every man who lives in a civilized state gets for nothing. What we pay for steam transportation is no tribe, but imagine, if you can, eastern Massachusetts cut off from steam connection with the rest of the world, turnpikes and sailing vessels remaining. The cost of food would rise so high that a quarter of the population would starve to death and another quarter would have to emigrate. To-day every man here gets an enormous advantage from the status of a society on a level of steam transportation, telegraph, and machinery, for which he pays nothing.

So far as I have yet spoken, we have before us the struggle of man with nature, but the social problems, strictly speaking, arise at the next step. Each man carries on the struggle to win his support for himself, but there are others by his side engaged in the same struggle. If the stores of nature were unlimited, or if the last unit of the supply she offers could be won as easily as the first, there would be no social problem. If a square mile of land could support an indefinite number of human beings, or if it cost only twice as much labor to get forty bushels of wheat from an acre as to get twenty, we should have no social problem. If a square mile of land could support millions, no one would ever emigrate and there would be no trade or commerce. If it cost only twice as much labor to get forty bushels as twenty, there would be no advance in the arts. The fact is far otherwise. So long as the population is low in proportion to the amount of land, on a given stage of the arts, life is easy and the competition of man with man is weak. When more persons are trying to live on a square mile than it can support, on the existing stage of the arts, life is hard and the competition of man with man is intense. In the former case, industry and prudence may be on a low grade; the penalties are not severe, or certain, or speedy. In the latter case, each individual needs to exert on his own behalf every force, original or acquired, which he can command. In the former case, the average condition will be one of comfort and the population will be all nearly on the average. In the latter case, the average condition will not be one of comfort, but the population will cover wide extremes of comfort and misery. Each will find his place according to his ability and his effort. The former society will be democratic; the latter will be aristocratic.

The constant tendency of population to outstrip the means of subsistence is the force which has distributed population over the world, and produced all advance in civilization. To this day the two means of escape for an overpopulated country are emigration and an advance in the arts. The former wins more land for the same people; the latter makes the same land support more persons. If, however, either of these means opens a chance for an increase of population, it is evident that the advantage so won may be speedily exhausted if the increase takes place. The social difficulty has only undergone a temporary amelioration, and when the conditions of pressure and competition are renewed, misery and poverty reappear. The victims of them are those who have inherited disease and depraved appetites, or have been brought up in vice and ignorance, or have themselves yielded to vice, extravagance, idleness, and imprudence. In the last analysis, therefore, we come back to vice, in its original and hereditary forms, as the correlative of misery and poverty.

The condition for the complete and regular action of the force of competition is liberty. Liberty means the security given to each man that, if he employs his energies to sustain the struggle on behalf of himself and those he cares for, he shall dispose of the product exclusively as he chooses. It is impossible to know whence any definition or criterion of justice can be derived, if it is not deduced from this view of things; or if it is not the definition of justice that each shall enjoy the fruit of his own labor and self-denial, and of injustice that the idle and the industrious, the self-indulgent and the self-denying, shall share equally in the product. Aside from the *a priori* speculations of philosophers who have tried to make equality an essential element in justice, the human race has recognized, from the earliest times, the above conception of justice as the true one, and has founded upon it the right of property. The right of property, with marriage and the family, gives the right of bequest.

Monogamic marriage, however, is the most exclusive of social institutions. It contains, as essential principles, preference, superiority, selection, devotion. It would not be at all what it is if it were not for these characteristic traits, and it always degenerates when these traits are not present. For instance, if a man should not have a distinct preference for the woman he married, and if he did not select her as superior to others, the marriage would be an imperfect one according to the standard of true monogamic marriage. The family under monogamy, also, is a closed group, having special interests and estimating privacy and reserve as valuable advantages for family development. We grant high prerogatives, in our society, to parents, although our observation teaches us that thousands of human beings are unfit to be parents or to be entrusted with the care of children. It follows, therefore, from the organization of marriage and the family, under monogamy, that great inequalities must exist in a society based on those institutions. The son of wise parents cannot start on a level with the son of foolish ones, and the man who has had no home discipline cannot be equal to the man who has had home discipline. If the contrary were true, we could rid ourselves at once of the wearing labor of inculcating sound morals and manners in our children.

Private property, also, which we have seen to be a feature of society organized in accordance with the natural conditions of the struggle for existence produces inequalities between men. The struggle for existence is aimed against nature. It is from her niggardly hand that we have to wrest the satisfactions for our needs, but our fellow-men are our competitors for the meager supply. Competition, therefore, is a law of nature. Nature is entirely neutral; she submits to him who most energetically and resolutely assails her. She grants her rewards to the fittest, therefore, without regard to other considerations of any kind. If, then, there be liberty, men get from her just in proportion to their works, and their having and enjoying are just in proportion to their being and their doing. Such is the system of nature. If we do not like it, and if we try to amend it, there is only one way in which we can do it. We can take from the better and give to the worse. We can deflect the penalties of those who have done ill and throw them on those who have done better. We can take the rewards from those who have done better and give them to those who have done worse. We shall thus lessen the inequalities. We shall favor the survival of the unfittest, and we shall accomplish this by destroying liberty. Let it be understood that we cannot go outside of this alternative: liberty, inequality, survival of the fittest; not-liberty, equality, survival of the unfittest. The former carries society forward and favors all its best members; the latter carries society downwards and favors all its worst members.

For three hundred years now men have been trying to understand and realize liberty. Liberty is not the right or chance to do what we choose; there is no such liberty as that on earth. No man can do as he chooses: the autocrat of Russia or the King of Dahomey has limits to his arbitrary will; the savage in the wilderness, whom some people think free, is the slave of routine, tradition, and superstitious fears; the civilized man must earn his living, or take care of his property, or concede his own will to the rights and claims of his parents, his wife, his children, and all the persons with whom he is connected by the ties and contracts of civilized life.

What we mean by liberty is civil liberty, or liberty under law; and this means the guarantees of law that a man shall not be interfered with while using his own powers for his own welfare. It is, therefore, a civil and political status; and that nation has the freest institutions in which the guarantees of peace for the laborer and security for the capitalist are the highest. Liberty, therefore, does not by any means do away with the struggle for existence. We might as well try to do away with the need of eating, for that would, in effect, be the same thing. What civil liberty does is to turn the competition of man with man from violence and brute force into an industrial competition under which men vie with one another for the acquisition of material goods by industry, energy, skill, frugality, prudence, temperance, and other industrial virtues. Under this changed order of things the inequalities are not done away with. Nature still grants her rewards of having and enjoying, according to our being and doing, but it is now the man of the highest training and not the man of the heaviest fist who gains the highest reward. It is impossible that the man with capital and the man without capital should be equal. To affirm that they are equal would be to say that a man who has no tool can get as much food out of the ground as the man who has a spade or a plough; or that the man who has no weapon can defend himself as well against hostile beasts or hostile men as the man who has a weapon. If that were so, none of us would work any more. We work and deny ourselves to get capital just because, other things being equal, the man who has it is superior, for attaining all the ends of life, to the man who has it not. Considering the eagerness with which we all seek capital and the estimate we put upon it, either in cherishing it if we have it, or envying others who have it while we have it not, it is very strange what platitudes pass current about it in our society so soon as we begin to generalize about it. If our young people really believed some of the teachings they hear, it would not be amiss to preach them a sermon once in a while to reassure them, setting forth that it is not wicked to be rich, nay even, that it is not wicked to be richer than your neighbor.

It follows from what we have observed that it is the utmost folly to denounce capital. To do so is to undermine civilization, for capital is the first requisite of every social gain, educational, ecclesiastical, political, aesthetic, or other.

It must also be noticed that the popular antithesis between persons and capital is very fallacious. Every law or institution which protects persons at the expense of capital makes it easier for persons to live and to increase the number of consumers of capital while lowering all the motives to prudence and frugality by which capital is created. Hence every such law or institution tends to produce a large population, sunk in misery. All poor laws and all eleemosynary institutions and expenditures have this tendency. On the contrary, all laws and institutions which give security to capital against the interests of other persons than its owners, restrict numbers while preserving the means of subsistence. Hence every such law or institution tends to produce a small society on a high stage of comfort and well-being. It follows that the antithesis commonly thought to exist between the protection of persons and the protection of property is in reality only an antithesis between numbers and quality.

I must stop to notice, in passing, one other fallacy which is rather scientific than popular. The notion is attributed to certain economists that economic forces are self-correcting. I do not know of any economists who hold this view, but what is intended probably is that many economists, of whom I venture to be one, hold that economic forces act compensatingly, and that whenever economic forces have so acted as to produce an unfavorable situation, other economic forces are brought into action which correct the evil and restore the equilibrium. For instance, in Ireland overpopulation and exclusive devotion to agriculture, both of which are plainly traceable to unwise statesmanship in the past, have produced a situation of distress. Steam navigation on the ocean has introduced the competition of cheaper land with Irish agriculture. The result is a social and industrial crisis. There are, however, millions of acres of fertile land on earth which are unoccupied and which are open to the Irish, and the economic forces are compelling the direct corrective of the old evils, in the way of emigration or recourse to urban occupations by unskilled labor. Any number of economic and legal nostrums have been proposed for this situation, all of which propose to leave the original causes untouched. We are told that economic causes do not correct themselves. That is true. We are told that when an economic situation becomes very grave it goes on from worse to worse and that there is no cycle through which it returns. That is not true, without further limitation. We are told that moral forces alone can elevate any such people again. But it is plain that a people which has sunk below the reach of the economic forces of self-interest has certainly sunk below the reach of moral forces, and that this objection is superficial and short-sighted. What is true is that economic forces always go before moral forces. Men feel self-interest long before they feel prudence, self-control, and temperance. They lose the moral forces long before they lose the economic forces. If they can be regenerated at all, it must be first by distress appealing to self-interest and forcing recourse to some expedient for relief. Emigration is certainly an economic force for the relief of Irish distress. It is a palliative only, when considered in itself, but the virtue of it is that it gives the non-emigrating population a chance to rise to a level on which the moral forces can act upon them. Now it is terribly true that only the better ones emigrate, and only the better ones among those who remain are capable of having their ambition and energy awakened, but for the rest the solution is famine and death, with a social regeneration through decay and the elimination of that part of the society which is not capable of being restored to health and life. As Mr. Huxley once said, the method of nature is not even a word and a blow, with the blow first. No explanation is vouchsafed. We are left to find out for ourselves why our ears are boxed. If we do not find out, and find out correctly, what the error is for which we are being punished, the blow is repeated and poverty, distress, disease, and death finally remove the incorrigible ones. It behooves us men to study these terrible illustrations of the penalties which follow on bad statesmanship, and of the sanctions by which social laws are enforced. The economic cycle does complete itself; it must do so, unless the social group is to sink in permanent barbarism. A law may be passed which shall force somebody to support the hopelessly degenerate members of a society, but such a law can only perpetuate the evil and entail it on future generations with new accumulations of distress.

The economic forces work with moral forces and are their handmaidens, but the economic forces are far more primitive, original, and universal. The glib generalities in which we sometimes hear people talk, as if you could set moral and economic forces separate from and in antithesis to each other, and discard the one to accept and work by the other, gravely misconstrue the realities of the social order.

We have now before us the facts of human life out of which the social problem springs. These facts are in many respects hard and stem. It is by strenuous exertion only that each one of us can sustain himself against the destructive forces and the ever recurring needs of life; and the higher the degree to which we seek to carry our development the greater is the proportionate cost of every step. For help in the struggle we can only look back to those in the previous generation who are responsible for our existence. In the competition of life the son of wise and prudent ancestors has immense advantages over the son of vicious and imprudent ones. The man who has capital possesses immeasurable advantages for the struggle of life over him who has none. The more we break down privileges of class, or industry, and establish liberty, the greater will be the inequalities and the more exclusively will the vicious bear the penalties. Poverty and misery will exist in society just so long as vice exists in human nature.

I now go on to notice some modes of trying to deal with this problem. There is a modern philosophy which has never been taught systematically, but which has won the faith of vast masses of people in the modem civilized world. For want of a better name it may be called the sentimental philosophy. It has colored all modern ideas and institutions in politics, religion, education, charity, and industry, and is widely taught in popular literature, novels, and poetry, and in the pulpit. The first proposition of this sentimental philosophy is that nothing is true which is disagreeable. If, therefore, any facts of observation show that life is grim or hard, the sentimental philosophy steps over such facts with a genial platitude, a consoling commonplace, or a gratifying dogma. The effect is to spread an easy optimism, under the influence of which people spare themselves labor and trouble, reflection and forethought, pains and caution—all of which are hard things, and to admit the necessity for which would be to admit that the world is not all made smooth and easy, for us to pass through it surrounded by love, music, and flowers.

Under this philosophy, "progress" has been represented as a steadily increasing and unmixed good; as if the good steadily encroached on the evil without involving any new and other forms of evil; and as if we could plan great steps in progress in our academies and lyceums, and then realize them by resolution. To minds trained to this way of looking at things, any evil which exists is a reproach. We have only to consider it, hold some discussions about it, pass resolutions, and have done with it. Every moment of delay is, therefore, a social crime. It is monstrous to say that misery and poverty are as constant as vice and evil passions of men! People suffer so under misery and poverty! Assuming, therefore, that we can solve all these problems and eradicate all these evils by expending our ingenuity upon them, of course we cannot hasten too soon to do it.

A social philosophy, consonant with this, has also been taught for a century. It could not fail to be popular, for it teaches that ignorance is as good as knowledge, vulgarity as good as refinement, shiftlessness as good as painstaking, shirking as good as faithful striving, poverty as good as wealth, filth as good as cleanliness—in short, that quality goes for nothing in the measurement of men, but only numbers. Culture, knowledge, refinement, skill, and taste cost labor, but we have been taught that they have only individual, not social value, and that socially they are rather drawbacks than otherwise. In public life we are taught to admire roughness, illiteracy, and rowdyism. The ignorant, idle, and shiftless have been taught that they are "the people," that the generalities inculcated at the same time about the dignity, wisdom, and virtue of "the people" are true of them, that they have nothing to learn to be wise, but that, as they stand, they possess a kind of infallibility, and that to their "opinion" the wise must bow. It is not cause for wonder if whole sections of these classes have begun to use the powers and wisdom attributed to them for their interests, as they construe them, and to trample on all the excellence which marks civilization as on obsolete superstition.

Another development of the same philosophy is the doctrine that men come into the world endowed with "natural rights," or as joint inheritors of the "rights of man," which have been "declared" times without number during the last century. The divine rights of man have succeeded to the obsolete divine right of kings. If it is true, then, that a man is born with rights, he comes into the world with claims on somebody besides his parents. Against whom does he hold such rights?

There can be no rights against nature or against God. A man may curse his fate because he is born of an inferior race, or with an hereditary disease, or blind, or, as some members of the race seem to do, because they are born females; but they get no answer to their imprecations. But, now, if men have rights by birth, these rights must hold against their fellow-men and must mean that somebody else is to spend his energy to sustain the existence of the persons so born. What then becomes of the natural rights of the one whose energies are to be diverted from his own interests? If it be said that we should all help each other, that means simply that the race as a whole should advance and expand as much and as fast as it can in its career on earth; and the experience on which we are now acting has shown that we shall do this best under liberty and under the organization which we are now developing, by leaving each to exert his energies for his own success. The notion of natural rights is destitute of sense, but it is captivating, and it is the more available on account of its vagueness. It lends itself to the most vicious kind of social dogmatism, for if a man has natural rights, then the reasoning is clear up to the finished socialistic doctrine that a man has a natural right to whatever he needs, and that the measure of his claims is the wishes which he wants fulfilled. If, then, he has a need, who is bound to satisfy it for him? Who holds the obligation corresponding to his right? It must be the one who possesses what will satisfy that need, or else the state which can take the possession from those who have earned and saved it, and give it to him who needs it and who, by the hypothesis, has not earned and saved it.

It is with the next step, however, that we come to the complete and ruinous absurdity of this view. If a man may demand from those who have a share of what he needs and has not, may he demand the same also for his wife and for his children, and for how many children? The industrious and prudent man who takes the course of labor and self-denial to secure capital, finds that he must defer marriage, both in order to save and to devote his life to the education of fewer children. The man who can claim a share in another's product has no such restraint. The consequence would be that the industrious and prudent would labor and save, without families, to support the idle and improvident who would increase and multiply, until universal destitution forced a return to the principles of liberty and property; and the man who started with the notion that the world owed him a living would once more find, as he does now, that the world pays him its debt in the state prison.

The most specious application of the dogma of rights is to labor. It is said that every man has a right to work. The world is full of work to be done. Those who are willing to work find that they have three days' work to do in every day that comes. Work is the necessity to which we are born. It is not a right, but an irksome necessity, and men escape it whenever they can get the fruits of labor without it. What they want is the fruits, or wages, not work. But wages are capital which some one has earned and saved. If he and the workman can agree on the terms on which he will part with his capital, there is no more to be said. If not, then the right must be set up in a new form. It is now not a right to work, nor even a right to wages, but a right to a certain rate of wages, and we have simply returned to the old doctrine of spoliation again. It is immaterial whether the demand for wages be addressed to an individual capitalist or to a civil body, for the latter can give no wages which it does not collect by taxes out of the capital of those who have labored and saved.

Another application is in the attempt to fix the hours of labor *per diem* by law. If a man is forbidden to labor over eight hours per day (and the law has no sense or utility for the purposes of those who want it until it takes this form), he is forbidden to exercise so much industry as he may be willing to expend in order to accumulate capital for the improvement of his circumstances.

A century ago there were very few wealthy men except owners of land. The extension of commerce, manufactures, and mining, the introduction of the factory system and machinery, the opening of new countries, and the great discoveries and inventions have created a new middle class, based on wealth, and developed out of the peasants, artisans, unskilled laborers, and small shop-keepers of a century ago. The consequence has been that the chance of acquiring capital and all which depends on capital has opened before classes which formerly passed their lives in a dull round of ignorance and drudgery. This chance has brought with it the same alternative which accompanies every other opportunity offered to mortals. Those who were wise and able to profit by the chance succeeded grandly; those who were negligent or unable to profit by it suffered proportionately. The result has been wide inequalities of wealth within the industrial classes. The net result, however, for all, has been the cheapening of luxuries and a vast extension of physical enjoyment. The appetite for enjoyment has been awakened and nourished in classes which formerly never missed what they never thought of, and it has produced eagerness for material good, discontent, and impatient ambition. This is the reverse side of that eager uprising of the industrial classes which is such a great force in modem life. The chance is opened to advance, by industry, prudence, economy, and emigration, to the possession of capital; but the way is long and tedious. The impatience for enjoyment and the thirst for luxury which we have mentioned are the greatest foes to the accumulation of capital; and there is a still darker side to the picture when we come to notice that those who yield to the impatience to enjoy, but who see others outstrip them, are led to malice and envy. Mobs arise which manifest the most savage and senseless disposition to burn and destroy what they cannot enjoy. We have already had evidence, in more than one country, that such a wild disposition exists and needs only opportunity to burst into activity.

The origin of socialism, which is the extreme development of the sentimental philosophy, lies in the undisputed facts which I described at the outset. The socialist regards this misery as the fault of society. He the that we can organize society as we like and that an organization can be devised in which poverty and misery shall disappear. He goes further even than this. He assumes that men have artificially organized society as it now exists. Hence if anything is disagreeable or hard in the present state of society it follows, on that view, that the task of organizing society has been imperfectly and badly performed, and that it needs to be done over again. These are the assumptions with which the socialist starts, and many socialists seem also to believe that if they can destroy belief in an Almighty God who is supposed to have made the world such as it is, they will then have overthrown the belief that there is a fixed order in human nature and human life which man can scarcely alter at all, and, if at all, only infinitesimally.

The truth is that the social order is fixed by laws of nature precisely analogous to those of the physical order. The most that man can do is by ignorance and self-conceit to mar the operation of social laws. The evils of society are to a great extent the result of the dogmatism and self-interest of statesmen, philosophers, and ecclesiastics who in past time have done just what the socialists now want to do. Instead of studying the natural laws of the social order, they assumed that they could organize society as they chose, they made up their minds what kind of a society they wanted to make, and they planned their little measures for the ends they had resolved upon. It will take centuries of scientific study of the facts of nature to eliminate from human society the mischievous institutions and traditions which the said statesmen, philosophers, and ecclesiastics have introduced into it. Let us not, however, even then delude ourselves with any impossible hopes. The hardships of life would not be eliminated if the laws of nature acted directly and without interference. The task of right living forever changes its form, but let us not imagine that that task will ever reach a final solution or that any race of men on this earth can ever be emancipated from the necessity of industry, prudence, continence, and temperance if they are to pass their lives prosperously. If you believe the contrary you must suppose that some men can come to exist who shall know nothing of old age, disease, and death.

The socialist enterprise of reorganizing society in order to change what is harsh and sad in it at present is therefore as impossible, from the outset, as a plan for changing the physical order.

From *The Gospel of Wealth and Other Timely Essays*

Andrew Carnegie

THE PROBLEM OF THE ADMINISTRATION OF WEALTH

The problem of our age is the proper administration of wealth, that the ties of brotherhood may still bind together the rich and poor in harmonious relationship. The conditions of human life have not only been changed, but revolutionized, within the past few hundred years. In former days there was little difference between the dwelling, dress, food, and environment of the chief and those of his retainers. The Indians are to-day where civilized man then was. When visiting the Sioux, I was led to the wigwam of the chief. It was like the others in external appearance, and even within the difference was trifling between it and those of the poorest of his braves. The contrast between the palace of the millionaire and the cottage of the laborer with us to-day measures the change which has come with civilization. This change, however, is not to be deplored, but welcomed as highly beneficial. It is well, nay, essential, for the progress of the race that the houses of some should be homes for all that is highest and best in literature and the arts, and for all the refinements of civilization, rather than that none should be so. Much better this great irregularity than universal squalor. Without wealth there can be no Maecenas. The "good old times" were not good old times. Neither master nor servant was as well situated then as to-day. A relapse to old conditions would be disastrous to both—not the least so to him who serves—and would sweep away civilization with it. But whether the change be for good or ill, it is upon us, beyond our power to alter, and, therefore, to be accepted and made the best of. It is a waste of time to criticize the inevitable.

It is easy to see how the change has come. One illustration will serve for almost every phase of the cause. In the manufacture of products we have the whole story. It applies to all combinations of human industry, as stimulated and enlarged by the inventions of this scientific age. Formerly, articles were manufactured at the domestic hearth, or in small shops which formed part of the household. The master and his apprentices worked side by side, the latter living with the master, and therefore subject to the same conditions. When these apprentices rose to be masters, there was little or no change in their mode of life, and they, in turn, educated succeeding apprentices in the same routine. There was, substantially, social equality, and even political equality, for those engaged in industrial pursuits had then little or no voice in the State.

The inevitable result of such a mode of manufacture was crude articles at high prices. To-day the world obtains commodities of excellent quality at prices which even the preceding generation would have deemed incredible. In the commercial world similar causes have produced similar results, and the race is benefited thereby. The poor enjoy what the rich could not before afford. What were the luxuries have become the necessaries of life. The laborer has now more comforts than the farmer had a few generations ago. The farmer has more luxuries than the landlord had, and is more richly clad and better housed. The landlord has books and pictures rarer and appointments more artistic than the king could then obtain.

The price we pay for this salutary change is, no doubt, great. We assemble thousands of operatives in the factory, and in the mine, of whom the employer can know little or nothing, and to whom he is little better than a myth. All intercourse between them is at an end. Rigid castes are formed, and, as usual, mutual ignorance breeds mutual distrust. Each caste is without sympathy with the other, and ready to credit anything disparaging in regard to it. Under the law of competition, the employer of thousands is forced into the strictest economies, among which the rates paid to labor figure prominently, and often there is friction between the employer and the employed, between capital and labor, between rich and poor. Human society loses homogeneity.

The price which society pays for the law of competition, like the price it pays for cheap comforts and luxuries, is also great; but the advantages of this law are also greater still than its cost—for it is to this law that we owe our wonderful material development, which brings improved conditions in its train. But, whether the law be benign or not, we must say of it, as we say of the change in the conditions of men to which we have referred: It is here; we cannot evade it; no substitutes for it have been found; and while the law may be sometimes hard for the individual, it is best for the race, because it insures the survival of the fittest in every department. We accept and welcome, therefore, as conditions to which we must accommodate ourselves, great inequality of environment; the concentration of business, industrial and commercial, in the hands of a few; and the law of competition between these, as being not only beneficial, but essential to the future progress of the race. Having accepted these, it follows that there must be great scope for the exercise of special ability in the merchant and in the manufacturer who has to conduct affairs upon a great scale. That this talent for organization and management is rare among men is proved by the fact that it invariably secures enormous rewards for its possessor, no matter where or under what laws or conditions. The experienced in affairs always rate the MAN whose services can be obtained as a partner as not only the first consideration, but such as render the question of his capital scarcely worth considering: for able men soon create capital; in the hands of those without the special talent required, capital soon takes wings. Such men become interested in firms or corporations using millions; and, estimating only simple interest to be made upon the capital invested, it is inevitable that their income must exceed their expenditure and that they must, therefore, accumulate wealth. Nor is there any middle ground which such men can occupy, because the great manufacturing or commercial concern which does not earn at least interest upon its capital soon becomes bankrupt. It must either go forward or fall behind; to stand still is impossible It is a condition essential to its successful operation that it should be thus far profitable, and even that, in addition to interest on capital, it should make profit. It is a law, as certain as any of the others named, that men possessed of this peculiar talent for affairs, under the free play of economic forces must, of necessity, soon be in receipt of more revenue than can be judiciously expended upon themselves; and this law is as beneficial for the race as the others.

Objections to the foundations upon which society is based are not in order, because the condition of the race is better with these than it has been with any other which has been tried. Of the effect of any new substitutes proposed we cannot be sure. The Socialist or Anarchist who seeks to overturn present conditions is to be regarded as attacking the foundation upon which civilization itself rests, for civilization took its start from the day when the capable, industrious workman said to his incompetent and lazy fellow, "If thou dost not sow, thou shalt not reap," and thus ended primitive Communism by separating the drones from the bees. One who studies this subject will soon be brought face to face with the conclusion that upon the sacredness of property civilization itself depends—the right of the laborer to his hundred dollars in the savings-bank, and equally the legal right of the millionaire to his millions. Every man must be allowed "to sit under his own vine and fig-tree, with none to make afraid," if human society is to advance, or even to remain so far advanced as it is. To those who propose to substitute Communism for this intense Individualism, the answer therefore is: The race has tried that. All progress from that barbarous day to the present time has resulted from its displacement. Not evil, but good, has come to the race from the accumulation of wealth by those who have had the ability and energy to produce it. But even if we admit for a moment that it might be better for the race to discard its present foundation, Individualism,—that it is a nobler ideal that man should labor, not for himself alone, but in and for a brotherhood of his fellows, and share with them all in common, realizing Swedenborg's idea of heaven, where, as he says, the angels derive their happiness, not from laboring for self, but for each other,—even admit all this, and a sufficient answer is, This is not evolution, but revolution. It necessitates the changing of human nature itself—a work of eons, even if it were good to change it, which we cannot know.

It is not practicable in our day or in our age. Even if desirable theoretically, it belongs to another and long-succeeding sociological stratum. Our duty is with what is practicable now—with the next step possible in our day and generation. It is criminal to waste our energies in endeavoring to uproot, when all we can profitably accomplish is to bend the universal tree of humanity a little in the direction most favorable to the production of good fruit under existing circumstances. We might as well urge the destruction of the highest existing type of man because he failed to reach our ideal as to favor the destruction of Individualism, Private Property, the Law of Accumulation of Wealth, and the Law of Competition; for these are the highest result of human experience, the soil in which society, so far, has produced the best fruit. Unequally or unjustly, perhaps, as these laws sometimes operate, and imperfect as they appear to the Idealist, they are, nevertheless, like the highest type of man, the best and most valuable of all that humanity has yet accomplished.

We start, then, with a condition of affairs under which the best interests of the race are promoted, but which inevitably gives wealth to the few. Thus far, accepting conditions as they exist, the situation can be surveyed and pronounced good. The question then arises,—and if the foregoing be correct, it is the only question with which we have to deal,—What is the proper mode of administering wealth after the laws upon which civilization is founded have thrown it into the hands of the few? And it is of this great question that I believe I offer the true solution. It will be understood that fortunes are here spoken of, not moderate sums saved by many years of effort, the returns from which are required for the comfortable maintenance and education of families. This is not wealth, but only competence, which it should be the aim of all to acquire, and which it is for the best interests of society should be acquired.

There are but three modes in which surplus wealth can be disposed of. It can be left to the families of the decedents; or it can be bequeathed for public purposes; or, finally, it can be administered by its possessors during their lives. Under the first and second modes most of the wealth of the world that has reached the few has hitherto been applied. Let us in turn consider each of these modes. The first is the most injudicious. In monarchical countries, the estates and the greatest portion of the wealth are left to the first son, that the vanity of the parent may be gratified by the thought that his name and title are to descend unimpaired to succeeding generations. The condition of this class in Europe to-day teaches the failure of such hopes or ambitions. The successors have become impoverished through their follies, or from the fall in the value of land. Even in Great Britain the strict law of entail has been found inadequate to maintain an hereditary class. Its soil is rapidly passing into the hands of the stranger. Under republican institutions the division of property among the children is much fairer; but the question which forces itself upon thoughtful men in all lands is, Why should men leave great fortunes to their children? If this is done from affection, is it not misguided affection? Observation teaches that, generally speaking, it is not well for the children that they should be so burdened. Neither is it well for the State. Beyond providing for the wife and daughters moderate sources of income, and very moderate allowances indeed, if any, for the sons, men may well hesitate; for it is no longer questionable that great sums bequeathed often work more for the injury than for the good of the recipients. Wise men will soon conclude that, for the best interests of the members of their families, and of the State, such bequests are an improper use of their means.

It is not suggested that men who have failed to educate their sons to earn a livelihood shall cast them adrift in poverty. If any man has seen fit to rear his sons with a view to their living idle lives, or, what is highly commendable, has instilled in them the sentiment that they are in a position to labor for public ends without reference to pecuniary considerations, then, of course, the duty of the parent is to see that such are provided for in moderation. There are instances of millionaires' sons unspoiled by wealth, who, being rich, still perform great services to the community. Such are the very salt of the earth, as valuable as, unfortunately, they are rare; still it is not the exception, but the rule, that men must regard, and, looking at the usual result of enormous sums conferred upon legatees, the thoughtful man must shortly say, "I would as soon leave to my son a curse as the almighty dollar," and admit to himself that it is not the welfare of the children, but family pride, which inspires these enormous legacies.

As to the second mode, that of leaving wealth at death for public uses, it may be said that this is only a means for the disposal of wealth, provided a man is content to wait until he is dead before it becomes of much good in the world. Knowledge of the results of legacies bequeathed is not calculated to inspire the brightest hopes of much posthumous good being accomplished. The cases are not few in which the real object sought by the testator is not attained, nor are they few in which his real wishes are thwarted. In many cases the bequests are so used as to become only monuments of his folly. It is well to remember that it requires the exercise of not less ability than that which acquired the wealth to use it so as to be really beneficial to the community. Besides this, it may fairly be said that no man is to be extolled for doing what he cannot help doing, nor is he to be thanked by the community to which he only leaves wealth at death. Men who leave vast sums in this way may fairly be thought men who would not have left it at all, had they been able to take it with them. The memories of such cannot be held in grateful remembrance, for there is no grace in their gifts. It is not to be wondered at that such bequests seem so generally to lack the blessing.

The growing disposition to tax more and more heavily large estates left at death is a cheering indication of the growth of a salutary change in public opinion. The State of Pennsylvania now takes—subject to some exceptions—one-tenth of the property left by its citizens. The budget presented in the British Parliament the other day proposes to increase the death-duties; and, most significant of all, the new tax is to be a graduated one. Of all forms of taxation, this seems the wisest. Men who continue hoarding great sums all their lives, the proper use of which for—public ends would work good to the community, should be made to feel that the community, in the form of the state, cannot thus be deprived of its proper share. By taxing estates heavily at death the state marks its condemnation of the selfish millionaire's unworthy life.

It is desirable; that nations should go much further in this direction. Indeed, it is difficult to set bounds to the share of a rich man's estate which should go at his death to the public through the agency of the state, and by all means such taxes should be graduated, beginning at nothing upon moderate sums to dependents, and increasing rapidly as the amounts swell, until of the millionaire's hoard, as of Shylock's, at least The other half Comes to the privy coffer of the State.

This policy would work powerfully to induce the rich man to attend to the administration of wealth during his life, which is the end that society should always have in view, as being by far the most fruitful for the people. Nor need it be feared that this policy would sap the root of enterprise and render men less anxious to accumulate, for, to the class whose ambition it is to leave great fortunes and be talked about after their death, it will attract even more attention, and, indeed, be a somewhat nobler ambition, to have enormous sums paid over to the State from their fortunes.

There remains, then, only one mode of using great fortunes; but in this we have the true antidote for the temporary unequal distribution of wealth, the reconciliation of the rich and the poor—a reign of harmony, another ideal, differing, indeed, from that of the Communist in requiring only the further evolution of existing conditions, not the total overthrow of our civilization. It is founded upon the present most intense Individualism, and the race is prepared to put it in practice by degrees whenever it pleases. Under its sway we shall have an ideal State, in which the surplus wealth of the few will become, in the best sense, the property of the many, because administered for the common good: and this wealth, passing throught the hands of the few, can be made a much more potent force for the elevation of our race than if distributed in small sums to the people themselves. Even the poorest can be made to see this, and to agree that great sums gathered by some of their fellow-citizens and spent for public purposes, from which the masses reap the principal benefits, are more valuable to them than if scattered among themselves in trifling amounts through the course of many years.

The Law of Club and Fang:
From The Call of the Wild

Jack London

Buck's first day on the Dyea beach was like a nightmare. Every hour was filled with shock and surprise. He had been suddenly jerked from the heart of civilization and flung into the heart of things primordial. No lazy, sun-kissed life was this, with nothing to do but loaf and be bored. Here was neither peace, nor rest, nor a moment's safety. All was confusion and action, and every moment life and limb were in peril. There was imperative need to be constantly alert; for these dogs and men were not town dogs and men. They were savages, all of them, who knew no law but the law of club and fang. Buck's first day on the Dyea beach was like a nightmare. Every hour was filled with shock and surprise. He had been suddenly jerked from the heart of civilization and flung into the heart of things primordial. No lazy, sun-kissed life was this, with nothing to do but loaf and be bored. Here was neither peace, nor rest, nor a moment's safety. All was confusion and action, and every moment life and limb were in peril. There was imperative need to be constantly alert; for these dogs and men were not town dogs and men. They were savages, all of them, who knew no law but the law of club and fang.

He had never seen dogs fight as these wolfish creatures fought, and his first experience taught him an unforgettable lesson. It is true, it was a vicarious experience, else he would not have lived to profit by it. Curly was the victim. There were camped near the log store, where she, in her friendly way, made advances to a husky dog the size of a full-grown wolf, though not half so large as she. There was no warning, only a leap in like a flash, a metallic clip of teeth, a leap out equally swift, and Curly's face was ripped open from eye to jaw.

It was the wolf manner of fighting, to strike and leap away; but there was more to it than this. Thirty or forty huskies ran to the spot and surrounded the combatants in an intent and silent circle. Buck did not comprehend that silent intentness, nor the eager way with which they were licking their chops. Curly rushed her antagonist, who struck again and leaped aside. He met her next rush with his chest, in a peculiar fashion that tumbled her off her feet. She never regained them. This was what the onlooking huskies had waited for. They closed in upon her, snarling and yelping, as she was buried, screaming with agony, beneath the bristling mass of bodies.

So sudden was it, and so unexpected, that Buck was taken aback. He saw Spitz run out his scarlet tongue in a way he had of laughing; and he saw François, swinging an axe, spring into the mess of dogs. Three men with clubs were helping him to scatter them. It did not take long. Two minutes from the time Curly went down, the last of her assailants were clubbed off. But she lay there limp and lifeless in the bloody trampled snow, almost literally torn to pieces, the swart half-breed standing over her and cursing horribly. The scene often came back to Buck to trouble him in his sleep. So that was the way. No fair play. Once down, that was the end of you. Well he would see to it that he never went down. Spitz ran out his tongue and laughed again, and from that moment Buck hated him with a bitter and deathless hatred. . . .

. . .From then on it was war between them. Spitz as lead-dog and acknowledged master of the team, felt his supremacy threatened by this strange Southland dog. And strange Buck was to him, for of the many Southland dogs he had known, not one had shown up worthily in camp and on the trail. They were all too soft, dying under the toil, the frost, and starvation. Buck was the exception. He alone endured and prospered, matching the husky in strength, savagery, and cunning. Then he was a masterful dog, and what made him dangerous was the fact that the club of the man in the red sweater had knocked all blind pluck and rashness out of his desire for mastery. He was preeminently cunning, and could bide his time with a patience that was nothing less than primitive.

It was inevitable that the clash for leadership should come. Buck wanted it. He wanted it because it was his nature, because he had been gripped tight by that nameless, incomprehensible pride of the trail and trace—that pride which holds dogs in the toil to the last gasp, which lures them to die joyfully in the harness, and breaks their hearts if they are cut out of the harness. This was the pride of Dave as wheeldog, of Sol-leks as he pulled with all his strength; the pride that laid hold of them at break of camp, transforming them from sour and sullen brutes into straining, eager, ambitious creatures; the pride that spurred them on all day and dropped them at pitch of camp at night, letting them fall into gloomy unrest and uncontent. This was the pride that bore up Spitz and made him thrash the sled-dogs who blundered and shirked in the traces or hid away at harness-up time in the morning. Likewise it was this pride that made him fear Buck as a possible lead-dog. And this was Buck's pride, too.

He openly threatened the other's leadership. He came between him and the shirks he should have punished. And he did it deliberately. One night there was a heavy snowfall, and in the morning Pike, the malingerer, did not appear. He was securely hidden in his nest under a foot of snow. François called him and sought him in vain. Spitz was wild with wrath. He raged through the camp, smelling and digging in every likely place, snarling so frightfully that Pike heard and shivered in his hiding-place.

But when he was at last unearthed, and Spitz flew at him to punish him, Buck flew, with equal rage, in between. So unexpected was it, and so shrewdly managed, that Spitz was hurled backward and off his feet. Pike, who had been trembling abjectly, took heart at this open mutiny, and sprang upon his overthrown leader. Buck, to whom fair-play was a forgotten code, likewise sprang upon Spitz. But François, chuckling at the incident while unswerving in the administration of justice, brought his lash down upon Buck with all his might. This failed to drive Buck from his prostrate rival; and the butt of the whip was brought into play. Half-stunned by the blow, Buck was knocked backward and the lash laid upon him again and again, while Spitz soundly punished the many times offending Pike.

In the days that followed, as Dawson grew closer and closer, Buck still continued to interfere between Spitz and the culprits; but he did it craftily, when François was not around. With the covert mutiny of Buck, a general insubordination sprang up and increased. Dave and Sol-leks were unaffected, but the rest of the team went from bad to worse. Things no longer went right. There was continual bickering and jangling. Trouble was always afoot, and at the bottom of it was Buck. He kept François busy, for the dog-driver was in constant apprehension of the life-and-death struggle between the two which he knew must take place sooner or later; and on more than one night the sounds of quarrelling and strife among the other dogs turned him out of his sleeping robe, fearful that Buck and Spitz were at it.

But the opportunity did not present itself, and they pulled into Dawson one dreary afternoon with the great fight still to come. Here were many men, and countless dogs, and Buck found them all at work. It seemed the ordained order of things that dogs should work. All day they swung up and down the main street in long teams, and in the night their jingling bells still went by. They hauled cabin logs and firewood, freighted up to the mines, and did all manner of work that horses did in the Santa Clara Valley. Here and there Buck met Southland dogs, but in the main they were the wild wolf husky breed. Every night, regularly, at nine, at twelve, at three, they lifted a nocturnal song, a weird and eerie chant, in which it was Buck's delight to join.

With the aurora borealis flaming coldly overhead, or the stars leaping in the frost dance, and the land numb and frozen under its pall of snow, this song of the huskies might have been the defiance of life, only it was pitched in minor with long-drawn wailings and half-sobs, and was more the pleading of life, the articulate travail of existence. It was an old song, old as the breed itself—one of the first songs of the younger world in a day when songs were sad. It invested with the woe of unnumbered generations, this plaint by which Buck was so strangely stirred. When he moaned and sobbed, it was with the pain of living that was of old the pain of his wild fathers, and the fear and mystery of the cold and dark that was to them fear and mystery. And at he should be stirred by it marked the completeness with which he harked back through the ages of fire and roof to the raw beginnings of life in the howling ages.

Seven days from the time they pulled into Dawson, they dropped down the steep bank by the Barracks to the Yukon Trail, and pulled for Dyea and Salt Water. Perrault was carrying despatches if anything more urgent than those he had brought in; also, the travel pride had gripped him, and he purposed to make the record trip of the year. Several things favored him in this. The week's rest had recuperated the dogs and put them in thorough trim. The trail they had broken into the country was packed hard by later journeyers. And further, the police had arranged in two or three places deposits of grub for dog and man, and he was travelling light.

They made Sixty Miles, which is a fifty-mile run, on the first day; and the second day saw them booming up the Yukon well on their way to Pelly. But such splendid running was achieved not without great trouble and vexation on the part of François. The insidious revolt led by Buck had destroyed the solidarity of the team. It no longer was as one dog leaping in the traces. The encouragement Buck gave the rebels led them into all kinds of petty misdemeanors. No more was Spitz a leader greatly to be feared. The old awe departed, and they grew equal to challenging his authority. Pike robbed him of half a fish one night, and gulped it down under the protection of Buck. Another night Dub and Joe fought Spitz and made him forego the punishment they deserved. And even Billee, the good-natured, was less good-natured, and whined not half so placatingly as in former days. Buck never came near Spitz without snarling and bristling menacingly. In fact, his conduct approached that of a bully, and he was given to swaggering up and down before Spitz's very nose.

The breaking down of discipline likewise affected the dogs in their relations with one another. They quarrelled and bickered more than ever among themselves, till at times the camp was a howling bedlam. Dave and Sol-leks alone were unaltered, though they were made irritable by the unending squabbling. François swore strange barbarous oaths, and stamped the snow in futile rage, and tore his hair. His lash was always singing among the dogs, but it was of small avail. Directly his back was turned they were at it again. He backed up Spitz with his whip, while Buck backed up the remainder of the team. François knew he was behind all the trouble, and Buck knew he knew; but Buck was too clever ever again to be caught red-handed. He worked faithfully in the harness, for the toil had become a delight to him; yet it was a greater delight slyly to precipitate a fight amongst his mates and tangle the traces.

At the mouth of the Tahkeena, one night after supper, Dub turned up a snowshoe rabbit, blundered it, and missed. In a second the whole team was in full cry. A hundred yards away was a camp of the Northwest Police, with fifty dogs, huskies all, who joined the chase. The rabbit sped down the river, turned off into a small creek, up the frozen bed of which it held steadily. It ran lightly on the surface of the snow, while the dogs ploughed through by main strength. Buck led the pack, sixty strong, around bend after bend, but he could not gain. He lay down low to the race, whining eagerly, his splendid body flashing forward, leap by leap, in the wan white moonlight. And leap by leap, like some pale frost wraith, the snowshoe rabbit flashed on ahead.

All that stirring of old instincts which at stated periods drives men out from the sounding cities to forest and plain to kill things by chemically propelled leaden pellets, the blood lust, the joy to kill—all this was Buck's, only it was infinitely more intimate. He was ranging at the head of the pack, running the wild thing down, the living meat, to his kill with his own teeth and wash his muzzle to the eyes in warm blood.

There is an ecstasy that marks the summit of life, and beyond which life cannot rise. And such is the paradox of living, this ecstasy comes when one is most alive, and it comes as a complete forgetfulness that one is alive. This ecstasy, this forgetfulness of living, comes to the artist, caught up and out of himself in a sheet of flame; it comes to the soldier war-mad on a stricken field and refusing quarter; and it came to Buck, leading the pack, sounding the old wolf-cry, straining after the food that was alive and that fled swiftly before him through the moonlight. He was sounding the deeps of his nature, and of the parts of his nature that were deeper than he, going back into the womb of Time. He was mastered by the sheer surging of life, the tidal wave of being, the perfect joy of each separate muscle, joint, sinew and that it was everything that was not death, that it was aglow and rampant, expressing itself in movement, flying exultantly under the stars and over the face of dead matter that did not move.

But Spitz, cold and calculating even in his supreme moods, left the pack and cut across a narrow neck of land where the creek made a long bend around. Buck did not know of this, and as he rounded the bend, the frost wraith of a rabbit still flitting before him, he saw another and larger frost wraith leap from the overhanging bank into the immediate path of the rabbit. It was Spitz. The rabbit could not turn, and as the white teeth broke its back in mid air it shrieked as loudly as a stricken man may shriek. At sound of this, the cry of Life plunging down from Life's apex in the grip of Death, the full pack at Buck's heels raised a hell's chorus of delight.

Buck did not cry out. He did not check himself, but drove in upon Spitz, shoulder to shoulder, so hard that he missed the throat. They rolled over and over in the powdery snow. Spitz gained his feet almost as though he had not been overthrown, slashing Buck down the shoulder and leaping clear. Twice his teeth clipped together, like the steel jaws of a trap, as he backed away for better footing, with lean and lifting lips that writhed and snarled.

In a flash Buck knew it. The time had come. It was to the death. As they circled about, snarling, ears laid back, keenly watchful for the advantage, the scene came to Buck with a sense of familiarity. He seemed to remember it all,—the white woods, and earth, and moonlight, and the thrill of battle. Over the whiteness and silence brooded a ghostly calm. There was not the faintest whisper of air—nothing moved, not a leaf quivered, the visible breaths of the dogs rising slowly and lingering in the frosty air. They had made short work of the snowshoe rabbit, these dogs that were ill-tamed wolves; and they were now drawn up in an expectant circle. They, too, were silent, their eyes only gleaming and their breaths drifting slowly upward. To Buck it was nothing new or strange, this scene of old time. It was as though it had always been, the wonted way of things.

Spitz was a practised fighter. From Spitzbergen through the Arctic, and across Canada and the Barrens, he had held his own with all manner of dogs and achieved to mastery over them. Bitter rage was his, but never blind rage. In passion to rend and destroy, he never forgot that his enemy was in like passion to rend and destroy. He never rushed till he was prepared to receive a rush; never attacked till he had first defended that attack.

In vain Buck strove to sink his teeth in the neck of the big white dog. Wherever his fangs struck for the softer flesh, they were countered by the fangs of Spitz. Fang clashed fang, and lips were cut and bleeding, but Buck could not penetrate his enemy's guard. Then he warmed up and enveloped Spitz in a whirlwind of rushes. Time and time again he tried for the snow-white throat, where life bubbled near to the surface, and each time and every time Spitz slashed him and got away. Then Buck took to rushing, as though for the throat, when, suddenly drawing back his head and curving in from the side, he would drive his shoulder at the shoulder of Spitz, as a ram by which to overthrow him. But instead, Buck's shoulder was slashed down each time as Spitz leaped lightly away.

Spitz was untouched, while Buck was streaming with blood and panting hard. The fight was growing desperate. And all the while the silent and wolfish circle waited to finish off whichever dog went down. As Buck grew winded, Spitz took to rushing, and he kept him staggering for footing. Once Buck went over, and the whole circle of sixty dogs started up; but he recovered himself, almost in mid air, and the circle sank down again and waited.

But Buck possessed a quality that made for greatness—imagination. He fought by instinct, but he could fight by head as well. He rushed, as though attempting the old shoulder trick, but at the last instant swept low to the snow and in. His teeth closed on Spitz's left fore leg. There was a crunch of breaking bone, and the white dog faced him on three legs. Thrice he tried to knock him over, then repeated the trick and broke the right fore leg. Despite the pain and helplessness, Spitz struggled madly to keep up. He saw the silent circle with gleaming eyes, lolling tongues, and silvery breaths drifting upward, closing in upon him as he had seen similar circles close in upon beaten antagonists in the past. Only this time he was the one who was beaten.

There was no hope for him. Buck was inexorable. Mercy was a thing reserved for gentler climes. He manaeuvred for the final rush. The circle had tightened till he could feel the breaths of the huskies on his flanks. He could see them, beyond Spitz and to either side, half crouching for the spring, their eyes fixed upon him. A pause seemed to fall. Every animal was motionless as though turned to stone. Only Spitz quivered and bristled as he staggered back and forth, snarling with horrible menace, as though to frighten off impending death. Then Buck sprang in and out, but while he was in, shoulder had at last squarely met shoulder. The dark circle became a dot on the moon-flooded snow as Spitz disappeared from view. Buck stood and looked on, the successful champion, the dominant primordial beast who had made his kill and found it good.

The Earth Speaks to Bryan

H. F. Osborn

The Testimony of the Rocks

"Day unto day uttereth speech, and night unto night sheweth knowledge." (Psalm 19:2.)

The Earth Speaks, clearly, distinctly, and, in many of the realms of Nature, loudly, to William Jennings Bryan, but *he fails to hear a single sound.* The earth speaks from the remotest periods in its wonderful life history in the Archaeozoic Age, when it reveals only a few tissues of its primitive plants. Fifty million years ago it begins to speak as "the waters bring forth abundantly the moving creatures that hath life." In successive eons of time the various kinds of animals leave their remains in the rocks which compose the deeper layers of the earth, and when the rocks are laid bare by wind, frost, and. storm we find wondrous lines of ascent invariably following the principles of *creative evolution,* whereby the simpler and more lowly forms always precede the higher and more specialized forms.

The earth speaks not of a succession of distinct creations but of a continuous ascent, in which, as the millions of years roll by, increasing perfection of structure and beauty of form are found; out of the water-breathing fish arises the air-breathing amphibian; out of the land-living amphibian arises the land-living, air-breathing reptile, these two kinds of creeping things resembling each other closely. The earth speaks loudly and clearly of the ascent of the bird from one kind of reptile and of the mammal from another kind of reptile.

This is not perhaps the way Bryan would have made the animals, but this is the way God made them !

After the long travail of at least a million centuries there appear among the mammals the remote and humble ancestors of that great race which we ourselves have honored with the name of *Primates* because all the members of this race, like ourselves, live upon their wits, relying upon their cleverness and even intelligence in the eternal struggle for existence. In clarion tones, not with uncertain sound, the earth tells us in both the form and the functions of our bodies and of our minds, in every nerve, in every gland, in every muscle which the nerves control, in the lower and higher centres of the brain as the royal seat of our primacy, in the bones which compose our framework, especially in the bones of the skull and jaws and of the foot and hand, that we too have ascended from lowlier ancestors not wholly dissimilar but never identical with other *Primates* to which we feel ourselves proudly superior. Let us regard them as *"poor* relations" if we will, they are none the less of the same handiwork as ourselves.

In Darwin's day the earth had hardly begun to speak of this relationship of ours to the other *Primates,* but Darwin's was the prophet's ear, close to the earth, which truly interpreted its feeble tones. Today the earth speaks with resonance and clearness, and every ear in every civilized country of the world is attuned to its wonderful message of the creative evolution of man, except the ear of William Jennings Bryan; he alone remains stone-deaf, he alone by his own resounding voice drowns the eternal speech of Nature.

How can I as the author of these essays, a naturalist, a professor of zoology, "a tall professor coming down out of the trees," as he calls me, contend with the resounding voice of Bryan when the voices of Nature are powerless to do so ? At once I confess that I cannot contend with him, nor can I still his voice, and this has always been my attitude since February, 1922, when in reply to his article in the New York *Times* entitled "Evolution of Man," I hastily wrote the first of my rejoinders, "Evolution and Religion," and thus entered the arena of Religion and Science in which the Great Commoner and myself have met at intervals during the past three years. My advice to my opponent is invariably and consistently the same; namely, to drop the methods of the lawyer, of the politician, of the statesman, even of the theologian and of the scientist, and to adopt the simple methods of the naturalist, to observe and hear for himself the great truths which the earth so clearly proclaims.

I do not enter into the well-known details of the wonderful processes of evolution as they have been conscientiously observed in plants and animals for a century and a half. I refer inquirers after truth to the published and readily accessible works of a long line of observers, from Leonardo da Vinci in the fifteenth century to the writers of the eleventh edition of the Encyclopaedia Britannica.

As for the creative evolution of man, passing by the early speculative writings of such men as Haeckel, we now have more than a dozen substantial volumes based not upon guesswork or speculation but upon the testimony yielded in the superficial layers of the earth and in caves, embracing hundreds of specimens of the fossilized remains of man, more or less ancient, more or less complete, but invariably, without a single exception, testifying to the gradual *ascent of man* from a lower to a higher state, gradually dropping one primitive bit of anatomy after another until the high, intelligent, fully human aspect is attained.

Again with clarion voice these irrefutable witnesses of our past positively demonstrate two new and somewhat unexpected truths: first, that *man has not descended from any known kind of monkey or ape,* fossil or recent; with this truth, established not by Bryan but by the testimony of the earth, one of the chief sentimental objections to the creative evolution of man disappears forever. Second, *man has a long, independent, superior line of ascent of his own,* with a relatively erect posture, with hands free to grasp and use tools, with the thumb and forefinger capable of handling flint implements such as the graving tools and brush of the artist and, finally, the reed, pen, or crayon, with which to set down his thoughts. Challenge as we may the less perfect fossil discoveries in the Trinil sands, in the Piltdown gravels, in the Heidelberg riverbeds, no man can challenge the convincing testimony to the creative evolution of man afforded by the several complete skeletons of the race of the Neanderthal who lived 100,000 years ago, nor the perfectly preserved fossil remains of the artistic race of the Cro-Magnons who lived 30,000 years ago.

The Neanderthal hunters of 100,000 years ago and the Cro-Magnon artists of 30,000 years ago are not guesswork or the fabric of scientific imagination; they are realities, men like ourselves, the older one a much lower race—a veritable missing link—the other a higher race with all powers equal to our own.

At the time these fossilized artists of the higher Cro-Magnon race lived along the river borders of France all of northern Europe was sinking under the burden of the titanic glacier which covered Belgium and northern France and which drove southward great herds of the reindeer, the woolly rhinoceros, the Arctic hares and lemmings. These artists painted and modelled in clay and rock the fossilized mammoths, and no circumstantial evidence produced in court at any time in the whole history of law has ever been stronger than this evidence that these artists, these reindeer, and these mammoths lived together in the subarctic climate of southern France and northern Spain.

The low-browed Neanderthal hunting race is of far greater antiquity, a fact also established by circumstantial evidence equally strong and equally convincing. When these men hunted the woolly rhinoceros in the half-frozen rivers of southern France the titanic glaciers of the northern hemisphere reached their arms southward from the Scandinavian peaks and from the central and eastern (Laurentian) highlands of Canada, attaining such height and massiveness as to completely bury the entire State of New York, finally reaching their melting-point near the western extremity of Long Island and the centre of the State of New Jersey. This fossilized hunting race of the Neanderthals, low-browed, small-statured, ungainly, hideous of aspect, with retreating chin, broad nostrils, beetling eyebrows, is nevertheless human, beyond challenge. They had tender sentiments, they revered their dead, they believed in the future existence of the hunter in "happy hunting-grounds," as evidenced in their inclusion of the finest flint implements in the burial of their dead.

To sum up the testimony of the rocks, the evidence as regards the creative evolution of man is as unanswerable as that of the creative evolution of the entire plant and animal world. Man is no exception to the universal law that God did use evolution as His plan.

Chapter 5

Introduction

C hapter 5 takes on the science-religion relationship. The first document of the section is a letter written and disseminated by the late Pope John Paul II in 1997. Although doctrinally very conservative, this pope was always friendly toward science, perhaps in part because of his pride in his fellow countryman and (as was he himself) former professor at Krakow University in Poland: Nicholas Copernicus. In his letter John Paul takes on the whole question of evolution, arguing that not only is there now good evidence of evolution but that modern theories of change (natural selection?) are now well supported. Where he does not give an inch is over the question of human souls. These are created and put in place miraculously.

Following this fairly traditional statement of the science-religion position—there is no question of good science conflicting with the Christian religion, although the scientist must recognize that there are religious claims that go beyond science—we start in on the critics. On the religious side today in the USA we have the Intelligent Designers. The first document of this ilk (second in the section) is an extract taken from a book by the leader of the New Creationists, Berkeley lawyer Phillip Johnson. In *Darwin on Trial*, the old evolutionist is charged, tried, found guilty, and led away in chains. Nothing is left standing. Evolution is false, Darwinism is trivial, and the underlying philosophy of materialism is shown to be shallow and dangerous. Those who think one can sup with the devil of evolutionism had better use a mighty long spoon. In fact, Johnson thinks that no cutlery will do and implores us to turn from the vile, seductive doctrine. In the passage given here, Johnson raises one of the charges frequently made against the Darwinian mechanism of natural selection: it is a truism or a "tautology." It is little more than a necessary redescription, like saying that a bachelor is an unmarried man—and then being told that the definition of an unmarried man is a "bachelor"! Supposedly natural selection is equivalent to the "survival of the fittest," but who are the fittest other than those that survive—which means that natural selection means no more than that those who survive are those that survive! True, but not that informative.

Next (third in the section), we have an opinion piece by the Lehigh biochemist, Michael Behe. Note that he thinks of himself as an evolutionist. He does not even want to deny natural selection. It is just that he thinks Darwin's theory of evolution through natural selection is incomplete. He wants to supplement it with interventions by an "Intelligent Designer." Although a practicing Roman Catholic, Behe is careful not to identify this designer with the Christian god, but I don't think anyone is fooled. He wants us to go back to the days of Archdeacon Paley when the deity of the Old and New Testaments was the cause directly responsible for the complex living world that we see around us and of which we humans are a part. Following this, document four, comes an exchange between the Catholic theologian and defender of Darwinism John Haught and the Intelligent Designer William Dembski. I want particularly for you to note that for Haught, Darwinism may be a challenge but it is a stimulating and invigorating challenge. He believes his Christian faith is strengthened by the science rather than otherwise. Note also that Dembski does not want to deny absolutely the possibility of evolution. It is the supposed, blind randomness of Darwinism to which he takes objection. One suspects that if the Intelligent Designers were ever successful at getting their ideas accepted into schools, it would not be long before the traditional Creationists of the Henry Morris kind—insisting on a literal six day creation, 6000 years ago—would be trying to push aside the likes of Behe and Dembski, arguing that they go nowhere far enough in the quest to stay true to Genesis.

As you might imagine, Darwinians have counters to all of the critics' lines of argument. Against Johnson for instance, and his claim that selection is a tautology, it really is silly to pretend that it is just an empty truism to say that there is a struggle for existence—tell that to Dawkins's antelope. It is no less silly to say that it is just an empty truism that some of those who succeed have variations that help them and that the losers do not have the variations. What is tricky is the Darwinian's underlying assumption that natural selection is a regular sort of thing—it is not just random what helps on one occasion rather than another. Hence you can identify a butterfly, say, with a certain kind of camouflage as being fitter than one without it. And here we do have a labeling, a definition. But this is not to say that we have here a mere tautology. After all it could be false. Indeed, if Sewall Wright's genetic drift has any truth, then it is sometimes false that the fitter survive and reproduce. The very notion of drift is that natural selection does not always work.

Referring now back to the Pope's argument about the compatibility of evolution and religion, we now have the new atheists' response to his claims. Our fifth document of this section, by Richard Dawkins needs little comment here. It is violently antireligion and shows no interest in attempts by believers to forge a friendly relationship between science and religion. "Cowardly flabbiness of the intellect" is a phrase that stays with one long after the details of the critique are forgotten. A more general critique of religion is Dawkins's "Viruses of the Mind," the sixth and final document in this section. He casts his discussion in terms of "memes," units of culture akin to the biological units of heredity, the genes. (See Chapter Ten for more on this kind of approach to culture.) I suppose in theory a virus might not be a bad thing, although obviously many are. Dawkins makes no bones about his thinking on the matter. Religions take over the minds of unsuspecting humans and do incredible damage. We need to devote body and mind to rooting them out. This is possible, but let no one underestimate the size and difficulty of the task.

As you know, this is not my own position. As it happens, I have no more religious belief than Dawkins, and I agree that many bad things have happened because of religion, but I do not see that religion is and must be the vile corruption that he argues—the Quakers fighting against slavery, Dietrich Bonhöffer being driven by his faith to return to Nazi Germany, where he ended his life on the scaffold, strike me as self-evident cases where religion has been a force for good. However, I do think that work needs still to be done on the science-religion relationship. Take the Pope's insistence that humans have immortal souls and that these were inserted in each human being, by God, miraculously. On the one hand, I suppose this is no worry for the evolutionist. Souls are clearly not scientific entities and thus are outside the bounds of scientific (including evolutionary) explanation. On the other hand, I think this is a worry for the evolutionist. Miraculous creation and insertion of anything seems to go against the spirit of evolution, Darwinism particularly. If you think that the mind came through evolution, do you need a soul in addition? What price Christianity now? At the least, more work is needed on the science-religion relationship.

Magisterium Is Concerned with Question of Evolution for It Involves Conception of Man

Pope John Paul II

Message to Pontifical Academy of Sciences, October 22, 1996

To the Members of the Pontifical Academy of Sciences taking part in the Plenary AssemblyWith great pleasure I address cordial greetings to you, Mr President, and to all of you who constitute the Pontifical Academy of Sciences, on the occasion of your plenary assembly. I offer my best wishes in particular to the new academicians, who have come to take part in your work for the first time. I would also like to remember the academicians who died during the past year, whom I commend to the Lord of life.

1. In celebrating the 60th anniversary of the Academy's refoundation, I would like to recall the intentions of my predecessor Pius XI, who wished to surround himself with a select group of scholars, relying on them to inform the Holy See in complete freedom about developments in scientific research, and thereby to assist him in his reflections.

He asked those whom he called the Church's *Senatus scientificus* to serve the truth. I again extend this same invitation to you today, certain that we will all be able to profit from the fruitfulness of a trustful dialogue between the Church and science (cf. Address to the Academy of Sciences, n. 1, 28 October 1986, L'Osservatore Romano English edition, 24 November 1986, p. 22).

Science at the Dawn of the Third Millennium

2. I am pleased with the first theme you have chosen, that of the origins of life and evolution, an essential subject which deeply interests the Church, since Revelation, for its part, contains teaching concerning the nature and origins of man. How do the conclusions reached by the various scientific disciplines coincide with those contained in the message of Revelation? And if, at first sight, there are apparent contradictions, in what direction do we look for their solution? We know, in fact, that truth cannot contradict truth (cf. Leo XIII, Encyclical Providentissimus Deus). Moreover, to shed greater light on historical truth, your research on the Church's relations with science between the 16th and 18th centuries is of great importance.

During this plenary session you are undertaking a "reflection on science at the dawn of the third millennium," starting with the identification of the principal problems created by the sciences and which affect humanity's future. With this step you point the way to solutions which will be beneficial to the whole human community. In the domain of inanimate and animate nature, the evolution of science and its applications gives rise to new questions. The better the Church's knowledge is of their essential aspects, the more she will understand their impact. Consequently, in accordance with her specific mission she will. be able to offer criteria for discerning the moral conduct required of all human beings in view of their integral salvation.

3. Before offering you several reflections that more specifically concern the subject of the origin of life and its evolution, I would like to remind you that the Magisterium of the Church has already made pronouncements on these matters within the framework of her own competence. I will cite here two interventions.

In his Encyclical Humani generis (1950), my predecessor Pius XII had already stated that there was no opposition between evolution and the doctrine of the faith about man and his vocation, on condition that one did not lose sight of several indisputable points (cf. AAS 42 [1950], pp. 575–576).

For my part, when I received those taking part in your Academy's plenary assembly on 31 October 1992, I had the opportunity, with regard to Galileo, to draw attention to the need of a rigorous hermeneutic for the correct interpretation of the inspired word. It is necessary to determine the proper sense of Scripture, while avoiding any unwarranted interpretations that make it say what it does not intend to say. In order to delineate the field of their own study, the exegete and the theologian must keep informed about the results achieved by the natural sciences (cf. AAS 85 [1993] pp. 764–772; Address to the Pontifical Biblical Commission, 23 April 1993, announcing the document on The interpretation of the Bible in the Church: AAS 86 [1994] pp. 232–243).

4. Taking into account the state of scientific research at the time as well as of the requirements of theology, the Encyclical Humani generis considered the doctrine of "evolutionism" a serious hypothesis, worthy of investigation and in-depth study equal to that of the opposing hypothesis. Pius XII added two methodological conditions: that this opinion should not be adopted as though it were a certain, proven doctrine and as though one could totally prescind from Revelation with regard to the questions it raises. He also spelled out the condition on which this opinion would be compatible with the Christian faith, a point to which I will return.

Today, almost half a century after the publication of the Encyclical, new knowledge has led to the recognition of more than one hypothesis in the theory of evolution. It is indeed remarkable that this theory has been progressively accepted by researchers, following a series of discoveries in various fields of knowledge. The convergence, neither sought nor fabricated, of the results of work that was conducted independently is in itself a significant argument in favour of this theory.

What is the significance of such a theory? To address this question is to enter the field of epistemology. A theory is a metascientific elaboration, distinct from the results of observation but consistent with them. By means of it a series of independent data and facts can be related and interpreted in a unified explanation. A theory's validity depends on whether or not it can be verified, it is constantly tested against the facts; wherever it can no longer explain the latter, it shows its limitations and unsuitability. It must then be rethought.

Furthermore, while the formulation of a theory like that of evolution complies with the need for consistency with the observed data, it borrows certain notions from natural philosophy.

And, to tell the truth, rather than the theory of evolution, we should speak of several theories of evolution. On the one hand, this plurality has to do with the different explanations advanced for the mechanism of evolution, and on the other, with the various philosophies on which it is based. Hence the existence of materialist, reductionist and spiritualist interpretations. What is to be decided here is the true role of philosophy and, beyond it, of theology.

5. The Church's Magisterium is directly concerned with the question of evolution, for it involves the conception of man: Revelation teaches us that he was created in the image and likeness of God (cf. Gn 1:27–29). The conciliar Constitution Gaudium et spes has magnificently explained this doctrine, which is pivotal to Christian thought. It recalled that man is :the only creature on earth that God has wanted for its own sake" (n. 24). In other terms, the human individual cannot be subordinated as a pure means or a pure instrument, either to the species or to society, he has value per se. He is a person. With his intellect and his will, he is capable of forming a relationship of communion, solidarity and self-giving with his peers. St Thomas observes that man's likeness to God resides especially in his speculative intellect for his relationship with the object of his knowledge resembles God's relationship with what he has created (Summa Theologica, I-II, q. 3, a.

5, ad 1). But even more, man is called to enter into a relationship of knowledge and love with God himself, a relationship which will find its complete fulfilment beyond time, in eternity. All the depth and grandeur of this vocation are revealed to us in the mystery of the risen Christ (cf. Gaudium et spes, n. 22). It is by virtue of his spiritual soul that the whole person possesses such a dignity even in his body. Pius XII stressed this essential point: if the human body takes its origin from pre-existent living matter the spiritual soul is immediately created by God ("animal enim a Deo immediate creari catholica fides nos retinere inhet"; Encyclical Humani generic, AAS 42 [1950], p. 575).

Consequently, theories of evolution which, in accordance with the philosophies inspiring them, consider the mind as emerging from the forces of living matter, or as a mere epiphenomenon of this matter, are incompatible with the truth about man. Nor are they able to ground the dignity of the person.

6. With man, then, we find ourselves in the presence of an ontological difference, an ontological leap, one could say. However, does not the posing of such ontological discontinuity run counter to that physical continuity which seems to be the main thread of research into evolution in the field of physics and chemistry? Consideration of the method used in the various branches of knowledge makes it possible to reconcile two points of view which would seem irreconcilable. The sciences of observation describe and measure the multiple manifestations of life with increasing precision and correlate them with the time line. The moment of transition into the spiritual cannot be the object of this kind of observation, which nevertheless can discover at the experimental level a series of very valuable signs indicating what is specific to the human being. But the experience of metaphysical knowledge, of self-awareness and self-reflection, of moral conscience, freedom, or again, of aesthetic and religious experience, falls within the competence of philosophical analysis and reflection while theology brings out its ultimate meaning according to the Creator's plans.

We Are Called to Enter Eternal Life

7. In conclusion, I would like to call to mind a Gospel truth which can shed a higher light on the horizon of your research into the origins and unfolding of living matter. The Bible in fact bears an extraordinary message of life. It gives us a wise vision of life inasmuch as it describes the loftiest forms of existence. This vision guided me in the Encyclical which I dedicated to respect for human life, and which I called precisely Evangelium vitae.

It is significant that in St John's Gospel life refers to the divine light which Christ communicates to us. We are called to enter into eternal life, that is to say, into the eternity of divine beatitude.

To warn us against the serious temptations threatening us, our Lord quotes the great saying of Deuteronomy: "Man shall not live by bread alone, but by every word that proceeds from the mouth of God" (Dt 8:3, cf. Mt 4:4).

Even more, "life" is one of the most beautiful titles which the Bible attributes to God. He is the living God.

I cordially invoke an abundance of divine blessings upon you and upon all who are close to you.

From the Vatican, 22 October 1996.

William Dembski and John Haught Spar on Intelligent Design

Rebecca Flietstra

Research News: *Would each of you summarize your position on origins and suggest a book or article that you recommend as a further introduction?*

John Haught: My position is that the Darwinian revolution is a great opportunity and a great gift for theology. I did a piece in Commonwealth Magazine on the Darwinian revolution, Evolution and God's Humility [Jan. 28, 2000] that summarizes God After Darwin as well as anything could in five or six pages. I agree with Dembski that Darwinism does not tell us everything, but it has uncovered things about the natural world we did not know about before. Even though at first sight Darwinism may seem to contradict things we knew theologically, if we look at them carefully from a theological point of view, they actually allow us to come to grips with the radical roots of the Christian tradition. In a way, a depth dimension in the world and in God has been opened that could not be seen before evolutionary science.

William Dembski: I look at a certain type of information that arises in contexts where we know intelligence to be operating. I call this information specified complexity, develop it formally, and then show that is indeed reliably correlated with the effects of intelligence, which is a source of a lot of controversy right now. I would refer people to my book *No Free Lunch*.

RN: *The subject of apologetics often comes up in these discussions. Could you reflect on the way your work is, or is not, apologetic.*

WD: My work certainly can be used for apologetics. People who are using it apologetically seem to be operating within a Christian evangelical framework. Intelli-

---

John Haught and William Dembski sat down at Oxford University with Rebecca Flietstra to talk about evolution and intelligent design for Research News. Haught is professor of theology at Georgetown University and Dembski is the founder of the International Society for Complexity, Information, and Design (ISCID). Fliestra teaches biology at Point Loma University.

gent design fits very nicely with this older theological tradition of examining the vestiges of creation where there is evidence of divine handiwork. Yes, that evidence is very limited in what it can tell you about God, but it still makes God's handiwork evident. That is where it is getting apologetic mileage.

There is also resistance to this. There is concern about anthropocentrism in the question, In what sense can the design that we see in a human context be extrapolated? Nevertheless, it is going in that direction.

I would like to further explore the apologetic movement, but that is not my main emphasis. I would prefer to see if the intelligent design movement could be developed as a scientific program. If intelligent design does not pan out as a scientific program, then any apologetic is going to come crashing down. I am sensitive to that.

JH: As a theologian I am interested in apologetics because there has to be an apologetic aspect to all Christian theology. I use apologetics in Paul Tillich's sense of the term as answering theology.

Tillich's point was that we should never make statements theologically that do not respond to actual questions that people are asking. That is the sense in which my theology is apologetic.

I think one question people ask today is an extension of a human question meaning.

As people look into the evolutionary picture, it can be very frightening. People want to ask, What is the meaning of this? What is going on here?

William James once wrote, I've finally concluded that something is going on in this universe, and that novelty is real. I want to emphasize both of those points. The universe may say something significant. Something momentous is happening, and responding to the question of meaning.

More than that, something new is constantly coming about. For me, the Darwinian, or evolutionary, picture of things allows the new creation to appear in a way that the pre-evolutionary universe does not.

RN: *You both draw on Polanyi's idea of contingency but come to radically different conclusions. How is this possible?*

JH: That is a very good question. I am a fond admirer of Polanyi, and I think we both would agree with Polanyi that Darwin does not give us the whole answer of things. I think, though, that Polanyi was more open to Darwinian biology than Bill thinks.

What I like about Polanyi is his approach to reductionism. It is very, very fertile and has not been drawn upon enough. He has answered the biological reductionism of people like Crick and others quite well. And I agree that he points to something like what Bill calls the fact of specified complexity, of informational sequencing that cannot be accounted for in terms of chemistry and physics as such. Even so, I think Polanyi is more open to an evolutionary view of things than Bill is.

WD: You are probably right that Polanyi would have accepted more of an evolutionary view than I do. However, there is a commonality between us because I am not for a static universe I do want to see the emergence of novelty. The question is how that happens in natural history. I do not think Darwinism is the whole show, though it is part of the show.

I am often called an anti-evolutionist, but I could be comfortable with common descent that can be squared with the Christian tradition. The problem is that common descent, or common ancestry, has been tied to the Darwinian mechanism. That mechanism is supposed to drive the whole of evolution.

If that mechanism is thrown into question, I question it; you [Haught] do not then how do we explain genealogical interrelatedness? Can the genealogical interrelatedness of all organisms instead be thrown into question? The evidence will go where it will on that, but I find the things that you are emphasizing about novelty and Whitehead's notion of beauty very congenial.

The Christian tradition that I am most comfortable with is Eastern Orthodox, which seems to have a commonality with these notions. Perhaps it is only over the adequacy of the Darwinian mechanism, or the extent to which it applies, that we disagree.

JH: Well, this is an interesting clarification! As I have talked to people who are critical of your work, they often say that the biggest problem with intelligent design is that it suppresses the massive amounts of evidence found in the fossil record, biogeographical distribution, radiometric dating, embryology, comparative anatomy and so forth. The thing that causes intelligent design to seem somewhat marginal to the scientific enterprise is that the data which scientists work to gather is not fully taken into account by the movement.

WD: I want to resist that. I recently read a strong case for common ancestry by a geneticist. However, I also see counter-evidence. Still, I do not want to dismiss the findings of scientists, because scientists sweat blood trying to understand nature's workings. I respect the fact that knowledge is hard to get, but the issue is how we put it all together. If intelligent-design researchers are on to something, science is going to require some fundamental rethinking. How far that rethinking is going to go is not clear at this point.

JH: Is there solidarity inside the intelligent design movement? For example, I attended a conference at Calvin College [Design, Self-Organization, and the Integrity of Creation] and listened to Jonathan Wells and his very vehement denial of the possibility of common descent. I wonder how you can get along with people within the movement who are apparently so much more anti-evolutionist than you claim to be.

WD: John Roche describes intelligent design as a big tent with many people under it from young-earth creationists to Michael Behe who accepts common descent. I am not sure that he would go to the mat for it, but he says there is good evidence. So there is a broad spectrum and that is just within the Christian world!

JH: I know that you have mentioned Muslims who have been attracted to intelligent design. I have a friend who is Islamic and a scientist, named Seyyed Hossein Nasr. When he writes about Darwin, he immediately sees materialism.

One of the things the intelligent design movement has been very sensitive to Phillip Johnson especially is the fact that so much evolutionary science is presented to the public wrapped in the blanket of metaphysical materialism. Many Muslim philosophers then, are very sensitive to that. They are attracted to intelligent design because it has really signaled this.

The problem both the intelligent design movement and Nasr have is the refusal to allow, in principle, a disengagement of the Darwinian data of the information that scientists are gathering about the fossil record from the metaphysical-philosophical overlay that is often put on it.

I wonder if we could make more progress in this discussion if the data of evolutionary science could, in principle, be dissociated from materialism and so-called naturalism?

WD: I want to think that through closely. Bruce Gordon has presented lectures stating that there is no problem squaring neo-Darwinism with Christian faith. I need to think about the neo-Darwinian mechanism that is said to be driven by chance mutations.

What metaphysical sense can be given to that? The role of chance? The way it is presented, especially by the materialistic neo-Darwinists, is that there is no direction to the mechanism driving evolution the mechanism is blind. What if there is a place for teleology in this scheme? Can the idea of multiple levels of explanation assist here?

It is evident to me that there are neo-Darwinists who are Christians, who subscribe to the creeds and can confess them with a straight face, but are they doing it coherently? In other words, is the metaphysic of materialism actually entering the theory of neo-Darwinism in some substantive way? I suspect it is in how you make sense of the random errors, namely, the chance.

JH: That is where I think good theological sense can be made of contingency of the undirected aspects of evolution in terms of a God who wants the world to become independent of God, so that a dialogical relationship with deity would be possible.

To try to imagine the alternative a universe which is directed in every respect is, logically speaking, simply an extension of God, rather than something over against God, capable of rebelling as I think creation does at times.

RN: *This is a strong point of contention. How much freedom does God grant the world? Is it correct to say that intelligent design has a very controlling God, and process theology has an ultra-permissive God? What would you see as a major point of disagreement in each other's positions?*

JH: I do not think that either Bill or I go to those extremes. I suspect our religious sense of God is much closer together than that. I think when we sit down and talk to each other that often we find ourselves closer than we might think.

If I could specify a sharp difference, it is a methodological one. When I talk to scientists about evolution, I do not want them to have to talk about intelligence. I expect them to give me what they have found through the old-fashioned scientific method scientific investigation. I do not mind that they also say as far as scientific explanation is concerned, natural selection is a perfectly good explanation.

What I do not want them to say is, That's the exhaustive explanation of this phenomenon of life. I want to make room alongside for theological explanation. The introduction of intelligence at the level of scientific explanation seems, from a point of view of a science-and-religion conversation, to be unwarranted.

WD: From my vantage point, science is calling for intelligence to be introduced! John is right about this being a sticking point, and it is going to make the science-religion dialogue at this point more difficult.

JH: Can I add another wrinkle? Intelligent design is often defended in very, very conservative political and religious journals. I wonder whether one of the reasons that the whole idea of evolution is frightening or distasteful to some people in our country and not just religious people is that the first thing that evolution implies is cumulative change over time. That means that things do not stay the same, whereas the political and religious right often wants things to stay the same. To what extent is this issue a political thing?

WD: That is a tough one. In a past life, I organized an intelligent-design think tank at Baylor. The idea was to do the science and to get away from political concerns. Unfortunately, the whole thing was politicized by people who were opposed to intelligent design (see *Research News*, December 2000). It is a problem that intelligent design is politicized by all sides. That is why my concern is to try to get a scientific program up and running. It is tough.

RN: *Do you have any concluding thoughts?*

JH: We are at a point of being inadequate in our understanding of everything especially life. I want to end on that point. There is so much room for further depth of understanding.

WD: One of the things I want to see happen in the intelligent design movement is to have more conversations with people like John Haught. Isolation has been one of my main concerns both personally and for the intellectual movement I represent.

We tend to work in isolated pockets. I remember when I was at Princeton Seminary; the science-religion dialogue was well represented there especially with Wentzel van Huyssteen and James Loder. If we can have more conversations, a lot of misconceptions can be cleared up.

Darwin Under the Microscope

Michael J. Behe

BETHLEHEM, PA Pope John Paul II's statement last week that evolution is "more than just a theory" is old news to a Roman Catholic scientist like myself.

I grew up in a Catholic family and have always believed in God. But beginning in parochial school I was taught that He could use natural processes to produce life. Contrary to conventional wisdom, religion has made room for science for a long time. But as biology uncovers startling complexity in life, the question becomes, can science make room for religion?

In his statement, the Pope was careful to point out that it is better to talk about "theories of evolution" rather than a single theory. The distinction is crucial. Indeed, until I completed my doctoral studies in biochemistry, I believed that Darwin's mechanism—random mutation paired with natural selection—was the correct explanation for the diversity of life. Yet I now find that theory incomplete.

In fact, the complex design of the cell has provoked me to stake out a distinctly minority view among scientists on the question of what caused evolution. I believe that Darwin's mechanism for evolution doesn't explain much of what is seen under a microscope. Cells are simply too complex to have evolved randomly; intelligence was required to produce them.

I want to be explicit about what I am, and am not, questioning. The word "evolution" carries many associations. Usually it means common descent—the idea that all organisms living and dead are related by common ancestry. I have no quarrel with the idea of common descent, and continue to think it explains similarities among species. By itself, however, common descent doesn't explain the vast differences among species.

That's where Darwin's mechanism comes in. "Evolution" also sometimes implies that random mutation and natural selection powered the changes in life. The idea is that just by chance an animal was born that was slightly faster or stron-

---

Michael J. Behe, associate professor of biochemistry at Lehigh University, is the author of "Darwin's Black Box: The Biochemical Challenge to Evolution."

ger than its siblings. Its descendants inherited the change and eventually won the contest of survival over the descendants of other members of the species. Over time, repetition of the process resulted in great changes—and, indeed, wholly different animals.

That's the theory. A practical difficulty, however, is that one can't test the theory from fossils. To really test the theory, one has to observe contemporary change in the wild, in the laboratory or at least reconstruct a detailed pathway that might have led to a certain adaptation.

Darwinian theory successfully accounts for a variety of modern changes. Scientists have shown that the average beak size of Galapagos finches changed in response to altered weather patterns. Likewise, the ratio of dark-to light-colored moths in England shifted when pollution made light-colored moths more visible to predators. Mutant bacteria survive when they become resistant to antibiotics. These are all clear examples of natural selection in action. But these examples involve only one or a few mutations, and the mutant organism is not much different from its ancestor. Yet to account for all of life, a series of mutations would have to produce very different types of creatures. That has not yet been demonstrated.

Darwin's theory encounters its greatest difficulties when it comes to explaining the development of the cell. Many cellular systems are what I term "irreducibly complex." That means the system needs several components before it can work properly. An everyday example of irreducible complexity is a mousetrap, built of several pieces (platform, hammer, spring and so on). Such a system probably cannot be put together in a Darwinian manner, gradually improving its function. You can't catch a mouse with just the platform and then catch a few more by adding the spring. All the pieces have to be in place before you catch any mice.

An example of an irreducibly complex cellular system is the bacterial flagellum: a rotary propeller, powered by a flow of acid, that bacteria use to swim. The flagellum requires a number of parts before it works—a rotor, stator and motor. Furthermore, genetic studies have shown that about 40 different kinds of proteins are needed to produce a working flagellum.

The intracellular transport system is also quite complex. Plant and animal cells are divided into many discrete compartments; supplies, including enzymes and proteins, have to be shipped between these compartments. Some supplies are packaged into molecular trucks, and each truck has a key that will fit only the lock of its particular cellular destination. Other proteins act as loading docks, opening the truck and letting the contents into the destination compartment.

Many other examples could be cited. The bottom line is that the cell—the very basis of life—is staggeringly complex. But doesn't science already have answers, or partial answers, for how these systems originated? No. As James Shapiro, a biochemist at the University of Chicago, wrote, "There are no detailed Darwinian accounts for the evolution of any fundamental biochemical or cellular system, only a variety of wishful speculations."

A few scientists have suggested non-Darwinian theories to account for the cell, but I don't find them persuasive. Instead, I think that the complex systems were designed—purposely arranged by an intelligent agent.

Whenever we see interactive systems (such as a mousetrap) in the everyday world, we assume that they are the products of intelligent activity. We should extend the reasoning to cellular systems. We know of no other mechanism, including Darwin's, which produces such complexity. Only intelligence does.

Of course, I could be proved wrong. If someone demonstrated that, say, a type of bacteria without a flagellum could gradually produce such a system, or produce any new, comparably complex structure, my idea would be neatly disproved. But I don't expect that to happen.

Intelligent design may mean that the ultimate explanation for life is beyond scientific explanation. That assessment is premature. But even if it is true, I would not be troubled. I don't want the best scientific explanation for the origins of life; I want the correct explanation.

Pope John Paul spoke of "theories of evolution." Right now it looks as if one of those theories involves intelligent design.

Obscurantism to the Rescue

Richard Dawkins

A cowardly flabbiness of the intellect afflicts otherwise rational people confronted with long-established religions (though, significantly, not in the face of younger traditions such as Scientology or the Moonies). S. J. Gould (1997), commenting in his *Natural History* column on the Pope's attitude to evolution, is representative of a dominant strain of conciliatory thought, among believers and non-believers alike:

> Science and religion are not in conflict, for their teachings occupy distinctly different domains (p 16).
> I believe, with all my heart, in a respectful, even *loving* concordat . . . (p 60; my emphasis).
> Well, what are these two distinctly different domains, these "nonoverlapping magisteria" that should snuggle up together in a respectful and loving concordat? Gould again:
> The net of science covers the empirical universe: what is it made of (fact) and why does it work this way (theory). The net of religion extends over questions of moral meaning and value (p 60).

Would that it were that tidy. In a moment I'll look at what the Pope actually says about evolution, and then at other claims of his church, to see if they really are so neatly distinct from the domain of science. First though, a brief aside on the claim that religion has some special expertise to offer us on moral questions. This is often blithely accepted even by the non-religious, presumably in the course of a civilized "bending over backwards" to concede the best point your opponent has to offer—however weak that best point may be.

The question, "What is right and what is wrong?" is a genuinely difficult question that science certainly cannot answer. Given a moral premise or a priori moral belief, the important and rigorous discipline of secular moral philosophy can pursue scientific or logical modes of reasoning to point up hidden implications of such beliefs, and hidden inconsistencies between them. But the absolute moral premises themselves must come from elsewhere, presumably from unargued conviction. Or, it might be hoped, from religion—meaning some combination of authority, revelation, tradition and scripture.

Unfortunately, the hope that religion might provide a bedrock, from which our otherwise sand-based morals can be derived, is a forlorn one. In practice no civilized person uses scripture as ultimate authority for moral reasoning. Instead, we pick and choose the nice bits of scripture (like the Sermon on the Mount) and blithely ignore the nasty bits (like the obligation to stone adulteresses, execute apostates and punish the grandchildren of offenders). The God of the Old Testament himself, with his pitilessly vengeful jealousy, his racism, sexism and terrifying bloodlust, will not be adopted as a literal role model by anybody you or I would wish to know. Yes, *of course* it is unfair to judge the customs of an earlier era by the enlightened standards of our own. But that is precisely my *point*! Evidently, we have some alternative source of ultimate moral conviction which overrides scripture when it suits us.

That alternative source seems to be some kind of liberal consensus of decency and natural justice, which changes over historical time, frequently under the influence of secular reformists. Admittedly, that doesn't sound like bedrock. But in practice we, including the religious among us, give it higher priority than scripture. In practice we more or less ignore scripture, quoting it when it supports our liberal consensus, quietly forgetting it when it doesn't. And, wherever that liberal consensus comes from, it is available to all of us, whether we are religious or not.

Similarly, great religious teachers like Jesus or Gautama Buddha may inspire us, by their good example, to adopt their personal moral convictions. But again we pick and choose among religious leaders, avoiding the bad examples of Jim Jones or Charles Manson, and we may choose good secular role models such as Jawaharlal Nehru or Nelson Mandela. Traditions too, however anciently followed, may be good or bad, and we use our secular judgment of decency and natural justice to decide which ones to follow, which to give up.

But that discussion of moral values was a digression. I now turn to my main topic of evolution, and whether the Pope lives up to the ideal of keeping off the scientific grass. His Message on Evolution to the Pontifical Academy of Sciences begins with some casuistical doubletalk designed to reconcile what John Paul is about to say with the previous, more equivocal pronouncements of Pius XII, whose acceptance of evolution was comparatively grudging and reluctant. Then the Pope comes to the hard task of reconciling scientific evidence with "revelation."

Revelation teaches us that [man] was created in the image and likeness of God . . . if the human body takes its origin from pre-existent living matter, the spiritual soul is immediately created by God. . . . Consequently, theories of evolution which, in accordance with the philosophies inspiring them, consider the mind as emerging from the forces of living matter, or as a mere epiphenomenon of this matter, are incompatible with the truth about man. . . . With man, then, we find ourselves in the presence of an ontological difference, an ontological leap, one could say (John Paul II 1996, this issue p 383).

To do the Pope credit, at this point he recognizes the essential contradiction between the two positions he is attempting to reconcile:

> However, does not the posing of such ontological discontinuity run counter to that physical continuity which seems to be the main thread of research into evolution in the field of physics and chemistry? (John Paul II 1996, this issue p 383).

Never fear. As so often in the past, obscurantism comes to the rescue:

> Consideration of the method used in the various branches of knowledge makes it possible to reconcile two points of view which would seem irreconcilable. The sciences of observation describe and measure the multiple manifestations of life with increasing precision and correlate them with the time line. The moment of transition to the spiritual cannot be the object of this kind of observation, which nevertheless can discover at the experimental level a series of very valuable signs indicating what is specific to the human being (John Paul II 1996, this issue p 383).

In plain language, there came a moment in the evolution of hominids when God intervened and injected a human soul into a previously animal lineage (When? A million years ago? Two million years ago? Between *Homo erectus* and *Homo sapiens*? Between "archaic" *Homo sapiens* and *H. sapiens sapiens*?). The sudden injection is necessary, of course, otherwise there would be no distinction upon which to base Catholic morality, which is speciesist to the core. You can kill adult animals for meat, but abortion and euthanasia are murder because human life is involved.

Catholicism's "net" is not limited to moral considerations, if only because Catholic morals have scientific implications. Catholic morality demands the presence of a great gulf between *Homo sapiens* and the rest of the animal kingdom. Such a gulf is fundamentally antievolutionary. The sudden injection of an immortal soul in the time-line is an antievolutionary intrusion into the domain of science.

More generally, it is completely unrealistic to claim, as Gould and many other do, that religion keeps itself away from science's turf, restricting itself to morals and values. A universe with a supernatural presence would be a fundamentally and qualitatively different kind of universe from one without. The difference is, inescapably, a scientific difference. Religions make existence claims, and this means scientific claims.

The same is true of many of the major doctrines of the Roman Catholic Church. The Virgin Birth, the bodily Assumption of the Blessed Virgin Mary, the Resurrection of Jesus, the survival of our own soul after death: these are all claims of a clearly scientific nature. Either Jesus had a corporeal father or he didn't. This is not a question of "values" or "morals"; it is a question of sober fact. WE may not have the evidence to answer it, but it is a scientific question, nevertheless. You may be sure that, if any evidence supporting the claim were discovered, the Vatican would not be reticent in promoting it.

Either Mary's body decayed when she died, or it was physically removed from this planet to Heaven. The official Roman Catholic doctrine of Assumption, promulgated as recently as 1950, implies that Heaven has a physical location and exists in the domain of physical reality—how else could the physical body of a woman go there? I am not, here, saying that the doctrine of the Assumption of the Virgin is necessarily false (although of course I think it is). I am simply rebutting the claim that it is outside the domain of science. On the contrary, the Assumption of the Virgin is transparently a scientific theory. So is the theory that our souls survive bodily death, and so are all stories of angelic visitations, Maryan manifestations and miracles of all types.

There is something dishonestly self-serving in the tactic of claiming that all religious beliefs are outside the domain of science. On the one hand miracle stories and the promise of life after death are used to impress simple people, win converts and swell congregations. It is precisely their scientific power that gives these stories their popular appeal. But at the same time it is considered below the belt to subject the same stories to the ordinary rigors of scientific criticism: these are religious matters and therefore outside the domain of science. But you cannot have it both ways. At least, religious theorists and apologists should not be allowed to get away with having it both ways. Unfortunately all too many of us, including nonreligious people, are unaccountably ready to let them get away with it.

I suppose it is gratifying to have the Pope as an ally in the struggle against fundamentalist creationism. It is certainly amusing to see the rug pulled out from under the feet of Catholic creationists such as Michael Behe. Even so, given a choice between honest to goodness fundamentalism on the one hand, and the obscurantist, disingenuous doublethink of the Roman Catholic Church on the other, I know which I prefer.

References

Gould, S. J. 1997. Nonoverlapping magisteria: science and religion are not in conflict, for their teachings occupy distinctly different domains. *Natural History* 3 March: 16.

John Paul II. 1996. Message to Pontifical Academy of Sciences, 22 October. *L'Osservatore Romano* 30 October: 3, 7.

Viruses of the Mind

Richard Dawkins

1991

The haven all memes depend on reaching is the human mind, but a human mind is itself an artifact created when memes restructure a human brain in order to make it a better habitat for memes. The avenues for entry and departure are modified to suit local conditions, and strengthened by various artificial devices that enhance fidelity and prolixity of replication: native Chinese minds differ dramatically from native French minds, and literate minds differ from illiterate minds. What memes provide in return to the organisms in which they reside is an incalculable store of advantages—with some Trojan horses thrown in for good measure. . .

Daniel Dennett, *Consciousness Explained*

1 Duplication Fodder

A beautiful child close to me, six and the apple of her father's eye, believes that Thomas the Tank Engine really exists. She believes in Father Christmas, and when she grows up her ambition is to be a tooth fairy. She and her school-friends believe the solemn word of respected adults that tooth fairies and Father Christmas really exist. This little girl is of an age to believe whatever you tell her. If you tell her about witches changing princes into frogs she will believe you. If you tell her that bad children roast forever in hell she will have nightmares. I have just discovered that without her father's consent this sweet, trusting, gullible six-year-old is being sent, for weekly instruction, to a Roman Catholic nun. What chance has she?

A human child is shaped by evolution to soak up the culture of her people. Most obviously, she learns the essentials of their language in a matter of months. A large dictionary of words to speak, an encyclopedia of information to speak about, complicated syntactic and semantic rules to order the speaking, are all transferred from older brains into hers well before she reaches half her adult size. When you are pre-programmed to absorb useful information at a high rate, it is hard to shut out pernicious or damaging information at the same time. With so many mindbytes to be downloaded, so many mental codons to be replicated, it is no wonder that child brains are gullible, open to almost any suggestion, vulnerable to subversion, easy prey to Moonies, Scientologists and nuns. Like immune-deficient patients, children are wide open to mental infections that adults might brush off without effort.

DNA, too, includes parasitic code. Cellular machinery is extremely good at copying DNA. Where DNA is concerned, it seems to have an eagerness to copy, seems eager to be copied. The cell nucleus is a paradise for DNA, humming with sophisticated, fast, and accurate duplicating machinery.

Cellular machinery is so friendly towards DNA duplication that it is small wonder cells play host to DNA parasites—viruses, viroids, plasmids and a riff-raff of other genetic fellow travelers. Parasitic DNA even gets itself spliced seamlessly into the chromosomes themselves. "Jumping genes" and stretches of "selfish DNA" cut or copy themselves out of chromosomes and paste themselves in elsewhere. Deadly oncogenes are almost impossible to distinguish from the legitimate genes between which they are spliced. In evolutionary time, there is probably a continual traffic from "straight" genes to "outlaw," and back again (Dawkins, 1982). DNA is just DNA. The only thing that distinguishes viral DNA from host DNA is its expected method of passing into future generations. "Legitimate" host DNA is just DNA that aspires to pass into the next generation via the orthodox route of sperm or egg. "Outlaw" or parasitic DNA is just DNA that looks to a quicker, less cooperative route to the future, via a squeezed droplet or a smear of blood, rather than via a sperm or egg.

For data on a floppy disc, a computer is a humming paradise just as cell nuclei hum with eagerness to duplicate DNA. Computers and their associated disc and tape readers are designed with high fidelity in mind. As with DNA molecules, magnetized bytes don't literally "want" to be faithfully copied. Nevertheless, you can write a computer program that takes steps to duplicate itself. Not just duplicate itself within one computer but spread itself to other computers. Computers are so good at copying bytes, and so good at faithfully obeying the instructions contained in those bytes, that they are sitting ducks to self-replicating programs: wide open to subversion by software parasites. Any cynic familiar with the theory of selfish genes and memes would have known that modern personal computers, with their promiscuous traffic of floppy discs and e-mail links, were just asking for trouble. The only surprising thing about the current epidemic of computer viruses is that it has been so long in coming.

2 Computer Viruses: a Model for an Informational Epidemiology

Computer viruses are pieces of code that graft themselves into existing, legitimate programs and subvert the normal actions of those programs. They may travel on exchanged floppy disks, or over networks. They are technically distinguished from "worms" which are whole programs in their own right, usually traveling over networks. Rather different are "Trojan horses," a third category of destructive programs, which are not in themselves self-replicating but rely on humans to replicate them because of their pornographic or otherwise appealing content. Both viruses and worms are programs that actually say, in computer language, "Duplicate me." Both may do other things that make their presence felt and perhaps satisfy the hole-in-corner vanity of their authors. These side-effects may be "humorous" (like the virus that makes the Macintosh's built-in loudspeaker enunciate the words "Don't panic," with predictably opposite effect); malicious (like the numerous IBM viruses that erase the hard disk after a sniggering screen-announcement of the impending disaster); political (like the Spanish Telecom and Beijing viruses that protest about telephone costs and massacred students respectively); or simply inadvertent (the programmer is incompetent to handle the low-level system calls required to write an effective virus or worm). The famous Internet Worm, which paralyzed much of the computing power of the United States on November 2, 1988, was not intended (very) maliciously but got out of control and, within 24 hours, had clogged around 6,000 computer memories with exponentially multiplying copies of itself.

"Memes now spread around the world at the speed of light, and replicate at rates that make even fruit flies and yeast cells look glacial in comparison. They leap promiscuously from vehicle to vehicle, and from medium to medium, and are proving to be virtually unquarantinable" (Dennett 1990, p.131). Viruses aren't limited to electronic media such as disks and data lines. On its way from one computer to another, a virus may pass through printing ink, light rays in a human lens, optic nerve impulses and finger muscle contractions. A computer fanciers' magazine that printed the text of a virus program for the interest of its readers has been widely condemned. Indeed, such is the appeal of the virus idea to a certain kind of puerile mentality (the masculine gender is used advisedly), that publication of any kind of "how to" information on designing virus programs is rightly seen as an irresponsible act.

I am not going to publish any virus code. But there are certain tricks of effective virus design that are sufficiently well known, even obvious, that it will do no harm to mention them, as I need to do to develop my theme. They all stem from the virus's need to evade detection while it is spreading.

A virus that clones itself too prolifically within one computer will soon be detected because the symptoms of clogging will become too obvious to ignore. For this reason many virus programs check, before infecting a system, to make sure that they are not already on that system. Incidentally, this opens the way for a defense against viruses that is analogous to immunization. In the days before a specific anti-virus program was available, I myself responded to an early infection of my own hard disk by means of a crude "vaccination." Instead of deleting the virus that I had detected, I simply disabled its coded instructions, leaving the "shell" of the virus with its characteristic external "signature" intact. In theory, subsequent members of the same virus species that arrived in my system should have recognized the signature of their own kind and refrained from trying to double-infect. I don't know whether this immunization really worked, but in those days it probably was worth while "gutting" a virus and leaving a shell like this, rather than simply removing it lock, stock and barrel. Nowadays it is better to hand the problem over to one of the professionally written anti-virus programs.

A virus that is too virulent will be rapidly detected and scotched. A virus that instantly and catastrophically sabotages every computer in which it finds itself will not find itself in many computers. It may have a most amusing effect on one computer—erase an entire doctoral thesis or something equally side-splitting—but it won't spread as an epidemic.

Some viruses, therefore, are designed to have an effect that is small enough to be difficult to detect, but which may nevertheless be extremely damaging. There is one type, which, instead of erasing disk sectors wholesale, attacks only spreadsheets, making a few random changes in the (usually financial) quantities entered in the rows and columns. Other viruses evade detection by being triggered probabilistically, for example erasing only one in 16 of the hard disks infected. Yet other viruses employ the time-bomb principle. Most modern computers are "aware" of the date, and viruses have been triggered to manifest themselves all around the world, on a particular date such as Friday 13th or April Fool's Day. From the parasitic point of view, it doesn't matter how catastrophic the eventual attack is, provided the virus has had plenty of opportunity to spread first (a disturbing analogy to the *Medawar/*Williams theory of ageing: we are the victims of lethal and sub-lethal genes that mature only after we have had plenty of time to reproduce (Williams, 1957)). In defense, some large companies go so far as to set aside one "miner's canary" among their fleet of computers, and advance its internal calendar a week so that any time-bomb viruses will reveal themselves prematurely before the big day.

Again predictably, the epidemic of computer viruses has triggered an arms race. Anti-viral software is doing a roaring trade. These antidote programs—"Interferon," "Vaccine," "Gatekeeper" and others—employ a diverse armory of tricks. Some are written with specific, known and named viruses in mind. Others intercept any attempt to meddle with sensitive system areas of memory and warn the user.

The virus principle could, in theory, be used for non-malicious, even beneficial purposes. Thimbleby (1991) coins the phrase "liveware" for his already-implemented use of the infection principle for keeping multiple copies of databases up to date. Every time a disk containing the database is plugged into a computer, it looks to see whether there is already another copy present on the local hard disk. If there is, each copy is updated in the light of the other. So, with a bit of luck, it doesn't matter which member of a circle of colleagues enters, say, a new bibliographical citation on his personal disk. His newly entered information will readily infect the disks of his colleagues (because the colleagues promiscuously insert their disks into one another's computers) and will spread like an epidemic around the circle. Thimbleby's liveware is not entirely virus-like: it could not spread to just anybody's computer and do damage. It spreads data only to already-existing copies of its own database; and you will not be infected by liveware unless you positively opt for infection.

Incidentally, Thimbleby, who is much concerned with the virus menace, points out that you can gain some protection by using computer systems that other people don't use. The usual justification for purchasing today's numerically dominant computer is simply and solely that it *is* numerically dominant. Almost every knowledgeable person agrees that, in terms of quality and especially user-friendliness, the rival, minority system is superior. Nevertheless, ubiquity is held to be good in itself, sufficient to outweigh sheer quality. Buy the same (albeit inferior) computer as your colleagues, the argument goes, and you'll be able to benefit from shared software, and from a generally large circulation of available software. The irony is that, with the advent of the virus plague, "benefit" is not all that you are likely to get. Not only should we all be very hesitant before we accept a disk from a colleague. We should also be aware that, if we join a large community of users of a particular make of computer, we are also joining a large community of viruses—even, it turns out, *disproportionately* larger.

Returning to possible uses of viruses for positive purposes, there are proposals to exploit the "poacher turned gamekeeper" principle, and "set a thief to catch a thief." A simple way would be to take any of the existing anti-viral programs and load it, as a "warhead," into a harmless self-replicating virus. From a "public health" point of view, a spreading epidemic of anti-viral software could be especially beneficial because the computers most vulnerable to malicious viruses—those whose owners are promiscuous in the exchange of pirated programs—will also be most vulnerable to infection by the healing anti-virus. A more penetrating anti-virus might—as in the immune system—"learn" or "evolve" an improved capacity to attack whatever viruses it encountered.

I can imagine other uses of the computer virus principle which, if not exactly altruistic, are at least constructive enough to escape the charge of pure vandalism. A computer company might wish to do market research on the habits of its customers, with a view to improving the design of future products. Do users like to choose files by pictorial icon, or do they opt to display them by textual name only? How deeply do people nest folders (directories) within one another? Do people settle down for a long session with only one program, say a word processors, or are they constantly switching back and forth, say between writing and drawing programs? Do people succeed in moving the mouse pointer straight to the target, or do they meander around in time-wasting hunting movements that could be rectified by a change in design?

The company could send out a questionnaire asking all these questions, but the customers that replied would be a biased sample and, in any case, their own assessment of their computer-using behavior might be inaccurate. A better solution would be a market-research computer program. Customers would be asked to load this program into their system where it would unobtrusively sit, quietly monitoring and tallying key-presses and mouse movements. At the end of a year, the customer would be asked to send in the disk file containing all the tallyings of the market-research program. But again, most people would not bother to cooperate and some might see it as an invasion of privacy and of their disk space.

The perfect solution, from the company's point of view, would be a virus. Like any other virus, it would be self-replicating and secretive. But it would not be destructive or facetious like an ordinary virus. Along with its self-replicating booster it would contain a market-research warhead. The virus would be released surreptitiously into the community of computer users. Just like an ordinary virus it would spread around, as people passed floppy disks and e-mail around the community. As the virus spread from computer to computer, it would build up statistics on users behavior, monitored secretly from deep within a succession of systems. Every now and again, a copy of the viruses would happen to find its way, by normal epidemic traffic, back into one of the company's own computers. There it would be debriefed and its data collated with data from other copies of the virus that had come "home."

Looking into the future, it is not fanciful to imagine a time when viruses, both bad and good, have become so ubiquitous that we could speak of an ecological community of viruses and legitimate programs coexisting in the silicosphere. At present, software is advertised as, say, "Compatible with System 7." In the future, products may be advertised as "Compatible with all viruses registered in the 1998 World Virus Census; immune to all listed virulent viruses; takes full advantage of the facilities offered by the following benign viruses if present..." Word-processing software, say, may hand over particular functions, such as word-counting and string-searches, to friendly viruses burrowing autonomously through the text.

Looking even further into the future, whole integrated software systems might grow, not by design, but by something like the growth of an ecological community such as a tropical rainforest. Gangs of mutually compatible viruses might grow up, in the same way as genomes can be regarded as gangs of mutually compatible genes (Dawkins, 1982). Indeed, I have even suggested that our genomes should be regarded as gigantic colonies of viruses (Dawkins, 1976). Genes cooperate with one another in genomes because natural selection has favored those genes that prosper in the presence of the other genes that happen to be common in the gene pool. Different gene pools may evolve towards different combinations of mutually compatible genes. I envisage a time when, in the same kind of way, computer viruses may evolve towards compatibility with other viruses, to form communities or gangs. But then again, perhaps not! At any rate, I find the speculation more alarming than exciting.

At present, computer viruses don't strictly evolve. They are invented by human programmers, and if they evolve they do so in the same weak sense as cars or aeroplanes evolve. Designers derive this year's car as a slight modification of last year's car, and then may, more or less consciously, continue a trend of the last few years—further flattening of the radiator grill or whatever it may be. Computer virus designers dream up ever more devious tricks for outwitting the programmers of anti-virus software. But computer viruses don't—so far—mutate and evolve by true natural selection. They may do so in the future. Whether they evolve by natural selection, or whether their evolution is steered by human designers, may not make much difference to their eventual performance. By either kind of evolution, we expect them to become better at concealment, and we expect them to become subtly compatible with other viruses that are at the same time prospering in the computer community.

DNA viruses and computer viruses spread for the same reason: an environment exists in which there is machinery well set up to duplicate and spread them around and to obey the instructions that the viruses embody. These two environments are, respectively, the environment of cellular physiology and the environment provided by a large community of computers and data-handling machinery. Are there any other environments like these, any other humming paradises of replication?

3 The Infected Mind

I have already alluded to the programmed-in gullibility of a child, so useful for learning language and traditional wisdom, and so easily subverted by nuns, Moonies and their ilk. More generally, we all exchange information with one another. We don't exactly plug floppy disks into slots in one another's skulls, but we exchange sentences, both through our ears and through our eyes. We notice each other's styles of moving and dressing and are influenced. We take in advertising jingles, and are presumably persuaded by them, otherwise hard-headed businessmen would not spend so much money polluting the air with them.

Think about the two qualities that a virus, or any sort of parasitic replicator, demands of a friendly medium,. the two qualities that make cellular machinery so friendly towards parasitic DNA, and that make computers so friendly towards computer viruses. These qualities are, firstly, a readiness to replicate information accurately, perhaps with some mistakes that are subsequently reproduced accurately; and, secondly, a readiness to obey instructions encoded in the information so replicated.

Cellular machinery and electronic computers excel in both these virus-friendly qualities. How do human brains match up? As faithful duplicators, they are certainly less perfect than either cells or electronic computers. Nevertheless, they are still pretty good, perhaps about as faithful as an RNA virus, though not as good as DNA with all its elaborate proofreading measures against textual degradation. Evidence of the fidelity of brains, especially child brains, as data duplicators is provided by language itself. Shaw's Professor Higgins was able by ear alone to place Londoners in the street where they grew up. Fiction is not evidence for anything, but everyone knows that Higgins's fictional skill is only an exaggeration of something we can all do. Any American can tell Deep South from Mid West, New England from Hillbilly. Any New Yorker can tell Bronx from Brooklyn. Equivalent claims could be substantiated for any country. What this phenomenon means is that human brains are capable of pretty accurate copying (otherwise the accents of, say, Newcastle would not be stable enough to be recognized) but with some mistakes (otherwise pronunciation would not evolve, and all speakers of a language would inherit identically the same accents from their remote ancestors). Language evolves, because it has both the great stability and the slight changeability that are prerequisites for any evolving system.

The second requirement of a virus-friendly environment—that it should obey a program of coded instructions—is again only quantitatively less true for brains than for cells or computers. We sometimes obey orders from one another, but also we sometimes don't. Nevertheless, it is a telling fact that, the world over, the vast majority of children follow the religion of their parents rather than any of the other available religions. Instructions to genuflect, to bow towards Mecca, to nod one's head rhythmically towards the wall, to shake like a maniac, to "speak in tongues"—the list of such arbitrary and pointless motor patterns offered by religion alone is extensive—are obeyed, if not slavishly, at least with some reasonably high statistical probability.

Less portentously, and again especially prominent in children, the "craze" is a striking example of behavior that owes more to epidemiology than to rational choice. Yo-yos, hula hoops and pogo sticks, with their associated behavioral fixed actions, sweep through schools, and more sporadically leap from school to school, in patterns that differ from a measles epidemic in no serious particular. Ten years ago, you could have traveled thousands of miles through the United States and never seen a baseball cap turned back to front. Today, the reverse baseball cap is ubiquitous. I do not know what the pattern of geographical spread of the reverse baseball cap precisely was, but epidemiology is certainly among the professions primarily qualified to study it. We don't have to get into arguments about "determinism"; we don't have to claim that children are compelled to imitate their fellows' hat fashions. It is enough that their hat-wearing behavior, as a matter of fact, *is* statistically affected by the hat-wearing behavior of their fellows.

Trivial though they are, crazes provide us with yet more circumstantial evidence that human minds, especially perhaps juvenile ones, have the qualities that we have singled out as desirable for an informational parasite. At the very least the mind is a plausible *candidate* for infection by something like a computer virus, even if it is not quite such a parasite's dream-environment as a cell nucleus or an electronic computer.

It is intriguing to wonder what it might feel like, from the inside, if one's mind were the victim of a "virus." This might be a deliberately designed parasite, like a present-day computer virus. Or it might be an inadvertently mutated and unconsciously evolved parasite. Either way, especially if the evolved parasite was the memic descendant of a long line of successful ancestors, we are entitled to expect the typical "mind virus" to be pretty good at its job of getting itself successfully replicated.

Progressive evolution of more effective mind-parasites will have two aspects. New "mutants" (either random or designed by humans) that are better at spreading will become more numerous. And there will be a ganging up of ideas that flourish in one another's presence, ideas that mutually support one another just as genes do and as I have speculated computer viruses may one day do. We expect that replicators will go around together from brain to brain in mutually compatible gangs. These gangs will come to constitute a package, which may be sufficiently stable to deserve a collective name such as Roman Catholicism or Voodoo. It doesn't too much matter whether we analogize the whole package to a single virus, or each one of the component parts to a single virus. The analogy is not that precise anyway, just as the distinction between a computer virus and a computer worm is nothing to get worked up about. What matters is that minds are friendly environments to parasitic, self-replicating ideas or information, and that minds are typically massively infected.

Like computer viruses, successful mind viruses will tend to be hard for their victims to detect. If you are the victim of one, the chances are that you won't know it, and may even vigorously deny it. Accepting that a virus might be difficult to detect in your own mind, what tell-tale signs might you look out for? I shall answer by imaging how a medical textbook might describe the typical symptoms of a sufferer (arbitrarily assumed to be male).

1. The patient typically finds himself impelled by some deep, inner conviction that something is true, or right, or virtuous: a conviction that doesn't seem to owe anything to evidence or reason, but which, nevertheless, he feels as totally compelling and convincing. We doctors refer to such a belief as "faith."

2. Patients typically make a positive virtue of faith's being strong and unshakable, *in spite of* not being based upon evidence. Indeed, they may feel that the less evidence there is, the more virtuous the belief (see below).

This paradoxical idea that lack of evidence is a positive virtue where faith is concerned has something of the quality of a program that is self-sustaining, because it is self-referential (see the chapter "On Viral Sentences and Self-Replicating Structures" in Hofstadter, 1985). Once the proposition is believed, it automatically undermines opposition to itself. The "lack of evidence is a virtue" idea could be an admirable sidekick, ganging up with faith itself in a clique of mutually supportive viral programs.

3. A related symptom, which a faith-sufferer may also present, is the conviction that "mystery," *per se,* is a good thing. It is not a virtue to solve mysteries. Rather we should enjoy them, even revel in their insolubility.

Any impulse to solve mysteries could be serious inimical to the spread of a mind virus. It would not, therefore, be surprising if the idea that "mysteries are better not solved" was a favored member of a mutually supporting gang of viruses. Take the "Mystery of Transubstantiation." It is easy and non-mysterious to believe that in some symbolic or metaphorical sense the eucharistic wine turns into the blood of Christ. The Roman Catholic doctrine of transubstantiation, however, claims far more. The "whole substance" of the wine is converted into the blood of Christ; the appearance of wine that remains is "merely accidental," "inhering in no substance" (Kenny, 1986, p. 72). Transubstantiation is colloquially taught as meaning that the wine "literally" turns into the blood of Christ. Whether in its obfuscatory Aristotelian or its franker colloquial form, the claim of transubstantiation can be made only if we do serious violence to the normal meanings of words like "substance" and "literally." Redefining words is not a sin, but, if we use words like "whole substance" and "literally" for this case, what word are we going to use when we really and truly *want* to say that something did actually happen? As Anthony Kenny observed of his own puzzlement as a young seminarian, "For all I could tell, my typewriter might be Benjamin Disraeli transubstantiated...."

Roman Catholics, whose belief in infallible authority compels them to accept that wine becomes physically transformed into blood despite all appearances, refer to the "mystery" of transubstantiation. Calling it a mystery makes everything OK, you see. At least, it works for a mind well prepared by background infection. Exactly the same trick is performed in the "mystery" of the Trinity. Mysteries are not meant to be solved, they are meant to strike awe. The "mystery is a virtue" idea comes to the aid of the Catholic, who would otherwise find intolerable the obligation to believe the obvious nonsense of the transubstantiation and the "three-in-one." Again, the belief that "mystery is a virtue" has a self-referential ring. As Hofstadter might put it, the very mysteriousness of the belief moves the believer to perpetuate the mystery.

An extreme symptom of "mystery is a virtue" infection is Tertullian's "*Certum est quia impossibile est*" (It is certain because it is impossible"). That way madness lies. One is tempted to quote Lewis Carroll's White Queen, who, in response to Alice's "One can't believe impossible things" retorted "I daresay you haven't had much practice... When I was your age, I always did it for half-an-hour a day. Why, sometimes I've believed as many as six impossible things before breakfast." Or Douglas Adams' Electric Monk, a labor-saving device programmed to do your believing for you, which was capable of "believing things they'd have difficulty believing in Salt Lake City" and which, at the moment of being introduced to the reader, believed, contrary to all the evidence, that everything in the world was a uniform shade of pink. But White Queens and Electric Monks become less funny when you realize that these virtuoso believers are indistinguishable from revered theologians in real life. "It is by all means to be believed, because it is absurd" (Tertullian again). Sir Thomas Browne (1635) quotes Tertullian with approval, and goes further: "Methinks there be not impossibilities enough in religion for an active faith." And "I desire to exercise my faith in the difficultest point; for to credit ordinary and visible objects is not faith, but perswasion [sic]."

I have the feeling that something more interesting is going on here than just plain insanity or surrealist nonsense, something akin to the admiration we feel when we watch a ten-ball juggler on a tightrope. It is as though the faithful gain prestige through managing to believe even more impossible things than their rivals succeed in believing. Are these people testing—exercising—their believing muscles, training themselves to believe impossible things so that they can take in their stride the merely improbable things that they are ordinarily called upon to believe?

While I was writing this, the *Guardian* (July 29, 1991) fortuitously carried a beautiful example. It came in an interview with a rabbi undertaking the bizarre task of vetting the kosher-purity of food products right back to the ultimate origins of their minutest ingredients. He was currently agonizing over whether to go all the way to China to scrutinize the menthol that goes into cough sweets. "Have you ever tried checking Chinese menthol... it was extremely difficult, especially since the first letter we sent received the reply in best Chinese English, 'The product contains no kosher'... China has only recently started opening up to kosher investigators. The menthol should be OK, but you can never be absolutely sure unless you visit." These kosher investigators run a telephone hot-line on which up-to-the-minute red-alerts of suspicion are recorded against chocolate bars and cod-liver oil. The rabbi sighs that the green-inspired trend away from artificial colors and flavors "makes life miserable in the kosher field because you have to follow all these things back." When the interviewer asks him why he bothers with this obviously pointless exercise, he makes it very clear that the point is precisely that there *is* no point:

That most of the Kashrut laws are divine ordinances without reason given is 100 per cent the point. It is very easy not to murder people. Very easy. It is a little bit harder not to steal because one is tempted occasionally. So that is no great proof that I believe in God or am fulfilling His will. But, if He tells me not to have a cup of coffee with milk in it with my mincemeat and peaces at lunchtime, that is a test. The only reason I am doing that is because I have been told to so do. It is something difficult.

Helena Cronin has suggested to me that there may be an analogy here to Zahavi's handicap theory of sexual selection and the evolution of signals (Zahavi, 1975). Long unfashionable, even ridiculed (Dawkins, 1976), Zahavi's theory has recently been cleverly rehabilitated (Grafen, 1990 a, b) and is now taken seriously by evolutionary biologists (Dawkins, 1989). Zahavi suggests that peacocks, for instance, evolve their absurdly burdensome fans with their ridiculously conspicuous (to predators) colors, precisely *because* they are burdensome and dangerous, and therefore impressive to females. The peacock is, in effect, saying: "Look how fit and strong I must be, since I can afford to carry around this preposterous tail."

To avoid misunderstanding of the subjective language in which Zahavi likes to make his points, I should add that the biologist's convention of personifying the unconscious actions of natural selection is taken for granted here. Grafen has translated the argument into an orthodox Darwinian mathematical model, and it works. No claim is here being made about the intentionality or awareness of peacocks and peahens. They can be as sphexish or as intentional as you please (Dennett, 1983, 1984). Moreover, Zahavi's theory is general enough not to depend upon a Darwinian underpinning. A flower advertising its nectar to a "skeptical" bee could benefit from the Zahavi principle. But so could a human salesman seeking to impress a client.

The premise of Zahavi's idea is that natural selection will favor skepticism among females (or among recipients of advertising messages generally). The only way for a male (or any advertiser) to authenticate his boast of strength (quality, or whatever is is) is to prove that it is true by shouldering a truly costly handicap—a handicap *that only a genuinely strong* (high quality, etc.) male could bear. It may be called the principle of costly authentication. And now to the point. Is it possible that some religious doctrines are favored not *in spite of* being ridiculous but precisely *because* they are ridiculous? Any wimp in religion could believe that bread *symbolically* represents the body of Christ, but it takes a real, red-blooded Catholic to believe something as daft as the transubstantiation. If you believe that you can believe anything, and (witness the story of Doubting Thomas) these people are trained to see that as a virtue.

Let us return to our list of symptoms that someone afflicted with the mental virus of faith, and its accompanying gang of secondary infections, may expect to experience.

4. The sufferer may find himself behaving intolerantly towards vectors of rival faiths, in extreme cases even killing them or advocating their deaths. He may be similarly violent in his disposition towards apostates (people who once held the faith but have renounced it); or towards heretics (people who espouse a different—often, perhaps significantly, only very slightly different—version of the faith). He may also feel hostile towards other modes of thought that are potentially inimical to his faith, such as the method of scientific reason which may function rather like a piece of anti-viral software.

The threat to kill the distinguished novelist Salman Rushdie is only the latest in a long line of sad examples. On the very day that I wrote this, the Japanese translator of *The Satanic Verses* was found murdered, a week after a near-fatal attack on the Italian translator of the same book. By the way, the apparently opposite symptom of "sympathy" for Muslim "hurt," voiced by the Archbishop of Canterbury and other Christian leaders (verging, in the case of the Vatican, on outright criminal complicity) is, of course, a manifestation of the symptom we discussed earlier: the delusion that faith, however obnoxious its results, has to be respected simply because it *is* faith.

Murder is an extreme, of course. But there is an even more extreme symptom, and that is suicide in the militant service of a faith. Like a soldier ant programmed to sacrifice her life for germ-line copies of the genes that did the programming, a young Arab or Japanese [??!] is taught that to die in a holy war is the quickest way to heaven. Whether the leaders who exploit him really believe this does not diminish the brutal power that the "suicide mission virus" wields on behalf of the faith. Of course suicide, like murder, is a mixed blessing: would-be converts may be repelled, or may treat with contempt a faith that is perceived as insecure enough to need such tactics.

More obviously, if too many individuals sacrifice themselves the supply of believers could run low. This was true of a notorious example of faith-inspired suicide, though in this case it was not "kamikaze" death in battle. The Peoples' Temple sect became extinct when its leader, the Reverend Jim Jones, led the bulk of his followers from the United States to the Promised Land of "Jonestown" in the Guyanan jungle where he persuaded more than 900 of them, children first, to drink cyanide. The macabre affair was fully investigated by a team from the *San Francisco Chronicle* (Kilduff and Javers, 1978).

Jones, "the Father," had called his flock together and told them it was time to depart for heaven.

"We're going to meet," he promised, "in another place."
The words kept coming over the camp's loudspeakers.
"There is great dignity in dying. It is a great demonstration for everyone to die."

Incidentally, it does not escape the trained mind of the alert sociobiologist that Jones, within his sect in earlier days, "proclaimed himself the only person permitted to have sex" (presumably his partners were also permitted). "A secretary would arrange for Jones's liaisons. She would call up and say, 'Father hates to do this, but he has this tremendous urge and could you please...?' " His victims were not only female. One 17-year-old male follower, from the days when Jones's community was still in San Francisco, told how he was taken for dirty weekends to a hotel where Jones received a "minister's discount for Rev. Jim Jones and son." The same boy said: "I was really in awe of him. He was more than a father. I would have killed my parents for him." What is remarkable about the Reverend Jim Jones is not his own self-serving behavior but the almost superhuman gullibility of his followers. Given such prodigious credulity, can anyone doubt that human minds are ripe for malignant infection?

Admittedly, the Reverend Jones conned only a few thousand people. But his case is an extreme, the tip of an iceberg. The same eagerness to be conned by religious leaders is widespread. Most of us would have been prepared to bet that nobody could get away with going on television and saying, in all but so many words, "Send me your money, so that I can use it to persuade other suckers to send me their money too." Yet today, in every major conurbation in the United States, you can find at least one television evangelist channel entirely devoted to this transparent confidence trick. And they get away with it in sackfuls. Faced with suckerdom on this awesome scale, it is hard not to feel a grudging sympathy with the shiny-suited conmen. Until you realize that not all the suckers are rich, and that it is often widows' mites on which the evangelists are growing fat. I have even heard one of them explicitly invoking the principle that I now identify with Zahavi's principle of costly authentication. God really appreciates a donation, he said with passionate sincerity, only when that donation is so large that it hurts. Elderly paupers were wheeled on to testify how much happier they felt since they had made over their little all to the Reverend whoever it was.

5. The patient may notice that the particular convictions that he holds, while having nothing to do with evidence, do seem to owe a great deal to epidemiology. Why, he may wonder, do I hold *this* set of convictions rather than *that* set? Is it because I surveyed all the world's faiths and chose the one whose claims seemed most convincing? Almost certainly not. If you have a faith, it is statistically overwhelmingly likely that it is the same faith as your parents and grandparents had. No doubt soaring cathedrals, stirring music, moving stories and parables, help a bit. But by far the most important variable determining your

religion is the accident of birth. The convictions that you so passionately believe would have been a completely different, and largely contradictory, set of convictions, if only you had happened to be born in a different place. Epidemiology, not evidence.

6. If the patient is one of the rare exceptions who follows a different religion from his parents, the explanation may still be epidemiological. To be sure, it is *possible* that he dispassionately surveyed the world's faiths and chose the most convincing one. But it is statistically more probable that he has been exposed to a particularly potent infective agent—a John Wesley, a Jim Jones or a St. Paul. Here we are talking about horizontal transmission, as in measles. Before, the epidemiology was that of vertical transmission, as in Huntington's Chorea.

7. The internal sensations of the patient may be startlingly reminiscent of those more ordinarily associated with sexual love. This is an extremely potent force in the brain, and it is not surprising that some viruses have evolved to exploit it. St. Teresa of Avila's famously orgasmic vision is too notorious to need quoting again. More seriously, and on a less crudely sensual plane, the philosopher Anthony Kenny provides moving testimony to the pure delight that awaits those that manage to believe in the mystery of transubstantiation. After describing his ordination as a Roman Catholic priest, empowered by laying on of hands to celebrate Mass, he goes on that he vividly recalls the exaltation of the first months during which I had the power to say Mass. Normally a slow and sluggish riser, I would leap early out of bed, fully awake and full of excitement at the thought of the momentous act I was privileged to perform. I rarely said the public Community Mass: most days I celebrated alone at a side altar with a junior member of the College to serve as acolyte and congregation. But that made no difference to the solemnity of the sacrifice or the validity of the consecration.

It was touching the body of Christ, the closeness of the priest to Jesus, which most enthralled me. I would gaze on the Host after the words of consecration, soft-eyed like a lover looking into the eyes of his beloved... Those early days as a priest remain in my memory as days of fulfilment and tremulous happiness; something precious, and yet too fragile to last, like a romantic love-affair brought up short by the reality of an ill-assorted marriage. (Kenny, 1986, pp. 101-2)

Dr. Kenny is affectingly believable that it felt to him, as a young priest, as though he was in love with the consecrated host. What a brilliantly successful virus! On the same page, incidentally, Kenny also shows us that the virus is transmitted contagiously—if not literally then at least in some sense—from the palm of the infecting bishop's hand through the top of the new priest's head:

If Catholic doctrine is true, every priest validly ordained derives his orders in an unbroken line of laying on of hands, through the bishop who ordains him, back to one of the twelve Apostles... there must be centuries-long, recorded chains of layings on of hands. It surprises me that priests never seem to trouble to trace their spiritual ancestry in this way, finding out who ordained their bishop, and who ordained him, and so on to Julius II or Celestine V or Hildebrand, or Gregory the Great, perhaps. (Kenny, 1986, p. 101)

It surprises me, too.

4 Is Science a Virus

No. Not unless all computer programs are viruses. Good, useful programs spread because people evaluate them, recommend them and pass them on. Computer viruses spread solely because they embody the coded instructions: "Spread me." Scientific ideas, like all memes, are subject to a kind of natural selection, and this might look superficially virus-like. But the selective forces that scrutinize scientific ideas are not arbitrary and capricious. They are exacting, well-honed rules, and they do not favor pointless self-serving behavior. They favor all the virtues laid out in textbooks of standard methodology: testability, evidential support, precision, quantifiability, consistency, intersubjectivity, repeatability, universality, progressiveness, independence of cultural milieu, and so on. Faith spreads despite a total lack of every single one of these virtues.

You may find elements of epidemiology in the spread of scientific ideas, but it will be largely descriptive epidemiology. The rapid spread of a good idea through the scientific community may even look like a description of a measles epidemic. But when you examine the underlying reasons you find that they are good ones, satisfying the demanding standards of scientific method. In the history of the spread of faith you will find little else but epidemiology, and causal epidemiology at that. The reason why person A believes one thing and B believes another is simply and solely that A was born on one continent and B on another. Testability, evidential support and the rest aren't even remotely considered. For scientific belief, epidemiology merely comes along afterwards and describes the history of its acceptance. For religious belief, epidemiology is the root cause.

5 Epilogue

Happily, viruses don't win every time. Many children emerge unscathed from the worst that nuns and mullahs can throw at them. Anthony Kenny's own story has a happy ending. He eventually renounced his orders because he could no longer tolerate the obvious contradictions within Catholic belief, and he is now a highly respected scholar. But one cannot help remarking that it must be a powerful infection indeed that took a man of his wisdom and intelligence—President of the British Academy, no less—three decades to fight off. Am I unduly alarmist to fear for the soul of my six-year-old innocent?

Acknowledgement

With thanks to Helena Cronin for detailed suggestion on content and style on every page.

References

Browne, Sir T. (1635) *Religio Medici,* I, 9

Dawkins, R. (1976) *The Selfish Gene.* Oxford: Oxford University Press.

Dawkins, R. (1982) *The Extended Phenotype.* Oxford: W. H. Freeman.

Dawkins, R. (1989) *The Selfish Gene,* 2nd edn. Oxford: Oxford University Press.

Dennett, D. C. (1983) Intentional systems in cognitive ethology: the "Panglossian paradigm" defended. *Behavioral and Brain Sciences,* 6, 343–90.

Dennett, D. C. (1984) *Elbow Room: The Varieties of Free Will Worth Wanting.* Oxford: Oxford University Press.

Dennett, D. C. (1990) Memes and the exploitation of imagination. *The Journal of Aesthetics and Art Criticism,* 48, 127–35.

Grafen, A. (1990a) Sexual selection unhandicapped by the Fischer process. *Journal of Theoretical Biology,* 144, 473–516.

Grafen, A. (1990b) Biological signals as handicaps. *Journal of Theoretical Biology,* 144, 517–46.

Hofstadter, D. R. (1985) *Metamagical Themas.* Harmondsworth: Penguin.

Kenny, A. (1986) *A Path from Rome* Oxford: Oxford University Press.

Kilduff, M. and Javers, R. (1978) *The Suicide Cult.* New York: Bantam.

Thimbleby, H. (1991) Can viruses ever be useful? *Computers and Security,* 10, 111–14.

Williams, G. C. (1957) Pleiotropy, natural selection, and the evolution of senescence. *Evolution,* 11, 398–411.

Zahavi, A. (1975) Mate selection—a selection for a handicap. *Journal of Theoretical Biology,* 53, 205–14.

Chapter 6

Introduction

Chapter 6 starts with the article written by Sewall Wright in which he introduced his metaphor of an adaptive landscape. This paper was given at an international congress of genetics in 1932. Dobzhansky saw it and at once realized that he had here the theory that he needed on which he could hang all of his own empirical research. He said, "I fell in love with Wright," and he was only half joking. Do not be worried if the article seems a bit technical and mathematical. Rest assured that none of this can be really important and that you can skip it. None of the leading synthetic theorists (with the exception of Simpson) could follow any mathematics whatsoever, so the details cannot have been essential! What matters are the pictures—the landscape and then the way in which populations are supposed to drift down from peaks into valleys, quite against the power of selection, and then shoot up the other sides under the influence of selection. I should say that this theory, which has been very influential—although there are now some evolutionists starting to doubt its truth entirely—has absolutely no basis in Darwin's *Origin*. However, as I said in the text, it is just the sort of thing that you would expect from a neo-Spencerian.

The second and third documents are letters. First I give you a letter written by Ernst Mayr, the ornithologist and systematist, when he was the founding editor of the journal *Evolution*. See how Mayr is trying to articulate the kind of science to which he thinks evolutionists should aspire. It should be experimental, factual, quantitative, and so forth. In short, all of the things that one finds in the best kinds of science, like physics. Above all else, one must avoid "philosophy and speculation"! One must get into "factors and causes." The second letter is by Dobzhansky, sent to an eminent historian of science, John Greene. Both men were practicing Christians and were trying to work out the relationship between science and religion, in particular between evolution and its ideological underpinnings (progress specifically).

Dobzhansky was an enthusiast for progress—remember that he was a Teilhard de Chardin supporter—and claimed that evolution itself supports his belief. Progress was not something written in but rather something that could be read out. What this letter shows is that although the synthetic theorists may have worked hard (and successfully) at professionalizing their subject, their motives and interests still overlapped strongly with those of earlier evolutionists in the Thomas

Henry Huxley mould. Say what you like, evolutionary theory was and remains more than just a theory.

The Roles of Mutation, Inbreeding, Crossbreeding and Selection in Evolution

Sewall Wright

The enormous importance of biparental reproduction as a factor in evolution was brought out a good many years ago by EAST. The observed properties of gene mutation—fortuitous in origin, infrequent in occurrence and deleterious when not negligible in effect—seem about as unfavorable as possible for an evolutionary process. Under biparental reproduction, however, a limited number of mutations which are not too injurious to be carried by the species furnish an almost infinite field of possible variations through which the species may work its way under natural selection.

Estimates of the total number of genes in the cells of higher organisms range from 1000 up. Some 400 loci have been reported as having mutated in Drosophila during a laboratory experience which is certainly very limited compared with the history of the species in nature. Presumably, allelomorphs of all type genes are present at all times in any reasonably numerous species. Judging from the frequency of multiple allelomorphs in those organisms which have been studied most, it is reasonably certain that many different allelomorphs of each gene are in existence at all times. With 10 allelomorphs in each of 1000 loci, the number of possible combinations is 10^{1000} which is a very large number. It has been estimated that the total number of electrons and protons in the whole visible universe is much less than 10^{100}.

However, not all of this field is easily available in an interbreeding population. Suppose that each type gene is manifested in 99 percent of the individuals, and that most of the remaining 1 percent have the most favorable of the other allelomorphs, which in general means one with only a slight differential effect. The average individual will show the effects of 1 percent of the 1000, or 10 deviations from the type, and since this average has a standard deviation of [square root]10 only a small proportion will exhibit more than 20 deviations from type where 1000 are possible. The population is thus confined to an infinitesimal portion of the field of possible gene combinations, yet this portion includes some 10^{40} homozygous combinations, on the above extremely conservative basis, enough so that there is no reasonable chance that any two individuals have exactly the same genetic constitution in a species of millions of millions of individuals persisting over millions of generations. There is no difficulty in accounting for the probable genetic uniqueness of each individual human being or other organism which is the product of biparental reproduction.

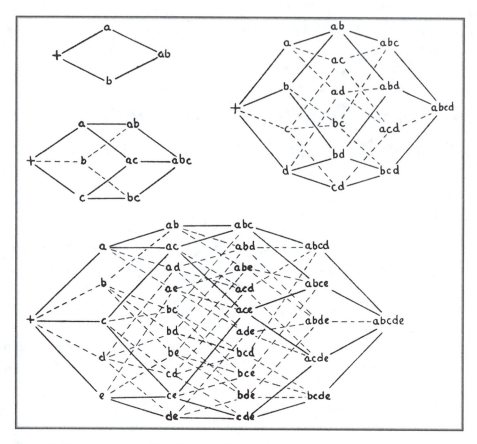

Figure 1: The combination of 2 to 5 paired allelomorphs.

If the entire field of possible gene combinations be graded with respect to adaptive value under a particular set of conditions, what would be its nature? Figure 1 shows the combinations in the cases of 2 to 5 paired allelomorphs. In the last case, each of the 32 homozygous combinations is at one remove from 5 others, at two removes from 10, etc. It would require 5 dimensions to represent these relations symmetrically; a sixth dimension is needed to represent level of adaptive value. The 32 combinations here compare with 10^{1000} in a species with 1000 loci each represented by 10 allelomorphs, and the 5 dimensions required for adequate representation compare with 9000. The two dimensions of figure 2 are a very inadequate representation of such a field. The contour lines are intended to represent the scale of adaptive value.

One possibility is that a particular combination gives maximum adaptation and that the adaptiveness of the other combinations falls off more or less regularly according to the number of removes. A species whose individuals are clustered about some combination other than the highest would move up the steepest gradient toward the peak, having reached which it would remain unchanged except for the rare occurrence of new favorable mutations.

But even in the two factor case (figure 1) it is possible that there may be two peaks, and the chance that this may be the case greatly increases with each additional locus. With something like 10^{1000} possibilities (figure 2) it may be taken as certain that there will be an enormous number of widely separated harmonious

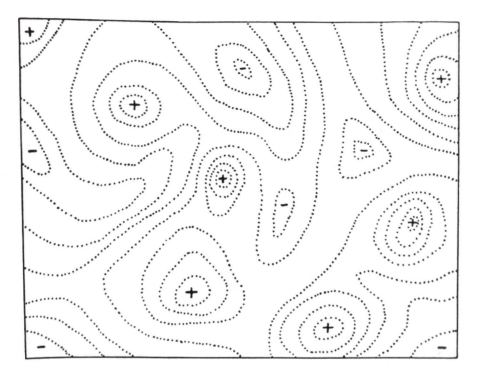

Figure 2: Diagrammatic representation of the field of gene combinations in two dimensions instead of many thousands. Dotted lines represent contours with respect to adaptiveness.

combinations. The chance that a random combination is as adaptive as those characteristic of the species may be as low as 10^{xe100} and still leave room for 10^{800} separate peaks, each surrounded by 10^{100} more or less similar combinations. In a rugged field of this character, selection will easily carry the species to the nearest peak, but there may be innumerable other peaks which are higher but which are separated by "valleys." The problem of evolution as I see it is that of a mechanism by which the species may continually find its way from lower to higher peaks in such a field. In order that this may occur, there must be some trial and error mechanism on a grand scale by which the species may explore the region surrounding the small portion of the field which it occupies. To evolve, the species must not be under strict control of natural selection. Is there such a trial and error mechanism?

At this point let us consider briefly the situation with respect to a single locus. In each graph in figure 3 the abscissas represent a scale of gone frequency, 0 percent of the type genes to the left, 100 percent to the right. The elementary evolutionary process is, of course, change of gene frequency, a practically continuous process. Owing to the symmetry of the Mendelian mechanism, any gene frequency tends to remain constant in the absence of disturbing factors. If the type gene mutates at a certain rate, its frequency tends to move to the left, but at a continually decreasing rate. The type gene would ultimately be lost from the population if there were no opposing factor. But the type gene is in general favored by selection. Under selection, its frequency tends to move to the right. The rate is greatest at some point near the middle of the range. At a certain gene frequency

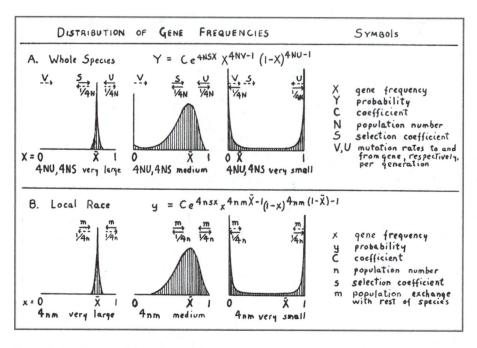

Figure 3: Random variability of a gene frequency under various specified conditions.

the opposing pressures are equal and opposite, and at this point there is consequently equilibrium. There are other mechanisms of equilibrium among evolutionary factors which need not be discussed here. Note that we have here a theory of the stability of species in spite of continuing mutation pressure, a continuing field of variability so extensive that no two individuals are ever genetically the same, and continuing selection.

If the population is not indefinitely large, another factor must be taken into account: the effects of accidents of sampling among those that survive and become parents in each generation and among the germ cells of these, in other words, the effects of inbreeding. Gene frequency in a given generation is in general a little different one way or the other from that in the preceding, merely by chance. In time, gene frequency may wander a long way from the position of equilibrium, although the farther it wanders the greater the pressure toward return. The result is a frequency distribution within which gene frequency moves at random. There is considerable spread even with very slight inbreeding and the form of distribution becomes U-shaped with close inbreeding. The rate of movement of gene frequency is very slow in the former case but is rapid in the latter (among unfixed genes). In this case, however, the tendency toward complete fixation of genes, practically irrespective of selection, leads in the end to extinction.

In a local race, subject to a small amount of crossbreeding with the rest of the species (figure 3, lower half), the tendency toward random fixation is balanced by immigration pressure instead of by mutation and selection. In a small sufficiently isolated group all gene frequencies can drift irregular back and forth about their mean values at a rapid rate, in terms of geologic time, without reaching fixation and giving the effects of close inbreeding. The resultant differentiation

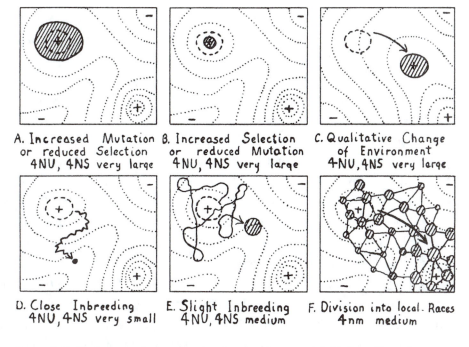

A. Increased Mutation or reduced Selection 4NU, 4NS very large

B. Increased Selection or reduced Mutation 4NU, 4NS very large

C. Qualitative Change of Environment 4NU, 4NS very large

D. Close Inbreeding 4NU, 4NS very small

E. Slight Inbreeding 4NU, 4NS medium

F. Division into local Races 4nm medium

Figure 4: Field of gene combinations occupied within the general field of possible combinations. Type of history under specified conditions indicated by relation to initial field (heavy broken contour) and arrow.

of races is of course increased by any local differences in the conditions of selection.

Let us return to the field of gene combinations (figure 4). In an indefinitely large but freely interbreeding species living under constant conditions, each gene will reach ultimately a certain equilibrium. The species will occupy a certain field of variation about a peak in our diagram (heavy broken contour in upper left of each figure). The field occupied remains constant although no two individuals are ever identical. Under the above conditions further evolution can occur only by the appearance of wholly new (instead of recurrent) mutations, and ones which happen to be favorable from the first. Such mutations would change the character of the field itself, increasing the elevation of the peak occupied by the species. Evolutionary progress through this mechanism is excessively slow since the chance of occurrence of such mutations is very small and, after occurrence, the time required for attainment of sufficient frequency to be subject to selection to an appreciable extent is enormous.

The general rate of mutation may conceivably increase for some reason. For example, certain authors have suggested an increased incidence of cosmic rays in this connection. The effect (figure 4A) will be as a rule a spreading of the field occupied by the species until a new equilibrium is reached. There will be an average lowering of the adaptive level of the species. On the other hand, there will be a speeding up of the process discussed above, elevation of the peak itself through appearance of novel favorable mutations. Another possibility of evolutionary advance is that the spreading of the field occupied may go so far as to include another and higher peak, in which case the species will move over and occupy the region about this. These mechanisms do not appear adequate to explain evolution to an important extent.

The effects of reduced mutation rate (figure 4B) are of course the opposite: a rise in average level, but reduced variability, less chance of novel favorable mutation, and less chance of capture of a neighboring peak.

The effect of increased severity of selection (also 4B) is, of course, to increase the average level of adaptation until a new equilibrium is reached. But again this is at the expense of the field of variation of the species and reduces the chance of capture of another adaptive peak. The only basis for continuing advance is the appearance of novel favorable mutations which are relatively rapidly utilized in this case. But at best the rate is extremely slow even in terms of geologic time, judging from the observed rates of mutation.

Relaxation of selection has of course the opposite effects and thus effects somewhat like those of increased mutation rate (figure 4A).

The environment, living and non-living, of any species is actually in continual change. In terms of our diagram this means that certain of the high places are gradually being depressed and certain of the low places are becoming higher (figure 4C). A species occupying a small field under influence of severe selection is likely to be left in a pit and become extinct, the victim of extreme specialization to conditions which have ceased, but if under sufficiently moderate selection to occupy a wide field, it will merely be kept continually on the move. Here we undoubtedly have an important evolutionary process and one which has been generally recognized. It consists largely of change without advance in adaptation. The mechanism is, however, one which shuffles the species about in the general field. Since the species will be shuffled out of low peaks more easily than high ones, it should gradually find its way to the higher general regions of the field as a whole.

Figure 4D illustrates the effect of reduction in size of population below a certain relation to the rate of mutation and severity of selection. There is fixation of one or another allelomorph in nearly every locus, largely irrespective of the direction favored by selection. The species moves down from its peak in an erratic fashion and comes to occupy a much smaller field. In other words there is the deterioration and homogeneity of a closely inbred population. After equilibrium has been reached in variability, movement becomes excessively slow, and, such as there is, is nonadaptive. The end can only be extinction. Extreme inbreeding is not a factor which is likely to give evolutionary advance.

With an intermediate relation between size of population and mutation rate, gene frequencies drift at random without reaching the complete fixation of close inbreeding (figure 4E). The species moves down from the extreme peak but continually wanders in the vicinity. There is some chance that it may encounter a gradient leading to another peak and shift its allegiance to this. Since it will escape relatively easily from low peaks as compared with high ones, there is here a trial and error mechanism by which in time the species may work its way to the highest peaks in the general field. The rate of progress, however, is extremely slow since change of gene frequency is of the order of the reciprocal of the effective population size and this reciprocal must be of the order of the mutation rate in order to meet the conditions for this case.

Finally (figure 4F), let us consider the case of a large species which is subdivided into many small local races, each breeding largely within itself but occasionally crossbreeding. The field of gene combinations occupied by each of these local races shifts continually in a nonadaptive fashion (except in so far as there are local differences in the conditions of selection). The rate of movement may be enormously greater than in the preceding case since the condition for such movement is that the reciprocal of the population number be of the order of the proportion of crossbreeding instead of the mutation rate. With many local races, each spreading over a considerable field and moving relatively rapidly in the more general field about the controlling peak, the chances are good that one at least will come under the influence of another peak. If a higher peak, this race will expand in numbers and by crossbreeding with the others will pull the whole species toward the new position. The average adaptiveness of the species thus advances under intergroup selection, an enormously more effective process than intragroup selection. The conclusion is that subdivision of a species into local races provides the most effective mechanism for trial and error in the field of gene combinations.

It need scarcely be pointed out that with such a mechanism complete isolation of a portion of a species should result relatively rapidly in specific differentiation, and one that is not necessarily adaptive. The effective intergroup competition leading to adaptive advance may be between species rather than races. Such isolation is doubtless usually geographic in character at the outset but may be clinched by the development of hybrid sterility. The usual difference of the chromosome complements of related species puts the importance of chromosome aberration as an evolutionary process beyond question, but, as I see it, this importance is not in the character differences which they bring (slight in balanced types), but rather in leading to the sterility of hybrids and thus making permanent the isolation of two groups.

How far do the observations of actual species and their subdivisions conform to this picture ? This is naturally too large a subject for more than a few suggestions.

That evolution involves nonadaptive differentiation to a large extent at the subspecies and even the species level is indicated by the kinds of differences by which such groups are actually distinguished by systematists. It is only at the subfamily and family levels that clear-cut adaptive differences become the rule (ROBSON, JACOT). The principal evolutionary mechanism in the origin of species must thus be an essentially nonadaptive one.

That natural species often are subdivided into numerous local races is indicated by many studies. The case of the human species is most familiar. Aside from the familiar racial differences recent studies indicate a distribution of frequencies relative to an apparently nonadaptive series of allelomorphs, that determining blood groups, of just the sort discussed above. I scarcely need to labor the point that changes in the average of mankind in the historic period have come about more by expansion of some types and decrease and absorption of others than by uniform evolutionary advance. During the recent period, no doubt, the phases of intergroup competition and crossbreeding have tended to overbalance the process of local differentiation, but it is probable that in the hundreds of thousands of years of prehistory, human evolution was determined by a balance between these factors.

Subdivision into numerous local races whose differences are largely non-adaptive has been recorded in other organisms wherever a sufficiently detailed study has been made. Among the land snails of the Hawaiian Islands, GULICK (sixty years ago) found that each mountain valley, often each grove of trees, had its own characteristic type, differing from others in "nonutilitarian" respects. GULICK attributed this differentiation to inbreeding. More recently CRAMPTON has found a similar situation in the land snails of Tahiti and has followed over a period of years evolutionary changes which seem to be of the type here discussed. I may also refer to the studies of fishes by DAVID STARR JORDAN, garter snakes by RUTHVEN, bird lice by KELLOGG, deer mice by OSGOOD, and gall wasps by KINSEY as others which indicate the role of local isolation as a differentiating factor. Many other cases are discussed by OSBORN and especially by RENSCH in recent summaries. Many of these authors insist on the nonadaptive character of most of the differences among local races. Others attribute all differences to the environment, but this seems to be more an expression of faith than a view based on tangible evidence.

An even more minute local differentiation has been revealed when the methods of statistical analysis have been applied. SCHMIDT demonstrated the existence of persistent mean differences at each collecting station in certain species of marine fish of the fjords of Denmark, and these differences were not related in any close way to the environment. That the differences were in part genetic was demonstrated in the laboratory. DAVID THOMPSON has found a correlation between water distance and degree of differentiation within certain fresh water species of fish of the streams of Illinois. SUMNER's extensive studies of subspecies of Peromyscus (deer mice) reveal genetic differentiations, often apparently nonadaptive, among local populations and demonstrate the genetic heterogeneity of each such group.

The modern breeds of livestock have come from selection among the products of local inbreeding and of crossbreeding between these, followed by renewed inbreeding, rather than from mass selection of species. The recent studies of the geographical distribution of particular genes in livestock and cultivated plants by SEREBROVSKY, PHILIPTSCHENKO and others are especially instructive with respect to the composition of such species.

The paleontologists present a picture which has been interpreted by some as irreconcilable with the Mendelian mechanism, but this seems to be due more to a failure to appreciate statistical consequences of this mechanism than to anything in the data. The horse has been the standard example of an orthogenetic evolutionary sequence preserved for us with an abundance of material. Yet MATHEW's interpretation as one in which evolution has proceeded by extensive differentiation of local races, intergroup selection, and crossbreeding is as close as possible to that required under the Mendelian theory.

Summing up: I have attempted to form a judgment as to the conditions for evolution based on the statistical consequences of Mendelian heredity. The most general conclusion is that evolution depends on a certain balance among its factors. There must be gene mutation, but an excessive rate gives an array of freaks, not evolution; there must be selection, but too severe a process destroys the field of variability, and thus the basis for further advance; prevalence of local inbreeding within a species has extremely important evolutionary consequences, but too close inbreeding leads merely to extinction. A certain amount of crossbreeding is favorable but not too much. In this dependence on balance the species is like a living organism. At all levels of organization life depends on the maintenance of a certain balance among its factors.

More specifically, trader biparental reproduction a very low rate of mutation balanced by moderate selection is enough to maintain a practically infinite field of possible gene combinations within the species. The field actually occupied is relatively small though sufficiently extensive that no two individuals have the same genetic constitution. The course of evolution through the general field is not controlled by direction of mutation and not directly by selection, except as conditions change, but by a trial and error mechanism consisting of a largely nonadaptive differentiation of local races (due to inbreeding balanced by occasional crossbreeding) and a determination of long time trend by intergroup selection. The splitting of species depends on the effects of more complete isolation, often made permanent by the accumulation of chromosome aberrations, usually of the balanced type. Studies of natural species indicate that the conditions for such an evolutionary process are often present.

Literature Cited

Crampton, H. E., 1925 Contemporaneous organic differentiation in the species of Partula living in Moorea, Society Islands. Amer. Nat. 59:5–35.

East, E. M., 1918 The role of reproduction in evolution. Amer. Nat. 52:273–289.

Gulick, J. T., 1905 Evolution, racial and habitudinal. Pub. Carnegie Instn. 25:1–269.

Jacot, a. P., 1932 The status of the species and the genus. Amer. Nat. 66:346–364.

Jordan, D. S., 1908 The law of geminate species. Amer. Nat. 42:73–80.

Kellogg, V. L., 1908 Darwinism, today. 403 pp. New York: Henry Holt and Co.

Kinsey, a. C., 1930 The gall wasp genus Cynips. Indiana Univ. Studies. 84–86:1–577.

Mathew, W. D., 1926 The evolution of the horse. A record and its interpretation. Quart. Rev. Biol. 1:139–185.

Osborn, H. F., 1927 The origin of species. V. Speciation and mutation. Amer. Nat. 49:193–239.

Osgood, W. H., 1909 Revision of the mice of the genus Peromyscus. North American Fauna 28:1–285.

Philiptschenko, J., 1927 Variabilit3t and Variation. 101 pp. Berlin.

Rensch, B., 1929 Das Prinzip geographischer Rassenkreise und das Problem der Artbildung. 206 pp. Berlin: Gebrüder Borntraeger.

Robson, G. C., 1928 The species problem. 283 pp. Edinburgh and London: Oliver and Boyd.

Ruthven, a. G., 1908 Variation and genetic relationships of the garter snakes. U. S. Nat. Mus. Bull. 61:1–301.

Schmidt, J., 1917 Statistical investigations with *Zoarces viviparus* L. J. Genet. 7:105–118.

Serebrovsky, a. S., 1929 Beitrag zur geographischen Genetic des Haushahns in Sowjet-Russland. Arch. f. Geflügelkunde Jahrgang 3: 161–169.

Sumner, F. B., 1932 Genetic, distributional, and evolutionary studies of the subspecies of deer mice (Peromyscus). Bibl. genet. 9:1–106.

Thompson, D. H., 1931 Variation in fishes as a function of distance. Trans. Il1. State Acad. of Sci. 23:276–281.

Wright, S., 1931 Evolution in Mendelian populations. Genetics 16:97–159.

Letter to G. F. Ferris

E. Mayr

March 29, 1948

Dr. G. F. Ferris
Natural History Museum
Stanford University
California

Dear Dr. Ferris,

I must apologize for not writing you earlier but there was some unexpected delay in getting your manuscript back from one of the readers.

Let me say right at the beginning that I read your paper with the greatest of pleasure. I think it does just what you intend, namely, stimulate the reader to think about a lot of things he normally takes for granted. Your paper is a fine example of a scientific stop-look-and-listen attitude in research and one of the unfortunately so rare cases of formulation of the problem before initiation of a research program (see later parts of this letter for a discussion of some of the points which you brought up). It pains me therefore to have to report to you that both readers of the Editorial Board say that the paper is not suitable for EVOLUTION in its present form. Reluctantly I have come to the conclusion that they may be right that the present paper is more a discussion of scientific methodology than one of the causes and factors of evolution. Much of your discussion appears to be particularly suitable as an introductory chapter of a book rather than as an article in a periodical.

However, in view of the obvious evolutionary importance of your studies on the morphology of the Annulata and the desirability to make them available to the readers of EVOLUTION, I was wondering whether it would not be possible for you to publish elsewhere that part of your manuscript that discusses scientific methods and to utilize the evolutionary working hypotheses of pages 15 and 16 as the introductory pages of one or two factual papers dealing with evolution in the Annulata. It has so far been the editorial policy of EVOLUTION to present concrete facts in every paper followed by the conclusions to be drawn from these facts. This policy was adopted deliberately because the prestige of evolutionary research has suffered in the past because of too much philosophy and speculation. A number of subjects from the domain of your present research have occurred to me which would be in harmony with our editorial policy. For example, it would be extremely interesting to have a paper on the single subject, The Origin of New Structures Among the Annulata. As you say with so much justification in Postulate Number 4, new structures originate extremely rarely. This, of course, is true in the vertebrates also. A paper on stable structures in an otherwise variable environment would also be very fascinating. This subject, incidentally, might permit a tie-in with certain of the problems of genetics. A third paper might be written on evolutionary trends in the morphology of the Annulata. A fourth one might be on parallelism, etc.

Your work appears so novel and important that I hope sincerely that the report of the readers of the Editorial Board does not discourage you because I am very much looking forward to receiving a manuscript from you along one of the above-mentioned lines.

Before closing, I would like to discuss one or two thoughts with you that came to my mind while reading your paper. I think you might say that there are four major evolutionary problems:

The fact of evolution
The material of evolution
The origin of discontinuities
The course of evolution.

As far as (1) is concerned, it requires no integration with genetics and is the proper field of phylogenetic paleontology, comparative embryology, and comparative anatomy. It is (2) that is the proper domain of genetics. It is (3) that is the principal domain of taxonomy since the origin of species is what you really meant by the origin of discontinuities. As far as (4) is concerned, it deals with the factors and causes of evolutionary change and it is here where the various fields concur and where the field of ecology enters the picture. Since there is so much agreement nowadays on (1), (2), and (3), the trend of evolutionary research has been very strongly toward (4). This is realized by the geneticists themselves, as most strongly expressed by Muller in some recent statements.

I was wondering as to the propriety of using the work "principle" as you have done. As Northrop says correctly, the first step of research is the formulation of a problem. The second step is the establishment of a working hypothesis while the final, ultimate answer is a Principle. I have a feeling that in a preliminary investigation one should not start with a set of principles. Rather, these principles should be the final conclusion of the investigation. However, this matter being one of terminology, no doubt everybody has his own subjective interpretation.

Let me conclude by saying how much I enjoyed reading your manuscript and I express the hope that you will be able to submit a paper along the lines suggested by me.

Very sincerely yours,
Ernst Mayr

Letter to John Greene

Theodosius Dobzhansky

New York, November 23, 1961

Dear Greene:

Let me assure you that I am grateful to you for your detailed reply, and that I also enjoy an opportunity to discuss matters of such vital concern to us with a man of your intellectual stature and your intellectual background. Even if we find ourselves in rather basic disagreement, let us drop the titles of "professors," and let us discuss things, even though this is very difficult in letters and we have not an opportunity to do so in any other way.

You say you do not understand where I stand. Let me remove all doubts about this. I am a Christian, hence I stand with my good friend Birch, and you, and Teilhard, and certainly not with Huxley, although his is very much a majority opinion among at least the natural scientists. I am not a Whiteheadian, and so differ from Birch who likes Whitehead, which to me is a rather meaningless philosophy.

It is hard to go much beyond these sweeping statements, but let me try. You and I will agree that the world is not a "devil's vaudeville" (Dostoyevsky's words), but is meaningful. Evolution (cosmic + biological + human) is going towards something, we hope some City of God. This belief is not imposed on us by our scientific discoveries, but if we wish (but not if we do not wish) we may see in nature manifestations of the Omega, or your creative ground (you spell this not with capitals?), or simply of God.

But let us face the problem of how the creative ground creates. I think it is here that your view is weak. You yourself say that if God was trying to make an Equus out of an Eohippus, then "He must be branded as a slow, clumsy, inefficient workman." Well, this is the crux of the whole matter. Nothing makes sense if you suppose that God makes special interventions to direct evolution. I refuse to believe in "direction" in any other sense than that the Alpha and the Omega[1] of

---

[1]These are Teilhard de Chardin's concepts.

evolution are simultaneously present in God's eyes (like to Laplace's universal intelligence). To direct evolution in any other sense, God must induce mutations, shuffle the nucleotides in DNA, and give from time to time little pushes to natural selection at critical moments. All this makes no sense to me (nor to Birch, Teilhard is unclear on this). I cannot believe that God becomes from time to time a particularly powerful enzyme.

I see no escape from thinking that God acts not in fits of miraculous interventions,—but in all significant and insignificant, spectacular and humdrum events. Panentheism, you may say? I do not think so, but if so then there is this much truth in panentheism. The really tough point is, of course, in what sense can God's action be seen in all that happens. I am not foolish enough to think that I can solve this. Perhaps Teilhard had a hint, very obscurely expressed. You are one of the few who could perhaps get on in trying to solve this. But look, if everything that happens is in some sense "trial and error," then it is merely redundant to say that "someone" is trying to accomplish something in particular in one place but not in another. But I refuse to abstain from talking about progress, improvement, and creativity. Why should I? Some extreme "scientists" would eliminate expressions such as that the eye is built so that the animal can see things. Perhaps extremes converge, and you would join these mechanistic purists? In evolution some organisms progressed and improved and stayed alive, others failed to do so and became extinct. Some adaptations are better than others—for the organisms having them; they are better for survival rather than for death. Yes, life is a value and a success, death is valueless and a failure. So, some evolutionary changes are better than others. Yes, life is trying hard to hang on and to produce more life. I see no need *at this point* to say that some creative ground is trying to get better adaptations and better values and more living substance. For one thing, this creative ground is only very slowly learning how to do it, and is even now clumsy and inefficient about it as you yourself admit! Your surprise at the fact that a process devoid of intelligence and will has produced beings which have intelligence and will is for me hard to understand, I cannot follow you into saying that this *proves* that this process itself is (and was?) *not* devoid of these things. Evolution, biological and especially human, is a process which generates novelties. Very remarkable indeed; but not a proof of the action of a creative ground shuffling genes. I feel really shocked—arguing with you I have to sound like a Huxley, something I do not like at all.

"The Death of Adam" showed clearly that to you evolution is something unwelcome though unavoidable. Your new book and your letter underscore this. To me, like to Birch and to Teilhard, evolution is a bright light. But it does not follow that evolution is a source of natural theology and a "proof" of the existence of God. I am driven to the view of such a conservative as [Karl] Barth which you quote on page 57 of your recent book (although otherwise I have no stomach for the neo-orthodoxy).[2] I am groping for a tolerable self-consistent Weltanschauung but do not claim having found one. Teilhard seemingly did, but perhaps not unexpectedly his book only hints at it and is unable to spell it out. Let me try to say this—evolution should be eventually understood *sub specie religionis* and religion *sub specie evolutionis*. But neither is deducible from the other. Both have to be somehow integrated in one's philosophy of "ultimate concern." I hope this is what you are struggling to do, but I submit that your attempt to view evolution as being actually propelled by a "creative ground is no more satisfactory than all finalistic theories of evolution (popularized or rather vulgarized by Lecomte du Noüy [1947]).

I fear the above sentences may be misunderstood—I do not doubt that at some level evolution, like everything in the world, is a manifestation of God's activity. All that I say is that *as a scientist* I do not observe anything that would prove this. In short, as scientists Laplace and myself "have no need of this hypothesis," but as a human being I do need this hypothesis! But I cannot follow your advice and put these things in water-tight compartments, and see only "change" and no "progress," only "change" and no "trial and error." For as a scientist I observe that evolution is on the whole progressive, its "creativeness" is increasing, and these findings I find fitting nicely into my general thinking, in which your "creative ground" is perfectly acceptable.

Well, this is a very long letter, and I fear still not succeeding to explain what I would like to explain, doubtless because my thinking is too hazy and unsatisfactory. But this is the best I can do in a letter. And I still hope that some day in a future not too remote we may find an opportunity to spend a good long evening or maybe a part of a night over glasses of wine if you use it or over cups of coffee if you don't, discussing these things and helping each other to see holes in our respective ideas.

With warmest personal regards,

Sincerely yours,

Th. Dobzhansky

P.S. I trust you would not object against sending your letter and mine to Charles Birch in Australia.

---

[2]" . . . it is not the existence of the world in its manifoldness, from which we are to read off the fact that God is its Creator. The world . . . gives us no information about God as the Creator.. . . But when God has been known and then known again in the world . . . , that is because He is to be sought and found by us in Jesus Christ (Barth 1959, 51. quoted in Greene 1961a, 57–58).

Chapter 7

Introduction

Now we come to the origin of life. The first document for this section is that paper in which J. B. S. Haldane puts forward his suggestion about life's origins. Haldane, I should explain, was the son of an eminent physiologist who was also something of a vitalist, J. S. Haldane. The son—Jack or J. B. S—defined himself with and against the father, on the one hand having imbibed science from birth but on the other hand being as outrageously reductionistic or materialistic (or whatever, as the mood suited him) as possible. After spending World War I happily killing Germans (typically, Haldane ended his life as a pacifistic, vegetarian Hindu), Haldane became not only a biochemist but also a brilliant population geneticist. At the same time he began writing for the general public (equally typically, the upper-class Haldane could be ferociously rude to underlings), and some of his most stimulating ideas appeared in more popular venues. This piece appeared in the *Rationalist Annual,* a place for nontechnical material by and for atheists and others.

It is clear from his piece, "The Origin of Life," that it was the virus that stimulated Haldane's imagination. He saw the virus as something halfway between being fully living and being totally inert. On the one hand, in the right conditions it can multiply at a terrific rate. On the other hand, the right conditions seem to be real, living organisms as its hosts and sources of food and energy generally. The virus therefore is the kind of bridge that Haldane thought must have come into existence at the beginning when life was first developing. Then, thanks to a lack of oxygen, primitive forms were able to survive and even reproduce, eventually leading to more sophisticated forms until one had full-blown organisms. The key therefore is a gradual development rather than a one-step spontaneous generation. At the end of his piece, Haldane could not forbear speculating on the origins of mind—a matter of great interest to his father. He comes to the somewhat double-edged conclusion that the biochemist's "ignorance disqualifies him no more than the historian or the geologist from attempting to solve a historical problem." I wonder what the older Haldane thought of that?

The second document is by the British chemist (who worked for many years in the USA) Leslie Orgel. He reviews the situation as it was known in the mid 1990s, showing how much more we know than did Haldane, but at the same time conceding that we do not yet know the full answer—indeed, we are probably still some way from this. The Creationists have had a field day with this article, pointing triumphantly to the fact that one of the leading origin of life researchers admits to great ignorance. Are they right in making this assumption? Orgel, who died recently, would certainly not have agreed. He never doubted that there is a scientific solution to be found, even if we have not yet succeeded. His attitude was that the success of science in the past when faced with difficult problems makes it reasonable to expect success in the future with today's difficult problems. Indeed, he would have agreed, that that is what makes science worthwhile. Not that the problems are easy, but that there are problems and we have the tools to tackle them. It is only a half century since we discovered the double helix. It is way too soon to be giving up yet!

The Origin of Life

J. B. S. Haldane

Until about 150 years ago it was generally believed that living beings were constantly arising out of dead matter. Maggots were supposed to be generated spontaneously in decaying meat. In 1668 Redi showed that this did not happen provided insects were carefully excluded. And in 1860 Pasteur extended the proof to the bacteria which he had shown were the cause of putrefaction. It seemed fairly clear that all the living beings known to us originate from other living beings. At the same time Darwin gave a new emotional interest to the problem. It had appeared unimportant that a few worms should originate from mud. But if man was descended from worms such spontaneous generation acquired a new significance. The origin of life on the Earth would have been as casual an affair as the evolution of monkeys into man. Even if the latter stages of man's history were due to natural causes, pride dung to a supernatural, or at least surprising, mode of origin for his ultimate ancestors. So it was with a sigh of relief that a good many men, whom Darwin's arguments had convinced, accepted the conclusion of Pasteur that life can originate only from life. It was possible either to suppose that life had been supernaturally created on Earth some millions of years ago, or that it had been brought to Earth by a meteorite or by microorganisms floating through interstellar space. But a large number, perhaps the majority, of biologists, believed, in spite of Pasteur, that at some time in the remote past life had originated on Earth from dead matter as the result of natural processes.

The more ardent materialists tried to fill in the details of this process, but without complete success. Oddly enough, the few scientific men who professed idealism agreed with them. For if one can find evidences of mind (in religious terminology the finger of God) in the most ordinary events, even those which go on in the chemical laboratory, one can without much difficulty believe in the origin of life from such processes. Pasteur's work therefore appealed most strongly to those who desired to stress the contrast between mind and matter. For a variety of obscure historical reasons, the Christian Churches have taken this latter point of view. But it should never be forgotten that the early Christians held many views which are now regarded as materialistic. They believed in the resurrection of the body, not the immortality of the soul. St Paul seems to have attributed consciousness and will to the body. He used a phrase translated in the revised version as 'the mind of the flesh', and credited the flesh with a capacity for hatred, wrath and other mental functions. Many modern physiologists hold similar beliefs. But, perhaps unfortunately for Christianity, the Church was captured by a group of very inferior Greek philosophers in the third and fourth centuries AD. Since that date views as to the relation between mind and body which St Paul, at least, did not hold, have been regarded as part of Christianity, and have retarded the progress of science.

It is hard to believe that any lapse of time will dim the glory of Pasteur's positive achievements. He published singularly few experimental results. It has even been suggested by a cynic that his entire work would not gain a Doctorate of Philosophy today! But every experiment was final. I have never heard of any one who has repeated any experiment of Pasteur's with a result different from that of the master. Yet his deductions from these experiments were sometimes too sweeping. It is perhaps not quite irrelevant that he worked in his latter years with half a brain. His right cerebral hemisphere had been extensively wrecked by the bursting of an artery when he was only forty-five years old; and the united brain-power of the microbiologists who succeeded him has barely compensated for that accident. Even during his lifetime some of the conclusions which he had drawn from his experimental work were disproved. He had said that alcoholic fermentation was impossible without life. Buchner obtained it with a cell-free and dead extract of yeast. And since his death the gap between life and matter has been greatly narrowed

When Darwin deduced the animal origin of man, a search began for a 'missing link' between ourselves and the apes. When Dubois found the bones of Pithecanthropus some comparative anatomists at once proclaimed that they were of animal origin, while others were equally convinced that they were parts of a human skeleton. It is now generally recognized that either party was right, according to the definition of humanity adopted. Pithecanthropus was a creature which might legitimately be described either as a man or an ape, and its existence showed that the distinction between the two was not absolute.

Now the recent study of ultramicroscopic beings has brought up at least one parallel case, that of the bacteriophage, discovered by d'Herelle, who had been to some extent anticipated by Twort. This is the case of a disease, or, at any rate, abnormality of bacteria. Before the size of the atom was known there was no reason to doubt that

Big fleas have little fleas
Upon their backs to bite 'em;
The little ones have lesser ones,
And so ad infinitum.

But we now know that this is impossible. Roughly speaking, from the point of view of size, the bacillus is the flea's flea, the bacteriophage of the bacillus' flea; but the bacteriophage's flea would be of the dimensions of an atom, and atoms do not behave like fleas. In other words, there are only about as many atoms in a cell as cells in a man. The link between living and dead matter is therefore somewhere between a cell and an atom.

D'Herelle found that certain cultures of bacteria began to swell up and burst until all had disappeared. If such cultures were passed through a filter fine enough to keep out all bacteria, the filtrate could infect fresh bacteria and so on indefinitely. Though the infective agents cannot be seen with a microscope, they *can* be counted as follows. If an active filtrate containing bacteriophage be poured over a colony of bacteria on a jelly, the bacteria will all, or almost all, disappear. If it be diluted many thousand times, a few islands of living bacteria survive for some time. If it be diluted about ten million fold, the bacteria are destroyed round only a few isolated spots, each representing a single particle of bacteriophage.

Since the bacteriophage multiplies, d'Herelle believes it to be a living organism. Bordet and others have taken an opposite view. It will survive heating and other insults which kill the large majority of organisms, and will multiply only in presence of living bacteria, though it can break up dead ones. Except perhaps in presence of bacteria, it does not use oxygen or display any other signs of life. Bordet and his school therefore regard it as a ferment which breaks up bacteria as our own digestive ferments break up our food, at the same time inducing the disintegrating bacteria to produce more of the same ferment. This is not as fantastic as it sounds, for most cells while dying liberate or activate ferments which digest themselves. But these ferments are certainly feeble when compared with the bacteriophage.

Clearly we are in doubt as to the proper criterion of life. D'Herelle says that the bacteriophage is alive, because, like the flea or the tiger, it can multiply indefinitely at the cost of living beings. His opponents say that it can multiply only as long as its food is alive, whereas the tiger certainly, and the flea probably, can live on dead products of life. They suggest that the bacteriophage is like a book or a work of art, which is constantly being copied by living beings, and is therefore only metaphorically alive, its real life being in its copiers.

The American geneticist Muller has, however, suggested an intermediate view. He compares the bacteriophage to a gene—that is to say, one of the units concerned in heredity. A fully coloured and a sported dog differ because the latter has in each of its cells one or two of a certain gene, which we know is too small for the microscope to see. Before a cell of a dog divides this gene divides also, so that each of the daughter-cells has one, two, or none according with the number in the parent cell. The ordinary spotted dog is healthy, but a gene common among German dogs causes a roan colour when one is present, while two make the dog nearly white, wall-eyed and generally deaf, blind or both. Most of such dogs die young, and the analogy to the bacteriophage is fairly close. The main difference between such a lethal gene, of which many are known, and the bacteriophage, is that the one is only known inside the cell, the other outside. In the present state of our ignorance we may regard the gene either as a tiny organism which can divide in the environment provided by the rest of the cell; or as a bit of machinery which the 'living' cell copies at each division. The truth is probably somewhere in between these two hypotheses.

Unless a living creature is a piece of dead matter plus a soul (a view which finds little support in modern biology) something of the following kind must be true. A simple organism must consist of parts A, B, C, D and so on, each of which can multiply only in presence of all, or almost all, of the others. Among these parts are genes, and the bacteriophage is such a part which has got loose. This hypothesis becomes more plausible if we believe in the work of Hauduroy, who finds that the ultramicroscopic particles into which the bacteria have been broken up, and which pass through filters that can stop the bacteria, occasionally grow up again into bacteria after a lapse of several months. He brings evidence to show that such fragments of bacteria may cause disease, and d'Herelle and Peyre claim to have found the ultramicroscopic form of a common staphylococcus, along with bacteriophage, in cancers, and suspects that this combination may be the cause of that disease.

On this view the bacteriophage is a cog, as it were, in the wheel of a life-cycle of many bacteria. The same bacteriophage can act on different species and is thus, so to say, a spare part which can be fitted into a number of different machines, just as a human diabetic can remain in health when provided with insulin manufactured by a pig. A great many kinds of molecule have been got from cells, and many of them are very efficient when removed from it. One can separate from yeast one of the many tools which it uses in alcoholic fermentation, an enzyme called invertase, and this will break up six times its weight of cane-sugar per second for an indefinite time without wearing out. As it does not form alcohol from the sugar, but only a sticky mixture of other sugars, its use is permitted in the United States in the manufacture of confectionery and cake-icing. But such fragments do not reproduce themselves, though they take part in the assimilation of food by the living cell. No one supposes that they are alive. The bacteriophage is a step beyond the enzyme on the road to life, but it is perhaps an exaggeration to call it fully alive. At about the same stage on the road are the viruses which cause such diseases as smallpox, herpes, and hydrophobia. They can multiply only in living tissue, and pass through filters which stop bacteria.

With these facts in mind we may, I think, legitimately speculate on the origin of life on this planet. Within a few thousand years from its origin it probably cooled down so far as to develop a fairly permanent solid crust. For a long time, however, this crust must have been above the boiling-point of water, which condensed only gradually. The primitive atmosphere probably contained little or no oxygen, for our present supply of that gas is only about enough to burn all the coal and other organic remains found below and on the Earth's surface. On the other hand, almost all the carbon of these organic substances, and much of the carbon now combined in chalk, lime stone, and dolomite, were in the atmosphere as carbon dioxide. Probably a good deal of the nitrogen now in the air was combined with metals as nitride in the Earth's crust, so that ammonia was constantly being formed by the action of water. The Sun was perhaps slightly brighter than it is now, and as there was no oxygen in the atmosphere the chemically active ultra-violet rays from the Sun were not, as they now are, mainly stopped by ozone (a modified form of oxygen) in the upper atmosphere, and oxygen itself lower down. They penetrated to the surface of the land and sea, or at least to the clouds.

Now, when ultra-violet light acts on a mixture of water, carbon dioxide, and ammonia, a vast variety of organic substances are made, including sugars and apparently some of the materials from which proteins are built up. This fact has been demonstrated in the laboratory by Baly of Liverpool and his colleagues. In this present world, such substances, if left about, decay—that is to say, they are destroyed by micro-organisms. But before the origin of life they must have accumulated till the primitive oceans reached the consistency of hot dilute soup. To-day an organism must trust to luck, skill, or strength to obtain its food. The first precursors of life found food available in considerable quantifies, and had no competitors in the struggle for existence. As the primitive atmosphere contained little or no oxygen, they must have obtained the energy which they needed for growth by some other process than oxidation—in fact, by fermentation, For, as Pasteur put it, fermentation is life without oxygen. If this was so, we should expect that high organisms like ourselves would start life as anaerobic beings, just as we start as single cells. This is the case. Embryo chicks for the first two or three days after fertilization use very little oxygen, but obtain the energy which they need for growth by fermenting sugar into lactic acid, like the bacteria which turns milk sour. So do various embryo mammals, and in all probability you and I lived mainly by fermentation during the first week of our pre-natal life. The cancer cell behaves in the same way. Warburg has shown that with its embryonic habit of unrestricted growth there goes an embryonic habit of fermentation.

The first living or half-living things were probably large molecules synthesized under the influence of the Sun's radiation, and only capable of reproduction in the particularly favourable medium in which they originated. Each presumably required a variety of highly specialized molecules before it could reproduce itself, and it depended on chance for a supply of them. This is the case today with most viruses, including the bacteriophage, which can grow only in presence of the complicated assortment of molecules found in a living cell.

The unicellular organisms, including bacteria, which were the simplest living things known a generation ago, are far more complicated. They are organisms— that is to say, systems whose parts co-operate. Each part is specialized to a particular chemical function, and prepares chemical molecules suitable for growth of the other parts. In consequence, the cell as a whole can usually subsist on a few types of molecule, which are transformed within it into the more complex substances needed for the growth of the parts.

The cell consists of numerous half-living chemical molecules suspended in water and enclosed in an oily film. When the whole sea was a vast chemical laboratory the conditions for the formation of such films must have been relatively favourable; but for all that life may have remained in the virus stage for many millions of years before a suitable assemblage of elementary units was brought together in the first cell. There must have been many failures, but the first successful cell had plenty of food, and an immense advantage over its competitors.

It is probable that all organisms now alive are descended from one ancestor, for the following reason. Most of our structural molecules are asymmetrical, as shown by the fact that they rotate the plane of polarized light, and often form asymmetrical crystals. But of the two possible types of any such molecule, related to one another like a right and left boot, only one is found throughout living nature. The apparent exceptions to this rule are all small molecules which are not used in the building of the large structures which display the phenomena of life. There is nothing, so far as we can see in the nature of things to prevent the existence of looking-glass organisms built from molecules which are, so to say, the mirror-images of those in our own bodies. Many of the requisite molecules have already been made in the laboratory. If life had originated independently on several occasions, such organisms would probably exist. As they do not, this event probably occurred only once, or, more probably, the descendants of the first living organism rapidly evolved far enough to overwhelm any later competitors when these arrived on the scene.

As the primitive organisms used up the foodstuffs available in the sea some of them began to perform in their own bodies the synthesis formerly performed haphazardly by the sunlight, thus ensuring a liberal supply of food. The first plants thus came into existence, living near the surface of the ocean, and making food with the aid of sunlight as do their descendants today. It is thought by many biologists that we animals are descended from them. Among the molecules in our own bodies are a number whose structure resembles that of chlorophyll, the green pigment with which the plants have harnessed the sunlight to their needs. We use them for other purposes than the plants—for example, for carrying oxygen—and we do not, of course, know whether they are, so to speak, descendants of chlorophyll or merely cousins. But since the oxygen liberated by the first plants must have killed off most of the other organisms, the former view is the more plausible.

The above conclusions are speculative. They will remain so until living creatures have been synthesized in the biochemical laboratory. We are a long way from that goal. It was only this year that Pictel for the first time made cane-sugar artificially. It is doubtful whether any enzyme has been obtained quite pure. Nevertheless I hope to live to see one made artificially. I do not think I shall behold the synthesis of anything so nearly alive as a bacteriophage or a virus, and I do not suppose that a self-contained organism will be made for centuries. Until that is done the origin of life will remain a subject for speculation. But such speculation is not idle, because it is susceptible of experimental proof or disproof.

Some people will consider it a sufficient refutation of the above theories to say that they are materialistic, and that materialism can be refuted on philosophical grounds. They are no doubt compatible with materialism, but also with other philosophical tenets. The facts are, after all, fairly plain. Just as we know of sight only in connection with a particular kind of material system called the eye, so we know only of life in connection with certain arrangements of matter, of which the biochemist can give a good, but far from complete, account. The question at issue is: 'How did the first such system on this planet originate ?' This is a historical problem to which I have given a very tentative answer on the not unreasonable hypothesis that a thousand million years ago matter obeyed the same laws that it does today.

This answer is compatible, for example, with the view that pre-existent mind or spirit can associate itself with certain kinds of matter. If so, we are left with the mystery as to why mind has so marked a preference for a particular type of colloidal organic substances. Personally I regard all attempts to describe the relation of mind to matter as rather clumsy metaphors. The biochemist knows no more, and no less, about this question than anyone else. His ignorance disqualifies him no more than the historian or the geologist from attempting to solve a historical problem.

The Origin of Life—A Review of Facts and Speculations

Leslie E. Orgel

Three popular hypotheses attempt to explain the origin of prebiotic molecules: synthesis in a reducing atmosphere, input in meteorites and synthesis on metal sulfides in deep-sea vents. It is not possible to decide which is correct. It is also unclear whether the RNA world was the first biological world or whether some simpler world preceded it.

The problem of the origin of life on the earth has much in common with a well-constructed detective story. There is no shortage of clues pointing to the way in which the crime, the contamination of the pristine environment of the early earth, was committed. On the contrary, there are far too many clues and far too many suspects. It would be hard to find two investigators who agree on even the broad outline of the events that occurred so long ago and made possible the subsequent evolution of life in all its variety. Here, I outline two of the main questions and some of the conflicting evidence that has been used in attempts to answer them. First, however, I summarize the few areas where there is fairly general agreement.

L. E. Orgel is at the Salk Institute for Biological Studies, 10010 N. Torrey Pines Road, La Jolla, CA 92037, USA. Email: orgel@salk.edu

The earth is slightly more than 4.5 billion years old. For the first half billion years or so after its formation, it was impacted by objects large enough to evaporate the oceans and sterilize the surface[1,2]. Well-preserved microfossils of organisms that have morphologies similar to those of modern blue-green algae, and date back about 3.5 billion years, have been found[3], and indirect but persuasive evidence supports the proposal that life was present 3.8 billion years ago[4]. Life, therefore, originated on or was transported to the earth at some point within a window of a few hundred million years that opened about four billion years ago. The majority of workers in the field reject the hypothesis that life was transported to the earth from somewhere else in the galaxy and take it for granted that life began *de novo* on the early earth.

The uniformity of biochemistry in all living organisms argues strongly that all modern organisms descend from a last-common ancestor (LCA). Detailed analysis of protein sequences suggests that the LCA had a complexity comparable to that of a simple modern bacterium and lived 3.2–3.8 billion years ago[5]. If we knew the stages by which the LCA evolved from abiotic components present on the primitive earth, we would have a complete account of the origin of life. In practice, the most ambitious studies of the origins of life address much simpler questions. Here, I discuss two of them. What were the sources of the small organic molecules that made up the first self-replicating systems? How did biological organization evolve from an abiotic supply of small organic molecules?

Abiotic synthesis of small organic molecules.

Miller, a graduate student who was working with Harold Urey, began the modern era in the study of the origin of life at a time when most people believed that the atmosphere of the early earth was strongly reducing. Miller[6] subjected a mixture of methane, ammonia and hydrogen to an electric discharge and led the products into liquid water. He showed that a substantial percentage of the carbon in the gas mixture was incorporated into a relatively small group of simple organic molecules and that several of the naturally occurring amino acids were prominent among these products. This was a surprising result; organic chemists would have expected a much-less- tractable product mixture. The Urey-Miller experiments were widely accepted as a model of prebiotic synthesis of amino acids by the action of lightning.

Miller and his co-workers went on to study electric-discharge synthesis of amino acids in greater detail[7,8]. Using more-powerful analytical techniques, they identified many more amino acids—some, but by no means all, of which occur in living organisms. They also showed that a major synthetic route to the amino acids is through the Strecker reaction—that is, from aldehydes, hydrogen cyanide and ammonia. Glycine, for example, is formed from formaldehyde, cyanide and ammonia—all of which can be detected among the products formed in the electric-discharge reaction.

In the years following the Urey-Miller experiments, the synthesis of biologically interesting molecules from products that could be obtained from a reducing gas mixture became the principle aim of prebiotic chemistry (Fig. 1). Remarkably, Oro and Kimble[9] were able to synthesize adenine from hydrogen cyanide and ammonia. Somewhat later, Sanchez, Ferris and I.[10] showed that cyanoacetylene is a major product of the action of an electric discharge on a mixture of methane and nitrogen and that cyanoacetylene is a plausible source of the pyrimidine bases uracil and cytosine. This new information, together with previous studies that showed that sugars are formed readily from formaldehyde.[11,12], convinced many students of the origins of life that they understood the first stage in the appearance of life on the earth: the formation of a prebiotic soup of biomonomers.

As in any good detective story, however, the principle suspect, the reducing atmosphere, has an alibi. Recent studies have convinced most workers concerned with the atmosphere of the early earth that it could never have been strongly reducing.[13] If this is true, Miller's experiments, and most other early studies of prebiotic chemistry, are irrelevant. I believe that the dismissal of the reducing atmosphere is premature, because we do not completely understand the early history of the earth's atmosphere. It is hard to believe that the ease with which sugars, amino acids, purines and pyrimidines are formed under reducing-atmosphere conditions is either a coincidence or a false clue planted by a malicious creator.

Many of those who dismiss the possibility of a reducing atmosphere believe that the crime was an outside job. A substantial proportion of the meteorites that fall on the earth belong to a class known as carbonaceous chondrites.[14] These are particularly interesting because they contain a significant amount of organic carbon and because some of the standard amino acids and nucleic-acid bases are present[8]. Could the prebiotic soup have originated in preformed organic material brought to the earth by meteorites and comets?

Supporters of the impact theory have argued convincingly that sufficient organic carbon must have been present in the meteorites and comets that reached the surface of the early earth to have stocked an abundant soup. However, would this material have survived the intense heating that accompanies the entry of large bodies into the atmosphere and their subsequent collisions with the surface of the earth? The results of theoretical calculations depend strongly on assumptions made about the composition and density of the atmosphere, the distribution of sizes of the impacting objects, etc.[15] The impact theory is probably the most popular at present, but nobody has proved that impacts were the most important sources of prebiotic organic compounds.

The newest suspects are the deep-sea vents, submarine cracks in the earth's surface where superheated water rich in transition-metal ions and hydrogen sulfide mixes abruptly with cold sea water. These vents are sites of abundant biological activity, much of it independent of solar energy. Wächtershäuser.[16,17] has proposed a scenario for the origin of life that might fit such an environment. He hypothesizes that the reaction between iron(II) sulfide and hydrogen sulfide [a reaction that yields pyrites (FeS^2) and hydrogen] could provide the free energy neces-

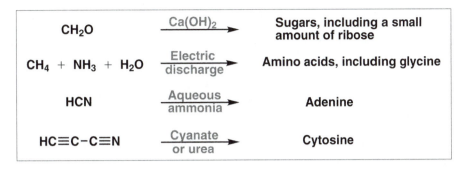

Figure 1: Early prebiotic syntheses of biomonomers.

sary for reduction of carbon dioxide to molecules capable of supporting the origin of life. He asserts that life originated on the surface of iron sulfides as a result of such chemistry. The assumptions that complex metabolic cycles self-organize on the surface and that the significant products never escape from the surface are essential parts of this theory; in Wächtershäuser's opinion, there never was a prebiotic soup!

Stetter and colleagues[18] have confirmed the novel suggestion that hydrogen sulfide, in the presence of iron(II) sulfide, acts as a reducing agent. They have reduced, for example, acetylene to ethane, and mercaptoacetic acid to acetic acid, but they have not reported reduction of CO^2. However, in a new study, Wächtershäuser and co-workers[19] have shown that FeS spiked with NiS reduces carbon monoxide. Given that carbon monoxide might well have been present in large amounts in the gases escaping from the vents, Wächtershäuser's findings could well prove important. If metal sulfides can be shown to catalyze the synthesis of a sufficient variety of organic molecules from carbon monoxide, the vent theory of the origins of biomonomers will become very attractive.

In summary, there are three main contending theories of the prebiotic origin of biomonomers (not to mention several other less-popular options). No theory is compelling, and none can be rejected out of hand. Perhaps it is time for a conspiracy theory; more than one of the sources of organic molecules discussed above may have collaborated to make possible the origin of life.

Self-organization

There is no general agreement about the source of prebiotic organic molecules on the early earth, but there are several plausible theories, each backed by some experimental data. The situation with regard to the evolution of a self-replicating system is less satisfactory; there are at least as many suspects, but there are virtually no experimental data.

The fairly general acceptance of the hypothesis that there was once an RNA world (i.e. a self-contained biological world in which RNA molecules functioned both as genetic materials and as enzyme-like catalysts) has changed the direction of research into the origins of life[20]. The central puzzle is now seen to be the origin of the RNA world. Two specific, but intertwined, questions are central to the debate. Was RNA the first genetic material or was it preceded by one or more simpler genetic materials? How much self-organization of reaction sequences is possible in the absence of a genetic material? I shall concentrate on the first question.

The assumption that a polymer that doubled as a genetic material and as a source of enzyme-like catalytic activity once existed profoundly changes the goals of prebiotic synthesis. The central issue becomes the synthesis of the first genetic monomers: nucleotides or whatever preceded them. The synthesis of amino acids, coenzymes, etc. becomes a side issue, because there is no reason to believe that they were ever synthesized abiotically; some or all of them might have been introduced as direct or indirect consequences of the enzyme-like activities of RNA or its precursor(s). Supporters of the hypothesis that RNA was the first genetic material must explain where the nucleotides came from and how they self-organized. Those who believe in a simpler precursor have the difficult task of identifying such a precursor, but they hope that explaining monomer synthesis will then be simpler.

Returning to the idiom of the detective story, accumulating evidence suggests that RNA, a prime suspect, could have completed the difficult task of organizing itself into a self-contained replicating system. It has proved possible to isolate sequences that catalyze a wide variety of organic reactions from pools of random RNA[21,22]. As regards the origin of the RNA world, the most important reactions are those in which a preformed template-RNA strand catalyzes the synthesis of its complement from monomers or short oligomers. Eklund and co-workers[23] have isolated catalysts for the ligation of short oligonucleotides surprisingly easily, and the catalysts carry out ligation with adequate specificity. These molecules are the RNA equivalents of the RNA and DNA ligases. Considerable progress has also been made in selecting RNA equivalents of RNA polymerases[24].

If the RNA world evolved *de novo*, it must have depended initially on an abiotic source of activated nucleotides. However, oxidation–reduction, methylation, oligosaccharide synthesis, etc., supported by nucleotide-containing coenzyme, probably became part of the chemistry of the RNA world before the invention of protein synthesis. Unfortunately, we cannot say just how complex the RNA world could have been until we know more about the range of reactions that can be catalyzed by ribozymes. It seems likely that RNA could have catalyzed most of the steps involved in the synthesis of nucleotides[25], and possibly the coupling of redox reactions to the synthesis of phosphodiesters and peptides, but this remains to be demonstrated experimentally.

The experiments on the selection of ribozymes that catalyze nucleic acid replication (discussed above) use as inputs pools of RNA molecules synthesized by enzymes. Recently, Ferris and coworkers[26,27] have made considerable progress in the assembly of RNA oligomers from monomers, using an abundant clay mineral, montmorillonite, as a catalyst. The substrates that they use, nucleoside 59-phosphorimidazolides, were probably not prebiotic molecules, but the experiments do indicate that the use of minerals as adsorbents and catalysts could allow the accumulation of long oligonucleotides once suitable activated monomers are available.

We have shown that, using activated monomers, non-enzymatic copying of a wide range of oligonucleotide sequences is possible[28] and have obtained similar, but less extensive, results for ligation of short oligomers.

An optimist could propose the following scenario. First, activated mononucleotides oligomerize on montmorillonite or an equivalent mineral. Next, copying of longer templates, using monomers or short oligomers as substrates, leads to the accumulation of a library of dsRNA molecules. Finally, an RNA double helix, one of whose strands has generalized RNA-polymerase activity, dissociates; the polymerase strand copies its complement to produce a second polymerase molecule, which copies the first to produce a second complement—and so on. The RNA world could therefore have arisen from a pool of activated nucleotides[29]. All that would have been needed is a pool of activated nucleotides!

Nucleotides are complicated molecules. The synthesis of sugars from formaldehyde gives a complex mixture, in which ribose is always a minor component. The formation of a nucleoside from a base and a sugar is not an easy reaction and, at least for pyrimidine nucleosides, has not been achieved under prebiotic conditions; the phosphorylation of nucleosides tends to give a complex mixture of products[30]. The inhibition of the template-directed reactions on Dtemplates by L-substrates is a further difficulty[31]. It is almost inconceivable that nucleic acid replication could have got started, unless there is a much simpler mechanism for the prebiotic synthesis of nucleotides. Eschenmoser and his colleagues[32] have had considerable success in generating ribose 2,4- diphosphate in a potentially prebiotic reaction from glycolaldehyde monophosphate and formaldehyde. Direct prebiotic synthesis of nucleotides by novel chemistry is therefore not hopeless. Nonetheless, it is more likely that some organized form of chemistry preceded the RNA world. This leads us to a discussion of genetic takeover.

Cairns-Smith[33], long before the argument became popular, emphasized how improbable it is that a molecule as high tech as RNA could have appeared *de novo* on the primitive earth. He proposed that the first form of life was a self-replicating clay. He suggested that the synthesis of organic molecules became part of the competitive strategy of the clay world and that the inorganic genome was taken over by one of its organic creations. Cairns-Smith's postulate of an inorganic life form has failed to gather any experimental support. The idea lives on in the limbo of uninvestigated hypotheses. However, Cairns-Smith also contemplated the possibility that RNA was preceded by one or more linear organic genomes[34]. This idea has taken root, but its implications have not always been appreciated.

If RNA was not the first genetic material, biochemistry might provide no clues to the origins of life. Presumably, the biological world that immediately preceded the RNA world already had the capacity to synthesize nucleotides. This should help us to formulate hypotheses about its chemical characteristics. However, if there were two or more worlds before the RNA world, the original chemistry might have left no trace in contemporary biochemistry. In that case, the chemistry of the origins of life is unlikely to be discovered without investigating in detail all the chemistry that might have occurred on the primitive earth—whether

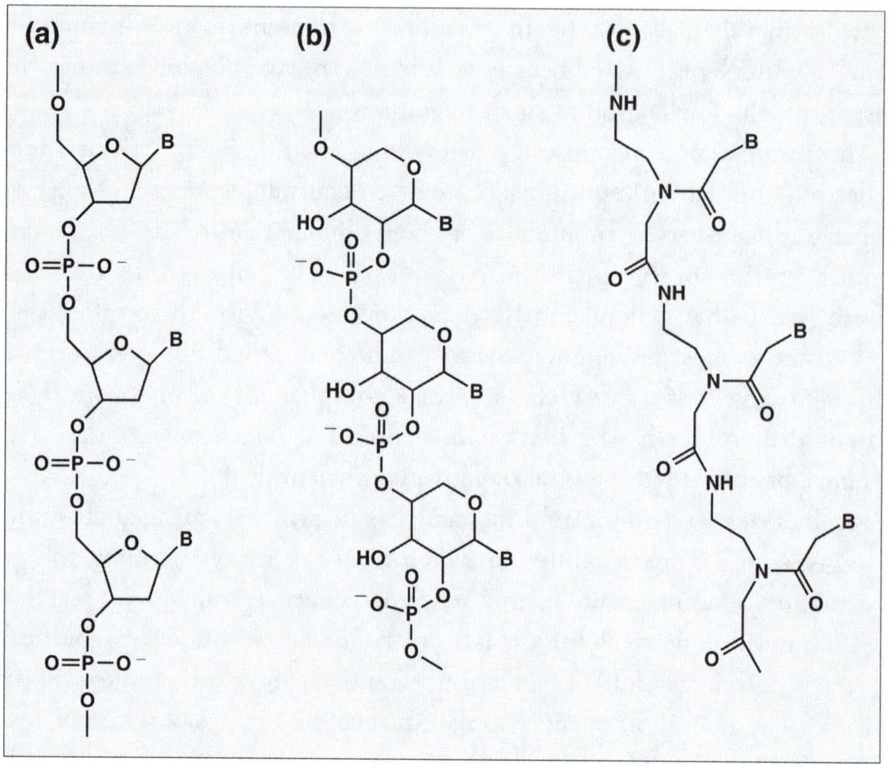

Figure 2: DNA and potentially informational oligonucleotide analogs. **(a)** DNA. **(b)** Pyranosyl analog of RNA. **(c)** Peptide nucleic acid.

or not that chemistry has any relation to biochemistry. This gloomy prospect has not prevented discussion of alternative genetic systems.

The only potentially informational systems, other than nucleic acids, that have been discovered are closely related to nucleic acids. Eschenmoser and his colleagues[35] have undertaken a systematic study of the properties of nucleic acid analogs in which ribose is replaced by another sugar or in which the furanose form of ribose is replaced by the pyranose form (Fig. 2b). Strikingly, polynucleotides based on the pyranosyl isomer of ribose (p-RNA) form Watson– Crick-paired double helices that are more stable than RNA, and p-RNAs are less likely than the corresponding RNAs to form multiple-strand competing structures[35]. Further- more, the helices twist much more gradually than those in the standard nucleic acids, which should make it easier to separate strands during replication. Pyrano- syl RNA seems to be an excellent choice as a genetic system; in some ways, it might be an improvement on the standard nucleic acids. However, prebiotic syn- thesis of pyranosyl nucleotides is not likely to prove much easier than synthesis of the standard isomers, although a route through ribose 2,4-diphosphate is being ex- plored by Eschenmoser and his colleagues.

Peptide nucleic acid (PNA) is another nucleic acid analog that has been stud- ied extensively (Fig. 2c). It was synthesized by Nielsen and colleagues[36] during work on antisense RNA. PNA is an uncharged, achiral analog of RNA or DNA; the ribose-phosphate backbone of the nucleic acid is replaced by a backbone held together by amide bonds. PNA forms very stable double helices with complementary RNA or DNA[36,37] We have shown that information can be transferred from PNA to RNA, and vice versa, in template-directed reactions[38,39] and that PNA–DNA chimeras form readily on either DNA or PNA templates.[40] Thus, a transition from a PNA world to an RNA world is possible. Nonetheless, I think it unlikely that PNA was ever important on the early earth, because PNA monomers cyclize when they are activated; this would make oligomer formation very difficult under prebiotic conditions.

The studies described above suggest that there are many ways of linking together nucleotide bases into chains that can form Watson–Crick double helices. Perhaps a structure of this kind will be discovered that can be synthesized easily under prebiotic conditions. If so, it would be a strong candidate for the very first genetic material. However, another possibility remains to be explored: the first genetic material might not have involved nucleoside bases. Two or more very simple molecules could have the pairing properties needed to form a genetic polymer—a positively charged and a negatively charged amino acid, for example. However, it is not clear that stable structures of this kind exist. RNA is clearly adapted to double-helix formation: its constrained backbone permits simultaneous base pairing and stacking. It is unlikely that much simpler molecules could substitute for the nucleotides. Perhaps some other interaction between the chains can stabilize a double helix in the absence of base stacking; binding to a mineral surface might supply the necessary constraints, but this remains to be demonstrated. In the absence of experimental evidence, little useful can be said.

The above discussion reveals a very large gap between the complexity of molecules that are readily synthesized in simulations of the chemistry of the early earth and the molecules that are known to form potentially replicating informational structures. Several authors have therefore proposed that metabolism came before genetics[41–43]. They have suggested that substantial organization of reaction sequences can occur in the absence of a genetic polymer and, hence, that the first genetic polymer probably appeared in an already-specialized biochemical environment. Because it is hard to envisage a chemical cycle that produces B-D-nucleotides, this theory would fit best if a simpler genetic system preceded RNA.

There is no agreement on the extent to which metabolism could develop independently of a genetic material. In my opinion, there is no basis in known chemistry for the belief that long sequences of reactions can organize spontaneously—and every reason to believe that they cannot. The problem of achieving sufficient specificity, whether in aqueous solution or on the surface of a mineral, is so severe that the chance of closing a cycle of reactions as complex as the reverse citric acid cycle, for example, is negligible. The same, I believe, is true for simpler cycles involving small molecules that might be relevant to the origins of life and also for peptide-based cycles.

Conclusion/outlook

In summary, there are several tenable theories about the origin of organic material on the primitive earth, but in no case is the supporting evidence compelling. Similarly, several alternative scenarios might account for the self-organization of a self-replicating entity from prebiotic organic material, but all of those that are well formulated are based on hypothetical chemical syntheses that are problematic. Returning to our detective story, we must conclude that we have identified some important suspects and, in each case, we have some ideas about the method they might have used. However, we are very far from knowing whodunit. The only certainty is that there will be a rational solution.

This review has necessarily been highly selective. I have neglected important aspects of prebiotic chemistry (e.g. the origin of chirality, the organic chemistry of solar bodies other than the earth, and the formation of membranes). The best source for such material is the journal *Origins of Life and Evolution of the Biosphere*, particularly those issues that contain the papers presented at meetings of the International Society for the Study of the Origin of Life.

Acknowledgements

This work was supported by NASA (grant number NAG5-4118) and NASA NSCORT/EXOBIOLOGY (grant number NAG5-4546). I thank Aubrey R. Hill, Jr for technical assistance and Bernice Walker for manuscript preparation.

References

1. Sleep, N. H., Zahnle, K. J., Kasting, J. F. and Morowitz, H. J. (1989) *Nature* 342, 139–142
2. Chyba, C. F. (1993) *Geochim. Cosmochim. Acta* 57, 3351–3358
3. Schopf, J. W. (1993) *The Earth's Earliest Biosphere: Its Origin and Evolution*, Princeton University Press
4. Mojzsis, S. J. *et al.* (1996) *Nature* 384, 55–59
5. Doolittle, R. F. (1997) *Proc. Natl. Acad. Sci. U. S. A.* 94, 13028–13033
6. Miller, S. L. (1953) *Science* 117, 528–529
7. Ring, D., Wolman, Y., Friedmann, N. and Miller, S. L. (1972) *Proc. Natl. Acad. Sci. U. S. A.* 69, 765–768
8. Wolman, Y., Haverland, H. and Miller, S. L. (1972) *Proc. Natl. Acad. Sci. U. S. A.* 69, 809–811
9. Oro, J. and Kimball, A. P. (1960) *Biochim. Biophys. Res. Commun.* 2, 407–412
10. Ferris, J. P., Sanchez, A. and Orgel, L. E. (1968) *J. Mol. Biol.* 33, 693–704
11. Butlerow, A. (1861) *Compt. Rend. Acad. Sci.* 53,145–147
12. Butlerow, A. (1861) *Liebig's Ann. Chem.*120, 295
13. Kasting, J. F. (1993) *Science* 259, 920–926
14. Cronin, J. R., *et al.* (1988) in *Meteorites and the Early Solar System* (Kerridge, J. F. and Matthew, M. S., eds), pp. 8191–1857 University of Arizona Press
15. Chyba, C. and Sagan, C. (1992) *Nature* 355, 125–132
16. Wächtershäuser, G. (1988) *Microbiol. Rev.* 52, 452–484
17. Wächtershäuser, G. (1992) *Prog. Biophys. Mol. Biol.* 58, 85–201

18. Blochl, E., Keller, M., Wächtershäuser, G. and Stetter, K. O. (1992) *Proc. Natl. Acad. Sci. U. S. A.* 89, 8117–8120

19. Huber, C. and Wächtershäuser, G. (1997) *Science* 276, 245–247

20. Gesteland, R. F. and Atkins, J. F. (1993) *The RNA World: The Nature of Modern RNA Suggests a Prebiotic World,* Cold Spring Harbor Laboratory Press

21. Pan, T. (1997) *Curr. Opin. Chem. Biol.* 1, 17–25

22. Breaker, R. R. (1997) *Curr. Opin. Chem. Biol.* 1, 26–31

23. Eklund, E. H., Szostak, J. W. and Bartel, D. P. (1995) *Science* 269, 364–370

24. Eklund, E. H. and Bartel, D. P. (1996) *Nature* 382, 373–376

25. Unrai, P. J. and Bartel, D. P. (1998) *Nature* 395, 260–263

26. Ferris, J. P. and Ertem, G. (1993) *J. Am. Chem. Soc.* 115, 12270–12275

27. Kawamura, K. and Ferris, J. P. (1994) *J. Am. Chem. Soc.* 116, 7564–7572

28. Hill, A. R., Jr, Wu, T. and Orgel, L. E. (1993) *Orig. Life Evol. Biosphere* 23, 285–290

29. Orgel, L. E. (1994) *Sci. Am.* 271, 52–61

30. Ferris, J. P. (1987) *Cold Spring Harbor Symp. Quant. Biol.* LII, 29–39

31. Joyce, G. F. et al. (1984) *Nature* 310, 602–604

32. Muller, D . et al. (1990) *Helv. Chim. Acta* 73, 1410–1468

33. Cairns-Smith, A. G. (1982) *Genetic Takeover and the Mineral Origins of Life*, Cambridge University Press

34. Cairns-Smith, A. G. and Davis, C. J. (1977) *in Encyclopaedia of Ignorance* (Duncan, R. and Weston-Smith, M., eds), pp. 397–403, Pergamon Press

35. Eschenmoser, A. (1997) *Orig. Life Evol. Biosphere* 27, 535–553

36. Egholm, M., Buchardt, O., Nielsen, P. E. and Berg, R. H. (1992) *J. Am. Chem. Soc.* 114, 1895–1897

37. Egholm, M. et al. (1993) *Nature* 365, 566–568

38. Schmidt, J. G., Christensen, L., Nielsen, P. E. and Orgel, L. E. (1997) *Nucleic Acids Res.* 25, 4792–4796

39. Schmidt, J. G., Nielsen, P. E. and Orgel, L. E. (1997) *Nucleic Acids Res.* 25, 4797–4802

40. Koppitz, M., Nielsen, P. E. and Orgel, L. E. (1998) *J. Am. Chem. Soc.* 120, 4563–4569

41. Kauffman, S. A. (1986) *J. Theor. Biol.* 119, 1–24

42. Wächtershäuser, G. (1988) *Microbiol. Rev.* 52, 452–484

43. De Duve, C. (1991) *Blueprint for a Cell: The Nature and Origin of Life*, Neil Patterson

Chapter 8

Introduction

C hapter 8 takes up the question of human evolution. The first document for this section is from Darwin and includes three extracts. We begin with the short passage in a private notebook that Darwin was keeping in the fall of 1838 (the exact date of the entry is November 27), where we have the first unambiguous reference to the mechanism he was to call "natural selection." Note that the mechanism is applied to humans, and most particularly to our brain and our instincts. Darwin was practically unique in that he never hesitated about the application of his evolutionary ideas to our species. The next reference is the full (and only) paragraph on humans that is given in the *Origin*. As I explain in the text, Darwin wanted to get the main ideas on evolution out before he turned to our own species. This he did in 1871 in the *Descent of Man,* and here I give the summary from the end of the book. Note that Darwin thinks that we humans are subject to the same laws as other animals and that sexual selection as well as natural selection was important in our origins. (Darwin also thought that Lamarckism, the inheritance of acquired characteristics, was important. Even the greatest genius makes mistakes.) He also makes it clear that he does not think the Adam and Eve story is compatible with evolutionism and that human intelligence and language, although very important, is part and parcel of the same naturalistic evolutionary package.

For the second document we jump almost exactly a hundred years. You have already been introduced to "Lucy," the wonderfully preserved specimen of *Australopithecus afarensis.* She was discovered by the American paleoanthropologist Donald Johanson, and in the extract reproduced here he tells of his discovery. In the text of this book (in Chapter 9), I quoted the passage from Edward O. Wilson's autobiography in which he remembers the excitement of reading and appreciating William Hamilton's model of kin selection as applied to hymenopteran sociality. I wanted there to give you a sense of the sheer thrill of ideas—ideas that simply sandbag you with their brilliance. Here, in parallel as it were, I want to give you a sense of the thrill of empirical discovery. Johanson is probably one of the most self-confident (and self-regarding) men in the whole of science—ego problems he does not have. But that is the way you have got to be if you are to spend days, weeks, months, traipsing through the hottest, driest, most hell-like places on earth, looking for those elusive bits of bone that point back to our evolutionary history. You have got to have faith in yourself and a brute determination to find what you seek. You must demand the answers of nature and refuse to quit until they are given. And what a reward when you succeed! Science—life—does not get any better than this.

From *Charles Darwin's Notebooks*

An habitual action must some way affect the brain in a manner which can be transmitted.—this is analogous to a blacksmith having children with strong arms.—The other principle of those children. which *chance?* produced with strong arms, outliving the weaker ones, may be applicable to the formation of instincts, independently of habits.—the limit of these two actions either on form or brain very hard to define.—Consider the acquirement of instinct by dogs, would show habit.—

From **The Origin of Species**

In the distant future I see open fields for far more important researches. Psychology will be based on a new foundation, that of the necessary acquirement of each mental power and capacity by gradation. Light will be thrown on the origin of man and his history.

From **The Descent of Man**

The main conclusion arrived at in this work, and now held by many naturalists who are well competent form a sound judgment, is that man is descended from some less highly organised form. The grounds upon which this conclusion rests will never be shaken, for the close similarity between man and the lower animals in embryonic development, as well as in innumerable points of structure and constitution, both of high and of the most trifling importance,—the rudiments which he retains, and the abnormal reversions to which he is occasionally liable,—are facts which cannot be disputed. They have long been known, but until recently they told us nothing with respect to the origin of man. Now when viewed by the light of our knowledge of the whole organic world, their meaning is unmistakable. The great principle of evolution stands up clear and firm, when these groups of facts are considered in connection with others, such as the mutual affinities of the members of the same group, their geographical distribution in past and present times and their geological succession. It is incredible that all these facts should speak falsely. He who is not content to look, like a savage, at the phenomena of nature as disconnected, cannot any longer believe that man is the work of a separate act of creation. He will be forced to admit that the close resemblance of the embryo of man to that, for instance, of a dog—the construction of his skull, limbs, and whole frame, independently of the uses to which the parts may be put, on the same plan with that of other mammals—the occasional reappearance of various structures, for instance of several distinct muscles, which man does not nor-really possess, but which are common to the Quadramana—and a crowd of analogous facts—all point in the plainest manner to the conclusion that man is the co-descendant with other mammals of a common progenitor.

We have seen that man incessantly presents individual differences in all parts of his body and in his mental faculties. These differences or variations seem to be induced by the same general causes, and to obey the same laws as with the lower animals. In both cases similar laws of inheritance prevail. Man tends to increase at a greater rate than his means of subsistence; consequently he is occasionally subject to a severestruggle for existence, and natural selection will have effected whatever lies within its scope. A succession of strongly-marked variations of a similar nature are by no means requisite; slight fluctuating differences in the individual suffice for the work of natural selection. We may feel assured that the inherited effects of the long-continued use or disuse of parts will have done much in the same direction with natural selection. Modifications formerly of importance, though no longer of any special use, will be long inherited. When one part is modified, other parts will change through the principle of correlation, of which we have instances in many curious cases of correlated monstrosities. Something may be attributed to the direct and definite—of the surrounding conditions of life, such as abundant food, heat, or moisture; and lastly, many characters of slight physiological importance, some indeed of considerable importance, have been gained through sexual selection.

No doubt man, as well as every other animal, presents structures, which as far as we can judge with our little knowledge, are not now of any service to him, nor have been so during any former period of his existence, either in relation to his general conditions of life or of one sex to the other. Such structures cannot be accounted for by any form of selection, or by the inherited effects of the use and disuse of parts. We know, however, that many strange and strongly-marked peculiarities of structure occasionally appear in our domesticated productions, and if the unknown causes which produce them were to act more uniformly, they would probably become common to all the individuals of the species. We may hope hereafter to understand something about the causes of such occasional modifications, especially through the study of monstrosities hence the labours of experimentalists, such as those of M. Camille Dareste, are full of promise for the future In the greater number of cases we can only say that the cause of each slight variation and of each monstrosity lies much more in the nature or constitution of the organism, than in the nature of the surrounding conditions; though new and changed conditions certainly play an important part in exciting organic changes of all kinds.

Through the means just specified, aided perhaps by others as yet undiscovered, man has been raised to his present state. But since he attained to the rank of manhood, he has diverged into distinct races, or as they may be more appropriately called sub-species. Some of these, for instance the Negro and European, are so distinct that, if specimens bad been brought to a naturalist without any further information, they would undoubtedly bare been considered by him as good and true species. Nevertheless all the races agree in so many unimportant details of structure and in so many mental peculiarities, that these can be accounted for only through inheritance from a common progenitor and a progenitor thus characterised would probably have deserved to rank as man.

It must not be supposed that the divergence of each race from the other races, and of all the races from a common stock, can bc traced back to any one pair of progenitors. On the contrary, at every stage in the process of modification, all the individuals which were in any way best fitted for their conditions of life, though in different degrees, would have survived in greater numbers than the less well fitted. The process would have been like that followed by man, when he does not intentionally select particular individuals, but breeds from all the superior and neglects all the inferior individuals. He thus slowly but surely modifies his stock, and unconsciously forms a new strain. So with respect to modifications, acquired independently of selection, and due to variations arising from the nature of the organism and the act/on of the surrounding conditions, or from changed habits of life, no single pair will have been modified in a much greater degree than the other pairs which inhabit the same country, for all will have been continually blended through free intercrossing.

By considering the embryological structure of man,—the homologies which he presents with the lower animals,—the rudiments which he retains,—and the perversions to which he is liable, we can partly recall in imagination the former condition of our early progenitors; and can approximately place them in their proper position in the zoological series. We thus learn that man is descended from a hairy quadruped, furnished with a tail and pointed cars, probably arboreal in its habits, and an inhabitant of the Old World. This creature, if its whole structure had been examined by a naturalist, would have been classed amongst the Quadrumana, as surely as would the common and still more ancient progenitor of the Old and New World monkeys. The Quadrumana and all the higher mammals are probably derived from an ancient marsupial animal, and this through a long line of diversified forms, either from some reptile-like or some amphibian-like creature, and this again from some fish-like animal. In the dim obscurity of the past we can see that the early progenitor of all the Vertebrata must have been an aquatic animal, provided with branchiae, with the two sexes united in the same individual, and with the most important organs of the body (such as the brain and heart) imperfectly developed. This animal seems to have been more like the larvae of our existing marine Ascidians than any other known form.

The greatest difficulty which presents itself, when we are driven to the above conclusion on the origin of man, is the high standard of intellectual power and of moral disposition which he has attained. But every one who admits the general principle of evolution, must see that the mental powers of the higher animals, which are the same in kind with those of mankind, though so different in degree, are capable of advancement. Thus the interval between the mental powers of one of the higher apes and of a fish, or between those of an ant and scale-insect, is immense. The development of these powers in animals does not offer any special difficulty; for with our domesticated animals, the mental faculties are certainly variable, and the variations are inherited. No one doubts that these faculties are of the utmost importance to animals in a state of nature. Therefore the conditions are favourable for their development through natural selection. The same conclusion may be extended to man; the intellect must have been all-important to him, even at a very remote period, enabling him to use language to invent and make weapons, tools, traps, &c.; by which means, in combination with his social habits, he long ago became the most dominant of all living creatures.

A great stride in the development of the intellect will have followed, as soon as, through a previous considerable advance, the half-art and half-instinct of language came into use; for the continued use of language will have reacted on the brain, and produced an inherited effect; and this again will have reacted on the improvement of language. The large size of the brain in man, in comparison with that of the lower animals, relatively to the size of their bodies, may be attributed in chief part, as Mr. Chauncey Wright has well remarked,[1] to the early use of some simple form of language,—that wonderful engine which affixes signs to all sorts of objects and qualities, and excites trains of thought which would never arise from the mere impression of the senses, and if they did arise could not be followed out. The higher intellectual powers of man, such as those of ratiocination, abstraction, sell; consciousness, &c., will have followed from the continued improvement of other mental faculties; but without considerable culture of the mind, both in the race and in the individual, it is doubtful whether these high powers would be exercised, and thus fully attained.

The development of the moral qualities is a more interesting and difficult problem. Their foundation lies in the social instincts, including in this term the family ties. These instincts are of a highly complex nature, and in the case of the lower animals give special tendencies towards certain definite actions; but the more important elements for us are love, and the distinct emotion of sympathy. Animals endowed with the social instincts take pleasure in each other's company, warn each other of danger, defend and aid each other in many ways. These instincts are not extended to all the individuals of the species, but only to those of the same community. As they are highly beneficial to the species, they have in all probability been acquired through natural selection.

A moral being is one who is capable of comparing his past and future actions and motives,—of approving of some and disapproving of others; and the fact that man is the one being who with certainty can be thus designated makes the greatest of all distinctions between him and the lower animals. But in our third chapter I have endeavoured to shew that the moral sense follows, firstly, from the enduring

---

[1]On the "Limits of Natural Selection," in the *North American Review*, Oct. 1870, p.295.

and always present nature of the social instincts, in which respect man agrees with the lower animals; and secondly, from his mental faculties being highly active and his impressions of past events extremely vivid, in which respects he differs from the lower animals. Owing to this condition of mind, man cannot avoid looking backwards and comparing the impressions of past events and actions. lie also continually looks forward. Hence, after some temporary desire or passion has mastered his social instincts, bc will reflect and compare the now weakened impression of such past impulses, with the ever present social instinct; and he will then feel that sense of dissatisfaction which all unsatisfied instincts leave behind them. Consequently he resolves to act differently for the future—and this is conscience. Any instinct which is permanently stronger or more enduring than another gives rise to a feeling which we express by saying that it ought to be obeyed. A pointer dog, if able to reflect on his past conduct, would say to himself, I ought (as indeed we say of him) to have pointed at that hare and not have yielded to the passing temptation of hunting it.

Social animals are partly impelled by a wish to aid the members of the same community in a general manner, but more commonly to perform certain definite actions. Man is impelled by the same general wish to aid his fellows, but has few or no special instincts. He differs also from the lower animals in being able to express his desires by words, which thus become the guide to the aid required and bestowed. The motive to give aid is likewise somewhat modified in man: it no longer consists solely of a blind instinctive impulse, but is largely influenced by the praise or blame of his fellow men. Both the appreciation and the bestowal of praise and blame rest on sympathy; and this emotion, as we have seen, is one of the most important elements of the social instincts. Sympathy, though gained as an instinct, is also much strengthened by exercise or habit. As all men desire their own happiness, praise or blame is bestowed on actions and motives, according as they lead to this end; and as happiness is an essential part of the general good, the greatest-happiness principle indirectly serves as a nearly safe standard of right and wrong. As the reasoning powers advance and experience is gained, the more remote effects of certain lines of conduct on the character of the individual, and on the general good, are perceived; and then the self-regarding virtues,. from coming within the scope of public opinion, receive praise, and their opposites receive blame. But with the less civilised nations reason often errs, and many bad customs and base superstitions come within the same scope, and consequently are esteemed as high virtues, and their breach as heavy crimes.

The moral faculties are generally esteemed, and with justice, as of higher value than the intellectual powers. But we should always bear in mind that the activity of the mind in vividly recalling past impressions is one of the fundamental though secondary bases of conscience. This fact affords the strongest argument for educating and stimulating in all possible ways the intellectual faculties of every human being. No doubt a man with a torpid mind, if his social affections and sympathies are well developed, will be led to good actions, and may have a fairly sensitive conscience. But whatever renders the imagination of men more vivid and strengthens the habit of recalling and comparing past impressions will make the conscience more sensitive, and may even compensate to a certain extent for weak social affections and sympathies.

The moral nature of man has reached the highest standard as yet attained, partly through the advancement of the reasoning powers and consequently of a just public opinion, but especially through the sympathies being rendered more tender and widely diffused through the effects of habit, example, instruction, and reflection. It is not improbable that virtuous tendencies may through long practice be inherited. With the more civilised races, the conviction of the existence of an all-seeing Deity has had a potent influence on the advancement of morality. Ultimately man no longer accepts the praise or blame of his fellows as his chief guide, though few escape this influence, but his habitual convictions controlled by reason afford him the safest rule. His conscience then becomes his supreme judge and monitor. Nevertheless the first foundation or origin of the moral sense lies in the social instincts including sympathy; and these instincts no doubt were primarily gained, as in the case of the lower animals, through natural selection.

The belief in God has often been advanced as not only the greatest, but the most complete of all the distinctions between man and the lower animals. It is however impossible, as we have seen, to maintain that this belief is innate or instinctive in man. On the other hand a belief in all-pervading spiritual agencies seems to be universal; and apparently follows from a considerable advance in the reasoning powers of man, and from a still greater advance in his faculties of imagination, curiosity and wonder. I am aware that the assumed instinctive belief in God has been used by many persons as an argument for His existence. But this is a rash argument, as we should thus be compelled to believe in the existence of many cruel and malignant possessing only a little more power than man; for the belief in them is far more general than of a beneficent deity. The idea of a universal and beneficent Creator of the universe does not seem to arise in the mind of man, until he has been elevated by long-continued culture.

He who believes in the advancement of man from some lowly-organized form, will naturally ask how does this bear on the belief in the immortality of the soul. the barbarous races of man, as Sir J. Lubbock has shewn, possess no clear belief of this kind; but arguments derived from the primeval beliefs of savages are, as we have just seen, of little or no avail. Few persons feel any anxiety from the impossibility of determining at what precise period in the development of the individual, from the first trace of the minute germinal—to the child either before or after birth, man becomes an immortal being; and there is no greater cause for anxiety because the period in the gradually ascending organic scale cannot possibly be determined.[2]

I am aware that the conclusions arrived at in this work will be denounced by some as highly irreligious; that he who thus denounces them is bound to shew why it is more irreligious to explain the origin of man as a distinct species by de-

---

[2]The Rev. J. A. Picton gives a discussion to this effect in his *New Theories and the Old Faith*, 1870.

scent from some lower form, through the laws of variation and natural selection than to explain the birth of the individual through the laws of ordinary reproduction. The birth both of the species and of the individual are equally parts of that grand sequence of events, which our minds refuse to accept as the result of blind chance. The understanding revolts at such a conclusion, whether or not we are able to believe that every slight variation of structure,—the union of each pair in marriage,—the dissemination of each seed,—and other such events, have all been ordained for some special purpose.

Lucy: *The Beginnings of Humankind*

Donald Johanson & Maitland Edey

On the morning of November 30, 1974, I woke, as I usually do on a field expedition, at daybreak. I was in Ethiopia, camped on the edge of a small muddy river, the Awash, at a place called Hadar, about a hundred miles northeast of Addis Ababa. I had been there for several weeks, acting as coleader of a group of scientists looking for fossils.

For a few minutes I lay in my tent, looking up at the canvas above me, black at first but quickly turning to green as the sun shot straight up beyond the rim of hills off to the east. Close to the Equator the sun does that; there is no long dawn as there is at home in the United States. It was still relatively cool, not more than 80 degrees. The air had the unmistakable crystalline smell of early morning on the desert, faintly touched with the smoke of cooking fires. Some of the Afar tribesmen who worked for the expedition had brought their families with them, and there was a small compound of dome-shaped huts made of sticks and grass mats about two hundred yards from the main camp. The Afar women had been up before daylight, tending their camels and goats, talking quietly.

For most of the Americans in camp this was the best part of the day. The rocks and boulders that littered the landscape had bled away most of their heat during the night and no longer felt like stoves when you stood next to one of them. I stepped out of the tent and took a look at the sky. Another cloudless day; another flawless morning on the desert that would turn to a crisper later on. I washed my face and got a cup of coffee from the camp cook, Kabete. Mornings are not my favorite time. I am a slow starter and much prefer evenings and nights. At Hadar I feel best just as the sun is going down. I like to walk up one of the exposed ridges near the camp, feel the first stirrings of evening air and watch the hills turn purple. There I can sit alone for a while, think about the work of the day just ended, plan the next, and ponder the larger questions that have brought me to Ethiopia. Dry silent places are intensifiers of thought, and have been known to be since early Christian anchorites went out into the desert to face God and their own souls.

Tom Gray joined me for coffee. Tom was an American graduate student who had come out to Hadar to study the fossil animals and plants of the region, to reconstruct as accurately as possible the kinds and frequencies and relationships of what had lived there at various times in the remote past and what the climate had been like. My own target—the reason for our expedition—was hominid fossils: the bones of extinct human ancestors and their close relatives. I was interested in the evidence for human evolution. But to understand that, to interpret any hominid fossils we might find, we had to have the supporting work of other specialists like Tom.

"So, what's up for today?" I asked.

Tom said he was busy marking fossil sites on a map.

"When are you going to mark in Locality 162?"

"I'm not sure where 162 is," he said.

"Then I guess I'll have to show you." I wasn't eager to go out with Gray that morning. I had a tremendous amount of work to catch up on. We had had a number of visitors to the camp recently. Richard and Mary Leakey, two well-known experts on hominid fossils from Kenya, had left only the day before. During their stay I had not done any paperwork, any cataloguing. I had not written any letters or done detailed descriptions of any fossils. I *should* have stayed in camp that morning—but I didn't. I felt a strong subconscious urge to go with Tom, and I obeyed it. I wrote a note to myself in my daily diary: *Nov. 30, 1974. To Locality 162 with Gray in am. Feel good.*

As a paleoanthropologist—one who studies the fossils of human ancestors—I am superstitious. Many of us are, because the work we do depends a great deal on luck. The fossils we study are extremely rare, and quite a few distinguished paleoanthropologists have gone a lifetime without finding a single one. I am one of the more fortunate. This was only my third year in the field at Hadar, and I had already found several. I know I am lucky, and I don't try to hide it. That is why I wrote "feel good" in my diary. When I got up that morning I felt it was one of those days when you should press your luck. One of those days when something terrific might happen.

Throughout most of that morning, nothing did. Gray and I got into one of the expedition's four Land Rovers and slowly jounced our way to Locality 162. This was one of several hundred sites that were in the process of being plotted on a master map of the Hadar area, with detailed information about geology and fossils being entered on it as fast as it was obtained. Although the spot we were headed for was only about four miles from camp, it took us half an hour to get there because of the rough terrain. When we arrived it was already beginning to get hot.

At Hadar, which is a wasteland of bare rock, gravel and sand, the fossils that one finds are almost all exposed on the surface of the ground. Hadar is in the center of the Afar desert, an ancient lake bed now dry and filled with sediments that record the history of past geological events. You can trace volcanic-ash falls there, deposits of mud and silt washed down from distant mountains, episodes of volcanic dust, more mud, and so on. Those events reveal themselves like layers in a slice of cake in the gullies of new young rivers that recently have cut through the lake bed here and there. It seldom rains at Hadar, but when it does it comes in an overpowering gush—six months' worth overnight. The soil, which is bare of vegetation, cannot hold all that water. It roars down the gullies, cutting back their sides and bringing more fossils into view.

Gray and I parked the Land Rover on the slope of one of those gullies. We were careful to face it in such a way that the canvas water bag that was hanging from the side mirror was in the shade. Gray plotted the locality on the map. Then we got out and began doing what most members of the expedition spent a great deal of their time doing: we began surveying, walking slowly about, looking for exposed fossils.

Some people are good at finding fossils. Others are hopelessly bad at it. It's a matter of practice, of training your eye to see what you need to see. I will never be as good as some of the Afar people. They spend all their time wandering around in the rocks and sand. They have to be sharp-eyed; their lives depend on it. Anything the least bit unusual they notice. One quick educated look at all those stones and pebbles, and they'll spot a couple of things a person not acquainted with the desert would miss.

Tom and I surveyed for a couple of hours. It was now close to noon, and the temperature was approaching 110. We hadn't found much: a few teeth of the small extinct horse *Hipparion;* part of the skull of an extinct pig; some antelope molars; a bit of a monkey jaw. We had large collections of all these things already, but Tom insisted on taking these also as added pieces in the overall jigsaw puzzle of what went where.

"I've had it," said Tom. "When do we head back to camp?"

"Right now, But let's go back this way and survey the bottom of that little gully over there."

The gully in question was just over the crest of the rise where we had been working all morning. It had been thoroughly checked out at least twice before by other workers, who had found nothing interesting. Nevertheless, conscious of the "lucky" feeling that had been with me since I woke, I decided to make that small final detour. There was virtually no bone in the gully. But as we turned to leave, I noticed something lying on the ground partway up the slope.

"That's a bit of a hominid arm," I said.

"Can't be. It's too small. Has to be a monkey of some kind."

We knelt to examine it.

"Much too small," said Gray again.

I shook my head. "Hominid."

"What makes you so sure?" he said.

"That piece right next to your hand. That's hominid too."

"Jesus Christ," said Gray. He picked it up. It was the back of a small skull. A few feet away was part of a femur: a thighbone. "Jesus Christ," he said again. We stood up, and began to see other bits of bone on the slope: a couple of vertebrae, part of a pelvis—all of them hominid. An unbelievable, impermissible thought flickered through my mind. Suppose all these fitted together? Could they be parts of a single, extremely primitive skeleton? No such skeleton had ever been found—anywhere.

"Look at that," said Gray. "Ribs."

A single individual?

"I can't believe it," I said. "I just can't believe it."

"By God, you'd better believe it!" shouted Gray. "Here it is. Right here!" His voice went up into a howl. I joined him. In that 110-degree heat we began jumping up and down. With nobody to share our feelings, we hugged each other, sweaty and smelly, howling and hugging in the heat-shimmering gravel, the small brown remains of what now seemed almost certain to be parts of a single hominid skeleton lying all around us.

"We've got to stop jumping around," I finally said. "We may step on something. Also, we've got to make sure."

"Aren't you sure, for Christ's sake?"

"I mean, suppose we find two left legs. There may be several individuals here, all mixed up. Let's play it cool until we can come back and make absolutely sure that it all fits together."

We collected a couple of pieces of jaw, marked the spot exactly and got into the blistering Land Rover for the run back to camp. On the way we picked up two expedition geologists who were loaded down with rock samples they had been gathering.

"Something big," Gray kept saying to them. "Something big. Something *big.*"

"Cool it," I said.

But about a quarter of a mile from camp, Gray could not cool it. He pressed his thumb on the Land Rover's horn, and the long blast brought a scurry of scientists who had been bathing in the river. "We've got it," he yelled. "Oh, Jesus, we've got it. We've got The Whole Thing!"

That afternoon everyone in camp was at the gully, sectioning off the site and preparing for a massive collecting job that ultimately took three weeks. When it was done, we had recovered several hundred pieces of bone (many of them fragments) representing about forty percent of the skeleton of a single individual. Tom's and my original hunch had been right. There was no bone duplication.

But a single individual of what? On preliminary examination it was very hard to say, for nothing quite like it had ever been discovered. The camp was rocking with excitement. That first night we never went to bed at all. We talked and talked. We drank beer after beer. There was a tape recorder in the camp, and a tape of the Beatles song "Lucy in the Sky with Diamonds" went belting out into the night sky, and was played at full volume over and over again out of sheer exuberance. At some point during that unforgettable evening—I no longer remember exactly when—the new fossil picked up the name of Lucy, and has been so known ever since, although its proper name—its acquisition number in the Hadar collection—is AL 288-1.

Chapter 9

Introduction

Chapter 9 takes us into the sociobiology controversy. There are two documents here. First we have a statement about the sociobiological program by Edward O. Wilson and a critique by a group of radical scientists who went under the somewhat ungainly name of the "Science for the People Sociobiology Study Group." (The group included many eminent scientists, among them Richard Lewontin and Stephen Jay Gould.) Of course, part of this debate is as scientific as you like. Wilson argues that we have now reached the point where Darwinian evolutionary theory can be extended out to social behavior and that this includes human social behaviour. The critics beg to differ. Part of the debate takes us beyond science. Wilson, as we see, and as we saw even more in the fifth chapter of the text, is trying to give us a vision of life for the modern post-Christian age. His critics think that he is simply giving us a fancy version of a bourgeois, middle-class, white American's view of life, dressed up with pseudoscientific language to sound like the objective truth.

Sociobiology—Another Biological Determinism

Sociobiology Study Group of Science for the People

Biological determinism represents the claim that the present states of human societies are the specific result of biological forces and the biological "nature" of the human species. Determinist theories all describe a particular model of society which corresponds to the socioeconomic prejudices of the writer. It is then asserted that this pattern has arisen out of human biology and that present human social arrangements are either unchangeable or if altered will demand continued conscious social control because these changed conditions will be "unnatural." Moreover, such determinism provides a direct justification for the status quo as "natural," although some determinists dissociate themselves from some of the consequences of their arguments. The issue, however, is not the motivation of individual creators of determinist theories, but the way these theories operate as powerful forms of legitimation of past and present social institutions such as aggression, competition, domination of women by men, defense of national territory, individualism, and the appearance of a status and wealth hierarchy.

The earlier forms of determinism in the current wave have now been pretty well discredited. The claims that there is a high heritability of IQ, which implies both the unchangeability of IQ and a genetic difference between races or between social classes, have now been thoroughly debunked.

The simplistic forms of the human nature argument given by Lorenz, Ardrey, Tiger and Fox, and others have no scientific credit and have been scorned as works of "advocacy" by E. O. Wilson, whose own book, *Sociobiology.' The New Synthesis,* is the manifesto of a new, more complex, version of biological determinism, no less a work of "advocacy" than its rejected predecessors. This book, whose first chapter is on "The Morality of the Gene," is intended to establish sociology as a branch of evolutionary biology, encompassing all human societies, past and present. Wilson believes that "sociology and the other social sciences, as well as the humanities, are the last branches of biology waiting to be included in the Modem Synthesis" (p. 4).

This is no mere academic exercise. For more than a century the idea that human social behavior is determined by evolutionary imperatives operating on inherited dispositions has been seized upon and widely entertained not so much for its alleged correspondence with reality as for its more obvious political value. Among the better known examples are Herbert Spencer's argument in *Social Statics* (1851) that poverty and starvation were natural agents cleansing society of the unfit, and Konrad Lorenz's call in 1940 in Germany for "the extermination of elements of the population loaded with dregs," based upon his ethological theories.

In order to make their case, determinists construct a selective picture of human history, ethnography, and social relations. They misuse the basic concepts and facts of genetics and evolutionary theory, asserting things to be true that are totally unknown, ignoring whole aspects of the evolutionary process, asserting that conclusions follow from premises when they do not. Finally, they invent ad hoc hypotheses to take care of the contradictions and carry on a form of "scientific reasoning" that is untestable and leads to unfalsifiable hypotheses. What follows is a general examination of these elements in sociobiological theory, especially as elaborated in E. O. Wilson's *Sociobiology.*

A Version of Human Nature

For the sociobiologist the first task is to delineate a model of human nature that is to be explained. Among Wilson's universal aspects of human nature are:

- territoriality and tribalism (pp. 564–565);
- indoctrinability—"Human beings are absurdly easy to indoctrinate—they *seek* it" (p. 562);
- spite and family chauvinism—"True spite is commonplace in human societies, undoubtedly because human beings are keenly aware of their own blood lines and have the intelligence to plot intrigue" (p. 119);

- reciprocal altruism (as opposed to true unselfishness)—"Human behavior abounds with reciprocal altruism," as for example, "aggressively moralistic behavior," "self-righteousness, gratitude and sympathy" (p. 120);
- blind faith—"Men would rather believe than know" (p. 561);
- warfare (p. 572) and genocide (p. 573)—"the most distinctive human qualities" emerged during the "autocatalytic phase of social evolution" which occurred through intertribal warfare, "genocide" and "genosorption."

The list is not exhaustive and is meant only to show how the outlines of human nature are viewed myopically, through the lens of modern Euro-American culture.

To construct such a view of human nature, Wilson must abstract himself totally from any historical or ethnographic perspective. His discussion of the economy of scarcity is an excellent example. An economy of relative scarcity and unequal distribution of rewards is stated to be an aspect of human nature:

> The members of human society sometimes cooperate closely in *insectan* fashion [our emphasis], but more frequently they compete for the limited resources allocated to their role sector. The best and the most entrepreneurial of the role-actors usually gain a disproportionate share of the rewards (p. 554).

There is a great deal of ethnographic and historical description entirely, contradicting this conception of social organization. It ignores, for example, the present and historical existence of societies not differentiated in any significant way by "role sectors"; without scarcities differentially induced by social institutions for different subpopulations of the society; not differentiated by lower and higher ranks and strata (Birket-Smith1959; Fried 1967; Harris 1968; Krader 1968).Realizing that history and ethnography do not support the universality of their description of human nature, sociobiologists claim that the exceptions are "temporary aberrations" or deviations. Thus, although genocidal warfare is (assertedly) universal, "it is to be expected that some isolated cultures will escape the process for generations at a time, in effect reverting temporarily to what ethnographers classify as a pacific state" (p. 574).

Another related ploy is the claim that ethnographers and historians have been too narrow in their definitions and have not realized that apparently contradictory evidence is really confirmatory.

> Anthropologists often discount territorial behavior as a general human attribute. This happens when the narrowest *concept* of the phenomenon is borrowed from zoology . . . it is necessary to define territory more broadly animals respond to their neighbors in a highly variable manner the scale may run from open hostility . . . to oblique forms of advertisement or *no territorial behavior at all* [our emphasis].If these qualifications are accepted it is reasonable to conclude that territoriality is a general trait of hunter-gatherer societies. (pp. 564–565)

Wilson's views of aggression and warfare are subject to this ploy of all-embracing definition on the one hand and erroneous historical-ethnographic data on the other. "Primitive" warfare is rarely lethal to more than one or at most a few individuals in an episode of warfare, virtually without significance genetically or demographically (Livingstone 1968). Genocide was virtually unknown until state-organized societies appeared in history (as far as can be made out from the archeological and documentary records).

We have given only examples of the general advocacy method employed by sociobiologists in a procedure involving definitions which exclude nothing and the laying of Western conceptual categories onto "primitive" societies.

Humans as Animals—the Meaning of Similarity

To support a biologistic explanation of human institutions it is useful to claim an evolutionary relationship between the nature of human social institutions and "social" behavior in other animals. Obviously sociobiologists would prefer to claim evolutionary homology, rather than simple analogy, as the basis for the similarity in behavior between humans and other animals; then they would have a prima facie case for genetic determination. In some sections of *Sociobiology,* Wilson attempts to do this by listing "universal" features of behavior in higher primates including humans. But claimed external similarity between humans and our closest relatives (which are by no means very close to us) does not imply genetic continuity. A behavior that may be genetically coded in a higher primate may be purely learned and widely spread among human cultures as a consequence of the enormous flexibility of our brain.

More often Wilson argues from evolutionary analogy. Such arguments operate on shaky grounds. They can never be used to assert genetic similarity, but they can serve as a plausibility argument for natural selection of human behavior by assuming that natural selection has operated on different genes in the two species but has produced convergent responses as independent adaptations to similar environments. The argument is not even worth considering unless the similarity is so precise that identical function cannot be reasonably denied, as in the classic case of evolutionary convergence—the eyes of vertebrates and octopuses. Here Wilson fails badly, for his favorite analogies arise by a twisted process of imposing human institutions on animals by metaphor, and then rederiving the human institutions as special cases of the more general phenomenon "discovered" in nature. In this way human institutions suddenly become "natural" and can be viewed as a product of evolution.

A classic example, long antedating *Sociobiology,* is "slavery" in ants. "Slave-making" species capture the immature stages of "slave" species and bring them back to their own nests. When the captured workers hatch, they perform housekeeping tasks with no compulsion as if they were members of the captor species. Why is this "slavemaking" instead of "domestication"? Human slavery involves members of one's own species under continued compulsion. It is an economic institution in societies producing an economic surplus, with both slave and products as commodities in exchange. It has nothing to do with ants except by weak and meaningless analogy. Wilson expands the realm of these weak analogies (chapter 27) to find barter, division of labor, role playing, culture, ritual, religion, magic, esthetics, and tribalism among nonhumans. But if we insist upon seeing animals in the mirror of our own social arrangements, we cannot fail to find any human institutions we want among them.

Genetic Bases of Behavior

We can dispense with the direct evidence for a genetic basis of various human social forms in a single word, "None." The genetics of normal human behavior is in a rudimentary state because of the impossibility of reproducing particular human genotypes over and over, or of experimentally manipulating the environments of individuals or groups. There is no evidence that meets the elementary requirements of experimental design, that such traits as xenophobia, religion, ethics, social dominance, hierarchy formation, slavemaking, etc., are in any way coded specifically in the genes of human beings.

And indeed, Wilson offers no such evidence. Instead, he makes confused and contradictory statements about what is an essential element in the argument. If there are no genes for parent-offspring conflict, then there is no sense in talking about natural selection for this phenomenon. Thus, he speaks of "genetically programmed sexual and parent-offspring conflict" (p. 563), yet there is the "considerable technical problem of distinguishing behavioral elements and combinations that emerge. . .independently of learning and those that are shaped at least to some extent by learning" (p. 159). In fact, it cannot be done.

Elsewhere, the *capacity* to learn is stated to be genetic in the species, so that "it does not matter whether aggression is wholly innate or acquired partly or wholly by learning" (p. 255). But it does matter. If all that is genetically programmed into people is that "genes promoting flexibility in social behavior are strongly selected" (p. 549) and if "genes have given away most of their sovereignty" (p. 550), then biology and evolution give no insight into the human condition except the most trivial one, that the *possibility* of social behavior is part of human biology. However, in the next phrase Wilson reasserts the sovereignty of the genes because they "maintain a certain amount of influence in at least the behavioral qualities that underly the variations between cultures." It is stated as *fact* that genetical differences underlie variations between cultures, when no evidence at all exists for this assertion and there is some considerable evidence against it.

Since sociobiologists can adduce no facts to support the genetic basis for human social behavior, they try two tacks. First, the suggestion of evolutionary homology between behavior in the human species and other animals, if correct, would imply a genetic basis in us. But the evidence for homology as opposed to analogy is very weak. Second, they postulate genes right and left and then go on to argue as if the genes were demonstrated facts. There are hypothetical altruist genes, conformer genes, spite genes, learning genes, homosexuality genes, and so on. An instance of the technique is on pages 554–555 of Wilson's book: "Dahlberg showed that *if* a single gene appears that is responsible for success and upward shift in status." and "Furthermore, *there are many* Dahlberg genes . . ." (our emphases throughout). Or on page 562: "*If we assume* for argument that indoctrinability evolves . . ." and "Societies containing higher frequencies of conformer genes replace those that disappear . . ."(our emphasis). Or consult nearly any page of Trivers (1971) for many more examples.

Geneticists long ago abandoned the naive notion that there are genes for toes, genes for ankles, genes for the lower leg, genes for the kneecap, or the like. Yet sociobiologists break the totality of human social phenomena into arbitrary units, which they reify as "organs of behavior," postulating particular genes for each.

Everything Is Adaptive

The next step in the sociobiological argument is to try to show that the hypothetical, genetically programmed behavior organs have evolved by natural selection. The assertion that all human behavior is or has been adaptive is an outdated expression of Darwinian evolutionary theory, characteristic of Darwin's 19th century defenders who felt it necessary to prove everything adaptive. It is a deeply conservative politics, not an understanding of modern evolutionary theory, that leads one to see the wonderful operation of adaptation in every feature of human social organization.

There is no hint in *Sociobiology* that at this very moment the scientific community of evolutionary geneticists is deeply split on the question of how important adaptive as opposed to random processes are in manifest evolution. More important, there is a strain in modern evolutionary thought, going back to Julian Huxley, that avoids much of the tortured logic required by extreme selectionism, by emphasizing allometry. Organs, not themselves under direct natural selection, may change because of their developmental links to other features that are under selection. Many aspects of human social organization, if not all, may be simply the consequence of increased plasticity of neurological response and cognitive capacity.

The major assertion of sociobiologists that human social structures exist because of their superior adaptive value is only an assumption for which no tests have even been proposed. The entire the0?Y is so constructed that *no tests are possible*. The mode of explanation involves three postulated-levels of the operation of natural selection: (1) classical individual selection to account for obviously self-serving behaviors: (2) kin selection to account for altruistic behaviors or submissive acts toward relatives; (3) reciprocal altruism to account for altruistic behaviors toward unrelated persons. All that remains is to make up a "just-so" story of adaptation with the appropriate form of selection acting. For some traits it is easy to invent a story. The "genes" for social dominance, aggression, entrepreneurship, successful deception, and so on will "obviously" be advantageous at the individual level. For example, evidence is presented (p. 288) that dominant males impregnate a disproportionate share of females in mice, baboons, and Yanamamo Indians. In fact, in the ethnographic literature there are numerous examples of groups whose political "leaders" do not have greater access to mates. In general it is hard to demonstrate a correlation of any of the sociobiologists' "adaptive" social behaviors with actual differential reproduction.

Other traits require more ingenuity. Homosexuality would seem to be at a reproductive disadvantage since "of course, homosexual men marry much less frequently and have far fewer children" (Dr. Kinsey disagreed, and what about homosexual women?). But a little ingenuity solves the problem: "The homosexual members of primitive societies may have functioned as helpers . . . [operated] with special efficiency in assisting close relatives" (p. 555). Kin selection saves the day when one's imagination for individual selection fails.

Only one more imaginative mechanism is needed to rationalize such phenomena as friendship, morality, patriotism, and submissiveness, even when the bonds do not involve relatives. The theory of reciprocal altruism (Trivers 1971) proposes that selection has operated such that risk taking and acts of kindness can be recognized and reciprocated so that the net fitness of both participants is increased.

The trouble with the whole system is that nothing is explained because everything is explained. If individuals are selfish, that is explained by simple individual selection. If, on the contrary, they are altruistic, it is kin selection or reciprocal altruism. If sexual identities are unambiguously heterosexual, individual fertility is increased. If, however, homosexuality is common, it is a result of kin selection. Sociobiologists give us no example that might conceivably contradict their scheme of perfect adaptation.

Variations of Cultures in Time and Space

There does exist one possibility of tests of sociobiological hypotheses when they make specific *quantitative* predictions about rates of change of characters in time and about the degree of differentiation between populations of a species. Population genetics makes specific predictions about rates of change, and there are hard data on the degree of genetic differentiation between human populations for biochemical traits. Both the theoretical rates of *genetic* change in time and the observed *genetic* differentiation between populations are too small to agree with the very rapid changes that have occurred in human *cultures* historically and the very large *cultural* differences observed among contemporaneous populations. So, for example, the rise of Islam after the 7th century to supreme cultural and political power in the West, to its subsequent rapid decline after the 13th century (a cycle occupying fewer than 30 generations) was too rapid by orders of magnitude for any large change by natural selection. The same problem arises for the immense cultural differences between contemporary groups, since we know from the study ' of enzyme-specifying genes that there is very little genetic differentiation between nations and races.

Wilson acknowledges and deals with both of these dilemmas by a bold stroke: He invents a new phenomenon. It is the "multiplier effect" (pp. 11–13, 569–572), which postulates that very small differences in the frequency of hypothetical genes for altruism, conformity, indoctrinability, etc., could move a whole society from one cultural pattern to another. The only evidence offered for this "multiplier effect" is a description of differences in behavior between closely related species of insects and of baboons. There is, however, no evidence about the amount of *genetic* difference between these closely related species nor how many tens or hundreds of thousands of generations separate the members of these species pairs since their divergence. The multiplier effect, by which any arbitrary but unknown genetic difference can be converted to any cultural difference you please, is a pure invention of convenience without any evidence to support it. It has been created out of whole cloth to seal off the last aperture through which the theory might have been tested against the real world.

An Alternative View

It is often stated by biological determinists that those who oppose them are "environmental determinists," who believe that the behavior of individuals is precisely determined by some sequence of environmental events in childhood. Such an assertion reveals the essential narrowness of viewpoint in determinist ideologies. First, they see the individual as the basic elements of determination and behavior, whereas society is simply the sum of all the individuals in it. But the truth is that the individual's social activity is to be understood only by first understanding social institutions. We cannot understand what it is to be a slave or a slave owner without first understanding the institution of slavery, which defines and creates both slave and owner.

Second, determinists assert that the evolution of societies is the result of changes in the frequencies of different sorts of individuals within them. But this confuses cause and effect. Societies evolve because social and economic activity alter the physical and social conditions in which these activities occur. Unique historical events, actions of some individuals, and the altering of consciousness of masses of people interact with social and economic forces to influence the timing, form, and even the possibility of particular changes; individuals are not totally autonomous units whose individual qualities determine the direction of social evolution. Feudal society did not pass away because some autonomous force increased the frequency of entrepreneurs. On the contary, the economic activity of Western feudal society itself resulted in a change in economic relations which made serfs into peasants and then into landless industrial workers with all the immense changes in social institutions that were the result.

Finally, determinists assert that the possibility of change in social institutions is limited by the biological constraints on individuals. But we know of no relevant constraints placed on social processes by human biology. There is no evidence from ethnography, archaeology, or history that would enable us to circumscribe the limits of possible human social organization. What history and ethnography do provide us with are the materials for building a theory that will itself be an instrument of social change.

References

Birket-Smith, K. (1959). *The Eskimos,* 2nd ed. (London: Methuen).

Fried, M. (1967). *The Evolution of Political Society.* (New York: Random House).

Harris, M. (1968). Law and Order in Egalitarian Societies. Pages 369–391 in *Culture. Man and Nature* (New York: Crowell).

Krader, L. (1968). "Government Without the State." In *Formation of the State* (Englewood Cliffs, N.J.: Prentice Hall) pp. 53–110.

Lorenz, K. (1940). "Durch Domestikation verursachte Störungen arteigenen Verhaltens." *Zeitschrift für angewandte Psychologie und Charakterkunde* 59, pp. 56–75. (As quoted in Cloud, W., 1973, "Winners and Sinners." *The Sciences* 13, pp. 16–21).

Livingstone, F. (1968). "The Effects of Warfare on the Biology of the Human Species." In M. Fried, M. Harris, and R. Murphy, eds. *War.' The Anthropology of Armed Conflict and Aggression* (Garden City: Natural History Press) pp. 3–15.

Spencer, H. (1851). *Social Statics;* (London: Chapman).

Trivers, R. (1971). "The Evolution of Reciprocal Altruism." *Q. Rev. Biol. 46,* pp. 35–57

Wilson, E. O. (1975). *Sociobiology: The New Synthesis* (Cambridge, Mass.: Harvard University Press).

Academic Vigilantism and the Political Significance of Sociobiology

Edward O. Wilson

The best response to a political attack of the kind exemplified by the preceding article, "Sociobiology—Another Biological Determinism," is perhaps no response at all. Some of my colleagues have offered that advice. But the problem is larger than the personal distress that this and earlier activities of the Science for the People group have caused me. The issue at hand, I submit, is vigilantism: the judgment of a work of science according to whether it conforms to the political convictions of the judges, who are self-appointed. The sentence for scientists found guilty is to be given a label and to be associated with past deeds that all decent persons will find repellent.

Thus, in a statement published earlier in *The New York Review of Books* (Allen et el. 1975), the Science for the People group characterized my book *Sociobiology: The New Synthesis* (Wilson 1975a) as the latest attempt to reinvigorate theories that in the past "provided an important basis for the enactment of sterilization laws and restrictive immigration laws by the United States between 1910 and 1930 and also for the eugenics policies which led to the establishment of gas chambers in Nazi Germany." To this malicious charge they added, "Wilson joins the long parade of biological determinists whose work has served to buttress the institutions of their society by exonerating them from responsibility for social problems." The tone of the present *BioScience* article is muted, but the innuendo is clear and remains the same.

This tactic, which has been employed by members of Science for the People against other scientists, throws the person criticized into the role of defendant and renders his ideas easier to discredit. Free and open discussion becomes difficult, as the critics continue to press their campaign, and the target struggles to clear his name. The problem is increased by difficulties in knowing with whom one is dealing. The statements are often published over long lists of names, shifts in committee membership occur through time, and the authors' names are withheld from some of the documents. (All have occurred during the present controversy.)

Despite the protean physical form taken by the Sociobiology Study Group of Science for the People, the belief system they promote is clear-cut and rigid. They postulate that human beings need only decide on the kind of society they wish, and then find the way to bring it into being. Such a vision can be justified if human social behavior proves to be infinitely malleable. In their earlier *New York Review* statement (Allen et al. 1975) the group therefore maintained that although eating, excreting, and sleeping may be genetically determined, social behavior is entirely learned; this belief has been developed further in the *BioScience* article. In contrast, and regardless of all they have said, I am ideologically indifferent to the degree of determinism in human behavior. If human beings proved infinitely malleable, as they hope, then one could justify any social or economic arrangement according to his personal value system. If on the other hand, human beings proved completely fixed, then the status quo could be justified as unavoidable.

Few reasonable persons take the first extreme position and none the second. On the basis of objective evidence the truth appears to lie somewhere in between, closer to the environmentalist than to the genetic pole. That was my wholly empirical conclusion in *Sociobiology: The New Synthesis* and continues to be in later writings. There is no reasonable way that this generalization can be construed as a support of the status quo and continued injustice, as the Science for the People group have now, on four painful occasions, claimed. I have personally argued the opposite conclusion, most fully and explicitly in my *New York Times Magazine* article of 12 October 1975 (Wilson 1975b). The Science for the People group have not found it convenient to mention this part of my writings.

With the exception of the Science for the People group, all of the biologists and social scientists whose reviews of *Sociobiology: The New Synthesis* I have seen understood the book correctly. None has read a reactionary political message into it, even though the reviewers represent a variety of personal political persuasions; and none has found my assessment of the degree of determinism in human social behavior out of line with the empirical evidence. The Science for the People group have utterly misrepresented the spirit and content of the portions of *Socio-*

biology devoted to human beings. They have done so, it would seem, in order to have a conspicuous straw man against which their views can be favorably pitted, and to obscure the valid points in *Sociobiology* which do indeed threaten their own extreme position. Let me document this interpretation with responses to the specific criticisms made by the 35 cosigners.

Response to Criticisms

First, it should be noted that *Sociobiology: The New Synthesis* is a large book, within which only chapter 27 and scattered paragraphs in earlier chapters refer to man. The main theses of sociobiology are based on studies of a myriad of animal species conducted by hundreds of investigators in various biological disciplines. It has been possible to derive propositions by the traditional postulational and deductive methods of theoretical science, and to test many of them rigorously by quantitative studies. Once can cite the work on kin selection in social Hymenoptera, the elaboration of caste systems in social insects, the economic functions of vertebrate territories, the ecological causes of ungulate social behavior, the repertory size and transmission characteristics of communication systems, and others. These ideas and data provide the main thrust of general sociobiology.

In my book human sociobiology was approached tentatively and in a taxonomic rather than a political spirit. The final chapter opens with the following passage: "Let us now consider man in the free spirit of natural history, as though we were zoologists from another planet completing a catalog of social species on Earth. In this macroscopic view the humanities and social sciences shrink to specialized branches of biology; history, biography, and fiction are the research protocols of human ethology; and anthropology and sociology together constitute the sociobiology of a single primate species."

It is the intellectually viable contention of the final chapter that the sociobiological methods which have proved effective in the study of animals can be extended to human beings, even though our vastly more complex, flexible behavior will make the application technically more difficult. The degree of success cannot yet be predicted. Chapter 27 was intended to be a beginning rather than a conclusion, and other reviewers have so interpreted it. In it I have characterized the distinctive human traits as best I could from the literature of the social sciences, and I have offered a set of hypotheses about the evolution of the traits stated in a way that seemed to make them most susceptible to analysis by sociobiological methods.

The Science for the People group ignore this main thrust of the book. They cite piece by piece incorrectly, or out of context, and then add their own commentary to furnish me with a political attitude I do not have and the book with a general conclusion that is not there. The following examples cover nearly all of their points.

Roles

The 35 cosigners have me saying that role sectors, and thus certain forms of economic role behavior associated with role sectors, are universal in man. On pages 552 and 554, the reader will find that I did not include role sectors among the widespread or universal traits. What I said was that when role sectors occur, certain economic features are associated with them.

Territory

It is now well known that animal territories commonly vary in size and quality of defense according to habitat, season, and population density. Under some circumstances many species show no territorial behavior, but it is necessary for them to display the behavior under other, specified circumstances in order to be called territorial—an obvious condition. This is the reason I have called the human species territorial. No contradiction in definitions exists; the cosigners have made it appear to exist by simply deleting three key pieces from the quoted statement. Most human societies are territorial most of the time.

Warfare

In Sociobiology I presented widespread lethal warfare in early human groups as a working hypothesis, not as a fact, contrary to what the cosigners suggest. And it is a hypothesis wholly consistent with the evidence: military activity and territorial expansion have been concomitants throughout history and at all levels of social organization (Otterbein 1970), and they can hardly fail to have had significant demographic and genetic consequences.

Slavery and Other Terms

The cosigners state that I claim to have found barter, religion, magic, and tribalism among nonhumans. I have made no such claim. The cosigners do not like to see terms such as slavery, division of labor, and ritual used in both zoology and the social sciences. Do they wish also to expunge communication, dominance, monogamy, and parental care from the vocabulary of zoology?

Genetic Bases of Behavior

The cosigners claim that no evidence exists for the genetic basis of particular forms of social behavior. Their statement indicates that they do not use the same criteria as other biologists. To postulate the existence of genes for the diagnostic human traits is not to imply that there exists one gene for spite, another for homosexuality, and so on, as one might envision the inheritance of flower color or seed texture in garden peas. The tendency to develop such behaviors, in a distinctively human form, is part of an immensely complex social repertory which is undoubtedly dependent on large numbers of genes.

My emphasis in *Sociobiology* was on the most, widespread, distinctive qualities of human behaviors—"human nature" if you wish—and the possible reasons why the underlying genes are different from those affecting social behavior in other species. Certain forms of human social behavior, such as the facial expressions used to convey the basic emotions, are relatively inflexible and transcultural. Human expressions, in fact, are so similar to those of the higher cercopithecoid primates as to suggest the possible existence of true homology (*Sociobiology*, pp. 227–228). Other kinds of response, including those under the categories of aggression, sexuality, and conformity, are of course subject to great variation through differences in experience. But as plastic as these latter behaviors might seem to us, they still form only a small subset of the many versions found in social species as a whole. It seems inconceivable that human beings could be socialized into the distinctive patterns of, say, ring-tailed lemurs, hamadryas baboons, or gibbons, vice versa. This is the ordinary criterion on which the expression "genetic control of human social behavior" in sociobiology is based. The main idea conveyed by the final chapter of my book is that such a comparison with other social species will place human behavior in a clearer evolutionary perspective.

With reference to genetic variation between human populations, there is no firm evidence. As usual, the cosigners misrepresent what I said. Here is their claim: "It is stated as a *fact* that genetical differences underlie variations between cultures, when no evidence at all exists for this assertion and there is some considerable evidence against it" (emphasis theirs). Here is what I really said, in the very sentences to which they allude (p. 550): "Even a small portion of this [genetic] variance invested in population differences *might* predispose societies toward cultural differences. At the very least, we should try to measure this amount. It is not valid to point to the absence of a behavioral trait in one or a few societies as conclusive evidence that the trait is environmentally induced and has no genetic disposition in man. The very opposite *could* be true" (italics newly added).

Adaptation versus Non-adaptation

The Science for the People group state that I believe all social behavior to be adaptive and hence "normal." This is so patently false that I am surprised the cosigners could bring themselves to say it. I have on the contrary discussed circumstances under which certain forms of animal social behavior become maladaptive, with examples and ways in which the deviations can be analyzed (pp. 33–34). With reference to human social behavior I have said (Wilson 1975b, an article well known to the cosigners): "When any genetic bias is demonstrated, it cannot be used to justify a continuing practice in present and future societies. Since most of us live in a radically new environment of our own making, the pursuit of such a practice would be bad biology; and like all bad biology, it would invite disaster." I then cited examples of maladaptive behavior in human beings. Furthermore, both

R.L. Trivers and I have provided varieties of adaptation hypotheses that compete with each other and against the non-adaptation hypothesis, contrary to the assertion of the Science for the People group (see, e.g., *Sociobiology:* pp. 123–124, 309–311, 326–327, 416–418).

Cultural Evolution

The cosigners propose that "sociobiological hypotheses" can be tested by seeing whether certain short-term episodes in history, such as the rise and decline of Islam, occurred too rapidly to be due to genetic change. They conclude that the theory of population genetics excludes that possibility. I agree, and that is why neither I nor any other sociobiologist of my acquaintance has ever proposed such hypotheses. The examples I used in *Sociobiology* to make the same point are the origin of the slave society of Jamaica, the decline of the Ik in Uganda, the alteration of Irish society following the potato famine, and the shift in the Japanese authority structure following World War II (pp. 548–550). I see no reason why the subject was even brought up. (A fuller discussion of the rates of cultural evolution and the complementarity of cultural to genetic evolution can be found in pages 168–175 and 555–562 of *Sociobiology.)*

Comments on the Debate

I now invite readers to check each of the pronouncements in the article by the 35 cosigners against the actual statements in my book, in the true context in which the statements were made. I suggest that they will encounter very little correspondence, and I am confident that they will be left with no doubt as to my true meaning.

How is it possible for the Science for the People group to misrepresent so consistently the content of a book, in contrast to all of the many other reviewers among their scientific colleagues? There is first the circumstance of the size and composition of the group. It has grown from 16, when it called itself The Genetic Engineering Group of Scientists and Engineers for Social and Political Action (in the magazine *Science for the People,* November 1975), to the present 35 now identified as the Sociobiology Study Group of Science for the People. The membership is heterogeneous: from the best count I can make there are eight professors in several fields of science in the Boston area; other members include at least one psychiatrist, a secondary school teacher, students and research assistants. Furthermore, in conformity with their political convictions the group really does believe in collective decision making and writing, so perhaps the result is not all that surprising. (In the issue of *Science for the People* just mentioned, the two main targets of criticism were myself, for biological determinism, and the Soviet Union, for revisionism.)

But the other, more important cause of the problem, and the reason I have not been able to find the matter as humorous as have some of my colleagues, is the remorseless zeal of the cosigners. By their own testimony they worked for months on the project. They appear to have been alarmed by the impact a critical success of the book might have on the acceptability of their own political views. One of the faculty members, in a *Harvard Crimson* interview on 3 November 1975, stated that the group was formed of persons who became interested "in breaking down the screen of approval" around the book. Clamorous denunciations followed during a closely packed series of lectures, work sessions, and release of printed statements. In October 1975 a second professorial member of the group drafted a 5,000-word position paper for *The New York Times* which characterized me as an ideologue and a privileged member of modern Western industrial society whose book attempts to preserve the status quo *(The New York Times,* 9 November 1975).' Later the same person (who shares the identical privileges at Harvard) startled me even more by declaring that "Sociobiology is not a racist doctrine" but "any kind of genetic determinism can and does feed other kinds, including the belief that some races are superior to others" *(Harvard Crimson,* 3 December 1975).

The latter argument is identical to that advanced simultaneously by student members of the Harvard-Radcliffe Committee against Racism, who, citing the Science for the People statement for authority, did not hesitate to label the book "dangerously racist" in leaflets distributed through the Boston area. Both the logic and the accusation were false and hurtful, and at this point the matter was close to getting out of hand.

On various occasions and with only limited success the Harvard faculty has attempted to protect itself from activities of this kind. During an earlier, similar episode 100 of its members published a statement that "In an academic community the substitution of personal harassment for reasoned inquiry is intolerable. The openminded search for truth cannot proceed in an atmosphere of political intimidation." This is the melancholy principle which has been confirmed by the exchange now extended to *BioScience.* In the Boston area at the present time it has become difficult to conduct an open forum on human sociobiology, or even general sociobiology, without falling into the role of either prosecutor or defendant.

The Political Significance of Sociobiology

Finally and briefly, let me express what I consider to be the real significance of human sociobiology for political and social thought. The question that science is now in a position to approach is the very origin and meaning of human values, from which all ethical pronouncements and much of political practice flow. Philosophers themselves have not explored the problem; traditional ethical philosophy begins with premises that are examined with reference to their consequences but not their origins. Thus, John Rawls open his celebrated *A Theory of Justice* (1971) with a proposition he regards as beyond dispute: "In a just society the liberties of equal citizenship are taken as settled; the rights secured by justice are not subject to political bargaining or to the calculus of social interests." Robert

Nozick launches his equally celebrated *Anarchy, State, and Utopia* (1974) with a similarly firm proposition: "Individuals have fights, and there are things no person or group may do to them (without violating their rights). So strong and far-reaching are these rights that they raise the question of what, if anything, the state and its officials may do."

These two premises are somewhat different in content, and they lead to radically different prescriptions. Rawls would allow rigid social control to secure as close an approach as possible to the equal distribution of society's rewards. Nozick sees the ideal society as one governed by a minimal state, empowered only to protect its citizens from force and fraud, and with unequal distribution of rewards wholly permissible Rawls rejects the meritocracy; Nozick accepts it as desirable except in those cases where local communities voluntarily decide to experiment with egalitarianism.

Whether in conflict or agreement, where do such fundamental premises come from? What lies behind the intuition on which they are based? Contemporary philosophers have progressed no further than Sophocles' Antigone, who said of moral imperatives, "They were not born today or yesterday; they die not, and none knoweth whence they sprung."

At this point the 35 members of the Science for the People group also come to a halt. At the close of their essay they imply the central issue to be a decision about the kind of the society we want to live in; humanity can then find the way to bring this society into being But which persons are the "we" who will decide, and whose moral precepts must thereby be validated? The group believe that all social behavior is learned and transmitted by culture. But if this is true, the value system by which "we" will decide social policy is created by the culture in which the most powerful decision makers were reared and hence must inevitably validate the status quo, the very condition which the Science for the People group reject. The solution to the conundrum must be that their premise of complete environmentalism is wrong.

The evidence that human nature is to some extent genetically influenced is in my opinion decisive. In the present space I can only suggest that the reader consider the facts presented in *Sociobiology* and in the very extensive primary literature on the subject, some of which is cited in this work. It follows that value systems are probably influenced, again to an unknown extent, by emotional responses programmed in the limbic system of the brain. The qualities that comprise human nature in the Mating of New Guinea as recognizably as they did in the Greeks at Troy are surely due in part to constraints within the unique human genotype. The challenge of human sociobiology, shared with the social sciences, is to measure the degree of these constraints and to infer their significance through the reconstruction of the evolutionary history of the mind. The enterprise is the logical complement to the continued study of cultural evolution.

Even if that formidable challenge is successfully met, however, it will still leave the ethical question: To what extent should the censors and motivators in the emotive centers of the brain be obeyed? Given that these controls deeply and unconsciously affect our moral decisions, how faithfully must they be consulted once they have been defined and assayed as a biological process? The answer must confront what appears to me to be the true human dilemma. We cannot follow the suggestions of the censors and motivators blindly. Although they are the source of our deepest and most compelling feelings, their genetic constraints evolved during the millions of years of prehistory, under conditions that to a large extent no longer exist. At some time in the future it will be necessary to decide how human we wish to remain, in this the ultimate biological sense, and to pick and choose consciously among the emotional guides we have inherited.

This dilemma should engender a sense of reserve about proposals for radical social change based on utopian intuition. To the extent that the biological interpretation noted here proves correct, men have rights that are innate, rooted in the ineradicable drives for survival and self-esteem, and these rights do not require the validation of ad hoc theoretical constructions produced by society. If culture is all that created human rights, as the extreme environmentalist position holds, then culture can equally well validate their removal. Even some philosophers of the radical left see this flaw in the position taken by Science for the People. Noam Chomsky, whose own linguistic research has provided evidence for the existence of genetic influence, considers extreme environmentalism to be a belief susceptible to dictatorships of both the left and the right:

> One can easily see why reformers and revolutionaries should become radical environmentalists, and there is no doubt that concepts of immutable human nature can be and have been employed to erect barriers against social change and to defend established privilege. But a deeper look will show that the concept of the "empty organism," plastic and unstructured, apart from being false, also serves naturally as the support for the most reactionary social doctrines. If people are, in fact, malleable and plastic beings with no essential psychological nature, then why should they not be controlled and coerced by those who claim authority, special knowledge, and a unique insight into what is best for those less enlightened? . . . The principle that human nature, in its psychological aspects, is nothing more than a product of history and given social relations removes all barriers to coercion and manipulation by the powerful. This too, I think, may be a reason for its appeal to intellectual ideologists, of whatever political persuasion (Chomsky 1975, p. 132).

Chomsky and I, not to mention Herbert Marcuse (who has a similar belief in the biological conservatism of human nature), can scarcely be accused of having linked arms to preserve the status quo, and yet that would seem to follow from the strange logic employed by the Science for the People group.

In their corybantic attentions to sociobiology, the Science for the People group have committed what can be usefully termed the Fallacy of the Political Consequent. This is the assumption that political belief systems can be mapped one-on-one onto biological or psychological generalizations. Another particularly ironic example is the response to B. F. Skinner's writings. Skinner is a radical environmentalist, whose conclusions about human behavior are essentially indistinguishable from those of the Science for the People group. Yet the particular political conclusions he has drawn are anathema to the radical left, who reject them as elitist, reactionary, and so forth. The cause of the Fallacy of the Political Consequent is the failure to appreciate adequately that scientific theories and political ideas are both complex and tenuously linked, and that political ideas are shaped in good part by personal judgments lying outside the domain of scientific evaluation.

All political proposals, radical and otherwise, should be seriously received and debated. But whatever direction we choose to take in the future, social progress can only be enhanced, not impeded, by the deeper investigation of the genetic constraints of human nature, which will steadily replace rumor and folklore with testable knowledge. Nothing is to be gained by a dogmatic denial of the existence of the constraints or attempts to discourage public discussion of them. Knowledge humanely acquired and widely shared, related to human needs but kept free of political censorship, is the real science for the people.

References

Allen, E. et at. (1975). "Against 'Sociobiology'," *The New York Review of Books*, November 13.

Chomsky, N. (1975). *Reflections on Language* (New York: Pantheon Books Random House).

Nozick, R. (1974). A*narchy, State and Utopia* (New York: Basic Books).

Otterbein, K. F. (1970). *The Evolution of War* (New Haven, Conn.: Human Relations Area Files Press).

Rawls, J. (1971). *A Theory of Justice* (Cambridge, Mass.: Harvard University Press).

Wilson, E. O. (1975a). *Sociobiology: The New Synthesis* (Cambridge, Mass.: Harvard University Press).

————. (1975b). "Human Decency Is Animal." *The New York Times Magazine*. October 12, pp. 38–50.

Chapter 10

Introduction

There are two documents in this section, both of which are referred to in the main text. The first is a piece by the philosopher Stephen Toulmin, written about forty years ago. Toulmin, an Englishman although now long resident in the United States, was one of a group of thinkers (the most notable representative being Thomas Kuhn, the author of The Structure of Scientific Revolutions) around 1960 who were very interested in the nature of science (in other words, with philosophical questions) who thought that the way to go was to immerse oneself in the great scientific texts of the past (in other words, with historical answers). This led him (as indeed it led others, including Kuhn) to a kind of evolutionary view of science, seeing continuity in the changes and the influence of the past upon the present. One could have left this as a rather loose metaphor, but Toulmin pushed the idea much harder, arguing that in fact the nature and change of science is remarkably like that of the nature and change of biological species. "The evolutionary development of natural science" is Toulmin's forthright exposition of this insight.

The second document is the piece that in the main text I highlight as what many take to be a dreadful example of **how not to do philosophy.** Edward O. Wilson and I start with the basic claim that moral philosophy, ethics, should be put on a naturalist basis and that means an evolutionary basis. Of course, what this means is not necessarily the same thing to the same people—consider how Christians interpret the Love Commandment (love your neighbor as yourself) in so many different ways. My strong suspicion is that Wilson and I never truly agreed on the answer to the question of foundations. For me, "turning moral philosophy into an applied science" means ultimately showing that there are no foundations. For Wilson, "turning moral philosophy into an applied science" means ultimately deriving the foundations of morality from the upwardly progressive nature of the evolutionary process. But we had a lot of fun writing the piece and learnt as much about our own thinking as we did about the thinking of our co-author. I hope you too will enjoy the piece and, in reacting against us, learn about your own thinking.

The Evolutionary Development of Natural Science

Stephen E. Toulmin

In the course of the first three centuries of modern science—from around 1600 A.D. until a generation ago—all aspects of the natural world in turn came under the scientist's scrutiny: the stars and the earth, living creatures and their fossil remains, atoms and cells, chickadees and chimpanzees, primitive societies and mental disorders. I say "all aspects," but it would be more exact to say "nearly all." For, throughout this period, one thing was generally exempted from the scope of scientific inquiry: although with the passage of time many aspects of human behavior came to be studied from different points of view—so giving rise to the new sciences of ethnology, anthropology, sociology, and abnormal psychology—the activities of *the scientist himself* were not normally considered a suitable object for scientific study and analysis. Right through the nineteenth century, any suggestion of a "science of science" would have struck men as a kind of *lèse-raison*. The business of science (it was thought) is to study the causes of natural phenomena; whereas science itself, as a rational activity, presumably operated on a higher level, and could not be thought of as a "natural phenomenon."

More recently, this self-denying ordinance has been somewhat relaxed. Twentieth-century science is less committed than the science of earlier centuries to explaining its phenomena in terms of rigid, mechanistic, cause-and-effect ideas. As a result, some of the restrictions earlier placed on scientific inquiries have been weakened, and the nature and working of science itself have been analyzed from various different points of view. Let me begin by reminding you about three of these lines of attack, which have up to now been followed largely independently.

(1) To begin with, the development of' natural science has been studied in a quantitative, statistical manner. For more than a century, since the pioneer work of Quetelet, statisticians have been developing techniques for describing and analyzing organic populations and growth-processes. As a result, it has become a commonplace that certain standard forms of growth-curve recur in a wide range of contexts, both biological and sociological: so that one and the same numerical pattern may be manifested equally in the growth of a bean-stalk, the spread of an infectious disease through a population, and the sales of domestic refrigerators. (The classic account of this general theory is to be found in D' Arcy Thompson's splendid treatise *On Growth and Form.*) Yet it was barely ten years ago that Professor Derek Price of Yale first demonstrated that these very same growth-patterns are discoverable also in the statistics of scientific activity [1]. If we provide ourselves with numerical indices for measuring the sheer quantity of scientific work being done at any time, we find

(Price showed) those very same "S-shaped" or "logistic" growth-curves which are already familiar in the case of organic activities of other kinds.

The activities of scientists, accordingly, can be subjected to numerical analysis as legitimately—at any rate—as the activities of other social groups and professions. For what they are worth, the resulting discoveries can be highly suggestive. Not that they tell us everything, by any means: the answers to which social statistics can lead us are limited by the questions which statistical method permits us to ask. The same kind of limitation is involved here as (for instance) in gas theory, where thermodynamics and statistical mechanics give a great deal of insight at a macroscopic level, but tell us only the very minimum about the individual molecules of different gases. (Much of the virtue of statistical mechanics, indeed, lies in the fact that it is neutral as between different gases. So too, the sociometrics of science is, inevitably, neutral as between scientific inquiries of different kinds.) The content and merit of different pieces of scientific work must be judged, first, by criteria drawn from outside the statistics of science: in the nature of the case, sociometric methods of inquiry give us only numerical answers to quantitative questions.

(2) Meanwhile, other scholars and scientists have been studying the development of science from a different, genetic point of view. Their concerns are with the internal development of the scientific tradition, and with the processes by which scientific ideas grow out of and displace one another. For them, the process of scientific development is to be thought of, not so much as a quantitative, organic growth-process, but rather as a dialectical sequence: problems lead to solutions, which in turn lead to new problems, whose solutions pose new problems again . . . investigations yield ideas, which provide material for new investigations, out of which emerge further ideas. . . and so on.

This genetic or problematic approach to the development of science can be considered from two somewhat different points of view: sociohistorical or logico-philosophical. One may study the problematic development of science in the hope of building up an historical understanding of the characteristic processes of intellectual change in natural science; or alternatively one may aim at producing a logical analysis (or "rational reconstruction") of the methods of inquiry and argument by which scientific progress is properly made. Either way, this approach to the study of scientific development also is subject to a certain self-limitation. It gives an account of scientific development in which factors *outside* the disciplinary procedures of the natural science in question are referred to only marginally, if at all. To use a biological metaphor: it studies the ontogeny or morphogenesis of a science in isolation from its ecological environment. Clearly, for many purposes, the resulting abstraction may be both legitimate and fruitful; but it too is, nevertheless, an abstraction.

(3) If the morphogenetic study of development abstracts a particular science from its wider environment, and considers its internal development in

isolation, the natural complement consists in a purely sociological approach to the development of science. And many people during the last half century have indeed been drawn toward a study of the external, environmental interactions between science—regarded as a social phenomenon—and the larger culture or society within which the scientist has to operate: its institutions, its social structure, politics, and economics.

Once again, many of the results have been profoundly interesting, and in some cases unexpected. The work of such men as Dean Don K. Price of Harvard has led us to understand in a new way the manner in which the different scientific subdisciplines have become organized into institutional "guilds," and the processes by which scientific work has acquired the new economic, political, and social impact characteristic of the last hundred years [2]. Meanwhile, there has been a perennial temptation to look for a "feed-back" from the social context into the actual content of scientific ideas: to speculate (for instance) that, in some manner or other, the development of thermodynamic theory in the first half of the nineteenth century *must* reflect in its structure contemporary developments in the technology of steam-locomotives. (Yet the actual *form* of this influence has up to now proved elusive.) Others have gone further, and hinted, e.g., that Darwin's theory of natural selection should be thought of as a reflection of contemporary beliefs about *laissez-faire* economics-at which point, most readers begin to feel that ingenuity has lapsed into implausibility [3]. In such a generalized sociology of science, as in the numerical statistics of scientific growth, one must feel that the processes by which the content of science develops slip through our intellectual sieve; up to now attempts to force answers about *content* out of questions about the *social ecology* of science seem only to have distorted that content.

Yet it is worth asking: "Can we not find a fresh standpoint, from which we can preserve the real virtues of all these three distinct approaches, within the framework of a single, coherent account of scientific development?" If each of the three approaches does have real merit, it must surely be possible to harmonize them. For, manifestly, the internal development of scientific thought does have a kind of rationality and method; even though accident, spontaneity, and in some cases inspired blundering, have had their parts to play. Manifestly, too, there are quite genuine interactions between scientific thought and its social environment, although these are more subtle than the naïver Marxists would imply. The question, therefore, is: "How are we to bring these approaches closer together? Can we look at all these questions from a standpoint which makes the nature of their convergence more evident?"

This problem will be our chief topic in all that follows. The task will be to argue our way to a provisional model for analyzing the process of scientific development—a model, in the sense of a theoretical pattern showing the inter-relations of different concepts and questions; but a merely provisional one, since on this occasion we can deal with the problem only to a first-order (perhaps even a "zeroth-order") approximation. Still, despite the crudity of this initial treatment, it will perhaps be a worthwhile achievement if we can simply establish that the three familiar approaches toward the study of scientific development *are* harmonizable within a larger, more integrated account.

II

We can usefully preface our analysis with a reminder, and with a truism. The initial reminder is the following. If we have any difficulty in relating our views about the internal development of natural science with our views about the external influences affecting the growth and development of science, this is partly because the contrast compels us necessarily—to oscillate between talking about the *ideas* of the natural sciences, and talking about the *men* who conceived, held, and/or rejected those ideas. A more comprehensive account of the development of science will require us to see how a history of ideas is to be related to a history of people: that is, how, within an evolving tradition of ideas, the actual content of the tradition affects and is affected by the activities of the human beings carrying the tradition.

However different these two aspects may appear, and however different the idioms in which we must describe them, the development of a system of ideas, and the intellectual activities of the people involved in that development, are two faces of a single coin; and the comprehensive account at which we are aiming must, at the very least, show how the life of ideas dovetails in with the lives of men. More specifically: the continuity and change which are characteristic of an evolving intellectual tradition must be related, in any such account, to the processes of transmission by which the ideas in question are passed on from one generation of human "carriers" to the next. (In this context, the word "carriers" is, quite deliberately, ambiguous: nor need the manifest implication—that scientific curiosity in general, and specific ideas in particular, spread through a population infectiously, like a disease—be regarded as derogatory: after all, the statistical evidence already hints at the possibility that the spread of ideas follows patterns familiar from epidemiology.)

The initial truism points in the same direction. For it is a commonplace to remark that an intellectual tradition is "scientific" *only* if the men who carry it in any particular generation regard the ideas to which their training exposes them in a sufficiently critical spirit—only (that is) if they are motivated by genuine, first-hand curiosity, by a spirit of innovation, by a desire to build up a more adequate, detailed, and/or elegant synthesis of the knowledge transmitted to them than that of their predecessors—only, in brief, if they are men with the "intellectual fidgets." In that case, we can ask: "How is it that the intellectual fidgets essential to scientific advance come to be infectious?" or, "How are the symptoms and after-effects characteristic of this intellectual state able to take permanent hold and establish themselves, within a population of inquirers, or tradition of ideas?" Again, to put the same questions more portentously, we may ask: "By what processes do intellectual innovations originate, spread, and establish themselves within a scientific tradition?"

III

With these two points in mind, we may now go back and take a second look at the process of scientific development. To begin with, we remarked that analyses of scientific development currently deal with two contrasted groups of questions—one group concerned with the internal development of scientific ideas, the other with the sociological and statistical aspects of scientific activities. Our two preliminary observations suggest, however, that this sharp contrast between the "internal" and "external" aspects of science should be replaced, rather, by a *spectrum* of questions—ranging from those which involve almost exclusively internal considerations, to those which are concerned predominantly with external (socio-historical, political, or economic) factors.

(1) To begin at the latter end of the spectrum: the social history of science has, as one of its central problems, the question, "What conditions must hold if there are to be any opportunities for scientific innovation at all?" Notice that this is not primarily a psychological question, since one may take it for granted that, within any population whatever, there will be a minority of human beings having the necessary innate curiosity. Essentially, it is a sociological question, arising out of the observation that different societies and cultures, at different stages in their history, provide different opportunities and/or incentives to intellectual innovation—or, more commonly, put different obstacles and/or disincentives in the way of intellectual heterodoxy. (If we ask, for instance, why cosmology and astronomy developed more slowly in China than in the West, we must bear in mind the tendency of eminent Chinese in the classical period to complain about the prevalence of unconventional ideas: where the cultural elite regards intellectual innovations as "dangerous thoughts," the institutions needed for the effective development of new scientific ideas can hardly flourish [4].

Indeed: whenever one turns to consider the development of science in any particular culture, nation, or epoch, one fruitful first question can be, "On whose back was Science riding at this stage?" Just because disinterested curiosity about the natural world, being in itself a "pure" form of intellectual activity, pays no particular dividends beyond the satisfactions of better understanding, it has never by itself given men a living. The fruitful development of science has always been contingent on other activities or institutions, which—inadvertently or by design—have provided occasions for men to pursue scientific investigations. In retrospect, it may be obvious that the development of natural science is one of the crucial achievements of human civilization; yet, sociologically speaking, scientific activities have hitherto been merely epiphenomena.

If men in earlier epochs and other cultures have "changed their minds" about Nature, this has happened always as a *by-product* of activities having more direct social, economic, or political functions. In the great days of Babylon, for instance, striking progress took place in computational astronomy; but the men concerned developed these techniques in their capacity as government servants—for purposes of official prognostication and calendrical computation. In medieval Islam, again, the natural sciences of Greek antiquity were kept alive, and developed further, at a time when they were languishing in Europe; but there, too, the men responsible earned their living by other means—in most cases, as court physicians. Among the men who established the Royal Society in seventeenth-century London, a few were scholars of independent means, yet many of them needed other sources of professional income; and the finance for the Royal Society itself was obtained from King Charles II by the Secretary of the Admiralty (Samuel Pepys, the diarist, who was also the first Secretary of the Royal Society), through the same concatenation of circumstances that led so much American research in the 1950's to be financed through the Office of Naval Research. In the next century, we find the Anglican and Dissenting Churches providing employment for educated men which left them enough surplus energy and resources to pursue significant scientific work as well . . . And so the story goes on; with the National Aeronautics and Space Administration as only one more in a long sequence of institutions which have provided extraneous occasions for the scientifically-minded to exercise their disinterested curiosities.

To sum up this first group of questions: what opportunities any culture provides for the development of heterodox ideas about nature, and what *volume of innovation* one finds there, are matters which depend predominantly on factors *external* to the scientific developments in question. Faced with problems concerning the volume of scientific work being done on a given subject within some particular society, we can reasonably enough cite social, economic, institutional, and similar factors as the major considerations bearing on such issues. Even here, one has to qualify the generalization by the use of such words as "predominant" and "major," for the "ripeness" or "unripeness" of a particular subject also serves to enhance or inhibit intellectual curiosity. (When a problem shows signs of yielding to investigation, a bandwagon effect frequently follows; and conversely, a recalcitrant field of inquiry will remain comparatively neglected despite otherwise favorable social and institutional conditions; but these qualifications are—arguably—second-order ones.)

(2) So much for the factors which determine how large a pool of scientific variants and novelties is under consideration at any particular place and time. But, when we turn our attention away from the sheer *size* of this pool, and start to ask questions about its *contents,* the picture begins to change. For why (we may ask) do scientists choose the particular new lines of thought (innovations, variants) they do? What considerations incline them to favor—say—"corpuscular," "fluid," or "field" theories of physical phenomena at any particular stage, and to ignore alternative

possibilities, even when experiment does not choose between them? Where a dominant *direction* of variation can be observed within any particular science, or where some particular direction of innovation appears to have been excessively neglected, a new type of issue arises. Within the total volume of intellectual variants under discussion, what factors determine which types of option are, and which are not pursued?

Questions of this kind are, perhaps, the most complicated that can arise for the historian of scientific thought. On the one hand, the considerations which incline scientists working in neurophysiology (say) or atomic physics or optics, at any particular stage, to take certain general types of hypothesis more seriously than others must undoubtedly be related to the intellectual situation within that branch of natural science at the moment in question. (Notice: we are here concerned with the "initial plausibility" attributed to certain classes of hypotheses, not with their "verification" or "establishment." We are asking how scientists come to take certain kinds of new suggestions seriously in the first place—considering them to be worthy of investigation at all—rather than with the standards they apply in deciding that those suggestions are in fact sound and acceptable.) So it is clear that the existence and continuity of certain "schools," "fashions," or "points of view" within, say, physical theory must be regarded essentially as an internal, professional matter; and will need to be analyzed and explained, substantially, in terms of the longer-term historical evolution of ideas within that particular area of science.

Even so, this will rarely be the whole story, and sometimes it may be only a small part of it. In plenty of cases, the justification for taking a particular kind of scientific hypothesis seriously has to be sought outside the intellectual content of that particular science. The influence of Platonist ideas on Johann Kepler, for instance, shows that any attempt to draw a hard and fast boundary around "astronomical" considerations would be vain. Likewise in nineteenth-century zoology: there too, a satisfactory story must bring in, e.g., the inhibiting influence of orthodox natural theology, on the one hand, and the positive influence of Malthus' theories of human population-growth on the other. When we are concerned with the content of the pool of intellectual variants, accordingly, rather than with its sheer volume, we have to consider this as the product—in varying proportions—of both "internal" and "external" factors.

(3) However, if we proceed still further along the spectrum of possible questions, we shall find the balance tilting sharply in the other direction. Consider the question: "What factors determine which of the intellectual variants circulating in any generation are selected out and incorporated into the tradition of scientific thought?" Evidently enough, the course of intellectual change within the sciences depends not merely on intellectual variation, but even more on the collective decisions by which certain new suggestions are generally accepted as "established" and transmitted to the next generations of scientists as "well-attested" results. The crucial factor in this selection-process is the set of criteria in

the light of which that choice is made. How do scientists determine this choice? Faced with that question, we must give a double answer—in part, one concerned with aspirations; in part, one which recognizes historical actualities.

Suppose we consider only aspirations: i.e., the explicit program to which natural scientists would subscribe as a question (so to speak) of ideology. As a matter of broad principle, scientists commonly take it for granted that their criteria of "truth," "verification," or "falsification" are stateable in absolute terms. In principle, that is, these criteria should be the same for scientists in all epochs, in all cultures, and should remain unaffected by such factors as political prejudice and theological conservatism. To formulate the criteria in explicit terms may be a taxing and contentious task, but at any rate (they believe) one is entitled to demand that any solution to this problem shall provide a satisfactory "demarcation criterion" for setting off irrelevant, "extrinsic" considerations from relevant, "intrinsic" considerations.

So much for theoretical aspirations; but, when we turn to look at historical actualities, the picture becomes slightly more complex. True: one may certainly argue that these selection-criteria are—and are rightly—determined *predominantly* by the professional values of the community of scientists in question. (This, as Michael Polanyi has argued, is one fundamental element in the political theory of the "Republic of Science") [5]. Yet there are reasons for wondering whether, in actual fact, this absolute independence of the selection-criteria from social and historical factors *has* ever been entirely realized; or, indeed, whether it ever *could* be. Many people will recall the passage in Pierre Duhem's book, *The Aim and Structure of Physical Theory,* in which he compares and contrasts the *styles* of theory found acceptable, respectively, by physical scientists in nineteenth-century Britain and France. French physicists writing about electricity and magnetism (he points out) demanded formal, axiomatized mathematical expositions, with all the assumptions and deductions set out clearly and unambiguously. British physicists working in the same area operated, rather, in terms of mechanical models: these were to a large extent intuitive rather than explicit, and they served their explanatory function by exploiting the power of analogy rather than the rigor of deduction. Duhem confesses himself to be, in this respect, an authentic Frenchman. Commenting on Oliver Lodge's new textbook of electrical theory, he remarks:

> In it there are nothing but strings which move around pullies, which roll around drums, which go through pearl beads, which carry weights; and tubes which pump water while others swell and contract; toothed wheels which are geared to one another and engage hooks. We thought we were entering the tranquil and neatly-ordered abode of reason, but we find ourselves in a factory.

Nor (Duhem argues) does this represent merely a temporary fad on the part of these particular English physicists. The habit of organizing physical ideas in terms of concrete analogies, rather than in abstract, mathematical form, is deeply rooted among English scientists, and represents the application within the scientific area of an even broader habit of mind, whose influence ranges over much larger regions of cultural and intellectual life. He compares this contrast between British and French patterns of thought in science with the contrast between Shakespeare and Racine, that between the *Code Napoléon* and the British tradition of Common Law, and that between the philosophies of Francis Bacon and René Descartes. The *espr'it géometrique* is a part of the French intellectual inheritance, in its widest terms; and this has served to influence the selection-criteria by which French scientists choose between rival hypotheses, just as it has served to influence so many other aspects of French intellectual life. Conversely, the habit of thinking in terms of particulars, and considering them in intuitive and imaginative terms—that *espr'i de finesse* which Pascal contrasted with *esprit géometrique*—has been equally characteristic of British habits of thought [6]. As a matter of historical fact, accordingly, the considerations bearing on the "establishment" of novel scientific hypotheses just cannot be stated in a form which will be *absolutely* invariant as between different epochs, different nations, and different cultural contexts. As an aspiration or ideal, such an absolute invariance may be something worth aiming at; but it has never been entirely realized in fact.

Nor is this solely a matter of historical fact. To go further: there are reasons for questioning whether such an ideal, absolute invariance is even attainable. For the processes of "proving," "establishing," "checking out," and/or "attempting to falsify" the novel ideas up for discussion within science at any time are *themselves* subject to a historical development of their own. In a, striking series of papers, Dr. Imre Lakatos has demonstrated that our concepts of "proof" and "refutation" have been subject to a slow, but definite and inescapable historical evolution *even within pure mathematics*. What counted as a proof or a refutation for Theaetetus or Euclid, for Wallis or Newton, for Euler or Gauss, for Dedekind or Weierstrass cannot be represented in terms of some unique, eternal, historically unchanging, logical pattern. On the contrary, throughout, the history of mathematical thought, the concepts of "proof" and "refutation" have themselves been slowly changing: more slowly (it is true) than the content of mathematics itself, but changing none the less [7]. And if this is true even within pure mathematics—which of all disciplines can most plausibly claim to illustrate the eternal virtues of a formalized logic—must we not suppose that the criteria of "verification," "establishment," and the like in natural science also have undergone a similar historical development?

IV

At this point we can make explicit the intellectual model toward which this discussion has been leading us. For, in the course of expounding all these considerations, we have fallen again and again—quite naturally—into the vocabulary of organic evolution. Science develops (we have said)as the outcome of a double process: at each stage, a pool of competing intellectual variants is in circulation, and in, each generation a selection process is going on, by which certain of these variants are accepted and incorporated into the science concerned, to be passed on to the next generation of workers as integral elements of the tradition.

Looked at in these terms, a particular scientific discipline—say, "atomic physics"–needs to be thought of, not as the contents of a textbook bearing any specific elate, but rather as a developing subject having a continuing identity through time, and characterized as much by its process of growth as by the content of any-one historical cross-section. Such a tradition will then display both elements of continuity and elements of variability. Why do we regard the atomic physics of 1960 as part of the "same" subject as the atomic physics of 1910, 1920, ... or 1950? Fifty years can transform the actual content of a subject beyond recognition; yet there remains a perfectly genuine continuity, both intellectual and institutional. This reflects both the master-pupil relationship, by which the tradition is passed on, and also the genealogical sequence of intellectual problems around which the men in question have focused their work. Moving from one historical cross-section to the next, the actual ideas transmitted display neither a complete breach at any point—the idea of absolute "scientific revolutions" involves an over-simplification [8]–nor perfect replication, either. The change from one cross-section to the next is an *evolutionary* one in this sense too: that later intellectual cross-sections of a tradition reproduce the content of their immediate predecessors, as modified by those particular intellectual novelties which were selected out in the meanwhile—in the light of the professional standards of the science of the time.

An "evolutionary" account of scientific change puts us in a position to re-interpret the spectrum of questions we constructed for ourselves in the preceding section. At one extreme, we saw, the *volume* of new intellectual innovations is highly sensitive to external factors: the relevant questions correspond, in the zoological sphere, to questions about the frequency of mutations within an organic population, and mutation-frequency too is highly sensitive to external influences such as cosmic rays. At the other extreme, the selective factors by which new ideas, or new organic forms are perpetuated for incorporation into the subsequent population, arise very much more from the detailed interaction between the variants and the immediate environment they face. At this level, considerations of an external kind—whether to do with cosmic rays, or with the social context—lose their earlier importance. Now the only question is "Do the new forms meet the detailed demands of the situation significantly better than their predecessors?" And those demands have to do predominantly with the narrower issues on which competitive survival depends.

Does the historical development of a science ever fit this evolutionary pattern perfectly? Can we use it as an instrument for analyzing scientific growth with any confidence? There is no point in making exaggerated claims for the model at this stage. Rather, we should explore its implications in a hypothetical way, to see whether it yields abstractions by which the patterns of scientific history can be more clearly described.

Suppose, then, that there *are* certain phases in the history of scientific thought which, for all practical purposes, do exemplify the evolutionary pattern expounded here. Suppose, that is, that there *are* certain periods of scientific development during which all significant changes in the content of a particular science were in fact the outcome of intellectual selections, made according to strictly professional criteria, from among pools of intellectual variants from a previous tradition of ideas. In such a case (to coin a word) we may speak of the scientific tradition in question as a *compact* tradition. Other traditions, which change in a less systematic way, can, by contrast, be referred to as more or less *diffuse*.

Evidently, to the extent that it presupposes our model, the concept of a "compact tradition" has the status of an intellectual ideal, having the same virtues and limitations as the concept of an "ideal gas" or "rigid body" or "inertial frame of reference." We are not obliged to demonstrate that *all* scientific changes whatever conform to this ideal, any more than we need demonstrate that *all* material bodies are "perfectly rigid," or all actual gases "ideal." Still, if we find as we go along that the notion of a "compact tradition" can be used to throw light on a variety of historical processes within the development of science; and if we find that the deviations from this pattern can, in their own ways, be explained quite as interestingly and illuminatingly as examples of conformity to it—if "diffuse" and "compact" changes are equally significant in their own ways—in that ease, we shall be entitled to conclude that the notion is justifying itself. A full discussion of this topic, however, will have to wait for another occasion [9].

V

Certainly, we must concede, there are clear instances in which the actual facts of scientific development *do not* fit our basic pattern at all accurately. Notoriously, the historical development of some natural sciences has included, e.g., cases in which the intellectual variants available for discussion at a given time were not adequately checked or tested, and for many years went—so to speak–"underground": a classic instance of this is Mendel's theory of genetical "factors." In a sense (one might say) Mendel's theory represented an intellectual variant available within the pool, but one which was overlooked and so failed to establish itself for more than 35 years. Yet, on second thought, one may inquire: "On its first presentation, was Mendel's novel theory really introduced into the general pool of available variants at all?" Was it (that is) put into effective circulation among professional, biologists in such a way that its virtues could be properly appraised? Arguably, this did not happen: the very limited contact between the Abbe Mendel and other theoretical biologists in his time shunted his variant off into a corner, where it could not demonstrate its merits in competition with its rivals [10].

Again, the development of scientific thought includes occasional phases of a kind which have no obvious analogy in the sphere of organic evolution. For instance, a kind of hybridization sometimes takes place between different branches of science, so giving rise to brand-new specialities, with subsequent genealogies and histories of their own: the most striking recent example of this was the emergence of molecular biology around 1950, through the cross-fertilization of crystallography and biochemistry. By itself, bur model of a *compact* tradition does not give us the means of analyzing or understanding such a hybridization.

Yet again, other fields of intellectual inquiry—known as "sciences" at any rate to their participants—develop in a way which scarcely exemplifies at all the orderly, cumulative pattern characteristic of a compact tradition. In sociology, for instance, the ideas of any one generation seem to have more in common with the ideas current two generations before than with those of the intervening generation. There is a kind of *pendulum-swing* in the ideas of the subject, by which, e.g., "historical-evolutionary" (or "diachronic") patterns of thought alternate with "functional" (or "synchronic") patterns of thought. The latest phases in the work of Talcott Parsons, for instance, thus recall the ideas of sociologists before 1900, rather than those of sociology during the interwar years [11] .

Once again, however, these criticisms may not represent so much *objections* to our model of a "compact tradition"; rather, they may indicate merely the need for further *refinements* to the model. After all, the very fact that the intellectual tradition of theoretical sociology lacks that compactness which one can find within (say) atomic physics is itself a significant fact. Perhaps there are quite genuine reasons, both intellectual and professional, why sociological theory should not yet have *acquired* the maturity required to guarantee such a compactness and continuity. And perhaps, in his own time, Mendel's ideas inevitably remained "recessive," just because their author was effectively isolated from the rest of professional biology. If that were so, the failure of genetics and sociology to conform to our ideal of a "compact" tradition would do as much to confirm the relevance of that ideal as the actual conformity of more mature and established sciences.

VI

The prime merit of the model expounded here is this: it focuses attention—in a way dispassionate and abstract accounts of the history of "scientific thought" tend not to do—on the questions, "*Who* carries the tradition of scientific thought? *Who* is responsible for the innovations by which this tradition changes? *Who* determines the manner in which the selection is made between these innovations? " And these questions lead one to examine the crucial relationship, within the larger process of scientific change, between the individual scientific innovator and the professional guild by which his ideas are judged. Just how far afield a study of this crucial relationship can lead us is another story, which we cannot go into here; but one point at any rate must be noted.

According to one widely accepted picture of science, the fundamental advances in our knowledge of Nature have all come about through Great Men *changing their* minds—having the honesty and candor to acknowledge the unexpectedness of certain phenomena, and the courage to modify their concepts in the light of these unforeseen observations. This picture of science as progressing through the successive discoveries of Great Men is an agreeable and engaging one, if what Science requires is folk-heroes to populate its Pantheon; yet a little reflection on the actual structure of scientific change may justify one in questioning its accuracy. Indeed, it is a matter for debate how far great scientists *do* in fact ever change their minds; and the actual historical development of Science would—arguably—have been very little different, even if no such "mind-changes" had ever taken place.

Consider, for instance, the work of Isaac Newton himself. We tend to think of Newton as the great intellectual innovator, yet it is worth reminding ourselves how little the basic framework of ideas within which he operated changed between the years of his youth and his old age. The final *Queries,* added to later editions of the *Opticks,* serve substantially to work out in more detail, and provide fresh illustrations of, ideas which had been present in rough, if embryonic, form even in his earliest speculations. True: for a while in middle life, having discovered how easily such hypotheses could generate bitter contention with his colleagues, Newton soft-pedaled his thoughts about the ether, and concentrated on less disputatious matters. Still, it is a closer approximation to the truth to represent Newton's intellectual development as comprising the progressive ramification of a fundamentally unchanging natural philosophy, than as involving a series of daring intellectual changes and reappraisals: The great and real change for which Isaac Newton is remembered was that between his own ideas and those which he inherited in his youth from his predecessors. The crucial change, that is to say, was a change *between* the generation of Newton's predecessors and Newton's own generation, rather than a change *within* the intellectual development of Newton himself [12].

The development of Max Planck's thought provides another interesting illustration. We tend to think of Planck as one of the conscious revolutionaries of science—as a man who helped to found twentieth-century quantum theory, through a daring breach with the work of his predecessors. Yet Planck saw his own work in quite a different perspective. He put forward his hypothesis that the emission of electromagnetic radiation by material bodies is "quantized," as a regrettable, but necessary refinement on Maxwell's classical theories, not as their abandonment. And he continued for some ten years to believe that the electromagnetic field in itself is something continuous, rather than characterized by discrete units. Indeed, the appearance of Einstein's theory of the "photon," in 1905, filled him initially with indignation: he found himself quite unable to accept it, since it struck him as involving a needless abandonment of Maxwell's electromagnetism, just at a moment when Maxwell's theory was finally establishing its credentials.

The younger generation of scientists, by contrast, had no hesitation in accepting both Planck's and Einstein's innovations; and they soon identified them as the joint pillars of the new "quantum" theory of radiation, on which any future work in optics and electromagnetism would have to be based. So much was this so that, in retrospect, we normally forget that there ever was a difference of opinion between the two men. Planck's skepticism was, in fact, almost universally disregarded. Whether or not he chose to accept Einstein's "photon" interpretation of the quantum hypothesis, scarcely mattered to his contemporaries: this became purely a biographical question about Max Planck himself. The general tradition—the theoreticians' consensus—moved at once beyond him. And, by the time Planck was finally reconciled to Einstein's position, after the Solvay conference of 1911, the development of the conceptual tradition within theoretical physics was rapidly leaving him behind.

These examples suggest an answer to one of our central questions: "If we regard a particular scientific discipline as a tradition, what should we think of as forming an historical cross-section of that tradition?" The answer toward which we are moving is the following. The carrier of scientific thought, at any particular stage, is the relevant "generation" of original young research workers. Each new generation *re-creates* for itself a vision of nature, which owes much to the ideas of its immediate masters and teachers, but in which the ideas of the preceding generation are never replicated exactly. (Perfect replication is the mark of "Scholasticism.") The operative question for any adequate philosophy or logic of science accordingly is: "What criteria does each new generation of scientists rely on, in deciding which aspects of their elders' theories to carry over into their own ideas about nature, and which to abandon in favor of current variants and innovations?"

VII

I shall end with a warning. In talking about the development of natural science as "evolutionary," I have not been employing a mere *façon de parler,* or analogy, or metaphor. The idea that the historical changes by which scientific thought develops frequently follow an "evolutionary" pattern needs to be taken quite seriously; and the implications of such a pattern of change can be, not merely suggestive, but explanatory.

To a philosopher of science, these implications are attractive for two reasons in particular. To begin with, they make more intelligible the justice of Karl Popper's central thesis: his insistence that "scientific method" depends on only two fundamental maxims—freedom of conjecture, and severity of criticism. For, if the fundamental mission of scientific thought in any human generation is to adapt itself better to the demands of the existing intellectual situation, these will be precisely the two cardinal virtues of science. Freedom of conjecture enlarges the available pool of variants: severity of criticism enhances the degree of selective pressure. Just as, in the organic world, adequate adaptation can be achieved only given a sufficient rate of mutation and a sufficient selective pressure, so, within the context of an evolutionary theory of scientific change, the double formula, "Conjectures and Refutations," makes perfect sense.

The present model has one other philosophical attraction. As we saw at the outset, the "ologies" of science, viz., the philosophy of science, the logic of inquiry, the sociology, history, and psychology of science, its politics and its economics—all of these disciplines have developed, hitherto, in more or less complete independence. Yet anyone who takes a serious interest in several of these sub-branches of the world of learning must feel a certain irritation at the necessity to switch categories every time he moves from one of these fields of inquiry to another. If the present argument has no other value, it does at any rate begin to show how reasonable and plausible connections could be established between the views of science as seen from all these different directions.

References

1. D. J. DE S. PRICE, *Little Science, Big Science* (New York Columbia U.P. 1962).
2. DON K. PRICE, *Government and Science* (New York U. Press, 1954), and *The Scientific Estate* (Cambridge, Mass.: Harvard U.P., 1965). .
3. See, for instance, J. D. BERNAL, *Science in History* (London: Watts and New York: Hawthorn, 1954) sec. 9.6. .
4. *Cf.* JOSEPH NEEDHAM, *Science and Civilisation in China,* Vol. 3 (Cambridge U.P., 1959) esp. sec. 20(c)(2), pp. 186 ff.
5. MICHAEL POLANYl, "The Republic of Science: Its Political and Economic Theory," *Minerva,* Vol. I, No.1 (Autumn, 1952), pp. 54-73.
6. PIERRE DUHEM, *The Aim and Structure of Physical Theory* (Eng. tr. P. P. Wiener, Princeton, U.P., 1954), ch. IV, pp. 55-104.
7. IMRE LAKATOS, "Proofs and Refutations," *British Journal for Philosophy of Science,* Vol. XIV (1963-4) pp. 1-25, 120-176, 221-264, and 296-342.
8. Even THOMAS S. KUHN, who argued so persuasively for the idea in *The Structure of Scientific Revolutions* (Chicago U.P., 1962), now seems to be retreating from the implications of his own earlier position: see, e.g., his paper "Logic of Discovery or Psychology of Research" in the forthcoming collection *The Philosophy of K. R. Popper* (ed. P. A. Schilpp), The Library of Living Philosophers Series.
9. Discussed In Part I of my forthcoming *New Inquiries into Human Understanding.*
10. E. B. GASKING, "Why Was Mendel's Work Ignored?", *Journal for the History of Ideas,* Vol. XX (1959), pp. 60-84.
11. *Cf.* J. W. BURROW, *Evolution and Society* (Cambridge U.P. 1967).
12. Newton's final account of his natural philosophy, as expounded in Query 31 of the third edition of Newton's *Opticks,* strikingly resembles ideas adumbrated in his earliest notebooks: on this point, see Isaac Newton, *Mathematical Principles of Natural Philosophy* (ed. F. Cajori, Berkeley: California U.P. (1934), esp. note 55, pp. 671-679.

Moral Philosophy as Applied Science

Michael Ruse and Edward O. Wilson

(1) For much of this century, moral philosophy has been constrained by the supposed absolute gap between *is* and *ought*, and the consequent belief that the facts of life cannot of themselves yield an ethical blueprint for future action. For this reason, ethics has sustained an eerie existence largely apart from science. Its most respected interpreters still believe that reasoning about right and wrong can be successful without a knowledge of the brain, the human organ where all the decisions about right and wrong are made. Ethical premises are typically treated

in the manner of mathematical propositions: directives supposedly independent of human evolution, with a claim to ideal, eternal truth.

While many substantial gains have been made in our understanding of the nature of moral thought and action, insufficient use has been made of knowledge of the brain and its evolution. Beliefs in extrasomatic moral truths and in an absolute is/ought barrier are wrong. Moral premises relate only to our physical nature and are the result of an idiosyncratic genetic history—a history which is nevertheless powerful and general enough within the human species to form working codes. The time has come to turn moral philosophy into an applied science because, as the geneticist Hermann J. Muller urged in 1959, 100 years without Darwin are enough.[1]

(2)The naturalistic approach to ethics, dating back through Darwin to earlier pre-evolutionary thinkers, has gained strength with each new advance in biology and the brain sciences. Its contemporary version can be expressed as follows:

Everything human, including the mind and culture, has a material base and originated during the evolution of the human genetic constitution and its interaction with the environment. To say this much is not to deny the great creative power of culture, or to minimize the fact that most causes of human thought and behaviour are still poorly understood. The important point is that modern biology can account for many of the unique properties of the species. Research on the subject is accelerating, quickly enough to lend plausibility to the belief that the human condition can eventually be understood to its foundation, including the sources of moral reasoning.

This accumulating empirical knowledge has profound consequences for moral philosophy. It renders increasingly less tenable the hypothesis that ethical truths are extrasomatic, in other words divinely placed within the brain or else outside the brain awaiting revelation. Of equal importance, there is no evidence to support the view—and a great deal to contravene it—that premises can be identified as global optima favouring the survival of any civilized species, in whatever form or on whatever planet it might appear. Hence external goals are unlikely to he articulated in this more pragmatic sense.

Yet biology shows that internal moral premises do exist and can be defined more precisely. They are immanent in the unique programmes of the brain that originated during evolution. Human mental development has proved to be far richer and more structured and idiosyncratic than previously suspected. The constraints on this development are the sources of our strongest feelings of right and wrong, and they are powerful enough to serve as a foundation for ethical codes. But the articulation of enduring codes will depend upon a more detailed knowledge of the mind and human evolution than we now possess. We suggest that it will prove possible to proceed from a knowledge of the material basis of moral feeling to generally accepted rules of conduct. To do so will be to escape—not a minute too soon—from the debilitating absolute distinction between *is* and *ought*.

---

[1]H. J. Muller is quoted by G. G. Simpson in *This View of Life* (New York: Harcourt, Brace & World, 1964), 36.

(3) All populations of organisms evolve through a law-bound causal process, as first described by Charles Darwin in his *Origin of Species*. The modern explanation of this process, known as natural selection, can be briefly summarized as follows. The members of each population vary hereditarily in virtually all traits of anatomy, physiology, and behaviour. Individuals possessing certain combinations of traits survive and reproduce better than those with other combinations. As a consequence, the units that specify physical traits—genes and chromosomes—increase in relative frequency within such populations, from one generation to the next. This change in different traits, which occurs at the level of the entire population, is the essential process of evolution. Although the agents of natural selection act directly on the outward traits and only rarely on the underlying genes and chromosomes, the shifts they cause in the latter have the most important lasting effects. New variation across each population arises through changes in the chemistry of the genes and their relative positions on the chromosomes. Nevertheless, these changes (broadly referred to as mutations) provide only the raw material of evolution. Natural selection, composed of the sum of differential survival and reproduction, for the most part determines the rate and direction of evolution.[2]

Although natural selection implies competition in an abstract sense between different forms of genes occupying the same chromosome positions or between different gene arrangements, pure competition, sometimes caricatured as 'nature red in tooth and claw', is but one of several means by which natural selection can operate on the outer traits. In fact, a few species are known whose members do not compete among themselves at all. Depending on circumstances, survival and reproduction can be promoted equally well through the avoidance of predators, more efficient breeding, and improved co-operation with others.[3]

In recent years there have been several much-publicized controversies over the pace of evolution and the universal occurrence of adaptation.[4] These uncertainties should not obscure the key facts about organic evolution: that it occurs as a universal process among all kinds of organisms thus far carefully examined, that the dominant driving force is natural selection, and that the observed major pat-

---

[2]See the following widely used textbooks: J. Roughgarden, *Theory of Population Genetics and Evolutionary Ecology: An Introduction* (New York: Macmillan, 1979); D. L. Hartl, *Principles of Population Genetics* (Sunderland, Mass.: Sinauer Associates, 1980); R. M. May (ed.), *Theoretical Ecology: Principles and Applications*, 2nd edn (Sunderland, Mass.: Sinauer Associates, 1981); J. R. Krebs and N. B. Davies (eds), *Behavioural Ecology: An Evolutionary Approach*, 2nd edn (Sunderland, Mass.: Sinauer Associates, 1984).

[3]Reviews of the various modes of selection, including forms that direct individuals away from competitive behaviour, can be found in E. 0. Wilson, *Sociobiology: The New Synthesis* (Cambridge, Mass.: Belknap Press of Harvard University Press, 1975); G. F. Oster and E. 0. Wilson, *Caste and Ecology in the Social Insects* (Princeton University Press, 1978), S. A. Boorman and P. R. Levitt, *The Genetics of Altruism* (New York: Academic Press, 1980); D. S. Wilson, *The Natural Selection of Populations and Communities* (Menlo Park, Calif.: Benjamin/Cummings, 1980).

[4]For example, the debate over 'punctuated equilibrium' versus 'gradualism' among palaeontologists and geneticists. For most biologists, the issue is not the mechanism of evolution but the conditions under which evolution sometimes proceeds rapidly and sometimes slows to a crawl. There is no difficulty in explaining the variation in rates. On the contrary, there is a surplus of plausible explanations, virtually all consistent with Neo-Darwinian theory, but insufficient data to choose among them. See, for example, S. J. Gould and N. Eldredge, 'Punctuated Equilibria: The Tempo and Mode of Evolution Reconsidered', *Paleobiology 3* (1977) 115-151, and J. R. G. Turner, '"The hypothesis that explains mimetic resemblance explains evolution the gradualist-saltationist schism', in M. Grene (ed.), *Dimensions of Darwinism* (Cambridge University Press, 1983), 129-169.

terns of change are consistent with the known principles of molecular biology and genetics. Such is the view held by the vast majority of the biologists who actually work on heredity and evolution.[5] To say that not all they facts have been explained, to point out that forces and patterns may yet be found that are inconsistent with the central theory—healthy doubts present in any scientific discipline—is by no means to call into question the prevailing explanation of evolution. Only a demonstration of fundamental inconsistency can accomplish that much, and nothing short of a rival explanation can bring the existing theory into full disarray.

There are no such crises. Even Motoo Kimura, the principal architect of the 'neutralist' theory of genetic diversity—which proposes that most evolution at the molecular level happens through random factors—allows that 'classical evolution theory has demonstrated beyond any doubt that the basic mechanism for adaptive evolution is natural selection acting on variations produced by changes in chromosomes and genes. Such considerations as population size and structure, availability of ecological opportunities, change of environment, life-cycle "strategies", interaction with other species, and in some situations kin or possibly group selection play a large role in our understanding of the process.'[6]

(4) Human evolution appears to conform entirely to the modern synthesis of evolutionary theory as just stated. We know now that human ancestors broke from a common line with the great apes as recently as six or seven million years ago, and that at the biochemical level we are today closer relatives of the chimpanzees than the chimpanzees are of gorillas.[7] Furthermore, all that we know about human fossil history, as well as variation in genes and chromosomes among individuals and the key events in the embryonic assembly of the nervous system, is consistent with the prevailing view that natural selection has served as the principal agent in the origin of humanity.

It is true that until recently information on the brain and human evolution was sparse. But knowledge is accelerating, at least as swiftly as the remainder of natural science, about a doubling every ten to fifteen years. Several key developments, made principally during the past twenty years, will prove important to our overall argument for a naturalistic ethic developed as an applied science.

The number of human genes identified by biochemical assay or pedigree analysis is at the time of writing 3,577, with approximately 600 placed to one or the other of the twenty-three pairs of chromosomes.[8] Because the rate at which this number has been accelerating (up from 1,200 in 1977), most of the entire complement of 100,000 or so structural genes may be characterized to some degree within three or four decades.

Hundreds of the known genes affect behaviour. The great majority do so simply by their effect on general processes of tissue development and metabolism,

[5]See footnote 2.

[6]M. Kimura, *The Neutral Theory of Molecular Evolution* (Cambridge University Press, 1983).

[7]C. G. Sibley and J. E. Ahiquist, 'The Phylogeny of the Hominoid Primates, as Indicated by DNA-DNA Hybridization', *Journal of Molecular Evolution* **20** (1984), 2-15.

[8]We are grateful to Victor A. McKusick for providing the counts of identified and inferred human genes up to 1984.

but a few have been implicated in more focused behavioural traits. For example, a single allele (a variant of one gene), prescribes the rare Lesch—Nyhan syndrome, in which people curse uncontrollably, strike out at others with no provocation, and tear at their own lips and fingers. Another allele at a different chromosome position reduces the ability to perform on certain standard spatial tests but not on the majority of such tests.[9] Still another allele, located tentatively on chromosome 15, induces a specific learning disability.[10]

These various alterations are of course strong and deviant enough to be considered pathological. But they are also precisely the kind usually discovered in the early stages of behavioural genetic analysis for any species. *Drosophila* genetics, for example, first passed through a wave of anatomical and physiological studies directed principally at chromosome structure and mechanics. As in present-day human genetics, the first behavioural mutants discovered were broadly acting and conspicuous, in other words those easiest to detect and characterize. When behavioural and biochemical studies grew more sophisticated, the cellular basis of gene action was elucidated in the case of a few behaviours, and the new field of *Drosophila* neurogenetics was born. The hereditary bases of subtle behaviours such as orientation to light and learning were discovered somewhat later.[11]

We can expect human behavioural genetics to travel along approximately the same course. Although the links between genes and behaviour in human beings are more numerous and the processes involving cognition and decision making far more complex, the whole is nevertheless conducted by cellular machinery precisely assembled under the direction of the human genome (that is, genes considered collectively as a unit). The techniques of gene identification, applied point by point along each of the twenty-three pairs of chromosomes, is beginning to make genetic dissection of human behaviour a reality.

Yet to speak of genetic dissection, a strongly reductionist procedure, is not to suggest that the whole of any trait is under the control of a single gene, nor does it deny substantial flexibility in the final product. Individual alleles (gene-variants) can of course affect a trait in striking ways. To take a humble example, the possession of a single allele rather than another on a certain point on one of the chromosome pairs causes the development of an attached earlobe as opposed to a pendulous earlobe. However, it is equally true that a great many alleles at different chromosome positions must work together to assemble the entire earlobe. In parallel fashion, one allele can shift the likelihood that one form of behaviour will develop as opposed to another, but many alleles are required to prescribe the ensemble of nerve cells, neurotransmitters, and muscle fibres that orchestrate the behaviour in the first place. Hence classical genetic analysis cannot by itself ex-

---

[9]G.C. Ashton, J. J. Polovina and S. G. Vandenberg, 'Segregation Analysis of Family Data for 15 Tests of Cognitive Ability', *Behaviour Genetics* **9** (1979), 329-347.

[10]S. D. Smith, W. J. Kimberling, B. F. Pennington and H. A. Lubs, 'Specific Reading Disability: Identification of an Inherited Form through Linkage Analysis', *Science* **219** (1982), 1345-1347.

[11]See J. C. Hall and R. J. Greenspan, 'Genetic Analysis of *Drosophila* Neurobiology', *Annual Review of Genetics* **13** (1979), 127-195.

plain all of the underpinnings of human behaviour, especially those that involve complex forms of cognition and decision making. For this reason behavioural development viewed as the interaction of genes and environment should also occupy centre stage in the discussion of human behaviour. The most important advances at this level are being made in the still relatively young field of cognitive psychology.[12]

(5) With this background, let us move at once to the central focus of our discussion: morality. Human beings, all human beings, have a sense of right and wrong, good and bad. Often, although not always, this 'moral awareness' is bound up with beliefs about deities, spirits, and other supersensible beings. What is distinctive about moral claims is that they are prescriptive; they lay upon us certain obligations to help and to co-operate with others in various ways. Furthermore, morality is taken to transcend mere personal wishes or desires. 'Killing is wrong' conveys more than merely 'I don't like killing'. For this reason, moral statements are thought to have an objective referent, whether the Will of a Supreme Being or eternal verities perceptible through intuition.

Darwinian biology is often taken as the antithesis of true morality. Something that begins with conflict and ends with personal reproduction seems to have little to do with right and wrong. But to reason along such lines is to ignore a great deal of the content of modern evolutionary biology. A number of causal mechanisms—already well confirmed in the animal world—can yield the kind of co-operation associated with moral behaviour. One is so-called 'kin selection'. Genes prescribing co-operation spread through the populations when self-sacrificing acts are directed at relatives, so that they (not the co-operators) are benefited, and the genes they share with the co-operators by common descent are increased in later generations. Another such co-operation-causing mechanism is 'reciprocal altruism'. As its name implies, this involves transactions (which can occur between non-relatives) in which aid given is offset by the expectation of aid received. Such mutual assistance can be extended to a whole group, whose individual members contribute to a general pool and (as needed) draw from the pool.[13]

Sociobiologists (evolutionists concerned with social behaviour) speak of acts mediated by such mechanisms as 'altruistic'. It must be recognized that this is now a technical biological term, and does not necessarily imply conscious free giving and receiving. Nevertheless, the empirical evidence suggests that co-operation between human beings was brought about by the same evolutionary mechanisms as those just cited. To include conscious, reflective beings is to go beyond the biological sense of altruism into the realm of genuine non-metaphorical altruism. We do not claim that people are either unthinking genetic robots or that they co-operate only when the expected genetic returns can be calculated in advance. Rather, human beings function better if they are deceived by their genes into thinking that there is a disinterested objective morality binding upon them, which

---

[12]See, for example, the recent analysis by J. R. Anderson, *The Architecture of Cognition* (Cambridge, Mass.: Harvard University Press, 1983).

[13]See footnote 3.

all should obey. We help others because it is 'right' to help them and because we know that they are inwardly compelled to reciprocate in equal measure. What Darwinian evolutionary theory shows is that this sense of 'right' and the corresponding sense of 'wrong', feelings we take to be above individual desire and in some fashion outside biology, are in fact brought about by ultimately biological processes. Such are the empirical claims. How exactly is biology supposed to exert its will on conscious, free beings? At one extreme, it is possible to conceive of a moral code produced entirely by the accidents of history. Cognition and moral sensitivity might evolve somewhere in some imaginary species in a wholly unbiased manner, creating the organic equivalent of an all-purpose computer. In such a blank-slate species, moral rules were contrived some time in the past, and the exact historical origin might now be lost in the mists of time. If proto-humans evolved in this manner, individuals that thought up and followed rules ensuring an ideal level of co-operation then survived and reproduced, and all others fell by the wayside.

However, before we consider the evidence, it is important to realize that any such even-handed device must also be completely gene-based and tightly controlled, because an exact genetic prescription is needed to produce perfect openness to any moral rule, whether successful or not. The human thinking organ must be indifferently open to a belief such as 'killing is wrong' or 'killing is right', as well as to any consequences arising from conformity or deviation. Both a very specialized prescription and an elaborate cellular machinery are needed to achieve this remarkable result. In fact, the blank-slate brain might require a cranial space many times that actually possessed by human beings. Even then a slight deviation in the many feedback loops and hierarchical controls would shift cognition and preference back into a biased state. In short, there appears to be no escape from the biological foundation of mind.

It can be stated with equal confidence that nothing like all-purpose cognition occurred during human evolution. The evidence from both genetic and cognitive studies demonstrates decisively that the human brain is not a tabula rasa. Conversely, neither is the brain (and the consequent ability to think) genetically determined in the strict sense. No genotype is known that dictates a single behaviour, precluding reflection and the capacity to choose from among alternative behaviours belonging to the same category. The human brain is something in-between: a swift and directed learner that picks up certain bits of information quickly and easily, steers around others, and leans toward a surprisingly few choices out of the vast array that can be imagined.

This quality can be made more explicit by saying that human thinking is under the influence of 'epigenetic rules', genetically based processes of development that predispose the individual to adopt one or a few forms of behaviours as opposed to others. The rules are rooted in the physiological processes leading from the genes to thought and action.[14] The empirical heart of our discussion is that we think morally because we are subject to appropriate epigenetic rules. These predispose us to think that certain courses of action are right and certain courses of action are wrong. The rules certainly do not lock people blindly into certain behaviours. But because they give the illusion of objectivity to morality, they lift us above immediate wants to actions which (unknown to us) ultimately serve our genetic best interests.

The full sequence in the origin of morality is therefore evidently the following: ensembles of genes have evolved through mutation and selection within an intensely social existence over tens of thousands of years; they prescribe epigenetic rules of mental development peculiar to the human species; under the influence of the rules certain choices are made from among those conceivable and available to the culture; and finally the choices are narrowed and hardened through contractual agreements and sanctification.

In a phrase, societies feel their way across the fields of culture with a rough biological map. Enduring codes are not created whole from absolute premises but inductively, in the manner of common law, with the aid of repeated experience, by emotion and consensus, through an expansion of knowledge and experience guided by the epigenetic rules of mental development, during which people sift the options and come to agree upon and to legitimate certain norms and directions.[15]

(6) Only recently have the epigenetic rules of mental development and their adaptive roles become accepted research topics for evolutionary biology. It should therefore not be surprising that to date the best understood examples of epigenetic rules are of little immediate concern to moral philosophers. Yet what such examples achieve is to draw us from the realm of speculative philosophy into the centre of ongoing scientific research. They provide the stepping stones to a more empirical basis of moral reasoning.

One of the most fully explored epigenetic rules concerns the constraint on colour vision that affects the cultural evolution of colour vocabularies. People see variation in the intensity of light (as opposed to colour) the way one might intuitively expect to see it. That is, if the level of illumination is raised gradually, from dark to brightly lit, the transition is perceived as gradual. But if the wavelength is changed gradually, from a monochromatic purple all across the visible spectrum to a monochromatic red, the shift is not perceived as a continuum. Rather, the full range is thought to comprise four basic colours (blue, green, yellow, red), each persisting across a broad band of wavelengths and giving way through ambiguous intermediate colour through narrow bands on either side. The physiological

---

[14]The evidence for biased epigenetic rules of mental development is summarized in C. J. Lumsden and E. 0. Wilson, *Genes, Mind, and Culture* (Cambridge, Mass.: Harvard University Press, 1981) and *Promethean Fire: Reflections on the Origin of Mind* (Cambridge, Mass.: Harvard University Press, 1983).

[15]A new discipline of decision-making is being developed in cognitive psychology based upon the natural means—one can correctly say the epigenetic rules—by which people choose among alternatives and reach agreements. See, for example, A. Tversky and D. Kahneman, 'The Framing of Decisions and the Psychology of Choice', *Science* **211** (1981), 453-458; and R. Axelrod, *The Evolution of Cooperation* (New York: Basic Books, 1984).

basis of this beautiful deception is partly known. There are three kinds of cones in the retina and four kinds of cells in the lateral geniculate nuclei of the visual pathways leading to the optical cortex. Although probably not wholly responsible, both sets of cells play a role in the coding of wavelength so that it is perceived in a discrete rather than continuous form. Also, some of the genetic basis of the cellular structure is known. Colour-blindness alleles on two positions in the X-chromosome cause particular deviations in wavelength perception.

The following experiment demonstrated the effect of this biological constraint on the formation of colour vocabularies. The native speakers of twenty languages from around the world were asked to place their colour terms in a standard chart that displays the full visible colour spectrum across varying shades of brightness. Despite the independent origins of many of the languages, which included Arabic, Ibidio, Thai, and Tzeltal, the terms placed together fall into four distinct clusters corresponding to the basic colours. Very few were located in the ambiguous intermediate zones.

A second experiment then revealed the force of the epigenetic rule governing this cultural convergence. Prior to European contact the Dani people of New Guinea possessed a very small colour vocabulary. One group of volunteers was taught a newly invented Dani-like set of colour terms placed variously on the four principal hue categories (blue, green, yellow, red). A second group was taught a similar vocabulary placed off centre, away from the main clusters formed by other languages. The first group of volunteers, those given the 'natural' vocabulary, learned about twice as quickly as those given the off-centre, less natural terms. Dani volunteers also selected these terms more readily when allowed to make a choice between the two sets.[16]

So far as we have been able to determine, all categories of cognition and behaviour investigated to the present time show developmental biases. More precisely, whenever development has been investigated with reference to choice under conditions as free as possible of purely experimental influence, subjects automatically favoured certain choices over others. Some of these epigenetic biases are moderate to very strong, as in the case of colour vocabulary. Others are relatively weak. But all are sufficiently marked to exert a detectable influence on cultural evolution.

Examples of such deep biases included the optimum degree of redundancy in geometric design; facial expressions used to denote the basic emotions of fear, loathing, anger, surprise, and happiness; descending degrees of preference for sucrose, fructose, and other sugars; the particular facial expressions used to respond to various distasteful substances; and various fears, including the fear-of-strangers response in children. One of the most instructive cases is provided by the phobias. These intense reactions are most readily acquired against snakes, spiders, high places, running water, tight enclosures, and other ancient perils of mankind for which epigenetic rules can be expected to evolve through natural selection. In

---

[16] E. Rosch, 'Natural Categories', *Cognitive Psychology* **4** (1973), 328-350.

contrast, phobias very rarely appear in response to automobiles, guns, electric sockets, and other truly dangerous objects in modern life, for which the human species has not yet had time to adapt through genetic change. Epigenetic rules have also been demonstrated in more complicated forms of mental development, including language acquisition, predication in logic, and the way in which objects are ordered and counted during the first steps in mathematical reasoning).[17]

We do not wish to exaggerate the current status of this area of cognitive science. The understanding of mental development is still rudimentary in comparison with that of most other aspects of human biology. But enough is known to see the broad outlines of complex processes. Moreover, new techniques are constantly being developed to explore the physical basis of mental activity. For example, arousal can be measured by the degree of alpha wave blockage, allowing comparisons of the impact of different visual designs. Electroencephalograms of an advanced design are used to monitor moment-by-moment activity over the entire surface of the brain. In a wholly different procedure, radioactive isotopes and tomography are combined to locate sites of enhanced metabolic activity. Such probes have revealed the areas of the brain used in specific mental operations, including the recall of melodies, the visualization of notes on a musical staff, and silent reading and counting.[18] There seems to be no theoretical reason why such techniques cannot be improved eventually to address emotions, more complex reasoning, and decision-making. There is similarly no reason why metabolic activity of the brain cannot be mapped in chimpanzees and other animals as they solve problems and initiate action, permitting the comparison of mental activity in human beings with that in lower species.

But what of morality? We have spoken of colour perception, phobias, and other less value-laden forms of cognition. We argue that moral reasoning is likewise moulded and constrained by epigenetic rules. Already biologists and behavioural scientists are moving directly into that area of human experience producing the dictates of right and wrong. Consider the avoidance of brother-sister incest, a negative choice made by the great majority of people around the world. By incest in this case is meant full sexual attraction and intercourse, and not merely exploratory play among children. When such rare matings do occur, lowered genetic fitness is the result. The level of homozygosity (a matching of like genes) in the children is much higher, and they suffer a correspondingly greater mortality and frequency of crippling syndromes due to the fact that some of the homozygous pairs of genes are defective. Yet this biological cause and effect is not widely perceived in most societies, especially those with little or no scientific knowledge of heredity. What causes the avoidance instead is a sensitive period between birth and approximately six years. When children this age are exposed to each other under conditions of close proximity (both 'use the same potty', as one anthropologist put it) they are unable to form strong sexual bonds during adolescence or

---

[17]The epigenetic rules of cognitive development analysed through the year 1980 are reviewed by C. J. Lumsden and E. 0. Wilson, op. cit.

[18]N. A. Lassen, D. H. Ingvar and E. Skinhøj, 'Brain Function and Blood Flow', *Scientific American* **239** (1978), 62-71.

later. The inhibition persists even when the pairs are biologically unrelated and encouraged to marry. Such a circumstance occurred, for example, when children from different families were raised together in Israeli kibbutzim and in Chinese households practising minor marriages.[19]

A widely accepted interpretation of the chain of causation in the case of brother-sister incest avoidance is as follows. Lowered genetic fitness due to inbreeding led to the evolution of the juvenile sensitive period by means of natural selection; the inhibition experienced at sexual maturity led to prohibitions and cautionary myths against incest or (in many societies) merely a shared feeling that the practice is inappropriate. Formal incest taboos are the cultural reinforcement of the automatic inhibition, an example of the way culture is shaped by biology. But these various surface manifestations need not be consulted in order to formulate a more robust technique of moral reasoning. What matters in this case is the juvenile inhibition: the measures of its strength and universality, and a deeper understanding of why it came into being during the genetic evolution of the brain.

Sibling incest is one of several such cases showing that a tight and formal connection can be made between biological evolution and cultural change. Models of sociobiology have now been extended to include the full co-evolutionary circuit, from genes affecting the direction of cultural change to natural selection shifting the frequencies of these genes, and back again to open new channels for cultural evolution. The models also predict the pattern of cultural diversity resulting from a given genotype distributed uniformly through the human species. It has just been seen how the avoidance of brother-sister incest arises from a strong negative bias and a relative indifference to the preferences of others. The quantitative models incorporating these parameters yield a narrow range of cultural diversity, with a single peak at or near complete rejection on the part of the members of most societies. A rapidly declining percentage of societies possess higher rates of acceptance. If the bias is made less in the model than the developmental data indicate, the mode of this frequency curve (that is, the frequency of societies whose members display different percentages of acceptance) shifts from one end of the acceptance scale towards its centre. If individuals are considerably more responsive to the preferences of others, the frequency curve breaks into two modes.[20]

Such simulations, employing the principles of population genetics as well as methods derived from statistical mechanics, are still necessarily crude and applicable only to the simplest forms of culture. But like behavioural genetics and the radionuclide-tomography mapping of brain activity, they give a fair idea of the kind of knowledge that is possible with increasing sophistication in theory and technique. The theory of the co-evolution of genes and culture can be used further to

[19]A. P. Wolf and C. S. Huang, *Marriage and Adoption in China*, 1845-1945 (Stanford University Press, 1980); J. Shepher, *Incest: A Biosocial View* (New York: Academic Press, 1983); P. L. van den Berghe, 'Human Inbreeding Avoidance: Culture in Nature', *The Behavioural and Brain Sciences* **6** (1983), 91-123.

[20]C. J. Lumsden and E. O. Wilson, op. cit. See also the précis of *Genes, Mind, and Culture* and commentaries on the book by twenty-three authors in *The Behavioural and Brain Sciences* **5** (1982), 1-37.

understand the origin and meaning of the epigenetic rules, including those that affect moral reasoning.

This completes the empirical case. To summarize, there is solid factual evidence for the existence of epigenetic rules—constraints rooted in our evolutionary biology that affect the way we think. The incest example shows that these rules, directly related to adaptive advantage, extend into the moral sphere. And the hypothesis of morality as a product of pure culture is refuted by the growing evidence of the co-evolution of genes and culture.

This perception of co-evolution is, of course, only a beginning. Prohibitions on intercourse with siblings hardly exhaust the human moral dimension. Philosophical reasoning based upon more empirical information is required to give a full evolutionary account of the phenomena of interest: philosophers' hands reaching down, as it were, to grasp the hands of biologists reaching up. Surely some of the moral premises articulated through ethical inquiry lie close to real epigenetic rules. For instance, the contractarians' emphasis on fairness and justice looks much like the result of rules brought about by reciprocal altruism, as indeed one distinguished supporter of that philosophy has already noted.[21]

(7) We believe that implicit in the scientific interpretation of moral behaviour is a conclusion of central importance to philosophy, namely that there can be no genuinely objective external ethical premises. Everything that we know about the evolutionary process indicates that no such extrasomatic guides exist. Let us define ethics in the ordinary sense, as the area of thought and action governed by a sense of obligation—a feeling that there are certain standards one ought to live up to. In order not to prejudge the issue, let us also make no further assumptions about content. It follows from what we understand in the most general way about organic evolution that ethical premises are likely to differ from one intelligent species to another. The reason is that choices are made on the basis of emotion and reason directed to these ends, and the ethical premises composed of emotion and reason arise from the epigenetic rules of mental development. These rules are in turn the idiosyncratic products of the genetic history of the species and as such were shaped by particular regimes of natural selection. For many generations—more than enough for evolutionary change to occur—they favoured the survival of individuals who practised them. Feelings of happiness, which stem from positive reinforcers of the brain and other elements that compose the epigenetic rules, are the enabling devices that led to such right action.

It is easy to conceive of an alien intelligent species evolving rules its members consider highly moral but which are repugnant to human beings, such as cannibalism, incest, the love of darkness and decay, parricide, and the mutual eating of faeces. Many animal species perform some or all of these things, with gusto and in order to survive. If human beings had evolved from a stock other than savanna-dwelling, bipedal, carnivorous man-apes we might do the same, feeling inwardly certain that such behaviours are natural and correct. In short, ethical premises are

---

[21]J. Rawls, *A Theory of Justice* (Cambridge, Mass.: Harvard University Press, 1971), 502-503.

the peculiar products of genetic history, and they can be understood solely as mechanisms that are adaptive for the species that possess them. It follows that the ethical code of one species cannot be translated into that of another. No abstract moral principles exist outside the particular nature of individual species.

It is thus entirely correct to say that ethical laws can be changed, at the deepest level, by genetic evolution. This is obviously quite inconsistent with the notion of morality as a set of objective, eternal verities. Morality is rooted in contingent human nature, through and through.

Nor is it possible to uphold the true objectivity of morality by believing in the existence of an ultimate code, such that what is considered right corresponds to what is truly right—that the thoughts produced by the epigenetic rules parallel external premises.[22] The evolutionary explanation makes the objective morality redundant, for even if external ethical premises did not exist, we would go on thinking about right and wrong in the way that we do. And surely, redundancy is the last predicate that an objective morality can possess. Furthermore, what reason is there to presume that our present state of evolution puts us in correspondence with ultimate truths? If there are genuine external ethical premises, perhaps cannibalism is obligatory.

(8) Thoughtful people often turn away from naturalistic ethics because of a belief that it takes the good will out of co-operation and reduces righteousness to a mechanical process. Biological 'altruism' supposedly can never yield genuine altruism. This concern is based on a half truth. True morality, in other words behaviour that most or all people can agree is moral, does consist in the readiness to do the 'right' thing even at some personal cost. As pointed out, human beings do not calculate the ultimate effect of every given act on the survival of their own genes or those of close relatives. They are more than just gene replicators. They define each problem, weigh the options, and act in a manner conforming to a well-defined set of beliefs—with integrity, we like to say, and honour, and decency. People are willing to suppress their own desires for a while in order to behave correctly.

That much is true, but to treat such qualifications as objections to naturalistic ethics is to miss the entire force of the empirical argument. There is every reason to believe that most human behaviour does protect the individual, as well as the family and the tribe and, ultimately, the genes common to all of these units. The advantage extends to acts generally considered to be moral and selfless. A person functions more efficiently in the social setting if he obeys the generally accepted moral code of his society than if he follows moment-by-moment egocentric calculations. This proposition has been well documented in the case of pre-literate societies, of the kind in which human beings lived during evolutionary time.

---

[22]This is the argument proposed by R. Nozick *in Philosophical Explanations* (Cambridge, Mass.: Belknap Press of Harvard University Press, 1981) in order to escape the implications of sociobiology.

While far from perfect, the correlation is close enough to support the biological view that the epigenetic rules evolved by natural selection.[23]

It should not be forgotten that altruistic behaviour is most often directed at close relatives, who possess many of the same genes as the altruist and perpetuate them through collateral descent. Beyond the circle of kinship, altruistic acts are typically reciprocal in nature, performed with the expectation of future reward either in this world or afterward. Note, however, that the expectation does not necessarily employ a crude demand for returns, which would be antithetical to true morality. Rather, I expect you (or God) to help me because it is right for you (or God) to help me, just as it was right for me to help you (or obey God). The reciprocation occurs in the name of morality. When people stop reciprocating, we tend to regard them as outside the moral framework. They are 'sociopathic' or 'no better than animals'.

The very concept of morality—as opposed to mere moral decisions taken from time to time—imparts efficiency to the adaptively correct action. Moral feeling is the shortcut taken by the mind to make the best choices quickly. So we select a certain action and not another because we feel that it is 'right', in other words, it satisfies the norms of our society or religion and thence, ultimately, the epigenetic rules and their prescribing genes. To recognize this linkage does not diminish the validity and robustness of the end result. Because moral consistency feeds mental coherence, it retains power even when understood to have a purely material basis.

For the same reason there is little to fear from moral relativism. A common argument raised against the materialist view of human nature is that if ethical premises are not objective and external to mankind, the individual is free to pick his own code of conduct regardless of the effect on others. Hence philosophy for the philosophers and religion for the rest, as in the Averrhoist doctrine. But our growing knowledge of evolution suggests that this is not at all the case. The epigenetic rules of mental development are relative only to the species. They are not relative to the individual. It is easy to imagine another form of intelligent life with non-human rules of mental development and therefore a radically different ethic. Human cultures, in contrast, tend to converge in their morality in the manner expected when a largely similar array of epigenetic rules meet a largely similar array of behavioural choices. This would not be the case if human beings differed greatly from one another in the genetic basis of their mental development.

Indeed, the materialist view of the origin of morality is probably less threatening to moral practice than a religious or otherwise non-materialistic view, for when moral beliefs are studied empirically, they are less likely to deceive. Bigotry declines because individuals cannot in any sense regard themselves as belonging to a chosen group or as the sole bearers of revealed truth. The quest for scientific understanding replaces the hajj and the holy grail. Will it acquire a similar pas-

---

[23]See footnote 16.

sion? That depends upon the value people place upon themselves, as opposed to their imagined rulers in the realms of the supernatural and the eternal.

Nevertheless, because ours is an empirical position, we do not exclude the possibility that some differences might exist between large groups in the epigenetic rules governing moral awareness. Already there is related work suggesting that the genes can cause broad social differences between groups—or, more precisely, that the frequency of genes affecting social behaviour can shift across geographic regions. An interesting example now being investigated is variation in alcohol consumption and the conventions of social drinking. Alcohol (ethanol) is broken down in two steps, first to acetaldehyde by the enzyme alcohol dehydrogenase and then to acetic acid by the enzyme acetaldehyde dehydrogenase. The reaction to alcohol depends substantially on the rate at which ethanol is converted into these two products. Acetaldehyde causes facial flushing, dizziness, slurring of words, and sometimes nausea. Hence the reaction to drinking depends substantially on the concentration of acetaldehyde in the blood, and this is determined by the efficiency of the two enzymes. The efficiency of the enzymes depends in turn on their chemical structure, which is prescribed by genes that vary within populations. In particular, two alleles (gene forms) are known for one of the loci (chromosome sites of the genes) encoding alcohol dehydrogenase, and two are known for a locus encoding acetaldehyde dehydrogenase. These various alleles produce enzymes that are either fast or slow in converting their target substances. Thus one combination of alleles causes a very slow conversion from ethanol to acetic acid, another the reverse, and so on through the four possibilities.

Independent evidence has suggested that the susceptibility to alcohol addiction is under partial genetic control. The tendency now appears to be substantially although not exclusively affected by the combination of genes determining the rates of ethanol and acetaldehyde conversion. Individuals who accumulate moderate levels of acetaldehyde are more likely to become addicted than those who sustain low levels. The propensity is especially marked in individuals who metabolize both ethanol and acetaldehyde rapidly and hence are more likely to consume large quantities to maintain a moderate acetaldehyde titre.

Differences among human populations also exist. Most caucasoids have slow ethanol and acetaldehyde conversion rates, and thus are able to sustain moderately high drinking levels while alone or in social gatherings. In contrast, most Chinese and Japanese convert ethanol rapidly and acetaldehyde slowly and thus built up acetaldehyde levels quickly. They reach intoxication levels with the consumption of a relatively small amount of alcohol.

Statistical differences in prevalent drinking habits are well known between the two cultures, with Europeans and North Americans favouring the consumption of relatively large amounts of alcohol dur ing informal gatherings and eastern Asiatics favouring the consumption of smaller amounts on chiefly ceremonial occasions. The divergence would now seem not to be wholly a matter of historical accident but to stem from biological differences as well. Of course a great deal remains to be learned Concerning the metabolism of alcohol and its effects on behaviour, but enough is known to illustrate the potential of the interaction of varying genetic material and the environment to create cultural diversity.[24]

It is likely that such genetic variation accounts for only a minute fraction of cultural diversity. It can be shown that a large amount of the diversity can arise purely from the statistical scatter due to differing choices made by genetically identical individuals, creating patterns that are at least partially predictable from a knowledge of the underlying universal bias.[25] We wish only to establish that, contrary to prevailing opinion in social theory but in concert with the findings of evolutionary biology, cultural diversity can in some cases be enhanced by genetic diversity. It is wrong to exclude a priori the possibility that biology plays a causal role in the differences in moral attitude among different societies. Yet even this complication gives no warrant for extreme moral relativism. Morality functions within groups and now increasingly across groups, and the similarities between all human beings appear to be far greater than any differences.

The last barrier against naturalistic ethics may well be a lingering belief in the absolute distinction between *is* and *ought*. Note that we say 'absolute'. There can be no question that *is* and *ought* differ in meaning, but this distinction in no way invalidates the evolutionary approach. We started with Hume's own belief that morality rests ultimately on sentiments and feelings. But then we used the evolutionary argument to discount the possibility of an objective, external reference for morality Moral codes are seen instead to be created by culture under the biasing influence of the epigenetic rules and legitimated by the illusion of objectivity. The more fully this process is understood, the sounder and more enduring can be the agreements.

Thus the explanation of a phenomenon such as biased colour vision or altruistic feelings does not lead automatically to the prescription of the phenomenon as an ethical guide. But this explanation, the *is* statement, underlies the reasoning used to create moral codes. Whether a behaviour is deeply ingrained in the epigenetic rules, whether it is adaptive or non-adaptive in modern societies, whether it is linked to other forms of behaviour under the influence of separate developmental rules: all these qualities can enter the foundation of the moral codes. Of equal importance, the means by which the codes are created, entailing the estimation of consequences and the settling upon contractual arrangements, are cognitive processes and real events no less than the more elementary elements they examine.

(9) No major subject is more important or relatively more neglected at the present time than moral philosophy. If viewed as a pure instrument of the humanities, it seems heavily worked, culminating a long and distinguished history. But if viewed as an applied science in addition to being a branch of philosophy, it is no better than rudimentary. This estimation is not meant to be derogatory. On the contrary, moral reasoning offers an exciting potential for empirical research and a

---

[24]E. Jones and C. Aoki, 'Genetic and Cultural Factors in Alcohol Use' (submitted to *Science*).
[25]C. J. Lumsden and E. 0. Wilson, op. cit., who show the way to predict cultural diversity caused by random choice patterns in different societies.

new understanding of human behaviour, providing biologists and psychologists join in its development. Diverse kinds of empirical information, best obtained through collaboration, are required to advance the subject significantly. As in twentieth-century science, the time of the solitary scholar pronouncing new systems in philosophy seems to have passed.

The very weakness of moral reasoning can be taken as a cause for optimism. By comparison with the financial support given other intellectual endeavours directly related to human welfare, moral philosophy is a starveling field. The current expenditure on health-related biology in the United States at the present time exceeds three billion dollars. Support has been sustained at that level or close to it for over two decades, with the result that the fundamental processes of heredity and much of the molecular machinery of the cell have been elucidated. And yet a huge amount remains to be done: the cause of cancer is only partly understood, while the mechanisms by which cells differentiate and assemble into tissues and organs are still largely unknown. In contrast, the current support of research on subjects directly related to moral reasoning, including the key issues in neurobiology, cognitive development, and sociobiology, is probably less than one per cent of that allocated to health-related biology. Given the complexities of the subject, it is not surprising that very little has been learned about the physical basis of morality—so little, in fact, that its entire validity can still be questioned by critics. We have argued that not only is the subject valid, but it offers what economists call increasing returns to scale. Small absolute increments in effort will yield large relative returns in concrete results. With this promise in mind, we will close with a brief characterization of several of the key problems of ethical studies as we see them.

First, only a few processes in mental development have been worked out in enough detail to measure the degree of bias in the epigenetic rules. The linkage from genes to cellular structure and thence to forms of social behaviour is understood only partially. In addition, a curious disproportion exists: the human traits regarded as most positive, including altruism and creativity, have been among the least analysed empirically. Perhaps they are protected by an unconscious taboo, causing them to be regarded as matters of the 'spirit' too sacred for material analysis.

Second, the interactive effects of cognition also remain largely unstudied. Among them are hierarchies in the expression of epigenetic rules. An extreme example is the suppression of preference in one cognitive category when another is activated. This is the equivalent of the phenomenon in heredity known as epistasis. We know in a very general way that certain desires and emotion-laden beliefs take precedence over others. Tribal loyalty can easily dominate other social bonds, especially when the group is threatened from the outside. Individual sacrifice becomes far more acceptable when it is believed to enhance future generations. The physical basis and relative quantitative strengths of such effects are almost entirely unknown.

Third, there is an equally enticing opportunity to create a comparative ethics, defined as the study of conceivable moral systems that might evolve in other intelligent species. Of course it is likely that even if such systems exist, we will never perceive them directly. But that is beside the point. Theoretical science, defined as the study of all conceivable worlds, imagines non-existent phenomena in order to classify more precisely those that do exist. So long as we confine ourselves to one rather aberrant primate species (our own), we will find it difficult to identify the qualities of ethical premises that can vary and thus provide more than a narrow perspective in moral studies. The goal is to locate human beings within the space of all possible moral systems, in order to gauge our strengths and weaknesses with greater precision.

Fourth, there are pressing issues arising from the fact that moral reasoning is dependent upon the scale of time. The trouble is that evolution gave us abilities to deal principally with short-term moral problems. ('Save that child!' 'Fight that enemy!') But, as we now know, short-term responses can easily lead to long-term catastrophes. What seems optional for the next ten years may be disastrous thereafter. Cutting forests and exhausting non-renewable energy sources can produce a healthy, vibrant population for one generation—and starvation for the next ten. Perfect solutions probably do not exist for the full range of time in most categories of behaviour. To choose what is best for the near future is relatively easy. To choose what is best for the distant future is also relatively easy, providing one is limited to broad generalities. But to choose what is best for both the near and distant futures is forbiddingly difficult, often drawing on internally contradictory sentiments. Only through study will we see how our short-term moral insights fail our long-term needs, and how correctives can be applied to formulate more enduring moral codes.

Chapter 11

Introduction

There are two batches of documents in this section. The first three, an opinion piece by me, a criticism by the developmental biologist Scott Gilbert, and a response by me, take on the question of natural selection in an age of evolutionary development theory. I argue that "evo-devo" is terrific new science, and were I an evolutionary biologist I would want to work in that area. But I don't think it threatens the Darwinian paradigm at all. It is all a question of natural selection working over the information thrown up by evo-devo. In his critique, Scott Gilbert argues that evo-devo now can be seen as providing the driving force of evolution and that basically natural selection simply has a mopping up function. I respond that Gilbert is just plain wrong. Embryology was a disaster for evolutionary studies in the years after the Origin. Let us not repeat the same mistakes all over again.

The second batch of documents focus on a recent paper by the philosopher Jerry Coyne in which he argues even more strongly than Gilbert that Darwinian theory, a theory focusing on natural selection, has had its day. Although Fodor's stance is a bit like that of Philip Johnson and of his critique of Darwinism—better at criticizing others than really providing an alternative of his own—it is clear that Fodor like Gilbert thinks that evo-devo might be the key to the evolutionism of the future. Expectedly Fodor upset a lot of orthodox evolutionists, both practicing biologists like Jerry Coyne and practicing philosophers like Daniel Dennett. I am not sure that anyone convinces anyone, but the issues are aired with vigor. Note the response by Steven Rose. He is a biologist and so does not want to trash his science to quite the extent that Fodor does, but he is also a Marxist throwback to the 1970s, when he was Britain's most ardent critic of Edward O. Wilson. Rose cannot resist the temptation to jab away at the regular Darwinians, even as he argues against Fodor. (Some of the references made by Rose will make more sense when we come to the next and final Chapter, dealing with Stephen Jay Gould.)

Concluding we have the delightful poem by Colin Boatman
It about sums things up perfectly.

Forty Years a Philosopher of Biology:
Why Evo Devo Makes Me Still Excited About My Subject

Michael Ruse

I have been a philosopher of biology now for forty years. When I started back around 1965, the biggest intellectual problem that we then faced was about the status of evolutionary theory. Was it a science like other sciences and, if so, was it as good as other sciences? General consensus was that either it was different—more historical or holistic or whatever—or that it really did not cut the mustard as a real science (aka physics). I think that I and others—most especially David Hull, but also including Ken Schaffner and Bill Wimsatt—did sterling work on this problem and that by the 1970s, no one could deny that evolutionary theory was good strong science and should apologize to no one, especially not physicists. There was still debate about the exact nature of evolutionary thinking and my suspicion is that many of us did think that there was something distinctive about biology. Speaking for myself, back then and still today, I argue that there is a teleological element to thinking about organisms—I do not think this a weakness but rather a reflection of the fact that natural selection has made organisms different from inanimate objects. Biology reflects this through the metaphor of design, and it is right and appropriate that it should do so.[i]

There were other related problems that fascinated us back then. One much-discussed topic was that of species—are they in some sense real in a way that other taxonomic groups are not? People like Ernst Mayr had made their biological reputations on taxonomy, and they insisted that this was a major issue, and we philosophers of biology agreed. Much effort was therefore put into trying to see how and why species function and what if anything makes them distinctive. My sense today is that although some do still write on the species problem, it does not have the immediacy that it had back then. So much has been said and written that we are all a bit exhausted. Also many—certainly this is true of me—feel that the stress on species is a distortion brought on by the fact that people like Mayr themselves worked with organisms (like birds) that form clearly defined biological

---

[i] Several years ago I published a little handbook to the philosophy of biology called *The Philosophy of Biology Today*. If anyone is interested in the ideas of the time and wants full references, I suggest they turn to this. If you want to see what I thought important back in the early days, look at my first little book, *The Philosophy of Biology*. My own recent attempt to put the teleology question in context is *Darwin and Design: Does Evolution have a Purpose?* I should say that this is the final work of a trilogy dealing with values in science, the earlier volumes being Monad to Man: The Concept of Progress in Evolutionary Biology and *Mystery of Mysteries: Is Evolution a Social Construction?* Although I can pride myself on being one of the founders of contemporary philosophy of biology, simply working within the paradigm has never really appealed to me and as it became more popular my own work turned more and more to issues in the history of ideas. In part, this is because I have the personality that likes working in new areas and in part because I came from the generation that took history of science very seriously. Look at Werner Callebaut's fascinating sociological study, *Taking the Naturalistic Turn*, for more on the background to me and others of my cohort.

species. What about plants and what about microorganisms? Is this a bit of a pseudo problem?

One topic that was not much discussed back then was human evolution—a little book I published in 1973 had no mention of the topic. That of course was to change drastically in the late 1970s, thanks to the sociobiology controversy. For a while it looked set to tear apart the now-growing philosophy of biology community, with people siding up against each other on the issue of whether we humans are indeed part and parcel of the biological world, and whether natural selection is the key to human understanding. Although tension still exists, I think wisely we philosophers realized that we have our problems, and they are not necessarily the problems of the scientists. If Harvard *primae donnae* want to quarrel that is their business and not ours. If immodestly I can take some credit for pouring oil on troubled waters, it is that in the 1980s I founded the journal *Biology and Philosophy*, and made it clear from Day One that all and any positions would find my pages open to them. No one could ever say that they were excluded on ideological grounds. Goodness, I even published Creationists!

What are the big issues today? What would I write on if in 2005, as in 1965, I were looking for a doctoral thesis topic? I pride myself on having a pretty good nose for a problem, and if I were going in the direction of straight philosophy of biology—as opposed to something that was going to bring in history, ever a fondness of mine—I would without hesitation go for evolutionary development, "evo-devo." I think some of the most incredible discoveries of recent years have come from this area—the amazing homologies between humans and fruitflies for starters. It almost makes me want to be an empirical scientist! (Not really. I love science and I love the history of science, but ultimately my tastes are metaphysical—perhaps even religious in a secular sense—and that is what makes scholarship so important for me.)[ii]

A big problem that I had when I started out doing the philosophy of biology—a big problem that my students always have—is learning how to do the subject. You read some exciting science and all you want to do is write it down, so you end with basically a *Scientific American* type of essay, talking about the science with enthusiasm but with no real philosophy. You have to keep asking: Why is what I am writing philosophy and not simply science journalism? So the question I now ask is: What are the philosophical problems I see raised by evo-devo? Let me list three, but if others disagree and have alternatives, then so much the better. (Psychologically, I am a huge Popperian. Get the ideas out on the table and let us fight like tigers to choose the best.)

First, there is the old question of the science itself. Since the time of the synthesis in the 1930s, the dominant paradigm has been Darwinism—natural selection "er alles. Does evo-devo, with the stress on development and not just genes on a string, threaten this? Some scientists think that it does. "The homologies of

---

[ii]Sean Carroll's new book, *Endless Forms Most Beautiful*, is a great place to start on the science. Follow it with Andrew Knoll's *Life on a Young Planet* to think of evo-devo in connection with paleontology, which I am about to suggest one should. I have myself written on evo-devo in *Darwin and Design* and in a new book, *Darwinism and its Discontents*.

process within morphogenetic fields provide some of the best evidence for evolution—just as skeletal and organ homologies did earlier. Thus, the evidence for evolution is better than ever. The role of natural selection in evolution, however, is seen to play less an important role. It is merely a filter for unsuccessful morphologies generated by development. Population genetics is destined to change if it is not to become as irrelevant to evolution as Newtonian mechanics is to contemporary physics" (Gilbert, Opitz, and Raff 1996, 368). Personally, I think this is hogwash. I think that Charles Darwin himself would be incredibly excited by the findings of evo-devo—he was ever fond of embryology—and argue correctly that evo-devo will complement natural selection not contradict it. But whether I am right or whether the scientists are right is a philosophical question and a terrific thesis topic.

Second, and perhaps this fits in with the ideas in the last paragraph, I see fascinating connections between evo-devo and paleontology. Stephen Jay Gould realized this for his last book, the monstrous *The Structure of Evolutionary Theory*, has a full exposition of the latest evo-devo findings and argues that changes in development can have direct and important effects on the overall nature of evolutionary change. I find it fascinating that we might have a direct link between change at the fastest individual level and change at the slowest group level. Of course, this was the presupposition of Haeckel's biogenetic law—ontogeny recapitulates phylogeny—but although there have been some recent attempts to resuscitate it, I doubt that anyone today really thinks that this law works. Now, however, I believe that a truer and more profitable connection may exist. This is a hunch and it calls for careful and detailed conceptual investigation. Philosophers, come forward

Third and finally, there is the human realm. What does evo-devo mean for human evolution, and particularly does it have implications for thought and culture and everything else that we associate with the human realm and distinctiveness. I fear too many people look desperately for something—anything—that will allow them to argue that humans are different, that we do not evolve as do others. They will turn to evo-devo to prove just this. This is not the sentiment leading me to ask for work in this area. I am a hard-line Darwinian. But hard-line Darwinians know that new ideas are challenges and opportunities not barriers or refutations. It seems to me that evo-devo is showing us terrifically interesting things about how the various parts of the body are put together and coordinated. Surely this is going to affect cognition as well as everything else. Do we learn, not necessarily in the most efficient way, but in a way that is dictated by our biology, which in turn means the order in which the genes that we have get switched on and off and so froth? I know that there are really important scientific discoveries waiting out there. The philosophers stay with them.

I doubt I shall be around in forty years, 2045, to see how my intuitions play out. But I wish I could be. And that makes me a lucky person. I got into a new field forty years ago, and today we still have truly exciting problems and possibilities. That is what makes life worth living.

Reference List

Callebaut, W. 1993. *Taking the Naturalistic Turn*. Chicago, Ill.: University of Chicago Press.

Carroll, S B. 2005. *Endless Forms Most Beautiful: The New Science of Evo Devo*. New York: Norton.

Gilbert, S F, J M Opitz, and R A Raff. 1996. Resynthesizing evolutionary and developmental biology. *Developmental Biology* 173: 357-72.

Gould, S J. 2002. *The Structure of Evolutionary Theory*. Cambridge, Mass.: Harvard University Press.

Knoll, A. 2003. *Life on a Young Planet: The First Three Billion Years of Evolution on Earth*. Princeton, N.J.: Princeton University Press.

Ruse, M. 1973. *The Philosophy of Biology*. London: Hutchinson.

————. 1988. *Philosophy of Biology Today*. Albany, NY: SUNY Press.

————. 1996. *Monad to Man: The Concept of Progress in Evolutionary Biology*. Cambridge, Mass.: Harvard University Press.

————. 1999. *Mystery of Mysteries: Is Evolution a Social Construction?* Cambridge, Mass.: Harvard University Press.

————. 2003. *Darwin and Design: Does Evolution have a Purpose?* Cambridge, Mass.: Harvard University Press.

————. 2006. *Darwinism and its Discontents*. Cambridge: Cambridge University Press.

The Generation of Novelty:
The Province of Developmental Biology

Scott F. Gilbert

In his op-ed piece, "Forty years a philosopher of biology: Why EvoDevo makes me still excited about my subject," Michael Ruse (2006) presents a tamed version of EvoDevo which will trouble no waters and which would integrate easily into the existing framework of evolution proposed by the population geneticists of the 1930s. In that paper, and even more explicitly at the conference "The Making Up of Organisms" (Ecole Normale Sup'erieure, Paris, June 8–10, 2006), Ruse opined that natural selection alone has the power to create evolutionary novelty. In both instances, he cited our 1996 paper (Gilbert et al. 1996) and quoted the following paragraph from it:

> The homologies of process within morphogenetic fields provides the best evidence for evolution—just as skeletal and organ homologiesflpar did earlier. Thus, the evidence for evolution is better than ever. The role of natural selection in evolution, however, is seen to play less an important role. It is merely a filter for unsuccessful morphologies generated by development. Population genetics is destined to change if it is not to become irrelevant to evolution as Newtonian mechanics is to contemporary physics.

Gilbert teaches in the Department of Biology, Swarthmore College, Swarthmore, PA, USA,
sgilbert1@cc.swarthmore.edu

Strong words. But I would contend that the past decade has proven those words remarkably accurate. Ruse, on the other hand, declares them to be "hogwash." "Hogwash" is a technical term in American rural philosophy, meaning "I don't have the data, but I know it to be wrong." Taking a leaf from the Creationists' instruction manual (e.g., Wells 2005), Ruse then portrays the EvoDevo statement as being anti-Darwinian, continuing, "I think that Charles Darwin himself would be incredibly excited by the findings of EvoDevo—he was ever fond of embryology—and argue that EvoDevo will complement natural selection, not contradict it." Michael, the supplementation of natural selection is precisely what EvoDevo is trying to do. Take for instance the question of how novelties of the arthropod body plan arose. Hughes and Kaufman (2002) begin their study, "To answer this question by invoking natural selection is correct—but insufficient. The fangs of a centipede . . . and the claws of a lobster accord these organisms a fitness advantage. However, the crux of the mystery is this: From what developmental genetic changes did these novelties arise in the first place?" Even in the 1996 paragraph quoted above, we merely thought to give natural selection a less important role, not abandon it. Similarly, in all of my writings on EvoDevo (e.g., Gilbert 2003, 2006), I have stressed the complementary nature of the population-genetic approach and the developmental-genetic approach. However, where we differ is that I think that natural selection has to relinquish its claim to being the sole (or even the major) mechanism for generating diversity. Natural selection oversteps its bounds when its advocates claim that it both generates and selects variation. Generating variation is the province of development.

The notion that natural selection could create variation exists because until recently the only genetics available to explain evolution was population genetics. Genetics was (as Kettlewell would claim), "Darwin's missing evidence." But both *population* genetics and *developmental* genetics have to be recognized. Darwin did not have a theory of variation. The genetics of the 20th century gave an inkling of what might be involved. Gray moths could become darkly peppered moths when exposed for generations to a darkened habitat. Those moths that had more cryptic coloration and could avoid predators survived to mate and their descendents had the more protective wings. Was natural selection creating novelty? Only by expanding the definition of natural selection *to include* development. Mutation and recombination were invoked as mechanisms by which genes could be altered to generate evolutionary innovations. But this really isn't "natural selection," it's more of a general statement about some unknown set of mechanisms active in development.

Darwin (1859) realized that selection could not act upon traits that had not yet appeared, noting that "characters may have originated from quite secondary sources, independently from natural selection." He continued this line of reasoning in his book on variation and domestication (Darwin 1883: p. 282), where he admits, "the external conditions of life are quite insignificant, in relationship to any particular variation, in comparison with the organization and constitution of the being which varies. We are thus driven to conclude that in most cases the conditions of life play a subordinate part in causing any particular modification." At best natural selection "creates" novelty by preparing a permissive environment for it. Thus, if variant A is more likely to arise from variant B than from variant C, then if the environment selects for B, the appearance of A is more probable. But this says nothing about the generation of A and why such generation is more likely from B. The mechanisms allowing B (but not C) to give rise to A are part of development (indeed, the "classic" area of developmental constraints).

Developmental genetics now has such a theory of evolutionary variation (reviewed in Carroll et al. 2005; Gilbert 2006). The tenets of these theories involve transcription factors and paracrine factors, concepts that were unknown to Darwin and to the architects of the Modern Synthesis. First, there are two major preconditions for developmental alterations that can generate morphological change. The first is *gene duplication* wherein genes can make copies of themselves and the sister genes mutate independently to assume different functions. Entire families of genes (*Hox* genes, *globin* genes, *cadherin* genes, *TGF-â* genes) have been produced this way. The second precondition is *modularity*. Modularity pervades development (Raff 1996; Schlosser and Wagner 2004). This means that a change can occur in one area of the body and need not affect another. Indeed, one of the most important aspects of EvoDevo is that not only are the *anatomical* units modular (such that one part of the body can develop differently than the others), but the DNA regions that form the *enhancers* of genes (telling the gene when, where, and how much it can be transcribed) are also modular. Thus, if a particular gene loses or gains a modular enhancer element, the organism containing that particular enhancer allele will express that gene in different places or at different times or different amounts than those organisms retaining the original allele.

These changes can cause different morphologies to develop (Sucena and Stern 2000; Shapiro et al. 2004; Maas and Fallon 2005). Modular units allow certain parts of the body to change without interfering with the functions of other parts.

The importance of enhancer modularity in evolution has been dramatically demonstrated in three-spine stickleback fishes. Freshwater sticklebacks evolved from marine sticklebacks about 12,000 years ago, as the marine populations colonized the newly formed freshwater lakes at the end of the last ice age. The marine sticklebacks have a pelvic spine that serves as protection against predation by other fish. It lacerates the mouths of those fish who would try to eat it. The freshwater sticklebacks, however, do not have these pelvic spines. This may be because they lack the predators that the marine fish have and the predators of the fresh water sticklebacks are invertebrates that capture them by grasping onto such spines. Thus, the freshwater populations of this species have evolved a pelvis without such lacerating appendages.

To determine which genes might be involved in this difference between marine and freshwater populations, David Kingsley's laboratory (Shapiro et al. 2004) mated individuals from certain marine populations (with pelvic spines) and freshwater populations (without spines). The resulting offspring were bred to each other and produced numerous progeny, some of which had pelvic spines and some of which didn't. Using molecular markers that could identify specific regions of the parental chromosomes, they found that nearly all the fish with pelvic spines had a portion of chromosome 7 from the marine parent, while nearly all the fish that lacked pelvic spines obtained this region from the freshwater parent. This genetic region contained the gene-encoding transcription factor Pitx1.

When they compared the amino acid sequences of the Pitx1 protein between marine and freshwater sticklebacks, there were no differences. However, there was a critically important difference when they compared the *expression patterns* of the *Pitx1* gene between these species. In both species, *Pitx1* was seen to be expressed in the precursors of the thymus, nose, and sensory neurons. In the marine species, *Pitx1* was also expressed in the pelvic region. But in the freshwater populations, the pelvic expression of *Pitx1* was absent or severely reduced. Since the coding region of *Pitx1* is not mutated (and since the gene involved in the pelvic spine differences maps to the site of the *Pitx1* gene, and the difference between the freshwater and marine species involves the expression of this gene at a particular site), it is reasonable to conclude that the *enhancer region* containing the information to express *Pitx1* in the pelvic area no longer functions in the freshwater fish. Thus, the modularity of the enhancer has enabled this particular expression domain to be lost, and with it the loss of the pelvic spine. No other function of *Pitx1* had to be disturbed.

In addition to the two preconditions for evolution by changing development, EvoDevo has also recognized four mechanisms of *bricolage* which are responsible for producing these variations (Arthur 2004; Gilbert 2006):

—heterotopy (change in location)
—heterochrony (change in time)
—heterotypy (change in kind)
—heterometry (change in amount).

Although these mechanisms can be employed at any level of development, I will focus on the level of transcription, since investigations have focused on this area and because it is the most gene-oriented. References to the papers here can be found in Gilbert (2006).

Heterotopy of gene expression involves changing the types of cells expressing a particular gene. Heterotopy of *Fgf10* expression in the turtle dermis may explain the formation of the carapace (Cebra-Thomas et al. 2005). Gremlin expression in the interdigital web of the duck hind limb (where it is not seen in the chicken or mouse) goes a long way to explaining how ducks got their webbed feet. Indeed, Gremlin inhibits the signal for programmed cell death, and if Gremlin protein is added to embryonic chick foot webbing, the chick foot becomes webbed, too. The different expression patterns of the *Ubx* and *Abd* genes between lobsters and shrimp explain the divergence of the animals in our seafood platters, and the difference in the epidermal expression of *BMP2* and *Shh* genes explains how feathers may have evolved from scales. Indeed, the proximate cause of the Genesis curse against snakes is the heterotopic expression of the *Hoxc-6* gene during snake embryonic development, where altered expression prevents limb development.

Heterochrony of gene expression involves the timing of gene expression. The origin of the vertebrate jaw comes, in part, from heterochronic gene expression (Shigetani et al. 2002), as does the elongation of the bat digits necessary to produce the wing (Sears et al. 2006). In this latter example, the gene encoding the paracrine factor BMP2 is expressed in the digital mesoderm for a longer period of time compared to that of other mammals. *Heterotypy* concerns changing the actual protein that is being made. Heterotypy of the gene encoding the Ultrabithorax (Ubx) transcription factor may explain why insects have just six legs, while other arthropod groups (think of spiders, millipedes, centipedes, and shrimp) have many more. The *Distal-less* gene in arthropods is essential for leg formation. Throughout most families of the arthropod lineage, Ubx protein does not inhibit the *Distal-less* gene. However, in the insect lineage, a mutation occurred in the *Ubx* gene wherein the original 3 end of the protein-coding region was replaced by a group of nucleotides encoding a stretch of about ten alanine residues. This polyalanine region functions as a repressor of *Distal-less* transcription. When a shrimp *Ubx* gene is experimentally modified to encode this polyalanine region, it, too, represses the *Distal-less* gene. The ability of insect Ubx protein to inhibit *Distal-less* thus appears to be the result of a gain-of-function mutation that characterizes the insect lineage.

Heterometry involves changing the amount of gene expression. Evolution only rarely proceeds by total loss of function. Rather, the alterations of the amount of function can give different phenotypes. One way of providing such variations is to alter the amount of gene transcription. Indeed, some of the best examples of heterometry in action are Darwin's celebrated finches. Systematists have shown that these species evolved in a particular manner, with one of the major separations being between the cactus finches and the ground finches. The ground finches have evolved deep, broad beaks that enable them to crack seeds open, whereas the cactus finches have evolved narrow pointed beaks that allow them to probe cactus flowers and fruits for insects and flower parts. Developmental research demonstrates that species differences in the beak pattern are caused by changes in the growth of the neural crest-derived mesenchyme of the frontonasal process (i.e., those cells that form the facial bones). Abzhanov and his colleagues (2004) found a remarkable correlation between the beak shape of the finches and timing and amount of BMP4 expression. No other paracrine factor showed such differences. The expression of BMP4 in ground finches started earlier and was much greater than the expression of BMP4 in cactus finch beaks. In all cases, the BMP4 expression pattern correlated with the broadness and deepness of the beak. Experimentally adding BMP4 will deepen chick beaks.

Another example of heterometric variation involves the evolution of the *IL4* gene in human populations. Most of human variation (both pathological and non-pathological) does not come from changes in the structural genes. Rather it arises from mutations in the *regulatory regions* of these genes (Rockman and Wray 2002;

Rockman et al. 2003). A single base pair mutation in the enhancer of the *IL4* gene creates a new binding site for transcription factor NFAT, a more rapid transcription of IL4 and higher levels of that protein. Moreover, population genetic studies show that this regulatory allele has been positively selected in particular populations and not others. Having this allele appears to be advantageous in those populations exposed to intestinal helminth parasites. However, this is not an exonic mutation in the actual protein; rather, it is an enhancer of the gene encoding this regulatory protein.

Recent research in developmental biology has also shown that in addition to producing new evolutionary variants, these four mechanisms also explain such evolutionary phenomena as parallel evolution (which has been used to justify the notion that natural selection is itself "creative"). Comparative developmental studies of the insect eye (Oakley and Cunningham 2002), stickleback fish armor plates and spines (Colosimo et al. 2004, 2005), as well as avian and *Drosophila* pigment patterns (Gompel et al. 2005; Mundy 2005) show that parallel evolution results from the independent recruitment of similar developmental pathways by different organisms. Thus, the loss of the pelvic spines in other stickleback species appears to be caused by independent losses of the Pitx1 expression domain mentioned earlier (Colosimo et al. 2004). Instead of extrinsic selection pressures being thought to play a dominant role in such phenomena, intrinsic developmental factors are now seen to play a critical role in producing these parallel variations (Hall 2003; Rudel and Sommer 2003;West-Eberhard 2003).[1]

What we see here is variation caused by developmental mechanisms. I have emphasized those involving gene transcription because these are the mechanisms closest to the genes themselves. These four mechanisms each involve changes in gene transcription during embryonic development. They each involve the signaling molecules whereby cell fates are determined—transcription factors and paracrine factors. They change the way the embryo is constructed and thereby change the phenotype in ways that natural selection can then test. Natural selection alone generates neither novelty nor variation. Development does. Natural selection can clear the area so that these new variants can spread through a population, and it can promote an environment permissive for such change. But the motor of evolutionary innovation is not natural selection; it is development. Biodiversity can be explained only when population genetics and developmental biology complement each other; but this can happen only if the proponents of natural selection allow developmental biology its proper place as an explanatory agent. Darwin originated much of evolutionary theory; but he lacked a theory of variation. His colleague Thomas Huxley (1878/1896) was more of an embryologist than Darwin, and he intuited that variation must be caused by inherited alterations of development. "Evolution is not a speculation but a fact;" he wrote, "and it takes place by epigenesis."

Note

1. Indeed, in some of these papers (especially Colosimo et al. 2004, and 2005 and Rockman et al. 2003) one sees precisely the critical importance of the population genetics of regulatory alleles, as mentioned in the paragraph that so offended Ruse.

References

Abzhanov A, Protas M, Grant BR, Grant PR, Tabin CJ (2004) *Bmp4* and morphological variation of beaks in Darwin's finches. Science 305: 1462–1465.

Arthur W (2004) Biased Embryos and Evolution. New York: Cambridge University Press.

Carroll SB, Grenier JK, Weatherbee SD (2005) From DNA to Diversity: Molecular Genetics and the Evolution of Animal Design. Malden, MA: Blackwell.

Cebra-Thomas J, Tan F, Sistla S, Estes E, Bender G, Kim C, Riccio P, Gilbert SF (2005) How the turtle forms its shell: A paracrine hypothesis of carapace formation. Journal of Experimental Zoology B (Molecular Development and Evolution) 304: 558–569.

Colosimo PF, Hosemann KE, Balabhadra S, Villarreal G Jr, Dickson M, Grimwood J, Schmutz J, Myers RM, Schluter D, Kingsley DM (2005) Widespread parallel evolution in sticklebacks by repeated fixation of ectodysplasin alleles. Science 307: 1928–1933.

Colosimo PF, Peichel CL, Nereng K, Blackman BK, ShapiroMD, Schluter D, Kingsley DM (2004) The genetic architecture of parallel armor plate reduction in three spine sticklebacks. PLoS Biology 2: 635–641.

Darwin C (1859) On the Origin of Species. London: John Murray.

Darwin C (1883) The Variation of Animals and Plants under Domestication, 2nd ed., Vol. 2. New York: D. Appleton.

Gilbert SF (2003) The morphogenesis of evolutionary developmental biology. International Journal of Developmental Biology 47: 467–477.

Gilbert, SF (2006) Developmental Biology, 8th ed. Sunderland, MA: Sinauer.

Gilbert SF, Opitz JM, Raff RA (1996) Resynthesizing evolutionary and developmental biology. Developmental Biology 173: 357–372. Gompel N, Prud'homme B, Wittkopp, PJ, Kassner VA, Carroll SB (2005) Chance caught on the wing: *Cis*-regulatory evolution and the origin of pigment patterns in *Drosophila*. Nature 433: 481–487.

Hall BK (2003) Descent with modification: The unity underlying homology and homoplasy as seen through an analysis of development and evolution. Biological Reviews 78: 409–433.

Hughes CL, Kaufman TC (2002) Hox genes and the evolution of the arthropod body plan. Evolution and Development 4: 459–499.

Huxley TH (1878/1896) Evolution in Biology [Reprint] Darwiniana: Collected Essays. New York: Appleton, p. 202 http://aleph0.clarku.edu/huxley/CE2/EvBio.html.

Maas SA, Fallon JF (2005) Single base pair change in the long-range Sonic hedgehog limb-specific enhancer is a genetic basis for preaxial polydactyly. Developmental Dynamics 232: 345–348.

Mundy NI (2005) A window on the genetics of evolution: MC1R and plumage colouration in birds. Proceedings: Biological Science 272: 1633–1640.

Oakley TH, Cunnigham CW (2002) Molecular phylogenetic evidence for the independent evolutionary origin of an arthropod compound eye. Proceedings of the National Academy of Sciences of the USA 99: 1426–1430.

Raff RA (1996) The Shape of Life: Genes, Development, and the Evolution of Animal Form. Chicago: University of Chicago Press.

Rockman MV, HahnMW, Soranzo N, Goldstein DB,WrayGA(2003) Positive selection on a human-specific transcription factor binding site regulating IL4 expression. Current Biology 13: 2118–2123.

Rockman MV, Wray GA (2002) Abundant raw material for *cis*-regulatory evolution in humans. Molecular Biology and Evolution 19: 1991–2004.

Rudel D, Sommer RJ (2003) The evolution of developmental mechanisms. Developmental Biology 264: 15–37.

Ruse M (2006) Forty years a philosopher of biology: Why EvoDevo makes me still excited about my subject. Biological Theory 1: 35–37.

Schlosser G, Wagner GP, eds (2004) Modularity in Development and Evolution. Chicago: University of Chicago Press.

Sears KE, Behringer RR, Rasweiler JJ 4th, Niswander LA (2006) Development of bat flight: Morphological and molecular evolution of bat wing digits. Proceedings of the National Academy of Sciences of the USA 103: 6581–6586.

Shapiro MD, Marks ME, Peichel CL, Blackman BK, Nereng KS, J'onsson B, Schluter D, Kingsley DM (2004) Genetic and developmental basis of evolutionary pelvic reduction in threespine sticklebacks. Nature 428: 717–723.

Shigetani Y, Sugahara F, Kawakami Y, Murakami Y, Hirano S, Kuratani S (2002) Heterotopic shift of epithelial-mesenchymal interactions in vertebrate jaw evolution. Science 296: 1316–1319.

Sucena E, Stern D (2000) Divergence of larval morphology between *Drosophila sechellia* and its sibling species caused by *cis*-regulatory evolution of *ovo/shaven-baby*. Proceedings of the National Academy of Sciences of the USA 97: 4530–4534.

Wells J (2005) Give me that old time evolution: A response to the New Republic. http://www.iconsofevolution.com/embedJonsArticles.php3?id=2933

West-Eberhard MJ (2003) Developmental Plasticity and Evolution. Oxford: Oxford University Press.

Bare-Knuckle Fighting:
Evo Devo versus Natural Selection

By Michael Ruse

Embryology played a very important role in the theory of evolution that Charles Darwin presented in his *Origin of Species*. He complained indeed that his friends did not realize how important. But for all that, the history of the relationship between Darwinian Theory and embryology has not been a happy one. Karl Ernst von Baer, the most important embryologist of the nineteenth century and a great influence on Darwin, rejected evolution through natural selection (Hull 1973). And after the *Origin* was published, the evolutionists of the day—starting with Ernst Haeckel but then backed by the English scientists including Thomas Henry Huxley and his students—slipped quickly and readily into a non-Darwinian mode. They showed the real influence on their thinking, namely *Naturphilosophie,* as they happily drew analogies between the development of the organism and the evolution of the group—most notoriously in Haeckel's biogenetic law, "ontogeny recapitulates phylogeny."

Not to put too fine a point on it, this was an absolute bloody disaster for evolutionary studies. Grotesquely speculative phylogentic hypothesizing replaced—

failed to allow—serious experimental work on the causes of evolution. At a time when physics was about to make the discoveries of the millennium, evolutionary biology was not serious science. It had become a quasi-religion to replace Christianity. It was good for the museum and little more. There was little wonder that by the end of the nineteenth century the good thinkers—men like Raphael Weldon and Edward Poulton, Edmund Wilson and William Bateson—were turning to other issues and other problems. Even into the twentieth century, embryology was a hindrance not a help. When the new science of genetics strove to find a place in universities, it was opposed by the embryologists and had to stay and thrive in the agricultural schools and institutes.

One is hardly surprised therefore, that when the synthesis of Darwin and Mendel came in the 1930s, the new breed of evolutionists wanted little to do with embryology. Even the one embryologist with a good claim to being in the synthesis—Gavin de Beer (1940)—spent most of his scientific life trying to show just how hopeless was the work of the previous embryologists who had presumed to meddle in things evolutionary. Of course, there was interest in aspects of development. I take it that Cyril Darlington's (1932) work, for one example, shows this. But it is true that empirically and conceptually embryology was shut out. The jump was from genotype to phenotype, with not much in between.

As Scott Gilbert (2006) has clearly shown, things have changed. In the past—let us say—thirty years, embryology has shot forward and now—in its modern incarnation, evo devo—is not only a terrific science in its own right, but has much to add to our picture of evolution. Let it be clearly understood that no one, certainly not I, disputes this. I certainly do not do so in the op ed piece by me (Ruse 2006) that has Gilbert so riled up. In my opinion, the saddest evolutionist still living is Richard C. Lewontin, who around 1975 basically gave up doing real science because he thought that evolutionary studies had reached an impasse. Since then, not only have we had three decades of wonderful work on the evolution of animal behaviour, but evolutionary development has come along and shown us so much. Who could not be thrilled at the discovery of the molecular homologies between fruitflies and humans? (Actually, I will answer that question for you. The Creationists! My jaw dropped when Gilbert linked me—me, the person who has done more in the past three decades than any other living being to fight the Creationists—with the Moonie, Jonathan Wells. Goodness, somebody is feeling insecure.)

So, let us be clear about where there are no disagreements. Evo devo is terrific science and I applaud the examples that Gilbert gives in his letter roasting my views on evolution. Where then is the disagreement? It is over the role of natural selection, past and present. Again, let us be clear about the disagreements. Nobody, certainly no Darwinian, has ever thought that selection generates variation. It is true that through such mechanisms as balanced superior heterozygote fitness, it is thought that selection can keep variation in populations, waiting as it were to be used by new challenges (this sounds teleological but it is not meant that way). But the variations themselves are things that selection works on. So again, when

Gilbert gives us all of the wonderful ways in which evo devo shows how development leads to variations, the Darwinian applauds. The person who would not applaud obviously is the person who thinks that all variation is a matter of single, new mutations. But such a person is not the Darwinian.

What then is the Darwinian claiming? That selection is the crucial mechanism bringing on adaptation, and that it is silly and misleading to speak of it as "merely a filter for unsuccessful morphologies generated by development." Gilbert himself shows how wrong it is to talk in this way, because he gives the example of the stickleback that has gone in two completely different ways with respect to morphology—the one getting spines to fight off predators and the other eschewing spines to fight off predators! Adaptation—"organized complexity" as we Darwinians call it—is the key phenomenon in the living world, and it is selection and selection alone that explains it. Nothing in evo devo pushes this to one side. Mrs Beeton took her hare to make her stew. We Darwinians take variation to make organized complexity. Mrs Beeton used her stove. We use natural selection.

This is not in any way to belittle the importance of evo devo. It is rather to point out its conceptual role. What I suspect is the hope and the big mistake at the back of the thinking of people like Gilbert is that some day development unaided will take over the job of creating adaptation. Selection then will truly be only a garbage can for failed morphologies. But to think that development really is going to get this really is to get you into teleology—it is precisely the position of the Intelligent Designers, who at least have God to help them in the crucial moves. Or it is to be in the position of Thomas Henry Huxley, whom Gilbert praises, a man who never took adaptation seriously. He denied that butterfly colors have any adaptive significance—and did this after Bates's brilliant work on mimicry. Huxley was a great Darwinian in the sense of pushing evolution. Huxley was a lousy Darwinian in the sense of getting the theory right.

This disagreement over evo devo and natural selection and their respective roles is no mere spat between two aging evolutionists—okay, I am aging, Scott is the Peter Pan of the evolutionary world. This is a really important matter. Darwinism is evolution through natural selection, working on the genes as they express themselves in phenotypes. Evo devo has come along and is going to show fantastic amounts about how this expression works, and I am sure going to show much about how selection can and cannot function and take effect. There are going to be revolutionary changes, just as there was when Mendelian genetics was replaced by molecular genetics. But all of this is going to be working within the framework of the existing theory. There is going to be no new theory, with selection and genetics replaced by embryology. The evolutionists tried that gambit after the *Origin* and it put back evolution studies for seventy five years. Please, please, as evolutionists let us learn from the past and not repeat the mistakes of our predecessors.

Reference List

Bowler, P. J. 1988. *The non-Darwinian Revolution: Reinterpreting a Historical Myth*. Baltimore, Md.: Johns Hopkins University Press.

———. 1996. *Life's Splendid Drama*. Chicago, Ill.: University of Chicago Press.

Darlington, C. D. 1932. Recent Advances in Cytology. With foreword by J. B. S. Haldane. London: Churchill.

de Beer, G. R. 1940. *Embryos and Ancestors*. Oxford: Oxford University Press.

Gilbert, S. 2006. The generation of novelty: The province of developmental biology. *Biological Theory* 1: 209-12.

Hull, D. L., Editor. 1973. *Darwin and His Critics*. Cambridge, Mass.: Harvard University Press.

Richards, R J. 1992. *The Meaning of Evolution: The Morphological Construction and Ideological Reconstruction of Darwin's Theory*. Chicago: University of Chicago Press.

Richards, R. J. 2003. *The Romantic Conception of Life: Science and Philosophy in the Age of Goethe*. Chicago: University of Chicago Press.

———. 2004. Michael Ruse's design for living. *Journal of the History of Biology* 37: 25-38.

Ruse, M. 1996. *Monad to Man: The Concept of Progress in Evolutionary Biology*. Cambridge, Mass.: Harvard University Press.

———. 2006. Forty years as a philosopher of biology: why evodevo makes me still excited about my subject. *Biological Theory* 1: 35-37.

Why Pigs Don't Have Wings

Jerry Fodor

Die Meistersinger is, by Wagner's standards, quite a cheerful opera. The action turns on comedy's staple, the marriage plot: get the hero and the heroine safely and truly wed with at least a presumption of happiness ever after. There are cross-currents and undercurrents that make *Meistersinger*'s libretto subtle in ways that the librettos of operas usually aren't. But for once Nietzsche is nowhere in sight and nobody dies; the territory is closer to T*he Barber of Seville* than to *The Ring*. Yet, in the first scene of Act 3, the avuncular Hans Sachs, whose benevolent interventions smooth the lovers' course, delivers an aria of bitter reflection on the human condition. It comes as rather a shock:

> Madness, Madness!
> Madness everywhere.
> Wherever I look . . . .
> People torment and flay each other
> In useless, foolish anger
> Till they draw blood.
> Driven to flight,
> They think they are hunting.
> They don't hear their own cry of pain . . . .
> When he digs into his own flesh,
> Each thinks he is giving himself pleasure.

So 'what got into Sachs?' is a well-known crux for Wagner fans, and one the opera doesn't resolve. (By Scene 2 of Act 3 Sachs is back on the job, arranging for Walther to get his Eva and vice versa.) Sachs isn't, of course, the first to wonder why we are so prone to making ourselves miserable, and the question continues to be pertinent. We have just seen the last of a terrible century with, quite possibly, worse to come. Why is it so hard for us to be good? Why is it so hard for us to be happy?

One thing, at least, has been pretty widely agreed: we can't expect much help from science. Science is about facts, not norms; it might tell us how we are, but it couldn't tell us what is wrong with how we are. There couldn't be a science of the human condition. Thus the received view ever since Hume taught that *ought* doesn't come from *is*. Of late, however, this Humean axiom has come under attack, and a new consensus appears to be emerging: Sachs was right to be worried; we are all a little crazy, and for reasons that Darwin's theory of evolution is alleged to reveal. What's wrong with us is that the kind of mind we have wasn't evolved to cope with the kind of world that we live in. Our kind of mind was selected to solve the sorts of problems that confronted our hunter-gatherer forebears thirty thousand years or so ago; problems that arise for small populations trying to make a living and to reproduce in an ecology of scarce resources. But, arguably, that kind of mind doesn't work very well in third millennium Lower Manhattan, where there's population to spare and a Starbucks on every block, but survival depends on dodging the traffic, finding a reliable investment broker and not having more children than you can afford to send to university. It's not that our problems are harder than our ancestors' were; by what measure, after all? It's rather that the mental equipment we've inherited from them isn't appropriate to what we're trying to do with it. No wonder it's driving us nuts.

This picture—that our minds were formed by processes of evolutionary adaptation, and that the environment they are adapted to isn't the one that we now inhabit—has had, of late, an extraordinarily favourable press. Darwinism has always been good copy because it has seemed closer to our core than most other branches of science: botany, say, or astronomy or hydrodynamics. But if this new line of thought is anywhere near right, it is closer than we had realised. What used to rile Darwin's critics most was his account of the phylogeny of our species. They didn't like our being just one branch among many in the evolutionary tree; and they liked still less having baboons among their family relations. The story of the consequent fracas is legendary, but that argument is over now. Except, perhaps, in remote backwaters of the American Midwest, the Darwinian account of our species' history is common ground in all civilised discussions, and so it should be. The evidence really is overwhelming.

But Darwin's theory of evolution has two parts. One is its familiar historical account of our phylogeny; the other is the theory of natural selection, which purports to characterise the mechanism not just of the formation of species, but of all evolutionary changes in the innate properties of organisms. According to selection theory, a creature's 'phenotype'—the inventory of its heritable traits, including, notably, its heritable mental traits—is an adaptation to the demands of its ecological situation. Adaptation is a name for the process by which environmental variables select among the creatures in a population the ones whose heritable properties are most fit for survival and reproduction. So environmental selection for fitness is (perhaps plus or minus a bit) the process par excellence that prunes the evolutionary tree.

More often than not, both halves of the Darwinian synthesis are uttered in the same breath; but it's important to see that the phylogeny could be true even if the adaptationism isn't. In principle at least, it could turn out that there are indeed baboons in our family tree, but that natural selection isn't how they got there. It's the adaptationism rather than the phylogeny that the Darwinist account of what ails us depends on. Our problem is said to be that the kind of mind we have is an anachronism; it was selected for by an ecology that no longer exists. Accordingly, if the theory of natural selection turned out not to be true, that would cut the ground from under the Darwinist diagnosis of our malaise. If phenotypes aren't selected at all, then there is, in particular, nothing that they are selected for. That applies to psychological phenotypes inter alia.

In fact, an appreciable number of perfectly reasonable biologists are coming to think that the theory of natural selection can no longer be taken for granted. This is, so far, mostly straws in the wind; but it's not out of the question that a scientific revolution—no less than a major revision of evolutionary theory—is in the offing. Unlike the story about our minds being anachronistic adaptations, this new twist doesn't seem to have been widely noticed outside professional circles. The ironic upshot is that at a time when the theory of natural selection has become an article of pop culture, it is faced with what may be the most serious challenge it has had so far. Darwinists have been known to say that adaptationism is the best idea that anybody has ever had. It would be a good joke if the best idea that anybody has ever had turned out not to be true. A lot of the history of science consists of the world playing that sort of joke on our most cherished theories.

Two kinds of consideration now threaten to displace natural selection from its position at the centre of evolutionary theory; one is more or less conceptual, the other is more or less empirical.

The conceptual issue. There is, arguably, an equivocation at the heart of selection theory; and slippage along the consequent faultline threatens to bring down the whole structure. Here's the problem: you can read adaptationism as saying that environments select creatures for their fitness; or you can read it as saying that environments select traits for their fitness. It looks like the theory must be read both ways if it's to do the work that it's intended to: on the one hand, forces of selection must act on individual creatures since it is individual creatures that live, struggle, reproduce and die. On the other hand, forces of selection must act on traits since it is phenotypes—bundles of heritable traits—whose evolution selection theory purports to explain. It isn't obvious, however, that the theory of selection can sustain both readings at once. Perhaps the consensus view among

Darwinists is that phenotypes evolve because fit individuals are selected for the traits that make them fit. This way of putting it avoids the ambiguity, but whether it's viable depends on whether adaptationism is able to provide the required notion of 'selection for'; and it seems, on reflection, that maybe it can't. Hence the current perplexity.

History might reasonably credit Stephen J. Gould and Richard Lewontin as the first to notice that something may be seriously wrong in this part of the wood. Their 1979 paper, 'The Spandrels of S. Marco and The Panglossian Paradigm: A Critique of the Adaptationist Programme', ignited an argument about the foundations of selection theory that still shows no signs of quieting. A spandrel is one of those more-or-less triangular spaces that you find at the junctures of the arches that hold up a dome. They are often highly decorated; painters competed in devising designs to fit them. Indeed (and this is Gould and Lewontin's main point), casual inspection might suggest that the spandrels are there because they provide the opportunity for decoration; that, an adaptationist might say, is what spandrels were selected for. But actually, according to Gould and Lewontin, that gets things backwards. In fact, spandrels are a by-product of an arch-and-dome architecture; decide on the latter and you get the former for better or worse. Arches were selected for holding up domes; spandrels just came along for the ride.

I assume that Gould and Lewontin got their architectural history right, but it doesn't really matter for the purposes at hand. What matters is that though spandrels survived and flourished, nothing at all follows about what, if anything, they were selected for. To a first approximation, you have spandrels if and only if you have a dome that's supported by arches; the two are, as logicians say, coextensive. Is it, then, that selection for arches explains why there are spandrels? Or is it that selection for spandrels explains why there are arches? It looks, so far, as though the story could go either way; so what tips the balance? Surely it's that domes and arches are designed objects. Somebody actually thought about, and decided on, the architecture of San Marco; and what he had in mind when he did so was that the arches should support the dome, not that they should form spandrels at their junctures. So that settles it: the spandrels weren't selected for anything at all; they're just part of the package. The question, however, is whether the same sort of reasoning can apply to the natural selection of the phenotypic traits of organisms, where there is, by assumption, no architect to do the deciding. If cathedrals weren't designed but grew in the wild, would the right evolutionary story be that they have arches because they were selected for having spandrels? Or would it be that they have spandrels because they were selected for having arches? Or neither? Or both?

It's a commonplace that Darwin constructed the theory of natural selection with an eye to what breeders do when they choose which creatures to encourage to reproduce. This reading of Darwin is by no means idiosyncratic. Darwin 'argues by example, not analogy,' Adam Gopnik wrote in the *New Yorker* in October last year. 'The point of the opening of "The Origin" isn't that something similar happens with domesticated breeds and natural species; the point is that the very same thing happens, albeit unplanned and over a much longer period.' It's true, of course, that breeding, like evolution, can alter phenotypes over time, with consequent effects on phylogenetic relations. But, on the face of it, the mechanisms by which breeding and evolution operate could hardly be more different. How could a studied decision to breed for one trait or another be 'the very same thing' as the adventitious culling of a population? Gopnik doesn't say.

The present worry is that the explication of natural selection by appeal to selective breeding is seriously misleading, and that it thoroughly misled Darwin. Because breeders have minds, there's a fact of the matter about what traits they breed for; if you want to know, just ask them. Natural selection, by contrast, is mindless; it acts without malice aforethought. That strains the analogy between natural selection and breeding, perhaps to the breaking point. What, then, is the intended interpretation when one speaks of natural selection? The question is wide open as of this writing.

The answers that have been suggested so far have not been convincing. In particular, though there is no end of it in popular accounts of adaptationism, it is a Very Bad Idea to try and save the bacon by indulging in metaphorical anthropomorphisms. It couldn't, for example, be literally true that the traits selected for are the ones Mother Nature has in mind when she does the selecting; nor can it be literally true that they are the traits one's selfish genes have in mind when they undertake to reproduce themselves. There is, after all, no Mother Nature, and genes don't have, or lack, personality defects. Metaphors are fine things; science probably couldn't be done without them. But they are supposed to be the sort of things that can, in a pinch, be cashed. Lacking a serious and literal construal of 'selection for', adaptationism founders on this methodological truism.

There are delicious ironies here. Getting minds in general, and God's mind in particular, out of biological explanations is a main goal of the adaptationist programme. I am, myself, all in favour of that; since I'm pretty sure that neither exists, I see nothing much to choose between God and Mother Nature. Maybe one can, after all, make sense of mindless environmental variables selecting for phenotypic traits. That is, maybe one can get away with claiming that phenotypes are like arches in that both are designed objects. The crucial test is whether one's pet theory can distinguish between selection for trait A and selection for trait B when A and B are coextensive: were polar bears selected for being white or for matching their environment? Search me; and search any kind of adaptationism I've heard of. Nor am I holding my breath till one comes along.

The empirical issue. It wouldn't be unreasonable for a biologist of the Darwinist persuasion to argue like this: 'Bother conceptual issues and bother those who raise them. We can't do without biology and biology can't do without Darwinism. So Darwinism must be true.' Darwinists do often argue this way; and the fear of hyperbole seems not to inhibit them. The biologist Theodosius Dobzhansky said that nothing in biology makes sense without Darwinism, and he is widely paraphrased. The philosopher Daniel Dennett says that 'in a single stroke, the idea of evolution by natural selection unifies the realm of life, meaning and purpose with the realm of space and time, cause and effect, mechanism and physical law.' (Phew!) Richard Dawkins says, 'If superior creatures from space ever visit earth, the first question they will ask, in order to assess the level of our civilisation, is: "Have they discovered evolution yet?"' Shake a stick at a Darwinist treatise and you're sure to find, usually in the first chapter, claims for the indispensability of adaptationism. Well, if adaptationism really is the only game in town, if the rest of biology really does presuppose it, we had better cleave to it warts and all. What is indispensable therefore cannot be dispensed with, as Wittgenstein might have said. The breaking news, however, is that serious alternatives to adaptationism have begun to emerge; ones that preserve the essential claim that phenotypes evolve, but depart to one degree or other from Darwin's theory that natural selection is the mechanism by which they do. There is now far more of this sort of thing around than I am able to survey. But an example or two may give the feel of it.

Adaptationism is a species of what one might call 'environmentalism' in biology. (It's not, by any means, the only species; Skinnerian learning theory is another prime example.) The basic idea is that where you find phenotypic structure, you can generally find corresponding structure in the environment that caused it. Phylogeny tells us that phenotypes don't occur at random; they form a more or less orderly taxonomic tree. Very well then, there must be nonrandomness in the environmental variables by which the taxonomic tree is shaped. Dennett has put this idea very nicely: 'Functioning structure carries implicit information about the environment in which its function "works". The wings of a seagull . . . imply that the creature whose wings they are is excellently adapted for flight in a medium having the specific density and viscosity of the atmosphere within a thousand metres or so of the surface of the Earth.' So, phenotypes carry information about the environment in which they evolved in something like the way that the size, shape, whatever, of a crater carries information about the size, shape, whatever, of the meteor that made it. Phenotypes aren't, in short, random collections of traits, and nonrandomness doesn't occur at random; the more nonrandomness there is, the less likely it is to have been brought about by chance. That's a tautology. So, if the nonrandomness of phenotypes isn't a reflection of the orderliness of God's mind, per haps it is a reflection of the orderliness of the environments in which the phenotypes evolved. That's the theory of natural selection in a nutshell.

But as soon as it's put that way, it's seen not to be the only possibility. External environments are structured in all sorts of ways, but so, too, are the insides of the creatures that inhabit them. So, in principle at least, there's an alternative to Darwin's idea that phenotypes 'carry implicit information about' the environments in which they evolve: namely, that they carry implicit information about the endogenous structure of the creatures whose phenotypes they are. This idea currently goes by the unfortunate soubriquet 'Evo-Devo' (short for 'evolutionary-developmental theory'). Everybody thinks evo-devo must be at least part of the truth, since nobody thinks that phenotypes are shaped directly by environmental variables. Even the hardest core Darwinists agree that environmental effects on a creature's phenotype are mediated by their effects on the creature's genes: its 'genome'. Indeed, in the typical case, the environment selects a phenotype by selecting a genome that the phenotype expresses. Once in place, this sort of reasoning spreads to other endogenous factors. Phenotypic structure carries information about genetic structure. And genotypic structure carries information about the biochemistry of genes. And the biochemical structure of genes carries information about their physical structure. And so on down to quantum mechanics for all I know. It is, in short, an entirely empirical question to what extent exogenous variables are what shape phenotypes; and it's entirely possible that adaptationism is the wrong answer.

One can think of the Darwinian account of evolution as prompted by the question: why are some phenotypes more similar than others? Darwin's answer was that phenotypic similarity is, pretty generally, explained by common ancestry; and the more similar two creature's phenotypes, the less remote is the nearest ancestor that they share. There are isolated examples to the contrary, but there's no serious doubt that this account is basically correct. And, if it's not the best idea anybody ever had, it's pretty good by any of the local standards. When you ask Darwin's question—why are phenotypes often similar?—you do indeed get Darwin's answer. But if you ask instead why it is that some phenotypes don't occur, an adaptationist explanation often sounds somewhere between implausible and preposterous. For example, nobody, not even the most ravening of adaptationists, would seek to explain the absence of winged pigs by claiming that, though there used to be some, the wings proved to be a liability so nature selected against them. Nobody expects to find fossils of a species of winged pig that has now gone extinct. Rather, pigs lack wings because there's no place on pigs to put them. To add wings to a pig, you'd also have to tinker with lots of other things. In fact, you'd have to rebuild the pig whole hog: less weight, appropriate musculature, an appropriate metabolism, an apparatus for navigating in three dimensions, a streamlined silhouette and god only knows what else; not to mention feathers. The moral is that if you want them to have wings, you will have to redesign pigs radically. But natural selection, since it is incremental and cumulative, can't do that sort of thing. Evolution by natural selection is inherently a conservative process, and once you're well along the evolutionary route to being a pig, your further options are considerably constrained; you can't, for example, go back and retrofit feathers.

That all seems reasonable on the face of it; but notice that this sort of 'channelling' imposes kinds of constraint on what phenotypes can evolve that aren't explained by natural selection. Winged pigs were never on the cards, so nature never had to select against them. How many such cases are there? How often does a phenotype carry information not about a creature's environment but about aspects of its endogenous structure? Nobody knows.

But it bears emphasis that, on this way of thinking about evolution, the mechanisms by which phenotypes are constructed may very well be numerous and heterogeneous. This is one of the important ways in which evo-devo differs from adaptationism. Darwinists generally hold that natural selection, even if it isn't all there is to evolution, is vastly the most important part. By contrast, channelling couldn't conceivably explain the structure of phenotypes all by itself. But that leaves it open that channelling might be one among many mechanisms by which phenotypes express endogenous structure, and which, taken together, account for (some? many? all of?) the facts of evolution. If, as I suggested, the notion of natural selection is conceptually flawed, such alternatives would be distinctly welcome.

Here's another kind of process that appears to explain some (very striking) facts about phenotype formation, but is quite different from either adaptation or channelling. In fact, it takes us back to spandrels. Gould and Lewontin say that spandrels are an artefact of selection for arches. Lacking arches, domes fall down; so arches are selected for supporting domes. But arches are linked to spandrels for reasons of geometry; so spandrels aren't selected for, they are 'free riders' on selection for arches. The moral is that phenotypic traits can carry information about linkages among the mechanisms that produce them. Free-riding is always suggestive of such linkages, and free-riding is ubiquitous in evolution.

There's a really lovely experiment that provides an example. The working hypothesis was succinctly summarised by Lyudmila Trut in *American Scientist* in 1999: 'Because behaviour is rooted in biology, selecting for tameness and against aggression means selecting for physiological changes in the systems that govern the body's hormones and neurochemicals. Those changes, in turn, could have had far-reaching effects on the development of the animals themselves, effects that might well explain why different animals would respond in similar ways when subjected to the same kinds of selective pressures.' In the vocabulary I've been using: one might expect a galaxy of other phenotypic traits to be endogenously linked to tameness, and hence to free-ride on selection for it. Such properties would co-evolve with tameness even if they have little or no systematic effect on fitness; in effect there would be evolution without adaptation. Moreover, insofar as the genetic and physiological mechanisms that link tameness to its free-riders hold across a range of species, one might expect that selecting for tameness will have similar phenotypic by-products in creatures of quite different kinds.

The experimental investigation of these hypotheses involved forty years of inbreeding for tameness in thirty or so generations of silver foxes. The results are impressive. On the one hand, foxes that were bred for tameness also tended to share a number of other phenotypic traits. Unlike their feral cousins, they tend to evolve floppy ears, brown moulting, grey hairs, short curly tails, short legs and piebald coloration (in particular, white flashes). Inbreeding for tameness also had characteristic effects on the reproductive cycles of the foxes and on the average size of their litters. And these are all traits that other domestic animals (dogs, cats, goats, cows) also tend to have. An adaptationist might well wonder what it is about dogs, cats etc that makes curly tails good for their fitness in an ecology of domestication. The answer, apparently, is 'nothing'. Curly tails aren't fitness enhancing, they just happen to be linked to tameness, so selection for the second willy-nilly selects the first.

This case is much like that of spandrels, but much worse from an adaptationist's point of view. You can explain the linkage between domes, arches and spandrels; the geometry and mechanics of the situation demands it. But the ancillary phenotypic effects of selection for tameness seem to be perfectly arbitrary. In p articular, they apparently aren't adaptations; there isn't any teleological explanation—any explanation in terms of fitness—as to why domesticated animals tend to have floppy ears. They just do. It's possible, of course, that channelling and free-riding are just flukes and that most or all of the other evolutionary determinants of phenotypic structure are exogenous. It's also possible that palaeontologists will someday dig up fossilised pigs with wings. But don't bet on it.

So what's the moral of all this? Most immediately, it's that the classical Darwinist account of evolution as primarily driven by natural selection is in trouble on both conceptual and empirical grounds. Darwin was too much an environmentalist. He seems to have been seduced by an analogy to selective breeding, with natural selection operating in place of the breeder. But this analogy is patently flawed; selective breeding is performed only by creatures with minds, and natural selection doesn't have one of those. The alternative possibility to Darwin's is that the direction of phenotypic change is very largely determined by endogenous variables. The current literature suggests that alterations in the timing of genetically controlled developmental processes is often the endogenous variable of choice; hence the 'devo' in 'evo-devo'.

But I think there's also a moral about what attitude we should take towards our science. The years after Darwin witnessed a remarkable proliferation of other theories, each seeking to co-opt natural selection for purposes of its own. Evolutionary psychology is currently the salient instance, but examples have been legion. They're to be found in more or less all of the behavioural sciences, to say nothing of epistemology, semantics, theology, the philosophy of history, ethics, sociology, political theory, eugenics and even aesthetics. What they have in common is that they attempt to explain why we are so-and-so by reference to what being so-and-so buys for us, or what it would have bought for our ancestors. 'We like telling stories because telling stories exercises the imagination and an imagination would have been a good thing for a hunter-gatherer to have.' 'We don't approve of eating grandmother because having her around to baby-sit was useful in the hunter-gatherer ecology.' 'We like music because singing together strengthened the bond between the hunters and the gatherers (and/or between the hunter-gatherer grownups and their hunter-gatherer offspring)'. 'We talk by making noises and not by waving our hands; that's because hunter-gatherers lived in the savannah and would have had trouble seeing one another in the tall grass.' 'We like to gossip because knowing who has been up to what is important when fitness depends on co-operation in small communities.' 'We don't all talk the same language because that would make us more likely to interbreed with foreigners (which would be bad because it would weaken the ties of hunter-gatherer communities).' 'We don't copulate with our siblings because that would decrease the likelihood of interbreeding with foreigners (which would be bad because, all else being equal, heterogeneity is good for the gene pool).' I'm not making this up, by the way. Versions of each of these theories can actually be found in the adaptationist literature. But, in point of logic, this sort of explanation has to stop somewhere. Not all of our traits can be explained instrumentally; there must be some that we have simply because that's the sort of creature we are. And perhaps it's unnecessary to remark that such explanations are inherently post hoc (Gould called them 'just so stories'); or that, except for the prestige they borrow from the theory of natural selection, there isn't much reason to believe that any of them is true.

The high tide of adaptationism floated a motley navy, but it may now be on the ebb. If it does turn out that natural selection isn't what drives evolution, a lot of loose speculations will be stranded high, dry and looking a little foolish. Induction over the history of science suggests that the best theories we have today will prove more or less untrue at the latest by tomorrow afternoon. In science, as elsewhere, 'hedge your bets' is generally good advice.

As for Sachs, I wouldn't think of arguing that we are either mostly happy or mostly good. But I doubt that's because of what our minds were selected for. Maybe the real trouble is that our neurones aren't hooked together quite right, or that some of our hormones aren't entirely reliable; with the effect, in either case, that getting some of the things we want isn't compatible with getting the others. Or that some of them we can't have at all. Anyhow, for what it's worth, I really would be surprised to find out that I was meant to be a hunter-gatherer since I don't feel the slightest nostalgia for that sort of life. I loathe the very idea of hunting, and I'm not all that keen on gathering either. Nor can I believe that living like a hunter-gatherer would make me happier or better. In fact, it sounds to me like absolute hell. No opera. And no plumbing.

Letters in Response to "Why Pigs Don't Have Wings"

From *Simon Blackburn, Department of Philosophy, University of Cambridge*

My colleague Jerry Fodor has added his name to the list of those who have taken themselves to have 'conceptual' objections to the idea of adaptation by natural selection (*LRB*, 18 October). His problem is fortunately quite easily solved. He takes from Stephen Jay Gould and Richard Lewontin the question: if two traits occur together, how do we know which was 'selected' for without appeal to the mind of a designer? Fodor urges that when we take away the designer, the question is unanswerable, unless we make a metaphorical and flat-footed appeal to Mother Nature. But this is not so. Two traits may be found together in nature, but one can play a causal role in producing a reproductive advantage, when the other does not. It may be that all and only vertebrates with eyes weigh a little bit extra because they carry various proteins (crystallins) around that go to making up eyeballs. But the sensitivity to light is what gives the advantage, not the little bit of extra weight due to carrying crystallin. Otherwise flatfish might as well have eyes on their undersides, and we might have turned out blind, but with devices for holding crystallin in our armpits. Similarly Fodor triumphantly asks whether it is being white or being the same colour as the environment that is good for polar bears. A brief look at the life of polar bears, and other bears, and animals such as ptarmigan or mountain hares that change colour with the seasons, forces just one answer. Camouflage helps across the board; being white only helps when it coincides with it.

From *Tim Lewens, History and Philosophy of Science, University of Cambridge*

When one is consciously designing something, it makes perfect sense to say that some features are there on purpose, others mere side-effects of intentional decisions. Jerry Fodor thinks that no parallel distinction is available in the mindless world of evolution, hence there is no way to say which organic traits are adaptations, and which are merely side-effects of selection going on somewhere else. This, he believes, means that the very ideas of adaptation and natural selection are incoherent.

Yet Fodor's comments later in his article suggest a perfectly good answer to a problem he says is insoluble. He tells us that 'curly tails aren't fitness-enhancing, they just happen to be linked to tameness, so selection for the second willy-nilly selects the first.' To be sure, he is discussing an example of an artificially selected trait. Even so, the conceptual resource he uses to distinguish between the trait that is selected for, and the trait that is merely linked to one that is selected for, is fitness enhancement, and there is nothing in this concept that draws on notions of what a designer intentionally chooses. If Fodor's test for adaptation works in the realm of artificial selection, it works in the realm of natural selection, too.

Further, Fodor suggests that most attempts to make adaptation respectable appeal to suspect metaphors of what Mother Nature is aiming at. Some do, but here is the philosopher of biology Elliott Sober's solution to the problem, which he gave in 1984, and which is basically the same as Fodor's own implicit proposal: "'Selection of" pertains to the effects of a selection process, whereas "selection for" describes its causes. To say there is selection for a given property means that having the property causes success in survival and reproduction.' If a property doesn't cause success in survival and reproduction, but is linked to one that does, then there is no selection for that property. This is precisely why Fodor thinks that although there is selection of curly tails, there is no selection for curly tails.

Finally, Fodor tells us that 'the crucial test is whether one's pet theory can distinguish between selection for trait A and selection for trait B when A and B are coextensive: were polar bears selected for being white or for matching their environment? Search me; and search any kind of adaptationism I've heard of.' What adaptationists need is a test that tells them, for example, whether there is selection for polar bears having white fur, having warm fur, or both. The Fodor/Sober test can tell us that: if we dye the fur of polar bears green and there is no impact on their survival or reproduction, then this provides evidence that there is selection for warm fur, and that whiteness simply follows along because whiteness and warmth are linked. But it is not necessary that our test tell us whether there is selection for whiteness or for matching the environment. If you dyed the fur of polar bears green, then they would also fail to match their environment. If we then observe that they do worse in terms of survival and reproduction, our test suggests that there is selection both for being white, and for matching the environment. But that is hardly surprising, because polar bears are camouflaged in virtue of being white. The fact that our test doesn't discriminate between selection favouring whiteness and selection favouring matching the background doesn't show that we have a test with no discriminatory power. It consequently fails to undermine the distinction between 'selection of' and 'selection for', it fails to show that the concept of adaptation is flawed, and it fails to make problems for natural selection.

From *Ian Cross, Faculty of Music, University of Cambridge*

There is a significant word missing from Jerry Fodor's entertaining dismissal of Darwinian theory: variation. Darwin starts *The Origin of Species* by ruminating on the causes of variation within species, particularly species that have been domesticated. Variation allows for differential chances of survival of members of a species through processes of natural selection; some, by virtue of being somewhat different from their conspecifics, will be better able to cope with environmental pressures and be more likely to survive, procreate and hence pass on their genes to the next generation. This is why, in Darwin's original formulation, evolution occurs through processes of natural variation and natural selection. What Fodor appears to be attacking is not so much natural selection but rather an extreme adaptationist view of the evolutionary process wherein each and every trait of an animal is held to arise as an adaptation to the environment. But it would be difficult to find any reasoned expression of such a view; as Fodor himself points out, pigs don't have wings not because it would not be evolutionarily advantageous for them to fly, but because they're just not built that way.

From *Jerry Coyne, Philip Kitcher, University of Chicago, Columbia University*

Jerry Fodor makes the striking claim that evolutionary biologists are abandoning natural selection as the principal, or even an important, cause of evolutionary change, and that 'it's not out of the question that a scientific revolution—no less than a major revision of evolutionary theory—is in the offing' (*LRB*, 18 October). This is news to us, and, we believe, will be news to most knowledgeable people as well. The idea of natural selection is, in fact, alive and well, and remains the only viable explanation of the apparent 'design' of organisms—the remarkable fit between them and their environments and lifestyles—that once was ascribed to the divine.

Fodor's 'conceptual' charge against natural selection is that the whole notion is incoherent. Breeders can select for features of organisms, because they can identify the traits they wish to develop. Unless you have some illicit personification—Mother Nature—who observes and chooses, natural selection doesn't work like that. So, to cite Fodor's example, we can't tell whether polar bears were selected for being white or for matching their environment. This is very odd reasoning. The concept of 'selecting for' characteristics is largely a philosopher's invention, one put to hefty work by philosophers of mind and language in particular as they strive to understand how psychological states can have content. Fodor knows all this, but he seems to know nothing about the way the notion of natural selection has been used in evolutionary explanations for the past 148 years.

Darwin would have seen the history of the polar bears along the following lines: some ancestors had different versions of the hereditary material that caused them to be paler than their fellows; this difference caused them to be less visible to their prey in their Arctic environment, and thus to have an edge when it came to hunting; that edge made them more successful in leaving descendants who inherited the fortunate variation. After Mendel, Thomas Morgan, Watson and Crick, we can do better: the ancestral bears had some difference in their DNA (perhaps a mutation or a gene rearrangement); that difference led to a difference in the type or expression of proteins affecting the biochemistry of hair follicles; that difference led to paler fur and a better match to the surroundings, producing greater prowess in hunting and increased reproductive success. Nobody has to decide if there was selection 'for' the modified DNA, or 'for' the protein differences, or 'for' the different organisation of the cells, or 'for' the whiteness, or 'for' the camouflage.

It is easy to see that natural selection makes sense of the important distinctions. Suppose, by some accident, that all and only the bears with the lucky variation were born on a Thursday. It would not follow that bears have been selected 'for' being born on Thursdays. This was an important insight underlying the work of Stephen Jay Gould and Richard Lewontin, cited by Fodor. In philosophical discussions, that insight has grown in an extraordinarily distorted fashion, so that philosophers struggle to develop a notion of 'selection for' that will discriminate finely among all traits. That is a mug's game, as Fodor correctly sees. It is a large leap, however, to suppose that the fact that you cannot make all distinctions means that you cannot make any. As the bear example illustrates, biologists can make the important distinctions. Whiteness and camouflage (along with protein balances and forms of genetic material) are candidates 'for' natural selection because they figure in the causal history of the changes in the bears; being a Thursday's cub isn't a candidate because it doesn't play a comparable causal role.

Fodor's second argument turns on an 'empirical' issue. Allegedly, 'serious alternatives to adaptationism have begun to emerge.' The rival mechanisms Fodor cites are supplements to natural selection, not replacements. Moreover, they are further articulations of ideas that have been evolutionary orthodoxy for generations. The first of Fodor's alleged alternatives is 'evo-devo', the field of evolutionary developmental biology. The remit of evo-devo is to explain how adaptive differences in animal form—say, the camouflage patterns on butterfly wings that protect them from predators—have resulted from the way the genes themselves behave (how particular genes deposit pigment in the right place on a wing). Evo-devo is not an alternative to adaptation; rather, it is a way to explain how the genes mechanistically produce adaptations. In fact, Sean Carroll, one of the most prominent 'evo-devotees', notes in his recent book, *Endless Forms Most Beautiful*, that evo-devo is completely consistent with the Darwinian theory of natural selection producing adaptations via cumulative genetic change. The constraints of development may tell us why an eye, for example, has a particular form (our retina lies behind the blood vessels and nerves that feed it because retinas evolved from everted portions of the brain), but they cannot tell us why eyes are there in the first place. They are there because the gradual acquisition of vision gave animals a leg up in the evolutionary struggle for existence.

Similarly, as Fodor notes, many features of organisms can be by-products of evolution rather than the direct objects of natural selection. Our blood is red, for example, not because it is good for blood to be a particular colour, but because the haemoglobin molecules that carry oxygen absorb light in such a way as to make them red. But the 'by-product' explanation cannot explain apparent design. Why are so many animals camouflaged to match their background? Can that be a result of evo-devo or a mere by-product of something else? Neither is likely. Experiments have shown that more camouflaged animals are eaten less often by predators. This is exactly what you'd expect if natural selection built such adaptations, and not what you'd predict if camouflage resulted simply from developmental constraints or was a by-product of something else. And how do Fodor's alternatives explain the sharp teeth of sharks or the ability of some Arctic fish to load their blood with 'antifreeze' proteins to keep them from freezing solid in cold waters? Adaptation is not a failed explanation: it is a testable hypothesis, and has been tested—and confirmed—many times over.

From *Daniel Dennett, Tufts University*

I love the style of Jerry Fodor's latest attempt to fend off the steady advance of evolutionary biology into the sciences of the mind. He tells us that 'an appreciable number of perfectly reasonable biologists' are thinking seriously of giving up on the half of Darwinism that concerns natural selection. Did you know that? I didn't. In fact, I wonder if the appreciable number is as high as one. Fodor gives no names so we'll just have to wait for more breaking news. He does provide two of his favourite foretastes, however: evo-devo and the famous case of the domesticated Russian foxes. These interesting developments both fit handsomely within our ever-growing understanding of how evolution by natural selection works. Briefly, evo-devo drives home the importance of the fact that in addition to the information in the genes (the 'recipes' for making offspring), there is information in the developmental processes (the 'readers' of the recipes), and both together need to be considered in a good explanation of the resulting phenotypes, since the interactions between them can be surprising. Of course the information in the developmental processes is itself all a product of earlier natural selection, not a gift from God or some otherwise inexplicable contribution. The foxes are a striking instance of how selection acting on one trait can bring other traits along with it—which may then be subject to further selection. It corrects the naive assumption that *everything* is directly evolvable—docile foxes with zebra stripes, or green foxes, or pigs with wings—but nobody makes that assumption, aside from the straw men constructed by some ideologues.

I won't bother correcting, one more time, Fodor's breezy misrepresentation of Gould and Lewontin's argument about 'spandrels', except to say that far from suggesting an alternative to adaptationism, the very concept of a spandrel depends on there being adaptations: the arches and domes are indeed selected for, and they bring spandrels along in their wake. No 'perfectly reasonable biologist' has claimed that the hugely various and exquisitely tuned sense organs of animals, or the superbly efficient water-conserving methods of desert plants, are spandrels, even if they spawn spandrels galore.

What could drive Fodor to hallucinate the pending demise of the theory of evolution by natural selection? A tell-tale passage provides the answer: 'Science is about facts, not norms; it might tell us how we are, but it couldn't tell us what is wrong with how we are. There couldn't be a science of the human condition.' There can indeed be a science of the human condition, but it won't tell us, directly, 'what is wrong with what we are'. It can, however, constrain our ultimately political exploration of what we think we ought to be by telling us what is open to us, given what we are. Fodor's mistake, which he is hardly alone in making, is to suppose that if our minds are scientifically explicable bio-mechanisms, then there could not be any room at all for values. That just does not follow, but if you believe it, and if you cherish—as of course you should—the world of values, then you have to stand firm against *any* physical science of the mind. It's admirable, in a way, if you like that kind of philosophy. But it is better to repair the mistake; then you can have a science of the mind and values too. And you don't have to misrepresent science out of fear of what it might be telling us.

From *Steven Rose, Open University, Milton Keynes*

Jerry Fodor's attack on ultra-Darwinian pan-adaptationism (and Flintstone evolutionary psychology) is spot on, but he does less than justice to Darwin, or to modern pluralistic evolutionary theory. Fodor argues that Darwin was unwise to draw analogies between the artificial selection employed by animal breeders and the mechanism of natural selection. But whether the selection pressure is provided by breeders choosing among pigeons for the most spectacular fantail, or a lion-rich environment selecting for faster-running antelopes, the analogy holds. The difference is that, far more than in artificial selection, the natural environment itself changes in response to the presence of the faster-running antelopes (more intensive grazing, reduction in lion population or whatever). The metaphor of selection is unfortunate as it implies that the 'selected' organisms are merely passive, whereas in fact organisms select environments just as environments select organisms.

Furthermore, Darwin was himself a pluralist; as he insisted in later editions of the *Origin of Species*, natural selection is only one of a number of motors of evolutionary change. Modern selection theory (in the hands of other than ultra-Darwinists) recognises multiple levels at which selection works: gene, genome, organism (phenotype), population and species. It also recognises that what evolves is not an adult phenotype but an entire developmental system (faster-running antelopes do not emerge fully grown). By contrast with pan-adaptationism, pluralistic evolutionary theory recognises the presence of spandrels (non-adaptive features of a phenotype, such as the red colour of blood) and exaptations: features originally selected with one function which then come to have another, such as feathers, which were a thermo-regulatory mechanism before they took on their role in flying birds.

From *Colin Tudge, Wolvercote, Oxfordshire*

Jerry Fodor tells us: 'There is no Mother Nature.' This is biology's common assumption (and was probably Darwin's), but it does not come out of science. It is a piece of metaphysical dogma. Many philosophers and scientists argue that 'mind' is part of the fabric of the universe, and this embedded intelligence might indeed be equated either with 'Mother Nature' or with God in such a way that imbues the universe with purpose. This is a perfectly reasonable position, and Fodor's denial is simply a decision, common to all atheists, not to take this position seriously. Darwin's idea of evolution by means of natural selection is perfectly compatible with the idea of God, as many theologians and quite a few scientists acknowledged as soon as *Origin* was published.

It is a long time now since I read Dobzhansky's essay of 1973, but it was not called 'Nothing in biology makes sense without Darwinism'. It was called 'Nothing in biology makes sense except in the light of evolution' (*American Biology Teacher*, Vol. 35). Since Fodor is at pains to point out that 'evolution' should not be conflated with 'Darwinian natural selection', this is a strange lapse. In fact, Dobzhansky admired Teilhard de Chardin, who came very close to saying that intelligence is embedded in the fabric of the universe.

From *Jerry Fodor, Rutgers University, New Jersey*

A perceptible flurry in the dovecote. Here are some replies to my critics. It seems to me that Simon Blackburn has comprehensively missed the point (Letters, 1 November). He takes the problem I raised to be epistemological: 'If two traits occur together, how do we know which was "selected" for?' But I don't do epistemology, and that isn't what I'm worried about (nor, by the way, is it what worried Gould and Lewontin). My question was: how can the operation of selection distinguish traits that are coextensive in a creature's ecology? Perhaps news about mountain hares and such tells us what colour was selected for in polar bears. But selection didn't consider mountain hares when it coloured polar bears. Nor, quite generally, did it consider such counterfactuals as 'what would happen to white bears if the colour of their environment changed?'

The same applies to Tim Lewens's line of thought. The selection of colour in polar bears can't be contingent on such counterfactuals as: 'what if one dyed their fur green?' In fact, it can't be contingent on any counterfactuals at all. We can apply the 'method of differences' to figure out what colour evolution made the polar bear; but selection can't apply the method of differences to figure out what colour to make them. That's because we have minds but it doesn't.

Some of my critics point out the importance of linkage as a mechanism that might explain why, for example, domesticated foxes have floppy ears. Quite so, but linkage is an endogenous trait, and adaptationism is committed to explaining phenotypes by reference to exogenous variables.

The same applies to the remarks by Steven Rose (Letters, 15 November). To give up on the idea that selection is determined by largely exogenous forces is to abandon adaptationism in all but name. No doubt, if we knew enough about the macro and microstructure of organisms (and of their ecologies) we would understand their evolution. If that's adaptationism, then I'm an adaptationist too (and so is every materialist since Lucretius).

Jerry Coyne and Philip Kitcher make the usual mistake. In fact, I am not worrying about whether we can tell if 'polar bears were selected for being white or for matching their environment'. I repeat: I don't do epistemology. Nor do I deny that we can often focus on different aspects of the causal history underlying an episode of selection. The problem is that it makes no sense at all to speak of the aspect of a causal history that selection focuses on; to say (as it might be) that selection focused on the whiteness of the polar bear rather than its match to the surround. Selection doesn't focus: it just happens.

Coyne and Kitcher then say that 'the concept of "selecting for" characteristics is largely a philosopher's invention.' I don't know who invented it, but that can't be right. If the theory of adaptation fails to explain what phenotypic traits were selected for, it won't generalise over possible-but-not-actual circumstances; it won't, for example, tell us whether purple polar bears would have survived in the ecology that supports ours. It will not be 'news to most knowledgeable people' that empirical theories are supposed to support relevant counterfactuals. If adaptationism doesn't, that *is* news.

Coyne and Kitcher suggest that evo-devo doesn't purport to be an alternative to adaptationism but rather is 'consistent with' natural selection. That's right but not relevant. Part of my point was that if adaptationism is independently incoherent (as, in fact, I believe it to be) then we're in want of an alternative. Evo-devo may reasonably be considered a step towards supplying one.

They also say that it doesn't matter whether selection can draw all the distinctions between traits so long as it can draw the important ones. I don't know how they tell which ones are important, but they ought to bear this in mind: selection is insensitive to the difference between any traits that are even *locally* confounded (i.e. that are confounded in a creature's actual history of causal interactions with its ecology). It can't, for example, distinguish encounters with big tails from encounters with colourful tails if all and only the big tails Miss Peacock has come across are colourful. (Of course, *we* can tell the difference between selecting for one and selecting for the other; that's because, unlike natural selection, we have minds.) If it isn't important (to, for example, ethology) whether it's big tails or colourful tails that lady peacocks like, then so much the worse for importance.

Finally, Coyne and Kitcher ask how anything but adaptationism can explain the match between a creature's phenotype and its ecology. This question is entirely pertinent. But they will have to read about it in Fodor and Piatelli-Palmarini (forthcoming).

Over the years, I've been finding it increasingly difficult to figure out which bits of Daniel Dennett's stuff are supposed to be the arguments and which are just rhetorical posturing. In the present case, I give up. I'll take it more or less paragraph by paragraph. Dennett speaks of the 'steady advance of evolutionary biology into the sciences of the mind'. He provides no examples, however, and surely he knows that there is a considerable body of literature to the contrary. (See, for example, David Buller's book *Adapting Minds*.) Even Dennett's fellow-critics of my piece express, in several cases, attitudes towards the evolutionary psychology programme ranging from scepticism to despair: it's a recurrent theme of theirs that Fodor is, of course, right about EP; but he's wrong about natural selection at large.

I cite the fox experiments and the literature on evo-devo as evidence of the importance of endogenous factors in directing the course of evolution. Dennett does not deny that lots of endogenous factors constrain the course of evolution; or that the cases I cited are instances; or that appeals to endogenous variables are alternatives to natural selection. 'Of course the information in the developmental processes is itself all a product of earlier natural selection.' What's the argument for that, I wonder. It appears, prima facie, simply to beg the question at issue.

Dennett can't be bothered to correct my 'breezy misrepresentation of Gould and Lewontin'. In fact, he can't even be bothered to say what it consists in. That being so, I can't be bothered to refute him.

'The very concept of a spandrel depends on there being adaptations.' This suggests that Dennett has utterly lost track of the argument. Of course the spandrels are free-riders on the architect's design for the arches and domes. But the question I wanted to raise was precisely whether this account of selection-for can be extended to cases where, by general consensus, there isn't any architect. In particular, I claim, Darwin overplayed the analogy between artificial selection (where there is somebody who does the selecting) and 'natural' selection (where there isn't). How could anybody who actually read my article have missed this?

I said that metaphors like 'evolution selects for what Mother Nature intends it to' have to be cashed. The rules of the game require respectable adaptationists to give an account of selection-for that doesn't appeal to agency. Suppose (what's not obvious) that explaining the scientific results really does require a notion of biological function (hence of selection-for). It simply doesn't follow that it requires a notion of biological function that is reconstructed in terms of selection history. Dennett must know that, de facto, there is no such notion. Biological function is itself an intentional concept, so appeals to it don't cash the Mother Nature metaphor; they just take out loans on its being cashed sooner or later. It seems that everybody understands this except Dennett.

Finally, Dennett says I am worried about preserving my values in the face of scientific reduction. Where on earth did he get that idea? I've spent more of my life than I like to think about arguing that ontological questions about reduction are neutral with respect to epistemological questions about intentional explanations. As a matter of fact . . .

But on second thoughts, to hell with it.

The reader may wonder whether there are any general morals to draw from all this. There are three: don't forget the importance of getting the counterfactuals right; don't confuse your ontology with your epistemology; and do try to keep your cool.

From *Simon Blackburn, Jerry Coyne, Philip Kitcher, Tim Lewens, Steven Rose, University of Cambridge, University of Chicago, Columbia University, University of Cambridge, Open University*

Jerry Fodor persists with two provocative claims: first, that natural selection explanations are incoherent; second, that there is some alternative explanation for adaptive phenomena such as camouflage or beak shape (*Letters*, 29 November).

To show the incoherence of anything, you have to address it in the form in which its professional expositors deploy it. In large numbers of articles and books, published from 1859 to the present, evolutionary biologists use the following style of explanation. A characteristic of an organism (the colour of an animal's coat, say) is as it is because of a historical process. In some ancestral population there was a variant type that differed from the rest in ways that enhanced reproductive success. (White polar bears, for example, more camouflaged than their brown confrères, were better at sneaking up on seals, were better fed and left more offspring.) If the variant has a genetic basis, its frequency increases in the next generation.

Is this incoherent? Nothing Fodor says bears on that question. Instead, he opposes a very particular way of presenting the explanation. Some people think we can talk of 'selection for' a characteristic, and identify rather precisely the traits that have been 'selected for'. Fodor tries to argue that this is wrong: that there is no single correct answer (whether we know it or not) to the question of whether it was the whiteness of polar bears or their blending in with their surroundings that was 'selected for'. Whether he is right is a philosophical issue about which people can disagree, but it has nothing to do with the coherence of Darwinian explanation. Natural selection proceeds if three elements are in place: variation in a trait, an effect of the variation on reproductive success, and some means by which the trait is inherited. Both the whiteness and the environmental blending emerged from the historical process that the selection explanation describes.

Although Fodor follows a long line of people, including Darwin himself, who recognise constraints on natural selection, he advocates something far more ambitious than his predecessors. He wants a replacement of natural selection, not supplements to it. Some of the signatories to this letter have emphasised the importance of constraints, and have written against the hyper-Darwinian practice of seeing adaptation everywhere. None of us has ever supposed that the appeal to constraints could eliminate all mention of selection.

Cases of convergent evolution are vivid illustrations of natural selection's importance. Ichthyosaurs, sharks and dolphins share a similar body form; marsupial and placental mammals have counterparts that are almost identical in form. In different lines of descent, similar traits emerge. Fodor would have us believe that natural selection plays no role whatsoever in explaining these facts. Indeed, he doesn't say how he thinks convergence—or any adaptation—should be explained, but merely tells us that he and a coauthor have something up their sleeve. The task they envisage is far more ambitious than that attempted by brilliant evolutionary theorists who have wanted to 'expand' Darwinism (for example, Stephen Jay Gould). Given the evidence that at least one of these would-be revolutionaries has little acquaintance with the biological theory he aspires to replace, we have little reason to think they will succeed.

Jerry Fodor writes: Blackburn et al have a number of complaints about what I wrote. The first is exegetical: they say that the kind of adaptationism I've attacked is not one that paradigm adaptationists endorse. I think that even a cursory glance at the relevant literature shows this is false. The standard current formulation has it that a main goal of evolutionary theory is to explain the distribution of phenotypic traits in populations of organisms, and that natural selection is the key to such explanations: organisms are selected for the ecological fitness of their phenotypes. Patently, any such theory is in want of a coherent account of what it is for a creature to be selected for some or other of its traits. But I don't propose to argue the exegetical point. Let those the shoe fits wear it. I'm content if what I wrote serves a cautionary function: if you find yourself tempted to espouse this sort of adaptationism, don't!

Their second claim is that there is no incoherence (or, anyhow, none of the sort that I alleged) in selection theory as correctly understood. They don't, however, say what the correct understanding is. Rather, they offer some potted polar bear history: 'White polar bears ... more camouflaged than their brown confrères, were better at sneaking up on seals, were better fed and left more offspring.' I don't know whether this story is true (neither, I imagine, do they), but let's suppose it is. They ask, rhetorically, whether I think it's incoherent. Well, of course I don't, but that's because they've somehow left out the Darwin bit. To get it back in, you have to add that the white bears were selected 'because of' their improved camouflage, and that the white bears were 'selected for' their improved camouflage: i.e. that the improved camouflage 'explains' why the white bears survived and flourished. But now we get the incoherence back too. What Darwin failed to notice (and what paradigm adaptationists continue to fail to notice) is that the theory of natural selection entails none of these. In fact, the theory of natural selection leaves it wide open what (if anything) the white bears were selected for. Here's the argument. Consider any trait X that was locally coextensive with being white in the polar bear's evolutionary ecology. Selection theory is indifferent between 'the bears were selected for being white' and 'the bears were selected for being X.' What's 'incoherent' is to admit that the theory of natural selection can't distinguish among locally coextensive properties while continuing to claim that natural selection explains why polar bears are white. Do not reply: 'But it's just obvious that, if the situation was as Blackburn et al describe, then it was the whiteness of the bears that mattered.' The question is not what is obvious to the theorist; the question is what follows from the theory. Why is it so hard to get this very rudimentary distinction across?

Having got all that wrong, Blackburn et al add that 'Fodor tries to argue that ... there is no single correct answer ... to the question of whether it was the whiteness of polar bears or their blending in with their surroundings that was "selected for".' But I don't argue anything of the sort. Since the hypotheses that the bears were selected for being white and that they were selected for matching their environments support different counterfactuals (what would have happened if their environment had been orange?) they can perfectly well be distinguished in (for example, experimental) environments in which one trait is instantiated and the other one isn't. I don't claim that locally coextensive properties are indistinguishable in principle. I claim that, since the theory of natural selection fails to distinguish them, there must be something wrong with the theory. (I also don't claim to have 'some alternative explanation for adaptive phenomena'; only that there had better be one sooner or later; and that it's a plausible guess that, when there is, it will explain adaptive phenomena largely by appeal to endogenous constraints on phenotypes.)

Finally, they say that whether I'm right about all this is 'a philosophical issue'. I don't know how they decide such things; maybe they think that philosophical issues are the ones that nobody else cares about (a masochistic metatheory that many philosophers apparently endorse). Anyhow, the kind of philosophy I do consists largely of minding other people's business. I am, to be sure, in danger of having insufficient 'acquaintance with the biological theory that [I aspire] to replace'; but I'm prepared to risk it. A blunder is a blunder for all that, and it doesn't take an ornithologist to tell a hawk from a handsaw. Tom Kuhn remarks that you can often guess when a scientific paradigm is ripe for a revolution: it's when people from outside start to stick their noses in.

The Boatman Poem

Colin Boatman

Professor Jerry A Fodor
makes claims as a Darwin de-coder
but his theories so muddled
leave readers befuddled
and impart a distinct fishy odour.

He avers that Darwin's adaptation
falls short as a good explanation
for the white of the hair
of the North-polar bear,
but he can't or he don't (if he can then he won't) tell us his own variation.

He hasn't much grounding in science
so has to place all his reliance
on a tricky disguise
philosophy-wise
to support his adaptive defiance.

He claims many experts agree
with his take on heredity—
but who these can be
is a deep mystery.
They're presented anonymously.

Professor Fodor is loath
to accept the bear's furry white growth
as selection for fitness
at the same time as witness
to environment—but why not BOTH?

Chapter 12

Introduction

Chapter 12 deals with the critics, particularly the ideas of Stephen Jay Gould and others working on problems of macroevolution. My first document is one of his justly celebrated essays, "The Episodic Nature of Evolutionary Change," from his monthly column in *Natural History*. Here Gould lays out his theory of punctuated equilibria in the clear prose for which he is justly celebrated. He argues that evolutionary change as revealed by the fossil record is much more jerky or spasmodic than is believed and claimed by classical Darwinism, arguing also that this is no artifact of the record—incomplete fossilization or whatever—but a reflection of reality. Evolution's history is one of inaction or lack of change, broken by periods when there is a rapid move from one form to another. Although Gould is careful to hedge his bets and certainly does come out foursquare for this philosophy, note the favorable reference to Marxism. As I explain in the text, I am inclined to think that Gould's real philosophical roots lie further back in the German idealism of the early nineteenth century, and I draw your attention to Gould's mention of the philosopher Hegel, who was himself an ardent proponent of *Naturphilosophie*.

My second document, countering Gould, is by John Maynard Smith. Asking "Did Darwin Get It Right?" Maynard Smith—a student of J. B. S. Haldane—answers his question with a ringing affirmative. As explained in the text, Maynard Smith feels that Gould misrepresents or caricatures the Darwinian position. No one ever said that everything must be adaptive all of the time. The point rather is that organic features are rooted in and related to adaptation. Having four limbs rather than six may not now be a necessity, but in the past, when vertebrates were aquatic, it was really important. Natural selection was all-important even if now it plays no direct role. You should judge Maynard Smith's arguments on their merits, but you might like to know that he too was a Marxist, breaking from that philosophy in the 1950s because of the treatment of geneticists when Lysenko was in power and because of political events, particularly the Hungarian Revolution in 1956, which was put down brutally by the Soviets. Do we have here at work a philosophical disagreement as much as anything purely scientific?

The Episodic Nature of Evolutionary Change

S. J. Gould

On November 23, 1859, the day before his revolutionary book hit the stands, Charles Darwin received an extraordinary letter from his friend Thomas Henry Huxley. It offered warm support in the coming conflict, even the supreme sacrifice: "I am prepared to go to the stake, if requisite . . . I am sharpening up my claws and beak in readiness." But it also contained a warning: "You have loaded yourself with an unnecessary difficulty in adopting *Natura non facit salturn* so unreservedly."

The Latin phrase, usually attributed to Linnaeus, states that "nature does not make leaps." Darwin was a strict adherent to this ancient motto. As a disciple of Charles Lyell, the apostle of gradualism in geology, Darwin portrayed evolution as a stately and orderly process, working at a speed so slow that no person could hope to observe it in a lifetime. Ancestors and descendants, Darwin argued, must be connected by "infinitely numerous transitional links" forming "the finest graduated steps." Only an immense span of time had permitted such a sluggish process to achieve so much.

Huxley felt that Darwin was digging a ditch for his own theory. Natural selection required no postulate about rates; it could operate just as well if evolution proceeded at a rapid pace. The road ahead was rocky enough; why harness the theory of natural selection to an assumption both unnecessary and probably false? The fossil record offered no support for gradual change: whole faunas had been wiped out during disarmingly short intervals. New species almost always appeared suddenly in the fossil record with no intermediate links to ancestors in older rocks of the same region. Evolution, Huxley believed, could proceed so rapidly that the slow and fitful process of sedimentation rarely caught it in the act.

The conflict between adherents of rapid and gradual change had been particularly intense in geological circles during the years of Darwin's apprenticeship in science. I do not know why Darwin chose to follow Lyell and the gradualists so strictly, but I am certain of one thing: preference for one view or the other had nothing to do with superior perception of empirical information. On this question, nature spoke (and continues to speak) in multifarious and muffled voices. Cultural and methodological preferences had as much influence upon any decision as the constraints of data.

On issues so fundamental as a general philosophy of change, science and society usually work hand in hand. The static systems of European monarchies won support from legions of scholars as the embodiment of natural law. Alexander Pope wrote:

> Order is Heaven's first law; and this confessed,
> Some are, and must be, greater than the rest.

As monarchies fell and as the eighteenth century ended in an age of revolution, scientists began to see change as a normal part of universal order, not as aberrant and exceptional. Scholars then transferred to nature the liberal program of slow and orderly change, that they advocated social transformation in human society. To many scientists, natural cataclysm seemed as threatening as the reign of terror that had taken their great colleague Lavoisier.

Yet the geologic record seemed to provide as much evidence for cataclysmic as for gradual change. Therefore, in defending gradualism as a nearly universal tempo, Darwin had to use Lyell's most characteristic method of argument—he had to reject literal appearance and common sense for an underlying "reality." (Contrary to popular myths, Darwin and Lyell were not the heros of true science, defending objectivity against the theological fantasies of such "catastrophists" as Cuvier and Buckland. Catastrophists were as committed to science as any gradualist; in fact, they adopted the more "objective" view that one should believe what one sees and not interpolate missing bits of a gradual record into a literal tale of rapid change.) In short, Darwin argued that the geologic record was exceedingly imperfect—a book with few remaining pages, few lines on each page, and few words on each line. We do not see slow evolutionary change in the fossil record because we study only one step in thousands. Change seems to be abrupt because the intermediate steps are missing.

The extreme rarity of transitional forms in the fossil record persists as the trade secret of paleontology. The evolutionary trees that adorn our textbooks have data only at the tips and nodes of their branches; the rest is inference, however reasonable, not the evidence of fossils. Yet Darwin was so wedded to gradualism that he wagered his entire theory on a denial of this literal record:

The geological record is extremely imperfect and this fact will to a large extent explain why we do not find interminable varieties, connecting together all the extinct and existing forms of life by the finest graduated steps. He who rejects these views on the nature of the geological record, will rightly reject my whole theory.

Darwin's argument still persists as the favored escape of most paleontologists from the embarrassment of a record that seems to show so little of evolution directly. In exposing its cultural and methodological roots, I wish in no way to impugn the potential validity of gradualism (for all general views have similar roots). I wish only to point out that it was never "seen" in the rocks.

Paleontologists have paid an exorbitant price for Darwin's argument. We fancy ourselves as the only true students of life's history, yet to preserve our favored account of evolution by natural selection we view our data as so bad that we almost never see the very process we profess to study.

For several years, Niles Eldredge of the American Museum of Natural History and I have been advocating a resolution of this uncomfortable paradox. We believe that Huxley was right in his warning. The modern theory of evolution does not require gradual change. In fact, the operation of Darwinian processes should yield exactly what we see in the fossil record. It is gradualism that we must reject, not Darwinism.

The history of most fossil species includes two features particularly inconsistent with gradualism:

> *Stasis*. Most species exhibit no directional change during their tenure on earth. They appear in the fossil record looking much the same as when they disappear; morphological change is usually limited and directionless.
>
> *Sudden appearance,* In any local area, a species does not arise gradually by the steady transformation of its ancestors; it appears all at once and "fully formed."

Evolution proceeds in two major modes. In the first, phyletic transformation, an entire population changes from one state to another. If all evolutionary change occurred in this mode, life would not persist for long. Phyletic evolution yields no increase in diversity, only a transformation of one thing into another. Since extinction (by extirpation, not by evolution into something else) is so common, a biota with no mechanism for increasing diversity would soon be wiped out. The second mode, speciation, replenishes the earth. New species branch off from a persisting parental stock.

Darwin, to be sure, acknowledged and discussed the process of speciation. But he cast his discussion of evolutionary change almost totally in the mold of phyletic transformation. in this context, the phenomena of stasis and sudden appearance could hardly be attributed to anything but imperfection of the record; for if new species arise by the transformation of entire ancestral populations, and if we almost never see the transformation (because species are essentially static through their range), then our record must be hopelessly incomplete.

Eldredge and I believe that speciation is responsible for almost all evolutionary change. Moreover, the way in which it occurs virtually guarantees that sudden appearance and stasis shall dominate the fossil record.

All major theories of speciation maintain that splitting takes place rapidly in very small populations. The theory of geographic, or allopatric, speciation is preferred by most evolutionists for most situations (allopatric means "in another place").* A new species can arise when a small segment of the ancestral population is isolated at the periphery of the ancestral range. Large, stable central populations exert a strong homogenizing influence. New and favorable mutations are diluted by the sheer bulk of the population through which they must spread. They may build slowly in frequency, but changing environments usually cancel their se-

---

*I wrote this essay in 1977. Since then, a major shift of opinion has been sweeping through evolutionary biology. The allopatric orthodoxy has been breaking down and several mechanisms of sympatric speciation have been gaining both legitimacy and examples. (In sympatric speciation, new forms arise within the geographic range of their ancestors.) These sympatric mechanisms are united in their insistence upon the two conditions that Eldredge and I require for our model of the fossil *record*—*rapid* origin in a *small* population. In fact, they generally advocate smaller groups and more rapid change than conventional allopatry envisages (primarily because groups in potential contact with their forebears must move quickly towards reproductive isolation, lest their favorable variants be diluted by breeding with the more numerous parental forms). See White (1978) for a thorough discussion of these sympatric models.

lective value long before they reach fixation. Thus, phyletic transformation in large populations should be very rare—as the fossil record proclaims.

But small, peripherally isolated groups are cut off from their parental stock. They live as tiny populations in geographic corners of the ancestral range. Selective pressures are usually intense because peripheries mark the edge of ecological tolerance for ancestral forms. Favorable variations spread quickly. Small, peripheral isolates are a laboratory of evolutionary change.

What should the fossil record include if most evolution occurs by speciation in peripheral isolates? Species should be static through their range because our fossils are the remains of large central populations. In any local area inhabited by ancestors, a descendent species should appear suddenly by migration from the peripheral region in which it evolved. In the peripheral region itself, we might find direct evidence of speciation, but such good fortune would be rare indeed because the event occurs so rapidly in such a small population. Thus, the fossil record is a faithful rendering of what evolutionary theory predicts, not a pitiful vestige of a once bountiful tale.

Eldredge and I refer to this scheme as the model *of punctuated equilibria.* Lineages change little during most of their history, but events of rapid speciation occasionally punctuate this tranquillity. Evolution is the differential survival and deployment of these punctuations. (In describing the speciation of peripheral isolates as very rapid, I speak as a geologist. The process may take hundreds, even thousands of years; you might see nothing if you stared at speciating bees on a tree for your entire lifetime. But a thousand years is a tiny fraction of one percent of the average duration for most fossil invertebrate species—5 to 10 million years. Geologists can rarely resolve so short an interval at all; we tend to treat it as a moment.)

If gradualism is more a product of Western thought than a fact of nature, then we should consider alternate philosophies of change to enlarge our realm of constraining prejudices. In the Soviet Union, for example, scientists are trained with a very different philosophy of change—the so-called dialectical laws, reformulated by Engels from Hegel's philosophy. The dialectical laws are explicitly punctuational. They speak, for example, of the "transformation of quantity into quality." This may sound like mumbo jumbo, but it suggests that change occurs in large leaps following a slow accumulation of stresses that a system resists until it reaches the breaking point. Heat water and it eventually boils. Oppress the workers more and more and bring on the revolution. Eldredge and I were fascinated to learn that many Russian paleontologists support a model similar to our punctuated equilibria.

I emphatically do not assert the general "truth" of this philosophy of punctuational change. Any attempt to support the exclusive validity of such a grandiose notion would border on the nonsensical. Gradualism sometimes works well. (I often fly over the folded Appalachians and marvel at the striking parallel ridges left standing by gradual erosion of the softer rocks surrounding them.) I make a simple plea for pluralism in guiding philosophies, and for the recognition that such philosophies, however hidden and unarticulated, constrain all our thought. The dialectical laws express an ideology quite openly; our Western preference for gradualism does the same thing more subtly.

Nonetheless, I will confess to a personal belief that a punctuational view may prove to map tempos of biological and geologic change more accurately and more often than any of its competitors—if only because complex *systems* in steady state are both common and highly resistant to change. As my colleague British geologist Derek V. Ager writes in supporting a punctuational view of geologic change: "The history of any one part of the earth, like the life of a soldier, consists of long periods of boredom and short periods of terror."

Did Darwin Get It Right?

John Maynard Smith

I think I can see what is breaking down in evolutionary theory—the strict construction of the modern synthesis with its belief in pervasive adaptation, gradualism and extrapolation by smooth continuity from causes of change in local populations to major trends and transitions in the history of life.

A new and general evolutionary theory will embody this notion of hierachy and stress a variety of themes either ignored or explicitly rejected by the modern synthesis.

These quotations come from a recent paper in *Palaeobiology* by Stephen Jay Gould. What is the new theory? Is it indeed likely to replace the currently orthodox "neo-Darwinian" *view*? Proponents of the new view make a minimum and a maximum claim. The minimum claim is an empirical one concerning the nature of the fossil record. It is that species, once they come into existence, persist with little or no change, often for millions of years ("stasis"), and that evolutionary change is concentrated into relatively brief periods ("punctuation"), these punctuational changes occurring at the moment when a single species splits into two. The maximal claim is a deduction from this, together with arguments drawn from the study of development: it is that evolutionary change, when it does occur, is not caused by natural selection operating on the genetic differences between members of populations, as Darwin argued and as most contemporary evolutionists would agree, but by some other process. I will discuss these claims in turn; as will be apparent, it would be possible to accept the first without being driven to accept the second.

The claim of stasis and punctuation will ultimately be settled by a study of the fossil record. I am not a palaeontologist, and it might therefore be wiser if I were to say merely that some palaeontologists assert that it is true, and others are vehemently denying it. There is something, however, that an outsider can say. it is that the matter can be settled only by a statistical analysis of measurements of fossil populations from different levels in the rocks, and not by an analysis of the lengths of time for which particular named species or genera persist in the fossil record. The trouble with the latter method is that one does not know whether one is studying the rates of evolution of real organisms, or merely the habits of the taxonomists who gave the names to the fossils. Suppose that in some lineage evolutionary change took place at a more or less steady rate, to such an extent that the earliest and latest forms are sufficiently different to warrant their being placed in different species. If there is at some point a gap in the record, because suitable deposits were not being laid down or have since been eroded, then there will be a gap in the sequence of forms, and taxonomists will give fossils before the gap one name and after it another. It follows that an analysis of named forms tells us little: measurements of populations, on the other hand, would reveal whether change was or was not occurring before and after the gap.

My reason for making this rather obvious point is that the only extended presentation of the punctuationist view—Stanley's book, *Macroevolution*—rests almost entirely on an analysis of the durations of named species and genera. When he does present population measurements, they tend to support the view that changes are gradual rather than sudden. ! think that at least some of the changes he presents as examples of sudden change will turn out on analysis to point the other way. I was unable to find any evidence in the book which supported, let alone established, the punctuationist view.

Of course, that is not to say that the punctuationist view is not correct. One study, based on a proper statistical analysis, which does support the minimal claim, but not the maximal one, is Williamson's study of the freshwater molluscs (snails and bivalves) of the Lake Turkana region of Africa over the last five million years. Of the 21 species studied, most showed no substantial evolutionary change during the whole period: "stasis" was a reality. The remaining six species were more interesting. They also showed little change for most of the period. There was, however, a time when the water table fell and the lake was isolated from the rest of the rift valley. When this occurred, these six species changed rather rapidly. Through a depth of deposit of about one meter, corresponding roughly to 50,000 years, successive populations show changes of shape great enough to justify placing the later forms in different species. Later, when the lake was again connected to the rest of the rift valley, these new forms disappear suddenly, and are replaced by the original forms, which presumably re-entered the lake from outside, where they had persisted unchanged.

This is a clear example of stasis and punctuation. However, it offers no support for the view that changes, when they do occur, are not the result of selection acting within populations. Williamson does have intermediate populations, so we know that the change did not depend on the occurrence of a "hopeful monster" (see below), or on the existence of an isolated population small enough to permit random changes to outweigh natural selection. The example is also interesting in showing how we may be misled if we study the fossil record only in one place. Suppose that, when the water table rose again, the new form had replaced the original one in the rest of the rift valley, instead of the other way round. Then, if we had examined the fossil record anywhere else but in Lake Turkana, we would have concluded, wrongly, that an effectively instantaneous evolutionary change had occurred.

Williamson's study suggests an easy resolution of the debate. Both sides are right, and the disagreement is purely semantic. A change taking 50,000 years is sudden to a palaeontologist but gradual to a population geneticist. My own guess is that there is not much more to the argument than that. However, the debate shows no signs of going away.

One question that arises is how far the new ideas are actually new. Much less so, I think, than their proponents would have us believe. They speak and write as if the orthodox view is that evolution occurs at a rate which is not only "gradual" but uniform. Yet George Gaylord Simpson, one of the main architects of the "modern synthesis" now under attack, wrote a book, *Tempo and Mode in Evolution,* devoted to emphasizing the great variability of evolutionary rates. It has never been part of the modern synthesis that evolutionary rates are uniform.

Yet there is a real point at issue. If it turns out to be the case that all, or most, evolutionary change is concentrated into brief periods, and associated with the splitting of lineages, that would require some serious rethinking. Oddly enough, it is not so much the sudden changes which would raise difficulties, but the intervening stasis. Why should a species remain unchanged for millions of years? The explanation favored by most punctuationists is that there are "developmental constraints" which must be overcome before a species cab change. The suggestion is that the members of a given species share a developmental pathway which can he modified so as to produce some kinds of change in adult structure rather easily, and other kinds of change only with great difficulty, or not at all. I do not doubt that this is true: indeed, in my book *The Theory of Evolution,* published in 1958 and intended as a popular account of the modern synthesis, I spent some time emphasizing that "the pattern of development of a given species is such that there are only a limited number of ways in which it can be altered without causing complete breakdown." Neo-Darwinists have never supposed that genetic mutation is equally likely to produce changes in adult structure in any direction: all that is assumed is that mutations do not, as a general rule, adapt organisms to withstand the agents which caused them, What is at issue, then, is not whether there are developmental constraints, because clearly there are, but whether such constraints can account for stasis in evolution.

I find it hard to accept such an explanation for stasis, for two reasons. The first is that artificial selection can and does produce dramatic morphological change: one has only to look at the breeds of dogs to appreciate that. The second is that species are not uniform in space. Most species with a wide geographical range show differences between regions. Often these differences are so great that one does not know whether the extreme forms would behave as a single species if they met. Occasionally we know that they would not. This requires that a ring of forms should arise, with the terminal links overlapping. The Herring Gull and Lesser Black-Backed Gull afford a familiar example. In Britain and Scandinavia they behave as distinct species, without hybridizing, but they are linked by a series of forms encircling the Arctic.

Stasis in time is, therefore, a puzzle, since it seems not to occur in space. The simplest explanation is that species remain constant in time if their environments remain constant. It is also worth remembering that the hard parts of marine invertebrates, on which most arguments for stasis are based, tell us relatively little about the animals within. There are on our beaches two species of periwinkle whose shells are indistinguishable, but which do not interbreed and of which one lays eggs and the other bears live young.

The question of stasis and punctuation will be settled by a statistical analysis of the fossil record. But what of the wider issues? Is mutation plus natural selection within populations sufficient to explain evolution on a large scale, or must new mechanisms be proposed?

It is helpful to start by asking why Darwin himself was a believer in gradual change. The reason lies, I believe, in the nature of the problem he was trying to solve. For Darwin, the outstanding characteristic of living organisms which called for an explanation was the detailed way in which they are adapted to their forms of life. He knew that "sports"—structural novelties of large extent—did arise from time to time, but felt that fine adaptation could not be explained by large changes of this kind: it would be like trying to perform a surgical operation with a mechanically-controlled scalpel which could only be moved a foot at a time. Gruber has suggested that Darwin's equating of gradual with natural and of sudden with supernatural was a permanent feature of this thinking, which predated his evolutionary views and his loss of religious faith. It may have originated with Archbishop Sumner's argument (on which Darwin made notes when a student at Cambridge) that Christ must have been a divine rather than a human teacher because of the suddenness with which his teachings were accepted. Darwin seems to have retained the conviction that sudden changes are supernatural long after he had rejected Sumner's application of the idea.

Whatever the source of Darwin's conviction, 1 think he was correct both in his emphasis on detailed adaptation as the phenomenon to be explained, and in his conviction that to achieve such adaptation requires large numbers of selective events. It does not, however, follow that all the steps had to be small. I have always had a soft spot for "hopeful monsters": new types arising by genetic mutation, strikingly different in some respects from their parents, and taking a first step in the direction of some new adaptation, which could then be perfected by further smaller changes. We know that mutations of large effect occur: our only problem is whether they are ever incorporated during evolution, or are always eliminated by selection. I see no *a priori* reason why such large steps should not occasionally happen in evolution. What genetic evidence we have points the other way, however. On the relatively few occasions when related species differing in some morphological feature have been analyzed genetically, it has turned out, as Darwin would have expected had he known of the possibility, that the difference is caused by a number of genes, each of small effect.

As I see it, a hopeful monster would still stand or fall by the test of natural selection. There is nothing here to call for radical rethinking. Perhaps the greatest weakness of the punctuationists is their failure to suggest a plausible alternative mechanism. The nearest they have come is the hypothesis of "species selection." The idea is that when a new species arises, it differs from its ancestral species in ways which are random relative to any long-term evolutionary trends. Species will differ, however, in their likelihood of going extinct, and of splitting again to form new species. Thus selection will operate between species, favoring those characteristics which make extinction unlikely and splitting likely. In "species selection," as compared to classical individual selection, the species replaces the individual organism, extinction replaces death, the splitting of species into two replaces birth, and mutation is replaced by punctuational changes at the time of splitting.

Some such process must take place. I have argued elsewhere that it may have been a relevant force in maintaining sexual reproduction in higher animals. It is, however, a weak force compared to typical Darwinian between-individual selection, basically because the origin and extinction of species are rare events compared to the birth and death of individuals. Some critics of Darwinism have argued that the perfection of adaptation is too great to be accounted for by the selection of random mutations. I think, on quantitative grounds, that they are mistaken. If however, they were to use the same argument to refute species selection as the major cause of evolutionary trends, they might well be right. For punctuationists, one way out of the difficulty would be to argue that adaptation is in fact less precise than biologists have supposed. Gould has recently tried this road. As it happens, ! think he is right to complain of some of the more fanciful adaptive explanations that have been offered, but I also think that he will find that the residue of genuine adaptive fit between structure and function is orders of magnitude too great to be explained by species selection.

One other extension of the punctuationist argument is worth discussing. As explained above, stasis has been explained by developmental constraints. This amounts to saying that the developmental processes are such that only certain kinds of animal are possible and viable. The extension is to apply the same idea to explain the existence of the major patterns of organization, or "bauplans," observable in the natural world. The existence of such bauplans is not at issue. For example, all vertebrates, whether swimming, flying, creeping or burrowing, have the same basic pattern of an internal jointed backbone with a hollow nerve cord above it and segmented body muscles on either side of it, and the vast majority have two pairs of fins, or of legs which are derived from fins (although a few have lost one or both pairs of appendages). Why should this be so?

Darwin's opinion is worth quoting. In The *Origin of Species,* he wrote:

> It is generally acknowledged that all organic beings have been formed on two laws—Unity of Type, and the Conditions of Existence. By unity of type is meant that fundamental agreement in structure which we see in organic beings of the same class, and which is quite independent of their habits of life. On my theory,

unity of type is explained by unity of descent. The expression of conditions of existence, so often insisted on by the illustrious Cuvier, is fully embraced by the principle of natural selection. For natural selection acts by either now adapting the varying parts of each being to its organic and inorganic conditions of life; or by having adapted them during the long-past periods of time ... Hence, in fact, the law of Conditions of Existence is the higher law; as it includes, through the inheritance of former adaptations, that of Unity of Type.

That is, we have two pairs of limbs because our remote ancestors had two pairs of fins, and they had two pairs of fins because that is an efficient number for a swimming animal to have.

I fully share Darwin's opinion. The basic vertebrate pattern arose in the first place as an adaptation for sinusoidal swimming. Early fish have two pairs of fins for the same reason that most early aeroplanes had wings and tailplane: two pairs of fins is the smallest number that can produce an upward or downward force through any point in the body. In the same vein, insects (which are descended from animals with many legs) have six legs because that is the smallest number which permits an insect to take haft its legs off the ground and not fall over.

The alternative view would be that there are (as yet unknown) laws of form or development which permit only certain kinds of organisms to exist—for example, organisms with internal skeletons, dorsal nerve cords and four legs, or with external skeletons, ventral nerve cords and six legs—and which forbid all others, in the same way that the laws of physics permit only elliptical planetary orbits, or the laws of chemistry permit only certain compounds. This view is a manifestation of the "physics envy" which still infects some biologists. 1 believe it to be mistaken. In some cases it is demonstrably false. For example, some of the earliest vertebrates had more than two pairs of fins (just as some early aeroplanes had a noseplane as well as a tailplane). Hence there is no general law forbidding such organisms.

What I have said about bauplans does not rule out the possibility that there may be a limited number of kinds of unit developmental process which occur, and which are linked together in various ways to produce adult structures. The discovery of such processes would be of profound importance for biology, and would no doubt influence our views about evolution.

One last word needs to be said about bauplans. They may, as Darwin thought, have arisen in the first place as adaptations to particular ways of life, but, once having arisen, they have proved to be far more conservative in evolution than the way of life which gave them birth. Apparently it has been easier for organisms to adapt to new ways of life by modifying existing structures than by scrapping them and starting afresh. It is for this reason that comparative anatomy is a good guide to relationship.

Punctuationist views will, I believe, prove to be a ripple rather than a revolution. Their most positive achievement may be to persuade more people to study populations of fossils with adequate statistical methods. In the meanwhile, those who would like to believe that Darwin is dead, whether because they are creationists, or because they dislike the apparently Thatcherite conclusions which have been drawn from his theory, or find the mathematics of population genetics too hard for them, would be well advised to be cautious: the reports of his death have been exaggerated.

Part Four:
BIOGRAPHIES

Biographies in this section were commissioned by Grey House Publishing and developed independently of author Michael Ruse.

Louis Agassiz

Louis Agassiz (1807–1873) was a Swiss-American scientist who worked primarily in the fields of zoology, glaciology, and geology. He was known for a variety of important scientific contributions including being the first person to propose that the Earth had gone through an ice age. He was also a prominent critic of Charles Darwin's theory of evolution.

Agassiz was born in Môtier, Switzerland. He chose initially medicine as his profession and left to study at the universities of Zurich, Heidelberg, and Munich. During this time, his interests broadened to include natural history and botany. He qualified as a doctor of medicine in Munich in 1830 and moved to Paris shortly after. While in Paris, Agassiz studied with the German naturalist Alexander von Humboldt, who introduced him to geology, and the French naturalist Georges Cuvier, who introduced him to zoology. Soon after, Agassiz was recruited to work on a project to cataloging a large number of new species of fish brought back from an expedition in Brazil. This project would be the beginning of a lifelong interest in the study of fish. In 1832, Agassiz returned to Switzerland where he was appointed professor of natural history at the University of Neuchâtel. His focus during this period was the study of fossil fish. The resulting work, *Recherches sur les poissons fossiles* (1833–1843), was one of his most important publications and would elevate him to worldwide prominence as a scientist.

Agassiz also gained attention in 1837 for being the first scientist to propose that that the Earth had gone through an ice age. He elaborated on this theory in a study of glaciers entitled, *Etudes sur les glaciers,* which was published in 1840. In 1846, Agassiz left Switzerland for the United States. He originally intended to remain there only a short time to give lectures and study the geology of North America, but he decided to stay because of the superior financial and scientific conditions that the country provided. He was appointed professor of zoology and geology at Harvard University in 1847 and founded Harvard's Museum of Comparative Zoology in 1859.

Agassiz was also known for his criticism of Darwin's theory of evolution. This was probably due to the influence of Cuvier. After *The Origin*, he debated evolutionary issues with Darwin's American supporter, the professor of botany at Harvard, Asa Gray. Agassiz was an advocate of polygenism, meaning that he believed that different races of human beings came from different origins and were each created individually. Specifically, he argued that different races were each created by God simultaneously, with different biological characteristics, and that they were placed on the continents where they were intended to dwell.

Michael Behe

Michael Behe (b. 1952) was born in Harrisburg, Pennsylvania. He is a biochemist who is well-known as an advocate for the theory of intelligent design, which asserts that some structures are too complex to be adequately attributed to evolutionary processes.

Behe argues that what he terms the "irreducible complexity" of essential cellular structures proves that an intelligent, or divine, being was responsible for creating the physical world. This claim strongly supported the intelligent design movement. Behe's views are strongly disputed by the scientific community, which argues that the idea of "irreducible complexity," a theory which has not been extensively tested and validated experimentally, cannot be used to support any scientific conclusions. Behe's work has been largely dismissed as "pseudoscience;" his testimony in the Dover lawsuit related to schools' teaching of evolution as scientific fact was cited by the judge as support for the idea of intelligent design as essentially religious, rather than scientific, in nature. Behe does not align himself with the full-blown creationist theories of the development of the earth, however. He accepts the idea that all humans descended from primates, as well as the scientific community's consensus on the age of the planet earth as being 4.6 billion years old. He insists, though, that increasing knowledge of the physical world allowed by modern instruments, which are increasingly more sensitive, uncovers a complexity which indicates that only an intelligent being could have created so many interlocking systems working in harmony. Behe also argues that, although his conclusions are not tested in the traditional sense, his theories are based on his interpretation of science which has been extensively peer reviewed. The "purposeful arrangement of parts," he says, allows confidence in the idea of involvement of a designer in life on earth.

Behe, at one time in his career, accepted a naturalistic theory of evolution, but his doubts about the theory of natural selection led him to seek alternate explanations. Proponents of intelligent design, like Behe, hope to overthrow science's current emphasis on finding only natural explanations. They believe that this "methodological atheism" has harmed the fabric of society, and that the theory of evolution hurts mankind by its implication that humans have no moral or spiritual nature. The movement seeks either to end the teaching of evolution in schools, or to have intelligent design taught in the classroom as science on an equal footing with evolution. Supporters of intelligent design emphasize disagreement among scientists in order to discredit naturalistic evolution. Scientists counter these arguments by saying that, although there is disagreement among individual scientists, evolution is one of the most robust and widely accepted principles in the modern scientific community.

Friedrich von Bernhardi

Friedrich von Bernhardi (1849–1930) was a Prussian general and military historian who was best known for using the theories of Charles Darwin to justify war and military aggression. He was a best-selling author prior to World War I and his writings were used in support of Germany's aggressive actions towards other countries. In his most famous book, *Deutschland und der Machete Krieg* (published in 1911 [*Germany and the Next War*]), Bernhardi used Darwin's theory of natural selection and the idea of "survival of the fittest" to argue that war is a natural, undeniable part of biological existence.

Bernhardi was born in St. Petersburg, Russia. His parents were Estonian-German and when Bernhardi was two years old his family moved to the German province of Silesia. During the Franco-Prussian war of 1870–1871, Bernhardi had his first experience with military service, as a cavalry lieutenant in the Prussian Army. After the war, he continued his military career and served as the army's chief war historian from 1898–1901. In 1909, Bernhardi was appointed the commanding general of the Seventh Army Corps. He served in World War I and was awarded the *Pour le Mérite* in 1916 for his role in the German defense against the Brusilov Offensive.

Bernhardi was known more for his writing than for his military career, though the two were closely linked. His main theories were outlined in *Germany and the Next War*, published three years before Germany entered into World War I. In the book Bernhardi described war as a "biological necessity" and claimed that any diplomatic attempt to avoid it was an attempt to go against nature. He viewed arbitration as weakness. Bernhardi cited the study of plant life as evidence of the existence of war in the natural world, and therefore, as justification for war between human populations. Bernhardi further used Darwin's ideas to argue that the question of war was a decision between the expansion, or death of a society; he invoked a sense of higher morality in stating that a country would either become a world power, or face certain decline; the two extremes equated to biological adaptation and survival or extinction in Darwin's theory.

Although Bernhardi's ideas were not the official policy of the German government, or even the German military leaders, they were consistent with the general thinking of extreme German nationalists of the time. Bernhardi had a profound influence on many high-ranking officials and his books were widely read. His application of the theory of evolution to military policy was an extreme example of Social Darwinism.

William Jennings Bryan

William Jennings Bryan (1860–1925) was a Democratic Party nominee for the presidency of the United States, one of the most popular public speakers of his time, and a noted opponent of Darwinian evolution.

In what became known as "The Scopes Monkey Trial," Bryan famously argued for the state of Tennessee in its prosecution of a teacher who attempted to teach evolution in his public school classroom. The case became a critical turning point in public perception, paving the way for the legitimization of evolution as a science. In May of 1925, John Scopes, a high school teacher, was arrested and charged with teaching evolution from a chapter in a textbook which described and endorsed ideas developed by Charles Darwin in his *Origin of the Species* (1859), in defiance of the state's so-called Butler Act, which made such instruction a crime. The American Civil Liberties Union provided Scope's defense, and prosecution was initially handled by two local attorneys. However, the prominent Bryan, who had not tried a case in 36 years, was invited to join the prosecution and quickly accepted. As the trial gained national prominence, Clarence Darrow then volunteered to lead the defense, setting the stage for a battle which was anxiously followed all across the country.

Bryan opposed Darwinism throughout his professional life because he believed that its account of the descent of man undermined the Bible and that its teaching would lead to widespread immorality. Some have argued that Bryan worried that Darwinism would encourage the influence of eugenics (controlling human mating to create "improved" races), but since he failed to mention such a theory at any time in public, this has not gained much support.

In what was billed as, "The Trial of the Century," with religion facing off against science, the ACLU had initially intended to oppose the Butler Act on the theory that it violated a teacher's constitutional rights and was therefore unconstitutional. Pushed to take a stronger stand on the science by Darrow, they changed their strategy. The defense instead proposed that there is no conflict between evolution and the creation account found in the Bible. The court chose to ignore Darrow's argument and instead narrowed the question, at Bryan's urging, to the question of whether or not Scopes had taught evolution without allowing other considerations. The judge also refused to permit many defense witnesses to testify.

After Bryan bragged to the court that he was "an expert in the Bible," the defense took the unusual step of calling Bryan to the stand. Darrow intended to demonstrate that belief in the historical accuracy of the Bible, and its many miracles, was illogical. Darrow questioned Bryan on the story of Eve having been created from Adam's rib and other incidents he felt proved that the stories of the Bible could not be scientific and should not be used as a base for teaching science. Bryan avoided most direct questions. Instead, he reframed the debate as an attempt by the defense to discredit and embarrass people who believed in the Bible. Darrow responded that he intended to prevent "bigots and ignoramuses from controlling the education of the United States." The judge eventually tired of the fighting and dismissed the witness and testimony as "irrelevant to the case;" each side claimed victory.

Scopes was convicted of teaching evolution after nine minutes of jury deliberation. His conviction was later overturned by a higher court based on a technicality. Supporters of evolution gained credibility from the controversy, however, the chilling effects of the trial led to the removal of evolutionary discussions from textbooks and the failure to teach evolution in American schools for many decades. Generally, his supporters thought he acquitted himself well in the courtroom. It should be noted that he was not a hard-line literalist, believing that the days of creation were over very long periods of time.

Bryan died in his sleep just five days after the conclusion of the trial, and the Christian liberal arts school, Bryan College, was founded in his memory.

Edward Drinker Cope

Edward Drinker Cope (1840–1897) was an influential American paleontologist and comparative anatomist who discovered almost 1,000 species of extinct mammals in the western United States. Although his most widely-known theories of evolution were later proved to be largely incorrect, he helped define the field of paleontology for generations to come.

Cope's most enduring legacy lies in his numerous publications describing genera and species, many of which are still used today. He was a prolific author, publishing about 1,400 scientific articles describing more than 1,200 previously verte-brate species. His massive volume, *Vertebrata of the Cretaceous Formations of the West*, was popularly known as "Cope's Bible." In 1877, Cope purchased half the rights to the journal "*American Naturalist*" in order to have an outlet for his many writ-ings. He was influential in discovering many varieties of dinosaurs; and the histori-cal surveys of the American West, for which he initially gained fame, were pre-cursors of the U.S. Geological Survey.

Together with fellow paleontologist Orthniel Marsh, Cope took part in the fa-mous "Bone Wars," a fierce competition to discover and describe fossilized re-mains in the West. The two were initially friends who traveled together, sharing discoveries. When the relationship soured, they began a near-frantic rivalry to in-troduce more discoveries, stopping only after the near-bankruptcy of both parties. When the competition began in 1877, only eighteen species of dinosaur were known. Together, the two men were responsible for naming 136 new species, al-though some of these were later discovered to be duplications. By 1892, when the competition is generally considered to have ended, both men were famous, but their public accusations and counter-accusations of fraud and carelessness had se-verely damaged the reputation of paleontology in the United States.

Cope opposed Darwin's theories of evolution based on natural selection, instead forming his work on the ideas of the Neo-Lamarckian school, of which he was one of the founders, and which promoted the law of use and disuse, that an animal will favor the use of certain anatomical parts and these parts, over time, will be-come stronger and larger. The giraffe's neck, for example, stretched to reach the tops of trees and this feature was passed directly onto its offspring. In contrast to a Darwinian belief in natural selection, Cope believed that most changes in a species occurred through the addition of developmental stages, a theory he termed "accel-erated growth." He also popularized the idea that mammals tend to become larger over time, which although influential for some time, has been widely discarded. Cope's personal beliefs tended toward racism, and his work was used by scientists for many years to justify imperialism.

Georges Cuvier

Georges Cuvier (1769–1832) was born in Montbéllard, a province close to Germany that was incorporated into France during his childhood. During the course of his life, he was a well-respected naturalist and zoologist whose work covered a range of subjects, including comparative anatomy, species classification, and extinction. Although, during his lifetime, he was a vocal critic of the then-current theories of evolution, much of his work was used in the evolutionary theories that would come later. His most significant contribution to both to the field of evolution and to science as a whole was definitively proving the possibility of species extinction. His most well known publication was *Règne animal distribué d'après son organisation* (1817), which was translated into English as *The Animal Kingdom*.

Cuvier attended the Karlsschule military academy in Stuttgart, Germany for four years. After completing his studies, he returned to France to accept a position as a tutor for a wealthy family in Normandy. During that time, Cuvier met the agriculturist A.H. Tessier, who introduced him to the prominent naturalist Étienne Geoffroy Saint-Hilaire. Cuvier was appointed an assistant at the Muséum National d'Hisoire Naturelle in Paris in 1795.

In 1796, Cuvier wrote a paper that compared the skeletal remains of both Indian and African elephants to mammoth fossils. By establishing definitive differences between the fossils and the living species, Cuvier was able to prove for the first time that the mammoths were extinct, and therefore, that extinction was possible. This was significant because, at the time, it was still widely believed that extinction was impossible, due to the perfection of God's creation. This work greatly enhanced Cuvier's reputation as a scientist. The studies were also seen as breakthroughs in the fields of paleontology and comparative anatomy.

Cuvier was a vocal critic of the evolutionary theories of his time, and of the scientists who supported them. Cuvier was able to explain his beliefs on extinction without accepting evolution because he believed in catastrophism, specifically the idea that one or more natural catastrophes caused the mass extinction of a large number of species. The leading evolutionary theory of the time was that of the French scientist, Jean Baptiste de Lamarck, supported by Geoffroy Saint-Hilaire. Cuvier's skepticism was based on the his belief in a correlation of parts within organisms which would render a species unable to survive if any single part was significantly changed independently of the whole. Cuvier was especially critical of Lamarck's theory that individual elements of organisms could change based on use or disuse. The nature and degree of Cuvier's criticism became a deterrent for other scientists working in the field of evolution at the time. However, despite his criticism, Cuvier's work on proving the possibility of species extinction, as well as his pioneering work in comparative anatomy and his discovering the progressive nature of the fossil record actually provided support for the evolutionary theories of later scientists.

Martin Daly

Martin Daly is a Canadian psychologist who has worked primarily on evolutionary psychology, specifically the evolutionary perspective on interpersonal violence. He is best known for his work, with partner Margo Wilson, on the "Cinderella effect," which refers to the phenomenon in which children are significantly more likely to be abused or neglected by their stepparents, than by their biological parents.

Daly is currently a professor of psychology at McMaster University in Ontario, Canada. He served as president of the Human Behavior and Evolution Society from 1991 to 1993. He was also coeditor, with his partner Margo Wilson, of the journal *Evolution and Human Behavior*. He has authored several books with Wilson including *Sex, Evolution, and Behavior* (1978), *Homicide* (1988), and *The Truth About Cinderella: A Darwinian View of Parental Love* (1998). In recognition of his contributions to the scientific community, Daly was made a fellow of the Royal Society of Canada in 1998.

Daly's work has included research on many topics in the field of evolutionary psychology. He has studied social diversity among related animal species, sex differences, parent-offspring relations, lethal violence, and the evolutionary consequences of uncertain paternity in animals with internal fertilization. Daly has also done collaborative work with Wilson on the behavioral ecology of desert rodents, as well as epidemiological studies of homicide.

The Cinderella effect seeks to explain the disproportionately high rates of child mistreatment and violence by stepparents, compared to those of biological parents. Some studies have shown, for example, that children are 100 times more likely to be beaten by their stepfathers than by their biological fathers. The concepts take its name from the fairy-tale character who is poorly treated by her stepmother. The evolutionary psychology explanation of the phenomenon proposed by Daly and Wilson is that natural selection has allowed humans to adapt their parental behavior to be most advantageous to the survival of their own offspring.

While biological parents are genetically inclined to devote the maximum resources possible to ensure the well being of their offspring, stepparents have no evolutionary benefit to gain from their stepchildren and therefore no biological incentive to protect them.

Clarence Darrow

Clarence Darrow (1857–1938) was born in Kinsman Township, Ohio. A prominent defense attorney and leading member of the American Civil Liberties Union (ACLU), he was immortalized on both stage and screen as the attorney for the defense in the Scopes "Monkey Trial" of 1925.

Billed as "The Trial of the Century," the prosecution of high school science teacher John T. Scopes for teaching evolutionary theories in a public school classroom focused the country's attention on the debate over evolution. Darrow, an atheist, refused to proceed with a theory of defense based on constitutionally granted individual rights. Instead, he framed the debate as religious superstition versus hard-headed scientific inquiry. Famed orator William Jennings Bryan, a well-known attorney and opponent of evolution, joined the prosecution's team in order to defend the Bible-based theory of scientific education.

In 1925, the state of Tennessee passed a law entitled the Butler act, which prohibited the teaching of "any theory that denies the story of the Divine Creation of man as taught in the Bible, and to teach instead that man has descended from the lower animals." After the act's passage, the ACLU let it be known that they would defend anyone arrested of violating the Butler Act. That same year, John T. Scopes, a substitute science teacher, agreed to be arrested and indicted for teaching a class using a textbook that referenced Charles Darwin's *The Origin of Species*. The arrest made national news and, in fact, Scopes's bail was paid by the editor of the *Baltimore Sun*.

The ACLU's original strategy in challenging the Butler Act was to show that the Act violated teachers' individual first amendment rights and was therefore unconstitutional. However, once Darrow became involved in the defense, the strategy shifted to a preliminary argument that evolution and the creation account in the Bible were in fact compatible. As the trial progressed, Darrow entirely abandoned the ACLU's first amendment strategy, instead attacking the logical underpinnings of a literal interpretation of the Bible.

In response, Bryan, who had not tried a case in thirty-six years, argued that the idea that man descended from "not even American monkeys, but from old world monkeys," was merely an excuse for immorality and sin. Mr. Bryan was followed by Dudley C. Malone for the defense, an international divorce lawyer who provided what many consider the seminal speech of the trial. In his speech, he argued that the Bible should be left to the realm of theology and morality and not inserted into a course of science. At the conclusion of his speech, the audience in the court burst into cheers.

A highlight of the trial was the examination of Bryan by Darrow. This strange circumstance came about after the judge deemed the defense's evidence on the Bible irrelevant and excludable. As legend has it, Darrow asked, "Where are we to find an expert on the Bible who is acceptable to the court?" In response, Bryan volunteered.

In his questioning, Darrow took a combative approach, stating at one point, "You insult every man of science and learning in the world because he does not believe in your fool religion." The two men's view of their opponent's goal in the trial is particularly enlightening. Bryan felt that Darrow's purpose was "to cast ridicule on everybody who believes in the Bible," while Darrow retorted that, "we have the purpose of preventing bigots and ignoramuses from controlling the education of the United States." Their exchange grew only more heated from there, until the judge adjourned the court and had Bryan's testimony expunged from the record.

Ultimately, the judge refused to hear testimony from all but one of the defense's witnesses and had limited the argument to whether or not Scopes had taught evolution in the classroom. The day after his exchange with Bryan, Darrow requested that the jury be brought into the courtroom to issue a guilty verdict so that the case could be appealed to a superior court, where the larger issue of the Butler Act's constitutionality would be addressed. After an eight day trial, it took the jury only nine minutes to find Scopes guilty. As punishment, Scopes was ordered to pay a $100.00 fine.

Scopes appealed the court's decision the Tennessee Supreme Court, which overturned the criminal conviction on procedural grounds. In the Court's opinion, however, the ACLU's arguments that the Butler Act violates the Establishment Clause of the First Amendment as well as an individual's constitutional rights to due process were explicitly rejected. The Court based its reasoning on the argument that the state has a right to regulate its employees, including public school teachers, and that because no religion used the tenets of evolution as its creation story, prohibiting the teaching of evolution did not favor one religion over another. It wasn't until 1968 that the United States Supreme Court found that bans on evolution violated the Establishment Clause due to their purpose being wholly religious. Ultimately, the result of the trial was a broad shift in public opinion against the idea of Creationism, based in large part upon Clarence Darrow's widely publicized examination of William Jennings Bryan.

Unfortunately, the trial did have a chilling effect on the teaching of evolution in schools. Discussion of evolution was removed from textbooks and did not reappear until American school science education was revamped in the late 1950s after the shock of Sputnik showed that the USA was falling behind Russia in scientific achievements.

Erasmus Darwin

Erasmus Darwin (1731–1802) was born in Nottinghamshire, England. He is perhaps best known for being the grandfather of Charles Darwin, but his early contributions to the theory of evolution are significant in their own right. During his time, he was noted as a physician, philosopher, botanist, naturalist, and poet. In fact, he often wrote poetry about his ideas on scientific topics, including evolution.

His early scientific endeavors included the founding of the Lichfield Botanical Society and the resulting publications, *A System of Vegetables*, (1783–1785) and *The Families of Plants* (1787). Both works were English translations of research done by Swedish botanist Carl Linnaeus. Darwin was also well known for connecting his poems with the topics of his scientific study. His poetry was said to have been admired by many well-known English writers including Samuel Taylor Coleridge and William Wordsworth. Two of Darwin's earlier works included the long poems "The Loves of the Plants" and "Economy of Vegetation," which were published together under the title *The Botanic Garden*. His most famous poem was "The Temple of Nature," which was published in 1803, one year after his death. Aside from being his most famous poem, it is also the most directly related to his ideas on evolution. Originally titled "The Origin of Society," the poem follows the gradual changes and adaptations that take place between primitive microorganisms and highly developed modern civilizations.

Darwin's most important scientific work was *Zoönomia* (1794–1796), which contained ideas on evolution that were very similar to those of the more well known scientists that would come after him. According to his grandson Charles Darwin, Erasmus' *Zoönomia* included concepts that anticipated the work of Jean-Baptiste Lamarck, the famous French evolutionist who wrote a decade later. The major similarity between Erasmus Darwin's work and the theory of evolution that would be formalized by his successors is the idea that all living creatures descended from a single common origin. Erasmus Darwin called this the "one living filament" in *Zoönomia* but it would become known as the "common ancestor" in modern evolutionary theory. (Lamarck, actually, believed in continuous creation of new life, which then independently, but in parallel, would ascend the progressive chain of being.)

Although his ideas on the subject were not as advanced as those of his famous grandson Charles, both shared the idea that evolution was a process in which species could acquire new traits and abilities to improve their ability to function and survive. Erasmus Darwin also made the basic connection between evolution and competition between members of the same species. One of Erasmus Darwin's beliefs that differs somewhat from modern evolutionary theory is that the process of evolution includes an active element of will or desire on the part of the species involved. This idea was also shared by Lamarck.

Nicholas Davies

Nicholas Davies (b. 1952) is a British ecologist. His most significant work relating to evolutionary theory is his behavioral ecology research on the conflicts that arise among animal populations as a result of behavioral adaptations stemming from natural selection.

Davies received his B.A. from Cambridge University in 1973 and his Ph.D. from Oxford University in 1976. After graduating, he stayed at Oxford as a demonstrator in the department of zoology. At the same time, he worked as a junior research fellow at Wolfson College. In 1979, Davies moved from Oxford University to Cambridge University, while maintaining the same position as demonstrator in the department of zoology. He was promoted to lecturer in 1984, reader in 1992, and professor of behavioral ecology in 1995—a position he still holds today. Davies has received many awards and honors in recognition of his work including the Scientific Medal in 1987, and the Frink Medal in 2001, both from the Zoological Society of London. Davies was also made fellow of the Royal Society in 1994.

Davies' work has focused on the field of behavioral ecology, which studies the ecological and evolutionary basis for animal behavior and behavioral adaptations. Much of the work in the field has focused on optimal behavioral strategies that animals adopt to increase their changes of survival, and those of their offspring. Since conditions often change over time, organisms must adapt their behavior accordingly, a process that takes place as a result of natural selection. Davies' work has focused on the conflicts that arise between animals as they undergo behavioral changes to meet the requirements of new living conditions.

One of Davies' primary areas of research has been the behavioral conflicts between male and females of the same species. His ground-breaking work, especially on dunnocks (hedge sparrows), has shown that gender-based conflicts can result in changes in a species' mating system, as well as its parental care. Davies' other research focus has been breeding conflicts between different species and the evolutionary adaptations that result at different stages of development. Davies has written a number of books on behavioral ecology including *Cuckoos, Cowbirds and Other Cheats* (2000) and two textbooks on the subject coauthored with British zoologist J.R. Krebs.

Richard Dawkins

Richard Dawkins (b. 1941) is a British ethologist, evolutionary biologist, and popular science writer. In addition to his work in biological sciences, he is known for his views on atheism, creationism, and religion and is an outspoken critic of intelligent design. Dawkins is an advocate of the idea that evolution is a gene-centered process. He is also known for coining the term "meme" to describe the equivalents of genes when evolutionary principals are applied to social phenomena.

Dawkins was born in Nairobi, Kenya. His father was a British soldier who had remained in Kenya after the Second World War. In 1949, when Richard was eight, his family moved back to England. Dawkins expressed doubts about the existence of God from an early age and also questioned the customs of the Church of England. He attended Oundle School from 1954 to 1959 and later studied zoology at Balliol College, Oxford. He pursued further study at the University of Oxford, where he received his M.A. and Ph.D. He remained at Oxford as a research assistant for year after graduating, focusing mainly on animal decision-making. From 1967 to 1969, Dawkins was an assistant professor of zoology at the University of California, Berkeley. He then returned to the University of Oxford in 1970 where he initially was a lecturer in zoology. He became a reader in 1990 and was appointed Simonyi Professor for the Public Understanding of Science in 1995.

Dawkins is a strong supporter of the view that evolution is a gene-centered process, which he explained his 1976 book, *The Selfish Gene*. He sees natural selection as a competition between replicators that attempt to out-propagate each other and considers genes to be the principal units of selection in evolution. He also coined the term "meme" as the cultural equivalent of genes when Darwinian ideas are extended into the social realm. In his book *The Extended Phenotype* (1982), Dawkins introduced the idea that all of the phenotypic, or observable, effects that genes have on the outside world may influence their chances of being replicated.

In his book, *The Blind Watchmaker* (1986), Dawkins criticized the watchmaker analogy frequently used in defense of intelligent design and gave a brilliant exposition of Darwinian evolutionary theory. Dawkins has been especially vocal in his criticism of teaching intelligent design in schools, calling it, "not a scientific argument at all, but a religious one." In his book, *The God Delusion* (2006), he argued against the existence of a supernatural creator. Dawkins has been called "Darwin's Rottweiler," which is a reference to Thomas Huxley, another Darwin supporter, who was known as "Darwin's Bulldog."

William Dembski

William Dembski (b. 1960) is a leading proponent of intelligent design and therefore rejects the modern theory of evolution. He is perhaps best known for his use of mathematics to support his views on intelligent design, specifically though the concept of specified complexity.

Dembski was born in Chicago, Illinois. His father was a college biology professor and he was raised Catholic. He attended an all-male preparatory school, finished a year early, and enrolled at the University of Chicago. Dembski initially struggled in college and dropped out. After working for a short time at his mother's art dealership—during which time he became an Evangelical Christian—Dembski returned to school and earned a number of advanced degrees, including: a B.A. in psychology (1981), an M.S. in statistics (1983), and a Ph.D. in philosophy (1996) from the University of Illinois at Chicago; a Ph.D. in mathematics (1988) from the University of Chicago; and M.Div. from the Princeton Theological Seminary (1996). While at Princeton, Dembski was involved with the founding of the Charles Hodge Society, which was interested in promoting the conservative ideas of the theologian Hodge, who had founded the *Princeton Theological Review* in 1825. He also held a post-doctoral fellowship from the National Science Foundation from 1988 to 1991. Dembski was associated with Baylor University from 1999 to 2005.

As a strong supporter of intelligent design Dembski has accused mainstream science, as part of its commitment to "atheistic" materialism and naturalism, of ruling out intelligent design *a priori*. The basis for his argument in favor of intelligent design is the concept of specified complexity. The concept attempts to formalize patterns found in nature that are both specified and complex, and therefore, indicative of an intelligent creator. Dembski's work on specified complexity was based partially on the biochemist Michael Behe's concept of irreducible complexity, which states that irreducibly complex systems cannot evolve gradually. Dembski has said that his knowledge of statistics led him to believe that the large amount of biological diversity in the world was unlikely to have been produced through natural selection.

Dembski originally presented his ideas on the subject in his 1991 paper, "Randomness by Design." He has also written several books including, *The Design Inference: Eliminating Chance through Small Probabilities* (1998), and *No Free Lunch: Why Specified Complexity Cannot Be Purchased without Intelligence* (2002). The scientific community has largely rejected intelligent design, calling specified complexity mathematically unsound.

Daniel Dennett

Daniel Dennett (b. 1942) is an American philosopher who has done work on the philosophy of the mind, the philosophy of science, and the philosophy of biology, as well cognitive science and evolutionary biology. His most significant contribution to evolutionary theory has been his application of evolutionary ideas to explaining human consciousness.

Dennett was born in Boston, Massachusetts, but spent part of his childhood in Beirut, Lebanon, where his father was a covert counter-intelligence agent during World War II. His family returned to the United States in 1947, when Dennett was five years old. He attended Phillips Exeter Academy in New Hampshire. Dennett went on to study at Harvard University, where he earned his B.A. in Philosophy in 1963. He then attended Christ Church in Oxford, England and earned his Ph.D. in 1965. He is currently the Austin B. Fletcher Professor of Philosophy and Co-Director of the Center for Cognitive Studies at Tufts University. Dennett's rise to prominence as a philosopher included presenting John Locke lectures at the University of Oxford in 1983, Gavin David Young lectures at Adelaide, Australia, in 1985, and the Tanner Lecture at the University of Michigan in 1986. He was also given the Jean Nicod Prize in Paris in 2001.

Dennett developed his basic philosophical view while at Oxford, under the influence of Gilbert Ryle. His primary area of interest is the philosophy of the mind and finding empirical evidence relating to consciousness. His dissertation, *Content and Consciousness*, proposed the need for separating the explanation of the mind into a theory of content and a theory of consciousness. In his 1992 book, *Consciousness Explained*, Dennett applied evolutionary theory to explain the content-producing aspects of consciousness. He has devoted much of his recent work to approaching his philosophical interests from an evolutionary point of view. His major work on evolution, *Darwin's Dangerous Idea* (1996), defends Darwinian selection against critics like Stephen Jay Gould. In his 1997 book, *Kinds of Minds*, Dennett proposed an evolutionary explanation of the differences between human minds and animal minds. He proposed a philosophy of how free will could be compatible with naturalism in *Freedom Evolves*, published in 2003. Most recently, Dennett applied evolutionary theory to propose possible reasons for religious adherence in his 2006 book, *Breaking the Spell*. He is an enthusiast for Richard Dawkins's theory of memes. Dennett is also an advocate of Neutral Darwinism, the theory of biologist Gerald Edelman that explains the adaptive development of the brain in anatomical terms.

Theodosius Dobzhansky

Theodosius Dobzhansky (1900–1975) was a Ukrainian-born Russian geneticist and evolutionary biologist. He was best known for his work in shaping the unifying modern evolutionary synthesis, which helped create a unified account of evolution by connecting the ideas of biologists working in different areas.

Dobzhansky was born in Nemyriv, Ukraine, when it was part of Imperial Russia. In 1910, when Dobzhansky was 10 years old, his family moved to Kiev. During high school, he decided to become a biologist and began collecting different species of butterflies. It was also during this time that Dobzhansky first read Charles Darwin's *On the Origin of Species*, which further heightened his interest in biology. He entered the University of Kiev in 1917 and although he completed all of the requirements for a degree in 1921, he never formally received a diploma. Shortly after, he accepted a position as an assistant to the faculty of agriculture at the Polytechnic Institute of Kiev. In 1924, he moved to Leningrad, Russia to study under the entomologist Yuri Filipchenko.

Dobzhansky immigrated to the United States in 1927 with the help of a scholarship from the International Educational Board of the Rockefeller Foundation. He worked with Thomas Morgan Hunt at Columbia University and then followed him to California, where they worked together at the California Institute of Technology from 1930 to 1940. Hunt had done pioneering work in using fruit flies in genetics experiments and Dobzhansky was credited with applying the research more directly outside of the laboratory. He returned to New York in 1940, where he worked at Columbia University until 1962, and then at Rockefeller University (then called the Rockefeller Institute) until he retired in 1971. He moved to the University of California at Davis, where he worked with his former student, Francisco Ayala.

Dobzhansky's most significant contribution to evolutionary science was his work on the modern evolutionary synthesis, which combined the ideas of evolutionary biology with those of genetics. His book on the subject, *Genetics and the Origin of Species*, deeply influenced by population geneticist Sewall Wright's "shifting balance theory of evolution," was published in 1937 and has been cited by many scientists as being influential to the formation of the modern theory of evolution. In the book, Dobzhansky defined evolution as "a change in the frequency of an allele within a gene pool." He was instrumental in promoting the idea that the natural selection took place through mutation in genes.

Dobzhansky was a communicant of the Eastern Orthodox Church and considered himself a religious man. For him, evolution and religion were not mutually exclusive. In his famous essay, "Nothing in Biology Makes Sense Except in the Light of Evolution," which he wrote in 1973, he defended his position on religion and evolution and criticized creationists for implying that God was deceitful, which he considered blasphemous.

Ronald Fisher

Ronald Fisher (1890–1962) was born in London, England. Although his enthusiasm for eugenic theories of human relations damaged his legacy, his achievements as a statistician, evolutionary biologist, and geneticist led to his reputation as one of the greatest of Charles Darwin's successors.

Fisher is famous for his work in both genetics and statistics. In *The Genetical Theory of Natural Selection* (1930), Fisher established that principles of heredity, rather than other contradicting evolutionary theories, actually prove the missing link in the theory of evolution by natural selection. He believed that dominance of a particular trait develops gradually, through selection, indicating that large mutations are not the driving force in evolution. The book also summarized his views on genes controlling dominant characteristics. He carried out experiments on breeding, developing sophisticated new mathematical models to demonstrate evolutionary theories.

Through these models, Fisher also made many contributions as a statistician, including the development of methods suitable for small samples, the discovery of the precise distributions of many sample statistics, and the invention of the analysis of variance. Fisher eventually became one of the leading proponents of something called "population genetics," which showed that evolution could produce evolutionary change without the help of forces outside the organism. The insights into hereditary diseases, such as sickle cell anemia, made great strides in treatment and understanding of what previously had been puzzling disorders. Genetic disorders, he argued, are actually a byproduct of natural selection.

Despite his accomplishments, including sexual selection, mimicry and the evolution of dominance, Fisher was associated with movements that were later discarded by scientists in the mainstream. Using census data for Britain, Fisher argued that a fall in the fertility of the upper classes would in turn lead to the decline and fall of civilizations. Between 1929 and 1934, he led the Eugenics Society in a campaign to promote a law allowing sterilization on eugenic grounds. The father of eight children, Fisher argued that the financial advantage granted to families with fewer children should be abolished through the granting of subsidies, or allowances, to the larger families, where parents are of superior stock. He claimed publicly that human beings "differ profoundly in their innate capacity for intellectual and emotional development." He also opposed the conclusion that smoking causes cancer, saying famously, "correlation does not equal causation."

Jerry Fodor

Jerry Fodor (b. 1935) was born in New York City, New York. He is a cognitive scientist and an outspoken opponent of the evolutionary theory of natural selection as determinative in the development of human cognition.

Fodor is most noted for his commitment to the theory of psychological nativism, which contrasts starkly with the widely accepted scientific idea of a "blank slate" at birth. Nativism is the belief that many cognitive functions and concepts are innate and biologically determined—that preexisting ideas are placed in the minds of all humans before birth by some being or process. According to Fodor, certain skills or abilities are "hard wired" into the brain and many abilities, such as language, would not be possible without some genetic pre-environment contribution. This indicates that mental states such as beliefs and desires are relations between individuals and mental representations. Language is structurally similar among many different and unrelated human groups, he says, which leads to the conclusion that humans understand language before birth. Scientists dispute this, but Fodor argues that cognitive science is "in the dark," throwing out theories at random to explain the human interactions and cognitive processes which they observe.

Dismissing adaptive ideas of mental development, Fodor revived the idea of a "modular" mind, which previously was associated only with pseudoscience such as phrenology. Modular theories propose that the structure of the brain determines its function. In phrenology, for example, practitioners claim that they can understand the nature of a person's intelligence based on the size and shape of the person's skull. More sophisticated modular theorists claim only that mental faculties can be associated with specific areas of the brain, and so at least in part are genetically determined. Darwinians believe that the mind develops in response to stimuli and environment, and so can change over time. Fodor dismisses these theories as they are unable to explain the relatively constant nature of concepts, such as "Bachelor." Fodor opposes reductive principles relied on by the majority of the mainstream scientific community, instead noting that since genes code for traits, it is likely that there is much that science currently does not understand about the brain and mental representations.

Critics of Fodor agree that, although he is correct that certain options are precluded from evolution, this does not prove that Darwin-influenced evolutionary theories are unsustainable. While some species will never develop certain traits, they will develop other traits to compensate. A bat, for example, will not develop feathers but will find another way to fly. Adaptation is not precluded by the existence of multiple options in a system.

Brian Goodwin

Brian Goodwin (b. 1931) was born in Montreal, Canada and is an influential figure in the "new biology," whose proponents believe that evolutionary variation is constrained by structural laws. He is considered a founder of a new branch of mathematical biology that uses the methodologies of mathematics and physics to understand the natural world and physical processes.

Goodwin focused his theories on morphogenesis and evolution, where he has sharpened scientists' understanding of the role of natural selection. According to Goodwin, all shape and form cannot be sufficiently explained by genetics. He believes that while genes will determine which molecules an organism can produce, the molecular composition does not determine the organism's form. Instead, the form develops according to structural laws.

Goodwin's interpretation rejects the traditional evolutionary metaphors of species regeneration based on conflict, competition, and selfish genes, for what he has called "evolution as a dance." His ideas contrast strongly with the Darwinian reading of organisms with adaptations furthering survival and reproduction. According to Goodwin, organisms form patterns, dynamically generated and repeated continuously, as they perpetually renew. Rather than simply fighting against each other, organisms are engaged in "self creation." As a result, what exists in the natural world is something more than an evolutionarily tested collection of random winners, but rather a universe created on orderly principles. Goodwin believes that something he calls "deterministic chaos" generates order in randomly created patterns. For example, snowflakes are each individual, but are still found always in patterns that Goodwin terms "generic forms." Goodwin believes such forms are characterized by the dynamic processes that generated the snowflakes, or other entities like organisms, and that scientists can discover the organizing principles underlying their emergence. Darwinism, then, according to Goodwin, has failed to explain the origin of the species because it has failed to explain the emergence of organisms.

Goodwin has argued that his work can be used to extend the domain of reliable scientific knowledge about reality to include theories of systems of coherent wholes. Western science has focused on discrete things which can be measured, while ignoring the larger ecosystem. Goodwin believes that a study of the "qualities of coherent wholes," such as the overall health of a body or an ecosystem, will bring further insights which provide important implications for modern societies. At the same time, he has pushed for a "devolutionized" theory of human activity, arguing that local activities, centered on an understanding of the peculiarities of a particular place, is a stronger model for human activity than the previous idea of "control and manipulation of nature." Homogenised cultures, he believes, are widely unsustainable models for growth, leading to "maximum entropy," and eventual self-destruction.

Goodwin has retired from the Open University in the UK; he currently teaches at the Schumacher College in Devon, UK.

Stephen Jay Gould

Stephen Jay Gould (1941–2002) was an American paleontologist and evolutionary biologist. His most significant contribution to the history of evolutionary theory was the concept of punctuated equilibrium, which he developed with fellow paleontologist Niles Eldredge.

Gould was born in the borough of Queens in New York City. He recalled being taken to the American Museum of Natural History in New York when he was five years old. Upon seeing the dinosaur exhibit, he decided to become a paleontologist. As an undergraduate, Gould studied geology at Antioch College and graduated in 1963. He then completed his graduate studies at Columbia University, where he earned his Ph.D. in paleontology in 1967. Immediately after graduating, Gould accepted a position at Harvard University, where he would remain for the rest of his life. In 1973, he was appointed professor of geology and curator of invertebrate paleontology at Harvard's Museum of Comparative Zoology. Then in 1982, he became the Alexander Agassiz Professor of Zoology. He also worked at the American Museum of Natural History. Gould received numerous awards and honors in recognition of his work. In 1983, he was made a fellow by the American Association of the Advancement of Science, and would later serve as the organization's president from 1999 to 2001. He also served as the president of the Paleontological Society from 1985 to 1986. In 2008, Gould was posthumously awarded the Darwin-Wallace Medal by the Linnean Society of London.

Gould's most significant contribution to evolutionary science was his theory of punctuated equilibrium, which he proposed with Niles Eldredge in 1972. The theory claimed that evolutionary changes take place in short, concentrated bursts, between long periods of evolutionary stability in which relative little activity occurs. During the periods of change, instances of branching evolution are possible. Punctuated equilibrium sharply contests the theory of phyletic gradualism, which claims that evolutionary development takes place through gradual, continuous changes.

Gould's contribution to the study of evolutionary developmental biology was significant, specifically his description of the concept of terminal addition, in which organisms evolve a final stage of individual development by shortening the earlier stages. The majority of his empirical research was done with the land snails Poecilozonite and Cerion. While Gould published a number of books on a wide range of evolutionary topics, two of his most significant were *Ontogeny and Phylogeny* (1977) and *The Structure of Evolutionary Theory* (2002). The latter summarized his views on modern evolutionary theory. Gould was a critic of both strict selectionism and sociobiology. He was also a vocal opponent of creationism and believed that religion and science were two distinct fields that should be kept separate from one another.

Peter and Rosemary Grant

Peter and Rosemary Grant are a British married couple of evolutionary biologists. They are best known for their work studying Darwin's Finches on the Galapagos Islands.

Peter earned his B.A. at Cambridge University in 1960 and his Ph.D. at the University of British Columbia in 1964. He taught at McGill University from 1965 to 1977. He then taught at the University of Michigan from 1977 to 1985 before transferring to Princeton University, where he is currently a professor. Peter's primary research interests include ecology, evolution, and behavior. In particular, he has focused on the origin of new species and the effect of ecological interactions on the persistence and extinction of species in various environments.

Rosemary earned her B.Sc. at Edinburgh University in 1960. Between 1960 and 1985, she worked as a research associate at the University of British Columbia, Yale University, McGill University, and the University of Michigan. She then earned her Ph.D. at Uppasala University in 1985. She is currently a lecturer in ecology and senior research biologist at Princeton University. Rosemary's primary research interests include the way in which natural selection produces genetic diversity, and the effect that the process has on speciation.

The Grants are known for their long-term study of Darwin's Finches on the island of Daphne Major in the Galapagos. Since 1973, they have spent six months out of every year on the Galapagos. Their work has included the tagging and taking blood samples from the finches for genetic comparisons and tracking. They have produced extensive evidence of evolutionary activity by showing the way in which changes in the food supply or severe environmental changes can, through natural selection, lead to rapid changes in body and beak size. They have clarified the process by which new species arise and by which genetic diversity is maintained within a population.

Their findings have revealed four primary evolutionary patterns: there is a large quantity of heritable variation within populations; variation can be subject to the effects of natural and sexual selection; the makeup of communities of finches is largely determined by variations in food supply; and different finch species are able to recognize each other by both their appearance and their song.

In recognition of their work, the Grants have received numerous honors and awards. In 2005 they were given the Balzan Prize for population biology. They are both fellows of the Royal Society. In 2008, the Grants were given the Darwin-Wallace Award of the Linnean Society of London.

Asa Gray

Asa Gray (1810–1888) was an American botanist and natural historian. He was known for his significant work in the identification and classification of North American plants. He was a friend of Charles Darwin and provided information that aided in the development of Darwin's theories. Gray was a vocal supporter of Darwin and was especially noted for his efforts to reconcile the differences between evolutionary theory and Christian belief.

Gray was born in Sauquoit, New York. He attended Fairfield academy before going on to study medicine at Bridgewater. He completed his medical studies in 1831, but decided shortly after to give up his practice and study botany. He studied with John Torrey, a botanist for the state of New York and together they published the comprehensive *Flora of North America* (1838–1843). The work was the first of Gray's many significant publications in the field. In 1842, Gray was appointed professor of natural history at Harvard University, where he would remain until 1873. Gray was instrumental in the development and expansion of the botany department at the university; he donated a large collection of books and plants and the Gray Herbarium was named for him in recognition. Gray's most famous publication was *The Manual of the Botany of the Northern United States, from New England to Wisconsin and South to Ohio and Pennsylvania Inclusive*, which became known simply as *Gray's Manual* (1848).

Gray corresponded often with Charles Darwin. He provided Darwin with information from his research that would give support for the theories in *On the Origin of Species*. Specifically, Gray's research on the close relationship between plants of East Asia and North America served as a key piece of evidence for Darwin's theory of evolution because it demonstrated that the flora of two different areas shared a common genetic origin.

Gray was one of the select few who knew about Darwin's theory before the publication of *The Origin*. When the work was published, he wrote an enthusiastic review in *The American Journal of Science*. Gray was Darwin's primary supporter in the United States. He debated publicly and in print with his fellow Harvard biologist, Louis Agassiz. Gray's most significant contribution to the debate on evolution was his attempt to reconcile the differences between Darwin's theory and the Biblical story of creation. The subject was controversial then, just as it is today and Gray was one of the first scientists to argue that it was possible for the two competing ideas to coexist. However, he did this by supposing that new variations are directed by God, an assumption that Darwin found unacceptable. The two men remained friends and allies, agreeing to disagree. While Gray acknowledged that Darwin's theory could be used to support atheistic views, he argued that the same was true of many scientific ideas and that it was not the only way to interpret the theory. Gray later published an influential collection of his writings on the subject, *Darwiniana*.

Ernst Haeckel

Ernst Haeckel (1834–1919) was a German biologist, naturalist, philosopher, physician, and illustrator. His major work in natural sciences were the classification and naming of thousands of new species and the creation a genealogical tree that connected all types of living creatures. He also coined a number of common biological terms. His contributions to evolutionary science were his controversial recapitulation theory and his promotion and popularization of Charles Darwin's work, especially in Germany.

Haeckel was born in Potsdam, Germany (then part of Prussia). He graduated from Cathedral High School in 1852, and then studied medicine with a number of physicians in Berlin. He earned his M.D. in 1857 and received his license to practice shortly after. He practiced medicine briefly before deciding to go back to school to continue his studies. He earned his Ph.D. in zoology from the University of Jena, where he studied under noted anatomist Carl Gegenbaur. In 1862,

Haeckel became a professor of comparative anatomy at the university, where he taught until he retired in 1909.

One of Haeckel's primary activities during the period was the naming and describing of species; he was credited with naming thousands of new species between 1859 and 1887. He was also a skilled illustrator and included drawings with many of his classifications. Haeckel was internationally recognized for his work in both the United States and New Zealand, where there is a Mount Haeckel named for him.

Haeckel was responsible for promoting the theory of evolution in Germany and elsewhere. His master work *Generelle Morphologie* (1866) was followed by *Natürliche Schöpfungsgeschichte*, intended for a wider audience and illustrated throughout. The book was published in Germany in 1868, where it became a bestseller, and then translated into English in 1876 as *The History of Creation*.

Despite his support for the theory of evolution, Haeckel's personal view on the subject varied slightly from Darwin's. Haeckel believed in a form of Lamarckism that held that racial characteristics were acquired though interactions with one's environment. He also considered the social sciences to be areas of applied biology, which has since been disputed. Haeckel's most significant contribution to the science of evolution, however, was the development of the recapitulation theory. The theory connected ontogeny (the development of form) to phylogeny (evolutionary descent). In essence, it proposed that as organisms developed before birth, they passed through stages in which they took the adult form of all of their preceding evolutionary ancestors. Haeckel supported the theory with detailed drawing of the various phases of development. The theory and its evidence have since been dismissed by many evolutionists as oversimplified and ignoring obvious indications to the contrary. Nevertheless, it was still a significant contribution to evolutionary theory and provided the basis for research in innovative directions.

John Burdon Sanderson Haldane

John Burdon Sanderson Haldane (1892–1964) was a British-born evolutionary biologist and geneticist. He was best known for being one of the founders of population genetics. He also proposed a concept connecting the physical size of an organism to the nature of its biological makeup, which became known as Haldane's principle.

Haldane was born in Oxford, England. His father was a physiologist and the young Haldane took interest in his father's work from a very young age. He attended both Eton and New College Oxford and also served in the British Army during World War One. In 1919, Haldane became a fellow of New College. Then in 1922, he moved to Cambridge to accept a Readership in Biochemistry at Trinity College. Haldane remained in Cambridge until 1932 focusing primarily on the study of enzymes and genetics, and in particular, on the mathematical aspects of genetics. He also wrote a large number of essays and published many of them in a collection in 1927 entitled *Possible Worlds*. In 1932, Haldane moved to London to accept a teaching position at University College, where he would spend the majority of his career. He was initially a professor of genetics, but was appointed Weldon Professor of Biometry after four years. Haldane received many awards in recognition of his work, including the Darwin Medal from the Royal Society in 1952 and the Huxley Memorial Medal from the Royal Anthropological Institute in 1956. At the end of his life, Haldane moved to India and took to wearing native dress. This act of Ghandi-like humility was balanced by his conviction that truly he belonged to the Brahmin caste.

Haldane was credited as one of the founding figures in the field of population genetics, along with R.A. Fisher and Sewall Wright. Specifically, Haldane contributed to the mathematical theory of natural selection. His series of papers on the subject were collectively entitled "A Mathematical Theory of Natural Selection and Artificial Selection." His work focused on showing the direction and rates of changes of gene frequencies. He also investigated the interaction of natural selection with both mutation and migration. Haldane summarized his findings on the subject in his 1932 book, *The Causes of Evolution*. Because it involved explaining the theory of evolution from the perspective of multiple fields of science, Haldane's work became a part of what was later known as the modern evolutionary synthesis. The synthesis acknowledged natural selection at the mechanism of evolution and explained the theory in light of Medelian genetics.

In his essay *On Being the Right Size*, Haldane asserted that the physical size of an organism often dictated the makeup of its biological structures. He used the example of insects not having oxygen-carrying bloodstreams because their small bodies were able to absorb all of the oxygen that they needed through simple diffusion of air; humans, by contrast, required more complex oxygen distribution systems partially due to their larger size. The concept became known as Haldane's principle and has been applied in the field of energy economics.

Haldane also wrote a provocative essay on the origin of life, published in 1929, about the same time that the Russian scientist A.I. Oparin was formulating such ideas. Haldane suggested that life evolved gradually in a "hot, dilute soup." (Darwin earlier had suggested a "warm, little pond.") Typically, Haldane's contribution did not appear in a professional journal but in the *Rationalist Annual*.

William D. Hamilton

William D. Hamilton (1936–2000) was a British evolutionary biologist and theorist. His work focused on forming the genetic basis of the concept of kin selection, and was a significant development in the gene-centric view of evolutionary theory. Hamilton's views was seen by many as a precursor to the field of sociobiology. He also did significant work on the roles of sex ratios and sexual reproduction in evolutionary theory.

Hamilton was born in Cairo, Egypt to New Zealand-born parents. Soon after Hamilton was born, his family moved to Kent, England. During the Second World War, he was evacuated to Edinburgh, where he developed an interest in natural history. While in Edinburgh, he spent much of his free time collecting butterflies and other insects. After the war, Hamilton returned to Kent, where he attended Tonbridge School. He remained at Tonbridge for an extra year so that he could take the Cambridge entrance exams. After traveling in France and completing two years of national service, Hamilton moved to Cambridge to attend St. John's College. As an undergraduate, he was disappointed by the fact that many biologists did not seem to be strong supporters of the theory of evolution. It was during this time that Hamilton discovered the work of evolutionary biologist and statistician Ronald Fisher. Though Fisher was not highly regarded at Cambridge, Hamilton was inspired by his work on developing a mathematical basis for the genetics of evolution.

Hamilton's increasingly varied interests led him to a course on human demographics at the London School of Economics. As his focus turned towards mathematics and genetics, he studied first under statistician John Hajnal, and then under geneticist Cedric Smith at University College London. Hamilton worked as a lecturer at Imperial College London from 1964 to 1978 before becoming a professor of evolutionary biology at the University of Michigan. He was elected a Fellow of the Royal Society in 1980 and in 1984 accepted an invitation to be the Royal Society Research Professor at Oxford University, a position he held for the rest of his life. He was also a visiting professor at both Harvard University and the University of São Paulo.

Hamilton's primary contribution to evolutionary genetics was his work on the mathematical basis for the idea that organisms could increase the fitness of their own genes by aiding their close relatives, even at their own expense. While he was not the first person to propose this theory, Hamilton developed an equation to explain the relationship between the cost of genetic fitness to the actor and the benefit to the recipient. The relationship became known as Hamilton's rule. He first proposed the concept in his 1964 publication entitled "The Genetical Evolution of Social Behavior." It has since become an important aspect of both evolutionary genetics and social evolution.

Hamilton was also responsible for other innovative ideas, including the insight that competition among siblings for mates might lead to the disruption of normal sex ratios. In the final years of his life, he championed the view that sex serves the end of protection from parasites (The juggling of gene ratios due to sex makes for moving targets, less easy to hit by fast-evolving attackers.)

Charles Hodge

Charles Hodge (1787–1878) was an American preacher, educator, writer, and systemic theologian. He was also one of the most prominent American supporters of Calvinist theology during the 19th century. He strongly opposed Charles Darwin's theory of evolution on the grounds that it was the equivalent of atheism because it denied the role of God in the process of biological design.

Hodge was born in Philadelphia, Pennsylvania. He attended the College of New Jersey (now Princeton University). Raised a Presbyterian, he made a public expression of his faith during his last year of college. After graduating, he entered the Princeton Theology Seminary in 1816. Hodge was licensed as a minister by the Presbytery of Philadelphia in 1819 and began preaching regularly in and around Philadelphia. In 1822, he was appointed professor of Biblical and Oriental literature at the Princeton Theology Seminary. In 1825, Hodge founded *Biblical Repertory*, a quarterly publication that would later become the *Princeton Review* in 1877. He served as the journal's chief editor and principal contributor until 1868, and remained at the seminary for the rest of his life, except to travel from 1826–1828, studying in Paris, Halle, and Berlin. In 1840, Hodge was made chair of exegetical and didactic theology.

Hodge was an outspoken critic of Charles Darwin's theory of evolution and wrote two books that discussed the subject. Hodge's primary criticism of the theory was that it excluded the intelligent design of God. He could neither believe, nor accept the idea that all of the complex organisms in existence were the result purely of natural laws.

Hodge's major contribution to scholarship was his magisterial *Systematic Theology* (1872). Basing his thinking on the book of Genesis, he argued that man and his soul were created with the direct, purposeful intervention of God. Hodge viewed the intelligent work of the mind as separate from the processes of nature and argued that a species as complex as human beings could not be created—as he saw it—by chance. He also challenged the idea that all species in existence shared a common ancestor, citing common sense as his rational. He concluded that if God had no place in Darwin's theory, than the theory equated to atheism.

The discussion of evolution in *Systematic Theology* was but a small part of a very large whole. Hodge's second book touching on the subject, *What is Darwinism?* (1874), was exclusively on the evolution question. Hodge restated many of his ideas from *Systematic Theology*, but focused primarily on the denial of the role of God in the theory. Despite his strong views on Darwin's theories, Hodge did not believe that all theories of evolution were inherently in conflict in with religion. He also openly praised Darwin for his intelligence and his methods.

Joseph Dalton Hooker

Joseph Dalton Hooker (1817–1911) was born in Suffolk, England. A noted botanist and explorer in his own right, he was also a close personal friend of Charles Darwin.

It's likely that Hooker was the first person to hear of Darwin's theory of evolution, as well as the first to publicly defend the theory. In an 1844 letter to Hooker, Darwin outlined his early theories on the transmutation of the species and natural selection. Their correspondence continued as Darwin developed his theory. Later in life, after enduring years of controversy over the theory of evolution, Darwin claimed that Hooker, "was the one living soul from whom I have received constant sympathy."

In addition to encouragement, Hooker provided practical support for Darwin's work. After Darwin received an essay in 1858 from the naturalist Alfred Russsel Wallace—an essay containing a full exposition of the idea of evolution through natural selection—Hooker, the well-connected son of a famous botanist, arranged to have pertinent writings by Darwin and Wallace published by the Linnean Society in the summer of 1858. Darwin recorded his thanks in the Introduction to *The Origin of Species*, (1859). In 1859 (just one month after the publication of Darwin's seminal work), Hooker published *The Introductory Essay to the Flora Tasmania*, in which he announced his support for the theory of evolution by natural selection. With this, Hooker became the first prominent scientist to support Darwin's theory and helped garner attention and further support for the theory of evolution. As president of the 1868 meeting of the British Association, Hooker used his address to the assembly as an opportunity to further propagate Darwinian theories. A member of the highly influential X-Club, he eventually became President of the Royal Society; he later received three of its medals. Throughout this career, he championed the idea of evolution, doing much to augment the credibility of proponents of the theory.

Hooker also took part in the famous 1860 debate on the validity of evolution, which galvanized Darwin's supporters and brought the theory to the attention of the public. According to many, it was Hooker, rather than the primary speaker Huxley, who most effectively defended evolution against Bishop Samuel Wilberforce, one of the most distinguished and respected public speakers of the time. Many said Hooker's reply to Wilberforce, "stunned him and left him no answer," as Hooker took the last word in the storied debate.

Through his many expeditions, including trips to Palestine, Morocco and the United States, Hooker built a respected scientific reputation in his own right. He eventually succeeded his father as director of the Royal Botanic Gardens, Kew, which attained world renown during his tenure. Hooker died in 1911; his widow refused an offer to have him buried next to Darwin.

Sarah Blaffer Hrdy

Sarah Blaffer Hrdy (b. 1946) is a noted U.S. biological anthropologist. Her discoveries of reproductive strategies used by male and female primates have been hugely influential on current evolutionary theories of gender and familial relationships.

Hrdy is most well-known for her theories involving mothers and children. Working with monkeys, she demonstrated that a mother may abort, abandon or even kill offspring that she lacks the resources to raise to maturity. This theory is backed by birth rates, which fall as women in developing countries gain access to birth control. Her controversial work was initially derided by her peers, but her theories have slowly gained widespread acceptance. Today she continues to press for increasingly popular ideas such as low-cost community daycare institutions, which she believes could serve as a substitution for the alliances tribal mothers formed to care for each other's children in earlier times.

Hrdy's theories, initially seen as very radical, developed from her fascination with a Harvard professor's comments on life among the langurs, small Asian monkeys. Harvard anthropologist Irven DeVore believed he had discovered a relationship between overcrowding and the killing of infants in the langur colonies. Hrdy developed her Ph.D. thesis to test the hypothesis that overcrowding causes infanticide. She concluded that infanticide occurred independent of overcrowding, and instead seemed to be an evolutionary tactic. Males and females appeared to be making strategic choices to increase their chances of passing on their genes. This overturned the classic idea that primates acted for the good of the group. When Hrdy first published her ground-breaking work on infanticide, the ideas of "maternal instinct" and "mother love" were taken as an article of faith by most scientists. Hrdy's view is that rather than follow some inborn instinct, mothers make trade offs, weighing the best possible outcomes for the offspring. Since human infants require so much care, Hrdy believes that mothers work to get assistance from the males and other members of her group.

Hrdy initially gained prominence with the publication of *The Langurs of Abu: Female and Male Strategies of Reproduction*, (1977) which posited her theories of evolutionary tactics used by mating primates. In 1981, she published *The Woman That Never Evolved*, which was one of the New York Times' Notable Books of 1981, as a continuation on the theme. She then co-edited *Infanticide: Comparative and Evolutionary Perspectives* (1984) which was selected as a 1984–1985 "Outstanding Academic Books," by "Choice," the Journal of the Association of College and Research Libraries. She followed this with *Mother Nature—Maternal Instincts and How They Shape the Human Species* (1999). She continues to publish and speak, promoting social causes and the field of sociobiology.

Julian Huxley

Julian Huxley (1887–1975) was a British evolutionary biologist. He was well known for his presentation and popularization of scientific topics through books, magazines, and television, and was a gifted communicator. Huxley's most significant contribution to the history of evolutionary theory was his prominent support for natural selection during the twentieth century. He also played an important role in the development of the evolutionary synthesis.

Huxley was born in London and grew up in Surrey, England. He was interested in nature from a very young age and was given science lessons by his grandfather, noted evolutionary biologist Thomas Henry Huxley. He attended Eton College as a King's Scholar, where he developed an interest in ornithology. Huxley won a scholarship in zoology and entered Balliol College, Oxford in 1906. While in Oxford, he focused his studies on embryology and protozoa. He graduated in 1909. After spending a year at the Naples Marine Biological Station, Huxley was appointed demonstrator in the department of zoology and comparative anatomy at Oxford University. He took a particular interest in bird behavior, the nascent science of ethology, and was to publish a major work on the courtship of the great crested grebe. Then in 1912, he accepted an invitation to help set up the biology department at the newly founded Rice Institute in Houston, Texas. In 1916, Huxley returned to Europe to contribute to the British cause in the First World War, working on intelligence in northern Italy. After the war, Huxley returned to Oxford University, where he was made fellow and senior demonstrator in the department of zoology. Then in 1925, he became professor of zoology at King's College London. He resigned from the position in 1927, much to the surprise of his colleagues, to work with H.G. Wells on *The Science of Life*. The project was a three-volume publication on all aspects of biology that was published between 1929 and 1930. Huxley also continued as an honorary lecturer at King's College from 1927 to 1931.

Huxley spent part of the 1930s traveling abroad. He visited east Africa to advise the British Colonial Office on education in the region. He also oversaw the creation of national parks in Kenya and other countries. He was a member of Lord Hailey's African survey committee from 1933 to 1938. From 1935 to 1942, Huxley was appointed secretary to the Zoological Society of London, overseeing the society, its gardens, and the London Zoo. He was also the first director of UNESCO, a position he held from 1946 to 1948. He became a major figure in British intellectual life, especially through his participation in a long-running weekly radio talk show The Braius Trust. (He was also a source of scientific ideas for his younger brother, the novelist Aldous Huxley.)

Huxley wrote extensively on development, especially on comparative rates of growth. Huxley's most important contribution to the history of evolutionary theory was his advocacy for natural selection and the evolutionary synthesis during the twentieth century. His major work was the overview *Evolution: The Modern Synthesis* (1942). He was also a skilled communicator who used his influence to popularize and promote the theory of evolution to a wide audience. He was influencial to scientists and the general public alike, travelling extensively and giving first-hand accounts to support the theories he promoted. For his contributions to the field, he was given the Darwin Medal of the Royal Society in 1956 and the Darwin-Wallace Medal of the Linnaean Society in 1958.

Julian Huxley was as much a humanist as a scientist. Although an atheist, he was always looking for a meaning to life. This he found in the concept of progress, the evolutionary development from the blob to humans. He thought that from this progress arises the ethical imperative to cherish humankind. His ideas were expressed in an influential book, *Religion Without Revelation* (1927).

Thomas Henry Huxley

Thomas Henry Huxley (1825–1895) was an English biologist and anatomist who was best known for his vocal support of the theory of evolution and the ideas of Charles Darwin. Although he never fully accepted the theory of natural selection on the basis that there was not enough observable evidence of its existence, he remained one of Darwin's strongest proponents and was largely responsible for the widespread approval of Darwin's ideas in the scientific community. These actions earned Huxley the nickname, "Darwin's Bulldog."

Huxley was born in Ealing, England. He received little formal schooling but possessed a natural drive to learn, and educated himself by reading extensively. As a young boy, he studied the works of Thomas Carlyle, James Hutton, and William Hamilton. He also taught himself German, which he would use later in life to translate texts for Charles Darwin, as well as Latin and Greek. As a young adult, he focused first on the study of invertebrates, and then vertebrates.

Huxley apprenticed for a number of medical practitioners before eventually receiving a scholarship to study at Charing Cross Hospital in London. When he found himself in need of money, he applied for the Royal Navy and was made assistant surgeon on the HMS *Rattlesnake*. The ship left England in 1846 on a scientific surveying voyage to New Guinea and Australia. Huxley used the trip as an opportunity to study marine invertebrates and sent writings on his discoveries back to England. One of the essays that Huxley published as a result of the trip was *On the anatomy and the affinities of the family of Medusae* (1849). In recognition of the value of his work, Huxley was elected a fellow of the Royal Society upon his return to England in 1850.

Prior to Charles Darwin, Huxley had been critical of the theories of other evolutionary scientists such as Robert Chambers and Jean Baptiste Lamarck. Huxley's skepticism, particularly with regards to Lamarck's theory of transmutation, was that there was insufficient observable evidence. After reading Darwin's *On the Origin of Species* in 1859 however, he became completely convinced of the existence of evolution, famous proclaiming, "How extremely stupid not to have thought of that!" Despite his unwavering support of evolution as a whole, Huxley never fully committed to the mechanism of natural selection proposed by Darwin.

The support of Darwin that Huxley became known for started with his positive reviews of *On the Origin of Species*, some written anonymously. He also gave a public lecture in support of the publication at the Royal Institution in 1860. Huxley's most significant action in support of Darwin was his performance at a debate on the topic against the Bishop Samuel Wilberforce. The debate was held at Oxford University Museum in 1860 and was considered a key moment in history for the acceptance of the theory of evolution. There is today considerable doubt as to whether the debate was really as dramatic as history records. Like many myths, however, it certainly played a role in making evolution secure.

Huxley was also a leading figure in late-Victorian education, sitting on the first London School Board, as well as being dean at the new science museum in South Kensington (now, Imperial College). He worked non-stop, except when brought down by periods of crushing depression, an affliction that also cursed his equally energetic grandson, Julian.

Phillip Johnson

Phillip Johnson (b. 1940) is a retired law professor at the University of California, Berkeley, who became the modern father of the intelligent design movement. After becoming a born-again Christian, Johnson founded the campaign to promote the teaching of intelligent design as an alternative to evolution, despite the fact that he has no background in biological science. Johnson has been strongly criticized by the scientific community. He is the author of several books on intelligent design, as well as criminal law textbooks.

Johnson was born in Aurora, Illinois in 1940. He attended Harvard University where he earned a bachelor's degree in English Literature in 1961. He studied law at the University of Chicago and later served as a law clerk for Earl Warren, Chief Justice of the US Supreme Court. He has taught at the Boalt School of Law at the University of California, Berkeley, where he was a faculty member from 1967–2000. After a divorce, Johnson became a born-again Christian and an elder in the Presbyterian Church. While on sabbatical in England, he had an epiphany and decided to devote his life to the promotion of intelligent design.

Johnson is responsible for popularizing the term "intelligent design" in his 1991 book, *Darwin on Trial* and is considered the modern father of the movement. He defines intelligent design as the notion that because of the complexity of the natural world, God must have had a direct role in its design and creation. Intelligent design explicitly rejects the naturalistic theory of evolution, especially that of

Charles Darwin. Johnson is also a critic of the overall concept of methodological naturalism in science, which is based on the principal that investigation of the natural causes must be limited to observable phenomena. Instead, Johnson advocates his own philosophy, which he calls theistic realism.

Johnson has received much criticism from the scientific community for his strong attacks on evolution, and on science itself. He has called the theory of evolution "atheistic," "falsified by all of the evidence," and has said that its "logic is terrible." In response, many scientists have judged intelligent design to be unscientific or pseudo-science. Johnson's critics have also accused him of being intellectually dishonest and equivocal, particularly in his use of the term naturalism.

Johnson was also the founder of the "wedge movement," which was the term initially used for the campaign to promote intelligent design. Johnson used a wedge as a metaphor for an aggressive public relations effort to create space within the scientific realm for his theistic agenda. He also advocates for the teaching of intelligent design in schools, as well as the theory of evolution. He promotes the "teach the controversy" approach for introducing his ideas in public schools. His books on the subject include: *Darwin on Trial, Defeating Darwinism by Opening Minds,* and *The Wedge of Truth: Splitting the Foundations of Naturalism.*

Stuart Kauffman

Stuart Kauffman (b. 1939) is a theoretical researcher whose ideas on the development of life on earth have challenged traditional Darwinian theories.

Kauffman's most famous work centers on his theory of self-organization, which he believes can extend the basic concepts of Darwinian evolution. Kauffman has called his theory, "order for free," and has said that it is as important as the theory of natural selection in producing the complexity of biological systems and organisms. This complexity, he believes, might result as much from the internal organization of a system as from outside forces such as natural selection. Natural selection, in Darwinian thought, is the process whereby favorable traits are passed on through successive generations while unfavorable traits become less common. Over time, this prompts the development of distinctive traits, and even new species. In other words, natural selection provides the mechanism for evolutionary development. Kauffman, in contrast, proposes that the dynamics of a system can, by itself, increase the inherent order of that system. Through the interaction of positive feedback, negative feedback, multiple interactions, and the balance of exploitation and exploration, a system creates an internal system of order.

The theory has implications for many fields, including social sciences, economics and anthropology. However, despite the intuitive simplicity of the theory that systems can self-regulate, the idea has not gained wide credibility. It has proved difficult to define formally or mathematically, and most scientists have withheld judgment as a result. Some criticize the application of self-organization theories as reductionism, or an attempt to reconcile and explain one field of study through reference to another. Most scientists agree that reductionism has limited value for complex systems. Kauffman, however, advocates the view that the field of complex systems poses few limits to reductionism. He has rejected the idea that one science is simply a variation of another, and looks for more holistic theories.

Kauffman continues to develop theories on the origins of life, believing that current evolutionary science is limited in its theories on "the essence of what makes something alive." He believes he may have a theory which explains this essence, something found in his understanding of the autonomous agent. An autonomous agent is something that can act on its own behalf in an environment, able to both reproduce itself and do at least one thermodynamic work cycle. This definition encompasses almost all free-living cells, leading Kauffman to conclude that, rather than passive participants in a system, cells "can actually build things." This process, he argues, can supplement natural selection. When a cell creates something beneficial for itself, the process of natural selection takes over and continues to renew that trait. He believes this work necessitates "a theory of organization to describe what the biosphere is doing."

Originally a medical doctor, Kauffman also developed a widely-used process which, in effect, artificially evolves pharmaceutical drugs to suit a particular purpose. His work, based on evolutionary principles, may have significant implications for the treatment of cancer and stem cell research.

Prince Petr Kropotkin

Prince Petr Kropotkin (1842–1921) was one of Russia's most prominent anarchists. Though born into a position of privilege, Kropotkin was a vocal advocate for anarchist communism, a form of communism free of central government control. He used his position of influence to promote his ideas on a wide range of topics including evolution. In his publication *Mutual Aid: A Factor of Evolution* (1902), Kropotkin argued that the social abilities of a species were a major factor in determining its possibility of survival, development, and prosperity.

Kropotkin was born in Moscow. His father owned large plots of land and over a thousand serfs. From a young age, Kropotkin expressed concern for the condition of the Russian peasants, a concern that grew as he aged. In 1857, at age 15, Kropotkin enrolled in the Corps of Pages in St. Petersburg, exclusive group primarily for sons of nobility, where he received a formal education similar to that of a military academy. In 1862, he was promoted from the Corps of Pages to the Russian army. Kropotkin had no interest in a military career so he initially chose an assignment were he thought he could do administrative work. He then accepted assignments for a number of geographical survey expeditions to Manchuria and the surrounding areas.

In 1867, Kropotkin resigned from the army and began studying at the university in St. Petersburg, while also serving as a secretary for the Russian Geographical Society. In 1871, Kropotkin explored glacial deposits in Finland and Sweden on behalf of the Geographical Society, and in 1873, he published an important geographical work, which proved that the maps in existence at the time were misrepresentative of the physical features of Asia. He was then offered a more prominent position within the society but turned it down in favor of pursuing social causes. Shortly after, he joined the revolutionary party. He was arrested for his political activities, escaped, and spent many years abroad, primarily in England. At the end of his life, after the fall of the Tsar, he returned to Russia, but was much disillusioned by the Bolsheviks and their grab for power.

Throughout his life, Kropotkin wrote on a wide variety of topics, both scientific and political, and leveraged his position of influence to promote his ideas. Because of his advocacy of anarchism, he was sometimes revered to as "the Anarchist Prince." His most well known publications were *The Conquest of Bread*, *Fields, Factories and Workshops*, and *Mutual Aid: A Factor of Evolution*. In *Mutual Aid*, Kropotkin offered a theory of the survival of species, which borrowed some ideas from Charles Darwin, but focused more on the views of the "Social Darwinists" of the time. While Kropotkin agreed that survival and adaptation were based on competition between species, he concentrated on social factors, both for determining a species' chances for survival and for evaluating the level of development and prosperity that a species had reached in its evolution. His primary assertion was that more sociable species—those able to utilize mutual aid to minimize individual struggle—were the most likely to survive and prosper. In arguing as he did, Kropotkin showed his Russian roots, where a mutually supportive battle against natures' elements was considered far more significant than a bloody struggle between individuals.

Jean-Baptiste de Lamarck

Jean-Baptiste de Lamarck (1744–1829) was born in Bazentin, Picardie, France. During his lifetime, Lamarck was a solider, naturalist, zoologist, and academic—in addition to being an evolutionary scientist. Although many of his ideas on evolution have been disputed or discredited by proponents of the modern theory of evolution, his pioneering work on the subject was significant and often cited by other scientists for its importance. Although he had supporters, he was also much criticized while he was alive and died in poverty. He was blind for the last eleven years of his life.

Early in his life Lamarck was a solider and fought in the Pomeranian war. It was while he was stationed in Monaco that he became interested in botany, natural history, and medicine. He decided then that he wanted to study medicine, but didn't dedicate himself to his studies until 1766, when he was injured and resigned from the army. Before working on evolutionary theory, he worked on general cell theory. He was also known as an authority on the subject of invertebrate zoology at a time when most scientists were not interested in invertebrates. Lamarck's most important early publications were *Flore française*—which helped him to gain membership to the French Academy of Sciences in 1779—and *Système des animaux sans vertèbres*, which was published in 1801. He first outlined his views on evolution in a lecture entitled Floreal, given in 1800. He later elaborated on his ideas on the subject in three published works: *Recherches sur l'organisation des corps vivants* (1802), *Philosophie Zoologique* (1809), and *Histoire naturelle des animaux sans vertèbres* (1815–1822).

Lamarck's beliefs regarding the theory of evolution are often referred to as "soft inheritance," "inheritance of acquired characteristics," or simply as "Lamarckism." Today, the term Lamarckism often carries an negative connotation because Lamarck's theories were discredited by Charles Darwin's theory of natural selection. Lamarckism is based on two principals: organisms evolve from less complex to more complex, and organisms are shaped by their environments. Lamarck also believed that organisms' structures evolved based on use, often referred to the use-or-disuse theory. He did not believe in common descent, but rather that new life forms are being spontaneously generated all of the time, and then start up independent, but parallel, paths on the progressive chain of being.

The aspect of Lamarckism that is most inconsistent with modern evolutionary theory is the idea that evolutionary adaptations acquired by adults could be passed on through inheritance. This concept was opposed by Charles Darwin's widely accepted belief that evolution occurred through natural selection. Lamarck also did not believe in extinction, but rather that inferior species simply evolved into more complex, more perfect species.

Despite the criticisms of his theories, Lamarck's beliefs have much in common with the modern theory of evolution. He was a proponent of the idea that evolution was the result of, and governed by, natural laws. He was credited not only by Charles Darwin but also by such prominent evolutionary scientists as Ernst Haeckel for his pioneering work in the field. Specifically, he is often cited as one of the first scientists to develop various ideas on the evolution into a complete, unified theory.

Richard Lewontin

Richard Lewontin (b. 1929) is an American evolutionary biologist and geneticist. He has been heavily involved in contributing to the mathematical basis of both population genetics and evolutionary theory. He has also been a pioneer in the application of techniques from molecular biology to study the problems of evolution and genetic variation.

Lewontin was born in New York City. He attended both Forest Hills High School and the École Libre des Hautes Études before going on to Harvard University. He received his bachelor's degree in biology from Harvard in 1951. He then returned to New York to pursue graduate studies at Columbia University, where he earned his masters in mathematical statistics in 1952, and his Ph.D. in zoology in 1954. At Columbia, he studied under the noted Russian geneticist Theodosius Dobzhansky. After graduating, Lewontin accepted a teaching position at North Carolina State University. He went on to teach at the University of Rochester and the University of Chicago before settling in at Harvard in 1973, where he has served as a professor of zoology and of biology. He was also appointed Alexander Agassiz Research Professor in 2003.

One of Lewontin's major contributions to evolutionary science is his work with fellow geneticist J.L Hubby on developing the field of molecular evolution. In 1966, the two men published a paper together in the journal *Genetics* that created the basis for nearly all of the work that has been done in the field ever since. The major finding of the paper, using the new technique of gel electrophoresis, was the discovery of high levels of molecular variability among species. They found that the highest levels of variation in humans—between 80 and 85%—were found within local geographic groups and that differences attributable to groups traditionally identified as racially different accounted for a relatively small percentage of human genetic variability. Lewontin's findings and ideas were gathered together in his major work, *The Genetic Basis of Evolutionary Change* (1974).

In his paper, "Organism and Environment," Lewontin made his case against the traditional Darwinian view that, in natural selection, organisms were passive recipients of the effects of their environments. He argued that it was more accurate to view each organism as an active creator of its own environment. Lewontin was also a supporter of the idea of a hierarchy of the levels of selection, a position that is not only anti-reductionistic, but a reflection of the Marxist philosophy he embraced during the Vietnam War.

Lewontin and the late biologist Stephen Jay Gould were responsible for the introduction of the term "spandrel," which was used to describe evolutionary characteristics that developed as a necessary consequence of another feature, and not for its own value. The concept was introduced in the pair's 1979 paper, "The spandrels of San Marco and the Panglossian paradigm: a critique of the adaptationist programme." Lewontin has also been a critic of the work of sociobiologists who have applied evolutionary theories to social sciences in an attempt to explain behavioral patterns. Again, as a Marxist, he thinks this is an overly reductionistic approach to understanding.

Konrad Lorenz

Konrad Lorenz (1903–1989) was born in Vienna. A noted zoologist and winner of the 1973 Nobel Prize in Medicine, he is regarded as one of the founders of modern ethology, which is the scientific study of animal behavior.

Ethologists are concerned with the evolution of behavior, and seek to develop an understanding of behavior in terms of natural selection. Charles Darwin is considered the first modern ethologist. Throughout his career, Darwin promoted the investigation of animal learning and intelligence as a way to supplement the scientific understanding of evolution. Ethology did not gain widespread influence, however, until the 20th century, when Lorenz identified fixed action patterns, or instinctive responses, that would occur reliably in the presence of stimuli. These patterns could then be compared across species, highlighting similarities and differences. Based on his observations of the nature of these fixed action patterns, Lorenz developed the idea of an innate releasing mechanism which would explain instinctive behaviors. Lorenz observed that a geese will roll a displaced egg near its nest back to the nest. The sight of the egg triggers this response. The geese will also attempt to maneuver egg-shaped objects, even objects which are much too large to be mistaken for a goose egg. Since this rigid behavior is not evolutionarily optimal, learning and complexity become crucial for survival. Lorenz discovered that young birds could be trained to follow their mothers, even if the egg was incubated artificially, as long as the training stimulus was presented during a critical early period, which he termed a "sensitive period," that continued for a few days after hatching.

Lorenz's work encouraged the strong development of ethology as a science in the years before World War II. Ethology is now a well-recognized scientific discipline with a number of journals covering the subject, including the *Ethology Journal*. The Konrad Lorenz Institute for Evolution and Cognition Research in Vienna supports the articulation, analysis, and integration of biological theories and the exploration of their wider scientific and cultural significance. The international institute supports theoretical research primarily in the areas of evolutionary developmental biology and evolutionary cognitive science.

Lorenz's reputation was marred by his participation in the Nazi party of Germany before and during World War II. Critics charged that his scientific work had been contaminated by his Nazi sympathies, noting that in 1938 he wrote in his application to join the Nazi party, "I am able to say that my whole work is devoted to the ideas of the National Socialists." In 1940, Lorenz published a work which included Nazi ideas of science; he apologized for this when accepting the Nobel Prize.

Charles Lyell

Charles Lyell (1797–1875) was one of the preeminent geologists of his time. He was a friend of Charles Darwin, one who not only influenced Darwin, but articulated and supported the concept of uniformitarianism which played a crucial part in shaping Darwin's theory of evolution.

Lyell was born in Kinnordy, Scotland. His father, a lawyer, was interested in science and introduced his son to the field at a very young age. Lyell went on to attend Exeter College in Oxford. Though he studied geology as an undergraduate, upon graduation Lyell initially took up law as a profession. At the same time, he began to tour England and observed the geological features of each region. When his eyesight began to fail, he decided to move to geology as his full-time profession. He wrote his first paper in the field in 1822 and gave up law altogether in 1827. He became a respected scientist and from the 1830s onward, his career as a geologist provided him both notoriety and substantial income.

Lyell wrote a large number of books on the geological characteristics of different areas, including two books on his North American travels, published in 1845 and 1849. However, his most famous work was the *Principles of Geology*, which was first published in three volumes between 1830 and 1833. Lyell continued to revise *Principles* throughout his life and update its content as his ideas changed. It was in the *Principles* that Lyell argued for uniformitarianism, which had a great influence on Darwin.

Uniformitarianism, originally proposed by another Scottish geologist, James Hutton, was explained by Lyell in the *Principles* as the belief that geological change took place gradually over a long period of time and was an accumulation of countless minute changes. Lyell also believed that the same process that created change over long periods of time in the past was still in existence and could be observed directly in the present. Uniformitarianism was in direct contrast to catastrophism, which stated that geological change was caused by abrupt changes of unknown origin.

Despite the tremendous influence his work would have on Darwin's theories, Lyell himself struggled with the acceptance of evolution and was sometimes criticized for the equivocal nature of his views on the subject. In the first edition of the *Principles*, Lyell explicitly rejected Jean Baptiste Lamarck's ideas on the transmutation of species and continued to dispute evolution in the following eight editions; he finally expressed support for the theories of evolution in the tenth edition, though it was far from enthusiastic. Lyell later admitted that his ambiguity on the subject had been intentional because he had trouble reconciling the theory of natural selection with his religious views as a devout believer in the existence of God. (Lyell was less of a Christian and more of a deist, a believer in a God who is an unmoved mover. He worshipped with the Unitarians. It was the special status of humans that Lyell was most keen to preserve.)

Trofim Lysenko

Trofim Lysenko (1898–1976) was born in the Ukraine. Despite little education or formal training, he became a prominent scientist in the Soviet Union. Today his work has been widely dismissed by mainstream scientists as being little more than the product of Soviet ideology.

Lysenko generally opposed both traditional scientific inquiry methods and Darwin's theories of evolution. He claimed that organisms do not compete, they "cooperate." This phrasing meshed well with current political theories, and Soviet officials condoned his efforts. Based on this theory, he ordered the planting of trees in small groups. Unfortunately, only small numbers of the trees survived, posing huge economic costs to the country. Similar experiments based on practices such as "cooling" grain before planting, led to widespread food shortages. Soviet officials, who had hired Lysenko to address the problem of food shortages, eventually tired of supporting him and allowed the scientist to fade into obscurity.

Although prior to World War II his theories were taught in the Soviet Union as, "Lysenkoism," today much of his agricultural experimentation and research is viewed as little more than fraud. His career is noted for its emphasis on false science, distortions of Darwinism, and disdain for scientific principles. Soviets hailed the uneducated scientist as the embodiment of a peasant genius and despite the eventual disproving of nearly all his theories, because of the support of Stalin (and then Khrushchev), he faced little criticism during his tenure as the Secretary of Agriculture for the Soviet Union. During Lysenko's tenure, Soviet scientists made many fantastic claims which were later unreproducible; these included the development of a variety of peas which flourished in the snow, and the ability to turn bacteria into other species.

Today his name is most widely associated with the "Lysenko Effect," the idea that politically popular theories can gain support even when they fail tests of scientific rigor.

Thomas Robert Malthus

Thomas Robert Malthus (1766–1834) was an English economist and philosopher who worked in many fields, including political economics and demography. His theories on the relationship between human population growth and increases in food production were influential on thinkers in many fields, including evolutionary science.

Malthus was born in Surrey, England and received his earliest education at home and at the Dissenting Academy in Warrington. He was admitted to Jesus College in Cambridge in 1784 where he majored in mathematics. He continued his studies at the college and earned a Masters degree in 1791, and then became a fellow in 1793. He began serving as an Anglican country parson in 1805 and a professor of political economics at the East India Company College in 1805. He was accepted as a fellow of the Royal Society in 1816.

Malthus' scholarly work was based around his most famous publication, *An Essay on the Principle of Population*, of which he published six editions between 1798 and 1826. He continuously updated the work both to incorporate newly available information, and to address past criticisms. The main idea of *Essay* was that, throughout human history, the resources needed to sustain human life (such as food) had grown at a slower rate than human population, a trend that Malthus predicted would continue, leaving him skeptical about the possibility of human society experiencing much of an improvement in status in the future. A mathematical context that Malthus gave to the situation was that unrestrained human population growth occurred at a geometric (or exponential) rate, while food supply grew at an arithmetic rate. The factors that Malthus listed as being able to control excessive population growth were natural causes, misery, and vice. His ideas became known collectively as Malthusian theory. A scenario in which population growth increases uncontrolled at a higher rate than food production has been dubbed a "Malthusian catastrophe."

Malthus' influenced many evolutionary scientists who saw his work as evidence of the need for competition between members of the same populations and the same species. Some saw the struggle for survival as a catalyst by which evolutionary change could take place through natural selection. Two of the more prominent evolutionary scientists who cited Malthus as an influence were Charles Darwin and Alfred Russel Wallace. Both scientists viewed Malthusian theory as removed from the natural theology debate and therefore were able to apply it to their work in its most basic form. However, Malthus himself did not share the view of the evolutionary scientists he influenced. Instead, he saw his Principal of Population as further evidence of the existence of God and was proud that William Paley and other leading natural theologians of the time had adopted his views. For Malthus, the tensions to which he was pointing were not evidence of God's inadequacy or cruelty, but of God's power and forethought in giving us a stimulus to get to work. In later editions of his work, he did allow that the struggle could be avoided through restraint.

Lynn Margulis

Lynn Margulis (b.1938) was born in Chicago. She is an American biologist best known for her theory of the origin of eukaryotic organelles and contributions to endosymbiotic theory. Although controversial at the time she proposed them, her ideas on how certain organelles were formed are now generally accepted. Her work led to the understanding that evolution is more flexible than scientists once believed.

Margulis opposes the traditional views of evolution, which center on competition to explain strategies of regeneration and development. Instead, she stresses the importance of symbiotic or cooperative relationships between species. Symbiotic theories based on zoological or paleontological observations were first put forward in the mid-19th century, but were widely dismissed until Margulis. Her endosymbiotic theory became the first to rely on direct microbiological observations. Margulis refused to accept early criticism; eventually she was widely respected for her tenacity in pushing her ideas, despite the initial skepticism of her peers. Although once dismissed by critics as unworthy of discussion in respectable scientific circles, her ideas are now taught to high school students.

The theory of endosymbiotic focuses on notions of interdependence and cooperative existence of multiple organisms. Scientists had long understood that over time, natural selection, acting on mutations, could generate new species. In some cases, one organism can engulf another, yet both survive and evolve over millions of years into eukaryotic cells. Scientists were unsure, however, if this resulted from branching off of older species, or if certain lines continued. Genetic variation has been popularly understood to result as a transfer between bacterial cells or viral in nature. Margulis's endosymbiotic theories prompted new ideas on the composition of human genomes. Significant portions of the human genome originate from bacterial or viral sources. However, some of these are ancient in origin, while others are more recent. The theory of symbiotic, or parasitic, relationships between organisms allows for new ideas on the forces promoting genetic changes in humans. It is now widely accepted that symbiotic events impact the organization and complexity of many forms of life. Algae have swallowed up bacterial partners, or were swallowed in turn and found within other single cells.

Margulis's theories are mostly compatible with Darwinism, but she has spoken out against some aspects of traditional evolutionary thought. Margulis opposes views of evolution which are solely oriented around competition-based theories of development, arguing that a cost-benefit analysis has been over-emphasized to the detriment of scientific understanding. Margulis also dismisses her peer's prior emphasis on random mutations in evolution. Rather, she believes that tissues, organs, and new species evolve primarily through the long-lasting intimacy of strangers. The fusion allowed by symbiosis, then followed by natural selection, leads to the increasingly complex individuality among a species.

In recent years, Lynn Margulis has become a great enthusiast for the Gaia hypothesis, first proposed by the English scientist James Lovelock. This sees the whole earth as one living organism and argues that, without care, we could put all out of balance and destroy our home. This hypothesis obviously fits well with Margulis's overall anti-reductionistic, holistic philosophy of life.

Othniel Charles Marsh

Othniel Charles Marsh (1831–1899) was born in Lockport, New York. He was a pre-eminent paleontologist of the late 19th century, whose work influenced our modern conception of fossils, particularly dinosaurs.

Although his prolific work with North American vertebrates over a period of many decades sealed his place in history, Marsh is perhaps most famed for his discovery of the first pterodactyl (flying reptile) found in the United States. Later, in part through an intense competition with fellow paleontologist Edward Drinker Cope, he also found the fossils of more flying reptiles, early horses, and toothed birds, as well as many examples of Cretaceous and Jurassic dinosaurs. During a period known as "The Bone Wars," of 1877 to 1892, Marsh and Cope were so competitive with each other that, between the two of them, they named 120 new species of dinosaur and led to the discovery and greater understanding of more than 142 new species of dinosaur. Judging by numbers alone, Marsh arguably came out ahead in the bone wars; he discovered a total of 80 new dinosaur species to Cope's 56. Science, however, was the true beneficiary. Prior to Marsh's efforts in the latter half of the 19th century, there were only nine named species of dinosaur in North America. Unfortunately, the intense competition led both men to attempt to discredit each other's work in the eyes of the American public. Their sniping eventually harmed both reputations, and had a negative impact on the reputation of American paleontology in Europe for decades.

Marsh's ideas on paleontological principles were also very influential among evolutionary theorists, and have held up well over time. One of his most well-known arguments, that birds are descended from dinosaurs, became a widely propagated understanding which later experiments have supported to this day. He also discovered very early horse fossils, much smaller than the modern horse, which he dubbed *Equus parvulus* (Now known as Protohippus). These fossils became one of the "missing links" which led to the genealogy of the modern horse. By the mid-1870s, Marsh had an exceptional collection of early mammals, many extinct, the later study of which led to the formation of early theories of evolutionary development. The larger, faster modern horses, when contrasted with the smaller, more delicate fossils, supported the idea that traits evolve through natural means. Scientists later extrapolated that larger horses were better able to survive in a changing environment, which supported the idea of natural selection, as well as survival of the fittest. By his death, Marsh was one of the most prominent paleontologists in the world, and his reputation as the "first professor of paleontology in America" guaranteed him a place in history.

Ernst Mayr

Ernst Mayr (1904–2005) was a German evolutionary biologist, taxonomist, ornithologist, science historian, and naturalist. His most significant contribution to evolutionary science was his work on modern evolutionary synthesis, which combined Mendelian genetics, systematics, and Darwinian evolution. Mayr was also responsible for the development of the biological species concept.

Mayr was born in Kempten, Germany. When he was a boy, his father often took the family on field trips and Mayr took an interest in natural history from a young age. His family then moved to Dresden where he became involved in the Saxony Ornithologists' Association during high school. He entered the University of Greifswald in 1923, where he initially studied medicine to satisfy the wishes of his family. He later transferred to the University of Berlin and in 1926, earned his Ph.D. in ornithology, which was his true interest. Immediately after graduating, Mayr accepted a position at the Berlin Museum. From 1927 to 1930, he undertook an expedition to New Guinea on behalf of the American Museum of Natural History and subsequently accepted a curatorial position at the museum in 1931. Mayr began to publish a large number of articles on bird taxonomy. Then in 1942, he published his first book, *Systematics and the Origin of Species*, which built on Darwin's theory of evolution and expanded it to form the modern evolutionary synthesis. Mayr joined the faculty of Harvard University in 1953 were he taught until he retired in 1975. He also served as director of the Harvard's Museum of Comparative Zoology from 1961 to 1970.

Mayr's work on the development of modern evolutionary synthesis was a significant contribution to evolutionary theory. He was one of a group of life scientists who gathered around Theodosius Dobzhansky and who worked to give the synthesis of Darwinian selection and Mendelian genetics new meaning. Mayr believed that the process of evolution worked on an entire organism, not on single genes. He argued that the effects of individual genes could vary based on which other genes were present and he advocated the study of the genome as a whole. Later in life, this view put him at odds with fellow naturalist Richard Dawkins, though the two maintained a polite, professional relationship.

Mayr also championed the concept of biological species, which addressed the issue of how multiple species could evolve from a single common ancestor, as well as how exactly to define what a species was. Mayr proposed that a species was not simply a group of organisms that were morphologically similar, but a group of organisms that could breed only among themselves. He also explained that new species result when groups of organisms become isolated and begin to differ as a result of genetic drift and natural selection. Mayr argued that, over time, these differences resulted in new species. He also explained that changes were generally more rapid among small populations in extremely isolated environments, such as on islands. He spoke of this as the "founder effect," and hypothesized that at such time a genetic revolution occurred.

Gregor Mendel

Gregor Mendel (1822–1884) was an Austrian Augustinian priest and scientist who is often referred to as the father of genetics for his study of inheritance in pea plants. He was influential in demonstrating that the inheritance of traits was subject to specific laws. Although he received little recognition while he was alive, his work was rediscovered at the beginning of the twentieth century and has been incorporated into modern evolutionary synthesis as the mechanism by which inheritance operates in natural selection.

Mendel was born into a German-speaking family in Heinzendorf in the Austrian Empire (now Hyncice, Czech Republic). He grew up on his family's farm where he worked as a gardener and a beekeeper. He attended the Philosophical Institute in Olomouc from 1840 to 1843 and entered the Augustinian Abbey of St. Thomas in Brno in 1843. He was sent to study at the University of Vienna in 1851 and returned to the abbey to teach physics in 1853. He began to study the plants in the monastery's garden where he cultivated approximately 29,000 pea plants between 1856 and 1863. His experiments focused on what we would now call "alleles," the variations of gene pairs that determine inheritance of traits. He discovered that one out of four plants had purebred recessive alleles, one out of four had purebred dominant alleles, and two out of four were hybrid. These findings led to the Law of Segregation and the Law of Independent Assortment, which collectively became known as Mendel's Laws of Inheritance. Mendel outlined the concepts in his paper, "Experiments on Plant Hybridization," which was published in 1866. Later, Mendel attempted to replicate his findings with animals by using bees, but he never attained any conclusive results, partially due to the difficulty of controlling the mating habits of queen bees.

At the time it was published, the significance of Mendel's work was not recognized and it was rejected entirely by many. At the time, Darwin's theory of pangenes was generally accepted as the mechanism by which inheritance took place. Modern evolutionary scientists have generally discredited pangenes—which claimed that every cell in an organism shed individual gemmules to determine which traits are inherited. It was not until Mendel's work was rediscovered around the beginning of the twentieth century and scientists began to replicate his experiments that the implications of his studies were full appreciated. Subsequently, a debate ensured between the biologists who supported Mendel's ideas, and the statisticians who supported the concept of biometrics (continuous variation). Both sides agreed about the existence of evolution and natural selection, but they differed on whether selection was a key, creative part of the evolutionary process, or merely mopping up after major genetic changes (causing new variations) had occurred. Eventually, the two approaches were combined into the modern evolutionary synthesis, which uses Mendel's Laws to explain inheritance.

Henry Morris

Henry Morris (1918–2006) was born in Dallas, Texas. His long fight to gain scientific credibility for the idea of a divinely created world led to his title as "The Father of Creation Science." Morris's ideas were rejected by most mainstream scientists, but they continue to set the terms of the public debate over evolutionary theories.

Morris coined the term "creation science" (now called "Creationism") to describe the idea that a divine being created the earth and all living beings in their present form. He wrote more than 60 books and founded the California-based Institute for Creation Research to popularize his teachings, which were based on a literal interpretation of the Protestant Christian Bible. Proponents use supposed, current understandings of physical science principles to explain, or support, Biblical descriptions of miracles and other physical events. Morris believed, with unbending certainty, that the Earth was less than 10,000 years old and was created during a period of six 24-hour days, as described in the Biblical book of Genesis. He argued that fossils, which would seem to disprove his theories, were animals that died during the biblical flood or were the result of mistaken assumptions about age relied on by mainstream scientists.

Scientists have objected to Morris's methods. In particular, many have complained that the emphasis on the literal truth of the Bible encourages creation scientists to begin with a conclusion and reject or misinterpret any evidence or facts that do not support that conclusion. Many called the movement "pseudoscience," complaining that Morris's books are largely unsupported by evidence because Morris omitted theories or facts which did not support his conclusions. Morris's efforts, however, have led to creationists having a large platform and increasing public credibility for their religion-based beliefs. While some of his ideas, such as the claim that craters in the moon were caused by a battle between Satan and the angels, were widely derided, he remained popular among religious groups and gained a certain amount of prominence with the public.

Morris, whose training as a scientist was limited to the field of hydraulic engineering, applied his knowledge of the movement of water to the biblical descriptions of Noah's flood. His book, *The Genesis Flood*, (1961), co-authored by Princeton-trained scholar John Whitcomb, provided a scientific explanation for a story that many accepted on faith alone. The book became known as the founding document of creationism, and remains in print to this day. Morris served as president of the Institute for Creation Research, which he also founded, until 1995, when his son John D. Morris took over. He continued to write and remained president emeritus of the institution until his death.

Simon Conway Morris

Simon Conway Morris (b. 1951) is a British paleontologist who has worked extensively in the fields of early evolution and paleobiology. He is best known for his work on the Burgess Shale fossil fauna found in the Canadian Rocky Mountains. Conway Morris is an advocate for the concept of evolutionary convergence.

Conway Morris was born in London. He entered the University of Bristol in 1969 and graduated in 1972 with a BSc in geology. He then attended St. John's College at the University of Cambridge, where he earned his Ph.D. in 1976. After graduating, he remained at St. John's as a research fellow. In 1983,

Conway Morris was appointed lecturer in the department of earth sciences at the University of Cambridge and was then promoted to reader in 1991. He is currently a professor of evolutionary palaeobiology in the department of earth sciences, a position he has held since 1995. In recognition of his work, Conway Morris has received numerous awards and honors. He was awarded the Walcott Medal of the National Academy of Sciences in 1987, elected fellow of the Royal Society in 1990, and received the Lyell Medal of the Geological Society of London in 1998.

Conway Morris' most significant contribution to the history of evolutionary theory has been his work on the Canadian fossil fauna, known as the Burgess Shale, and other specimens of early evolution found in Greenland and China. He published a popular revision of his work on the fossils in his 1998 book, *The Crucible of Creation*. His work on the Burgess Shale, as well as Stephen Jay Gould's response to his findings, led Conway Morris to investigate further the concept of evolutionary convergence. He first published his work on the subject in his 2003 book, *Life's Solution: Inevitable Humans in a Lonely Universe*. He has also been involved with a website project to create a simple introduction to the concept of convergence and present its thousands of known examples.

Conway Morris is known as a skilled communicator, capable of conveying complex scientific ideas to a wide audience. He is also a Christian and regularly involved in various debates on science and religion. He is a strong critic of intelligent design, materialism, and reductionism. A strong supporter of the evolutionary convergence, Conway Morris has been critical of the widely accepted view that evolution is a process governed by the contingencies of circumstance. He has supported his view by citing the large extent to which evolutionary is highly predictable. Conway Morris also believes that humans have passed a threshold in which we have transcended our animal origins, but are far from the pinnacle of evolutionary development.

Hermann J. Muller

Hermann J. Muller (1890–1967) was an American geneticist and proponent of the idea that genetic mutations were the basis for natural selection. He did significant work in Thomas Morgan Hunt's Drosophila lab, but was best known for his discovery of the physiological and genetic effects of exposure to radiation. The discovery, which became known as x-ray mutagenesis, earned Muller a Nobel Prize.

Muller was born in New York City. He attend public schools were he excelled academically. He entered Columbia College at age 16 where he immediately became interested in biology. He was a supporter of the Mendelian-chromosome theory of heredity and of the concept that genetic mutations were the mechanism by which natural selection worked. He earned in his undergraduate degree in 1910 and remained at Columbia for graduate studies, partially because of his interest in Thomas Hunt Morgan's lab, which was doing groundbreaking genetics work with Drosophila, or fruit flies. Muller studied metabolism at Cornell University from 1911 to 1912, but he remained involved with the activity at Columbia. In 1912, Muller officially joined Morgan's lab, after two years of informal participation. In 1914, Muller was offered a position at the recently founded William Marsh Rice Institute in Houston. He was frustrated by the way that credit was assigned for the work in the lab so he quickly finished his Ph.D. and moved to Texas in time for the beginning of the 1915 academic year. At Rice, Muller taught biology and continued the Drosophila work on his own.

Morgan found himself shorthanded as a result of many of his students and assistants being drafted into World War I, and in 1918 convinced Muller to return to his lab. Muller did not stay long, however, leaving in 1920 to accept a position at the University of Texas. It was there that he made his breakthrough discovery that exposure to x-ray radiation led to the mutation of genes. He had begun exposing Drosophila to radiation as early as 1923, but initially had difficulty obtaining quantifiable results because the process made the flies sterile. It wasn't until 1926 that Muller released his findings in paper entitled, "The Problem of Genetic Modification."

In 1932, Muller moved to Berlin to work with geneticist Nicolay Timofeeff-Ressovsky. He intended the trip only as a brief sabbatical, but it turned into an eight-year stay that would take him to five countries. Very left-wing, he moved to work in the Soviet Union. He returned to the United States after Soviet leader Joseph Stalin found some of his work objectionable.

By the time Muller returned to America in 1940, his work on the effects of radiation exposure had gained him recognition, and he was awarded the Nobel Prize in 1946. After the bombings of Hiroshima and Nagasaki, the risks of radiation and nuclear fallout became a public issue. The work that made Muller famous also provided a significant contribution to the field of evolutionary genetics by helping to identify the causes of the mutation of genes.

John Henry Newman

John Henry Newman (1801–1890) was born in London, England. The greatest English churchman since the Reformation—with the possible exception of John Wesley—Newman moved from evangelicalism in his childhood, to the leadership of the High Church Anglican movement of the 1830s (the Oxford Movement) and then over to Rome, which many years later made him a Cardinal. Newman was first and foremost a theologian, second a preacher, and third an educator. He always had an educated interest in science and this extended to evolutionary ideas.

Newman was most famous for defending Darwin as "not necessarily atheistic." He argued that, rather than a denial of God's existence, evolutionary theories could be suggestive of "a larger theory of divine providence and skill." Where many church leaders saw cause for alarm and dismay, Newman saw the possibility for divine intervention and creation through natural processes. Even the Bible, he argued, could be considered the result of evolutionary forces, which refine and develop teachings and stories through a revelationary process. "It is not that first one truth is told, then another," he wrote in *An Essay on the Development of Christian Doctrine*, published in 1845, "but the whole truth or large portions of it are told at once, yet only in their rudiments or miniature, and they are expanded and finished in their parts, as the course of the revelation proceeds."

Newman did not believe that evidence of divine design could be extrapolated directly from scientific observation; he never claimed, like some intelligent design proponents, that a divine mind or designer had to have directly prompted natural phenomena. Rather, he objected to any theological inquiry that claimed divine design in nature existed apart from religious experience. In a letter about his seminal work, *A Grammar of Assent* (1870), he wrote, "I believe in design because I believe in God; not in a God because I see design." The natural world, he believed, could inspire a Christian to greater religious wonder and religious experience, without requiring adherence to dogmatic interpretations of scientific data. An understanding of science could allow the use of categories such as "natural selection," as a substitute for "divine guidance," or "chance," rather than "divine intervention." Instead of viewing Darwin's ideas as flying in the face of Christian tradition, Newman saw these ideas as compatible with the conception of a world where accidents occur, but where a divine presence is still available to comfort the faithful. Newman believed that God can only be proved to exist through an interior, unreasoned individual conviction, outside proof of divine existence being irrelevant. His theories allowed religious adherents to accept the validity of both evolution and religious teachings.

Newman's thoughtful observations made him lastingly popular as an inspirational writer, and Newman Centers have been established in his honor throughout the world to provide pastoral services and ministries to Catholics at non-Catholic universities. In 1991, Newman was declared "venerable," an early step in the process of nominating him for canonization as a saint in the Catholic Church. (It is true, however, that in line with the general downplaying of natural selection in the years after *The Origin*, Newman preferred the more directed evolutionism of the Catholic biologist St. George Mivart.)

Alexander Oparin

Alexander Oparin (1894–1980) was a Russian biochemist best known for his work on the origin of life. He studied the material processing and enzyme reactions in plants that demonstrated that many food-production processes involved biocatalysts. He also contributed to the founding of industrial biochemistry in the Soviet Union.

Oparin was born in Uglich, Russia, near Moscow. He attended the Moscow State University and graduated in 1917. He initially studied panspermia theories, which focused on the chemical precursors that formed the first microorganisms when life on earth began. Oparin then proposed his own theory of the beginning of life, which he presented to the Russian Botanical Society in 1922. The theory was then published in Russia his 1924 book, *The Origin of Life*, though it did not reach the west until the 1930s. In 1935, he founded the Biochemistry Institute at the Soviet Academy of Sciences with fellow biochemist Aleksei Bakh. Oparin became a corresponding member of the Academy of Sciences in 1939 and a full member in 1946. He was also elected president of the International Society for the Study of the Origins of Life in 1970. He received many awards in recognition of his work; he became a Hero of Socialist Labor in 1969, received the Lenin Prize in 1974, and was awarded the Lomonov Gold Medal in 1979 for outstanding achievements in biochemistry.

Oparin's most significant work focused on the question of how life on earth began. Though others had worked on the subject in the past, Oparin's work was considered by many to be the first modern appreciation of the specifics of the problem. He asserted that there was no fundamental difference between the composition of living organisms and non-living matter and proposed a theory of the way in which basic organic chemicals might have formed microscopic localized systems when the earth was still young. These systems were viewed as the precursors of cells, from which all living creatures evolved. Oparin suggested that a variety of coacervates formed in the "primordial ocean" that was present when the earth was still young, and that these elements eventually led to life through a process of a competitive struggle for existence, and later through natural selection. Oparin was influenced by the recent discovery of methane in the atmospheres of Jupiter. He theorized that Earth initially possessed an atmosphere rich in methane, ammonia hydrogen, and water vapor, which he considered to be the essential raw materials required for the development and evolution of life. Although Oparin was not able to perform experiments to support many of his theories, other scientists later did, and continued research based on his work. Among them was Stanley Miller, whose experiments in 1953 supported many of Oparin's earlier claims.

There has long been discussion about the extent to which Oparin's thinking was influenced by Marxism, specifically the anti-reductionistic philosophy expounded by Friedrich Engels in his *True Dialectics of Nature*. The answer is probably mixed. Oparin's early work owed little to dialectical meterialism; however, he found the philosophy congenial and, in his later writings, strove to present his ideas in terms of that philosophy.

Richard Owen

Richard Owen (1804–1892) was an English biologist, comparative anatomist and paleontologist. He was one of the main proponents for the founding of the British Museum of Natural History. Owen was best known for his vocal opposition to Charles Darwin's theory evolution, as well as for the resulting personal conflicts with Darwin and his supporters.

Owen was born in Lancaster, England. As a boy, he attended Lancaster Royal Grammar School. When he was 16, Owen was apprenticed to a local surgeon and in 1824, he entered the University of Edinburgh as a medical student. He left the university after one year and decided to complete his medical studies at St. Bartholomew's Hospital in London. After briefly contemplating a career in medicine, Owen decided to pursue academic research instead and accepted a position as assistant to the conservator of the museum of the Royal College of Surgeons. He found that his medical studies gave him a strong knowledge of comparative anatomy. In 1836, Owen was appointed Hunterian professor at the Royal College of Surgeons. He then became the conservator of the college's museum in 1849. In 1856, Owen left the college to become the superintendent of the natural history department of the British Museum. Using his position of influence, Owen was instrumental in the creation of the British Museum of Natural History. The museum eventually opened in South Kensington in 1881 and contains the naturally history collection previously located at the British Museum. Owen was also known for coining the term dinosauria (which then became dinosaur), meaning "terrible reptile."

Owen was an outspoken critic of Darwin's theory of evolution and in particular, of natural selection. He instead favored his own theories, which were influenced by German physiologist Johannes Peter Müller and based on the concept of an organizing energy that controlled tissue growth and determined a species' lifespan.

Owen was also a great enthusiast for the ideas of the German anatomist, the naturphilosopher Lorenz Oben. Since Oben was an evolutionist, the question is whether Owen was an evolutionist also. By the 1860s, he certainly was. There is some suspicion that he was an evolutionist as early as the 1830s, when (after the *Beagle* voyage) he and Darwin were intimate. However, it would always have been evolution of an idealistic kind, stressing form over function. And Owen, unlike Darwin, dependent on the patronage of others, would have had to keep silent. It could be that this led to a jealously of Darwin that was the real cause of his public disagreements.

When Darwin published *On the Origin of Species* in 1859, he attempted to convince Owen that the work was based on verifiable scientific laws, but Owen remained skeptical, especially of the concept of transmutation. Owen's criticism of Darwin's theories led to conflicts with other scientists, including Thomas Henry Huxley, and this led to a very public dispute over the relationship of humans to the great apes. Owen claimed that humans uniquely have a part of the brain known as the hippocampus minov. With glee, Huxley pointed out that Owen was working from badly pickled specimens, and the apes do indeed have the part. This point was a major plank in Huxley's evolutionary work *Man's Place in Nature* (1863). Owen's opposition to evolution also created conflicts with his work at the British Museum; when he bought a Archaeopteryx fossil for the museum in 1863, Owen claimed it was simply a bird, but others (Huxley especially) claimed that the fossil was the proto-bird with unfused wing fingers that Darwin had predicted would be discovered.

Archdeacon William Paley

Archdeacon William Paley (1743–1805) was born in Peterborough, England. He was a Christian apologist, philosopher and utilitarian. Although he died four years before the birth of Charles Darwin, his work in the field of natural theology continues to be frequently cited in religious opposition to Dawin's theory of natural selection. Paradoxically, the work also influenced Darwin greatly.

Paley attended Christ's College in Cambridge (the college that Darwin later attended) and graduated in 1763. He then became a fellow at Christ's College in 1766, and a tutor in 1768. He lectured at the college on a variety of subjects, including divinity and Greek testament, and taught a systematic course on moral philosophy that would later become the basis for much of his work. He served a number of different parishes in the area and eventually became the Archdeacon of Carlisle in 1782. In 1785, Paley published a collection of his lectures under the title *The Principles of Moral and Political Philosophy*, which was a great success. It was one of most influential philosophical texts in England of the era; it even became the ethical textbook of the University of Cambridge and a number of other schools. In 1789, Paley wrote the *View of the Evidences of Christianity*, which was an essay on the divine origin of Christianity.

Paley's *Natural Theology* was published in 1802. It would be his last book and the reason for which he is associated with the subject of evolution. Paley stated that his main purpose in writing the book was to establish, as an irrefutable truth, that the world was created and sustained by God in an intelligent, planned fashion. With the exception of one chapter on astronomy, nearly the entire book is based on examples in medicine and natural history; Paley used the complexity of human anatomy to argue that human beings could not exist if not for the input of an intelligent design by God. Paley also used what became known as the "Watchmaker analogy" to explain why he believed that God must have created humans as they are today. This analogy uses a watch as an example of a complex system that could not be created by chance and therefore must have had an intelligent creator. Though it had been used numerous times before to explain divine intervention in the design of other complex systems, Paley is often credited with connecting the Watchmaker analogy with biological development and the creation of human beings. Charles Darwin mentioned *Natural Theology* in his autobiography, saying that it profoundly influenced him when he read it as a student at Cambridge, though his views on the work changed significantly between that time and when he introduced his own theory of natural selection. Paley's work remains today as one of the primary sources cited by modern proponents of intelligent design.

The importance of Paley for Darwin was that, although Darwin may have rejected a miracle-intervening designer, he always accepted Paley's premise that the organic world is "as if" designed. Natural selection is intended to speak to this. Critics of Darwinism, like Stephen Jay Gould, attack the claim that the organic world is generally "as if" designed, and claim that Darwinians are hung up on an outmoded Christian natural theology.

Geoffrey Parker

Geoffrey Parker (b. 1944) is a British evolutionary biologist. He is best known for introducing the concept of sperm competition and for his work in applying game theory to biology. Parker has also done work on the evolution of competitive mate searching, animal distributions, animal fighting, coercion, interfamilial conflict, and complex life cycles.

As a child, Parker attended Lymm Grammar School in Cheshire County, England. He later enrolled in the University of Bristol, where he received his BSc in 1965 and his Ph.D. in 1969. While at Bristol, he studied under the late British entomologist H.E. Hinton. Parker's dissertation, *The Reproductive Behaviour and the Nature of Sexual Selection in Scatophaga Stercoraria L*, provided a quantitative test of Darwin's theory of sexual selection. The work was also an early application of optimality theory in biology, which would become one of Parker's principal areas of study.

After completing his studies at Bristol, Parker accepted a position as lecturer of zoology at the University of Liverpool. In 1978, he accepted a research fellowship at King's College in Cambridge, but returned to the University of Liverpool in 1979. Parker was elected to the Royal Society in 1989 and became the Derby Chair of Zoology in 1996.

Parker is best known for his concept of sperm competition, which he introduced in 1970. The concept is centered around the idea that multiple sperm from male organisms compete for the fertilization of the female organism's ova, and that the more resources that the male dedicates to the production and spread of his sperm, the more likely he is to inseminate the female. However, because the male's biological resources are finite, he must adjust the distribution of his resources to meet the requirements of a given situation. Parker's work on the subject has been influential in generating support within the scientific community for the gene-centered view of evolution, as well as providing a basis for the beginnings of behavioral ecology.

When sperm competition was applied to game theory in mathematics, the optimum allocation of a male's resources was understood in terms of the concept of an evolutionarily stable strategy (or ESS). An ESS is a stable situation in a population of different types (genes) where a member cannot benefit by changing type. If, for instance, one has a population of hawks and doves, it would not benefit an individual hawk (dove) to change to a dove (hawk). British evolutionary biologist and geneticist John Maynard Smith did pioneering work on the notion of an ESS and he and Parker collaborated on developing this idea. It has also been proposed that sperm competition may lead to evolutionary changes in organisms to increase their sperm production, such as larger testes.

Parker has also done work on various other evolutionary subjects relating to sexual species. In 1972, he collaborated with fellow scientists R.R. Baker and V.G.F. Smith to propose a theory of the evolution of anisogamy and the two sexes (Anisogamy is where one sex has much bigger sex cells than the other; human sperm, for example, is much smaller then the ovum). Parker has also done theoretical analysis of sexual conflict in evolution.

Louis Pasteur

Louis Pasteur (1822–1895) was a French chemist and microbiologist He was one of the founders of microbiology and did groundbreaking work in the field. He was best known for his work on germ theory and for his process for reducing the spread of disease through liquids, which became known as pasteurization. His experiments also disproved the concept of spontaneous generation, which was a significant aspect of many early, pre-Darwinian evolutionary theories.

Pasteur was born in Dole, France and grew up in Arbois. In school, his academic talents were recognized by his headmaster who recommended that he apply to a university. Following this advice, Pasteur attended the École Normale Supérieure in Paris. After graduating, he served briefly as professor of physics at Dijon Lycée before becoming professor of chemistry at Strasbourg University in 1848.

Pasteur's most important work was based on his dispelling of spontaneous generation, which led to his support of both biogenesis and germ theory. His findings also made him a strong critic of the theory of evolution, although this opposition was linked also to Pasteur's very conservative political beliefs. During the 18th century, it was widely believed that life could form from nonliving matter at a microscopic level. Pasteur disproved this theory with an experiment that used boiled chicken broth in an airtight apparatus to demonstrate that new organisms could not form in the absence of living matter. The result of Pasteur's discovery was the concept of biogenesis, which stated simply that life arose from life, and more specifically, that an organism could only come from the same type of organism.

The practical applications of Pasteur's work were extremely important to the advancement of medical sciences and for the prevention of disease transmission. The work that Pasteur became best known for was the development of pasteurization, a process in which liquids such as milk were heated to kill bacteria. His work also supported the concept of germ theory, which stated that microorganisms were the direct cause of many diseases. The theory, also known as the pathogenic theory of medicine, has become the cornerstone of modern medicine. The development was responsible for many life-saving concepts, including antiseptics used in surgical procedures and, later, antibiotics.

The implications of Pasteur's work for the theory of evolution were significant because spontaneous generation was taken by many to be an essential part of Darwin's theory. However, the full story is a little more complex. It is true that many pre-Darwinian theories (Lamarck's for instance) supposed spontaneous generation, but because by Darwin's time the idea was under attack, in *The Origin* Darwin said nothing at all about ultimate origins. A decade later, in a private letter, Darwin hypothesized that life began when the earth was still young in an environment that became known as the "primordial soup." He theorized that the conditions on earth several billion years ago that were necessary for the beginning of life no longer existed. Probably Darwin was not thinking in terms of spontaneous generation, but more of the idea of gradual development, which became popular in the 20th century.

Alvin Plantinga

Alvin Plantinga (born 1932) is an American philosopher who is best known for his work in the fields of epistemology, metaphysics, and the philosophy of religion. He is considered by many to be a central figure in the effort to gain respectability for the belief of God in academic philosophy. Plantinga believes that evolution implies naturalism, that naturalism is incoherent, and hence one should (rationally) believe in God.

Plantinga was born in Ann Arbor, Michigan. He left high school one year early and entered Jamestown College in 1949. Following his family to Grand Rapids, Michigan, Plantinga enrolled at Calvin College, where his father was a professor. Plantinga earned a scholarship during his first semester at Calvin and enrolled at Harvard University, where he spent two semesters from 1950 to 1951. He then returned to Calvin after being impressed by the philosophy professor William Harry Jellema, whom he heard speak during a school break. After finishing his undergraduate studies at Calvin, Plantinga started graduated studies at the University of Michigan in 1954. He then transferred to Yale University in 1955, where he earned his Ph.D. in 1958.

In 1958, Plantinga accepted a position as professor of philosophy at Wayne State University. He then began teaching at Calvin College in 1963, replacing his old professor, Jellema, who was retiring. Plantinga taught at Calvin for 19 years before accepting a position at the University of Notre Dame, where he is currently the John A. O'Brien Professor of Philosophy.

Plantinga is well known for his use of the concept free will in defense of the logical problem of the existence of evil. In his work on the subject, Plantinga makes a distinction between a defense, which looks for logical reasons for God permitting the existence of evil, and a theodicy, which attempts to justify evil by explaining why God permits the existence of evil. While not claiming that God permits evil for the reason of free will, Plantinga argues in his defense that is logically possible that God could not have created a world with good but no evil. Plantinga published his work on the subject in his 1974 book entitled *God, Freedom, and Evil*.

Plantinga's view on evolution encompasses a strong stance against naturalism. He believes that evolutionary thinking presupposes naturalism (everything working by law), that the naturalist should be an evolutionist, and that this connection makes naturalism incoherent. He argues that, because the naturalistic view of evolution states that cognitive function developed for the purpose of survival, there is no way to know that the beliefs produced by that function are true; therefore there is reason to doubt the products of that function, including both naturalism and evolution. Plantinga argues instead that if God created man, either through evolution or other means, that there is no reason to doubt the reliability of our cognitive facilities. Plantinga's views do not suggest or deny a correlation between true beliefs and survival. (Although Plantinga does not deny common descent, he favors intelligent design theory. Hence, his position would seem to be somewhat akin to that of Michael Behe.)

Felix Pouchet

Felix Pouchet (1800–1872) was a French naturalist, biologist, and science writer. He became known primarily as a proponent of the theory of spontaneous generation and the resulting debate with microbiologist Louis Pasteur. Though his other accomplishments were often overshadowed by this controversy, he contributed to a wide variety of scientific fields, including zoology, physiology, botany and microbiology. Pouchet's support for spontaneous generation stemmed from his enthusiasm for ideas of evolution in which he thought the concept played a crucial part.

Pouchet was born in Rouen, France. He demonstrated an interest in biology from a very young age. He attended the University of Rouen and graduated in 1927. He then became the director of the Muséum d'Histoire Naturelle de Rouen in 1828, where he would remain for the rest of his life. In 1838, he also took a position as professor of biology in the school of medicine at the University of Rouen. He received a number of awards in recognition of his service, including the Legion of Honor in 1843. He published an encyclopedia of science in 1865 entitled, *The Universe*, which was understandable by the layperson and used a large number of illustrations. In it, however, Pouchet mocked the work of Louis Pasteur, who was the leading critic of the theory of spontaneous generation at the time.

Pouchet became known for his support of the theory of spontaneous generation, and for his connection with Pasteur's disproval of the theory. Pouchet was a supporter of Darwin and he thought that the underlying implications of spontaneous generation were significant because the concept was a key part of how life was thought to have begun in the theory of evolution. The debate also had religious implications because Pasteur was a devout Catholic and France was a Catholic country. Pouchet's support of the theory added fuel to what was already a highly contentious environment. The controversy centered on whether or not microorganisms could generate on their own. Pouchet believed that three elements were necessary for the generation of life: decaying organic matter, air, and water. He also claimed that electricity and sunlight could encourage the process. Pasteur became famous for disproving the theory of spontaneous generation in 1861 by using an airtight apparatus to demonstrate that new organisms could not arise in the absence of living matter. Though Pouchet continued to argue his views, Pasteur dealt him a decisive blow by gaining the support of the French Academy of Science in 1864. After Pouchet died, the remaining support for spontaneous generation gradually declined. While not supporting spontaneous generation (the belief that life is generated instantaneously from non-life), many evolutionary scientists agree with Darwin's view that the process was natural and only possible in the unique conditions that were present when the earth was still young, billions of years ago.

Étienne Geoffroy Saint-Hilaire

Étienne Geoffroy Saint-Hilaire (1772–1844), a naturalist, was born in Étampes, France. He supported and expanded upon the evolutionary theories of his colleague Jean-Baptiste Lamarck, using the concepts of unity in organism design and species transmutation as a base for his work. His research included work in the fields of zoology, embryology, paleontology, and comparative anatomy. He is best known for his theory that all species stem from the same basic structural plan, which he explained in the publication, *Philosophie anatomique* (1818–1822).

Geoffroy studied at the Collège de Navarre in Paris. In 1793, he was appointed as professor of vertebrate zoology at the Musée National d'Histoire Naturelle. Shortly after, he became acquainted with the young scientist Georges Cuvier. The two exchanged letters until Geoffroy invited Cuvier to Paris in 1795 and Cuvier was appointed as an assistant at the Musée National d'Histoire Naturelle. The two scientists worked together on several projects, including the 1795 publication of *Histoire de Makis, ou singes de Madagascar*, in which Geoffrey presented his concept of the unity of organic composition. Despite their early collaboration, the views of Geoffrey and Cuvier diverged significantly with time—so much so, that they would eventually become adversaries later in life.

In 1798, Geoffrey participated in Napoleon's famous expedition to Egypt. Geoffrey was one of 167 scientists and artists selected for the trip. Shortly after returning from Egypt in 1808, Geoffroy was made a professor of zoology at the University of Paris. It was during this time that he began to devote himself more exclusively to the study of anatomy. In the years that followed, he developed his theory of unity in organic composition. Although developed independently, it bore strong resemblances to the archetypal views of the German Naturophilosopher Lorenz Oben and later, Richard Owen. The theory stated that all animals are made up of the same elements, in the same numbers, connected in the same way. These elements would later come to be known as homologous structures. He also believed that there was cohesion between the growth and development of an organism's different elements that created a natural balance within the animal as a whole. This theory was the basis of Geoffroy's conflict with Cuvier. While Cuvier accepted the existence of a natural balance between the structures within an organism, he was strongly opposed to the idea of structural unity between members of different species. Cuvier believed instead in the variation of elements within species based on their environments.

Geoffroy believed in the ability of a organism's environment to directly cause organic change. He took a great deal of interest in monsters and how they can be produced by environmental changes. This idea differed from those of the leading evolutionary scientists of the time, including Lamarck, who believed that a species' environment could affect its development only indirectly, by causing behavioral modifications. Geoffroy's views on the direct influence of an organism's environment are not shared by any proponents of the modern theory of evolution. Despite this, his work on structural unity would be important for scientists who came later, as many, including Charles Darwin, used homologous structures as a way of proving their evolutionary theories.

John J. Sepkoski Jr.

John J. Sepkoski Jr. (1948–1999) was an American paleontologist who studied the fossil records and diversity of life on Earth. He was best known for his work on mapping the diversity of life through the ages and for proposing that mass extinction events occur in cycles.

Sepkoski was born in Presque Isle, Maine. Growing up, he collected bones and fossils and decided to become a paleontologist at a very young age. As an undergraduate, he attended the University of Notre Dame where he earned his B.S. in 1970. He then went on to study at Harvard University where he earned his Ph.D. in geological sciences in 1977. Sepkoski did research for his Ph.D. at the Black Hills in South Dakota. He then taught briefly at the University of Rochester before accepting a position at the University of Chicago in 1978. He started as an assistant professor and was promoted to associate professor in 1982, and professor in 1986. Sepkoski also began working as a research associate at the Field Museum of Natural History in Chicago in 1980. He was a visiting professor at the California Institute of Technology in 1986, and at Harvard University from 1990 to 1991. In recognition of his work, Sepkoski was given the Charles Schuchert Award by the Paleontological Society in 1983.

Sepkoski's most significant contribution to our understanding of evolution was his work on mass extinction and extinction event cycles. He and his colleague David Raup developed a theory based on statistical analysis of fossil records that proposed that mass extinctions of marine animals have occurred approximately every 26 million years for the past 250 million years. According to the theory, these extinctions also account for the disappearance of the dinosaurs 65 million years ago. The previously accepted view in the scientific community was that mass extinctions occurred at random and not as part of a cycle. Sepkoski, by accumulating huge data sets, also did extensive work studying marine animal families and genera. In 1981, he identified the three evolutionary faunas of the marine animal fossil record: the Cambrian, the Paleozoic and the Modern. He then modeled the faunas through three coupled logistical functions using ideas he gleaned from the MacArthur-Wilson theory of island biogeography. Sepkoski's work and data continue to motivate a considerable amount of research in the field of paleobiology.

George Gaylord Simpson

George Gaylord Simpson (1902–1984) was an American paleontologist who contributed to the modern evolutionary synthesis. He was an expert on extinct mammals and did pioneering work on the migratory patters of the American fauna.

Simpson was born in Chicago but grew up in Denver. He entered the University of Colorado at Boulder in 1918, where he initially wanted to study creative writing, but soon switched to geology. He transferred to Yale University during his senior year after having been told that it was the best place to study geology and paleontology, and graduated in 1923. Simpson remained at Yale as a graduate student and earned his Ph.D. in 1926. He then moved to London to study primitive mammals by examining the specimens at the British Museum of Natural History. When he returned from England in 1927, he took a position as assistant curator of fossil vertebrates at the American Museum of Natural History in New York.

In the 1930s and early 1940s, Simpson's work became more theoretical as he began to focus more on the general subject of evolution, and less on extinct mammals and fossils. In 1944, he published *Tempo and Mode in Evolution*, which was significant in the development of the modern evolutionary synthesis concept. He was much influenced by the population geneticist Sewall Wright and the Russian-born fruit fly geneticist Theodosius Dobzhansky. Particularly innovative was Simpson's (mainly successful) attempt to interpret Wright's gene-based adaptive landscape in terms of fossil taxa as found in the geological record. From 1945 to 1959, Simpson was a professor of zoology at Columbia University as well as curator of the department of geology and paleontology at the American Museum of Natural History. He became the curator of the Museum of Comparative Zoology at Harvard University in 1959, and then professor of geosciences at the University of Arizona in 1970, where he taught until he retired in 1982.

Simpson was a strong supporter of Charles Darwin's theory of natural selection and provided support for evolution through his expertise in interpreting the fossil record. Specifically, he theorized that organisms evolved in three ways: speciation, phylectic evolution, and quantum evolution. The differences between the three were based on whether or not an entire species evolved, such as with phylectic evolution, or only individual members of a species evolved as a result of having been isolated from the main population, such as with quantum evolution. Simpson explained this concept in *Tempo and Mode in Evolution*, in which he described the way in which evolutionary change could be categorized by tempo, rate of change and mode. He was also one of the first scientists to apply concepts from mathematics and genetics to paleontology. His work in connecting the findings of a variety of scientific fields helped him with his work on the modern evolutionary synthesis.

By his own admission, Simpson was not an easy man with whom to work. However, he did build some close friendships—with Julian Huxley, who had been kind to the young Simpson on the first visit to London, and with the Jesuit paleontologist Pierre Teilhard de Cuardin, whose vision of a progressive evolutionary movement up to the Omega Point (Jesus Christ) Simpson could admire, but not share. He was a man of great integrity, who, although above age, served in very dangerous work in WW II. Typically, Simpson engaged in a horrific row with General George Patton who wanted Simpson to shave his beard. Also typically, Simpson found an obscure regulation, which let him keep it.

John Maynard Smith

John Maynard Smith (1920–2004) was a British theoretical evolutionary biologist and geneticist. His most significant contribution to the history of evolutionary theory was his application of the mathematical concept of game theory to evolution and the subsequent formalization of the evolutionary stable strategy. Maynard Smith also did work on the evolution of sex and signaling theory.

Maynard Smith was born in London. After his father's death in 1928, his family moved to Exmoor, where Maynard Smith developed an interest in natural history. He attended Eton College, but he was disappointed by the lack of formal science education available there, so he nurtured his growing interests in mathematics and Darwin's theory of evolution by reading books from the school's library. After Eton, Maynard Smith entered Trinity College, where he studied engineering. When the Second World War broke out in 1939, he volunteered for service but was rejected due to poor eyesight. He then returned to Trinity and finished his engineering degree in 1941. From 1942 to 1947, Maynard Smith applied his engineering background to the design of military aircraft. Deciding to change careers, Maynard Smith enrolled at University College London to study fruit fly genetics under noted geneticist J.B.S. Haldane. After graduating in 1952, Maynard Smith remained at the university where he lectured in zoology until 1965. During that time, he also directed the Drosophila lab and performed research on population genetics. In 1962, Maynard Smith became one of the founding members of the University of Sussex, and he served as dean of science from 1965 to 1985.

Maynard Smith's most significant contribution to evolutionary science was his formalization of a concept of game theory in mathematics. His thinking was focused on what was called an evolutionarily stable strategy (or ESS) and was based on a concept proposed by American geneticist George R. Price, going back before that to ideas developed by John Nash. Game theory refers to the field of study that models, in mathematical terms, behavior in strategic situations in which successful strategy is affected by the choices of others. Maynard Smith's ESS was a strategy that, once adopted by a population, was evolutionarily stable and could not be invaded by alternative strategies. His work on the subject was ultimately published in his 1982 book *Evolution and the Theory of Games*.

Maynard Smith published a wide variety of books on evolutionary topics, including a popular overview of evolution entitled *The Theory of Evolution* (1958) and his final book, *Animal Signals* (2003), on signaling theory. He received a number of honors and awards in recognition of his significant scientific contributions. He was elected a Fellow of the Royal Society in 1977 and was given the Darwin Medal in 1986.

Herbert Spencer

Herbert Spencer (1820–1903) was an English philosopher, political theorist, and sociological theorist. He was best known for developing an all-encompassing concept of evolution in which the physical world, biological organisms, human intelligence, culture and societies were all connected and governed by the same rules of development. He also coined the term "survival of the fittest" to describe Charles Darwin's theory of natural selection, though Spencer himself never fully agreed with the theory. His work encompassed a wide range of disciplines, including ethics, religion, politics, philosophy, biology, sociology, and psychology.

Spencer was born in Derby, England. He received most of his education at home from his father and uncle. Members of the Derby Philosophical Society also played a role in introducing him to the pre-Darwin theories of evolution of Erasmus Darwin and Jean Baptiste Lamarck. In other areas, Spencer was self-taught. As a young man, he worked as a civil engineer for the railroad industry while writing often for radical provincial journals. He served as an editor for *The Economist*, a free-trade journal, from 1848–1853. In 1851, he published his first book, *Social Statics*, in which he argued that the role of government would eventually diminish as people became adapted to the requirements of living in modern society. Spencer's most famous publication, *The System of Synthetic Philosophy*, was an ambitious project started in 1858 and which and took him the greater part of the rest of his life to complete. Spencer's goal for the publication was to prove that the basic concept of evolution could be applied to biology, psychology, sociology, and morality.

Although he came independently to the Darwinian notion of natural selection (several years before *The Origin* was published), Spencer himself agreed more with Jean Baptiste Lamarck's concepts of use-disuse and soft inheritance. He incorporated this into a progressive view of life history, where groups (organisms, humans, societies) get disturbed, and then strive upwardly to reach a new point of balance. Spencer referred to this as "dynamic equilibrium." Spencer's application of the theory of evolution to modern society, sociology and ethics became known as Social Darwinism, and is the idea for which Spencer is perhaps the most remembered. Social Darwinism was a major influence on many sociologists, political theorists, and others. It should be noted that, as with many broad moral systems (like Christianity), there was a notable lack of unanimity with respect to the actual ways in which social Darwinian principles should be enacted. Some were for capitalism, some for socialism, some for war, some for pacifism, some for chauvinism, some for feminism. It was perhaps because of this undue flexibility, some would say flabbiness, that Spencer's influence collapsed, never to be revived.

George Ledyard Stebbins

George Ledyard Stebbins (1906–2000) was an American botanist, geneticist, and evolutionary biologist. He was known for his work to incorporate genetics and Charles Darwin's theory of natural selection into a comprehensive synthesis of plant evolution. His work helped form the basis for the general concept of modern evolutionary synthesis.

Stebbins was born in Lawrence, New York. His parents were real estate financiers who developed Seal Harbor, Maine and encouraged Stebbins' interest in natural history during family trips to the area. After high school, Stebbins entered Harvard University. He originally intended to study political science, but switched his major to botany during his third year. He continued at Harvard for graduate school, first studying flowering plant taxonomy, and then switching to the cytology of plant reproductive processes. He completed his Ph.D. in botany in 1931.

In 1932, Stebbins accepted a teaching position at Colgate University and continued to work in cytogenics. Then in 1935, he accepted a position at the University of California, Berkeley, where he did research with noted geneticist E.B. Babcock. It was during this time that he became involved with a group of scientists known as the Bay Area Biosystematists, who worked on problems related to evolutionary biology. In 1939, Stebbins was made full professor in the department of genetics at UC Berkeley and was instrumental in shaping the development of the department. In 1946, he also became involved with the Society for the Study of Evolution, and was one of the few botanists associated with the organization.

In 1946 he gave a series of lectures on plants that combined the fields of genetics, ecology, systematics, cytology, and paleontology. The lectures, which were published in 1950 as *Variation and Evolution in Plants*, combined the theory of natural selection and genetics to explain plant speciation. It was the first book to provide a comprehensive explanation of how evolution functioned in plants at the genetic level and it provided the basis for future biological research in the field of plant evolutionary. Because of the book's scope of incorporating different scientific disciplines, it was considered a significant publication in the development of the modern evolutionary synthesis as well. The work thus completed the project started by Theodosius Dobzhansky in his *Genetics and the Origin of Species* (1937), and continued by Ernst Mayr in *Systematics and the Origin of Species* (1942) and George Gaylord Simpson in *Tempo and Mode in Evolution* (1944).

Stebbins was a strong proponent that evolution needed to be studied as a dynamic topic. He was responsible for developing evolution-based science programs for California high schools. He was also a member of the National Academy of Sciences.

Chris Stringer

Chris Stringer (b. 1947) is a British anthropologist and one of the leading supporters of the recent single-origin hypothesis, also known as the "Out of Africa" theory. The concept claims that modern humans originated in Africa approximately 200,000 years ago and migrated out of Africa sometime within the last 50,000 years (or somewhat earlier). The theory also claims that humans began to replace related hominid species once they left Africa.

Stringer studied anthropology at University College London and received both a Ph.D. and D.Sc. in anatomical science from Bristol University. He has worked at the Natural History Museum in London since 1973, where he is currently a research leader in the paleontology department. He also leads the Ancient Human Occupation of Britain project (AHOB), which aims to augment the existing knowledge of the spread of modern humans throughout the globe, as well as to explain the disappearance of the Neanderthals. His work has included collaboration with a wide variety of scientists in different disciplines of palaeoanthropology. He is also a Fellow of the Royal Society. His recent books include *The Complete World of Human Evolution* (2005) and *Homo Britannicus* (2006).

Stringer developed the Out of Africa model with fellow scientist Peter Andrews. The two men first published their work on the theory in a 1988 article entitled, "Genetic and Fossil Evidence of the Origin of Modern Humans." The theory stated that modern humans, or *Homo sapiens*, evolved into their current forms in east Africa around 200,000 years ago. According to the Stringer and Andrews, *Homo sapiens* migrated out of Africa approximately 50,000 years ago and, at that time, began to replace many of the related species in the hominidae family. The other hominids that existed at the time, and have since become extinct include *Homo erectus, Homo habilis, Homo antecessor*, and the Neanderthals. The model has been further substantiated by research using mitochondrial DNA, as well as by physical anthropological evidence from archaic specimens. Stringer's work is important to the history of evolutionary theory because it explains the earliest origins of humans, as well as provides evidence of the evolution from our closest genetic ancestors.

Other paleontologists have proposed a number of alternate competing theories of the origins of the first humans. Some theories (that of Milford Wolpoff) claim that humans originated in multiple geographic regions as a result of interbreeding between different early hominids. Other theories are similar to Stringer's hypothesis, but claim that humans left Africa as many as two million years ago. Though some debate on the issue still exists within the scientific community, Stringer's ideas have become widely accepted.

D'Arcy Wentworth Thompson

D'Arcy Wentworth Thompson (1860–1948) was a Scottish biologist and mathematician who was among the first to do work in the field of mathematical biology. He used mathematical descriptions to explain the development of species and proposed structuralism as an alternative evolutionary mechanism to natural selection.

Thompson was born in Edinburgh, Scotland. He entered the University of Edinburgh in 1878 with the intention of studying medicine, but two years into his studies he transferred to Cambridge University to study natural science. He received his B.A. in 1883 and accepted a position as professor of biology at University College, Dundee in 1884. He was appointed Chair of Natural History at St. Andrews University in 1917. In recognition of his work, Thompson was elected fellow of the Royal Society in 1916 and awarded the Darwin Medal in 1946.

Thompson was best known for proposing that the laws of physics and mechanics played a significant role in determining of the structure and form of developing organisms. He believed that biologists overemphasized the influence of natural selection in relation to the form of species. To support his view, Thompson presented numerous examples in which biological forms closely resembled mechanical phenomena. These examples included the similarities in form between jellyfish and drops of liquid falling into viscous liquid and between the structural form of the hollow bones of birds and those of engineering truss designs. Another example that became synonymous with Thompson and his ideas was the observable relationship between spiral structures in plants and the Fibonacci sequence in mathematics. He also described the way in which structural differences between related species could be explained through mathematical transformations.

Thompson published his views on the subject in his 1917 book entitled *On Growth and Form*. The book was extensively illustrated throughout and contained a large number of examples to support its ideas, but it was never fully accepted by the mainstream biological community. Many scientists saw it as purely descriptive and lacking a unifying thesis, criticisms that Thompson himself acknowledged. The book also did not contain an experimental hypothesis that could be tested. Nevertheless, Thompson's ideas were significant to the history of evolutionary theory because they proposed an alternative view on explaining the structure and form of the development of species. In recent years, Thompson's ideas have been championed by formalists such as Stephen Jay Gould and Brian Goodwin.

Alfred Russel Wallace

Alfred Russel Wallace (1823–1913) was a British naturalist, explorer, geographer, anthropologist and biologist. He is best known for developing his own theory of natural selection independently of Charles Darwin's. Their ideas were publicly presented, jointly, in the summer of 1858.

Born in Llanbadoc, Wales, Wallace moved to Herford, England with his family when he was five years old. He briefly attended grammar school but was forced to withdraw when his family ran into financial trouble. In 1837, at the age of 14, Wallace moved in with his older brother William and worked for him doing surveying work for the next six years. When difficult economic conditions forced him to leave his brother's business, he found work teaching drawing, mapmaking and surveying at the Collegiate School in Leicester. During this period, he spent most of his free time in the library reading the works of Thomas Robert Malthus, Charles Darwin, and Charles Lyell. He probably became an evolutionist around this time through reading *The Vestiges of the Natural History of Creation* (anonymously authored) by the Scottish publisher Robert Chambers.

Wallace was inspired by stories of the epic voyages of other scientists and decided that he too wanted to travel overseas to do research. In 1848, Wallace and fellow naturalist Henry Bates left England for Brazil, looking for insects in the Amazon Rainforest to sell to British collectors, as well as evidence to support the transmutation of species theory. Wallace returned to London in 1852, departing again in 1854 on another expedition to the Malay Archipelago (now Malaysia and Indonesia).

Wallace's observation during his trip to the Malay Archipelago further convinced him of the existence of evolution. He then combined his observations with Robert

Malthus' concept of competition among species as population control and arrived at his own theory of natural selection. Wallace was also in correspondence with Darwin at the time. In 1857, Wallace sent to Darwin his paper, "On the Law that has Regulated the Introduction of New Species," which he had written in 1855. Darwin's response acknowledged that the two men had both been working on the same concepts independently, but Darwin believed that his version of the theory predated Wallace's and was further developed. Wallace respected and trusted Darwin and sent him another paper in 1858, entitled, "On the Tendency of Varieties to Depart Indefinitely From the Original Type." It was this paper that had the idea of selection (although not by that name). Darwin was impressed by the work and passed it on to Charles Lyell and Joseph Hooker, who eventually had it published with material on natural selection written by Darwin. Wallace was satisfied by the arrangement and acknowledged that his ideas probably would not have been given much regard without the support of a respected scientist such as Darwin. In 1859, when Darwin published his landmark work, *On the Origin of Species*, Wallace was one of its strongest supporters.

Wallace returned to England in the early 1860s and went on writing about evolutionary topics. However, he became enamoured with spiritualism and decided that human evolution could not be purely natural. In response, Darwin penned *The Descent of Man* (1871), arguing that sexual selection was a major, hitherto-overlooked cause of our arrival here on earth. Wallace continued to write on science, including major work on biogeography. Increasingly, he turned to social issues, like land reform, where his socialist convictions, dating from his youth, could get his full attention.

Samuel Wilberforce

Samuel Wilberforce (1805–1873), the son of William Wilberforce, of slave-trade abolition fame, was an English bishop in the Church of England. He was noted for his skill as a public speaker, as well as for his vocal criticism of evolution. He became known for his 1860 debate with Thomas Henry Huxley in which the two clashed over Charles Darwin's theory of evolution. Because of his fluidity of speech, he was known as "Soapy Sam."

Wilberforce was born at Clapham Common, London. In 1823, he entered Oriel College in Oxford. He joined the United Debating Society and gained a reputation as a strong supporter of liberalism. He studied mathematics and classics and graduated in 1826. After spending a year traveling throughout Europe, Wilberforce was ordained and appointed curate at Checkendon. In 1830, he was presented by the Bishop of Winchester to the rectory of Brichstone in the Isle of Wight. In 1841, as a result of his public speaking skills, Wilberforce was selected as a Bampton lecturer at the University of Oxford. The selection gained him considerable recognition and, as a result, he was appointed chaplain to Prince Albert shortly after. Then in 1843, he was appointed sub-almoner to the Queen by the archbishop of York. He became Bishop of Oxford in 1845 (and was appointed to the more senior bishopric of Winchester in 1870). In 1854, he opened a theological college in Cuddesdon, which is now known as Ripon College. He was a leader of the High Church in the Church of England.

In 1860, Wilberforce took part in the debate on the theory of evolution for which he would become known. The debate took place at the Oxford University Museum of Natural History and pitted Wilberforce against Thomas Henry Huxley and other prominent supporters of Charles Darwin's theory; Darwin himself was not present. Wilberforce argued against evolution on largely scientific grounds, citing the fact that many prominent scientists of the day opposed the theory. However, the most memorable exchange of the debate was when he asked Huxley if he claimed decent from a monkey through his grandmother. It was generally believed by those present that Huxley presented the more convincing argument. Because of the prominence of the people involved, the debate was seen as a significant event in the history of the theory of evolution that helped to gain wider recognition and acceptance for the theory.

Historians have subsequently thrown doubt on whether the Huxley-Wilberforce confrontation was quite as dramatic as legend would have it. Certainly, although negative, the review of *The Origin* by Wilberforce in the *Quarterly Review* was respectful and admiring of Darwin as a scientist and a man. (Little known is the fact that Wilberforce was a fellow of the Royal Society.) However, true or not, the story took on mythical proportions and served for generations as a dreadful warning about the threat religion poses to science.

Edward O. Wilson

Edward O. Wilson (b. 1929) is an American biologist, researcher, theorist, and naturalist, who specializes in myrmecology (the study of ants). He is known for establishing the field of sociobiology and for his groundbreaking work on the behavior of insects. His sociobiological concepts apply Darwin's evolutionary theories to sociology by linking behavior to inherited biological traits.

Wilson was born in Birmingham, Alabama and grew up in both rural Alabama, and outside Washington D.C. He showed an interest in natural history from an early age. When he was 16, Wilson decided that he wanted to be an entomologist and began collecting insects. He initially started by collecting flies, but World War II led to a shortage of insect pins with which to store them, so he switched to ants, which could be stored in vials. With the encouragement of a myrmecologist at the National Museum of Natural History, Wilson began a comprehensive survey of the ants of Alabama. He then went on to earn his B.S. and M.S. from the University of Alabama, and later his Ph.D. from Harvard University, where he is now a professor emeritus.

Wilson is one of the founders of sociobiology, a field that studies social behavior on the basis of biological factors. His major work is *Sociobiology: The New Synthesis* (1975), followed by the Pulitzer Prize winning work on our own species, *On Human Nature* (1978). Wilson argues that evolutionary principals can be applied to behavioral study. Specifically, he claims that all animal behavior, including human behavior, is the direct result of inherited genetic traits, environmental stimuli, and past experiences. He therefore concludes that the idea of free will is an illusion. He has referred to the biological factors that influence social behavior as a "genetic leash." His theories, especially those that question the free will of humans, have been both influential and controversial.

Many of Wilson's views are based on his research with ants. He asserts that the social habits of ants are closely tied to the fact that they share similar genes, and their limited mating structure; since worker ants are sterile, the colony must rely on the queen to survive. He compares the structured, role-based society of ants to the more flexible human societies and argues that inherited biological traits have given humans more sophisticated reproductive capabilities that have a direct effect on human behavior. Wilson's systematic study of ants and ant behavior, which he completed with Bert Hölldobler, is entitled, *The Ants*, and was published in 1990.

Wilson has advocated for ways in which different fields of study can work more closely together, a concept which he terms "consilience." He presented the idea in his 1998 book, *Consilience: The Unity of Knowledge*. The concept promotes interdisciplinary study and argues that many phenomena that were previously only studied in the fields such as psychology, sociology, and anthropology can benefit from research using scientific methods.

An indefatigable worker, Wilson has also contributed to other areas, notably biogeography. He and theoretical biologist Robert MacArthur devised a formula showing the numbers of species on islands reach an equilibrium that is a function of the size of the island, the rate of extinction and the distance from the mainland (thus governing the rate of immigration). With student Dan Simberloff, Wilson performed famous experiments on islets in the Florida Keys, killing off the denizens, and then recording the rates of restocking.

Margo Wilson

Margo Wilson is a Canadian psychologist who has worked primarily on evolutionary psychology, specifically the evolutionary perspective on interpersonal violence. She is best known for her work, with partner Martin Daly, on the "Cinderella effect," which refers to the phenomenon in which children are significantly more likely to be abused or neglected by their stepparents, than by their biological parents.

Wilson is currently a professor of psychology at McMaster University in Ontario, Canada. She was also coeditor, with Daly, of the journal *Evolution and Human Behavior*. She and Daly have authored several books including *Sex, Evolution, and Behavior* (1978), *Homicide* (1988), and *The truth about Cinderella: A Darwinian view of parental love* (1998). In recognition of her contributions to the scientific community, Wilson was made a fellow of the Royal Society of Canada in 1998.

Wilson's work has included research on many topics in the field of evolutionary psychology. She has collaborated with Daly on the epidemiological analyses of patterns of risk of lethal and nonlethal violence in different categories of relationships, especially, marital and parent-offspring relationships. Wilson has also worked on the multi-disciplinary Ecowise project, which studies ecosystems near McMaster University from various perspectives.

Wilson is best known for her and Daly's work on the Cinderella effect. The concept seeks to explain the disproportionately high rates of child mistreatment and violence by stepparents, compared to those of biological parents. Some studies have shown, for example, that children are 100 times more likely to be beaten by their stepfathers than by their biological fathers. The concept takes its name from the fairy-tale character who is poorly treated by her stepmother. The evolutionary psychology explanation of the phenomenon proposed by Wilson and Daly was that natural selection has led humans adapt their parental behavior to be most advantageous to the survival of their own offspring. While biological parents are genetically inclined to devote the maximum resources possible to ensure the well being of their offspring, stepparents have no evolutionary benefit to gain from their stepchildren and therefore no biological incentive to protect them. Wilson and Daly first published their work on the subject in a 1996 article entitled "Violence Against Stepchildren."

Milford Wolpoff

Milford Wolpoff (b. 1942) is an American paleoanthropologist. He is the leading proponent of the multiregional evolution hypothesis that attempts to explain human evolution through local evolutionary events that took place across the world and rejects views of other theories that propose that human evolution was the result of speciation and the decline of competing species. Wolpoff's work is important to the history of evolutionary theory because it proposes an alternative hypothesis for the evolution of modern humans.

Wolpoff was born in Chicago. He attended the University of Illinois where he received first his A.B. and then, in 1964, his Ph.D.—both in anthropology. His formal education also included training in physics, evolutionary biology, and ecology. Since 1977, Wolpoff has been a professor of anthropology at the University of Michigan, as well as adjunct associate research scientist at the university's Museum of Anthropology. In addition to the multiregional evolutionary hypothesis, his primary areas of research have been the role of culture in early hominid evolution, allometry, robust australopithecine evolution, sexual dimorphism, taxonomy, and the role of genetics in paleoanthropological research.

Wolpoff is the author of many significant books in his field. He published the first and second editions of *Paleoanthropology* in 1980 and 1999, respectively. The work presents a history of known human fossil records, along with an analysis of human evolutionary patterns. Wolpoff also coauthored *Race and Human Evolution* (1997) with Rachel Caspari, which reviews the conflicting theories regarding human evolution and its their implications for views on race.

Wolpoff's primary work has been on the multiregional evolutionary hypothesis, which states that modern humans had origins in Africa, but migrated out of the continent before evolving to their current state through a series of local evolutionary events in various locations throughout the globe. As part of the hypothesis, Wolpoff advocates the idea that human evolution and the emergence of *Homo sapiens*, or modern-day humans, was a process of evolutionary change within the hominid populations of the time. The theory proposes that modern humans did not split into a separate species and replace other hominid species, but rather that humans evolved from earlier hominid species directly. Wolpoff's theory opposes the "Out of Africa" model of Chris Stringer, which claims that modern humans originated in Africa where they evolved to their current state before spreading throughout the globe.

Sewall Wright

Sewall Wright (1889–1988) was born in Melrose, Massachusetts. An American geneticist, his work informed and expanded evolutionary theories. He was among those who established population genetics as a science, and he became a key figure in the NeoDarwinian synthesis, which brought together genetics and Darwinian selection. This is considered to be one of the most important developments in evolutionary biology after Darwin's.

Wright is perhaps most famous for his papers on inbreeding, mating systems, and genetic drift, which established him as a principle founder of theoretical population genetics. The work he helped develop became the origin of modern evolutionary synthesis. Evolutionary geneticists attempt to account for evolution in terms of changes in genes and genotype frequencies within populations. Wright used guinea pigs to study the effects of genes on coat and eye color. Working in parallel with scientists J.B.S. Haldane and R.A. Fisher, he developed a mathematical basis for modern evolutionary theory, based on statistical techniques. Through the use of mathematical models, Wright developed methods to assess the degrees of inbreeding and its effects (his own parents were, in fact, first cousins). His work was initially intended for practical application in livestock breeding, but soon became much more significant.

Wright's "shifting balance theory of evolution," inspired by Herbert Spencer's ideas about dynamic equilibrium, and given empirical content by studies of the pedigrees of Shorthorn cattle (while he was working for the U.S. Department of Agriculture), supposed that populations of organisms break into small groups, that the vagaries of breeding ("genetic drift") bring about changes irrespective of selection, and then, when groups recombine, selection can work on the most useful. Wright backed this with the powerful metaphor of an adaptive landscape, showing organisms (thanks to drift) able to cross over valleys and climb ever-higher peaks on the other side.

Wright pioneered the use of the inbreeding coefficient and F-statistics, which became standard tools in the field of population genetics, and made major contributions to the fields of mammalian genetics and biochemical genetics. His graphical models are still widely used in social science. He was ahead of his time, noting early in the 20th century that genes act by controlling enzymes. Many of Wright's students, at both the University of Chicago and the University of Wisconsin-Madison, became significant contributors to the development of mammalian genetics, extending Wright's influence further. Wright was also a noted reviewer of manuscripts, which allowed him to increase his reputation through close association with influential works of other scientists. Wright also collaborated extensively with the more empirically minded Theodosius Dobzhansky, and jointly they produced several very important papers in the early years of the synthetic theory.

Wright always had philosophical interests, being attracted to a "pan-psychic monism," the belief that paralleling material objects at all levels are minds. He believed in a hierarchy of being, with ultimately all things coming together in one super mind. This tied in with the progressivism underlying his root beliefs in dynamic equilibrium.

Part Five:
APPENDICES

Agassiz, E. C., ed. 1885. *Louis Agassiz: His Life and Correspondence*. Boston: Houghton Mifflin.

Agassiz, L. 1859. *Essay on Classification*. London: Longman, Brown, Green, Longmans, and Roberts and Trubner.

Allan, J. M. 1869. On the real differences in the minds of men and women. *Journal of Anthropology* 7.

Allen, E., and others. 1977. Sociobiology: a new biological determinism. In *Biology as a Social Weapon*. Edited by Sociobiology Study Group of Boston. Minneapolis, Minn.: Burgess.

Allen, G. E. 1978. *Life Science in the Twentieth Century*. Cambridge: Cambridge University Press.

Alvarez, L. W., W. Alvarez, F. Asaro, and H. V. Michel. 1980. Extraterrestrial cause for the Cretaceous-Tertiary extinction. *Science* 208: 1095–1098.

Appel, T. A. 1987. *The Cuvier-Geoffroy Debate: French Biology in the Decades Before Darwin*. New York: Oxford University Press.

Bannister, R. 1979. *Social Darwinism: Science and Myth in Anglo-American Social Thought*. Philadelphia: Temple University Press.

Barrett, P. H., P. J. Gautrey, S. Herbert, D. Kohn, and S. Smith, eds. 1987. *Charles Darwin's Notebooks, 1836–1844*. Ithaca, N.Y.: Cornell University Press.

Barth, K. [1949] 1959. *Dogmatics in Outline*. New York: Harper and Row.

Bateson, B. 1928. *William Bateson, F.R.S., Naturalist: His Essays and Addresses together with a Short Account of His Life*. Cambridge: Cambridge University Press.

Beatty, J. 1987. Dobzhansky and Drift: Facts, Values and Chance in Evolutionary Biology. In *The Probabilistic Revolution*. Edited by L Kruger. Cambridge: MIT Press.

Beecher, H. W. 1885. *Evolution and Religion*. New York: Fords, Howard, and Hulbert.

Behe, M. 1996. *Darwin's Black Box: The Biochemical Challenge to Evolution*. New York: Free Press.

Bergson, H. 1907. *L'évolution créatrice*. Paris: Alcan.

Bowler, P. J. 1976. *Fossils and Progress*. New York: Science History Publications.

———. 1996. *Life's Splendid Drama*. Chicago: University of Chicago Press.

Brandon, R. M., and M. D. Rauscher. 1996. Testing adaptationism: a comment on Orzack and Sober. *American Naturalist* 148: 189–201.

Brewster, D. 1844. Vestiges. *North British Review* 3, no. 470–515.

Brown, P., T. Sutikna, M. J. Morewood, R. P. Soejono, E. Jatmiko, E. Wayhu Saptomo, and Rokus Awe Due. 2004. A new small-bodied hominin from the Late Pleistocene of Flores, Indonesia. *Nature* 431, no. 1055–1061.

Browne, J. 1995. *Charles Darwin: Voyaging. Volume 1 of a Biography*. New York: Knopf.

———. 2002. *Charles Darwin: The Power of Place. Volume II of a Biography*. New York: Knopf.

Brunet, M., F. Guy, D. Pilbeam, and others. 2002. A new hominid from the Upper Miocene of Chad Central Africa. *Nature* 418: 145–51.

Buber, M. 1937. *I and Thou*. Edinburgh: T. and T. Clark.

Buffon, G. L. L. 1749–1767. *Histoire naturelle, générale et particulaire*. 15 vols. Paris.

Bultmann, R. 1958. *Jesus Christ and Mythology*. New York: Charles Scribner's Sons.

Burchfield, J. D. 1975. *Lord Kelvin and the Age of the Earth*. New York: Science History Publications.

Burdach, C. F. 1832. *Die Physiologie als Erfahrungswissenschaft*. Leipzig: Leopold Voss.

Burkhardt, R. W. 1977. *The Spirit of System: Lamarck and Evolutionary Biology*. Cambridge: Harvard University Press.

Cain, J. A. 1993. Common problems and cooperative solutions: organizational activity in evolutionary studies 1936–1947. *Isis* 84: 1–25.

Cairns-Smith, A. G. 1982. *Genetic Takeover and the Mineral Origins of Life*. Cambridge: Cambridge University Press.

———. 1986. *Clay Minerals and the Origin of Life*. Cambridge: Cambridge University Press.

Calvin, J. 1847–1850. *Commentaries on the First Book of Moses Called Genesis*. J. King, translator. Edinburgh: Calvin Translation Society.

Cann, R. L., M. Stoneking, and A. C. Wilson. 1987. Mitochondrial DNA and human evolution. *Nature* 325: 31–36.

Cannon, W. B. 1931. *The Wisdom of the Body*. Cambridge: Harvard University Press.

Chalmers, D. J. 1996. *The Conscious Mind*. New York: Oxford University Press.

———. 1997. Facing up to the problem of consciousness. *Explaining Consciousness—The 'Hard Problem'*. Edited by J. Shear, 9–32. Cambridge: MIT Press.

Chambers, R. 1844. *Vestiges of the Natural History of Creation*. London: Churchill.

———. 1846. *Vestiges of the Natural History of Creation*. 5th ed. London: J. Churchill.

Coleman, W. 1964. *Georges Cuvier Zoologist. A Study in the History of Evolution Theory*. Cambridge: Harvard University Press.

Cosmides, L. 1989. The logic of social exchange: has natural selection shaped how humans reason? Studies with the Wason selection task. *Cognition* 31, no. 187–276.

Crook, P. 1994. *Darwinism: War and History*. Cambridge: University of Cambridge Press.

Cuvier, G. 1810. *Rapport Historique sur les progr_s des sciences naturelles*. Paris.

———. 1813. *Essay on the Theory of the Earth*. Robert Kerr, translator. Edinburgh: W. Blackwood.

———. 1817. *Le règne animal distribué d'aprés son organisation, pour servir de base à l'histoire naturelle des animaux et d'introduction à l'anatomie comparée*. Paris.

Daly, M., and M. Wilson. 1988. *Homicide*. New York: De Gruyter.

Darlington, C. D. 1932. Recent Advances in Cytology. Foreword by J. B. S. Haldane. London: Churchill.

Darwin, C. 1859. *On the Origin of Species by Means of Natural Selection, or the Preservation of Favoured Races in the Struggle for Life*. London: John Murray.

———. 1871. *The Descent of Man, and Selection in Relation to Sex*. London: John Murray.

Darwin, E. 1794–1796. *Zoonomia; or, The Laws of Organic Life*. London: J. Johnson.

———. 1801. Zoonomia; or, The Laws of Organic Life. 3rd ed. London: J. Johnson.

————. 1803. *The Temple of Nature*. London: J. Johnson.

Darwin, F. 1887. *The Life and Letters of Charles Darwin, Including an Autobiographical Chapter*. London: Murray.

Darwin, F., and A. C. Seward, eds. 1903. *More Letters of Charles Darwin*. 2 vols. London: John Murray.

Davies, N. B. 1992. *Dunnock Behaviour and Social Evolution*. Oxford: Oxford University Press.

Dawkins, R. 1976. *The Selfish Gene*. Oxford: Oxford University Press.

————. 1983. Universal Darwinism. *Molecules to Men*. D. S. Bendall, ed. Cambridge: University of Cambridge Press.

————. 1986. *The Blind Watchmaker*. New York: Norton.

————. 1995a. Richard Dawkins: A survival machine. *The Third Culture*. J. Brockman, ed. New York: Simon and Schuster.

————. 1995b. *A River Out of Eden*. New York: Basic Books.

————. 1997a. Human chauvinism: Review of Full House by Stephen Jay Gould. *Evolution* 51, no. 3: 1015–20.

————. 1997b. Obscurantism to the rescue. *Quarterly Review of Biology* 72: 397–99.

————. 1997c. Religion is a virus. *Mother Jones* November/December: 60–61.

————. 2006. *The God Delusion*. New York: Houghton Mifflin.

Dembski, W. A. 1998. *The Design Inference: Eliminating Chance through Small Probabilities*. Cambridge: Cambridge University Press.

Dennett, D. 2007. Letter to the editor. *London Review of Books* 29 (22) 15 November.

Dennett, D. C. 2006. *Breaking the Spell: Religion as a Natural Phenomenon*. New York: Viking.

Desmond, A. 1994. *Huxley, the Devil's Disciple*. London: Michael Joseph.

————. 1997. *Huxley, Evolution's High Priest*. London: Michael Joseph.

Desmond, A., and J. Moore. 1992. *Darwin: The Life of a Tormented Evolutionist*. New York: Warner.

Dobzhansky, T. 1937. *Genetics and the Origin of Species*. New York: Columbia University Press.

Dobzhansky, T., and H. Levene. 1955. Developmental Homeostasis in Natural Populations of Drosophila pseudoobscura XXIV. *Genetics* 40: 797–808.

Dobzhansky, T., and B. Wallace. 1953. The genetics of homeostasis in Drosophila. *Proceedings of the National Academy of Sciences* 39: 162–71.

Douady, S., and Y. Couder. 1992. Phyllotaxis as a physical self-organised growth process. *Physical Review Letters* 68: 2098–101.

Duarte, C., J. Mauricio, P. B. Petitt, P. Souto, E. Trinkaus, H. van der Plicht, and J. Zilhao. 1999. The early Upper Paleolithic human skeleton from the Abrigo do Lagar Velho (Portugal) and modern human emergence in Iberia. *Proceedings of the National Academy of Sciences (USA)* 96: 7604–9.

Duncan, D., ed. 1908. *Life and Letters of Herbert Spencer*. London: Williams and Norgate.

Dupree, A. H. 1959. *Asa Gray 1810–1888*. Cambridge: Harvard University Press.

Edwards, A. W. F. 2003. Human genetic diversity: Lewontin's fallacy. *BioEssays* 25: 798–801.

Eldredge, N., and S. J. Gould. 1972. Punctuated equilibria: an alternative to phyletic gradualism. *Models in Paleobiology*. T. J. M. Schopf, ed. 82–115. San Francisco: Freeman, Cooper.

Ellegård, A. 1958. *Darwin and the General Reader*. Goteborg: Goteborgs Universitets Arsskrift.

Elliot Smith, G. 1924. *Evolution of Man*. Oxford: Oxford University Press.

Erskine, F. 1995. "The Origin of Species" and the science of female inferiority. *Charles Darwin's "The Origin of Species": New Interdisciplinary Essays*, 95–121. Manchester: Manchester University Press.

Falk, D. 2004. *Braindance: New Discoveries about Human Origins and Brain Evolution*. Gainsville, Fla.: University of Florida Press.

Falk, D., C. Hildebolt, K. Smith, M. J. Morwood, T. Sutikna, P. Brown, [first initial of name?] Jatmiko, E. Wayhu Saptomo, B. Brunsden, and F. Prior. 2005. The Brain of LB1, *Homo floresiensis*. *Science* 308: 242.

Farley, J. 1977. *The Spontaneous Generation Controversy from Descartes to Oparin*. Baltimore: Johns Hopkins University Press.

Feduccia, A. 1996. *The Origin and Evolution of Birds*. New York: Yale University Press.

Fodor, J. 2007. Why pigs don't have wings. The case against natural selection. *London Review of Books* 29 (20) 18 October.

———. 2008. Letter to editor. *London Review of Books* 30 (1) January 3.

Fox, S. W. 1988. *The Emergence of Life: Darwinian Evolution from the Inside*. New York: Basic Books.

Friedlander, S. 1997. *Nazi Germany and the Jews: The Years of Persecution 1933–39*. London: Weidenfeld and Nicolson.

Gasman, D. 1971. *The Scientific Origins of National Socialism: Social Darwinism in Ernst Haeckel and the Monist League*. New York: Elsevier.

Gee, H. 1996. Box of Bones 'Clinches' Identity of Piltdown Palaeontology Hoaxer. *Nature* 382: 261–62.

Geoffroy Saint-Hilaire, E. 1818. *Philosophie anatomique*. Paris: Mequignon-Marvis.

Gilbert, S. F., J. M. Opitz, and R. A. Raff. 1996. Resynthesizing evolutionary and developmental biology. *Developmental Biology* 173: 357–72.

Gilkey, L. B. 1985. *Creationism on Trial: Evolution and God at Little Rock*. Minneapolis: Winston Press.

Gish, D. 1973. *Evolution: The Fossils Say No!* San Diego: Creation-Life.

Goodwin, B. 2001. *How the Leopard Changed its Spots,* 2nd ed. Princeton: Princeton University Press.

Gould, S. J. 1966. Allometry and size in ontogeny and phylogeny. *Biological Reviews of the Cambridge Philosophical Society* 41: 587–640.

———. 1971. D'Arcy Thompson and the science of form. *New Literary History* 2: 229–58.

———. 1977a. *Ever Since Darwin*. New York: Norton.

———. 1977b. *Ontogeny and Phylogeny*. Cambridge: Belknap Press.

———. 1980. *The Panda's Thumb*. New York: Norton.

———. 1980a. Is a new and general theory of evolution emerging? *Paleobiology* 6: 119–30.

———. 1980b. The Piltdown conspiracy. *Natural History* 89, no. August: 8–28.

————. 1981. *The Mismeasure of Man*. New York: Norton.

————. 1982. Darwinism and the expansion of evolutionary theory. *Science* 216: 380–7.

————. 1983. Irrelevance, submission, partnership: the changing role of paleontology in Darwin's three centennials and a modest proposal for macroevolution. *Evolution from Molecules to Men*. D. S. Bendall, ed. 347–66. Cambridge: Cambridge University Press.

————. 1984. Morphological channeling by structural constraint: Convergence in styles of dwarfing and gigantism in Cerion, with a description of two new fossil species and a report on the discovery of the largest Cerion. *Paleobiology* 10: 172–94.

————. 1985. *The Flamingo's Smile: Reflections in Natural History*. New York: Norton.

————. 1989. *Wonderful Life: The Burgess Shale and the Nature of History*. New York: W. W. Norton Co.

————. 1991. *Bully for Brontosaurus: Reflections in Natural History*. New York: Norton.

————. 1996. *Full House: The Spread of Excellence from Plato to Darwin*. New York: Paragon.

————. 1997. Darwinian Fundamentalism. *The New York Review*, no. 12 June: 34–37.

————. 1999. *Rocks of Ages: Science and Religion in the Fullness of Life*. New York: Ballantine.

————. 2002. *The Structure of Evolutionary Theory*. Cambridge: Harvard University Press.

Gould, S. J., and C. B. Calloway. 1980. Clams and brachiopods–ships that pass in the night. *Paleobiology* 6: 383–96.

Gould, S. J., and N. Eldredge. 1977. Punctuated equilibria: the tempo and mode of evolution reconsidered. *Paleobiology* 3: 115–51.

Gould, S. J., and R. C. Lewontin. 1979. The spandrels of San Marco and the Panglossian paradigm: a critique of the adaptationist programme. *Proceedings of the Royal Society of London, Series B: Biological Sciences* 205: 581–98.

Gould, S. J., and E. S. Vrba. 1982. Exaptation—a missing term in the science of form. *Paleobiology* 8: 4–15.

Graham, L. 1987. *Science, Philosophy, and Human Behavior in the Soviet Union*. New York: Columbia University Press.

Grant, M. 1916. *The Passing of the Great Race, or The Racial Basis of European History*. New York: Charles Scribner's Sons.

Grant, P. R. 1986. *Ecology and Evolution of Darwin's Finches*. Princeton: Princeton University Press.

Grant, P. R., and Grant B. R. 1995. Predicting microevolutionary responses to directional selection on heritable variation. *Evolution* 49: 241–51.

Grant, B. R., and P. R. Grant. 1989. *Evolutionary Dynamics of a Natural Population: The Large Cactus Finch of the Galapagos*. Chicago: University of Chicago Press.

Gray, A. 1876. *Darwiniana*. New York: D. Appleton. Notes: Reprinted ed. A.H. Dupree, 1963, Cambridge: Harvard University Press

————. 1881. *Structural Botany*. 6th ed. London: Macmillan.

Gray, J. L. 1894. *Letters of Asa Gray*. Boston: Houghton Mifflin.

Green, R. E., and et al. 206. Analysis of one million base pairs of Neanderthal DNA. *Nature* 444: 330–336.

Greene, J. C. 1959. *The Death of Adam: Evolution and its Impact on Western Thought*. Ames, Iowa: Iowa State University Press.

Haeckel E. 1866. *Generelle Morphologie der Organismen*. Berlin: Georg Reimer.

———. 1898. *The Last Link: Our Present Knowledge of the Descent of Man*. London: Black.

Haldane, J. B. S. 1929. The origin of life. *Rationalist Annual*: 1–10.

Hamilton, W. D. 1964a. The genetical evolution of social behaviour I. *Journal of Theoretical Biology* 7: 1–16.

———. 1964b. The genetical evolution of social behaviour II. *Journal of Theoretical Biology* 7: 17–32.

———. 2001. *Narrow Roads of Gene Land: The Collected Papers of W. D. Hamilton. Volume 2. Evolution of Sex*. Oxford: Oxford University Press.

Harris, S. 2004. *The End of Faith: Religion, Terror, and the Future of Reason*. New York: Free Press.

Haught, J. F. 1995. *Science and Religion: From Conflict to Conversation*. New York: Paulist Press.

Hegel, G. W. F. [1817] 1970. *Philosophy of Nature*. Oxford: Oxford University Press.

Hennig, W. 1966. *Phylogenetic Systematics*. Urbana, Ill.: University of Illinois Press.

Hitchens, C. 2007. *God is not Great: How Religion Poisons Everything*. New York: Hachette.

Hitler, A. 1925. *Mein Kampf*. London: Secker and Warburg.

Hodge, C. 1872. *Systematic Theology*. London and Edinburgh: Nelson.

Hrdy, S. B. 1981. *The Woman that Never Evolved*. Cambridge: Harvard University Press.

———. 1999. *Mother Nature: A History of Mothers, Infants, and Natural Selection*. New York: Pantheon Books.

Hull, D. L. 1988. *Science as a Process*. Chicago: University of Chicago Press.

Hume, D. [1779] 1947. *Dialogues Concerning Natural Religion*. N. K. Smith, ed. Indianapolis, Ind.: Bobbs-Merrill Co.

———. 1978. *A Treatise of Human Nature*. Oxford: Oxford University Press.

Hunter, G. 1914. *A Civic Biology: Presented in Problems*. New York: American Book Company.

Huxley, J. S. 1912. *The Individual in the Animal Kingdom*. Cambridge: Cambridge University Press.

———. 1927. *Religion Without Revelation*. London: Ernest Benn.

———. 1931. *What Dare I Think? The Challenge of Modern Science to Human Action and Belief*. London: Chatto.

———. 1934. *If I Were Dictator*. New York and London: Harper and Brothers.

———. 1942. *Evolution: The Modern Synthesis*. London: Allen and Unwin.

ditto 1943. *TVA: Adventure in Planning*. London: Scientific Book Club.

———. 1948. *UNESCO: Its Purpose and Its Philosophy*. Washington, D.C.: Public Affairs Press.

———. S. 1959. Introduction to Teilhard de Chardin's The Phenomenon of Man. 11–28. London: Collins.

Huxley, L. 1900. *The Life and Letters of Thomas Henry Huxley*. London: Macmillan.

Huxley, T. H. [1868]1898. On the animals which are most nearly intermediate between birds and reptiles, *Geological Magazine*, V, 357–65. In *The Scientific Memoirs of Thomas Henry Huxley*. M.

Foster, and E. R. Lankester, ed. 303–13. London: Macmillan. Notes: Scientific Memoirs, 3: 303–13

———. 1863. *Evidence as to Man's Place in Nature*. London: Williams and Norgate.

———. 1868. On some organisms living at great depths in the North Atlantic Ocean. *Quarterly Journal of Microscopial Science* 8: 203–12.

———. 1871. Administrative nihilism. Reprinted in *Methods and Results*, 251–89. London: Macmillan.

Huxley, T. H., and Huxley J. S. 1947. *Evolution and Ethics 1893–1943*. London: Pilot.

Hyatt, A. 1857. Alpheus Hyatt's Travel Book. Hyatt Papers, Syracuse University Library, Syracuse, New York.

———. 1889. Genesis of the Arietidae. *Bulletin of the Museum of Comparative Zoology* 16, no. 3: 238.

Jerison, H. 1973. *Evolution of the Brain and Intelligence*. New York: Academic Press.

Johanson, D., and M. Edey. 1981. *Lucy: The Beginnings of Humankind*. New York: Simon and Schuster.

John Paul II. 1997. The Pope's message on evolution. *Quarterly Review of Biology* 72: 377–83.

Johnson, P. E. 1993. *Darwin on Trial* 2nd ed. Washington, D.C.: Regnery Gateway.

———. 1995. *Reason in the Balance: The Case Against Naturalism in Science, Law and Education*. Downers Grove, Ill: InterVarsity Press.

Jones, G. 1980. *Social Darwinism and English Thought*. Brighton: Harvester.

Joravsky, D. 1970. *The Lysenko Affair*. Cambridge: Harvard University Press.

Kauffman, S. A. 1995. *At Home in the Universe: The Search for the Laws of Self-Organization and Complexity*. New York: Oxford University Press.

Kelly, A. 1981. *The Descent of Darwin: The Popularization of Darwinism in Germany, 1860–1914*. Chapel Hill: University of North Carolina Press.

Kimura, M. 1983. *Neutral Theory of Molecular Evolution*. Cambridge: Cambridge University Press.

King-Hele, D. 1963. *Erasmus Darwin: Grandfather of Charles Darwin*. New York: Scribners.

Krings, M., A. Stone, R. W. Schmitz, H. Krainitzki, M. Stoneking, and S. Pääbo. 1997. Neanderthal DNA sequences and the origin of modern humans. *Cell* 90: 19–30.

Kropotkin, P. 1902. *Mutual Aid*. Boston: Extending Horizons Books.

Kuhn, T. 1962. *The Structure of Scientific Revolutions*. Chicago: University of Chicago Press.

Lamarck, J. B. 1809. *Philosophie zoologique*. Paris: Dentu.

Landau, M. 1991. *Narratives of Human Evolution*. New Haven: Yale University Press.

Larson, E. J. 1997. *Summer for the Gods: The Scopes Trial and America's Continuing Debate over Science and Religion*. New York: Basic Books.

LeVay, S. 1993. *The Sexual Brain*. Cambridge: M.I.T. Press.

Lewin, R. 1989. *Human Evolution: An Illustrated Introduction*. 2nd ed. Oxford: Blackwell Scientific.

Lewontin, R. C. 1972. The apportionment of human diversity. *Evolutionary Biology* 6: 381–98. Notes: Lewontin, R. C. 1972. "The Apportionment of Human Diversity." *Evolutionary Biology* 6: 381–398.

———. 1982. *Human Diversity*. New York: Scientific American Library.

Lewontin, R. C., and J. L. Hubby. 1966. A molecular approach to the study of genic heterozygosity in natural populations. II. Amount of variation and degree of heterozygosity in natural populations of 'Drosophila pseudoobscura'. *Genetics* 54: 595–609.

Lewontin, R. C., J. A. Moore, W. B. Provine, and B. Wallace, eds. 1981. *Dobzhansky's Genetics of Natural Populations I-XLIII.* New York: Columbia University Press.

Linnaeus, C. 1758. *Systema naturae per regna tria naturae, secundum classes, ordines, genera, species, cum characteribus, differentiis, synonymis, locis.* London: British Museum (Natural History).

Lorenz, K. 1966. *On Aggression.* London: Methuen.

Lovejoy, A. O. 1936. *The Great Chain of Being.* Cambridge: Harvard University Press.

Lurie, E. 1960. *Louis Agassiz: A Life in Science.* Chicago: Chicago University Press.

Lyell, C. 1830–1833. *Principles of Geology: Being an Attempt to Explain the Former Changes in the Earth's Surface by Reference to Causes now in Operation.* London: John Murray.

———. 1863. *The Antiquity of Man.* London: Murray.

Malthus, T. R. [1826] 1914. *An Essay on the Principle of Population* 6th ed. London: Everyman.

Marchant, J, ed. 1916. *Alfred Russel Wallace: Letters and Reminiscences.* London: Cassell and Company, Ltd.

Margulis, L. 1970. *Origin of Eukaryotic Cells: Evidence and Research Implications for a Theory of the Origin and Evolution of Microbial, Plant, and Animal Cells on the Precambrian Earth.* New Haven: Yale University Press.

Maynard Smith, J. 1978. Optimization theory in evolution. *Annual Review of Ecology and Systematics* 9: 31–56.

———. 1981. Did Darwin get it right? *London Review of Books* 3, no. 11: 10–11.

———. 1982. *Evolution and the Theory of Games.* Cambridge: Cambridge University Press.

———. 1995. Genes, memes, and minds. *New York Review of Books* 42, no. 19: 46–48.

Mayr, E. 1942. *Systematics and the Origin of Species.* New York: Columbia University Press.

———. 1954. Change of genetic environment and evolution. *Evolution as a Process,* 157–80. London: Allen and Unwin.

———. 1963. *Animal Species and Evolution.* Cambridg: Harvard University Press.

McCosh, J. 1882. *The Method of Divine Government, Physical and Moral.* London: Macmillan.

McDonald, J. H., and M. Kreitman. 1991. Adaptive protein evolution at the Adh locus in *Drosophila. Nature* 351: 652–54.

McMullin, E., ed. 1985. *Evolution and Creation.* Notre Dame: University of Notre Dame Press.

McNeil, M. 1987. *Under the Banner of Science: Erasmus Darwin and His Age.* Manchester: Manchester University Press.

Medawar, P. [1961] 1967. Review of The Phenomenon of Man. In *The Art of the Soluble.* P Medawar, ed. London: Methuen and Co. Ltd. Originally in Mind 1961, 70: 99–106

Miller, K. 1999. *Finding Darwin's God.* New York: Harper and Row.

Mitchison, G J. 1977. Phyllotaxis and the Fibonacci series. *Science* 196: 270–275.

Mithen, S. 1996. *The Prehistory of the Mind.* London: Thames and Hudson.

Mitman, G. 1992. *The State of Nature: Ecology, Community, and American Social Thought, 1900–1950.* Chicago: University of Chicago Press.

Mivart, St. G. 1871. *Genesis of Species*. London: Macmillan.

Moore, A. 1890. The Christian doctrine of God. *Lux Mundi*. C. Gore, ed. London: John Murray.

Moore, J. 1979. *The Post-Darwinian Controversies: A Study of the Protestant Struggle to come to terms with Darwin in Great Britain and America, 1870–1900*. Cambridge: Cambridge University Press.

Morwood, M., and P. Van Oosterzee. 2007. *A New Human: The Startling Discovery and Strange Story of the "Hobbits" of Flores, Indonesia*. London: Collins.

Morwood, M. J., P. Brown, S. Y. Jatmiko, T. Sutikna, E. Wahyu Saptomo, K. E. Westaway, Rokus Awe Due, R. G. Roberts, T. Maeda, S. Wasisto, and T. Djubiantono. 2005. Further evidence for small-bodied hominins from the Late Pleistocene of Flores, Indonesia. *Nature* 437: 1012–17.

Morwood, M. J., Soejono, R. P. Roberts, T. R. G. Sutikna, C. S. M. Turney, K. E. Westaway, W. J. Rink, J. X. Zhao, G. D. van den Bergh, Rokus Awe Due, D. R. Hobbs, M. W. Moore, M. I. Bird, and L. K. Fifield. 2004. Archaeology and age of a new hominin from Flores in eastern Indonesia. *Nature* 431: 1087–91.

Muller, H. J. 1949. The Darwinian and Modern Conceptions of Natural Selection. *Proceedings of the American Philosophical Society* 93: 459–70.

Newman, J. H. 1971. *The Letters and Diaries of John Henry Newman, XXI*. C. S. Dessain and T. Gornall, eds. Edinburgh: Thomas Nelson.

———. 1973. *The Letters and Diaries of John Henry Newman, XXV*. C. S. Dessain and T. Gornall, ed. Oxford: Clarendon Press.

Niklas, K. J. 1988. The role of phyllotactic pattern as a 'developmental constraint' on the interception of light by leaf surfaces. *Evolution* 42: 1–16.

Noll, M. 2002. *America's God: From Jonathan Edwards to Abraham Lincoln*. New York: Oxford University Press.

Noonan, J. P., G. Coop, S. Kudaravalli, D. Smith, J. Krause, J. Alessi, F. Chen, D. Platt, S. Pääbo, J. Pritchard, and E. Rubin. 2006. Sequencing and Analysis of Neanderthal Genomic DNA. *Science* 314: 1113–18.

Numbers, R. L. 2006. *The Creationists: From Scientific Creationism to Intelligent Design*. Standard ed. Cambridge: Harvard University Press.

O'Brien, C. F. 1970. Eozoon canadense: The dawn animal of Canada. *ISIS* 61: 206–23.

Oakley, K. P. 1964. The Problem of Man's Antiquity. *Bulletin of the British Museum (Natural History)*, Geological Series 9 (5).

Oparin, A. [1924] 1967. The origin of life (Originally published as Proishkhozhdenie zhizni [1928]). *The Origin of Life*. A. Synge, translator. 199–234. Cleveland: World.

Orgel, L. E. The origin of life—a review of facts and speculations. *Trends in Biochemical Sciences*. 1995; 23:491–500.

Orzack, S. H., and E. Sober. 1994. Optimality models and the test of adaptationism. *American Naturalist* 143: 361–80.

———. , eds. 2001. *Adaptationism and Optimality*. Cambridge: Cambridge University Press.

Osborn, H. F. 1931. *Cope: Master Naturalist: The Life and Writings of Edward Drinker Cope*. Princeton: Princeton University Press.

Oster, G., and E. O. Wilson. 1978. *Caste and Ecology in the Social Insects*. Princeton: Princeton University Press.

Outram, D. 1984. *Georges Cuvier: Vocation, Science and Authority in Post Revolutionary France*. Manchester: Manchester University Press.

Paley, W. [1802]1819. *Natural Theology* (Collected Works: IV). London: Rivington.

Parker, G. 1978. Searching for mates. *Behavioural Ecology: An Evolutionary Approach*. Sunderland, Mass.: Sinauer.

Parker, G. A. 1970. The reproductive behaviour and the nature of sexual selection in Scatophaga stercovaria L. (Diptera: Scatophagidae)—VIII. The origin and evolution of the passive phase. *Evolution* 24: 744–88.

Pasteur, L. 1883. La dissymétrie moléculaire. *Oeuvres de Pasteur*, Vol. 1. Pasteur Vallery-Radot, ed. Paris: Masson.

Pilbeam, D. 1984. The descent of Hominoids and Hominids. *Scientific American* 250, no. 3: 84–97.

Pinker, S. 1994. *The Language Instinct: How the Mind Creates Language*. New York: William Morrow.

———. 1997. *How the Mind Works*. New York: Norton.

Pittenger, M. 1993. *American Socialists and Evolutionary Thought, 1870–1920*. Madison: University of Wisconsin Press.

Plantinga, A. 1991. An evolutionary argument against naturalism. *Logos* 12: 27–49.

———. 1991. When faith and reason clash: evolution and the Bible. *Christian Scholar's Review* 21, no. 1: 8–32 [Reprinted in D. Hull and M. Ruse, eds. *The Philosophy of Biology*, Oxford: Oxford University Press, 1998; 674–697].

Popper, K., and J. Eccles. 1977. *The Self and Its Brain*. Berlin: Springer International.

Popper, K. R. 1972. *Objective Knowledge*. Oxford: Oxford University Press.

Quine, W. V. O. 1969. *Ontological Relativity and Other Essays*. New York: Columbia University Press.

Raff, R. 1996. *The Shape of Life: Genes, Development, and the Evolution of Animal Form*. Chicago: University of Chicago Press.

Rainger, R. 1991. *An Agenda for Antiquity: Henry Fairfield Osborn and Vertebrate Paleontology at the American Museum of Natural History, 1890–1935*. Tuscaloosa: University of Alabama Press.

Rawls, J. 1971. *A Theory of Justice*. Cambridge: Harvard University Press.

Ray, J. Wisdom of God, Manifested in the Words of Creation. [pub info?]

Redi, F. 1688. *Experiments in the Generation of Insects*. M. Bigelow, translator. Chicago: Open Court Publishing.

Reeve, H. K., and P. W. Sherman. 1993. Adaptation and the goals of evolutionary research. *Quarterly Review of Biology* 68: 1–32.

Reichenbach, B. R. 1976. Natural evils and natural laws: a theodicy for natural evil. *International Philosophical Quarterly* 16: 179–96.

Richards, R. J. 1987. *Darwin and the Emergence of Evolutionary Theories of Mind and Behavior*. Chicago: University of Chicago Press.

———. 1992. *The Meaning of Evolution: The Morphological Construction and Ideological Reconstruction of Darwin's Theory*. Chicago: University of Chicago Press.

————. 2008. *The Tragic Sense of Life: Ernst Haeckel and the Struggle over Evolutionary Thought*. Chicago: University of Chicago Press.

Roger, J. 1997. *Buffon: A Life in Natural History*. S. L. Bonnefoi, translator. Ithaca, N.Y.: Cornell University Press.

Rosenberg, N. A., J. K. Pritchard, J. L. Weber, H. M. Cann, K. K. Kidd, L. A. Zhivotovsky, and M. Feldman. 2002. Genetic structure of human populations. *Science* 298: 2381–85.

Roth, G., J. Blanke, and D. B. Wake. 1994. Cell size predicts morphological complexity in the brains of frogs and salamanders. *Proceedings of the National Academy of the Sciences, USA* 91: 4796–800.

Rupke, N. A. 1994. *Richard Owen: Victorian Naturalist*. New Haven: Yale University Press.

Ruse, M. 1979 [1999]. *The Darwinian Revolution: Science Red in Tooth and Claw*. 2nd ed. Chicago: University of Chicago Press.

————. 1980. Charles Darwin and group selection. *Annals of Science* 37: 615–30.

————. 1982. *Darwinism Defended: A Guide to the Evolution Controversies*. Reading, Mass.: Benjamin/Cummings Pub. Co.

————. 1984. Is there a limit to our knowledge of evolution? *BioScience* 34, no. 2: 100–104.

————. 1996. *Monad to Man: The Concept of Progress in Evolutionary Biology*. Cambridge: Harvard University Press.

————. 1999. *Mystery of Mysteries: Is Evolution a Social Construction?* Cambridge: Harvard University Press.

————. 2001. *Can a Darwinian be a Christian? The Relationship between Science and Religion*. Cambridge: Cambridge University Press.

Ruse, M., and E. O. Wilson. 1985. The evolution of morality. *New Scientist* 1478: 108–28.

Russell, E. S. 1916. *Form and Function: A Contribution to the History of Animal Morphology*. London: John Murray.

Russett, C. E. 1976. *Darwin in America: The Intellectual Response. 1865–1912*. San Francisco: Freeman.

Sahlins, M. 1976. *The Use and Abuse of Biology*. Ann Arbor: University of Michigan.

Schiff, M., and R. C. Lewontin. 1986. *Education and Class: The Irrelevance of I Q Studies*. Oxford: Oxford University Press.

Sedgwick, A. 1845. Vestiges. *Edinburgh Review* 82: 1–85.

————. 1850. *Discourse on the Studies at the University of Cambridge*. 5th ed. Cambridge: Cambridge University Press.

Sepkoski Jr., J. J. 1976. Species diversity in the Phanerozoic—species-area effects. *Paleobiology* 2, no. 4: 298–303.

————. 1979. A kinetic model of Phanerozoic taxonomic diversity. II Early Paleozoic families and multiple equilibria. *Paleobiology* 5: 222–52.

————. 1984. A kinetic model of Phanerozoic taxonomic diversity. III Post-Paleozoic families and mass extinctions. *Paleobiology* 10: 246–67.

Settle, M. L. 1972. *The Scopes Trial: The State of Tennessee v. John Thomas Scopes*. New York: Franklin Watts.

Shor, E. N. 1974. *The Fossil Feud between E.D. Cope and O.C. Marsh*. Hicksville, N.Y.: Exposition Press.

Simpson, G. G. 1944. *Tempo and Mode in Evolution*. New York: Columbia University Press.

———. 1949. *The Meaning of Evolution*. New Haven: Yale University Press.

———. 1953. *The Major Features of Evolution*. New York: Columbia University Press.

Singer, P. 2005. Ethics and intuitions. *Journal of Ethics* 9: 331–52.

Sober, E., ed. 1984. *Conceptual Issues in Evolutionary Biology*. Cambridge: M.I.T. Press, 2nd ed. 1993.

Spencer, H. 1851. *Social Statics; Or the Conditions Essential to Human Happiness Specified and the First of them Developed*. London: J. Chapman.

———. 1852. The development hypothesis. In *Essays: Scientific, Political and Speculative*. H Spencer, 377–83. London: Williams and Norgate.

———. 1864. *Principles of Biology*. London: Williams and Norgate.

———. 1873. *The Study of Sociology*. Ann Arbor: University of Michigan Press.

———. 1877–1896. *Principles of Sociology*. London: Williams.

———. 1904. *Autobiography*. London: Williams and Norgate.

Stebbins, G. L. 1969. *The Basis of Progressive Evolution*. Chapel Hill: University of North Carolina Press.

Stebbins, G. L., and F. J. Ayala. 1981. Is a new evolutionary synthesis necessary? *Science* 213: 967–71.

Stringer C. 2002. Modern human origins—progress and prospects. *Philosophical Transactions of the Royal Society, London (B)* 357: 563–79.

———. 2003. Human Evolution: Out of Ethiopia. *Nature* 423: 692–95.

Sulloway, F. 1979. *Freud: Biologist of the Mind*. New York: Basic Books.

Sumner, W. G. 1914. *The Challenge of Facts and Other Essays*. New Haven: Yale University Press.

Thompson, D. W. 1917. *On Growth and Form*. Cambridge: Cambridge University Press.

———. 1948. *On Growth and Form*, 2nd ed. Cambridge: Cambridge University Press.

Tooby, J., L. Cosmides, and H. C. Barrett. 2005. Resolving the debate on innate ideas: Learnability constraints and the evolved interpenetration of motivational and conceptual functions. *The Innate Mind: Structure and Content*. P. Carruthers, S. Laurence, and S. Stich, eds. New York: Oxford University Press.

Toulmin, S. 1967. The evolutionary development of science. *American Scientist* 57: 456–71.

Trinkaus, E. 2006. Modern human versus Neandertal evolutionary distinctiveness. *Current Anthropology* 47: 597–620.

Trivers, R. L. 1971. The evolution of reciprocal altruism. *Quarterly Review of Biology* 46: 35–57.

Trivers, R. L., and D. E. Willard. 1973. Natural selection of parental ability to vary the sex ratio of offspring. *Science* 179: 90–92.

von Bernhardi, F. 1912. *Germany and the Next War*. London: Edward Arnold.

Wallace, A. R. 1858. On the tendency of varieties to depart indefinitely from the original type. *Journal of the Proceedings of the Linnean Society, Zoology* 3: 53–62.

————. 1870. *Contributions to the Theory of Natural Selection: A Series of Essays*. London: Macmillan.

————. 1900. *Studies: Scientific and Social*. London: Macmillan.

————. 1905. *My Life: A Record of Events and Opinions*. London: Chapman and Hall.

Watson, J. D., and F. H. C. Crick. 1953. Molecular structure of nucleic acids. *Nature* 171, no. 737.

Weikart, R. 2004. *From Darwin to Hitler: Evolutionary Ethics, Eugenics, and Racism in Germany*. New York: Palgrave Macmillan.

Westfall, R. S. 1980. *Never at Rest: A Biography of Isaac Newton*. Cambridge: University of Cambridge Press.

Whitcomb, J. C., and H. M. Morris. 1961. *The Genesis Flood: The Biblical Record and its Scientific Implications*. Philadelphia: Presbyterian and Reformed Publishing Company.

Wilberforce, S. 1860. Review of 'Origin of Species'. *Quarterly Review*.

Williams, G. C. 1966. *Adaptation and Natural Selection*. Princeton, N.J.: Princeton University Press.

————. 1975. *Sex and Evolution*. Princeton, N.J.: Princeton University Press.

Williamson, P. 1985. Punctuated equilibrium, morphological stasis and the paleontological documentation of speciation. *Biological Journal of the Linnaean Society of London* 26: 307–24.

Wilson, E. O. 1975. *Sociobiology: The New Synthesis*. Cambridge: Harvard University Press.

————. 1978. *On Human Nature*. Cambridge: Cambridge University Press.

————. 1980a. Caste and division of labor in leaf cutter ants (hymenoptera formicidae, Atta) I The overall pattern in Atta sexdens. *Behavioral Ecology and Sociobiology* 7: 143–56.

————. 1980b. Caste and division of labor in leaf cutter ants (hymenoptera formicidae, Atta). II The ergonomic optimization of leaf cutting. *Behavioral Ecology and Sociobiology* 7: 157–65.

————. 1994. *Naturalist*. Washington, D.C.: Island Books/Shearwater Books.

————. 2002. *The Future of Life*. New York: Vintage Books.

Winsor, M. P. 1991. *Reading the Shape of Nature: Comparative Zoology at the Agassiz Museum*. Chicago: University of Chicago Press.

Wolpoff, M., and R. Caspari. 1997. *Race and Human Evolution*. Boulder, Colorado: Westview.

Wright, G. F. 1882. *Studies in Science and Religion*. Andover, Mass.: Draper.

Wright, S. 1932. The roles of mutation, inbreeding, crossbreeding and selection in evolution. *Proceedings of the Sixth International Congress of Genetics* 1: 356–66.

Wynne-Edwards, V. C. 1962. *Animal Dispersion in Relation to Social Behaviour*. Edinburgh: Oliver and Boyd.

Young, R. M. 1985. *Darwin's Metaphor: Nature's Place in Victorian Culture*. Cambridge: Cambridge University Press.

Zihlman, A. 1981. Women as shapers of the human adaptation. *Woman the Gatherer*. F. Dahlberg, ed. 75–120. New Haven: Yale University Press.

Chronology

1912	Discovery of Piltdown Man	1964	W. D. Hamilton proposes theory of kin selection
1914–1918	World War I		
1925	Raymond Dart discovers Taung Baby	1966	Richard Lewontin uses gel electrophoretic techniques
	Scopes Monkey Trial	1972	Niles Eldredge and Stephen Jay Gould announce punctuated equilibria theory
1928	*The origin of life* by A. I. Oparin		
1929	*The origin of life* by J. B. S. Haldane	1974	Don Johanson discovers Lucy
	Wall Street crash	1975	*Sociobiology: The New Synthesis* by Edward O. Wilson
1930	*The Genetical Theory of Natural Selection* by R. A. Fisher	1976	*The Selfish Gene* by Richard Dawkins
1932	Adaptive landscape metaphor of Sewall Wright	1981	Arkansas Creation Trial
1933	Adolf Hitler comes to power	1995	Nobel Prize for Physiology or Medicine awarded to *Edward B. Lewis*, *Christiane Nüsslein-Volhard*, and *Eric F. Wieschaus* "for their discoveries concerning the genetic control of early *embryonic development*"
1937	*Genetics and the Origin of Species* by Theodosius Dobzhansky		
1939–1945	World War II		
1942	*Evolution: The Modern Synthesis* by Julian Huxley		
	Systematics and the Origin of Species by Ernst Mayr	1997	John Paul II issues papal letter accepting modern evolutionary theory
1944	*Tempo and Mode in Evolution* by George Gaylord Simpson	1999	State of Kansas takes evolution out of school curricula. (In the years following, Kansas moved back and forward on the evolution issue)
1947	Journal *Evolution* founded		
1950	*Variation and Evolution in Plants* by G. Ledyard Stebbins		
1953	Stanley Miller and Harold Urey simulate the natural production of amino acids	2001	Terrorists attack and destroy the World Trade Center in New York City, on September 11, killing nearly three thousand people
	Discovery of double helix by James Watson and Francis Crick	2005	Dover intelligent design trial
1955	*Le phénomène humaine* by Pierre Teilhard de Chardin	2006	*The God Delusion* by Richard Dawkins
1957	Sputnik	2009	200th anniversary of the birth of Charles Darwin and the 150th anniversary of the publication of *The Origin of Species*
	Syntactic Structures by Noam Chomsky		
1961	*Genesis Flood* by John Whitmore and Henry Morris		

Glossary

ABIOGENESIS: the natural development of life from nonliving materials (traditionally applied to the spontaneous generation of life from inorganic, never-living material).

ADAPTATION: any characteristic that aids its possessor to survive and reproduce.

ADAPTATIONISM: the belief that all organic characters are indeed adaptive.

ADAPTIVE LANDSCAPE: a metaphor introduced by Sewall Wright claiming that the fitness of organisms can be mapped as if on a hilly terrain.

AGNOSTICISM: the belief that one cannot know whether or not God exists.

ALLELE: any one of a number of forms of a gene that can occupy the same place (locus) on a chromosome.

ALLOMETRY: the study of the relative growth of some parts of an organism in comparison with other parts or the whole.

ALTRUISM: help given by one organism to another, at some biological cost, for the donor's long-term reproductive advantage.

AMINO ACID: the major complex organic molecules that serve as the building blocks of proteins.

ARACHNID: an arthropod (an invertebrate with segmented body, jointed limbs, and an external skeleton) with eight legs; includes spiders and scorpions.

ARCHAEOPTERYX: an ancient bird with many reptilian features (a missing link).

ARCHETYPE: the basic building plan of a group of animals such as vertebrates.

ARMS RACE (BIOLOGICAL): members of two lines competing against each other and developing evermore sophisticated adaptations.

ASTROLOGY: the system claiming that our destinies are controlled by the configurations of the heavens.

ATHEISM: the belief that God does not exist.

AUSTRALOPITHECUS: a genus that gave rise immediately to our genus, *Homo,* consisting of animals intermediate between apelike forms and humans.

AUTOCATALYTIC: becoming ever more powerful thanks to positive feedback mechanisms.

BALANCE HYPOTHESIS: the belief that natural selection holds many different alleles or genes in a balance or equilibrium within a population.

BALANCED SUPERIOR HETEROZYGOTE FITNESS: the claim that natural selection keeps different alleles in balance or equilibrium within a population, because the heterozygote is fitter than either homozygote.

BAUPLAN: an archetype.

BIOGENESIS: the natural development of life from living materials (as in normal generation).

BIOGENETIC LAW: the claim that ontogeny, individual development, recapitulates phylogeny, the evolution of the group.

BIOGEOGRAPHY: the study of the distribution of organisms.

BIOMETRICS/BIOMETRICIAN: the school of biologists at the beginning of the century committed to the belief that evolution could be studied through detailed quantification and statistical analysis.

CALVINISM: a branch of Protestantism which follows the sixteenth-century reformer John Calvin, marked especially by a belief in predestination (that is, that our fates are known to God from the first).

CAMBRIAN EXPLOSION: the rapid evolution and diversification of life forms in the Cambrian period, occurring between five and six hundred million years ago.

CASTE SYSTEM: the different forms or morphs found within species in the hymenoptera.

CATASTROPHISM: a geological theory of periodic, violent, earth upheavals, endorsed by Georges Cuvier, believed responsible for major geological formations.

CELL: the building blocks of complex organisms, being membrane-contained units containing within them the genes and other components necessary for life.

CENANCESTOR: the most recent, jointly shared ancestor of all living things.

CHAIN OF BEING: a doctrine popular in the Middle Ages claiming that all organisms can be put in a continuous line from the very simple to the very complex.

CHROMOSOME: a string-like entity, in the cell, that carries the genes.

CLADISM: a system of classification based on phylogeny and not on similarity.

CLASSICAL HYPOTHESIS: the claim that there is little genetic variation within a population thanks to the cleansing effect of natural selection.

CONSTRAINTS: the physical factors that keep an organism developing along certain fixed limited paths.

CREATION SCIENCE: the claim that the early chapters of Genesis can be given good scientific backing.

CREATIONIST: one who believes in the literal truth of the Bible, especially the early chapters of Genesis.

CYTOLOGY: the systematic study of cell structure.

DARWINIAN: a person who accepts Charles Darwin's theory of evolution through natural selection.

DARWINISM: the theory of evolution through natural selection.

DEEP STRUCTURE: the claim, first made by Noam Chomsky, that all languages have a fundamental, underlying similarity.

DEISM: the belief in God as unmoved mover.

DEOXYRIBONUCLEIC ACID (DNA): the macromolecule that transmits genetic information (the molecular gene).

DIVISION OF LABOR: the process of breaking down an activity into specialized tasks so that it can be performed much more efficiently.

DOMINANT: a gene (allele) whose effects mask the effects of its paired opposite at the same locus.

DROSOPHILA: a fruitfly, a popular organism for study by geneticists.

DUALIST/DUALISM: the belief that the mind is a substance (usually thought of as a thinking or spiritual substance) corresponding to the substance of the body (usually thought of as material substance).

DYNAMIC EQUILIBRIUM: a balance between opposing forces that is in constant motion, usually upwards.

ÉLAN VITAL/VITALISM: the force the vitalists believe animates living bodies.

EMBRANCHEMENT: one of the four divisions of the animal kingdom as supposed by Georges Cuvier (vertebrates, mollusks, articulata, and radiata).

EMBRYOLOGY: the study of developing multicellular organisms, particularly those in early development.

EPISTEMOLOGY: the theory of knowledge ("what can I know?").

ETHICS: the theory of morality ("what should I do?").

EUKARYOTE: an organism (like a mammal) where the DNA is on the chromosomes within the nucleus (see also, prokaryote).

EUGENICS: the movement that aimed to improve humankind through selective breeding.

EVANGELICAL CHRISTIANITY: a form of Protestantism that puts a major emphasis on personal commitment to Jesus as Lord, stressing the significance of the Bible taken fairly literally.

EVOLUTION (CAUSE): the mechanism or force behind evolutionary change.

EVOLUTION (PATH): the particular track that organisms have taken through history, usually known as phylogeny.

EVOLUTION (FACT): the belief that organisms living and dead are the end results of a natural process of development from one or a few forms.

EVOLUTIONARILY STABLE STRATEGY (ESS): a genetically programmed path or strategy taken by members of a group such that no other strategy or path can dislodge it.

EVOLUTIONARY DEVELOPMENT (EVO-DEVO): the study of development from an evolutionary perspective (the successor to classical embryology, now done from a molecular perspective).

EVOLUTIONARY PSYCHOLOGY: human sociobiology.

EXAPTATION: an organic feature without adaptive function.

FITNESS: the comparative ability of an organism to survive and reproduce and pass on its genes.

FORM: the particular shape of features of an organism, usually distinguished from function.

FOUNDER PRINCIPLE: the claim by the systematist Ernst Mayr that the making of a new species involves just a few organisms isolated from the main group.

FUNCTION: a thing that a characteristic does, believed by Darwinians to have been produced by natural selection and hence adaptive.

FUNDAMENTALIST/FUNDAMENTALISM: a form of American Protestant evangelicalism popular at the beginning of this century, committed to the literal truth of the Bible.

GEL ELECTROPHORESIS: a technique for detecting variations in molecules (and hence genes) by tracking their progress through a gel under the influence of an electric field.

GENE POOL: the collective genes of a population of organisms.

GENE: the ultimate unit of heredity, believed today to consist of ribonucleic acid (usually DNA).

GENETIC DRIFT: the claim by Sewall Wright that sometimes characteristics and their genes have so little effect that they escape the effects of natural selection and hence (proportionately) move randomly up or down within a population.

GENETICS: the theory of heredity dating back to the nineteenth-century monk Gregor Mendel.

GENOTYPE: the collective genes of any particular organism, to be contrasted with the phenotype, which refers to the collective physical characteristics of an organism. Also known as the "genome."

GROUP SELECTION: the belief that sometimes natural selection can work for the benefit of the group against the interests of some individuals.

HETEROGENESIS: the belief that life is created naturally from non-organic matter (traditionally applied to the spontaneous generation of life from formerly living materials, as maggots from meat).

HETEROSIS: superior heterozygote fitness.

HETEROZYGOTE: an organism that has different alleles occupying the paired loci of some chromosome.

HIERARCHY THEORY: the claim that supposes that different evolutionary forces operate at different levels, from micro-evolution through to macro-evolution (hence, antireductionistic).

HOBBIT: *Homo floresiensis*, the little human-like creature discovered on an Indonesian island.

HOLISTIC/HOLISM: the belief that in order to understand the individual one must look at the whole functioning organism and not just at the parts; to be contrasted with reductionism, the claim that higher level macrofunctions of organisms can be explained fully in terms of lower level microprocesses.

HOMEOBOX GENES: the DNA sequences that control the patterns of development.

HOMEOSTASIS: the state of balance or equilibrium brought on by different forces within an individual or a population.

HOMINIDS: humans and their relatively recent ancestors and relations (technically, members of the family Hominidae).

HOMO: the genus of organisms that includes *Homo sapiens* as one of its members.

HOMOLOGY/HOMOLOGIES: the isomorphisms between the parts of different organisms of different species, believed today to be a result of common ancestry.

HOMOZYGOTE: an organism with two identical alleles at some locus on the chromosomes.

HOX GENES: a subset of homeobox genes that control development along the main axis of animals, common to (homologous in) fruitflies and humans.

HYMENOPTERA: the ants, the bees, and the wasps.

IDENTITY THEORY: the philosophy of mind that takes mind and brain to be different aspects of the same substance.

INDIVIDUAL SELECTION: the claim that selection works for, and only for, the individual as opposed to the group.

INFANTICIDE: deliberate killing of the newborn.

INTELLIGENT DESIGN THEORY (IDT): a position arguing that the complexity of organisms is such that the only possible causal explanation is some form of intelligence.

IRREDUCIBLE COMPLEXITY: organic complexity too great to have come in a gradual fashion through a blind-law mechanism like natural selection.

"JUST SO STORIES": fantastical adaptive scenarios that could not possibly be true (a term borrowed from the tales of Rudyard Kipling).

KIN SELECTION: the action of a selective force that promotes the well being of close relatives as well as of the individual; a form of biological altruism.

KINETIC MODEL: a theory about movement.

LAISSEZ-FAIRE: a sociopolitical economic doctrine which claims that, for the most good to be achieved, there should be no state interference in the workings of individual firms or businesses.

LAMARCKISM: the belief that acquired characters can be inherited.

LAW: a statement about some particular regularity of nature as in Mendel's Laws or the Hardy-Weinberg Law.

LOCUS: a particular point on a chromosome matched by a corresponding point on the other member of paired chromosomes; can be occupied by the various different alleles in the set peculiar to a species.

MACROEVOLUTION: organic change over large quantities of time.

MEME: a unit of culture, analogous to the gene, a unit of heredity.

MENDELIAN(ISM): the genetical theory going back to Gregor Mendel, and stressing that the units of heredity, the genes, remain unchanged from generation to generation (a particulate theory).

METAPHYSICAL NATURALISM: the belief that physical nature is all that exists, hence excluding any deity or other supernatural entity.

METHODOLOGICAL NATURALISM: the assumption that physical nature is all that exists, taken in order to promote the success of science, but not making any ultimate metaphysical assumptions.

MICROEVOLUTION: organic change over short times.

MIRACLE: a special intervention by the Deity.

MITOCHONDRIA/MITOCHONDRIAL DNA: bodies within the cell, that are separate from the central nucleus, but carry genes.

MITOCHONDRIAL EVE: the female supposedly ancestral to all living humans, dating from about 140,000 years ago.

MODERNIST/MODERNISM: a movement, particularly in religion at about the beginning of the last century, which tried to accommodate and harmonize with the science of the day.

MOLECULAR CODE: the pattern governing the ordering of the DNA molecule's components, yielding information needed in the production of other cellular parts, notably amino acids.

MOLECULAR DRIFT: the belief that there is random change in populations at the molecular level, below the forces of natural selection (see also genetic drift, of which molecular drift is a special case).

MONAD: the most primitive (and presumably earliest) of all organisms.

MONOMORPHIC: having only one particular form, as opposed to dimorphic or polymorphic.

MORPHOLOGY: the study of the structure or form of organisms.

MUTATION: the spontaneous change in a gene, which may lead to new variations or characteristics.

NATURAL SELECTION: Charles Darwin's mechanism of evolution supposing that, since more organisms are born than can survive and reproduce, fitter organisms will be successful in the struggle for existence and hence permanent change will be effected.

NATURAL RELIGION/THEOLOGY: the belief that through reason one can get to know of the existence and nature of the Deity.

NATURALISM: a belief that one can explain everything through unbroken law.

NATURPHILOSOPHIE/NATURPHILOSOPHEN: an idealistic German morphological movement at the beginning of the nineteenth century that stressed form over function.

NEANDERTHAL: a subspecies of *Homo sapiens* that lived about 100,000 years ago.

NUCLEIC ACID: a chainlike, macromolecule found in cells, either DNA which (usually) carries the information of heredity, or RNA which reads the information from the DNA (in some viruses, RNA carries the information).

NUCLEUS/NUCLEI: the central part of a cell that contains chromosomes carrying genes.

ONTOGENY: the individual development of an organism.

OPTIMALITY MODEL: a theory or hypothesis trying to show how selection has made certain specified characteristics the most efficient possible for survival and reproduction.

ORDER FOR FREE: self organization.

ORIGINAL SIN: a Christian doctrine claiming that all humans are tainted with sin, due in some sense to the original misdeeds of Adam and Eve.

PALEOANTHROPOLOGY: the study of the fossil evidence for and evolutionary history of humankind.

PALEONTOLOGY: the study of life's past as revealed particularly through the fossils.

PANGLOSS(IAN): an extreme form of adaptationism, unjustifiably seeing selection as having produced everything (usually acting exclusively for the good of the individual), often marked by undue use of just-so stories (taken from Voltaire's *Candide*, where the Leibnizian philosopher Dr. Pangloss sees value in everything, including the most extreme disasters).

PARADIGM: a world picture in science, such as Darwin's theory of evolution through natural selection.

PHENOTYPE: the collective physical characteristics of an organism, contrasted with the genotype.

PHOTOSYNTHESIS: the capture of the sun's energy by green plants for use in the manufacture of carbon compounds.

PHRENOLOGY: a nineteenth-century pseudoscience which claimed that you could read character from the shape of the skull.

PHYLETIC GRADUALISM: Stephen Jay Gould's term for the Darwinian commitment to the gradual (as opposed to jerky) nature of the evolutionary process.

PHYLLOTAXIS: the spiral pattern shown by the flowers and fruits of many plants.

PHYLOGENY: the path of evolution.

PLEIOTROPY: the production of different characteristics by the same gene.

POLYMORPHIC: having different forms within the same population.

POPULATION GENETICS: the generalization of Mendelian genetics to deal with groups, thus focussing on the spread and change of genetic variation within a population.

PROBLEM OF EVIL: the theological problem of reconciling a Christian God who is both all good and all powerful with the existence of evil and pain in the world.

PROGRESS (BIOLOGICAL): the belief that the course of evolution has been from the simple to the complex.

PROGRESS (CULTURAL): the belief that humankind can and is improving its lot.

PROKARYOTE: a single-celled organism, like a bacterium, where the DNA floats free (as opposed to a eukaryote).

PROTEIN: a chain-like macromolecule, one of the building blocks of the cell, composed of amino acids as produced by the translation of the RNA coded from the DNA.

PROVIDENTIALISM: the belief that God is always present (immanent) in the universe and ready to intercede on our behalf.

PUNCTUATED EQUILIBRIA: the theory of Stephen Jay Gould that the course of evolution is one of lack of action (stasis) followed by short, sharp breaks or jumps.

RECESSIVE: an allele whose effects are masked by the other allele at the same locus (see dominant).

RECIPROCAL ALTRUISM: help given by one organism to another with the expectation that such help will be returned.

REDUCTIONISTIC/REDUCTIONISM: the claim that higher-level macro-functions of organisms can be explained fully in terms of lower-level micro-processes (to be contrasted with holism).

REVEALED RELIGION/THEOLOGY: that area of religious belief dealing with claims based on faith or dogma or authority; to be contrasted with natural theology.

REVERSE ENGINEERING: the process of discovering an organism's function by working backwards from its form to its possible design.

RIBONUCLEIC ACID: RNA the kind of nucleic acid that carries the information from the DNA to produce the amino acids that make up proteins; in some viruses, RNA acts as the carrier of information itself.

RIBOSOMES: particles within cells where RNA makes proteins.

SALTATION(ISM): the evolutionary theory claiming that organic change goes in jumps, usually through macromutations.

SELF ORGANIZATION: the belief that the laws of physics and chemistry unaided (by natural selection) can be responsible for the complexity of organisms.

SELFISH GENE: a metaphor produced by Richard Dawkins to emphasize that selection ultimately works for the benefit of the individual, perhaps even the individual gene, rather than the group.

SEXUAL SELECTION: a secondary form of selection posited by Charles Darwin and involving competition within species for mates.

SOCIOBIOLOGY: that area of evolutionary thought dealing with the development and maintenance of social behavior.

SPANDRELS: Stephen Gould's term for the nonfunctional byproducts of adaptations, taken from the triangular mosaic-covered areas to be seen at the tops of the columns in St. Mark's in Venice (properly these areas are called pendentives).

SPONTANEOUS GENERATION: the belief that life appears in one move, naturally from non-living matter.

STASIS: the periods of stability or nonchange as supposed in Stephen Gould's theory of punctuated equilibria.

SYNTHETIC THEORY: the theory of evolution combining Darwinian selection with Mendelian or post-Mendelian particulate genetics; sometimes called neo-Darwinism.

SYSTEMATICS: the area of evolutionary biology that classifies and explains the relationships between organisms.

TELEOLOGY: the claim that entities, organisms particularly, can and should be considered with reference to their ends or functions and not simply by the causes that brought them about.

THEIST: someone who believes in an immanent god, that is a god who is prepared to intervene in the creation: a term traditionally reserved for the followers of the major religions of Judaism, Christianity, and Islam.

TRANSUBSTANTIATION: the Catholic miracle where the bread and wine in the mass are changed into the body and blood of Jesus Christ.

TROLLEY PROBLEM: a moral paradox stemming from our willingness to sacrifice the individual in favor of the group when it is a theoretical decision but not when it involves personal action against the individual.

UNIFORMITARIANISM: the geological theory associated with Charles Lyell that supposes that all geological change takes place gradually through causes now in operation; to be contrasted with catastrophism.

UNITY OF TYPE: the fact that organisms exhibit significant isomorphisms or homologies, linked through the sharing of a common archetype or Bauplan, and explained by evolutionists as a function of shared descent.

UTILITY FUNCTION: the purpose or function supposed by engineers of machines and thus the thing to be discovered through reverse engineering.

VITALISM: the belief that there are unseen life forces controlling evolution.

YOUNG EARTH CREATIONISM: the belief that the world was literally created in six days, six thousand years ago, followed by a worldwide flood.

Illustration and Document Credits and Permissions

Chapter One

Illustrations

Chapter Two

Illustrations

Documents

Chapter Three

Illustrations

Documents

Chapter Four

Illustrations

Documents

Chapter Five

Illustrations

Documents

437   "William Dembski and John Haught Spar on Intelligent Design," by Rebecca Flietstra in *Research News and Opportunities in Science and Theology*, 2 (9):14-15 (2002) (*Science & Theology News*). Reprinted with permission.

442   "Darwin Under the Microscope," Michael Behe, *New York Times*, October 29, 1996. © 1996, The New York Times. Reprinted by permission.

444   R. Dawkins, "Obscurantism to the Rescue." *Quarterly Review of Biology* 72 (1997): 397–399. Reprinted courtesy of the University of Chicago Press.

448   "Viruses of the Mind," Richard Dawkins, From *Dennett and his Critics: Demystifying Mind*, Bo Dahlbom, ed., copyright © 1995 Wiley-Blackwell. Reproduced with permission of Blackwell Publishing Ltd.

Chapter Six

Illustrations

156   The Granger Collection, New York

157   The Granger Collection, New York

158   From W. Bateson, *Mendel's Principles of Heredity* (Cambridge: Cambridge University Press, 1913)

159   Courtesy of T. D. Lysenko

161   Courtesy of Sewall Wright

163   Courtesy Columbia University Press

165   G.G. Simpson, *Tempo and Mode in Evolution* (New York: Columbia University Press, 1994, p. 92)

166   Natural History Museum, London

176   Photograph: Camera Press Ltd. Source: Retna Ltd..I.

177   From the Author's Collection

Documents

465   Sewall Wright, "The Roles of Mutation, Inbreeding, Crossbreeding and Selection in Evolution," *Proceedings of the Sixth International Congress of Genetics* 1932, 356–366. Reprinted with permission of the Genetics Society of America.

474   Ernst Mayr to G. F. Ferris, March 29, 1948. Reprinted with the permission of the American Philosophical Society.

476   Theodosius Dobzhansky to John Greene, November 23, 1961. Reprinted with the permission of the American Philosophical Society.

Chapter Seven

Illustrations

182   Library of Congress, Prints & Photographs Division, LC-USZ62-3497

184   The Granger Collection, New York

190   From Sir C. Wyville Thomson, *The Depths of the Sea: An Account of the General Results of the Dredging Cruises of H.M.Ss. 'Porcupine' and 'Lightning' during the Summers of 1868, 1869, and 1870, under the Scientific Direction of Dr. Carpenter, F.R.S., J. Gwyn Jeffreys, F.R.S., and Dr. Wyville Thomson, F.R.S.* (New York, London: Macmillan, 1873)

191   From the Author's Collection

193   Royal College of Surgeons

194   Credit: James A. Sugar. Source: http://web99.arc.nasa.gov/~astrochm/Miller/

199   From the Author's Collection / Illustrator Martin Young

200   From M. Glaessner and M. Wade, "The late precambrian fossils from Ediacara, South Australia." *Paleontology* 9 (1966): 599–628. Used by permission of The Paleontological Association.

201   From J. William Schopf, "The Evolution of the Earliest Cells." *Scientific American* 239:3 (September 1978): 110–138

Chapter Eight

Illustrations

Documents

Chapter Nine

Illustrations

Chapter Ten

Chapter Eleven

Chapter Twelve

Illustrations

Documents

Biographies

615   Georges Cuvier
www.ucmp.berkeley.edu/history/cuvier.html
www.victorianweb.org/science/cuvier.html
Illustration: Library of Congress, Prints & Photographs Division, LC-USZ62-134030

616   Martin Daly
http://psych.mcmaster.ca/dalywilson/martin.html
www.science.mcmaster.ca/psychology/md.html
http://psych.mcmaster.ca/dalywilson/
http://psych.mcmaster.ca/dalywilson/margo.html
www.science.mcmaster.ca/psychology/margo.html

617   Clarence Darrow
www.law.umkc.edu/faculty/projects/ftrials/Darrow.htm
www.spartacus.schoolnet.co.uk/USAdarrow.htm
Illustration: Library of Congress, Prints & Photographs Division, LC-DIG-ggbain-06468

619   Erasmus Darwin
www.ucmp.berkeley.edu/history/Edarwin.html
www.victorianweb.org/science/edarwin.html

620   Nicholas Davies
www.zoo.cam.ac.uk/zoostaff/bbe/Davies/Nick1.htm
http://hcr3.isiknowledge.com/author.cgi?id=2751&cb=61
Illustration: From the Author's Collection

621   Richard Dawkins
www.richarddawkins.com/
www.time.com/time/specials/2007/time100/article/0,28804,1595326_1595329_
1616137,00.html
Illustration: From the Author's Collection

622   William Dembski
www.designinference.com/biosketch.htm

623   Daniel Dennett
http://ase.tufts.edu/cogstud/incbios/dennettd/dennettd.htm
www.edge.org/3rd_culture/bios/dennett.html
www.ted.com/index.php/speakers/dan_dennett.html

624   Theodosius Dobzhansky
www.mnsu.edu/emuseum/information/biography/abcde/dobzhansky_theodosius.html
www.stephenjaygould.org/people/theodosius_dobzhansky.html
Illustration: Courtesy Columbia University Press

625   Ronald Fisher
www.answers.com/topic/ronald-fisher
http://evolution.berkeley.edu/evolibrary/article/_0/history_19

626   Jerry Fodor
"Natural Selection Has Gone Bust," by Jerry Fodor, Uncommon Descent, October 13, 2007
Washington Post Obituary, March 1, 2006
University of California Museum of Paleontology
www.christiananswers.com
Suite 101.com
Illustration: Courtesy of Pedro Alcocer

628   Brian Goodwin
Katarxis No. 3, "A Conversation with Three Scientists." found at www.katarxis3.com/Three_
Scientists.htm
Norfolk Generation Network, "An Interview with Professor Brian Goodwin," by David King
Illustration: Courtesy of Brian Goodwin

629   Stephen Jay Gould
www.stephenjaygould.org/people/john_haldane.html

630   Peter and Rosemary Grant
www.eeb.princeton.edu/FACULTY/Grant_R/Grant_BR.html
www.eeb.princeton.edu/FACULTY/Grant_P/grantPeter.html
http://explore-evolution.unl.edu/grant.html

631   Asa Gray
www.answers.com/topic/asa-gray
www.asa3.org/aSA/PSCF/2001/PSCF9-01Miles.html
www.nceas.ucsb.edu/~alroy/lefa/Gray.html
www.famousamericans.net/asagray/

632   Ernst Haeckel
www.ucmp.berkeley.edu/history/haeckel.html
www.answersingenesis.org/creation/v18/i2/haeckel.asp
www.gennet.org/facts/haeckel.html
Illustration: Library of Congress, Prints & Photographs Division, LC-DIG-ggbain-05698

634   John Burdon Sanderson Haldane
Illustration: From the Author's Collection

635   William D. Hamilton
www.mnsu.edu/emuseum/information/biography/fghij/hamilton_william.html
www.stephenjaygould.org/people/william_hamilton.html
Illustration: From the Author's Collection

636   Charles Hodge
www.ccel.org/h/hodge/
www.theropps.com/papers/Winter1997/CharlesHodge.htm
www.theopedia.com/Charles_Hodge

638   Sarah Blaffer Hrdy
Claudia Glenn Dowling, "The Hardy Sarah Blaffer Hrdy," *Discover Magazine*, March 1, 2003
www.citrona.com
www.msnu.edu/emuseum/information/biography/fghij/hrdy_sarah.html

639   Julian Huxley
www.nndb.com/people/009/000100706/
http://encarta.msn.com/encyclopedia_761576201/julian_huxley.html

641   Thomas Henry Huxley
www.ucmp.berkeley.edu/history/thuxley.html
www.iep.utm.edu/h/huxley.htm
www.bbc.co.uk/history/historic_figures/huxley_thomas_henry.shtml

642   Phillip Johnson
www.law.umkc.edu/faculty/projects/ftrials/conlaw/johnsonp.html

643   Stuart Kauffman
www.edge.org/3rd_culture/kauffman03/kauffman_index.html

644   Prince Petr Kropotkin
www.uh.edu/engines/epi720.htm
Illustration: Library of Congress, Prints & Photographs Division, LC-DIG-ggbain-50403

646   Jean Baptiste de Lamarck
www.ucmp.berkeley.edu/history/lamarck.html
www.mnsu.edu/emuseum/information/biography/klmno/lemarck_jean.html

647   Richard Lewontin
http://authors.library.caltech.edu/5456/1/hrst.mit.edu/hrs/evolution/public/profiles/lewontin.html

648   Konrad Lorenz
http://nobelprize.org/nobel_prizes/medicine/laureates/1973/lorenz-autobio.html
Illustration: From the Author's Collection, photo by Jane Clelland

649   Charles Lyell
www.mnsu.edu/emuseum/information/biography/klmno/lyell_charles.html
www.victorianweb.org/science/lyell.html
www.gennet.org/facts/lyell.html
http://evolution.berkeley.edu/evolibrary/article/_0_0/history_12

651   Trofim Lysenko
Illustration: Courtesy of T.D. Lysenko

652   Thomas Robert Malthus
http://cepa.newschool.edu/het/profiles/malthus.htm
www.blupete.com/Literature/Biographies/Philosophy/Malthus.htm
www.usp.nus.edu.sg/victorian/economics/malthus.html
www.bbc.co.uk/history/historic_figures/malthus_thomas.shtml
www.econlib.org/library/Enc/bios/Malthus.html

653   Lynn Margulis
www.geo.umass.edu/faculty/margulis
www.evolution.berkely.edu/evolibrary/article/_0/history_24

654   Othniel Charles Marsh
"Fossil Horses and Othniel Charles Marsh," found at
www.geocities.com/ResearchTriangle/Lab/3773/OC_Marsh.html
Illustration: Library of Congress, Prints & Photographs Division, LC-DIG-cwpbh-04124

655   Ernst Mayr
www.pbs.org/wgbh/evolution/library/06/2/l_062_01.html
www.achievement.org/autodoc/page/may1bio-1
www.edge.org/3rd_culture/bios/mayr.html

656   Gregor Mendel
www.accessexcellence.org/RC/AB/BC/Gregor_Mendel.php
www.accessexcellence.org/RC/AB/BC/Gregor_Mendel.php
http://oz.plymouth.edu/~biology/history/mendel.html

658   Simon Conway Morris
www.enlightennext.org/magazine/bios/simon-conway-morris.asp
www.nndb.com/people/790/000031697/
www.christianpost.com/article/20070425/christian-origins-expert-promotes-evolution-at-texas-universities.htm

659   Hermann J. Muller
http://nobelprize.org/nobel_prizes/medicine/laureates/1946/muller-bio.html

660   John Henry Newman
Darwin Correspondence Project, www.darwinproject.ac.uk/content/view/110/114
www.newmanreader.org

662   Alexander (Aleksandr) Oparin
www.daviddarling.info/encyclopedia/O/Oparin.html
www.britannica.com/EBchecked/topic/429565/Aleksandr-Oparin

664   Archdeacon William Paley
www.wmcarey.edu/carey/paley/paley.htm

666   Geoffrey Parker
Illustration: Courtesy of Geoffrey Parker

667   Louis Pasteur
www.accessexcellence.org/RC/AB/BC/Louis_Pasteur.php
www.lucidcafe.com/library/95dec/pasteur.html
Illustration: Library of Congress, Prints & Photographs Division, LC-USZ62-3497

668   Alvin Plantinga
http://philosophy.nd.edu/people/all/profiles/plantinga-alvin/
http://philofreligion.homestead.com/plantingapage.html

669   Felix Pouchet
www.bookrags.com/biography/felix-archimede-pouchet-wog/

670   Etienne Geoffroy Saint-Hilaire
www.ucmp.berkeley.edu/history/hilaire.html
www.infoplease.com/ce6/people/A0820525.html

672   John J. Sepkoski Jr.
http://neo.jpl.nasa.gov/news/news016.htm
Illustration: Courtesy of John J. Sepkoski

673   George Gaylord Simpson
http://people.ucsc.edu/~laporte/simpson/Biography.html
www.pbs.org/wgbh/evolution/library/06/2/l_062_02.html
www.mnsu.edu/emuseum/information/biography/pqrst/simpson_george_gaylord.html

674   John Maynard Smith
www.edge.org/3rd_culture/maynard_smith04/maynard_smith04_index.html
http://homepage.ntlworld.com/marek.kohn/jms.html
www.muskingum.edu/~psych/psycweb/history/smith.htm
Illustration: From the Author's Collection

675   Herbert Spencer
www.iep.utm.edu/s/spencer.htm
www.victorianweb.org/philosophy/spencer/spencer.html

676   George Ledyard Stebbins
       http://biosci.ucdavis.edu/alumni/newsletter/spring00/stebbins.html
       www.ncbi.nlm.nih.gov/pubmed/11700300

677   Chris Stringer
       www.nationmaster.com/encyclopedia/Chris-Stringer
       http://archaeology.about.com/od/archaeologistss/g/stringerc.htm

678   D'Arcy Wentworth Thompson
       www-groups.dcs.st-and.ac.uk/~history/Mathematicians/Thompson_D'Arcy.html
       www.daviddarling.info/encyclopedia/T/Thompson.html

679   Alfred Russel Wallace
       www.mnsu.edu/emuseum/information/biography/uvwxyz/wallace_alfred_russel.html
       www.answers.com/topic/alfred-russel-wallace
       www.victorianweb.org/science/wallace/wallace1.htm
       www.iol.ie/~spice/alfred.htm

681   Edward O. Wilson
       www-museum.unl.edu/research/entomology/workers/EWilson.htm
       http://your.kingcounty.gov/solidwaste/NaturalConnections/edward_wilson_bio.htm

684   Milford Wolpoff
       www.lsa.umich.edu/anthro/faculty_staff/wolpoff.html
       http://discovermagazine.com/2001/jun/breakdialogue
       http://archaeology.about.com/od/archaeologistsw/g/wolpoffm.htm
       www.mnsu.edu/emuseum/information/biography/uvwxyz/wolpoff_milford.html
       Illustration: Courtesy of Milford Wolpoff

       General
       www.pbs.org/wgbh/evolution/library/02/3/l_023_01.html
       http://evolution.berkeley.edu/evolibrary/article/_0/history_09
       http://anthro.palomar.edu/evolve/evolve_1.htm
       NationMaster Encyclopedia
       Encyclopedia Britannica
       Wikipedia

Dr. Michael Ruse is a philosopher of science, who focuses on the philosophy of biology. He is well known for his work on the argument between creationism and evolutionary biology. He taught at the University of Guelph, in Ontario, Canada for 35 years before moving to Florida State University in 2000. As Lucyle T. Werkmeister Professor of Philosophy, he developed the program in The History and Philosophy of Science. He also founded the journal *Biology and Philosophy* and has published dozens of books and articles, including *The Philosophy of Biology; Monad to Man: The Concept of Progress in Evolutionary Biology; Mystery of Mysteries: Is Evolution a Social Construction?; Darwin and Design: Does Nature have a Purpose?;* and *The Evolution-Creation Struggle.*

In 1981, he was an expert witness for the American Civil Liberties Union in an ultimately successful defeat of a bill that mandated the teaching of Genesis in the State's biology classrooms. In response to this controversy, he later published *But is it Science? The Philosophical Question in the Creation-Evolution Controversy.*

Dr. Ruse is the editor of *The Cambridge Companion to the Philosophy of Biology, The Cambridge Companion to the Origin of Species, The Oxford Handbook to the Philosophy of Biology,* and the *Harvard Companion to Evolution* (known as *Evolution: The First Four Billion Years*).

More recently, Ruse has taken to arguing that science and religion can be harmonized. This is the topic of his 2001 book, *Can a Darwinian be a Christian? The Relationship between Science and Religion,* and the forthcoming *Making Room for Faith: Christianity in an Age of Science.* This last work draws on Ruse's experience as a Gifford Lecturer in 2001, in Glasgow, Scotland, on the topic of natural theology, which is God, the world, and the ways in which reason can and cannot throw light on their relationship.

He lives in Florida with his wife; he has five children.

Part Six:
INDEX

Index

Business Information ✦ Ratings Guides ✦ General Reference ✦ Education ✦
Statistics ✦ Demographics ✦ Health Information ✦ Canadian Information

Grey House Publishing

The Directory of Business Information Resources, 2009

With 100% verification, over 1,000 new listings and more than 12,000 updates, *The Directory of Business Information Resources* is the most up-to-date source for contacts in over 98 business areas – from advertising and agriculture to utilities and wholesalers. This carefully researched volume details: the Associations representing each industry; the Newsletters that keep members current; the Magazines and Journals - with their "Special Issues" - that are important to the trade, the Conventions that are "must attends," Databases, Directories and Industry Web Sites that provide access to must-have marketing resources. Includes contact names, phone & fax numbers, web sites and e-mail addresses. This one-volume resource is a gold mine of information and would be a welcome addition to any reference collection.

"This is a most useful and easy-to-use addition to any researcher's library." –The Information Professionals Institute

Softcover ISBN 978-1-59237-399-4, 2,500 pages, $195.00 | Online Database: http://gold.greyhouse.com Call (800) 562-2139 for quote

Hudson's Washington News Media Contacts Directory, 2009

With 100% verification of data, *Hudson's Washington News Media Contacts Directory* is the most accurate, most up-to-date source for media contacts in our nation's capital. With the largest concentration of news media in the world, having access to Washington's news media will get your message heard by these key media outlets. Published for over 40 years, Hudson's Washington News Media Contacts Directory brings you immediate access to: News Services & Newspapers, News Service Syndicates, DC Newspapers, Foreign Newspapers, Radio & TV, Magazines & Newsletters, and Freelance Writers & Photographers. The easy-to-read entries include contact names, phone & fax numbers, web sites and e-mail and more. For easy navigation, Hudson's Washington News Media Contacts Directory contains two indexes: Entry Index and Executive Index. This kind of comprehensive and up-to-date information would cost thousands of dollars to replicate or countless hours of searching to find. Don't miss this opportunity to have this important resource in your collection, and start saving time and money today. Hudson's Washington News Media Contacts Directory is the perfect research tool for Public Relations, Marketing, Networking and so much more. This resource is a gold mine of information and would be a welcome addition to any reference collection.

Softcover ISBN 978-1-59237-407-6, 800 pages, $289.00 | Online Database: http://gold.greyhouse.com Call (800) 562-2139 for quote

Nations of the World, 2009 A Political, Economic and Business Handbook

This completely revised edition covers all the nations of the world in an easy-to-use, single volume. Each nation is profiled in a single chapter that includes Key Facts, Political & Economic Issues, a Country Profile and Business Information. In this fast-changing world, it is extremely important to make sure that the most up-to-date information is included in your reference collection. This edition is just the answer. Each of the 200+ country chapters have been carefully reviewed by a political expert to make sure that the text reflects the most current information on Politics, Travel Advisories, Economics and more. You'll find such vital information as a Country Map, Population Characteristics, Inflation, Agricultural Production, Foreign Debt, Political History, Foreign Policy, Regional Insecurity, Economics, Trade & Tourism, Historical Profile, Political Systems, Ethnicity, Languages, Media, Climate, Hotels, Chambers of Commerce, Banking, Travel Information and more. Five Regional Chapters follow the main text and include a Regional Map, an Introductory Article, Key Indicators and Currencies for the Region. As an added bonus, an all-inclusive CD-ROM is available as a companion to the printed text. Noted for its sophisticated, up-to-date and reliable compilation of political, economic and business information, this brand new edition will be an important acquisition to any public, academic or special library reference collection.

"A useful addition to both general reference collections and business collections." –RUSQ

Softcover ISBN 978-1-59237-273-7, 1,700 pages, $175.00

The Directory of Venture Capital & Private Equity Firms, 2009

This edition has been extensively updated and broadly expanded to offer direct access to over 2,800 Domestic and International Venture Capital Firms, including address, phone & fax numbers, e-mail addresses and web sites for both primary and branch locations. Entries include details on the firm's Mission Statement, Industry Group Preferences, Geographic Preferences, Average and Minimum Investments and Investment Criteria. You'll also find details that are available nowhere else, including the Firm's Portfolio Companies and extensive information on each of the firm's Managing Partners, such as Education, Professional Background and Directorships held, along with the Partner's E-mail Address. *The Directory of Venture Capital & Private Equity Firms* offers five important indexes: Geographic Index, Executive Name Index, Portfolio Company Index, Industry Preference Index and College & University Index. With its comprehensive coverage and detailed, extensive information on each company, The Directory of Venture Capital & Private Equity Firms is an important addition to any finance collection.

"The sheer number of listings, the descriptive information and the outstanding indexing make this directory a better value than ...Pratt's Guide to Venture Capital Sources. Recommended for business collections in large public, academic and business libraries." –Choice

Softcover ISBN 978-1-59237-398-7, 1,300 pages, $565/$450 Lib | Online DB: http://gold.greyhouse.com Call (800) 562-2139 for quote

Business Information ♦ Ratings Guides ♦ General Reference ♦ Education ♦
Statistics ♦ Demographics ♦ Health Information ♦ Canadian Information

Grey House Publishing

The Encyclopedia of Emerging Industries

*Published under an exclusive license from the Gale Group, Inc.

The fifth edition of the *Encyclopedia of Emerging Industries* details the inception, emergence, and current status of nearly 120 flourishing U.S. industries and industry segments. These focused essays unearth for users a wealth of relevant, current, factual data previously accessible only through a diverse variety of sources. This volume provides broad-based, highly-readable, industry information under such headings as Industry Snapshot, Organization & Structure, Background & Development, Industry Leaders, Current Conditions, America and the World, Pioneers, and Research & Technology. Essays in this new edition, arranged alphabetically for easy use, have been completely revised, with updated statistics and the most current information on industry trends and developments. In addition, there are new essays on some of the most interesting and influential new business fields, including Application Service Providers, Concierge Services, Entrepreneurial Training, Fuel Cells, Logistics Outsourcing Services, Pharmacogenomics, and Tissue Engineering. Two indexes, General and Industry, provide immediate access to this wealth of information. Plus, two conversion tables for SIC and NAICS codes, along with Suggested Further Readings, are provided to aid the user. *The Encyclopedia of Emerging Industries* pinpoints emerging industries while they are still in the spotlight. This important resource will be an important acquisition to any business reference collection.

"This well-designed source…should become another standard business source, nicely complementing Standard & Poor's Industry Surveys. It contains more information on each industry than Hoover's Handbook of Emerging Companies, is broader in scope than The Almanac of American Employers 1998-1999, but is less expansive than the Encyclopedia of Careers & Vocational Guidance. Highly recommended for all academic libraries and specialized business collections." –Library Journal

Hardcover ISBN 978-1-59237-242-3, 1,400 pages, $325.00

Encyclopedia of American Industries

*Published under an exclusive license from the Gale Group, Inc.

The Encyclopedia of American Industries is a major business reference tool that provides detailed, comprehensive information on a wide range of industries in every realm of American business. A two volume set, Volume I provides separate coverage of nearly 500 manufacturing industries, while Volume II presents nearly 600 essays covering the vast array of services and other non-manufacturing industries in the United States. Combined, these two volumes provide individual essays on every industry recognized by the U.S. Standard Industrial Classification (SIC) system. Both volumes are arranged numerically by SIC code, for easy use. Additionally, each entry includes the corresponding NAICS code(s). The *Encyclopedia's* business coverage includes information on historical events of consequence, as well as current trends and statistics. Essays include an Industry Snapshot, Organization & Structure, Background & Development, Current Conditions, Industry Leaders, Workforce, America and the World, Research & Technology along with Suggested Further Readings. Both SIC and NAICS code conversion tables and an all-encompassing Subject Index, with cross-references, complete the text. With its detailed, comprehensive information on a wide range of industries, this resource will be an important tool for both the industry newcomer and the seasoned professional.

"Encyclopedia of American Industries contains detailed, signed essays on virtually every industry in contemporary society. ... Highly recommended for all but the smallest libraries." -American Reference Books Annual

Two Volumes, Hardcover ISBN 978-1-59237-244-7, 3,000 pages, $650.00

Encyclopedia of Global Industries

*Published under an exclusive license from the Gale Group, Inc.

This fourth edition of the acclaimed *Encyclopedia of Global Industries* presents a thoroughly revised and expanded look at more than 125 business sectors of global significance. Detailed, insightful articles discuss the origins, development, trends, key statistics and current international character of the world's most lucrative, dynamic and widely researched industries – including hundreds of profiles of leading international corporations. Beginning researchers will gain from this book a solid understanding of how each industry operates and which countries and companies are significant participants, while experienced researchers will glean current and historical figures for comparison and analysis. The industries profiled in previous editions have been updated, and in some cases, expanded to reflect recent industry trends. Additionally, this edition provides both SIC and NAICS codes for all industries profiled. As in the original volumes, *The Encyclopedia of Global Industries* offers thorough studies of some of the biggest and most frequently researched industry sectors, including Aircraft, Biotechnology, Computers, Internet Services, Motor Vehicles, Pharmaceuticals, Semiconductors, Software and Telecommunications. An SIC and NAICS conversion table and an all-encompassing Subject Index, with cross-references, are provided to ensure easy access to this wealth of information. These and many others make the *Encyclopedia of Global Industries* the authoritative reference for studies of international industries.

"Provides detailed coverage of the history, development, and current status of 115 of "the world's most lucrative and high-profile industries." It far surpasses the Department of Commerce's U.S. Global Trade Outlook 1995-2000 (GPO, 1995) in scope and coverage. Recommended for comprehensive public and academic library business collections." -Booklist

Hardcover ISBN 978-1-59237-243-0, 1,400 pages, $495.00

Business Information ✦ **Ratings Guides** ✦ **General Reference** ✦ **Education** ✦
Statistics ✦ **Demographics** ✦ **Health Information** ✦ **Canadian Information**

The Directory of Mail Order Catalogs, 2009

Published since 1981, *The Directory of Mail Order Catalogs* is the premier source of information on the mail order catalog industry. It is the source that business professionals and librarians have come to rely on for the thousands of catalog companies in the US. Since the 2007 edition, *The Directory of Mail Order Catalogs* has been combined with its companion volume, *The Directory of Business to Business Catalogs*, to offer all 13,000 catalog companies in one easy-to-use volume. Section I: Consumer Catalogs, covers over 9,000 consumer catalog companies in 44 different product chapters from Animals to Toys & Games. Section II: Business to Business Catalogs, details 5,000 business catalogs, everything from computers to laboratory supplies, building construction and much more. Listings contain detailed contact information including mailing address, phone & fax numbers, web sites, e-mail addresses and key contacts along with important business details such as product descriptions, employee size, years in business, sales volume, catalog size, number of catalogs mailed and more. *The Directory of Mail Order Catalogs*, now with its expanded business to business catalogs, is the largest and most comprehensive resource covering this billion-dollar industry. It is the standard in its field. This important resource is a useful tool for entrepreneurs searching for catalogs to pick up their product, vendors looking to expand their customer base in the catalog industry, market researchers, small businesses investigating new supply vendors, along with the library patron who is exploring the available catalogs in their areas of interest.

"This is a godsend for those looking for information." –Reference Book Review

Softcover ISBN 978-1-59237-396-3, 1,700 pages, $350/$250 Lib | Online DB: http://gold.greyhouse.com Call (800) 562-2139 for quote

Sports Market Place Directory, 2008

For over 20 years, this comprehensive, up-to-date directory has offered direct access to the Who, What, When & Where of the Sports Industry. With over 20,000 updates and enhancements, the *Sports Market Place Directory* is the most detailed, comprehensive and current sports business reference source available. In 1,800 information-packed pages, *Sports Market Place Directory* profiles contact information and key executives for: Single Sport Organizations, Professional Leagues, Multi-Sport Organizations, Disabled Sports, High School & Youth Sports, Military Sports, Olympic Organizations, Media, Sponsors, Sponsorship & Marketing Event Agencies, Event & Meeting Calendars, Professional Services, College Sports, Manufacturers & Retailers, Facilities and much more. The Sports Market Place Directory provides organization's contact information with detailed descriptions including: Key Contacts, physical, mailing, email and web addresses plus phone and fax numbers. *Sports Market Place Directory* provides a one-stop resources for this billion-dollar industry. This will be an important resource for large public libraries, university libraries, university athletic programs, career services or job placement organizations, and is a must for anyone doing research on or marketing to the US and Canadian sports industry.

"Grey House is the new publisher and has produced an excellent edition...highly recommended for public libraries and academic libraries with sports management programs or strong interest in athletics." -Booklist

Softcover ISBN 978-1-59237-348-2, 1,800 pages, $225.00 | Online Database: http://gold.greyhouse.com Call (800) 562-2139 for quote

Food and Beverage Market Place, 2009

Food and Beverage Market Place is bigger and better than ever with thousands of new companies, thousands of updates to existing companies and two revised and enhanced product category indexes. This comprehensive directory profiles over 18,000 Food & Beverage Manufacturers, 12,000 Equipment & Supply Companies, 2,200 Transportation & Warehouse Companies, 2,000 Brokers & Wholesalers, 8,000 Importers & Exporters, 900 Industry Resources and hundreds of Mail Order Catalogs. Listings include detailed Contact Information, Sales Volumes, Key Contacts, Brand & Product Information, Packaging Details and much more. *Food and Beverage Market Place* is available as a three-volume printed set, a subscription-based Online Database via the Internet, on CD-ROM, as well as mailing lists and a licensable database.

"An essential purchase for those in the food industry but will also be useful in public libraries where needed. Much of the information will be difficult and time consuming to locate without this handy three-volume ready-reference source." –ARBA

3 Vol Set, Softcover ISBN 978-1-59237-361-1, 8,500 pages, $595 | Online DB: http://gold.greyhouse.com Call (800) 562-2139 for quote

The Grey House Performing Arts Directory, 2009

The Grey House Performing Arts Directory is the most comprehensive resource covering the Performing Arts. This important directory provides current information on over 8,500 Dance Companies, Instrumental Music Programs, Opera Companies, Choral Groups, Theater Companies, Performing Arts Series and Performing Arts Facilities. Plus, this edition now contains a brand new section on Artist Management Groups. In addition to mailing address, phone & fax numbers, e-mail addresses and web sites, dozens of other fields of available information include mission statement, key contacts, facilities, seating capacity, season, attendance and more. This directory also provides an important Information Resources section that covers hundreds of Performing Arts Associations, Magazines, Newsletters, Trade Shows, Directories, Databases and Industry Web Sites. Five indexes provide immediate access to this wealth of information: Entry Name, Executive Name, Performance Facilities, Geographic and Information Resources. *The Grey House Performing Arts Directory* pulls together thousands of Performing Arts Organizations, Facilities and Information Resources into an easy-to-use source – this kind of comprehensiveness and extensive detail is not available in any resource on the market place today.

"Immensely useful and user-friendly … recommended for public, academic and certain special library reference collections." –Booklist

Softcover ISBN 978-1-59237-376-5, 1,500 pages, $185.00 | Online Database: http://gold.greyhouse.com Call (800) 562-2139 for quote

To preview any of our Directories Risk-Free for 30 days, call (800) 562-2139 or fax (518) 789-0556
www.greyhouse.com books@greyhouse.com

Business Information ◆ Ratings Guides ◆ General Reference ◆ Education ◆ Statistics ◆ Demographics ◆ Health Information ◆ Canadian Information

The Environmental Resource Handbook, 2008/09

The Environmental Resource Handbook is the most up-to-date and comprehensive source for Environmental Resources and Statistics. Section I: Resources provides detailed contact information for thousands of information sources, including Associations & Organizations, Awards & Honors, Conferences, Foundations & Grants, Environmental Health, Government Agencies, National Parks & Wildlife Refuges, Publications, Research Centers, Educational Programs, Green Product Catalogs, Consultants and much more. Section II: Statistics, provides statistics and rankings on hundreds of important topics, including Children's Environmental Index, Municipal Finances, Toxic Chemicals, Recycling, Climate, Air & Water Quality and more. This kind of up-to-date environmental data, all in one place, is not available anywhere else on the market place today. This vast compilation of resources and statistics is a must-have for all public and academic libraries as well as any organization with a primary focus on the environment.

> *"...the intrinsic value of the information make it worth consideration by libraries with environmental collections and environmentally concerned users."* –Booklist

Softcover ISBN 978-1-59237-195-2, 1,000 pages, $155.00 | Online Database: http://gold.greyhouse.com Call (800) 562-2139 for quote

New York State Directory, 2008/09

The New York State Directory, published annually since 1983, is a comprehensive and easy-to-use guide to accessing public officials and private sector organizations and individuals who influence public policy in the state of New York. *The New York State Directory* includes important information on all New York state legislators and congressional representatives, including biographies and key committee assignments. It also includes staff rosters for all branches of New York state government and for federal agencies and departments that impact the state policy process. Following the state government section are 25 chapters covering policy areas from agriculture through veterans' affairs. Each chapter identifies the state, local and federal agencies and officials that formulate or implement policy. In addition, each chapter contains a roster of private sector experts and advocates who influence the policy process. The directory also offers appendices that include statewide party officials; chambers of commerce; lobbying organizations; public and private universities and colleges; television, radio and print media; and local government agencies and officials.

> *"This comprehensive directory covers not only New York State government offices and key personnel but pertinent U.S. government agencies and non-governmental entities. This directory is all encompassing... recommended."* -Choice

New York State Directory - Softcover ISBN 978-1-59237-358-1, 800 pages, $145.00
Online Database: http://gold.greyhouse.com Call (800) 562-2139 for quote
New York State Directory with *Profiles of New York* – 2 Volumes, Softcover ISBN 978-1-59237-359-8, 1,600 pages, $225.00

The Grey House Homeland Security Directory, 2008

This updated edition features the latest contact information for government and private organizations involved with Homeland Security along with the latest product information and provides detailed profiles of nearly 1,000 Federal & State Organizations & Agencies and over 3,000 Officials and Key Executives involved with Homeland Security. These listings are incredibly detailed and include Mailing Address, Phone & Fax Numbers, Email Addresses & Web Sites, a complete Description of the Agency and a complete list of the Officials and Key Executives associated with the Agency. Next, *The Grey House Homeland Security Directory* provides the go-to source for Homeland Security Products & Services. This section features over 2,000 Companies that provide Consulting, Products or Services. With this Buyer's Guide at their fingertips, users can locate suppliers of everything from Training Materials to Access Controls, from Perimeter Security to BioTerrorism Countermeasures and everything in between – complete with contact information and product descriptions. A handy Product Locator Index is provided to quickly and easily locate suppliers of a particular product. This comprehensive, information-packed resource will be a welcome tool for any company or agency that is in need of Homeland Security information and will be a necessary acquisition for the reference collection of all public libraries and large school districts.

> *"Compiles this information in one place and is discerning in content. A useful purchase for public and academic libraries."* –Booklist

Softcover ISBN 978-1-59237-196-6, 800 pages, $195.00 | Online Database: http://gold.greyhouse.com Call (800) 562-2139 for quote

The Grey House Safety & Security Directory, 2009

The Grey House Safety & Security Directory is the most comprehensive reference tool and buyer's guide for the safety and security industry. Arranged by safety topic, each chapter begins with OSHA regulations for the topic, followed by Training Articles written by top professionals in the field and Self-Inspection Checklists. Next, each topic contains Buyer's Guide sections that feature related products and services. Topics include Administration, Insurance, Loss Control & Consulting, Protective Equipment & Apparel, Noise & Vibration, Facilities Monitoring & Maintenance, Employee Health Maintenance & Ergonomics, Retail Food Services, Machine Guards, Process Guidelines & Tool Handling, Ordinary Materials Handling, Hazardous Materials Handling, Workplace Preparation & Maintenance, Electrical Lighting & Safety, Fire & Rescue and Security. Six important indexes make finding information and product manufacturers quick and easy: Geographical Index of Manufacturers and Distributors, Company Profile Index, Brand Name Index, Product Index, Index of Web Sites and Index of Advertisers. This comprehensive, up-to-date reference will provide every tool necessary to make sure a business is in compliance with OSHA regulations and locate the products and services needed to meet those regulations.

"Presents industrial safety information for engineers, plant managers, risk managers,
and construction site supervisors…" –Choice

Softcover ISBN 978-1-59237-375-8, 1,500 pages, $165.00

The Grey House Transportation Security Directory & Handbook

This is the only reference of its kind that brings together current data on Transportation Security. With information on everything from Regulatory Authorities to Security Equipment, this top-flight database brings together the relevant information necessary for creating and maintaining a security plan for a wide range of transportation facilities. With this current, comprehensive directory at the ready you'll have immediate access to: Regulatory Authorities & Legislation; Information Resources; Sample Security Plans & Checklists; Contact Data for Major Airports, Seaports, Railroads, Trucking Companies and Oil Pipelines; Security Service Providers; Recommended Equipment & Product Information and more. Using the *Grey House Transportation Security Directory & Handbook*, managers will be able to quickly and easily assess their current security plans; develop contacts to create and maintain new security procedures; and source the products and services necessary to adequately maintain a secure environment. This valuable resource is a must for all Security Managers at Airports, Seaports, Railroads, Trucking Companies and Oil Pipelines.

"Highly recommended. Library collections that support all levels of readers, including professionals/practitioners; and
schools/organizations offering education and training in transportation security." -Choice

Softcover ISBN 978-1-59237-075-7, 800 pages, $195.00

The Grey House Biometric Information Directory

This edition offers a complete, current overview of biometric companies and products – one of the fastest growing industries in today's economy. Detailed profiles of manufacturers of the latest biometric technology, including Finger, Voice, Face, Hand, Signature, Iris, Vein and Palm Identification systems. Data on the companies include key executives, company size and a detailed, indexed description of their product line. Information in the directory includes: Editorial on Advancements in Biometrics; Profiles of 700+ companies listed with contact information; Organizations, Trade & Educational Associations, Publications, Conferences, Trade Shows and Expositions Worldwide; Web Site Index; Biometric & Vendors Services Index by Types of Biometrics; and a Glossary of Biometric Terms. This resource will be an important source for anyone who is considering the use of a biometric product, investing in the development of biometric technology, support existing marketing and sales efforts and will be an important acquisition for the business reference collection for large public and business libraries.

"This book should prove useful to agencies or businesses seeking companies that deal with biometric technology. Summing Up:
Recommended. Specialized collections serving researchers/faculty and professionals/practitioners." -Choice

Softcover ISBN 978-1-59237-121-1, 800 pages, $225.00

Business Information ✦ Ratings Guides ✦ General Reference ✦ Education ✦
Statistics ✦ Demographics ✦ Health Information ✦ Canadian Information

The Rauch Guide to the US Adhesives & Sealants, Cosmetics & Toiletries, Ink, Paint, Plastics, Pulp & Paper and Rubber Industries

The Rauch Guides save time and money by organizing widely scattered information and providing estimates for important business decisions, some of which are available nowhere else. Within each Guide, after a brief introduction, the ECONOMICS section provides data on industry shipments; long-term growth and forecasts; prices; company performance; employment, expenditures, and productivity; transportation and geographical patterns; packaging; foreign trade; and government regulations. Next, TECHNOLOGY & RAW MATERIALS provide market, technical, and raw material information for chemicals, equipment and related materials, including market size and leading suppliers, prices, end uses, and trends. PRODUCTS & MARKETS provide information for each major industry product, including market size and historical trends, leading suppliers, five-year forecasts, industry structure, and major end uses. Next, the COMPANY DIRECTORY profiles major industry companies, both public and private. Information includes complete contact information, web address, estimated total and domestic sales, product description, and recent mergers and acquisitions. *The Rauch Guides* will prove to be an invaluable source of market information, company data, trends and forecasts that anyone in these fast-paced industries.

"An invaluable and affordable publication. The comprehensive nature of the data and text offers considerable insights into the industry, market sizes, company activities, and applications of the products of the industry. The additions that have been made have certainly enhanced the value of the Guide." –Adhesives & Sealants Newsletter of the Rauch Guide to the US Adhesives & Sealants Industry

Paint Industry: Softcover ISBN 978-1-59237-127-3 $595 | Plastics Industry: Softcover ISBN 978-1-59237-128-0 $595 | Adhesives and Sealants Industry: Softcover ISBN 978-1-59237-129-7 $595 | Ink Industry: Softcover ISBN 978-1-59237-126-6 $595 | Rubber Industry: Softcover ISBN 978-1-59237-130-3 $595 | Pulp and Paper Industry: Softcover ISBN 978-1-59237-131-0 $595 | Cosmetic & Toiletries Industry: Softcover ISBN 978-1-59237-132-7 $895

Research Services Directory: Commercial & Corporate Research Centers

This ninth edition provides access to well over 8,000 independent Commercial Research Firms, Corporate Research Centers and Laboratories offering contract services for hands-on, basic or applied research. Research Services Directory covers the thousands of types of research companies, including Biotechnology & Pharmaceutical Developers, Consumer Product Research, Defense Contractors, Electronics & Software Engineers, Think Tanks, Forensic Investigators, Independent Commercial Laboratories, Information Brokers, Market & Survey Research Companies, Medical Diagnostic Facilities, Product Research & Development Firms and more. Each entry provides the company's name, mailing address, phone & fax numbers, key contacts, web site, e-mail address, as well as a company description and research and technical fields served. Four indexes provide immediate access to this wealth of information: Research Firms Index, Geographic Index, Personnel Name Index and Subject Index.

"An important source for organizations in need of information about laboratories, individuals and other facilities." –ARBA

Softcover ISBN 978-1-59237-003-0, 1,400 pages, $465.00

International Business and Trade Directories

Completely updated, the Third Edition of *International Business and Trade Directories* now contains more than 10,000 entries, over 2,000 more than the last edition, making this directory the most comprehensive resource of the worlds business and trade directories. Entries include content descriptions, price, publisher's name and address, web site and e-mail addresses, phone and fax numbers and editorial staff. Organized by industry group, and then by region, this resource puts over 10,000 industry-specific business and trade directories at the reader's fingertips. Three indexes are included for quick access to information: Geographic Index, Publisher Index and Title Index. Public, college and corporate libraries, as well as individuals and corporations seeking critical market information will want to add this directory to their marketing collection.

"Reasonably priced for a work of this type, this directory should appeal to larger academic, public and corporate libraries with an international focus." –Library Journal

Softcover ISBN 978-1-930956-63-6, 1,800 pages, $225.00

Business Information ◆ **Ratings Guides** ◆ General Reference ◆ Education ◆
Statistics ◆ Demographics ◆ Health Information ◆ Canadian Information

TheStreet.com Ratings Guide to Health Insurers

TheStreet.com Ratings Guide to Health Insurers is the first and only source to cover the financial stability of the nation's health care system, rating the financial safety of more than 6,000 health insurance providers, health maintenance organizations (HMOs) and all of the Blue Cross Blue Shield plans – updated quarterly to ensure the most accurate information. The Guide also provides a complete listing of all the major health insurers, including all Long-Term Care and Medigap insurers. Our *Guide to Health Insurers* includes comprehensive, timely coverage on the financial stability of HMOs and health insurers; the most accurate insurance company ratings available–the same quality ratings heralded by the U.S. General Accounting Office; separate listings for those companies offering Medigap and long-term care policies; the number of serious consumer complaints filed against most HMOs so you can see who is actually providing the best (or worst) service and more. The easy-to-use layout gives you a one-line summary analysis for each company that we track, followed by an in-depth, detailed analysis of all HMOs and the largest health insurers. The guide also includes a list of TheStreet.com Ratings Recommended Companies with information on how to contact them, and the reasoning behind any rating upgrades or downgrades.

> *"With 20 years behind its insurance-advocacy research [the rating guide] continues to offer a wealth of information that helps consumers weigh their healthcare options now and in the future." -Today's Librarian*

Issues published quarterly, Softcover, 550 pages, $499.00 for four quarterly issues, $249.00 for a single issue

TheStreet.com Ratings Guide to Life & Annuity Insurers

TheStreet.com Safety Ratings are the most reliable source for evaluating an insurer's financial solvency risk. Consequently, policyholders have come to rely on TheStreet.com's flagship publication, *TheStreet.com Ratings Guide to Life & Annuity Insurers*, to help them identify the safest companies to do business with. Each easy-to-use edition delivers TheStreet.com's independent ratings and analyses on more than 1,100 insurers, updated every quarter. Plus, your patrons will find a complete list of TheStreet.com Recommended Companies, including contact information, and the reasoning behind any rating upgrades or downgrades. This guide is perfect for those who are considering the purchase of a life insurance policy, placing money in an annuity, or advising clients about insurance and annuities. A life or health insurance policy or annuity is only as secure as the insurance company issuing it. Therefore, make sure your patrons have what they need to periodically monitor the financial condition of the companies with whom they have an investment. The TheStreet.com Ratings product line is designed to help them in their evaluations.

> *"Weiss has an excellent reputation and this title is held by hundreds of libraries. This guide is recommended for public and academic libraries." -ARBA*

Issues published quarterly, Softcover, 360 pages, $499.00 for four quarterly issues, $249.00 for a single issue

TheStreet.com Ratings Guide to Property & Casualty Insurers

TheStreet.com Ratings Guide to Property and Casualty Insurers provides the most extensive coverage of insurers writing policies, helping consumers and businesses avoid financial headaches. Updated quarterly, this easy-to-use publication delivers the independent, unbiased TheStreet.com Safety Ratings and supporting analyses on more than 2,800 U.S. insurance companies, offering auto & homeowners insurance, business insurance, worker's compensation insurance, product liability insurance, medical malpractice and other professional liability insurance. Each edition includes a list of TheStreet.com Recommended Companies by type of insurance, including a contact number, plus helpful information about the coverage provided by the State Guarantee Associations.

> *"In contrast to the other major insurance rating agencies...Weiss does not have a financial relationship worth the companies it rates. A GAO study found that Weiss identified financial vulnerability earlier than the other rating agencies." -ARBA*

Issues published quarterly, Softcover, 455 pages, $499.00 for four quarterly issues, $249.00 for a single issue

TheStreet.com Ratings Consumer Box Set

Deliver the critical information your patrons need to safeguard their personal finances with *TheStreet.com Ratings' Consumer Guide Box Set*. Each of the eight guides is packed with accurate, unbiased information and recommendations to help your patrons make sound financial decisions. TheStreet.com Ratings Consumer Guide Box Set provides your patrons with easy to understand guidance on important personal finance topics, including: *Consumer Guide to Variable Annuities, Consumer Guide to Medicare Supplement Insurance, Consumer Guide to Elder Care Choices, Consumer Guide to Automobile Insurance, Consumer Guide to Long-Term Care Insurance, Consumer Guide to Homeowners Insurance, Consumer Guide to Term Life Insurance,* and *Consumer Guide to Medicare Prescription Drug Coverage*. Each guide provides an easy-to-read overview of the topic, what to look out for when selecting a company or insurance plan to do business with, who are the recommended companies to work with and how to navigate through these often-times difficult decisions. Custom worksheets and step-by-step directions make these resources accessible to all types of users. Packaged in a handy custom display box, these helpful guides will prove to be a much-used addition to any reference collection.

Issues published twice per year, Softcover, 600 pages, $499.00 for two biennial issues

Business Information ✦ **Ratings Guides** ✦ General Reference ✦ Education ✦
Statistics ✦ Demographics ✦ Health Information ✦ Canadian Information

TheStreet.com Ratings Guide to Stock Mutual Funds

TheStreet.com Ratings Guide to Stock Mutual Funds offers ratings and analyses on more than 8,800 equity mutual funds – more than any other publication. The exclusive TheStreet.com Investment Ratings combine an objective evaluation of each fund's performance and risk to provide a single, user-friendly, composite rating, giving your patrons a better handle on a mutual fund's risk-adjusted performance. Each edition identifies the top-performing mutual funds based on risk category, type of fund, and overall risk-adjusted performance. TheStreet.com's unique investment rating system makes it easy to see exactly which stocks are on the rise and which ones should be avoided. For those investors looking to tailor their mutual fund selections based on age, income, and tolerance for risk, we've also assigned two component ratings to each fund: a performance rating and a risk rating. With these, you can identify those funds that are best suited to meet your - or your client's – individual needs and goals. Plus, we include a handy Risk Profile Quiz to help you assess your personal tolerance for risk. So whether you're an investing novice or professional, the *Guide to Stock Mutual Funds* gives you everything you need to find a mutual fund that is right for you.

> *"There is tremendous need for information such as that provided by this Weiss publication. This reasonably priced guide is recommended for public and academic libraries serving investors." -ARBA*

Issues published quarterly, Softcover, 655 pages, $499 for four quarterly issues, $249 for a single issue

TheStreet.com Ratings Guide to Exchange-Traded Funds

TheStreet.com Ratings editors analyze hundreds of mutual funds each quarter, condensing all of the available data into a single composite opinion of each fund's risk-adjusted performance. The intuitive, consumer-friendly ratings allow investors to instantly identify those funds that have historically done well and those that have under-performed the market. Each quarterly edition identifies the top-performing exchange-traded funds based on risk category, type of fund, and overall risk-adjusted performance. The rating scale, A through F, gives you a better handle on an exchange-traded fund's risk-adjusted performance. Other features include Top & Bottom 200 Exchange-Traded Funds; Performance and Risk: 100 Best and Worst Exchange- Traded Funds; Investor Profile Quiz; Performance Benchmarks and Fund Type Descriptions. With the growing popularity of mutual fund investing, consumers need a reliable source to help them track and evaluate the performance of their mutual fund holdings. Plus, they need a way of identifying and monitoring other funds as potential new investments. Unfortunately, the hundreds of performance and risk measures available, multiplied by the vast number of mutual fund investments on the market today, can make this a daunting task for even the most sophisticated investor. This Guide will serve as a useful tool for both the first-time and seasoned investor.

Editions published quarterly, Softcover, 440 pages, $499.00 for four quarterly issues, $249.00 for a single issue

TheStreet.com Ratings Guide to Bond & Money Market Mutual Funds

TheStreet.com Ratings Guide to Bond & Money Market Mutual Funds has everything your patrons need to easily identify the top-performing fixed income funds on the market today. Each quarterly edition contains TheStreet.com's independent ratings and analyses on more than 4,600 fixed income funds – more than any other publication, including corporate bond funds, high-yield bond funds, municipal bond funds, mortgage security funds, money market funds, global bond funds and government bond funds. In addition, the fund's risk rating is combined with its three-year performance rating to get an overall picture of the fund's risk-adjusted performance. The resulting TheStreet.com Investment Rating gives a single, user-friendly, objective evaluation that makes it easy to compare one fund to another and select the right fund based on the level of risk tolerance. Most investors think of fixed income mutual funds as "safe" investments. That's not always the case, however, depending on the credit risk, interest rate risk, and prepayment risk of the securities owned by the fund. TheStreet.com Ratings assesses each of these risks and assigns each fund a risk rating to help investors quickly evaluate the fund's risk component. Plus, we include a handy Risk Profile Quiz to help you assess your personal tolerance for risk. So whether you're an investing novice or professional, the *Guide to Bond and Money Market Mutual Funds* gives you everything you need to find a mutual fund that is right for you.

> *"Comprehensive... It is easy to use and consumer-oriented, and can be recommended for larger public and academic libraries." -ARBA*

Issues published quarterly, Softcover, 470 pages, $499.00 for four quarterly issues, $249.00 for a single issue

TheStreet.com Ratings Guide to Banks & Thrifts

Updated quarterly, for the most up-to-date information, *TheStreet.com Ratings Guide to Banks and Thrifts* offers accurate, intuitive safety ratings your patrons can trust; supporting ratios and analyses that show an institution's strong & weak points; identification of the TheStreet.com Recommended Companies with branches in your area; a complete list of institutions receiving upgrades/downgrades; and comprehensive coverage of every bank and thrift in the nation – more than 9,000. TheStreet.com Safety Ratings are then based on the analysts' review of publicly available information collected by the federal banking regulators. The easy-to-use layout gives you: the institution's TheStreet.com Safety Rating for the last 3 years; the five key indexes used to evaluate each institution; along with the primary ratios and statistics used in determining the company's rating. *TheStreet.com Ratings Guide to Banks & Thrifts* will be a must for individuals who are concerned about the safety of their CD or savings account; need to be sure that an existing line of credit will be there when they need it; or simply want to avoid the hassles of dealing with a failing or troubled institution.

> *"Large public and academic libraries most definitely need to acquire the work. Likewise, special libraries in large corporations will find this title indispensable." -ARBA*

Issues published quarterly, Softcover, 370 pages, $499.00 for four quarterly issues, $249.00 for a single issue

---

Business Information ✦ **Ratings Guides** ✦ General Reference ✦ Education ✦
Statistics ✦ Demographics ✦ Health Information ✦ Canadian Information

TheStreet.com Ratings Guide to Common Stocks

TheStreet.com Ratings Guide to Common Stocks gives your patrons reliable insight into the risk-adjusted performance of common stocks listed on the NYSE, AMEX, and Nasdaq – over 5,800 stocks in all – more than any other publication. TheStreet.com's unique investment rating system makes it easy to see exactly which stocks are on the rise and which ones should be avoided. In addition, your patrons also get supporting analysis showing growth trends, profitability, debt levels, valuation levels, the top-rated stocks within each industry, and more. Plus, each stock is ranked with the easy-to-use buy-hold-sell equivalents commonly used by Wall Street. Whether they're selecting their own investments or checking up on a broker's recommendation, TheStreet.com Ratings can help them in their evaluations.

"Users... will find the information succinct and the explanations readable, easy to understand, and helpful to a novice." -Library Journal

Issues published quarterly, Softcover, 440 pages, $499.00 for four quarterly issues, $249.00 for a single issue

TheStreet.com Ratings Ultimate Guided Tour of Stock Investing

This important reference guide from TheStreet.com Ratings is just what librarians around the country have asked for: a step-by-step introduction to stock investing for the beginning to intermediate investor. This easy-to-navigate guide explores the basics of stock investing and includes the intuitive TheStreet.com Investment Rating on more than 5,800 stocks, complete with real-world investing information that can be put to use immediately with stocks that fit the concepts discussed in the guide; informative charts, graphs and worksheets; easy-to-understand explanations on topics like P/E, compound interest, marked indices, diversifications, brokers, and much more; along with financial safety ratings for every stock on the NYSE, American Stock Exchange and the Nasdaq. This consumer-friendly guide offers complete how-to information on stock investing that can be put to use right away; a friendly format complete with our "Wise Guide" who leads the reader on a safari to learn about the investing jungle; helpful charts, graphs and simple worksheets; the intuitive TheStreet.com Investment rating on over 6,000 stocks — every stock found on the NYSE, American Stock Exchange and the NASDAQ; and much more.

"Provides investors with an alternative to stock broker recommendations, which recently have been tarnished by conflicts of interest. In summary, the guide serves as a welcome addition for all public library collections." -ARBA

Issues published quarterly, Softcover, 370 pages, $499.00 for four quarterly issues, $249.00 for a single issue

TheStreet.com Ratings' Reports & Services

- Ratings Online — An on-line summary covering an individual company's TheStreet.com Financial Strength Rating or an investment's unique TheStreet.com Investment Rating with the factors contributing to that rating; available 24 hours a day by visiting www.thestreet.com/tscratings or calling (800) 289-9222.
- Unlimited Ratings Research — The ultimate research tool providing fast, easy online access to the very latest TheStreet.com Financial Strength Ratings and Investment Ratings. Price: $559 per industry.

Contact TheStreet.com for more information about Reports & Services at www.thestreet.com/tscratings or call (800) 289-9222

TheStreet.com Ratings' Custom Reports

TheStreet.com Ratings is pleased to offer two customized options for receiving ratings data. Each taps into TheStreet.com's vast data repositories and is designed to provide exactly the data you need. Choose from a variety of industries, companies, data variables, and delivery formats including print, Excel, SQL, Text or Access.
- Customized Reports - get right to the heart of your company's research and data needs with a report customized to your specifications.
- Complete Database Download – TheStreet.com will design and deliver the database; from there you can sort it, recalculate it, and format your results to suit your specific needs.

Contact TheStreet.com for more information about Custom Reports at www.thestreet.com/tscratings or call (800) 289-9222

The Value of a Dollar 1600-1859, The Colonial Era to The Civil War

Following the format of the widely acclaimed, *The Value of a Dollar, 1860-2004*, *The Value of a Dollar 1600-1859, The Colonial Era to The Civil War* records the actual prices of thousands of items that consumers purchased from the Colonial Era to the Civil War. Our editorial department had been flooded with requests from users of our *Value of a Dollar* for the same type of information, just from an earlier time period. This new volume is just the answer – with pricing data from 1600 to 1859. Arranged into five-year chapters, each 5-year chapter includes a Historical Snapshot, Consumer Expenditures, Investments, Selected Income, Income/Standard Jobs, Food Basket, Standard Prices and Miscellany. There is also a section on Trends. This informative section charts the change in price over time and provides added detail on the reasons prices changed within the time period, including industry developments, changes in consumer attitudes and important historical facts. This fascinating survey will serve a wide range of research needs and will be useful in all high school, public and academic library reference collections.

"The Value of a Dollar: Colonial Era to the Civil War, 1600-1865 will find a happy audience among students, researchers, and general browsers. It offers a fascinating and detailed look at early American history from the viewpoint of everyday people trying to make ends meet. This title and the earlier publication, The Value of a Dollar, 1860-2004, complement each other very well, and readers will appreciate finding them side-by-side on the shelf." -Booklist

Hardcover ISBN 978-1-59237-094-8, 600 pages, $145.00 | Ebook ISBN 978-1-59237-169-3 www.greyhouse.com/ebooks.htm

The Value of a Dollar 1860-2009, Fourth Edition

A guide to practical economy, *The Value of a Dollar* records the actual prices of thousands of items that consumers purchased from the Civil War to the present, along with facts about investment options and income opportunities. This brand new Third Edition boasts a brand new addition to each five-year chapter, a section on Trends. This informative section charts the change in price over time and provides added detail on the reasons prices changed within the time period, including industry developments, changes in consumer attitudes and important historical facts. Plus, a brand new chapter for 2005-2009 has been added. Each 5-year chapter includes a Historical Snapshot, Consumer Expenditures, Investments, Selected Income, Income/Standard Jobs, Food Basket, Standard Prices and Miscellany. This interesting and useful publication will be widely used in any reference collection.

"Business historians, reporters, writers and students will find this source... very helpful for historical research. Libraries will want to purchase it." –ARBA

Hardcover ISBN 978-1-59237-403-8, 600 pages, $145.00 | Ebook ISBN 978-1-59237-173-0 www.greyhouse.com/ebooks.htm

Working Americans 1880-1999
Volume I: The Working Class, Volume II: The Middle Class, Volume III: The Upper Class

Each of the volumes in the *Working Americans* series focuses on a particular class of Americans, The Working Class, The Middle Class and The Upper Class over the last 120 years. Chapters in each volume focus on one decade and profile three to five families. Family Profiles include real data on Income & Job Descriptions, Selected Prices of the Times, Annual Income, Annual Budgets, Family Finances, Life at Work, Life at Home, Life in the Community, Working Conditions, Cost of Living, Amusements and much more. Each chapter also contains an Economic Profile with Average Wages of other Professions, a selection of Typical Pricing, Key Events & Inventions, News Profiles, Articles from Local Media and Illustrations. The *Working Americans* series captures the lifestyles of each of the classes from the last twelve decades, covers a vast array of occupations and ethnic backgrounds and travels the entire nation. These interesting and useful compilations of portraits of the American Working, Middle and Upper Classes during the last 120 years will be an important addition to any high school, public or academic library reference collection.

"These interesting, unique compilations of economic and social facts, figures and graphs will support multiple research needs. They will engage and enlighten patrons in high school, public and academic library collections." –Booklist

Volume I: The Working Class Hardcover ISBN 978-1-891482-81-6, 558 pages, $145.00 | Volume II: The Middle Class Hardcover ISBN 978-1-891482-72-4, 591 pages, $145.00 | Volume III: The Upper Class Hardcover ISBN 978-1-930956-38-4, 567 pages, $145.00 | www.greyhouse.com/ebooks.htm

Working Americans 1880-1999 Volume IV: Their Children

This Fourth Volume in the highly successful *Working Americans* series focuses on American children, decade by decade from 1880 to 1999. This interesting and useful volume introduces the reader to three children in each decade, one from each of the Working, Middle and Upper classes. Like the first three volumes in the series, the individual profiles are created from interviews, diaries, statistical studies, biographies and news reports. Profiles cover a broad range of ethnic backgrounds, geographic area and lifestyles – everything from an orphan in Memphis in 1882, following the Yellow Fever epidemic of 1878 to an eleven-year-old nephew of a beer baron and owner of the New York Yankees in New York City in 1921. Chapters also contain important supplementary materials including News Features as well as information on everything from Schools to Parks, Infectious Diseases to Childhood Fears along with Entertainment, Family Life and much more to provide an informative overview of the lifestyles of children from each decade. This interesting account of what life was like for Children in the Working, Middle and Upper Classes will be a welcome addition to the reference collection of any high school, public or academic library.

Hardcover ISBN 978-1-930956-35-3, 600 pages, $145.00 | Ebook ISBN 978-1-59237-166-2 www.greyhouse.com/ebooks.htm

To preview any of our Directories Risk-Free for 30 days, call (800) 562-2139 or fax (518) 789-0556
www.greyhouse.com books@greyhouse.com

Business Information ✦ Ratings Guides ✦ General Reference ✦ **Education** ✦
Statistics ✦ Demographics ✦ Health Information ✦ Canadian Information

Grey House Publishing

Working Americans 1880-2003 Volume V: Americans At War

Working Americans 1880-2003 Volume V: Americans At War is divided into 11 chapters, each covering a decade from 1880-2003 and examines the lives of Americans during the time of war, including declared conflicts, one-time military actions, protests, and preparations for war. Each decade includes several personal profiles, whether on the battlefield or on the homefront, that tell the stories of civilians, soldiers, and officers during the decade. The profiles examine: Life at Home; Life at Work; and Life in the Community. Each decade also includes an Economic Profile with statistical comparisons, a Historical Snapshot, News Profiles, local News Articles, and Illustrations that provide a solid historical background to the decade being examined. Profiles range widely not only geographically, but also emotionally, from that of a girl whose leg was torn off in a blast during WWI, to the boredom of being stationed in the Dakotas as the Indian Wars were drawing to a close. As in previous volumes of the *Working Americans* series, information is presented in narrative form, but hard facts and real-life situations back up each story. The basis of the profiles come from diaries, private print books, personal interviews, family histories, estate documents and magazine articles. For easy reference, *Working Americans 1880-2003 Volume V: Americans At War* includes an in-depth Subject Index. The Working Americans series has become an important reference for public libraries, academic libraries and high school libraries. This fifth volume will be a welcome addition to all of these types of reference collections.

Hardcover ISBN 978-1-59237-024-5, 600 pages, $145.00 | Ebook ISBN 978-1-59237-167-9 www.greyhouse.com/ebooks.htm

Working Americans 1880-2005 Volume VI: Women at Work

Unlike any other volume in the *Working Americans* series, this Sixth Volume, is the first to focus on a particular gender of Americans. *Volume VI: Women at Work*, traces what life was like for working women from the 1860's to the present time. Beginning with the life of a maid in 1890 and a store clerk in 1900 and ending with the life and times of the modern working women, this text captures the struggle, strengths and changing perception of the American woman at work. Each chapter focuses on one decade and profiles three to five women with real data on Income & Job Descriptions, Selected Prices of the Times, Annual Income, Annual Budgets, Family Finances, Life at Work, Life at Home, Life in the Community, Working Conditions, Cost of Living, Amusements and much more. For even broader access to the events, economics and attitude towards women throughout the past 130 years, each chapter is supplemented with News Profiles, Articles from Local Media, Illustrations, Economic Profiles, Typical Pricing, Key Events, Inventions and more. This important volume illustrates what life was like for working women over time and allows the reader to develop an understanding of the changing role of women at work. These interesting and useful compilations of portraits of women at work will be an important addition to any high school, public or academic library reference collection.

Hardcover ISBN 978-1-59237-063-4, 600 pages, $145.00 | Ebook ISBN 978-1-59237-168-6 www.greyhouse.com/ebooks.htm

Working Americans 1880-2005 Volume VII: Social Movements

Working Americans series, Volume VII: Social Movements explores how Americans sought and fought for change from the 1880s to the present time. Following the format of previous volumes in the Working Americans series, the text examines the lives of 34 individuals who have worked -- often behind the scenes --- to bring about change. Issues include topics as diverse as the Anti-smoking movement of 1901 to efforts by Native Americans to reassert their long lost rights. Along the way, the book will profile individuals brave enough to demand suffrage for Kansas women in 1912 or demand an end to lynching during a March on Washington in 1923. Each profile is enriched with real data on Income & Job Descriptions, Selected Prices of the Times, Annual Incomes & Budgets, Life at Work, Life at Home, Life in the Community, along with News Features, Key Events, and Illustrations. The depth of information contained in each profile allow the user to explore the private, financial and public lives of these subjects, deepening our understanding of how calls for change took place in our society. A must-purchase for the reference collections of high school libraries, public libraries and academic libraries.

Hardcover ISBN 978-1-59237-101-3, 600 pages, $145.00 | Ebook ISBN 978-1-59237-174-7 www.gale.com/gvrl/partners/grey.htm

Working Americans 1880-2005 Volume VIII: Immigrants

Working Americans 1880-2007 Volume VIII: Immigrants illustrates what life was like for families leaving their homeland and creating a new life in the United States. Each chapter covers one decade and introduces the reader to three immigrant families. Family profiles cover what life was like in their homeland, in their community in the United States, their home life, working conditions and so much more. As the reader moves through these pages, the families and individuals come to life, painting a picture of why they left their homeland, their experiences in setting roots in a new country, their struggles and triumphs, stretching from the 1800s to the present time. Profiles include a seven-year-old Swedish girl who meets her father for the first time at Ellis Island; a Chinese photographer's assistant; an Armenian who flees the genocide of his country to build Ford automobiles in Detroit; a 38-year-old German bachelor cigar maker who settles in Newark NJ, but contemplates tobacco farming in Virginia; a 19-year-old Irish domestic servant who is amazed at the easy life of American dogs; a 19-year-old Filipino who came to Hawaii against his parent's wishes to farm sugar cane; a French-Canadian who finds success as a boxer in Maine and many more. As in previous volumes, information is presented in narrative form, but hard facts and real-life situations back up each story. With the topic of immigration being so hotly debated in this country, this timely resource will prove to be a useful source for students, researchers, historians and library patrons to discover the issues facing immigrants in the United States. This title will be a useful addition to reference collections of public libraries, university libraries and high schools.

Hardcover ISBN 978-1-59237-197-6, 600 pages, $145.00 | Ebook ISBN 978-1-59237-232-4 www.greyhouse.com/ebooks.htm

Business Information ◆ Ratings Guides ◆ General Reference ◆ **Education** ◆
Statistics ◆ Demographics ◆ Health Information ◆ Canadian Information

Working Americans 1770-1896 Volume IX: From the Revolutionary War to the Civil War

Working Americans 1770-1869: From the Revolutionary War to the Civil War examines what life was like for the earliest of Americans. Like previous volumes in the successful Working Americans series, each chapter introduces the reader to three individuals or families. These profiles illustrate what life was like for that individual, at home, in the community and at work. The profiles are supplemented with information on current events, community issues, pricing of the times and news articles to give the reader a broader understanding of what was happening in that individual's world and how it shaped their life. Profiles extend through all walks of life, from farmers to merchants, the rich and poor, men, women and children. In these information-packed, fun-to-explore pages, the reader will be introduced to Ezra Stiles, a preacher and college president from 1776; Colonel Israel Angell, a continental officer from 1778; Thomas Vernon, a loyalist in 1776, Anna Green Winslow, a school girl in 1771; Sarah Pierce, a school teacher in 1792; Edward Hooker, an attorney in 1805; Jeremiah Greenman, a common soldier in 1775 and many others. Using these informationfilled profiles, the reader can develop an understanding of what life was like for all types of Americans in these interesting and changing times. This new edition will be an important acquisition for high school, public and academic libraries as well as history reference collections.

Hardcover ISBN 978-1-59237-371-0, 660 pages, $145.00

The Encyclopedia of Warrior Peoples & Fighting Groups

Many military groups throughout the world have excelled in their craft either by fortuitous circumstances, outstanding leadership, or intense training. This new second edition of *The Encyclopedia of Warrior Peoples and Fighting Groups* explores the origins and leadership of these outstanding combat forces, chronicles their conquests and accomplishments, examines the circumstances surrounding their decline or disbanding, and assesses their influence on the groups and methods of warfare that followed. Readers will encounter ferocious tribes, charismatic leaders, and daring militias, from ancient times to the present, including Amazons, Buffalo Soldiers, Green Berets, Iron Brigade, Kamikazes, Peoples of the Sea, Polish Winged Hussars, Teutonic Knights, and Texas Rangers. With over 100 alphabetical entries, numerous cross-references and illustrations, a comprehensive bibliography, and index, the *Encyclopedia of Warrior Peoples and Fighting Groups* is a valuable resource for readers seeking insight into the bold history of distinguished fighting forces.

"Especially useful for high school students, undergraduates, and general readers with an interest in military history." –Library Journal

Hardcover ISBN 978-1-59237-116-7, 660 pages, $135.00 | Ebook ISBN 978-1-59237-172-3 www.greyhouse.com/ebooks.htm

The Encyclopedia of Invasions & Conquests, From the Ancient Times to the Present

This second edition of the popular *Encyclopedia of Invasions & Conquests*, a comprehensive guide to over 150 invasions, conquests, battles and occupations from ancient times to the present, takes readers on a journey that includes the Roman conquest of Britain, the Portuguese colonization of Brazil, and the Iraqi invasion of Kuwait, to name a few. New articles will explore the late 20th and 21st centuries, with a specific focus on recent conflicts in Afghanistan, Kuwait, Iraq, Yugoslavia, Grenada and Chechnya. In addition to covering the military aspects of invasions and conquests, entries cover some of the political, economic, and cultural aspects, for example, the effects of a conquest on the invade country's political and monetary system and in its language and religion. The entries on leaders – among them Sargon, Alexander the Great, William the Conqueror, and Adolf Hitler – deal with the people who sought to gain control, expand power, or exert religious or political influence over others through military means. Revised and updated for this second edition, entries are arranged alphabetically within historical periods. Each chapter provides a map to help readers locate key areas and geographical features, and bibliographical references appear at the end of each entry. Other useful features include cross-references, a cumulative bibliography and a comprehensive subject index. This authoritative, well-organized, lucidly written volume will prove invaluable for a variety of readers, including high school students, military historians, members of the armed forces, history buffs and hobbyists.

"Engaging writing, sensible organization, nice illustrations, interesting and obscure facts, and useful maps make this book a pleasure to read." –ARBA

Hardcover ISBN 978-1-59237-114-3, 598 pages, $135.00 | Ebook ISBN 978-1-59237-171-6 www.gale.com/gvrl/partners/grey.htm

Encyclopedia of Prisoners of War & Internment

This authoritative second edition provides a valuable overview of the history of prisoners of war and interned civilians, from earliest times to the present. Written by an international team of experts in the field of POW studies, this fascinating and thought-provoking volume includes entries on a wide range of subjects including the Crusades, Plains Indian Warfare, concentration camps, the two world wars, and famous POWs throughout history, as well as atrocities, escapes, and much more. Written in a clear and easily understandable style, this informative reference details over 350 entries, 30% larger than the first edition, that survey the history of prisoners of war and interned civilians from the earliest times to the present, with emphasis on the 19th and 20th centuries. Medical conditions, international law, exchanges of prisoners, organizations working on behalf of POWs, and trials associated with the treatment of captives are just some of the themes explored. Entries are arranged alphabetically, plus illustrations and maps are provided for easy reference. The text also includes an introduction, bibliography, appendix of selected documents, and end-of-entry reading suggestions. This one-of-a-kind reference will be a helpful addition to the reference collections of all public libraries, high schools, and university libraries and will prove invaluable to historians and military enthusiasts.

"Thorough and detailed yet accessible to the lay reader. Of special interest to subject specialists and historians; recommended for public and academic libraries." - Library Journal

Hardcover ISBN 978-1-59237-120-4, 676 pages, $135.00 | Ebook ISBN 978-1-59237-170-9 www.greyhouse.com/ebooks.htm

Business Information ✦ Ratings Guides ✦ General Reference ✦ **Education** ✦
Statistics ✦ Demographics ✦ Health Information ✦ Canadian Information

The Encyclopedia of Rural America: the Land & People

History, sociology, anthropology, and public policy are combined to deliver the encyclopedia destined to become the standard reference work in American rural studies. From irrigation and marriage to games and mental health, this encyclopedia is the first to explore the contemporary landscape of rural America, placed in historical perspective. With over 300 articles prepared by leading experts from across the nation, this timely encyclopedia documents and explains the major themes, concepts, industries, concerns, and everyday life of the people and land who make up rural America. Entries range from the industrial sector and government policy to arts and humanities and social and family concerns. Articles explore every aspect of life in rural America. *Encyclopedia of Rural America*, with its broad range of coverage, will appeal to high school and college students as well as graduate students, faculty, scholars, and people whose work pertains to rural areas.

"This exemplary encyclopedia is guaranteed to educate our highly urban society about the uniqueness of rural America. Recommended for public and academic libraries." -Library Journal

Two Volumes, Hardcover, ISBN 978-1-59237-115-0, 800 pages, $250.00

The Religious Right, A Reference Handbook

Timely and unbiased, this third edition updates and expands its examination of the religious right and its influence on our government, citizens, society, and politics. From the fight to outlaw the teaching of Darwin's theory of evolution to the struggle to outlaw abortion, the religious right is continually exerting an influence on public policy. This text explores the influence of religion on legislation and society, while examining the alignment of the religious right with the political right. A historical survey of the movement highlights the shift to "hands-on" approach to politics and the struggle to present a unified front. The coverage offers a critical historical survey of the religious right movement, focusing on its increased involvement in the political arena, attempts to forge coalitions, and notable successes and failures. The text offers complete coverage of biographies of the men and women who have advanced the cause and an up to date chronology illuminate the movement's goals, including their accomplishments and failures. This edition offers an extensive update to all sections along with several brand new entries. Two new sections complement this third edition, a chapter on legal issues and court decisions and a chapter on demographic statistics and electoral patterns. To aid in further research, *The Religious Right*, offers an entire section of annotated listings of print and non-print resources, as well as organizations affiliated with the religious right, and those opposing it. Comprehensive in its scope, this work offers easy-to-read, pertinent information for those seeking to understand the religious right and its evolving role in American society. A must for libraries of all sizes, university religion departments, activists, high schools and for those interested in the evolving role of the religious right.

" Recommended for all public and academic libraries." - Library Journal

Hardcover ISBN 978-1-59237-113-6, 600 pages, $135.00 | Ebook ISBN 978-1-59237-226-3 www.greyhouse.com/ebooks.htm

From Suffrage to the Senate, America's Political Women

From Suffrage to the Senate is a comprehensive and valuable compendium of biographies of leading women in U.S. politics, past and present, and an examination of the wide range of women's movements. Up to date through 2006, this dynamically illustrated reference work explores American women's path to political power and social equality from the struggle for the right to vote and the abolition of slavery to the first African American woman in the U.S. Senate and beyond. This new edition includes over 150 new entries and a brand new section on trends and demographics of women in politics. The in-depth coverage also traces the political heritage of the abolition, labor, suffrage, temperance, and reproductive rights movements. The alphabetically arranged entries include biographies of every woman from across the political spectrum who has served in the U.S. House and Senate, along with women in the Judiciary and the U.S. Cabinet and, new to this edition, biographies of activists and political consultants. Bibliographical references follow each entry. For easy reference, a handy chronology is provided detailing 150 years of women's history. This up-to-date reference will be a must-purchase for women's studies departments, high schools and public libraries and will be a handy resource for those researching the key players in women's politics, past and present.

"An engaging tool that would be useful in high school, public, and academic libraries looking for an overview of the political history of women in the US." –Booklist

Two Volumes, Hardcover ISBN 978-1-59237-117-4, 1,160 pages, $195.00 | Ebook ISBN 978-1-59237-227-0 www.gale.com/gvrl/partners/grey.htm

Business Information ✦ Ratings Guides ✦ General Reference ✦ **Education** ✦
Statistics ✦ Demographics ✦ Health Information ✦ Canadian Information

An African Biographical Dictionary

This landmark second edition is the only biographical dictionary to bring together, in one volume, cultural, social and political leaders – both historical and contemporary – of the sub-Saharan region. Over 800 biographical sketches of prominent Africans, as well as foreigners who have affected the continent's history, are featured, 150 more than the previous edition. The wide spectrum of leaders includes religious figures, writers, politicians, scientists, entertainers, sports personalities and more. Access to these fascinating individuals is provided in a user-friendly format. The biographies are arranged alphabetically, cross-referenced and indexed. Entries include the country or countries in which the person was significant and the commonly accepted dates of birth and death. Each biographical sketch is chronologically written; entries for cultural personalities add an evaluation of their work. This information is followed by a selection of references often found in university and public libraries, including autobiographies and principal biographical works. Appendixes list each individual by country and by field of accomplishment – rulers, musicians, explorers, missionaries, businessmen, physicists – nearly thirty categories in all. Another convenient appendix lists heads of state since independence by country. Up-to-date and representative of African societies as a whole, An African Biographical Dictionary provides a wealth of vital information for students of African culture and is an indispensable reference guide for anyone interested in African affairs.

"An unquestionable convenience to have these concise, informative biographies gathered into one source, indexed, and analyzed by appendixes listing entrants by nation and occupational field." –Wilson Library Bulletin

Hardcover ISBN 978-1-59237-112-9, 667 pages, $135.00 | Ebook ISBN 978-1-59237-229-4 www.greyhouse.com/ebooks.htm

American Environmental Leaders, From Colonial Times to the Present

A comprehensive and diverse award winning collection of biographies of the most important figures in American environmentalism. Few subjects arouse the passions the way the environment does. How will we feed an ever-increasing population and how can that food be made safe for consumption? Who decides how land is developed? How can environmental policies be made fair for everyone, including multiethnic groups, women, children, and the poor? *American Environmental Leaders* presents more than 350 biographies of men and women who have devoted their lives to studying, debating, and organizing these and other controversial issues over the last 200 years. In addition to the scientists who have analyzed how human actions affect nature, we are introduced to poets, landscape architects, presidents, painters, activists, even sanitation engineers, and others who have forever altered how we think about the environment. The easy to use A–Z format provides instant access to these fascinating individuals, and frequent cross references indicate others with whom individuals worked (and sometimes clashed). End of entry references provide users with a starting point for further research.

"Highly recommended for high school, academic, and public libraries needing environmental biographical information." –Library Journal/Starred Review

Two Volumes, Hardcover ISBN 978-1-59237-119-8, 900 pages $195.00 | Ebook ISBN 978-1-59237-230-0 www.greyhouse.com/ebooks.htm

World Cultural Leaders of the Twentieth & Twenty-First Centuries

World Cultural Leaders of the Twentieth & Twenty-First Centuries is a window into the arts, performances, movements, and music that shaped the world's cultural development since 1900. A remarkable around-the-world look at one-hundred-plus years of cultural development through the eyes of those that set the stage and stayed to play. This second edition offers over 120 new biographies along with a complete update of existing biographies. To further aid the reader, a handy fold-out timeline traces important events in all six cultural categories from 1900 through the present time. Plus, a new section of detailed material and resources for 100 selected individuals is also new to this edition, with further data on museums, homesteads, websites, artwork and more. This remarkable compilation will answer a wide range of questions. Who was the originator of the term "documentary"? Which poet married the daughter of the famed novelist Thomas Mann in order to help her escape Nazi Germany? Which British writer served as an agent in Russia against the Bolsheviks before the 1917 revolution? A handy two-volume set that makes it easy to look up 450 worldwide cultural icons: novelists, poets, playwrights, painters, sculptors, architects, dancers, choreographers, actors, directors, filmmakers, singers, composers, and musicians. *World Cultural Leaders of the Twentieth & Twenty-First Centuries* provides entries (many of them illustrated) covering the person's works, achievements, and professional career in a thorough essay and offers interesting facts and statistics. Entries are fully cross-referenced so that readers can learn how various individuals influenced others. An index of leaders by occupation, a useful glossary and a thorough general index complete the coverage. This remarkable resource will be an important acquisition for the reference collections of public libraries, university libraries and high schools.

"Fills a need for handy, concise information on a wide array of international cultural figures."-ARBA

Two Volumes, Hardcover ISBN 978-1-59237-118-1, 900 pages, $195.00 | Ebook ISBN 978-1-59237-231-7 www.greyhouse.com/ebooks.htm

---

Business Information ♦ Ratings Guides ♦ General Reference ♦ **Education** ♦
Statistics ♦ Demographics ♦ Health Information ♦ Canadian Information

Political Corruption in America: An Encyclopedia of Scandals, Power, and Greed

The complete scandal-filled history of American political corruption, focusing on the infamous people and cases, as well as society's electoral and judicial reactions. Since colonial times, there has been no shortage of politicians willing to take a bribe, skirt campaign finance laws, or act in their own interests. Corruption like the Whiskey Ring, Watergate, and Whitewater cases dominate American life, making political scandal a leading U.S. industry. From judges to senators, presidents to mayors, *Political Corruption in America* discusses the infamous people throughout history who have been accused of and implicated in crooked behavior. In this new second edition, more than 250 A–Z entries explore the people, crimes, investigations, and court cases behind 200 years of American political scandals. This unbiased volume also delves into the issues surrounding Koreagate, the Chinese campaign scandal, and other ethical lapses. Relevant statutes and terms, including the Independent Counsel Statute and impeachment as a tool of political punishment, are examined as well. Students, scholars, and other readers interested in American history, political science, and ethics will appreciate this survey of a wide range of corrupting influences. This title focuses on how politicians from all parties have fallen because of their greed and hubris, and how society has used electoral and judicial means against those who tested the accepted standards of political conduct. A full range of illustrations including political cartoons, photos of key figures such as Abe Fortas and Archibald Cox, graphs of presidential pardons, and tables showing the number of expulsions and censures in both the House and Senate round out the text. In addition, a comprehensive chronology of major political scandals in U.S. history from colonial times until the present. For further reading, an extensive bibliography lists sources including archival letters, newspapers, and private manuscript collections from the United States and Great Britain. With its comprehensive coverage of this interesting topic, *Political Corruption in America: An Encyclopedia of Scandals, Power, and Greed* will prove to be a useful addition to the reference collections of all public libraries, university libraries, history collections, political science collections and high schools.

> *"...this encyclopedia is a useful contribution to the field. Highly recommended."* - CHOICE
> "Political Corruption should be useful in most academic, high school, and public libraries." Booklist

Two Volumes, Hardcover ISBN 978-1-59237-297-3, 500 pages, $195.00 | Ebook ISBN 978-1-59237-308-6
www.greyhouse.com/ebooks.htm

Religion and Law: A Dictionary

This informative, easy-to-use reference work covers a wide range of legal issues that affect the roles of religion and law in American society. Extensive A–Z entries provide coverage of key court decisions, case studies, concepts, individuals, religious groups, organizations, and agencies shaping religion and law in today's society. This *Dictionary* focuses on topics involved with the constitutional theory and interpretation of religion and the law; terms providing a historical explanation of the ways in which America's ever increasing ethnic and religious diversity contributed to our current understanding of the mandates of the First and Fourteenth Amendments; terms and concepts describing the development of religion clause jurisprudence; an analytical examination of the distinct vocabulary used in this area of the law; the means by which American courts have attempted to balance religious liberty against other important individual and social interests in a wide variety of physical and regulatory environments, including the classroom, the workplace, the courtroom, religious group organization and structure, taxation, the clash of "secular" and "religious" values, and the relationship of the generalized idea of individual autonomy of the specific concept of religious liberty. Important legislation and legal cases affecting religion and society are thoroughly covered in this timely volume, including a detailed Table of Cases and Table of Statutes for more detailed research. A guide to further reading and an index are also included. This useful resource will be an important acquisition for the reference collections of all public libraries, university libraries, religion reference collections and high schools.

Two Volumes, Hardcover ISBN 978-1-59237-298-0, 500 pages, $195.00 | Ebook ISBN 978-1-59237-309-3
www.greyhouse.com/ebooks.htm

Human Rights in the United States: A Dictionary and Documents

This two volume set offers easy to grasp explanations of the basic concepts, laws, and case law in the field, with emphasis on human rights in the historical, political, and legal experience of the United States. Human rights is a term not fully understood by many Americans. Addressing this gap, the new second edition of *Human Rights in the United States: A Dictionary and Documents* offers a comprehensive introduction that places the history of human rights in the United States in an international context. It surveys the legal protection of human dignity in the United States, examines the sources of human rights norms, cites key legal cases, explains the role of international governmental and non-governmental organizations, and charts global, regional, and U.N. human rights measures. Over 240 dictionary entries of human rights terms are detailed—ranging from asylum and cultural relativism to hate crimes and torture. Each entry discusses the significance of the term, gives examples, and cites appropriate documents and court decisions. In addition, a Documents section is provided that contains 59 conventions, treaties, and protocols related to the most up to date international action on ethnic cleansing; freedom of expression and religion; violence against women; and much more. A bibliography, extensive glossary, and comprehensive index round out this indispensable volume. This comprehensive, timely volume is a must for large public libraries, university libraries and social science departments, along with high school libraries.

> *"...invaluable for anyone interested in human rights issues ... highly recommended for all reference collections."*
> - American Reference Books Annual

Two Volumes, Hardcover ISBN 978-1-59237-290-4, 750 pages, $225.00 | Ebook ISBN 978-1-59237-301-7
www.greyhouse.com/ebooks.htm

Business Information ◆ Ratings Guides ◆ General Reference ◆ **Education** ◆
Statistics ◆ Demographics ◆ Health Information ◆ Canadian Information

The Comparative Guide to American Elementary & Secondary Schools, 2008

The only guide of its kind, this award winning compilation offers a snapshot profile of every public school district in the United States serving 1,500 or more students – more than 5,900 districts are covered. Organized alphabetically by district within state, each chapter begins with a Statistical Overview of the state. Each district listing includes contact information (name, address, phone number and web site) plus Grades Served, the Numbers of Students and Teachers and the Number of Regular, Special Education, Alternative and Vocational Schools in the district along with statistics on Student/Classroom Teacher Ratios, Drop Out Rates, Ethnicity, the Numbers of Librarians and Guidance Counselors and District Expenditures per student. As an added bonus, *The Comparative Guide to American Elementary and Secondary Schools* provides important ranking tables, both by state and nationally, for each data element. For easy navigation through this wealth of information, this handbook contains a useful City Index that lists all districts that operate schools within a city. These important comparative statistics are necessary for anyone considering relocation or doing comparative research on their own district and would be a perfect acquisition for any public library or school district library.

> *"This straightforward guide is an easy way to find general information.*
> *Valuable for academic and large public library collections." –ARBA*

Softcover ISBN 978-1-59237-223-2, 2,400 pages, $125.00 | Ebook ISBN 978-1-59237-238-6 www.greyhouse.com/ebooks.htm

The Complete Learning Disabilities Directory, 2009

The Complete Learning Disabilities Directory is the most comprehensive database of Programs, Services, Curriculum Materials, Professional Meetings & Resources, Camps, Newsletters and Support Groups for teachers, students and families concerned with learning disabilities. This information-packed directory includes information about Associations & Organizations, Schools, Colleges & Testing Materials, Government Agencies, Legal Resources and much more. For quick, easy access to information, this directory contains four indexes: Entry Name Index, Subject Index and Geographic Index. With every passing year, the field of learning disabilities attracts more attention and the network of caring, committed and knowledgeable professionals grows every day. This directory is an invaluable research tool for these parents, students and professionals.

> *"Due to its wealth and depth of coverage, parents, teachers and others... should find this an invaluable resource." -Booklist*

Softcover ISBN 978-1-59237-368-0, 900 pages, $145.00 | Online Database $195.00 | Online Database & Directory Combo $280.00

Educators Resource Directory, 2007/08

Educators Resource Directory is a comprehensive resource that provides the educational professional with thousands of resources and statistical data for professional development. This directory saves hours of research time by providing immediate access to Associations & Organizations, Conferences & Trade Shows, Educational Research Centers, Employment Opportunities & Teaching Abroad, School Library Services, Scholarships, Financial Resources, Professional Consultants, Computer Software & Testing Resources and much more. Plus, this comprehensive directory also includes a section on Statistics and Rankings with over 100 tables, including statistics on Average Teacher Salaries, SAT/ACT scores, Revenues & Expenditures and more. These important statistics will allow the user to see how their school rates among others, make relocation decisions and so much more. For quick access to information, this directory contains four indexes: Entry & Publisher Index, Geographic Index, a Subject & Grade Index and Web Sites Index. *Educators Resource Directory* will be a well-used addition to the reference collection of any school district, education department or public library.

> *"Recommended for all collections that serve elementary and secondary school professionals." –Choice*

Softcover ISBN 978-1-59237-179-2, 800 pages, $145.00 | Online Database $195.00 | Online Database & Directory Combo $280.00

Profiles of New York | Profiles of Florida | Profiles of Texas | Profiles of Illinois | Profiles of Michigan | Profiles of Ohio | Profiles of New Jersey | Profiles of Massachusetts | Profiles of Pennsylvania | Profiles of Wisconsin | Profiles of Connecticut & Rhode Island | Profiles of Indiana | Profiles of North Carolina & South Carolina | Profiles of Virginia | Profiles of California

The careful layout gives the user an easy-to-read snapshot of every single place and county in the state, from the biggest metropolis to the smallest unincorporated hamlet. The richness of each place or county profile is astounding in its depth, from history to weather, all packed in an easy-to-navigate, compact format. Each profile contains data on History, Geography, Climate, Population, Vital Statistics, Economy, Income, Taxes, Education, Housing, Health & Environment, Public Safety, Newspapers, Transportation, Presidential Election Results, Information Contacts and Chambers of Commerce. As an added bonus, there is a section on Selected Statistics, where data from the 100 largest towns and cities is arranged into easy-to-use charts. Each of 22 different data points has its own two-page spread with the cities listed in alpha order so researchers can easily compare and rank cities. A remarkable compilation that offers overviews and insights into each corner of the state, each volume goes beyond Census statistics, beyond metro area coverage, beyond the 100 best places to live. Drawn from official census information, other government statistics and original research, you will have at your fingertips data that's available nowhere else in one single source.

"The publisher claims that this is the 'most comprehensive portrait of the state of Florida ever published,' and this reviewer is inclined to believe it...Recommended. All levels." –Choice on Profiles of Florida

Each Profiles of... title ranges from 400-800 pages, priced at $149.00 each

America's Top-Rated Cities, 2008

America's Top-Rated Cities provides current, comprehensive statistical information and other essential data in one easy-to-use source on the 100 "top" cities that have been cited as the best for business and living in the U.S. This handbook allows readers to see, at a glance, a concise social, business, economic, demographic and environmental profile of each city, including brief evaluative comments. In addition to detailed data on Cost of Living, Finances, Real Estate, Education, Major Employers, Media, Crime and Climate, city reports now include Housing Vacancies, Tax Audits, Bankruptcy, Presidential Election Results and more. This outstanding source of information will be widely used in any reference collection.

"The only source of its kind that brings together all of this information into one easy-to-use source. It will be beneficial to many business and public libraries." –ARBA

Four Volumes, Softcover ISBN 978-1-59237-349-9, 2,500 pages, $195.00 | Ebook ISBN 978-1-59237-233-1
www.greyhouse.com/ebooks.htm

America's Top-Rated Smaller Cities, 2008/09

A perfect companion to *America's Top-Rated Cities, America's Top-Rated Smaller Cities* provides current, comprehensive business and living profiles of smaller cities (population 25,000-99,999) that have been cited as the best for business and living in the United States. Sixty cities make up this 2004 edition of America's Top-Rated Smaller Cities, all are top-ranked by Population Growth, Median Income, Unemployment Rate and Crime Rate. City reports reflect the most current data available on a wide-range of statistics, including Employment & Earnings, Household Income, Unemployment Rate, Population Characteristics, Taxes, Cost of Living, Education, Health Care, Public Safety, Recreation, Media, Air & Water Quality and much more. Plus, each city report contains a Background of the City, and an Overview of the State Finances. *America's Top-Rated Smaller Cities* offers a reliable, one-stop source for statistical data that, before now, could only be found scattered in hundreds of sources. This volume is designed for a wide range of readers: individuals considering relocating a residence or business; professionals considering expanding their business or changing careers; general and market researchers; real estate consultants; human resource personnel; urban planners and investors.

*"Provides current, comprehensive statistical information in one easy-to-use source…
Recommended for public and academic libraries and specialized collections." –Library Journal*

Two Volumes, Softcover ISBN 978-1-59237-284-3, 1,100 pages, $195.00 | Ebook ISBN 978-1-59237-234-8
www.greyhouse.com/ebooks.htm

Profiles of America: Facts, Figures & Statistics for Every Populated Place in the United States

Profiles of America is the only source that pulls together, in one place, statistical, historical and descriptive information about every place in the United States in an easy-to-use format. This award winning reference set, now in its second edition, compiles statistics and data from over 20 different sources – the latest census information has been included along with more than nine brand new statistical topics. This Four-Volume Set details over 40,000 places, from the biggest metropolis to the smallest unincorporated hamlet, and provides statistical details and information on over 50 different topics including Geography, Climate, Population, Vital Statistics, Economy, Income, Taxes, Education, Housing, Health & Environment, Public Safety, Newspapers, Transportation, Presidential Election Results and Information Contacts or Chambers of Commerce. Profiles are arranged, for ease-of-use, by state and then by county. Each county begins with a County-Wide Overview and is followed by information for each Community in that particular county. The Community Profiles within the county are arranged alphabetically. *Profiles of America* is a virtual snapshot of America at your fingertips and a unique compilation of information that will be widely used in any reference collection.

A Library Journal Best Reference Book "An outstanding compilation." –Library Journal

Four Volumes, Softcover ISBN 978-1-891482-80-9, 10,000 pages, $595.00

Business Information ◆ Ratings Guides ◆ General Reference ◆ Education ◆
Statistics ◆ **Demographics** ◆ Health Information ◆ Canadian Information

Grey House
Publishing

The Comparative Guide to American Suburbs, 2007/08

The Comparative Guide to American Suburbs is a one-stop source for Statistics on the 2,000+ suburban communities surrounding the 50 largest metropolitan areas – their population characteristics, income levels, economy, school system and important data on how they compare to one another. Organized into 50 Metropolitan Area chapters, each chapter contains an overview of the Metropolitan Area, a detailed Map followed by a comprehensive Statistical Profile of each Suburban Community, including Contact Information, Physical Characteristics, Population Characteristics, Income, Economy, Unemployment Rate, Cost of Living, Education, Chambers of Commerce and more. Next, statistical data is sorted into Ranking Tables that rank the suburbs by twenty different criteria, including Population, Per Capita Income, Unemployment Rate, Crime Rate, Cost of Living and more. *The Comparative Guide to American Suburbs* is the best source for locating data on suburbs. Those looking to relocate, as well as those doing preliminary market research, will find this an invaluable timesaving resource.

"Public and academic libraries will find this compilation useful…The work draws together figures from many sources and will be especially helpful for job relocation decisions." – Booklist

Softcover ISBN 978-1-59237-180-8, 1,700 pages, $130.00 | Ebook ISBN 978-1-59237-235-5 www.greyhouse.com/ebooks.htm

The American Tally: Statistics & Comparative Rankings for U.S. Cities with Populations over 10,000

This important statistical handbook compiles, all in one place, comparative statistics on all U.S. cities and towns with a 10,000+ population. *The American Tally* provides statistical details on over 4,000 cities and towns and profiles how they compare with one another in Population Characteristics, Education, Language & Immigration, Income & Employment and Housing. Each section begins with an alphabetical listing of cities by state, allowing for quick access to both the statistics and relative rankings of any city. Next, the highest and lowest cities are listed in each statistic. These important, informative lists provide quick reference to which cities are at both extremes of the spectrum for each statistic. Unlike any other reference, *The American Tally* provides quick, easy access to comparative statistics – a must-have for any reference collection.

"A solid library reference." -Bookwatch

Softcover ISBN 978-1-930956-29-2, 500 pages, $125.00 | Ebook ISBN 978-1-59237-241-6 www.greyhouse.com/ebooks.htm

The Asian Databook: Statistics for all US Counties & Cities with Over 10,000 Population

This is the first-ever resource that compiles statistics and rankings on the US Asian population. *The Asian Databook* presents over 20 statistical data points for each city and county, arranged alphabetically by state, then alphabetically by place name. Data reported for each place includes Population, Languages Spoken at Home, Foreign-Born, Educational Attainment, Income Figures, Poverty Status, Homeownership, Home Values & Rent, and more. Next, in the Rankings Section, the top 75 places are listed for each data element. These easy-to-access ranking tables allow the user to quickly determine trends and population characteristics. This kind of comparative data can not be found elsewhere, in print or on the web, in a format that's as easy-to-use or more concise. A useful resource for those searching for demographics data, career search and relocation information and also for market research. With data ranging from Ancestry to Education, *The Asian Databook* presents a useful compilation of information that will be a much-needed resource in the reference collection of any public or academic library along with the marketing collection of any company whose primary focus in on the Asian population.

"This useful resource will help those searching for demographics data, and market research or relocation information… Accurate and clearly laid out, the publication is recommended for large public library and research collections." -Booklist

Softcover ISBN 978-1-59237-044-3, 1,000 pages, $150.00

The Hispanic Databook: Statistics for all US Counties & Cities with Over 10,000 Population

Previously published by Toucan Valley Publications, this second edition has been completely updated with figures from the latest census and has been broadly expanded to include dozens of new data elements and a brand new Rankings section. The Hispanic population in the United States has increased over 42% in the last 10 years and accounts for 12.5% of the total US population. For ease-of-use, *The Hispanic Databook* presents over 20 statistical data points for each city and county, arranged alphabetically by state, then alphabetically by place name. Data reported for each place includes Population, Languages Spoken at Home, Foreign-Born, Educational Attainment, Income Figures, Poverty Status, Homeownership, Home Values & Rent, and more. Next, in the Rankings Section, the top 75 places are listed for each data element. These easy-to-access ranking tables allow the user to quickly determine trends and population characteristics. This kind of comparative data can not be found elsewhere, in print or on the web, in a format that's as easy-to-use or more concise. A useful resource for those searching for demographics data, career search and relocation information and also for market research. With data ranging from Ancestry to Education, *The Hispanic Databook* presents a useful compilation of information that will be a much-needed resource in the reference collection of any public or academic library along with the marketing collection of any company whose primary focus in on the Hispanic population.

"This accurate, clearly presented volume of selected Hispanic demographics is recommended for large public libraries and research collections."-Library Journal

Softcover ISBN 978-1-59237-008-5, 1,000 pages, $150.00

Business Information ◆ Ratings Guides ◆ General Reference ◆ Education ◆
Statistics ◆ **Demographics** ◆ Health Information ◆ Canadian Information

Ancestry in America: A Comparative Guide to Over 200 Ethnic Backgrounds

This brand new reference work pulls together thousands of comparative statistics on the Ethnic Backgrounds of all populated places in the United States with populations over 10,000. Never before has this kind of information been reported in a single volume. Section One, Statistics by Place, is made up of a list of over 200 ancestry and race categories arranged alphabetically by each of the 5,000 different places with populations over 10,000. The population number of the ancestry group in that city or town is provided along with the percent that group represents of the total population. This informative city-by-city section allows the user to quickly and easily explore the ethnic makeup of all major population bases in the United States. Section Two, Comparative Rankings, contains three tables for each ethnicity and race. In the first table, the top 150 populated places are ranked by population number for that particular ancestry group, regardless of population. In the second table, the top 150 populated places are ranked by the percent of the total population for that ancestry group. In the third table, those top 150 populated places with 10,000 population are ranked by population number for each ancestry group. These easy-to-navigate tables allow users to see ancestry population patterns and make city-by-city comparisons as well. This brand new, information-packed resource will serve a wide-range or research requests for demographics, population characteristics, relocation information and much more. *Ancestry in America: A Comparative Guide to Over 200 Ethnic Backgrounds* will be an important acquisition to all reference collections.

"This compilation will serve a wide range of research requests for population characteristics … it offers much more detail than other sources." –Booklist

Softcover ISBN 978-1-59237-029-0, 1,500 pages, $225.00

Weather America, A Thirty-Year Summary of Statistical Weather Data and Rankings

This valuable resource provides extensive climatological data for over 4,000 National and Cooperative Weather Stations throughout the United States. Weather America begins with a new Major Storms section that details major storm events of the nation and a National Rankings section that details rankings for several data elements, such as Maximum Temperature and Precipitation. The main body of Weather America is organized into 50 state sections. Each section provides a Data Table on each Weather Station, organized alphabetically, that provides statistics on Maximum and Minimum Temperatures, Precipitation, Snowfall, Extreme Temperatures, Foggy Days, Humidity and more. State sections contain two brand new features in this edition – a City Index and a narrative Description of the climatic conditions of the state. Each section also includes a revised Map of the State that includes not only weather stations, but cities and towns.

"Best Reference Book of the Year." –Library Journal

Softcover ISBN 978-1-891482-29-8, 2,013 pages, $175.00 | Ebook ISBN 978-1-59237-237-9 www.greyhouse.com/ebooks.htm

Crime in America's Top-Rated Cities

This volume includes over 20 years of crime statistics in all major crime categories: violent crimes, property crimes and total crime. *Crime in America's Top-Rated Cities* is conveniently arranged by city and covers 76 top-rated cities. Crime in America's Top-Rated Cities offers details that compare the number of crimes and crime rates for the city, suburbs and metro area along with national crime trends for violent, property and total crimes. Also, this handbook contains important information and statistics on Anti-Crime Programs, Crime Risk, Hate Crimes, Illegal Drugs, Law Enforcement, Correctional Facilities, Death Penalty Laws and much more. A much-needed resource for people who are relocating, business professionals, general researchers, the press, law enforcement officials and students of criminal justice.

"Data is easy to access and will save hours of searching." –Global Enforcement Review

Softcover ISBN 978-1-891482-84-7, 832 pages, $155.00

Business Information ♦ Ratings Guides ♦ General Reference ♦ Education ♦
Statistics ♦ Demographics ♦ <u>Health Information</u> ♦ Canadian Information

The Complete Directory for People with Disabilities, 2009

A wealth of information, now in one comprehensive sourcebook. Completely updated, this edition contains more information than ever before, including thousands of new entries and enhancements to existing entries and thousands of additional web sites and e-mail addresses. This up-to-date directory is the most comprehensive resource available for people with disabilities, detailing Independent Living Centers, Rehabilitation Facilities, State & Federal Agencies, Associations, Support Groups, Periodicals & Books, Assistive Devices, Employment & Education Programs, Camps and Travel Groups. Each year, more libraries, schools, colleges, hospitals, rehabilitation centers and individuals add *The Complete Directory for People with Disabilities* to their collections, making sure that this information is readily available to the families, individuals and professionals who can benefit most from the amazing wealth of resources cataloged here.

"No other reference tool exists to meet the special needs of the disabled in one convenient resource for information." –Library Journal

Softcover ISBN 978-1-59237-367-3, 1,200 pages, $165.00 | Online Database: http://gold.greyhouse.com Call (800) 562-2139 for quote

The Complete Learning Disabilities Directory, 2009

The Complete Learning Disabilities Directory is the most comprehensive database of Programs, Services, Curriculum Materials, Professional Meetings & Resources, Camps, Newsletters and Support Groups for teachers, students and families concerned with learning disabilities. This information-packed directory includes information about Associations & Organizations, Schools, Colleges & Testing Materials, Government Agencies, Legal Resources and much more. For quick, easy access to information, this directory contains four indexes: Entry Name Index, Subject Index and Geographic Index. With every passing year, the field of learning disabilities attracts more attention and the network of caring, committed and knowledgeable professionals grows every day. This directory is an invaluable research tool for these parents, students and professionals.

"Due to its wealth and depth of coverage, parents, teachers and others… should find this an invaluable resource." -Booklist

Softcover ISBN 978-1-59237-368-0, 900 pages, $145.00 | Online Database: http://gold.greyhouse.com Call (800) 562-2139 for quote

The Complete Directory for People with Chronic Illness, 2007/08

Thousands of hours of research have gone into this completely updated edition – several new chapters have been added along with thousands of new entries and enhancements to existing entries. Plus, each chronic illness chapter has been reviewed by a medical expert in the field. This widely-hailed directory is structured around the 90 most prevalent chronic illnesses – from Asthma to Cancer to Wilson's Disease – and provides a comprehensive overview of the support services and information resources available for people diagnosed with a chronic illness. Each chronic illness has its own chapter and contains a brief description in layman's language, followed by important resources for National & Local Organizations, State Agencies, Newsletters, Books & Periodicals, Libraries & Research Centers, Support Groups & Hotlines, Web Sites and much more. This directory is an important resource for health care professionals, the collections of hospital and health care libraries, as well as an invaluable tool for people with a chronic illness and their support network.

"A must purchase for all hospital and health care libraries and is strongly recommended for all public library reference departments." –ARBA

Softcover ISBN 978-1-59237-183-9, 1,200 pages, $165.00 | Online Database: http://gold.greyhouse.com Call (800) 562-2139 for quote

The Complete Mental Health Directory, 2008/09

This is the most comprehensive resource covering the field of behavioral health, with critical information for both the layman and the mental health professional. For the layman, this directory offers understandable descriptions of 25 Mental Health Disorders as well as detailed information on Associations, Media, Support Groups and Mental Health Facilities. For the professional, The Complete Mental Health Directory offers critical and comprehensive information on Managed Care Organizations, Information Systems, Government Agencies and Provider Organizations. This comprehensive volume of needed information will be widely used in any reference collection.

"… the strength of this directory is that it consolidates widely dispersed information into a single volume." –Booklist

Softcover ISBN 978-1-59237-285-0, 800 pages, $165.00 | Online Database: http://gold.greyhouse.com Call (800) 562-2139 for quote

Grey House
Publishing

The Comparative Guide to American Hospitals, Second Edition

This new second edition compares all of the nation's hospitals by 24 measures of quality in the treatment of heart attack, heart failure, pneumonia, and, new to this edition, surgical procedures and pregnancy care. Plus, this second edition is now available in regional volumes, to make locating information about hospitals in your area quicker and easier than ever before. The Comparative Guide to American Hospitals provides a snapshot profile of each of the nations 4,200+ hospitals. These informative profiles illustrate how the hospital rates when providing 24 different treatments within four broad categories: Heart Attack Care, Heart Failure Care, Surgical Infection Prevention (NEW), and Pregnancy Care measures (NEW). Each profile includes the raw percentage for that hospital, the state average, the US average and data on the top hospital. For easy access to contact information, each profile includes the hospital's address, phone and fax numbers, email and web addresses, type and accreditation along with 5 top key administrations. These profiles will allow the user to quickly identify the quality of the hospital and have the necessary information at their fingertips to make contact with that hospital. Most importantly, *The Comparative Guide to American Hospitals* provides easy-to-use Regional State by State Statistical Summary Tables for each of the data elements to allow the user to quickly locate hospitals with the best level of service. Plus, a new 30-Day Mortality Chart, Glossary of Terms and Regional Hospital Profile Index make this a must-have source. This new, expanded edition will be a must for the reference collection at all public, medical and academic libraries.

"These data will help those with heart conditions and pneumonia make informed decisions about their healthcare and encourage hospitals to improve the quality of care they provide. Large medical, hospital, and public libraries are most likely to benefit from this weighty resource."-Library Journal

Four Volumes Softcover ISBN 978-1-59237-182-2, 3,500 pages, $325.00 | Regional Volumes $135.00 |
Ebook ISBN 978-1-59237-239-3 www.greyhouse.com/ebooks.htm

Older Americans Information Directory, 2008

Completely updated for 2008, this sixth edition has been completely revised and now contains 1,000 new listings, over 8,000 updates to existing listings and over 3,000 brand new e-mail addresses and web sites. You'll find important resources for Older Americans including National, Regional, State & Local Organizations, Government Agencies, Research Centers, Libraries & Information Centers, Legal Resources, Discount Travel Information, Continuing Education Programs, Disability Aids & Assistive Devices, Health, Print Media and Electronic Media. Three indexes: Entry Index, Subject Index and Geographic Index make it easy to find just the right source of information. This comprehensive guide to resources for Older Americans will be a welcome addition to any reference collection.

"Highly recommended for academic, public, health science and consumer libraries…" –Choice

1,200 pages; Softcover ISBN 978-1-59237-357-4, $165.00 | Online Database: http://gold.greyhouse.com Call (800) 562-2139 for quote

The Complete Directory for Pediatric Disorders, 2008

This important directory provides parents and caregivers with information about Pediatric Conditions, Disorders, Diseases and Disabilities, including Blood Disorders, Bone & Spinal Disorders, Brain Defects & Abnormalities, Chromosomal Disorders, Congenital Heart Defects, Movement Disorders, Neuromuscular Disorders and Pediatric Tumors & Cancers. This carefully written directory offers: understandable Descriptions of 15 major bodily systems; Descriptions of more than 200 Disorders and a Resources Section, detailing National Agencies & Associations, State Associations, Online Services, Libraries & Resource Centers, Research Centers, Support Groups & Hotlines, Camps, Books and Periodicals. This resource will provide immediate access to information crucial to families and caregivers when coping with children's illnesses.

"Recommended for public and consumer health libraries." –Library Journal

Softcover ISBN 978-1-59237-150-1, 1,200 pages, $165.00 | Online Database: http://gold.greyhouse.com Call (800) 562-2139 for quote

The Directory of Drug & Alcohol Residential Rehabilitation Facilities

This brand new directory is the first-ever resource to bring together, all in one place, data on the thousands of drug and alcohol residential rehabilitation facilities in the United States. The Directory of Drug & Alcohol Residential Rehabilitation Facilities covers over 1,000 facilities, with detailed contact information for each one, including mailing address, phone and fax numbers, email addresses and web sites, mission statement, type of treatment programs, cost, average length of stay, numbers of residents and counselors, accreditation, insurance plans accepted, type of environment, religious affiliation, education components and much more. It also contains a helpful chapter on General Resources that provides contact information for Associations, Print & Electronic Media, Support Groups and Conferences. Multiple indexes allow the user to pinpoint the facilities that meet very specific criteria. This time-saving tool is what so many counselors, parents and medical professionals have been asking for. *The Directory of Drug & Alcohol Residential Rehabilitation Facilities* will be a helpful tool in locating the right source for treatment for a wide range of individuals. This comprehensive directory will be an important acquisition for all reference collections: public and academic libraries, case managers, social workers, state agencies and many more.

"This is an excellent, much needed directory that fills an important gap…" –Booklist

Softcover ISBN 978-1-59237-031-3, 300 pages, $135.00

To preview any of our Directories Risk-Free for 30 days, call (800) 562-2139 or fax (518) 789-0556
www.greyhouse.com books@greyhouse.com

Business Information ◆ Ratings Guides ◆ General Reference ◆ Education ◆
Statistics ◆ Demographics ◆ <ins>Health Information</ins> ◆ Canadian Information

The Directory of Hospital Personnel, 2009

The Directory of Hospital Personnel is the best resource you can have at your fingertips when researching or marketing a product or service to the hospital market. A "Who's Who" of the hospital universe, this directory puts you in touch with over 150,000 key decision-makers. With 100% verification of data you can rest assured that you will reach the right person with just one call. Every hospital in the U.S. is profiled, listed alphabetically by city within state. Plus, three easy-to-use, cross-referenced indexes put the facts at your fingertips faster and more easily than any other directory: Hospital Name Index, Bed Size Index and Personnel Index. *The Directory of Hospital Personnel* is the only complete source for key hospital decision-makers by name. Whether you want to define or restructure sales territories… locate hospitals with the purchasing power to accept your proposals… keep track of important contacts or colleagues… or find information on which insurance plans are accepted, *The Directory of Hospital Personnel* gives you the information you need – easily, efficiently, effectively and accurately.

"Recommended for college, university and medical libraries." -ARBA

Softcover ISBN 978-1-59237-402-1, 2,500 pages, $325.00 | Online Database: http://gold.greyhouse.com Call (800) 562-2139 for quote

The HMO/PPO Directory, 2009

The HMO/PPO Directory is a comprehensive source that provides detailed information about Health Maintenance Organizations and Preferred Provider Organizations nationwide. This comprehensive directory details more information about more managed health care organizations than ever before. Over 1,100 HMOs, PPOs, Medicare Advantage Plans and affiliated companies are listed, arranged alphabetically by state. Detailed listings include Key Contact Information, Prescription Drug Benefits, Enrollment, Geographical Areas served, Affiliated Physicians & Hospitals, Federal Qualifications, Status, Year Founded, Managed Care Partners, Employer References, Fees & Payment Information and more. Plus, five years of historical information is included related to Revenues, Net Income, Medical Loss Ratios, Membership Enrollment and Number of Patient Complaints. Five easy-to-use, cross-referenced indexes will put this vast array of information at your fingertips immediately: HMO Index, PPO Index, Other Providers Index, Personnel Index and Enrollment Index. *The HMO/PPO Directory* provides the most comprehensive data on the most companies available on the market place today.

"Helpful to individuals requesting certain HMO/PPO issues such as co-payment costs, subscription costs and patient complaints. Individuals concerned (or those with questions) about their insurance may find this text to be of use to them." -ARBA

Softcover ISBN 978-1-59237-369-7, 600 pages, $325.00 | Online Database: http://gold.greyhouse.com Call (800) 562-2139 for quote

Medical Device Register, 2009

The only one-stop resource of every medical supplier licensed to sell products in the US. This award-winning directory offers immediate access to over 13,000 companies - and more than 65,000 products – in two information-packed volumes. This comprehensive resource saves hours of time and trouble when searching for medical equipment and supplies and the manufacturers who provide them. Volume I: The Product Directory, provides essential information for purchasing or specifying medical supplies for every medical device, supply, and diagnostic available in the US. Listings provide FDA codes & Federal Procurement Eligibility, Contact information for every manufacturer of the product along with Prices and Product Specifications. Volume 2 - Supplier Profiles, offers the most complete and important data about Suppliers, Manufacturers and Distributors. Company Profiles detail the number of employees, ownership, method of distribution, sales volume, net income, key executives detailed contact information medical products the company supplies, plus the medical specialties they cover. Four indexes provide immediate access to this wealth of information: Keyword Index, Trade Name Index, Supplier Geographical Index and OEM (Original Equipment Manufacturer) Index. *Medical Device Register* is the only one-stop source for locating suppliers and products; looking for new manufacturers or hard-to-find medical devices; comparing products and companies; know who's selling what and who to buy from cost effectively. This directory has become the standard in its field and will be a welcome addition to the reference collection of any medical library, large public library, university library along with the collections that serve the medical community.

"A wealth of information on medical devices, medical device companies… and key personnel in the industry is provide in this comprehensive reference work... A valuable reference work, one of the best hardcopy compilations available." -Doody Publishing

Two Volumes, Hardcover ISBN 978-1-59237-373-4, 3,000 pages, $325.00

The Directory of Health Care Group Purchasing Organizations, 2008

This comprehensive directory provides the important data you need to get in touch with over 800 Group Purchasing Organizations. By providing in-depth information on this growing market and its members, *The Directory of Health Care Group Purchasing Organizations* fills a major need for the most accurate and comprehensive information on over 800 GPOs – Mailing Address, Phone & Fax Numbers, E-mail Addresses, Key Contacts, Purchasing Agents, Group Descriptions, Membership Categorization, Standard Vendor Proposal Requirements, Membership Fees & Terms, Expanded Services, Total Member Beds & Outpatient Visits represented and more. Five Indexes provide a number of ways to locate the right GPO: Alphabetical Index, Expanded Services Index, Organization Type Index, Geographic Index and Member Institution Index. With its comprehensive and detailed information on each purchasing organization, *The Directory of Health Care Group Purchasing Organizations* is the go-to source for anyone looking to target this market.

"The information is clearly arranged and easy to access…recommended for those needing this very specialized information." –ARBA

1,000 pages; Softcover ISBN 978-1-59237-287-4, $325.00 | Online Database: http://gold.greyhouse.com Call (800) 562-2139 for quote

Business Information ◆ Ratings Guides ◆ General Reference ◆ Education ◆
Statistics ◆ Demographics ◆ Health Information ◆ **Canadian Information**

Canadian Almanac & Directory, 2009

The Canadian Almanac & Directory contains sixteen directories in one – giving you all the facts and figures you will ever need about Canada. No other single source provides users with the quality and depth of up-to-date information for all types of research. This national directory and guide gives you access to statistics, images and over 100,000 names and addresses for everything from Airlines to Zoos - updated every year. It's Ten Directories in One! Each section is a directory in itself, providing robust information on business and finance, communications, government, associations, arts and culture (museums, zoos, libraries, etc.), health, transportation, law, education, and more. Government information includes federal, provincial and territorial - and includes an easy-to-use quick index to find key information. A separate municipal government section includes every municipality in Canada, with full profiles of Canada's largest urban centers. A complete legal directory lists judges and judicial officials, court locations and law firms across the country. A wealth of general information, the *Canadian Almanac & Directory* also includes national statistics on population, employment, imports and exports, and more. National awards and honors are presented, along with forms of address, Commonwealth information and full color photos of Canadian symbols. Postal information, weights, measures, distances and other useful charts are also incorporated. Complete almanac information includes perpetual calendars, five-year holiday planners and astronomical information. Published continuously for 160 years, *The Canadian Almanac & Directory* is the best single reference source for business executives, managers and assistants; government and public affairs executives; lawyers; marketing, sales and advertising executives; researchers, editors and journalists.

Hardcover ISBN 978-1-59237-370-3, 1,600 pages, $325.00

Associations Canada, 2009

The Most Powerful Fact-Finder to Business, Trade, Professional and Consumer Organizations
Associations Canada covers Canadian organizations and international groups including industry, commercial and professional associations, registered charities, special interest and common interest organizations. This annually revised compendium provides detailed listings and abstracts for nearly 20,000 regional, national and international organizations. This popular volume provides the most comprehensive picture of Canada's non-profit sector. Detailed listings enable users to identify an organization's budget, founding date, scope of activity, licensing body, sources of funding, executive information, full address and complete contact information, just to name a few. Powerful indexes help researchers find information quickly and easily. The following indexes are included: subject, acronym, geographic, budget, executive name, conferences & conventions, mailing list, defunct and unreachable associations and registered charitable organizations. In addition to annual spending of over $1 billion on transportation and conventions alone, Canadian associations account for many millions more in pursuit of membership interests. *Associations Canada* provides complete access to this highly lucrative market. *Associations Canada* is a strong source of prospects for sales and marketing executives, tourism and convention officials, researchers, government officials - anyone who wants to locate non-profit interest groups and trade associations.

Hardcover ISBN 978-1-59237-401-4, 1,600 pages, $325.00

Financial Services Canada, 2008/09

Financial Services Canada is the only master file of current contacts and information that serves the needs of the entire financial services industry in Canada. With over 18,000 organizations and hard-to-find business information, Financial Services Canada is the most up-to-date source for names and contact numbers of industry professionals, senior executives, portfolio managers, financial advisors, agency bureaucrats and elected representatives. Financial Services Canada incorporates the latest changes in the industry to provide you with the most current details on each company, including: name, title, organization, telephone and fax numbers, e-mail and web addresses. *Financial Services Canada* also includes private company listings never before compiled, government agencies, association and consultant services - to ensure that you'll never miss a client or a contact. Current listings include: banks and branches, non-depository institutions, stock exchanges and brokers, investment management firms, insurance companies, major accounting and law firms, government agencies and financial associations. Powerful indexes assist researchers with locating the vital financial information they need. The following indexes are included: alphabetic, geographic, executive name, corporate web site/e-mail, government quick reference and subject. *Financial Services Canada* is a valuable resource for financial executives, bankers, financial planners, sales and marketing professionals, lawyers and chartered accountants, government officials, investment dealers, journalists, librarians and reference specialists.

Hardcover ISBN 978-1-59237-278-2, 900 pages, $315.00

Directory of Libraries in Canada, 2008/09

The Directory of Libraries in Canada brings together almost 7,000 listings including libraries and their branches, information resource centers, archives and library associations and learning centers. The directory offers complete and comprehensive information on Canadian libraries, resource centers, business information centers, professional associations, regional library systems, archives, library schools and library technical programs. *The Directory of Libraries in Canada* includes important features of each library and service, including library information; personnel details, including contact names and e-mail addresses; collection information; services available to users; acquisitions budgets; and computers and automated systems. Useful information on each library's electronic access is also included, such as Internet browser, connectivity and public Internet/CD-ROM/subscription database access. The directory also provides powerful indexes for subject, location, personal name and Web site/e-mail to assist researchers with locating the crucial information they need. *The Directory of Libraries in Canada* is a vital reference tool for publishers, advocacy groups, students, research institutions, computer hardware suppliers, and other diverse groups that provide products and services to this unique market.

Hardcover ISBN 978-1-59237-279-9, 850 pages, $315.00

Canadian Environmental Directory, 2009

The Canadian Environmental Directory is Canada's most complete and only national listing of environmental associations and organizations, government regulators and purchasing groups, product and service companies, special libraries, and more! The extensive Products and Services section provides detailed listings enabling users to identify the company name, address, phone, fax, e-mail, Web address, firm type, contact names (and titles), product and service information, affiliations, trade information, branch and affiliate data. The Government section gives you all the contact information you need at every government level – federal, provincial and municipal. We also include descriptions of current environmental initiatives, programs and agreements, names of environment-related acts administered by each ministry or department PLUS information and tips on who to contact and how to sell to governments in Canada. The Associations section provides complete contact information and a brief description of activities. Included are Canadian environmental organizations and international groups including industry, commercial and professional associations, registered charities, special interest and common interest organizations. All the Information you need about the Canadian environmental industry: directory of products and services, special libraries and resource, conferences, seminars and tradeshows, chronology of environmental events, law firms and major Canadian companies, *The Canadian Environmental Directory* is ideal for business, government, engineers and anyone conducting research on the environment.

Softcover ISBN 978-1-59237-374-1, 900 pages, $325.00

Canadian Parliamentary Guide, 2008

An indispensable guide to government in Canada, the annual *Canadian Parliamentary Guide* provides information on both federal and provincial governments, courts, and their elected and appointed members. The Guide is completely bilingual, with each record appearing both in English and then in French. The Guide contains biographical sketches of members of the Governor General's Household, the Privy Council, members of Canadian legislatures (federal, including both the House of Commons and the Senate, provincial and territorial), members of the federal superior courts (Supreme, Federal, Federal Appeal, Court Martial Appeal and Tax Courts) and the senior staff for these institutions. Biographies cover personal data, political career, private career and contact information. In addition, the Guide provides descriptions of each of the institutions, including brief historical information in text and chart format and significant facts (i.e. number of members and their salaries). The Guide covers the results of all federal general elections and by-elections from Confederations to the present and the results of the most recent provincial elections. A complete name index rounds out the text, making information easy to find. No other resources presents a more up-to-date, more complete picture of Canadian government and her political leaders. A must-have resource for all Canadian reference collections.

Hardcover ISBN 978-1-59237-310-9, 800 pages, $184.00